T0271281

Programming in Ada 2022

Programming in Ada 2022

JOHN BARNES

Shaftesbury Road, Cambridge CB2 8EA, United Kingdom

One Liberty Plaza, 20th Floor, New York, NY 10006, USA

477 Williamstown Road, Port Melbourne, VIC 3207, Australia

314–321, 3rd Floor, Plot 3, Splendor Forum, Jasola District Centre, New Delhi – 110025, India

103 Penang Road, #05-06/07, Visioncrest Commercial, Singapore 238467

Cambridge University Press is part of Cambridge University Press & Assessment, a department of the University of Cambridge.

We share the University's mission to contribute to society through the pursuit of education, learning and research at the highest international levels of excellence.

www.cambridge.org
Information on this title: www.cambridge.org/9781009564779

DOI: 10.1017/9781009564809

When citing this work, please include a reference to the DOI 10.1017/9781009564809

First published 2024

A catalogue record for this publication is available from the British Library

A Cataloging-in-Publication data record for this book is available from the Library of Congress

ISBN 978-1-009-56477-9 Paperback

Additional resources for this publication at www.cambridge.org/barnes22.

To Barbara

Contents

Foreword xv
Preface xvii

Part 1 An Overview 1

1 Introduction 3

1.1 Standard development 3
1.2 Software engineering 4
1.3 Evolution and abstraction 6
1.4 Structure and objectives of this book 8
1.5 References 10

2 Simple Concepts 11

2.1 Key goals 11
2.2 Overall structure 12
2.3 The scalar type model 16
2.4 Arrays and records 19
2.5 Access types 21
2.6 Errors and exceptions 23
2.7 Terminology 25

3 Abstraction 27

3.1 Packages and private types 27
3.2 Objects and inheritance 29
3.3 Classes and polymorphism 34
3.4 Genericity 39
3.5 Object oriented terminology 41
3.6 Tasking 42

4 Programs and Libraries 47

4.1 The hierarchical library 47
4.2 Input–output 49

4.3 Numeric library 52
4.4 Running a program 54

Program 1 Magic Moments 59

Part 2 Algorithmic Aspects 63

5 Lexical Style 65

5.1 Syntax notation 65
5.2 Lexical elements 66
5.3 Identifiers 67
5.4 Numbers 68
5.5 Comments 70
5.6 Pragmas and aspects 71

6 Scalar Types 73

6.1 Object declarations and assignments 73
6.2 Blocks and scopes 75
6.3 Types 77
6.4 Subtypes 79
6.5 Simple numeric types 81
6.6 Enumeration types 87
6.7 The type Boolean 90
6.8 Categories of types 93
6.9 Expression summary 95

7 Control Structures 101

7.1 If statements 101
7.2 Case statements 105
7.3 Loop statements 108
7.4 Goto statements and labels 114
7.5 Statement classification 115

8 Arrays and Records 117

8.1 Arrays 117
8.2 Array types 122
8.3 Array aggregates 127
8.4 Characters and strings 132
8.5 Arrays of arrays and slices 135
8.6 One-dimensional array operations 138
8.7 Records 143
8.8 General aggregates 147

9 Expression Structures 151

9.1 Membership tests 151
9.2 If expressions 153

9.3 Case expressions 157
9.4 Quantified expressions 159
9.5 Declare expressions 161
9.6 Reduction expressions 162

10 Subprograms 165
10.1 Functions 165
10.2 Operators 173
10.3 Procedures 175
10.4 Aliasing 181
10.5 Named and default parameters 183
10.6 Overloading 185
10.7 Declarations, scopes, and visibility 186

11 Access Types 193
11.1 Flexibility versus integrity 193
11.2 Access types and allocators 195
11.3 Null exclusion and constraints 202
11.4 Aliased objects 204
11.5 Accessibility 208
11.6 Access parameters 210
11.7 Anonymous access types 214
11.8 Access to subprograms 218
11.9 Storage pools 224

Program 2 Sylvan Sorter 227

Part 3 The Big Picture 231

12 Packages and Private Types 233
12.1 Packages 233
12.2 Private types 238
12.3 Primitive operations and derived types 245
12.4 Equality 251
12.5 Limited types 255
12.6 Resource management 261

13 Overall Structure 267
13.1 Library units 267
13.2 Subunits 270
13.3 Child library units 272
13.4 Private child units 276
13.5 Mutually dependent units 283
13.6 Scope, visibility, and accessibility 287
13.7 Renaming 291
13.8 Programs, partitions, and elaboration 296

Program 3 Rational Reckoner 301

14 Object Oriented Programming 305

14.1 Type extension 305
14.2 Polymorphism 311
14.3 Abstract types and interfaces 319
14.4 Primitive operations and tags 322
14.5 Views and redispatching 332
14.6 Private types and extensions 338
14.7 Controlled types 346
14.8 Multiple inheritance 351
14.9 Multiple implementations 357

15 Exceptions 365

15.1 Handling exceptions 365
15.2 Declaring and raising exceptions 368
15.3 Checking and exceptions 374
15.4 Exception occurrences 376
15.5 Exception pragmas and aspects 380
15.6 Scope of exceptions 385

16 Contracts 389

16.1 Aspect specifications 389
16.2 Preconditions and postconditions 392
16.3 Type invariants 404
16.4 Subtype predicates 410
16.5 Global state 418
16.6 Messages 422

17 Numeric Types 427

17.1 Signed integer types 427
17.2 Modular types 433
17.3 Real types 435
17.4 Floating point types 437
17.5 Fixed point types 440
17.6 Decimal types 446

18 Parameterized Types 449

18.1 Discriminated record types 449
18.2 Default discriminants 453
18.3 Variant parts 459
18.4 Discriminants and derived types 463
18.5 Access types and discriminants 466
18.6 Private types and discriminants 473
18.7 Access discriminants 475

19 Generics 479

19.1 Declarations and instantiations 479
19.2 Type parameters 485
19.3 Subprogram parameters 495
19.4 Package parameters 502
19.5 Generic library units 508

20 Tasking 511

20.1 Parallelism 511
20.2 The rendezvous 513
20.3 Timing and scheduling 518
20.4 Protected objects 524
20.5 Simple select statements 531
20.6 Timed and conditional calls 535
20.7 Concurrent types and activation 537
20.8 Termination, exceptions, and ATC 544
20.9 Signalling and scheduling 550
20.10 Summary of structure 557

21 Object Oriented Techniques 561

21.1 Extension and composition 561
21.2 Using interfaces 564
21.3 Mixin inheritance 570
21.4 Linked structures 572
21.5 Iterators 575
21.6 Generalized iteration 580
21.7 Procedural iterators 587
21.8 Object factories 589
21.9 Controlling abstraction 593

22 Tasking Techniques 599

22.1 Dynamic tasks 599
22.2 Multiprocessors 602
22.3 Synchronized interfaces 610
22.4 Discriminants 621
22.5 Task termination 626
22.6 Clocks and timers 629
22.7 Profiles 638
22.8 Parallel blocks and loops 639
22.9 Conflict checking 642

Program 4 Super Sieve 645

Part 4 Completing the Story 649

23 Predefined Library 651

23.1 The package Standard 651
23.2 The package Ada 655
23.3 Characters and strings 658
23.4 Text buffers and images 677
23.5 Numerics 681
23.6 Big numbers 685
23.7 Input and output 692
23.8 Text input–output 698
23.9 Streams 708
23.10 Environment commands 715

Program 5 Wild Words 727

24 Container Library 731

24.1 Organization of library 732
24.2 Doubly linked lists 734
24.3 Vectors 748
24.4 Maps 754
24.5 Sets 767
24.6 Trees 780
24.7 Holders 792
24.8 Queues 794
24.9 Bounded containers 803
24.10 Indefinite containers 807
24.11 Sorting 813
24.12 Summary table 815

25 Interfacing 817

25.1 Representations 817
25.2 Unchecked programming 822
25.3 The package System 824
25.4 Storage pools and subpools 826
25.5 Other languages 834

Program 6 Playing Pools 839

26 The Specialized Annexes 843

26.1 Systems Programming 843
26.2 Real-Time Systems 847
26.3 Distributed Systems 851
26.4 Information Systems 852
26.5 Numerics 852
26.6 High Integrity Systems 858

27 Finale 861

27.1 Names and expressions 861
27.2 Type equivalence 865
27.3 Overall program structure 868
27.4 Portability 872
27.5 Penultimate thoughts 874
27.6 SPARK 839

Appendices 889

A1 Reserved Words, etc. 889

A1.1 Reserved words 889
A1.2 Predefined attributes 890
A1.3 Predefined aspects 898
A1.4 Predefined pragmas 902
A1.5 Predefined restrictions 904

A2 Glossary 907

Answers to Exercises 913
Bibliography 917
Index 919

Foreword

Ada 22 has arrived! The wheels of international standardization turn slowly even when there are no technical issues, so it was actually published in May 2023.

We are at a point in time where many core processors are becoming ubiquitous and need to be used efficiently and reliably. Ada 2022 will help with this admirably. It also fills in some gaps in what should, with the benefit of hindsight, have already been in earlier versions of Ada.

And so the Ada programming language continues to evolve in order to better support the development of software applications with high requirements for reliability and system integrity. The 4th major revision of the language, Ada 2022, includes improvements in many different areas, but with particular focus on a few topics:

- allowing software developers to easily and safely take advantage of the parallel execution capabilities of multi-core and multi-threaded architectures;
- allowing developers to more precisely express their intent regarding a program's structure and logic via improved contracts and other forms of assertions;
- providing improved support for containers;
- providing "creature comforts" to ease the use of common programming idioms.

The introduction of parallel loops and parallel block statements allows users to express the possibility of parallel execution of constructs that also have well-defined sequential semantics. The Global and Nonblocking aspects can be used to enable static detection of unsafe concurrent access to variables.

Contracts are improved along two different axes: contracts are supported in more contexts (e.g., the Default_Initial_Condition aspect and support for precondition/postcondition specifications for access-to-subprogram types and for generic formal subprograms), and more expressive forms of expressions are defined in order to make it possible to write more precise contracts (e.g., declare expressions, reduction expressions, delta aggregates, and calls to static functions). There is a sometimes-unappreciated point about some of these expression forms that may seem like just syntactic sugar (another example is Ada 2012's quantified expressions). It is true that the same value could be computed in a function body without too much trouble, but then what would the postcondition be for this function? How would the declaration of the function describe its result? So these new forms do enable more expressive contracts.

Support is added for iterating over containers, and for constructing container values using aggregates. The mechanisms used to accomplish this can also be used in user-defined container implementations; the implementation-defined containers have no special "magic" that is unavailable to users who want to "roll their own". Support is also provided for literals of private types, for indexing operations on a private type, for implicit dereferencing of certain non-access types; all of these are intended to make it easier to hide the implementation of a private type without taking useful functionality away from clients. And the private type in question may, of course, be a container type.

"Creature comforts" include the introduction of some features that have long been available in other programming languages. A name ("@") is defined for the target of an assignment statement, as in:

```
Function_Call.Some_Long_Name(Some_Index_Expression) := @ + 1;
```

The Image attribute is now available for almost all types and objects, not just for scalars; support for (in effect) user-specified Image functions is provided. Restrictions on aggregates are relaxed (index values can now be used in array aggregate component expressions); a discriminant value in an aggregate can be non-static in more cases. Predefined support for big numbers (integer and real) is added. Internationalization support is improved via the various new Wide_File_Names and Wide_Wide_File_Names packages.

And there are many, many other improvements to the language as well. The Jorvik profile, the System.Atomic_Operations package, the Object_Size attribute, iterator filters, and lots of other stuff.

And who better to guide a user through all of this material than John Barnes? John has been involved in Ada programming language design efforts since shortly after the birth of Ada Lovelace (or so it seems). He combines his detailed knowledge of the subject with a special talent for exposition. He has an unusual ability to anticipate, and correct for, the ways that a reader might be confused by whatever corner of the language he is describing. His choices of examples are invariably both enlightening and entertaining. His writing combines technical correctness with accessibility and humor.

For folks who want to write software that works correctly (admittedly, a small audience), there is value in learning about Ada and, in particular, about Ada 2022. And for anyone who wants to learn about Ada, there is both value and fun in learning about it from John's fine book.

Enjoy!

Steve Baird & Jeff Cousins,
Current and Past Chairs of the
ISO/IEC JTC1/SC22/WG9
Ada Rapporteur Group (ARG)

Preface

Welcome to *Programming in Ada 2022* which describes the Ada language as presented to ISO in 2022 and formally approved in early 2023.

The original language, devised in the 1980s, is known as Ada 83 and was followed by Ada 95, Ada 2005, Ada 2012, and now Ada 2022.

Ada has gained a reputation as being the language of choice when software needs to be correct. And as software pervades into more areas of society so that ever more software is safety critical or security critical, it is clear that the future for Ada is bright. One observes, for example, the growth in use of SPARK, the Ada based high integrity language widely used in areas such as avionics and signalling.

Ada 83 was a relatively simple but highly reliable language with emphasis on abstraction and information hiding. It was also notable for being the first practical language to include multitasking within the language itself.

Ada 95 added extra flexibility in the form of the full dynamic features of Object Oriented Programming (OOP) and was the first such language to become an ISO standard. Ada 95 also made important structural enhancements to visibility control and greatly improved multitasking by the addition of protected types.

Ada 2005 then added more flexibility in the OOP area by the addition of multiple inheritance via interfaces and it also added more facilities in the real-time area concerning scheduling algorithms, timing and so on. It also added further facilities to the standard library such as the introduction of the container library covering the manipulation of lists, vectors, sets, and maps.

Ada 2012 made further enhancements. Perhaps the most important was the addition of features for contracts such as pre- and postconditions and type invariants. These in turn showed the need for more flexible expressions and so conditional expressions, case expressions, and quantified expressions were also added. Tasking facilities were enhanced to recognize multicore architectures. The container library was also enhanced to include multiway trees and task-safe queues. Bounded forms of all containers which are important for high integrity systems where dynamic storage management is often not permitted were also added.

Ada 2022 makes important improvements in two key areas: reliability and efficiency. Reliability is improved by contracts which are strengthened by the addition of annotations describing the manipulation of global state. The form of contracts as pre- and postconditions which take the form of expressions has stimulated the enhancement of the power of expressions. So Ada is being nudged in the direction of expression/functional languages (but not too much). The other area of concern is efficiency and this is addressed by the inclusion of lightweight parallel features to enable the straightforward use of multiprocessors. Both reliability and efficiency are addressed in the container library by the concept of stable state which enables many common operations to be performed more efficiently but still with full reliability.

The main body of the book consists of 27 chapters grouped into four parts as follows

- Chapters 1 to 4 provide an overview which should give the reader an understanding of the overall scope of the language as well as the ability to run significant programs as examples – this part is particularly for newcomers to Ada.
- Chapters 5 to 11 cover the small-scale aspects such as the lexical details, scalar, array and simple record types, control and expression structures, subprograms and access types.
- Chapters 12 to 22 discuss the large-scale aspects including packages and private types, contracts, separate compilation, abstraction, OOP and tasking as well as exceptions and the details of numerics.
- Chapters 23 to 27 complete the story by discussing the predefined library, interfacing to the outside world, the extensive container library and the specialized annexes; there is also a finale concluding with some ruminations over correctness and a brief introduction to SPARK.

New sections within the existing chapters describe topics such as more elaborate aggregates, declare expressions, reduction expressions, global state, procedural iterators, parallel blocks and loops, and big numbers.

Perhaps the introduction of big numbers is my favourite enhancement. It enables me to add some examples of the work of the mathematician, Mersenne the Monk, and also the very important topic of encryption in an uncertain world.

The finale in Chapter 27 includes a fantasy customer in a shop trying to buy reusable software components and whose dream now seems as far away or indeed as near at hand as it did many years ago when I first toiled at this book. The discussion continues to take a galactic view of life and perhaps echoes the cover of the book which depicts two galaxies interacting.

There are just two appendices. The first summarises the reserved words, aspects, attributes, pragmas, and restrictions. The second is a glossary.

In order to avoid the book becoming unwieldy, the full details of the syntax for Ada 2022 and an updated table summarising the containers are now provided on the website. The answers to all exercises are also on the website but those for the introductory Part 1 are also included in the book for the convenience of readers encountering Ada for the first time. More details of the website will be found below.

Revisions to the language are described by Ada Issues (AIs) and a number of these are mentioned in the Index. For example AI05-220 refers to a change to Ada 2005, AI12-187 refers to a change to Ada 2012 and AI22-78 refers to a change to Ada 2022 following standardization (it concerns big decimal types).

And now I must thank all those who have helped with this new book. Many thanks therefore (in alphabetical order) to Steve Baird, Janet Barnes, Randy Brukardt, Jeff Cousins, Bob Duff, and Tucker Taft. Their much valued comments enabled me to improve the presentation and to eliminate a number of errors.

I must give special thanks to Jeff Cousins for his help in improving and testing the big number programs, and for writing an excellent introduction to Ada 2022 which goes by the enchanting name of the Language Enhancement Guide (or *LEG*). I must also give special thanks to Randy Brukardt without whose eternal energy as editor of the Ada standard, I would find writing about Ada a very difficult task indeed. Note that the Ada Reference Manual (or *ARM*) goes well with the *LEG*.

Finally, many thanks to my wife Barbara for help in typesetting and proof-reading and to friends at Cambridge University Press for their continued guidance and help.

John Barnes
Caversham, England
March 2024

Notes on the website

The website for this book is www.cambridge.org/barnes22. As well as the full syntax for Ada 2022 and a table giving a summary of all container operations in Ada 2022, it contains the answers to all the exercises, plus some obscure or obsolete material on exceptions, discriminants, and iterators which were in previous versions of the book.

It also contains additional material on the six sample programs in the body of the book and several further programs illustrating new features in Ada 2022 involving big numbers.

Each example in the book commences with some remarks about its purpose and overall structure. This is followed by the text of the program and then some notes on specific details. A desire to keep the program text short means that comments are at a minimum. However, the corresponding source text on the website includes much additional commentary. The website also includes further discussion and explanation and suggestions for enhancement. In general the programs use only those features of the language explained in detail by that point in the book.

The first program, Magic Moments, illustrates type extension and dispatching. It shows how the existence of common components and common operations enable dispatching to compute various geometrical properties almost by magic.

The Sylvan Sorter is an exercise in access types and basic algorithmic techniques including recursion.

The Rational Reckoner provides two examples of abstract data types – the rational numbers themselves and the stack which is the basis of the calculator part of the program.

The Super Sieve illustrates multitasking and communication between tasks both directly through entry calls and indirectly through protected objects. For added interest it is made generic so that more general primes than those from the familiar domain of integers may be found. This provides the opportunity to use a discriminated record type and a modular type to represent binary polynomials.

The program Wild Words is probably the hardest to follow because it is not based on any particular example described in the preceding chapters. It illustrates many of the facilities of the character and string handling packages as well as the generation of random numbers.

The program Playing Pools shows how users might write their own storage allocation package for the control of storage pools. The example shown enables the user to monitor the state of the pool and it is exercised by running the familiar Tower of Hanoi program which moves a tower of discs between three poles. Variety is provided by implementing the stack structures representing the three poles (and defined by an interface) in two different ways and dispatching to the particular implementation. The website also includes an extended version which uses three different ways.

The other sample programs on the website at the time of writing illustrate the use of big numbers which are described in Chapter 23.

The first program, which we can refer to as Clever Coding, illustrates the coding algorithm known as RSA after its inventors Ronald Rivest, Adi Shamir, and Leonard Adleman in 1977. The algorithm is outlined in Section 23.6. The program asks the user for the values of the parameters and whether coding of a number or letter sequence is required. The program then encodes the item and when requested decodes it thereby showing the original value once more.

The second program, which we can refer to as Mersenne Magic, illustrates Mersenne numbers and the amazing algorithm devised by Lucas and Lehmer that determines whether a particular Mersenne number is prime or not. The largest known prime number is a Mersenne prime. See Section 22.8.

The last program, which we can refer to as Fermat Factors, uses a method devised by Fermat for finding the factors of a number. There are in fact three versions of this program with increasing levels of complexity so altogether there are five new programs.

Information on many aspects of Ada such as vendors, standards, books and so on can be obtained from the websites listed in the Bibliography.

Part 1

An Overview

Chapter 1	**Introduction**	3
Chapter 2	**Simple Concepts**	11
Chapter 3	**Abstraction**	27
Chapter 4	**Programs and Libraries**	47
Program 1	**Magic Moments**	59

This first part covers the background to the Ada language and an overview of most of its key features. Enough material is presented here to enable a programmer to write complete simple programs.

Chapter 1 contains a short historical account of the origins and development of various versions of Ada from Ada 83 through Ada 95, Ada 2005, and Ada 2012 leading to Ada 2022 which is the topic of this book. There is also a general discussion of how the evolution of abstraction has been an important key to the development of programming languages in general.

Broadly speaking, the other three chapters in this part provide an overview of the material in the corresponding other parts of the book. Thus Chapter 2 describes the simple concepts familiar from languages such as C and Pascal and which form the subject of the seven chapters which comprise Part 2. Similarly Chapter 3 covers the important topic of abstraction which is the main theme of Part 3. And then Chapter 4 rounds off the overview by showing how a complete program is put together and corresponds to Part 4 which covers material such as the predefined library.

Chapter 3 includes a discussion of the popular topic of object oriented programming and illustrates the key concepts of classes as groups of related types, of type extension and inheritance as well as static polymorphism (genericity) and dynamic polymorphism leading to dynamic binding. It also includes a brief comparison between the terminology used by Ada and that used by some other languages. This chapter concludes with an introduction to tasking which is a very important aspect of Ada and is a topic not addressed by most programming languages at all.

This part concludes with the first of a number of complete programs designed to give the reader a better understanding of the way the various components of the language fit together. This particular program illustrates a number of aspects of type extension and polymorphism.

Those not familiar with Ada will find that this part will give them a fair idea of Ada's capabilities and lays the foundation for understanding the details presented in the remainder of the book.

1 Introduction

1.1 Standard development
1.2 Software engineering
1.3 Evolution and abstraction
1.4 Structure and objectives of this book
1.5 References

Ada 2022 is a comprehensive high level programming language especially suited for the professional development of large or critical programs for which correctness and robustness are major considerations. In this introductory chapter we briefly trace the development of Ada 2022 (and its predecessors), its place in the overall language scene and the general structure of the remainder of this book.

1.1 Standard development

Ada 2022 is a direct descendant of Ada 83 which was originally sponsored by the US Department of Defense for use in the embedded system application area. (An embedded system is one in which the computer is an integral part of a larger system such as a chemical plant, missile, or dishwasher.)

The story of Ada goes back to about 1974 when the United States Department of Defense realized that it was spending far too much on software, especially in the embedded systems area. To cut a long story short, the DoD sponsored the new language through a number of phases of definition of requirements, competitive and parallel development and evaluation which culminated in the issue of the ANSI standard for Ada in 1983[1]. The team that developed Ada was based at CII Honeywell Bull in France under the leadership of Jean D Ichbiah.

The language was named after Augusta Ada Byron, Countess of Lovelace (1815–52). Ada, the daughter of the poet Lord Byron, was the assistant and patron of Charles Babbage and worked on his mechanical analytical engine. In a very real sense she was therefore the world's first programmer. Note that the 1983 standard was published on Ada's birthday (December 12) and the standard number 1815 happens to be the year of her birth!

Ada 83 became ISO standard 8652 in 1987 and, following normal ISO practice, work leading to a revised standard commenced in 1988. The DoD, as the agent of

ANSI, the original proposers of the standard to ISO, established the Ada project in 1988 under the management of Christine M Anderson. The revised language design was contracted to Intermetrics Inc. under the technical leadership of S Tucker Taft. The revised ISO standard was published in February 1995 and so became Ada 95[2].

The maintenance of the language is performed by the Ada Rapporteur Group (ARG) of ISO/IEC committee SC22/WG9. The ARG under the leadership of Erhard Plödereder identified the need for some corrections and these were published as a Corrigendum on 1 June 2001[3].

Experience with Ada 95 and other modern languages such as Java indicated that further improvements would be very useful. The changes needed were not so large as the step from Ada 83 to Ada 95 and so an Amendment was deemed appropriate rather than a Revised standard. The ARG then developed the Amended language known as Ada 2005 under the leadership of Pascal Leroy[4].

Further experience with Ada 2005 showed that additions especially in the area of contracts and multiprocessor support would be appropriate. It was decided that this time a consolidated Edition should be produced. This was accordingly done under the leadership of Ed Schonberg and resulted in the version known as Ada 2012 which became an ISO standard towards the end of 2012[5]. Maintenance then continued under the leadership of Jeff Cousins. A number of corrections were made and resulted in the Corrigendum 1 version of Ada 2012 published in 2016[6].

It was clear that further development was required in order to take full advantage of the contract features and multiprocessor support. This progressed under the leadership of Steve Baird with the editor (as ever) being the seemingly indefatigable Randy Brukardt and resulted in the ISO standard for Ada 2022 which was published early in 2023[7].

1.2 Software engineering

It should not be thought that Ada is just another programming language. Ada is about Software Engineering, and by analogy with other branches of engineering it can be seen that there are two main problems with the development of software: the need to reuse software components as much as possible and the need to establish disciplined ways of working.

As a language, Ada largely solves the problem of writing reusable software components – or at least through its excellent ability to prescribe interfaces, it provides an enabling technology in which reusable software can be written.

The establishment of a disciplined way of working seems to be a Holy Grail which continues to be sought. One of the problems is that development environments change so rapidly that the stability necessary to establish discipline can be elusive.

But Ada is stable and encourages a style of programming which is conducive to disciplined thought. Experience with Ada shows that a well designed language can reduce the cost of both the initial development of software and its later maintenance.

The main reason for this is simply reliability. The strong typing and related features ensure that programs contain few surprises; most errors are detected at compile time and of those that remain, many are detected by run-time constraints. Moreover, the compile-time checking extends across compilation unit boundaries.

This aspect of Ada considerably reduces the costs and risks of program development compared for example with C and its derivatives such as C++. Moreover, an Ada compilation system includes the facilities found in separate tools such as 'lint' and 'make' for C. Even if Ada is seen as just another programming language, it reaches parts of the software development process that other languages do not reach.

Ada 95 added extra flexibility to the inherent reliability of Ada 83. That is, it kept the Software Engineering but allowed more flexibility. The features of Ada 95 which contributed to this more flexible feel are the extended or tagged types, the hierarchical library facility and the greater ability to manipulate pointers or references. Another innovation in Ada 95 was the introduction of protected types to the tasking model.

As a consequence, Ada 95 incorporated the benefits of object oriented languages without incurring the pervasive overheads of languages such as Smalltalk or the insecurity brought by the weak C foundation in the case of C++. Ada 95 remained a very strongly typed language but provided the benefits of the object oriented paradigm.

Ada 2005 added yet further improvements to the object model by adding interfaces in the style of Java and providing constructor functions and also extending the object model to incorporate tasking. Experience has shown that a standard library is important and accordingly Ada 2005 had a much larger library including predefined facilities for containers.

The Ada 2005 containers were doubly linked lists (so one could traverse a list in both directions), vectors (like lists but indexed as well, and not like vectors in mechanics), maps (for getting from a key to an element), and sets (collections of elements). Both sets and maps have ordered and hashed forms.

Ada has long been renowned as the flagship language for multitasking applications. (Multitasking is often known as multithreading.) This position was strengthened by the addition of further standard paradigms for scheduling and timing and the incorporation of the Ravenscar profile into Ada 2005. The Ravenscar profile enables the development of real-time programs with predictable behaviour.

Further improvements which resulted in Ada 2012 were in three main areas. First there was the introduction of material for 'programming by contract' such as pre- and postconditions somewhat on the lines of those found in Eiffel. There were also additions to the tasking model including facilities for mapping tasks onto multiprocessors. Other important extensions were additional facilities in the container library enabling further structures (such as trees and queues) to be addressed; other improvements also simplified many operations on the existing container structures.

Ada 2022 builds on the experience with Ada 2012 in a number of areas. The use of the contract model based on pre- and postconditions showed that their use would be simplified and encouraged by the addition of more expression features. However, although Ada 2022 has features enabling objects to be declared within expressions, Ada is not moving towards being an expression language of the style historically introduced by LISP with the attendant difficulties of legibility. Other proof related features such as the ability to describe the effect of a subprogram on the overall global state are also added. Major improvements are made to the container library enabling greater efficiency of operation whilst retaining security. And lightweight tasking is added to aid the simpler use of multiple processors.

Two kinds of application stand out where Ada is particularly relevant. The very large and the very critical.

Very large applications, which inevitably have a long lifetime, require the cooperative effort of large teams. The information hiding properties of Ada and especially the way in which integrity is maintained across compilation unit boundaries are invaluable in enabling such developments to progress smoothly. Furthermore, if and when the requirements change and the program has to be modified, the structure and especially the readability of Ada enable rapid understanding of the original program even if it is modified by a different team.

Very critical applications are those that just have to be correct otherwise people or the environment get damaged. Obvious examples occur in avionics, railway signalling, process control, and medical applications. Such programs may not be large but have to be very well understood and often mathematically proven to be correct. The full flexibility of Ada is not appropriate in this case but the intrinsic reliability of the strongly typed kernel of the language is exactly what is required. Indeed many certification agencies dictate the properties of acceptable languages and whereas they do not always explicitly demand a subset of Ada, nevertheless the properties are not provided by any other practically available language. The SPARK language which is based around a kernel subset of Ada illustrates how special tools can provide extra support for developing high integrity systems. A SPARK program includes additional information regarding data flow, state, and proof. In the original version of SPARK this was done in the form of Ada comments. However, in later versions of SPARK, information is presented using features of Ada including pre- and postconditions and additional assertions. This information is processed by the SPARK tools and can be used to show that a program meets certain criteria such as not raising any predefined exceptions. Much progress has been made in the area of proof in recent years. A brief introduction to SPARK will be found in Chapter 27.

1.3 Evolution and abstraction

The evolution of programming languages has apparently occurred in a rather *ad hoc* fashion but with hindsight it is now possible to see a number of major advances. Each advance seems to be associated with the introduction of a level of abstraction which removes unnecessary and harmful detail from the program.

The first advance occurred in the early 1950s with high level languages such as Fortran and Autocode which introduced 'expression abstraction'. It thus became possible to write statements such as

```
X = A + B(I)
```

so that the use of the machine registers to evaluate the expression was completely hidden from the programmer. In these early languages the expression abstraction was not perfect since there were somewhat arbitrary constraints on the complexity of expressions; subscripts had to take a particularly simple form for instance. Later languages such as Algol 60 removed such constraints and completed the abstraction.

The second advance concerned 'control abstraction'. The prime example was Algol 60 which took a remarkable step forward; no language since then has made such an impact on later developments. The point about control abstraction is that the flow of control is structured and individual control points do not have to be named or numbered. Thus we write

if X = Y **then** P := Q **else** A := B

and the compiler generates the gotos and labels which would have to be explicitly used in early versions of languages such as Fortran. The imperfection of early expression abstraction was repeated with control abstraction. In this case the obvious flaw was the horrid Algol 60 switch which has now been replaced by the case statement of later languages.

The third advance was 'data abstraction'. This means separating the details of the representation of data from the abstract operations defined upon the data.

Older languages take a very simple view of data types. In all cases the data is directly described in numerical terms. Thus if the data to be manipulated is not really numerical (it could be traffic light colours) then some mapping of the abstract type must be made by the programmer into a numerical type (usually integer). This mapping is purely in the mind of the programmer and does not appear in the written program except perhaps as a comment.

Pascal introduced a certain amount of data abstraction as instanced by the enumeration type. Enumeration types allow us to talk about the traffic light colours in their own terms without our having to know how they are represented in the computer. Moreover, they prevent us from making an important class of programming errors – accidentally mixing traffic lights with other abstract types such as the names of fish. When all such types are described in the program as numerical types, such errors can occur.

Another form of data abstraction concerns visibility. It has long been recognized that the traditional block structure of Algol and Pascal is not adequate. For example, it is not possible in Pascal to write two procedures to operate on some common data and make the procedures accessible without also making the data directly accessible. Many languages have provided control of visibility through separate compilation; this technique is adequate for medium-sized systems, but since the separate compilation facility usually depends upon some external system, total control of visibility is not gained. The module of Modula is an example of an appropriate construction.

Ada was probably the first practical language to bring together these various forms of data abstraction.

Another language which made an important contribution to the development of data abstraction is Simula 67 with its concept of class. This leads us into the paradigm now known as object oriented programming. There seems to be no precise definition of OOP, but its essence is a flexible form of data abstraction providing the ability to define new data abstractions in terms of old ones and allowing dynamic selection of types.

All types in Ada 83 were static and thus Ada 83 was not classed as a truly object oriented language but as an object based language. However, all later versions of Ada include the essential functionality associated with OOP such as type extension and dynamic polymorphism.

We are, as ever, probably too close to the current scene to achieve a proper perspective. Data abstraction in Ada 83 seems to have been not quite perfect, just as Fortran expression abstraction and Algol 60 control abstraction were imperfect in their day. It remains to be seen just how well Ada now provides what we might call 'object abstraction'. Indeed it might well be that inheritance and other aspects of OOP turn out to be unsatisfactory by obscuring the details of types although not hiding them completely; this could be argued to be an abstraction leak making the problems of program maintenance even harder.

A brief survey of how Ada relates to other languages would not be complete without mention of C and C++. These have a completely different evolutionary trail to the classic Algol–Pascal–Ada route.

The origin of C can be traced back to the CPL language devised by Strachey, Barron and others in the early 1960s. This was intended to be used on major new hardware at Cambridge and London universities but proved hard to implement. From it emerged the simple system programming language BCPL and from that B and then C. The essence of BCPL was the array and pointer model which abandoned any hope of strong typing and (with hindsight) a proper mathematical model of the mapping of the program onto a computing engine. Even the use of := for assignment was lost in this evolution which reverted to the confusing use of = as in Fortran. Having hijacked = for assignment, C uses == for equality thereby conflicting with several hundred years of mathematical usage. About the only feature of the elegant CPL remaining in C is the unfortunate braces {} and the associated compound statement structure which was abandoned by many other languages in favour of the more reliable bracketed form originally proposed by Algol 68. It is again tragic to observe that Java has used the familiar but awful C style. The very practical problems with the C notation are briefly discussed in Chapter 2.

Of course there is a need for a low level systems language with functionality like C. It is, however, unfortunate that the interesting structural ideas in C++ have been grafted onto the fragile C foundation. As a consequence although C++ has many important capabilities for data abstraction, including inheritance and polymorphism, it is all too easy to break these abstractions and create programs that violently misbehave or are exceedingly hard to understand and maintain. Java is free from most of these flaws but persists with anarchic syntax.

The designers of Ada 95 incorporated the positive dynamic facilities of the kind found in C++ onto the firm foundation provided by Ada 83. The designers of Ada 2005 and Ada 2012 have added further appropriate good ideas from Java. But the most important step taken by Ada 2012 was to include facilities for 'programming by contract' which in a sense is the ultimate form of abstraction. Ada 2022 adds further features of this nature especially concerning the manipulation of global state.

Ada thus continues to advance along the evolution of abstraction. It incorporates full object abstraction in a way that is highly reliable without incurring excessive run-time costs.

1.4 Structure and objectives of this book

Learning a programming language is a bit like learning to drive a car. Certain key things have to be learnt before any real progress is possible. Although we need not know how to use the cruise control, nevertheless we must at least be able to start the engine, steer, and brake. So it is with programming languages. We do not need to know all about Ada before we can write useful programs but quite a lot must be learnt. Moreover, many virtues of Ada become apparent only when writing large programs just as many virtues of a Rolls-Royce are not apparent if we only use it to drive to the local store.

This book is not an introduction to programming but an overall description of programming in Ada. It is assumed that the reader will have significant experience of programming in some other language such as Pascal, C or Java (or earlier versions of Ada). But a specific knowledge of any particular language is not assumed.

This book is in four main parts. The first part, Chapters 1 to 4, is an overview of much of the language and covers enough material to enable a wide variety of programs to be written; it also lays the foundation for understanding the rest of the material.

The second part, Chapters 5 to 11, covers the traditional algorithmic parts of the language and roughly corresponds to the domain addressed by Pascal or C although the detail is much richer. The third part, Chapters 12 to 22, covers modern and exciting material associated with data abstraction, programming in the large, OOP, contracts, and parallel processing.

Finally, the fourth part, Chapters 23 to 27, completes the story by discussing the predefined environment, interfacing to the outside world and the specialized annexes; the concluding chapter also pulls together a number of threads that are slightly dispersed in the earlier chapters and finishes with an introduction to SPARK.

There are also six complete program examples which are interspersed at various points. The first follows Chapter 4 and illustrates various aspects of OOP. The other programs illustrate access types, data abstraction, generics and tasking, string handling, and storage pools respectively. These examples illustrate how the various components provided by Ada can be fitted together to make a complete program. The full text of these programs including additional comments and other explanatory material will be found on the associated website. Two new programs concerning the use of big numbers for topics such as encryption will also be found on the website.

Most sections contain exercises. They are an integral part of the discussion and later sections often use the results of earlier exercises. Solutions to all exercises will be found on the website.

Most chapters end with a short checklist of key points and a brief summary of what is new in Ada 2022.

Two appendices are provided in order to make this book reasonably self-contained. The first covers matters such as reserved words, attributes, aspects, pragmas, and restrictions. The second is a glossary of terms. The book concludes with a Bibliography and Index.

Some material such as the complete syntax and a summary of operations on containers will be found on the website. Not only does this keep the size of the book under control but it also enables the reader to print out copies of this material to provide handy supplements which can be referred to in parallel with the main text.

This book includes references to Ada Issues. These are the reports of the Ada Rapporteur Group (ARG) which analyses technical queries and drafts the new Ada standards from time to time.

This book covers all aspects of Ada but does not explore every pathological situation. Its purpose is to describe the effect of and intended use of the features of

Ada. In a few areas the discussion is incomplete; these are areas such as system dependent programming, input–output, and the specialized annexes. System dependent programming (as its name implies) is so dependent upon the particular implementation that only a brief overview seems appropriate. Input–output, although important, does not introduce new concepts but is rather a mass of detail; again a simple overview is presented. And, as their name implies, the specialized annexes address the very specific needs of certain communities; to cover them in detail would make this book excessively long and so only an overview is provided.

Further details can be found in the *Ada Reference Manual* (the *ARM*) which is referred to from time to time. There is also an extended form of the reference manual known as the *Annotated Ada Reference Manual* (the *AARM*) – this includes much embedded commentary and is aimed at the language lawyer and compiler writer. An introduction to Ada 2022 known as the *Language Enhancement Guide* (the *LEG*) is also available. These documents will all be found on the Ada website as described in the Bibliography.

1.5 References

The references given here are to the formal standardization documents describing the various versions of the Ada Programming Language. References to other documents will be found in the Bibliography.

1 United States Department of Defense. *Reference Manual for the Ada Programming Language* (ANSI/MIL-STD-1815A). Washington DC, 1983.

2 International Organization for Standardization. *Information technology – Programming languages – Ada. Ada Reference Manual.* ISO/IEC 8652:1995(E).

3 International Organization for Standardization. *Information technology – Programming languages – Ada. Technical Corrigendum 1.* ISO/IEC 8652:1995/COR.1:2001.

4 International Organization for Standardization. *Information technology – Programming languages – Ada. Amendment 1.* ISO/IEC 8652:1995/AMD.1:2006.

5 International Organization for Standardization. *Information technology – Programming languages – Ada. Ada Reference Manual.* ISO/IEC 8652:2012(E).

6 International Organization for Standardization. *Information technology – Programming languages – Ada. Technical Corrigendum 1.* ISO/IEC 8652:2012(E)/COR.1:2016.

7 International Organization for Standardization. *Information technology – Programming languages – Ada. Ada Reference Manual.* ISO/IEC 8652:2023(E).

2 Simple Concepts

2.1 Key goals
2.2 Overall structure
2.3 The scalar type model
2.4 Arrays and records
2.5 Access types
2.6 Errors and exceptions
2.7 Terminology

This is the first of three chapters covering in outline the main goals, concepts, and features of Ada. Enough material is given in these chapters to enable the reader to write significant programs. It also allows the reader to create a framework in which the exercises and other fragments of program can be executed, if desired, before all the required topics are discussed in depth.

The material covered in this chapter corresponds approximately to that in simple languages such as Pascal and C.

2.1 Key goals

Ada is a large language since it addresses many important issues relevant to the programming of practical systems in the real world. It is much larger than Pascal which is really only suitable for training purposes and for small personal programs. Ada is similarly much larger than C although perhaps of the same order of size as C++. But a big difference is the stress which Ada places on integrity and readability. Some of the key issues in Ada are

- Readability – professional programs are read much more often than they are written. So it is important to avoid a cryptic notation such as in APL which, although allowing a program to be written down quickly, makes it almost impossible to be read except perhaps by the original author soon after it was written.
- Strong typing – this ensures that each object has a clearly defined set of values and prevents confusion between logically distinct concepts. As a consequence many errors are detected by the compiler which in other languages (such as C) can lead to an executable but incorrect program.

- Programming in the large – mechanisms for encapsulation, separate compilation and library management are necessary for the writing of portable and maintainable programs of any size.
- Exception handling – it is a fact of life that programs of consequence are rarely perfect. It is necessary to provide a means whereby a program can be constructed in a layered and partitioned way so that the consequences of unusual events in one part can be contained.
- Data abstraction – extra portability and maintainability can be obtained if the details of the representation of data are kept separate from the specifications of the logical operations on the data.
- Object oriented programming – in order to promote the reuse of tested code, the type flexibility associated with OOP is important. Type extension (inheritance), polymorphism and late binding are all desirable especially when achieved without loss of type integrity.
- Tasking – for many applications it is important to conceive a program as a series of parallel activities rather than just as a single sequence of actions. Building appropriate facilities into a language rather than adding them via calls to an operating system gives better portability and reliability.
- Generic units – in many cases the logic of part of a program is independent of the types of the values being manipulated. A mechanism is therefore necessary for the creation of related pieces of program from a single template. This is particularly useful for the creation of libraries.
- Communication – programs do not live in isolation and it is important to be able to communicate with systems possibly written in other languages.

An overall theme in the design of Ada was concern for the programming process as a human activity. An important aspect of this is enabling errors to be detected early in the overall process. For example a single typographical error in Ada usually results in a program that does not compile rather than a program that still compiles but does the wrong thing. We shall see some examples of this in Section 2.6.

2.2 Overall structure

An important objective of software engineering is to reuse existing pieces of program so that detailed new coding is kept to a minimum. The concept of a library of program components naturally emerges and an important aspect of a programming language is therefore its ability to access the items in a library.

Ada recognizes this situation and introduces the concept of library units. A complete Ada program is conceived as a main subprogram (itself a library unit) which calls upon the services of other library units. These library units can be thought of as forming the outermost lexical layer of the total program.

The main subprogram takes the form of a procedure of an appropriate name. The service library units can be subprograms (procedures or functions) but they are more likely to be packages. A package is a group of related items such as subprograms but may contain other entities as well.

Suppose we wish to write a program to print out the square root of some number. We can expect various library units to be available to provide us with a means of computing square roots and performing input and output. Our job is merely to write a main subprogram to use these services as we wish.

We will suppose that the square root can be obtained by calling a function in our library whose name is Sqrt. We will also suppose that our library includes a package called Simple_IO containing various simple input–output facilities. These facilities might include procedures for reading numbers, printing numbers, printing strings of characters and so on.

Our program might look like

```
with Sqrt, Simple_IO;                  -- with clause
procedure Print_Root is
   use Simple_IO;                      -- declarations here
   X: Float;
begin
   Get(X);                             -- statements here
   Put(Sqrt(X));
end Print_Root;
```

The program is written as a procedure called Print_Root preceded by a with clause giving the names of the library units which it wishes to use. Between **is** and **begin** we can write declarations, and between **begin** and **end** we write statements. Broadly speaking, declarations introduce the entities we wish to manipulate and statements indicate the actions to be performed. Note also the comments starting with a double hyphen.

We have introduced a variable X of type Float which is a predefined language type. Values of this type are a set of floating point numbers and the declaration of X indicates that X can have values only from this set. A value is assigned to X by calling the procedure Get which is in the package Simple_IO.

Writing

```
use Simple_IO;
```

gives us immediate access to the facilities in the package Simple_IO. If we had omitted this use clause we would have had to write

```
Simple_IO.Get(X);
```

in order to indicate where Get was to be found.

The program then calls the procedure Put in the package Simple_IO with a parameter which in turn is the result of calling the function Sqrt with the parameter X.

Some small-scale details should be noted. Statements and declarations all terminate with a semicolon; this is like C but unlike Pascal where semicolons are separators rather than terminators. The program contains various words such as **procedure**, Put and X. These fall into two categories. A few (74 in Ada 2022) such as **procedure** and **is** are used to indicate the structure of the program; they are reserved words and can be used for no other purpose. All others, such as Put and X, can be used as identifiers for whatever purpose we desire. Some of these, such as Float in the example, have a predefined meaning but we can nevertheless reuse them although it might be confusing to do so. For clarity in this book we write the reserved words in lower case bold and use leading capitals for all others. This is a matter of style; the language rules do not distinguish the two cases except when we consider the manipulation of characters themselves. Note also how the underline character is used to break up long identifiers to increase readability.

Finally, observe that the name of the procedure, Print_Root, is repeated between the final **end** and the terminating semicolon. This is optional but is recommended so as to clarify the overall structure although this is obvious in a small example.

Our program is very simple; it might be more useful to enable it to cater for a whole series of numbers and print out each answer on a separate line. We could stop the program somewhat arbitrarily by giving it a value of zero.

```
with Sqrt, Simple_IO;
procedure Print_Roots is
   use Simple_IO;
   X: Float;
begin
   Put("Roots of various numbers");
   New_Line(2);
   loop
      Get(X);
      exit when X = 0.0;
      Put(" Root of "); Put(X); Put(" is ");
      if X < 0.0 then
         Put("not calculable");
      else
         Put(Sqrt(X));
      end if;
      New_Line;
   end loop;
   New_Line;
   Put("Program finished");
   New_Line;
end Print_Roots;
```

The output has been enhanced by calling further procedures New_Line and Put in the package Simple_IO. A call of New_Line outputs the number of new lines specified by the parameter (of the predefined type Integer); the procedure New_Line has been written in such a way that if no parameter is supplied then a default value of 1 is assumed. There are also calls of Put with a string as argument. This is a different procedure from the one that prints the number X. The compiler distinguishes them by the different types of parameters. Having more than one subprogram with the same name is known as overloading. Note also the form of strings; this is a situation where the case of the letters does matter.

Various new control structures are also introduced. The statements between **loop** and **end loop** are repeated until the condition X = 0.0 in the **exit** statement is found to be true; when this is so the loop is finished and we immediately carry on after **end loop**. We also check that X is not negative; if it is we output the message 'not calculable' rather than attempting to call Sqrt. This is done by the if statement;

if the condition between **if** and **then** is true, then the statements between **then** and **else** are executed, otherwise those between **else** and **end if** are executed.

The general bracketing structure should be observed; **loop** is matched by **end loop** and **if** by **end if**. All the control structures of Ada have this closed form rather than the open form of Pascal and C which can lead to poorly structured and incorrect programs.

We will now consider in outline the possible general form of the function Sqrt and the package Simple_IO that we have been using.

The function Sqrt will have a structure similar to that of our main subprogram; the major difference will be the existence of parameters.

```
function Sqrt(F: Float) return Float is
   R: Float;
begin
   ...      -- compute value of Sqrt(F) in R
   return R;
end Sqrt;
```

We see here the description of the formal parameters (in this case only one) and the type of the result. The details of the calculation are represented by the comment. The return statement is the means by which the result of the function is indicated. Note the distinction between a function which returns a result and is called as part of an expression, and a procedure which does not have a result and is called as a single statement.

The package Simple_IO will be in two parts: the specification which describes its interface to the outside world, and the body which contains the details of how it is implemented. If it just contained the procedures that we have used, its specification might be

```
package Simple_IO is
   procedure Get(F: out Float);
   procedure Put(F: in Float);
   procedure Put(S: in String);
   procedure New_Line(N: in Integer := 1);
end Simple_IO;
```

The parameter of Get is an **out** parameter because the effect of a call such as Get(X); is to transmit a value out from the procedure to the actual parameter X. The other parameters are all **in** parameters because the value goes in to the procedures. The parameter mode can be omitted and is then taken to be **in** by default. There is a third mode **in out** which enables the value to go both ways. Both functions and procedures can have parameters of all modes. We usually give the mode **in** for procedures explicitly but omit it for functions.

Only a part of the procedures occurs in the package specification; this part is known as the procedure specification and just gives enough information to enable the procedures to be called.

We see also the two overloaded specifications of Put, one with a parameter of type Float and the other with a parameter of type String. Finally, note how the default value of 1 for the parameter of New_Line is indicated.

The package body for Simple_IO will contain the full procedure bodies plus any other supporting material needed for their implementation and is naturally hidden from the outside user. In vague outline it might look like

```
with Ada.Text_IO;
package body Simple_IO is

   procedure Get(F: out Float) is
      ...
   end Get;
   ...        -- other procedures similarly
end Simple_IO;
```

The with clause shows that the implementation of the procedures in Simple_IO uses the more general package Ada.Text_IO. The notation indicates that Text_IO is a *child* package of the package Ada. It should be noted how the full body of Get repeats the procedure specification which was given in the corresponding package specification. (The procedure specification is the bit up to but not including **is**.) Note that the package Ada.Text_IO really exists whereas Simple_IO is a figment of our imagination made up for the purpose of this example. We will say more about Ada.Text_IO in Chapter 4.

The example in this section has briefly revealed some of the overall structure and control statements of Ada. One purpose of this section has been to stress that the idea of packages is one of the most important concepts in Ada. A program should be conceived as a number of components which provide services to and receive services from each other.

Finally, note that there is a special package Standard which exists in every implementation and contains the declarations of all the predefined identifiers such as Float and Integer. Access to Standard is automatic and it does not have to be mentioned in a with clause. It is discussed in detail in Chapter 23.

Exercise 2.2

1 Suppose that the function Sqrt is not on its own but in a package Simple_Maths along with other mathematical functions Log, Ln, Exp, Sin, and Cos. By analogy with the specification of Simple_IO, write the specification of such a package. How would the program Print_Roots need to be changed?

2.3 The scalar type model

We have said that one of the key benefits of Ada is its strong typing. This is well illustrated by the enumeration type. Consider

```
declare
   type Colour is (Red, Amber, Green);
   type Fish is (Cod, Hake, Salmon);
   X, Y: Colour;
   A, B: Fish;
begin
```

```
    X := Red;          -- ok
    A := Hake;         -- ok
    B := X;            -- illegal
    ...
end;
```

This fragment of program is a block which declares two enumeration types Colour and Fish, and two variables of each type, and then performs various assignments. The declarations of the types give the allowed values of the types. Thus the variable X can only take one of the three values Red, Amber, or Green. The fundamental rule of strong typing is that we cannot assign a value of one type to a variable of a different type. So we cannot mix up colours and fishes and thus our (presumably accidental) attempt to assign the value of X to B is illegal and will be detected during compilation.

Four enumeration types are predefined in the package Standard. One is

```
type Boolean is (False, True);
```

which plays a key role in control flow. Thus the predefined relational operators such as < produce a result of this type and such a value follows **if** as in

```
if X < 0.0 then
```

The other predefined enumeration types are Character, Wide_Character, and Wide_Wide_Character whose values are the 8-bit ISO Latin-1, the 16-bit ISO Basic Multilingual Plane and the full 32-bit ISO 10646:2020 characters; these types naturally play an important role in input–output. Their literal values include the printable characters and these are represented by placing them in single quotes thus 'X' or 'a' or indeed '''.

The other fundamental types are the numeric types. One way or another, all other data types are built out of enumeration types and numeric types. The two major classes of numeric types are the integer types and floating point types (there are also fixed point types which are rather obscure and deserve no further mention in this overview). The integer types are further subdivided into signed integer types (such as Integer) and unsigned or modular types. All implementations have the types Integer and Float. An implementation may also have other predefined numeric types, Long_Integer, Long_Float, Short_Float and so on. There will also be specific integer types for an implementation depending upon the supported word lengths such as Integer_16 and unsigned types such as Unsigned_16.

One of the problems of numeric types is how to obtain both portability and efficiency in the face of variation in machine architecture. In order to explain how this is done in Ada it is convenient to introduce the concept of a derived type. (We will deal with derived types in more detail in the next chapter when we come to object oriented programming.)

The simplest form of derived type introduces a new type which is almost identical to an existing type except that it is logically distinct. If we write

```
type Light is new Colour;
```

then Light will, like Colour, be an enumeration type with literals Red, Amber, and Green. However, values of the two types cannot be arbitrarily mixed since they are logically distinct. Nevertheless, in recognition of the close relationship, a value of one type can be converted to the other by explicitly using the destination type name. So we can write

```
declare
   type Light is new Colour;
   C: Colour;
   L: Light;
begin
   L := Amber;                  -- the light amber, not the colour
   C := Colour(L);              -- explicit conversion
   ...
end;
```

whereas a direct assignment

```
   C := L;                      -- illegal
```

would violate the strong typing rule and this violation would be detected during compilation.

Returning now to the numeric types, if we write

```
type My_Float is new Float;
```

then My_Float will have all the operations (+, – etc.) of Float and in general can be considered as equivalent. Now suppose we transfer the program to a different computer on which the predefined type Float is not so accurate and that Long_Float is necessary. Assuming that the program has been written using My_Float rather than Float then replacing the declaration of My_Float by

```
type My_Float is new Long_Float;
```

is the only change necessary. We can actually do better than this by directly stating the precision that we require, thus

```
type My_Float is digits 7;
```

will cause My_Float to be based on the smallest predefined type with at least 7 decimal digits of accuracy.

A similar approach is possible with integer types so that rather than using the predefined types we can give the range of values required thus

```
type My_Integer is range –1000_000 .. +1000_000;
```

The point of all this is that it is not good practice to use the predefined numeric types directly when writing professional programs which may need to be portable. However, for simplicity, we will generally use the types Integer and Float in examples in most of this book. We will say no more about numeric types for the moment except that all the expected operations apply to all integer and floating point types.

2.4 Arrays and records

Ada naturally enables the creation of composite array and record types. Arrays may actually be declared without giving a name to the underlying type (the type is then said to be anonymous) but records always have a type name.

As an example of the use of arrays suppose we wish to compute the successive rows of Pascal's triangle. This is usually represented as shown in Figure 2.1. The reader will recall that the rows are the coefficients in the expansion of $(1 + x)^n$ and that a neat way of computing the values is to note that each one is the sum of the two diagonal neighbours in the row above.

Suppose that we are interested in the first ten rows. We can declare an array Pascal to hold such a row and a variable N to hold the row number by

```
Pascal: array (0 .. 10) of Integer;
N: Integer;
```

If the current values of the array Pascal correspond to row $n-1$, with the component Pascal(0) being 1, then the next row can be computed in a similar array Next by

```
Next(0) := 1;
for I in 1 .. N-1 loop
   Next(I) := Pascal(I-1) + Pascal(I);
end loop;
Next(N) := 1;
```

and then the array Next could be copied into the array Pascal.

This illustrates another form of loop statement where a controlled variable I takes successive values from a range; the variable is automatically declared to be of the type of the range which in this case is Integer. Note that the intermediate array Next could be avoided by iterating backwards over the array; we indicate this by writing **reverse** in front of the range thus

```
Pascal(N) := 1;
for I in reverse 1 .. N-1 loop
   Pascal(I) := Pascal(I-1) + Pascal(I);
end loop;
```

We can also declare arrays of several dimensions. So if we wanted to keep all the rows of the triangle we might declare

```
          1
        1   1
      1   2   1
    1   3   3   1
  1   4   6   4   1
1   5  10  10   5   1
```

Figure 2.1 Pascal's triangle.

```
Pascal2: array (0 .. 10, 0 .. 10) of Integer;
```

and then the loop for computing row *n* would be

```
Pascal2(N, 0) := 1;
for I in 1 .. N-1 loop
   Pascal2(N, I) := Pascal2(N-1, I-1) + Pascal2(N-1, I);
end loop;
Pascal2(N, N) := 1;
```

We have declared the arrays without giving a name to their type. We could alternatively have written

```
type Row is array (0 .. Size) of Integer;
Pascal, Next: Row;
```

where we have given the name Row to the type and then declared the two arrays Pascal and Next. There are advantages to this approach as we shall see later. Incidentally, the bounds of an array do not have to be a constant, they could be any computed values such as the value of some variable Size.

We conclude this brief discussion of arrays by noting that the type String which we encountered in Section 2.2 is in fact an array whose components are of the enumeration type Character. Its declaration (in the package Standard) is

```
type String is array (Positive range <>) of Character;
```

and this illustrates a form of type declaration which is said to be indefinite because it does not give the bounds of the array; these have to be supplied when an object is declared

```
A_Buffer: String(1 .. 80);
```

The identifier Positive in the declaration of the type String denotes what is known as a subtype of Integer; values of the subtype Positive are the positive integers and so the bounds of all arrays of type String must also be positive – the lower bound is of course typically 1 but need not be.

A record is an object comprising a number of named components typically of different types. We always have to give a name to a record type. If we were manipulating a number of buffers then it would be convenient to declare a record type containing the buffer and an indication of the start and finish of that part of the buffer actually containing useful data.

```
type Buffer is
   record
      Data: String(1 .. 80);
      Start, Finish: Integer;
   end record;
```

An individual buffer could then be declared by

```
My_Buffer: Buffer;
```

and the components of the buffer can then be manipulated using a dotted notation to select the individual components

```
My_Buffer.Start := 1;
My_Buffer.Finish := 3;
My_Buffer.Data(1 .. 3) := "XYZ";
```

Note that the last statement assigns values to the first three components of the array My_Buffer.Data using a so-called slice.

Whole array and record values can be created using aggregates which are simply a set of values in parentheses separated by commas.

Thus we could assign appropriate values to Pascal and to My_Buffer by

```
Pascal(0 .. 4) := (1, 4, 6, 4, 1);
My_Buffer := (Data => ('X', 'Y', 'Z', others => ' '), Start => 1, Finish => 3);
```

where in the latter case we have in fact assigned all 80 values to the array My_Buffer.Data and used **others** so that after the three useful characters the remainder of the array is padded with spaces. Note the nesting of parentheses and the optional use of named notation for the record components.

In Ada 2022 a record declaration can optionally repeat the name of the record type after **end record**, much as the name of a subprogram can follow **end**.

This concludes our brief discussion on simple arrays and records. In the next chapter we will show how record types can be extended.

Exercise 2.4

1. Write statements to copy the array Next into the array Pascal.
2. Write a nested loop to compute all the rows of Pascal's triangle in the two-dimensional array Pascal2.
3. Declare a type Month_Name and then declare a type Date with components giving the day, month, and year. Then declare a variable Today and assign Queen Victoria's date of birth to it (or your own).

2.5 Access types

The previous section showed how the scalar types (numeric and enumeration types) may be composed into arrays and records. The other vital means for creating structures is through the use of access types (the Ada name for pointer types); access types allow list processing and are typically used with record types.

The explicit manipulation of pointers or references has been an important feature of most languages since Algol 68. References rather dominated Algol 68 and caused problems and the corresponding pointer facility in Pascal is rather austere. The pointer facility in C on the other hand provides raw flexibility which is open to abuse and quite insecure and thus the cause of many wrong programs.

Ada provides both a high degree of reliability and flexibility through access types. A full description will be found in Chapter 11 but the following brief description will be useful for discussing polymorphism in the next chapter.

Ada access types must explicitly indicate the type of data to which they refer. The most general form of access types can refer to any data of the type concerned but we will restrict ourselves in this overview to applications which just refer to data declared in a storage pool (the Ada term for a heap).

For example suppose we wanted to declare various buffers of the type Buffer in the previous section. We might write

```
Handle: access Buffer;
...
Handle := new Buffer;
```

This allocates a buffer in the storage pool and sets a reference to it into the variable Handle. We can then refer to the various components of the buffer indirectly using the variable Handle

```
Handle.Start := 1;
Handle.Finish := 3;
```

and we can refer to the complete record as Handle.**all**. Note that Handle.Start is strictly an abbreviation for Handle.**all**.Start.

Access types are of particular value for list processing where one record structure contains an access value to another record structure. The classic example which we will encounter in many forms is typified by

```
type Cell is
   record
      Next: access Cell;
      Value: Data;
   end record;
```

The type Cell is a record containing a component Next which can refer to another similar record plus a component of some type Data. An example of the use of this sort of construction will be found in the next chapter.

Sometimes it is important to give a name to an access type. We can rewrite the above example so that the component Next is of a named type by first using an incomplete type thus

```
type Cell;                                 -- incomplete declaration
type Cell_Ptr is access Cell;

type Cell is                               -- the completion
   record
      Next: Cell_Ptr;
      Value: Data;
   end record;
```

Using this two stage approach and naming the access type is necessary if we are doing our own control of storage as described in Section 25.4.

Access types often refer to record types as in these examples but can refer to any type. Access types may also be used to refer to subprograms and this is particularly important for certain repetitive operations and when communicating with programs in other languages.

2.6 Errors and exceptions

We introduce this topic by considering what would have happened in the example in Section 2.2 if we had not tested for a negative value of X and consequently called Sqrt with a negative argument. Assuming that Sqrt has itself been written in an appropriate manner then it clearly cannot deliver a value to be used as the parameter of Put. Instead an exception will be raised. The raising of an exception indicates that something unusual has happened and the normal sequence of execution is broken. In our case the exception might be Constraint_Error which is a predefined exception declared in the package Standard. If we did nothing to cope with this possibility then our program would be terminated and no doubt the Ada Run Time System will give us a rude message saying that our program has failed and why. We can, however, look out for an exception and take remedial action if it occurs. In fact we could replace the conditional statement

```
if X < 0.0 then
   Put("not calculable");
else
   Put(Sqrt(X));
end if;
```

by the block

```
begin
   Put(Sqrt(X));
exception
   when Constraint_Error =>
      Put("not calculable");
end;
```

If an exception is raised by the sequence of statements between **begin** and **exception**, then control immediately passes to the one or more statements following the handler for that exception and these are obeyed instead. If there were no handler for the exception (it might be another exception such as Storage_Error) then control passes up the nested sequence of calls until we come to an appropriate handler or fall out of the main subprogram, which then becomes terminated as we mentioned with a message from the Run Time System.

The above example is not a good illustration of the use of exceptions since the event we are guarding against can easily be tested for directly. Nevertheless it does show the general idea of how we can look out for unexpected events, and leads us into a brief consideration of errors in general.

From the linguistic viewpoint, an Ada program may be incorrect for various reasons. There are four categories according to how they are detected.

- Many errors are detected by the compiler – these include simple punctuation mistakes such as leaving out a semicolon or attempting to violate the type rules such as mixing up colours and fishes. In these cases the program is said to be illegal and will not be executed.

- Other errors are detected when the program is executed. An attempt to find the square root of a negative number or divide by zero are examples of such errors. In these cases an exception is raised as we have just seen.
- There are also certain situations where the program breaks the language rules but there is no simple way in which this violation can be detected. For example a program should not use a variable before a value is assigned to it. In cases like this the behaviour is not predictable but will nevertheless lie within certain bounds. Such errors are called bounded errors.
- In more extreme situations there are some kinds of errors which can lead to quite unpredictable behaviour. In these (quite rare) cases we say that the behaviour is erroneous.

Finally, there are situations where, for implementation reasons, the language does not prescribe the order in which things are to be done. For example, the order in which the parameters of a procedure call are evaluated is not specified. If the behaviour of a program does depend on such an order then it is not considered to be incorrect but just not portable.

We mentioned earlier that an overall theme in the design of Ada was concern for correctness and that errors should be detected early in the programming process. As a simple example consider a fragment of program controlling the crossing gates on a railroad. First we have an enumeration type describing the state of a signal

```
type Signal is (Danger, Caution, Clear);
```

and then perhaps

```
if The_Signal = Clear then
   Open_Gates;
   Start_Train;
end if;
```

It is instructive to consider how this might be written in C and then to consider the consequences of various simple programming errors. Enumeration types in C are not strongly typed and essentially just provide names for integer (int) constants with the values 0, 1, 2 representing the three states. This has potential for errors because there is nothing in C that can prevent us from assigning a silly value such as 4 to a signal whereas it is not even possible to attempt such a thing when using the Ada enumeration type.

The corresponding text in C might be

```
if (the_signal == clear)
{
   open_gates();
   start_train();
}
```

It is interesting to consider what would happen in the two languages if we make various typographical errors. Suppose first that we accidentally type an extra semicolon at the end of the first line. The Ada program then fails to compile and the error is immediately drawn to our attention; the C program however still compiles and the condition is ignored (since it then controls no statements). The C program consequently always opens the gates and starts the train irrespective of the state of the signal!

Another possibility is that one of the = signs might be omitted in C. The equality then becomes an assignment and also returns the result as the argument for the test. The program still compiles, the signal is set clear, the condition is true (since clear is not 0) and so the gates are always opened and the train started on its perilous journey. The corresponding error in Ada might be to write := instead of = and then the program will not compile.

Of course, many errors cannot be detected at compile time. For example using My_Buffer and a variable Index of type Integer, we might write

```
Index := 81;
...
My_Buffer.Data(Index) := 'x';
```

which attempts to write to the 81st component of the array which does not exist. Such assignments are checked in Ada at run time and Constraint_Error would be raised so that the integrity of the program is not violated.

The corresponding instructions in C would undoubtedly overwrite an adjacent piece of storage and probably corrupt the value in My_Buffer.Start.

It often happens that variables such as Index can only sensibly have a certain range of values; this can be indicated by introducing a subtype

```
subtype Buffer_Index is Integer range 1 .. 80;
Index: Buffer_Index;
```

or by indicating the constraint directly

```
Index: Integer range 1 .. 80;
```

Applying a constraint to Index has the advantage that the attempt to assign 81 to it is itself checked and prevented so that the error is detected even earlier.

The reader may feel that such checks will make the program slower. This is not usually the case as we shall see in Chapter 15.

2.7 Terminology

We conclude this first introductory chapter with a few remarks on terminology. Every subject has its own terminology or jargon and Ada is no exception. (Indeed in Ada an exception is a kind of error as we have seen!) A useful glossary of terms will be found in Appendix 2.

Terminology will generally be introduced as required but before starting off with the detailed description of Ada it is convenient to mention a few concepts which will occur from time to time.

The term *static* refers to things that can be determined at compilation, whereas *dynamic* refers to things determined during execution. Thus a static expression is one whose value can be determined by the compiler such as

```
2 + 3
```

and a statically constrained array is one whose bounds are known at compilation.

The term *real* comes up from time to time in the context of numeric types. The floating point types and fixed point types are collectively known as real types. (Fixed point types are rather specialized and not discussed until Chapter 17.) Literals such as 2.5 are known as real literals since they can be used to denote values of both floating and fixed point types. Other uses of the term real will occur in due course.

The terminology used with exceptions in Ada is that an exception is *raised* and then *handled*. Some languages say that an exception is thrown and then caught.

We talk about statements being *executed* and expressions being *evaluated*. Moreover, declarations can also require processing and this is called being *elaborated*.

Object oriented programming has its own rather specialized terminology and a section is devoted to this in the next chapter. One particular term that seems to be overused is *interface*. We use interface in a very general sense to mean the description of a means of communication. As we noted in Section 2.2, the specification of a package describes its interface to the outside world. But interface also has a highly technical meaning in OOP as we shall see in the next chapter. As usual in any language, the context should clarify the intended meaning.

3 Abstraction

3.1 Packages and private types
3.2 Objects and inheritance
3.3 Classes and polymorphism
3.4 Genericity
3.5 Object oriented terminology
3.6 Tasking

As mentioned in Chapter 1, abstraction in various forms seems to be the key to the development of programming languages. In this chapter we survey various aspects of abstraction with particular emphasis on the object oriented paradigm.

3.1 Packages and private types

In the previous chapter we declared a type for the manipulation of a buffer as follows

```
type Buffer is
   record
      Data: String(1 .. 80);
      Start: Integer;
      Finish: Integer;
   end record;
```

in which the component Data actually holds the characters in the buffer and Start and Finish index the ends of the part of the buffer containing useful information. We also saw how the various components might be updated and read using normal assignment.

However, such direct assignment is often unwise since the user could inadvertently set inconsistent values into the components or read nonsense components of the array.

A much better approach is to create an Abstract Data Type (ADT) so that the user cannot see the internal details of the type but can only access it through various subprogram calls which define an appropriate protocol.

This can be done using a package containing a private type. Let us suppose that the protocol allows us to reload the buffer (possibly not completely full) and to read one character at a time. Consider the following

```
package Buffer_System is                    -- visible part

   type Buffer is private;
   procedure Load(B: out Buffer; S: in String);
   procedure Get(B: in out Buffer; C: out Character);

private                                     -- private part
   Max: constant Integer := 80;
   type Buffer is
      record
         Data: String(1 .. Max);
         Start: Integer := 1;
         Finish: Integer := 0;
      end record;
end Buffer_System;

package body Buffer_System is               -- package body
   procedure Load(B: out Buffer; S: in String) is
   begin
      B.Start := 1;
      B.Finish := S'Length;
      B.Data(B.Start .. B.Finish) := S;
   end Load;

   procedure Get(B: in out Buffer; C: out Character) is
   begin
      C := B.Data(B.Start);
      B.Start := B.Start + 1;
   end Get;
end Buffer_System;
```

Note how the package comes in two parts, the specification and the body. Basically, the specification describes the interface to other parts of the program and the body gives implementation details.

With this formulation the client can only access the information in the visible part of the specification which is the bit before the reserved word **private**. In this visible part the declaration of the type Buffer merely says that it is private and the full declaration then occurs in the private part. There are thus two views of the type Buffer; the external client just sees the partial view whereas within the package the code of the server subprograms can see the full view. The specifications of the server subprograms are naturally also declared in the visible part and the full bodies which give their implementation details are in the package body.

The net effect is that the user can declare and manipulate a buffer by simply writing

```
My_Buffer: Buffer;
...
```

```
Load(My_Buffer, Some_String);
...
Get(My_Buffer, A_Character);
```

but the internal structure is quite hidden. There are two advantages: one is that the user cannot inadvertently misuse the buffer and the second is that the internal structure of the private type could be rearranged if necessary and provided that the protocol is maintained the user program would not need to be changed.

This hiding of information and consequent separation of concerns is very important and illustrates the benefit of data abstraction. The design of appropriate interface protocols is the key to the development and subsequent maintenance of large programs.

The astute reader will note that we have not bothered to ensure that the buffer is not loaded when there is still unread data in it or the string is too long to fit, nor read from when it is empty. We could rectify this by declaring our own exception called perhaps Buffer_Error in the visible part of the specification thus

```
Buffer_Error: exception;
```

and then check within Load by for example

```
if S'Length > Max or B.Start <= B.Finish then
   raise Buffer_Error;
end if;
```

This causes our own exception to be raised if an attempt is made to overwrite existing data or if the string is too long.

As a minor point note the use of the constant Max so that the literal 80 only appears in one place. Note also the attribute Length which applies to any array and gives the number of its components. The upper and lower bounds of an array S are incidentally given by S'First and S'Last.

Another point is that the parameter B of Get is marked as **in out** because the procedure both reads the initial value of B and updates it.

Finally, note that the components Start and Finish of the record have initial values in the declaration of the record type; these ensure that when a buffer is declared these components are assigned sensible values and thereby indicate that the buffer is empty. An alternative would of course be to provide a procedure Reset but the user might forget to call it.

Exercise 3.1

1 Extend the package Buffer_System to include the exception Buffer_Error in its specification and appropriate checks in the subprogram bodies. Also add a function Is_Empty visible to the user.

3.2 Objects and inheritance

The term object oriented programming is well established. A precise definition is hard and is made worse by a lack of agreement on appropriate terminology. Ada uses carefully considered terminology which is rather different from that in some languages such as C++ but avoids the ambiguities which can arise from the confusing overuse of terms such as class.

As its name suggests, object oriented programming concerns the idea of programming around objects. A good example of an object in this sense is the variable My_Buffer of the type Buffer in the previous section. We conceive of the object such as a buffer as a coordinated whole and not just as the sum of its components (corresponding to a holistic rather than a reductionist view using terminology from Quantum Mechanics). Indeed the external user cannot see the components at all but can only manipulate the buffer through the various subprograms associated with the type.

Certain operations upon a type are called the primitive operations of the type. In the case of the type Buffer they are the subprograms declared in the package specification along with the type itself and which have parameters or a result of the type. In the case of a type such as Integer, the primitive operations are those such as + and - which are predefined for the type (and indeed they are declared in Standard along with the type Integer itself and so fit the same model). Such operations are called methods in some languages.

Other important ideas in OOP are

- the ability to define one type in terms of another and especially as an extension of another; this is type extension,
- the ability for such a derived type to inherit the primitive operations of its parent and also to add to and replace such operations; this is inheritance,
- the ability to distinguish the specific type of an object at run time from among several related types and in particular to select an operation according to the specific type; this is (dynamic) polymorphism.

In the previous chapter we showed how the type Light was derived from Colour and also showed how numeric portability could be aided by deriving a numeric type such as My_Float from one of the predefined types. These were very simple forms of inheritance; the new types inherited the primitive operations of the parent; however, the types were not extended in any way and underneath were really the same type. The main benefit of such derivation is simply to provide a different name and thereby to distinguish the different uses of the same underlying type in order to prevent us from inadvertently using a Light when we meant to use a Colour.

The more general case is where we wish to extend a type in some way and also distinguish objects of different types at run time. The most natural form of type for the purposes of extension is of course a record where we can consider extension as simply the addition of further components. The other point is that if we need to distinguish the type at run time then the object must contain an indication of its type. This is provided by a hidden component called the tag. Type extension in Ada is thus naturally carried out using tagged record types.

As a simple example suppose we wish to manipulate various kinds of geometrical objects. We can imagine that the kinds of objects form a hierarchy as shown in Figure 3.1. Thus we have points, polygons, circles, and shapes; and polygons are further subdivided into quadrilaterals and pentagons.

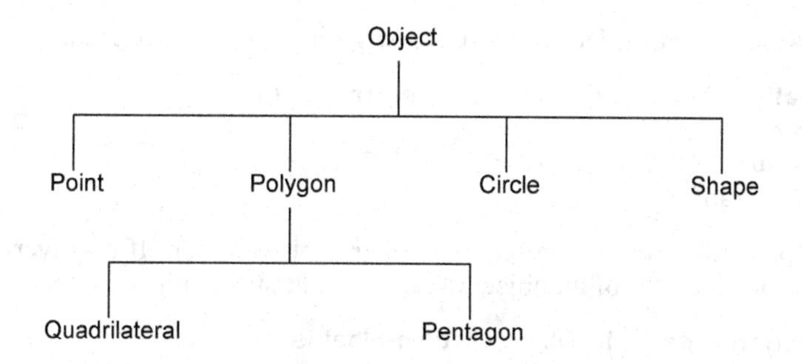

Figure 3.1 A hierarchy of geometrical objects.

All objects will have a position given by their *x*- and *y*-coordinates. So we declare the root of the hierarchy as

```
type Object is tagged
  record
    X_Coord: Float;
    Y_Coord: Float;
  end record;
```

Note carefully the introduction of the reserved word **tagged**. This indicates that values of the type carry a tag at run time and that the type can be extended. The other types of geometrical objects will be derived (directly or indirectly) from this type. For example we could have

```
type Circle is new Object with
  record
    Radius: Float;
  end record;
```

and the type Circle then has the three components X_Coord, Y_Coord, and Radius. It inherits the two coordinates from the type Object and the additional component Radius is added explicitly.

Sometimes it is convenient to derive a new type without adding any further components. For example

```
type Point is new Object with null record;
```

In this last case we have derived Point from Object but not added any new components. However, since we are dealing with tagged types we explicitly add **with null record**; to indicate that we did not want any new components. This means that it is clear from a declaration whether a type is tagged or not since it will have either **tagged** or **with** (or **interface** as we shall see later) if it is tagged.

The primitive operations of a type are those declared in the same package specification as the type and that have parameters or result of the type. On derivation these operations are inherited by the new type. They can be overridden by new versions and new operations can be added and these then become primitive operations of the new type and are themselves inherited by any further derived type.

Thus we might have declared a function giving the distance from the origin

```
function Distance(O: in Object) return Float is
begin
   return Sqrt(O.X_Coord**2 + O.Y_Coord**2);
end Distance;
```

The type Circle would then sensibly inherit this function. If however, we were concerned with the area of an object then we might start with

```
function Area(O: in Object) return Float is
begin
   return 0.0;
end Area;
```

which returns zero since a raw object has no area. The abstract concept of an area applies also to a circle and so it is appropriate that a function Area be defined for the type Circle. However, to inherit the function from the type Object is clearly inappropriate and so we explicitly declare

```
function Area(C: in Circle) return Float is
begin
   return π * C.Radius**2;
end Area;
```

which then overrides the inherited operation. (We assume that π is an appropriate constant such as that declared in Ada.Numerics; see Section 4.3.)

We can summarize these ideas by saying that the specification is always inherited whereas the implementation may be inherited but can be replaced.

It is important to remember that new primitive operations must be declared in the same package specification as the type itself. A derived type might be declared in the same package as its parent or it might be in a different package. So in the former case the overall structure might be

```
package Geometry is

   type Object is tagged ... ;

   function Distance(O: in Object) return Float;
   function Area(O: in Object) return Float;

   type Circle is new Object with ... ;

   function Area(C: in Circle) return Float;

   type Point is new Object with null record;

end Geometry;

package body Geometry is
   ...      -- bodies of function Distance and two functions Area
end Geometry;
```

Another approach is to declare each type in a distinct library package. Child packages with names of the form Geometry.Circle have a number of advantages such as minimizing the need for with clauses as we shall see in Program 1. Mixed strategies are also possible as illustrated in the exercise at the end of this section.

It is possible to convert a value from the type Circle to Object and vice versa. From circle to object is straightforward, we simply write

```
O: Object := (1.0, 0.5);
C: Circle := (0.0, 0.0, 34.7);
...
O := Object(C);
```

which effectively ignores the radius component. However, conversion in the other direction requires the provision of a value for the extra component and this is done by an extension aggregate thus

```
C := (O with Radius => 41.2);
```

where the expression O is extended after **with** by values for the extra components (in this case only the radius) written just as in a normal aggregate.

We conclude by noting that a private type can also be marked as tagged

```
type Shape is tagged private;
```

and the full type declaration must then (ultimately) be a tagged record

```
type Shape is tagged
   record ...
```

or derived from some tagged record type. On the other hand we might wish to make visible the fact that the type Shape is derived from Object and yet keep the additional components hidden. In this case we might write

```
with Geometry;
package Hidden_Shape is
   type Shape is new Geometry.Object with private;  -- client view
   ...
private
   type Shape is new Geometry.Object with            -- server view
      record
         ...   -- the private components
      end record;
end Hidden_Shape;
```

Note that it is not necessary for the full declaration of Shape to be derived directly from the type Object. It might be derived from Circle; all that matters is that Shape is ultimately derived from Object.

Exercise 3.2

1 Declare the type Object and the functions Distance and Area in a package Objects. Then declare a package Shapes containing the types Circle and Point and also a type Triangle with sides A, B, and C and appropriate functions returning the area. (Assume that Sqrt and π are somehow directly visible.)

3.3 Classes and polymorphism

In the previous section we showed how to declare a hierarchy of types derived from the type Object. We saw that on derivation further components and operations could be added and that operations could be replaced.

However, it is very important to note that an operation cannot be taken away nor can a component be removed. As a consequence we are guaranteed that all the types derived from a common ancestor will have all the components and operations of that ancestor.

So in the case of the type Object, all types in the hierarchy derived from Object will have the common components such as their coordinates X_Coord and Y_Coord and the common operations such as Distance and Area. Since they have these common properties it is natural that we should be able to manipulate a value of any type in the hierarchy without knowing exactly which type it is provided that we only use the common properties. Such general manipulation is done through the concept of a class.

Ada carefully distinguishes between an individual type such as Object on the one hand and the set of types such as Object plus all its derivatives on the other hand. A set of such types is known as a class. Associated with each class is a type called the class wide type which for the set rooted at Object is denoted by Object'Class. The type Object is referred to as a specific type when we need to distinguish it from a class wide type.

We can of course have subclasses, for example Polygon'Class represents the set of all types derived from and including Polygon. This is a subset of the class Object'Class. All the properties of Object'Class will also apply to Polygon'Class but not vice versa. For example, although we have not shown it, the type Polygon will presumably contain a component giving the length of the sides. Such a component will belong to all types of the class Polygon'Class but not to Object'Class.

As a simple example of the use of a class wide type consider the following function

```
function Moment(OC: Object'Class) return Float is
begin
   return OC.X_Coord * Area(OC);
end Moment;
```

Those who recall their school mechanics will remember that the moment of a force about a fulcrum is the product of the force by the distance from the fulcrum. So in our example, taking the *x*-axis as horizontal and the *y*-axis as vertical, the moment of the force of gravity on an object about the origin is the *x*-coordinate multiplied by its weight which we can take as being proportional to its area (and for simplicity we have taken to be just its area).

This function has a formal parameter of the class wide type Object'Class. This means it can be called with an actual parameter whose type is any specific type in the class comprising the set of all types derived from Object. Thus we could write

```
C: Circle ...
M: Float;
...
M := Moment(C);
```

Within the function Moment we can naturally refer to the specific object as OC. Since we know that the object must be of a specific type in the Object class, we are guaranteed that it will have a component OC.X_Coord. Similarly, we are guaranteed that the function Area will exist for the type of the object since it is a primitive operation of the type Object and will have been inherited by (and possibly overridden for) every type derived from Object. So the appropriate function Area is called and the result multiplied by the *x*-coordinate and returned as the result of the function Moment.

Note carefully that the particular function Area to be called is not known until the program executes. The choice depends upon the specific type of the parameter and this is determined by the tag of the object passed as the actual parameter; remember that the tag is a sort of hidden component of the tagged type. This selection of the particular subprogram according to the tag is known as dispatching and is a vital aspect of the dynamic behaviour provided by polymorphism.

Dispatching only occurs when the actual parameter is of a class wide type; if we call Area with an object of a specific type such as C of type Circle then the choice is made at compile time. Dispatching is often called late binding because the call is only bound to the called subprogram late in the compile–link–execute process. The binding to a call of Area with the parameter C of type Circle is called static binding because the subprogram to be called is determined at compile time.

Observe that the function Moment is not a primitive operation of any type; it is just an operation of Object'Class and it happens that a value of any specific type derived from Object is implicitly converted to the class wide type when passed as a parameter. Class wide types do not have primitive operations and so no inheritance is involved.

It is interesting to consider what would have happened if we had written

```
function Moment(O: Object) return Float is
begin
   return O.X_Coord * Area(O);
end Moment;
```

where the formal parameter is of the specific type Object. This always returns zero because the function Area for an Object always returns zero. If this function Moment were declared in the same package specification as Object then it would be a primitive operation of Object and thus inherited by the type Circle. However, the internal call would still be to the function Area for the type Object and not to the type Circle and so the answer would still be zero. This is because the binding is static and inheritance simply passes on the same code. The code mechanically works on a Circle because it only uses the Object part of the circle (we say that it sees the Object view of the Circle); but unfortunately it is not what we want. We could of course override the inherited operation by writing

```
function Moment(C: Circle) return Float is
begin
   return C.X_Coord * Area(C);
end Moment;
```

but this is both tedious and causes unnecessary duplication of similar code. The proper approach for such general situations is to use the original class wide version with its internal dispatching; this can be shared by all types without duplication and always calls the appropriate function Area.

A major advantage of using a class wide operation such as Moment is that a system using it can be written, compiled, and tested without knowing all the specific types to which it is to be applied. Moreover, we can then add further types to the system without recompilation of the existing tested system.

For example we could add a further type

```
type Pentagon is new Object with ...
function Area(P: Pentagon) return Float;
...
Star: Pentagon := ...
...
Put("Moment of star is ");
Put(Moment(Star));
```

and then the old existing tried and tested Moment will call the new Area for the Pentagon without being recompiled. (It will of course have to be relinked.)

This works because of the mechanism used for dynamic binding; the essence of the idea is that the class wide code has dynamic links into the new code and this is accessed via the tag of the type. This creates a very flexible and extensible interface ideal for building up a system from reusable components. Details of how this apparent magic might be implemented will be outlined when we discuss OOP in detail in Chapter 14.

One difficulty with the flexibility provided by class wide types is that we cannot know how much space might be occupied by an arbitrary object of the type because the type might be extended. So although we can declare an object of a class wide type it has to be initialized and thereafter that object can only be of the specific type of that initial value. Note that a formal parameter of a class wide type such as in Moment is allowed because it is similarly initialized by the actual parameter.

Another similar restriction is that we cannot have an array of class wide components (even if initialized) because the components might be of different specific types and so probably of different sizes and impossible to index efficiently.

One consequence of these necessary restrictions is that it is very natural to use access types with object oriented programming since there is no problem with pointing to objects of different sizes at different times.

Thus suppose we wanted to manipulate a series of geometrical objects; it is very natural to declare these in free storage as required. They can then be chained together on a list for processing. Consider

```
type Pointer is access Object'Class;

type Cell is
   record
      Next: access Cell;
      Element: Pointer;
   end record;
...
type List is access Cell;
A_List: List;
```

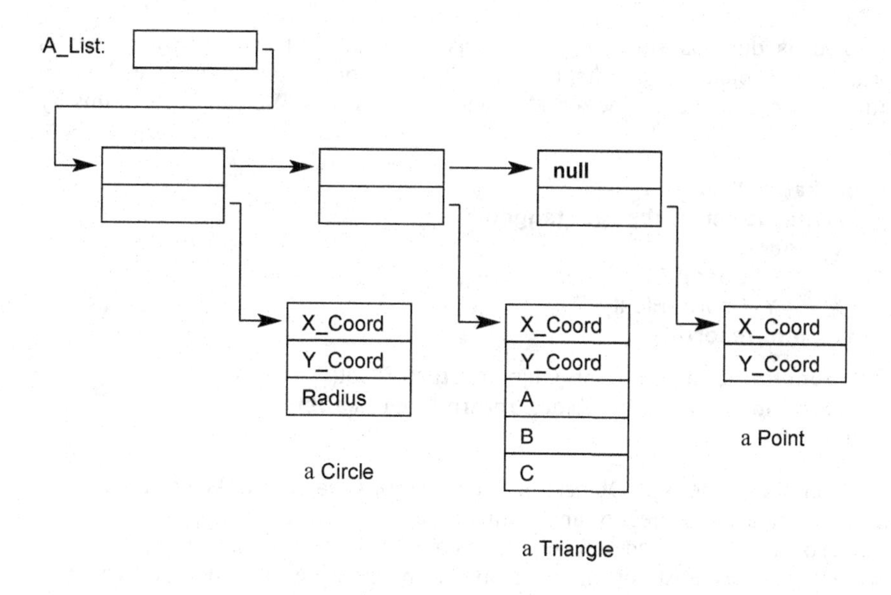

Figure 3.2 A chain of objects.

which enables us to create cells which can be linked together; each cell has an element which is a pointer to any geometrical object. We have also declared a named access type List and a variable of that type. Several objects can then be created and linked together to form a list as in Figure 3.2.

We can now easily process the objects on the list and might for example compute the total moment of the set of objects by calling the following function

```
function Total_Moment(The_List: List) return Float is
   Local: access Cell := The_List;
   Result: Float := 0.0;
begin
   loop
      if Local = null then                    -- end of list
         return Result;
      end if;
      Result := Result + Moment(Local.Element.all);
      Local := Local.Next;
   end loop;
end Total_Moment;
```

We conclude this brief survey of the OOP facilities in Ada by considering abstract types. It is sometimes the case that we would like to declare a type as the foundation for a class of types with certain common properties but without allowing objects of the original type to be declared. For example we probably would not want to declare an object of the raw type Object. If we wanted an object without any area then it would be appropriate to declare a Point. Moreover, the function Area for the type Object is dubious since it usually has to be overridden anyway. But it is important to be able to ensure that all types derived from Object do have an Area so that the dispatching in the function Moment always works. We can achieve this by writing

```
package Objects is
   type Object is abstract tagged
      record
         X_Coord: Float;
         Y_Coord: Float;
      end record;

   function Distance(O: in Object) return Float;
   function Area(O: in Object) return Float is abstract;
end Objects;
```

In this formulation the type Object and the function Area are marked as abstract. It is illegal to declare an object of an abstract type and an abstract subprogram has no body and so cannot be called. On deriving a concrete (that is, nonabstract) type from an abstract type any abstract inherited operations must be overridden by concrete operations. Note that we have declared the function Distance as not abstract; this is largely because we know that it will be appropriate anyway.

This approach has a number of advantages; we cannot declare a raw Object by mistake, we cannot inherit the silly body of Area and we cannot make the mistake of declaring the function Moment for the specific type Object (why?).

But despite the type being abstract we can declare the function Moment for the class wide type Object'Class. This always works because of the rule that we cannot declare an object of an abstract type; any actual object passed as a parameter must be of a concrete type and will have an appropriate function Area to which it can dispatch.

Note that it would be very sensible for the function Distance also to take a class wide parameter. This is because Distance is inherently the same for all types in the class and so cannot need to be overridden. So we would write

```
function Distance(OC: in Object'Class) return Float is
begin
   return Sqrt(OC.X_Coord**2 + OC.Y_Coord**2);
end Distance;
```

If an abstract type has no concrete operations at all and has no components then it can be declared as an interface. Interfaces are important for multiple inheritance where a type inherits properties from more than one ancestor. Thus if we decided that the type Object should not have any components (so that they would have to be added by each type such as Circle) then we could write

```
package Objects is
   type Object is interface;

   function Distance(OC: in Object'Class) return Float;
   function Area(O: in Object) return Float is abstract;
   function X_Coord(O: in Object) return Float is abstract;
```

```
    function Y_Coord(O: in Object) return Float is abstract;
end Objects;
```

We have also added functions to return the values of the coordinates since these are needed by the class wide function Distance.

The normal way of calling a subprogram in classical languages is to give its name and then a list of parameters. However, in OOP it is often better to consider the object as dominant and to give that first followed by the subprogram name and any other parameters. This is also possible in Ada for operations of tagged types provided always that the first parameter denotes the object. Thus we can write either of

```
X := Area(C);          -- classical notation
X := C.Area;           -- prefixed notation
```

There are a number of advantages of the prefixed notation which will be explained in detail in Chapter 14. An immediately obvious one is that it unifies access to a component and to a corresponding function. Thus we could write

```
function Moment(OC: in Object'Class) return Float is
begin
   return OC.X_Coord * OC.Area;
end Moment;
```

and this treats the function Area and the component X_Coord in the same manner. Moreover, if we later decided to use the interface style where the component value is returned by a function X_Coord then the text of the function Moment remains the same. And of course the function Distance remains the same as well.

Exercise 3.3

1 Declare a procedure Add_To_List which takes a list of type List and (an access to) any object of the type Object'Class and adds it to the list.

2 Write the body of the package Objects for the version with the abstract type.

3 How would the package Shapes of Exercise 3.2(1) need to be modified using the package Objects with the abstract type?

4 Why could we not declare the function Moment for the abstract type Object?

5 The moment of inertia of an object about the origin M_O is equal to its moment of inertia about its centre of gravity M_I plus MR^2 where M is its mass and R its distance from the origin. Assuming that we have functions MI for each specific type declare a class wide function MO returning the moment of inertia about the origin. Assume the mass is simply equal to the area.

3.4 Genericity

We have seen how class wide types provide us with dynamic polymorphism. This means that we can manipulate several types with a single construction and that the specific type is determined dynamically, that is, at run time. In this section we introduce the complementary concept of static polymorphism where again we have a single construction for the manipulation of several types but in this case the choice of type is made statically at compile time.

At the beginning of Section 2.2 we said that an important objective of software engineering is to reuse existing software components. However, the strong typing model of Ada (even with class wide types) sometimes gets in the way unless we have a method of writing software components which can be used for various different types. For example, the program to do a sort is largely independent of what it is sorting – all it needs is a rule for comparing the values to be sorted.

So we need a means of writing pieces of software that can be parameterized as required for different types. In Ada this is done by the generic mechanism. We can make a package or subprogram generic with respect to one or more parameters which can include types. Such a generic unit provides a template from which we can create genuine packages and subprograms by so-called instantiation. The full details of Ada generics are quite extensive and will be dealt with in Chapter 18. However, in the next chapter we will be discussing input–output and other aspects of the predefined library which make significant use of the generic mechanism and so a brief introduction is appropriate.

The standard package for the input and output of floating point values in text form is generic with respect to the actual floating type. This is because we want a single package to cope with all the possible floating types such as the underlying machine types Float and Long_Float as well as the portable type My_Float. Its specification is

```
generic
  type Num is digits <>;
package Float_IO is
  ...
  procedure Get(Item: out Num; ... );
  procedure Put(Item: in Num; ... );
  ...
end Float_IO;
```

where we have omitted various details relating to the format. The one generic parameter is Num and the notation **digits** <> indicates that it must be a floating point type and echoes the declaration of My_Float using **digits** 7 that we briefly mentioned in Section 2.3.

In order to create an actual package to manipulate values of the type My_Float, we write an instantiation thus

```
package My_Float_IO is new Float_IO(My_Float);
```

This creates a package with the name My_Float_IO where the formal type Num has been replaced throughout with our actual type My_Float. As a consequence, procedures Get and Put taking parameters of the type My_Float are created and we can then call these as required. But we are straying into the next chapter.

The kind of parameterization provided by genericity is similar but rather different from that provided through class wide types. In both cases the parameterization is over a related set of types with a set of common properties. Such a related set of types is termed a category of types.

A common form of category is a derivation class where all the types are derived from a common ancestor such as the type Object. Tagged derivation classes form the basis of dynamic polymorphism as we have seen.

But there are also broader forms of categories such as the set of all floating point types which are allowed as actual parameters for the generic package Float_IO. The parameters of generic units can use these broader categories. For example, a very broad category is the set of all types having assignment. Such a category would be a suitable basis for writing a generic sort routine.

In due course we shall see that the two forms of polymorphism work together; a common form of generic package is one which takes as a parameter a type from a tagged derivation class.

3.5 Object oriented terminology

It is perhaps convenient at this point to compare the Ada terminology with that used by other object oriented languages such as Smalltalk, C++ and Java. Those not familiar with such languages could skip this section before they get confused.

At least one term in common is inheritance. Ada 83 had inheritance although not type extension. Ada 95 introduced type extension and dynamic polymorphism with tagged types. Untagged record types are called structs in some languages.

Ada actually uses the term *object* to denote variables and constants in general whereas in the OO sense an object is an instance of an Abstract Data Type (ADT).

Many languages use *class* to denote what Ada calls a specific tagged type (or more strictly an ADT consisting of a tagged type plus its primitive operations). Ada (like the functional language Haskell) uses the word class to refer to a group of related types and not to a single type. The Ada approach clarifies the distinction between the group of types and a single type and strengthens that clarification by introducing class wide types as such. Many languages use the term class for both specific types and the group of types with much resulting confusion both in terms of description but also in understanding the behaviour of the program and keeping track of the real nature of an object. (Logically the distinction is between a set and the members of the set; we confuse these in casual human language but it's best to avoid the confusion if possible and Ada does just that.)

Primitive operations of Ada tagged types are often called methods or virtual functions. The call of a primitive operation in Ada is bound statically or dynamically according to whether the actual parameter is of a specific type or a class wide type. The rules in C++ are more complex and depend upon whether the parameter is a pointer and also whether the call is prefixed by its class name and whether it is virtual or not. In Ada, an operation with class wide formal parameters is always bound statically although it applies to all types of the class.

Dispatching is the Ada term for calling a primitive operation with dynamic binding and this provides dynamic polymorphism. Subprogram calls via access to subprogram types are also a form of dynamic binding.

Abstract types correspond to abstract classes in many languages. Abstract subprograms are pure virtual member functions in C++.

Ancestor or parent type and descendant or derived type become superclass and subclass. The Ada concept of subtype has no correspondence in languages which do not have range checks and has no relationship to subclass. An Ada subtype can never have more values than its base type, whereas a descendant type (subclass in other languages) can never have fewer values than its parent type.

Generic units in Ada provide static polymorphism and are similar to templates in C++ and generics in C# and Java. In Ada, the contract model is stronger than in C++ so that more checking is carried out on the generic template whereas C++ leaves many checks to the instance which produces less helpful diagnostics.

Ada includes interfaces which are very similar to interfaces in Java, C#, and CORBA and avoid the complexities of multiple inheritance in C++.

Encapsulation is important. We can distinguish module encapsulation (for information hiding) and object encapsulation (for controlled access). Ada packages provide module encapsulation whereas tasks and protected objects provide object encapsulation. C++ and Java really only provide module encapsulation mainly with classes and (more weakly) with their namespaces and packages. In Ada the package has other purposes unrelated to type extension and inheritance and the private type. Note that the effect of private and protected operations in C++ is provided in Ada by a combination of private types and child packages; the latter are kinds of friends.

3.6 Tasking

No survey of abstraction in Ada would be complete without a brief mention of tasking. It is often necessary to write a program as a set of parallel activities rather than just as one sequential program.

Most programming languages do not address this issue at all. Some argue that the underlying operating system provides the necessary mechanisms and that they are therefore unnecessary in a programming language. Such arguments do not stand up to careful examination for three main reasons

- Built-in syntactic constructions provide a degree of reliability which cannot be obtained through a series of individual operating system calls.
- Portability between systems is difficult if system calls are used directly.
- General purpose operating systems do not provide the degree of control and timing required by many applications.

An Ada program can be written as a series of interacting tasks. There are two main ways in which tasks can communicate: directly by sending messages to each other and indirectly by accessing shared data.

Direct communication between Ada tasks is achieved by one task calling an entry in another task. The calling (client) task waits while the called (server) task executes an accept statement in response to the call; the two tasks are closely coupled during this interaction which is called a rendezvous.

Controlled access to shared data is vital in tasking applications if interference is to be avoided. For example, returning to the character buffer example in Section 3.1, it would be a disaster if a task started to read the buffer while another task was updating it with further information since the component B.Start could be changed and a component of the Data array read by Get before the buffer had been correctly updated by Load. Ada prevents such interference by a construction known as a protected object.

The syntactic form of both tasks and protected objects is similar to that of a package. They have a specification part describing the interface presented to the client, a private part containing hidden details of the interface, and a body stating what they actually do.

The general client–server model can thus be expressed as

```
task Server is
   entry Some_Service(Formal: in out Data);
end;

task body Server is
begin
   ...
   accept Some_Service(Formal: in out Data) do
      ...      -- statements providing the service
   end;
   ...
end Server;

task Client;

task body Client is
   Actual: Data;
begin
   ...
   Server.Some_Service(Actual);
   ...
end Client;
```

A good example of the form of a protected object is given by the buffer example which could be rewritten as follows

```
protected type Buffer(Max: Integer) is           -- visible part
   procedure Load(S: in String);
   procedure Get(C: out Character);
private                                          -- private part
   Data: String(1 .. Max);
   Start: Integer := 1;
   Finish: Integer := 0;
end Buffer;

protected body Buffer is                         -- body
   procedure Load(S: in String) is
   begin
      Start := 1;
      Finish := S'Length;
      Data(Start .. Finish) := S;
   end Load;
```

```
   procedure Get(C: out Character) is
   begin
      C := Data(Start);
      Start := Start + 1;
   end Get;
end Buffer;
```

This construction uses a slightly different style from the package. It is a type in its own right whereas the package exported the type. Calls of the procedures use the dotted form of notation where the name of the particular protected object acts as a prefix to the call rather than being passed as a parameter. (This is similar to the prefixed notation for calling operations of tagged types mentioned in Section 3.3.) Another point is that within the bodies of the procedures, the references to the private data are naturally taken to refer to the current instance.

Note also that the type is parameterized by the discriminant Max. This enables us to give the actual size of a particular buffer when it is declared thus

```
B: Buffer(80);
```

Statements using the protected object B might then look like

```
B.Load(Some_String);
...
B.Get(A_Character);
```

The rules regarding protected objects ensure that several tasks cannot be executing subprograms Load and Get simultaneously (this is implemented by the use of so-called locks which are not accessible to the programmer). This formulation therefore prevents disastrous interference between several clients but nevertheless it does not prevent a call of Load from overwriting unread data. As before we could insert tests and raise an exception. But the proper approach is to cause the tasks to wait if circumstances are not appropriate. This can be done through the use of entries and barriers. The protected type might then be

```
protected type Buffer(Max: Integer) is
   entry Load(S: in String);
   entry Get(C: out Character);
private
   Data: String(1 .. Max);
   Start: Integer := 1;
   Finish: Integer := 0;
end Buffer;

protected body Buffer is
   entry Load(S: in String) when Start > Finish is
   begin
      Start := 1;
      Finish := S'Length;
      Data(Start .. Finish) := S;
   end Load;
```

```
      entry Get(C: out Character) when Start <= Finish is
      begin
         C := Data(Start);
         Start := Start + 1;
      end Get;
   end Buffer;
```

In this formulation, the procedures are replaced by entries and each entry body has a barrier condition. A task calling an entry is queued if the barrier is false and only allowed to proceed when the barrier becomes true. This construction is very efficient because task switching is minimized.

Tasking is discussed in detail in Chapters 20 and 22.

4 Programs and Libraries

4.1 The hierarchical library
4.2 Input–output
4.3 Numeric library
4.4 Running a program

In this final introductory chapter we consider the important topic of putting together a complete program. Such a program will inevitably use predefined material such as that for input and output and we therefore also briefly survey the structure and contents of the extensive predefined library.

4.1 The hierarchical library

A complete program is put together out of various separately compiled units. In developing a very large program it is inevitable that it will be conceived as a number of subsystems themselves each composed out of a number of separately compiled units.

We see at once the risk of name clashes between the various parts of the total system. It would be very easy for the designers of different parts to reuse popular package names such as Error_Messages or Debug_Info and so on.

In order to overcome this and other related problems, Ada has a hierarchical naming scheme at the library level. Thus a package Parent may have a child package with the name Parent.Child.

We immediately see that if our total system breaks down into a number of major parts such as acquisition, analysis, and report then name clashes will be avoided if it is mapped into three corresponding library packages plus appropriate child units. There is then no risk of a clash between the names Analysis.Debug_Info and Report.Debug_Info since they are now quite distinct.

The naming is hierarchical and can continue to any depth. A good example of the use of this hierarchical naming scheme is found in the standard libraries which are provided by every implementation of Ada.

We have already mentioned the package Standard which is an intrinsic part of the language. All library units can be considered to be children of Standard and it should never (well, hardly ever) be necessary to explicitly mention Standard at all (unless you do something crazy like redefine Integer to mean something else; the original could then still be referred to as Standard.Integer).

In order to reduce the risk of clashes with users' own names the predefined library comprises just three packages each of which has a number of children. The three packages are System, which is concerned with the control of storage and similar implementation matters; Interfaces, which is concerned with interfaces to other languages and the intrinsic hardware types; and finally Ada which contains the bulk of the predefined library.

The packages System and Interfaces are described in detail in Chapter 25. In this chapter we will briefly survey the main package Ada and some important child packages. A fuller description will be found in Chapters 23 and 24.

The package Ada itself (herself?) was traditionally written as

```
package Ada is
   pragma Pure(Ada);        -- as white as driven snow!
end Ada;
```

The pragma Pure indicates that Ada has no variable state (this concept is important for sharing in distributed systems, a topic outlined in Section 26.3). But the current style is to use an aspect clause as discussed in Section 5.6.

Important child packages of Ada are

Numerics – this contains the mathematical library providing the elementary functions, random number generators, facilities for complex numbers and big numbers, and vector and matrix manipulation.

Characters – this contains packages for classifying and manipulating characters as well as the names of all the characters in the Latin-1 set.

Strings – this contains packages for the manipulation of strings of various kinds: fixed length, bounded, and unbounded.

Containers – this contains packages for manipulating various data structures such as vectors, lists, trees, maps, sets, and queues.

Text_IO, Sequential_IO, and Direct_IO – these and other packages provide a variety of input–output facilities.

There are many other children of Ada and these will be mentioned as required.

Most library units are packages and it is easy to think that all library units must be packages; indeed only a package can have child units. But of course the main subprogram is a library subprogram and any library package can have child subprograms. A library unit can also be a generic package or subprogram and even an instantiation of a generic package or subprogram; this latter fact is often overlooked.

We have introduced the hierarchical library as simply a naming mechanism. It also has important information hiding and sharing properties which will be dealt with in detail in Chapter 13. An important example is that a child unit can access the information in the private part of its parent; but of course other units cannot see into the private part of a package. This and related facilities enable a group of units to share private information while keeping the information hidden from external clients.

It should also be noted that a child package does not need to have a with clause or use clause for its parent; this emphasizes the close relationship of the hierarchical structure and parallels the fact that we never need a with clause for Standard because all library units are children of Standard.

4.2 Input–output

The Ada language is defined in such a way that all input and output is performed in terms of other language features. There are no special intrinsic features just for input and output. In fact input–output is just a service required by a program and so is provided by one or more Ada packages. This approach runs the attendant risk that different implementations will provide different packages and program portability will be compromised. In order to avoid this, the *ARM* describes certain standard packages that will be available in all implementations. Other, more elaborate, packages may be appropriate in special circumstances and the language does not prevent this. Indeed, simple packages such as the illustrative Simple_IO may also be appropriate. Full consideration of input and output is deferred until Chapter 23. However, we will now describe how to use some of the features so that the reader will be able to run some simple exercises. We restrict ourselves to the input and output of simple text such as numbers, characters, and strings.

Text input–output is performed through the use of the standard package Ada.Text_IO. Unless we specify otherwise, all communication will be through two standard files, one for input and one for output, and we will assume that (as is likely for most implementations) these are such that input is from the keyboard and output is to the screen. The full details of Ada.Text_IO cannot be described here but if we restrict ourselves to just a few useful facilities it looks a bit like

```
with Ada.IO_Exceptions;
package Ada.Text_IO is
   type Count is ...          -- an integer type
   ...
   procedure New_Line(Spacing: in Count := 1);
   procedure Set_Col(To: in Count);
   function Col return Count;
   ...
   procedure Get(Item: out Character);
   procedure Get(Item: out String);
   procedure Put(Item: in Character);
   procedure Put(Item: in String);
   procedure Put_Line(Item: in String);
   ...
   ...                     -- package Float_IO as outlined in Section 3.4
   ...                     -- plus a similar package Integer_IO and so on
   ...
end Ada.Text_IO;
```

This package commences with a with clause for Ada.IO_Exceptions. This further package contains the declaration of a number of different exceptions relating to a variety of things which can go wrong with input–output. For toy programs the most likely to arise is Data_Error which would occur for example if we tried to read a number from the keyboard but then accidentally typed something that was not a number at all or was in the wrong format.

Note the outline declaration of the type Count. This is an integer type having similar properties to the type Integer and almost inevitably with the same implementation (just as the type My_Integer might be based on Integer). The parameter of New_Line is of the type Count rather than plain Integer, although since the parameter will often be a literal such as 2 (or be omitted so that the default of 1 applies) this will not be particularly evident.

The procedure Set_Col and function Col are useful for tabulation. The character positions along a line of output are numbered starting at 1. So if we write (assuming **use** Ada.Text_IO;)

```
Set_Col(10);
```

then the next character output will go at position 10. A call of New_Line naturally sets the current position to 1 so that output commences at the beginning of the line. The function Col returns the current position and so

```
Set_Col(Col + 10);
```

will move the position on by 10 and thereby leave 10 spaces. Note that Col is an example of a function that has no parameters.

A single character can be output by for example

```
Put('A');
```

and a string of characters by

```
Put("This Is a string of characters");
```

There is also a useful procedure Put_Line which takes a string as parameter and has the effect of Put with the string followed by a call of New_Line.

A value of the type My_Float can be output in various formats. But first we have to instantiate the package Float_IO mentioned in Section 3.4 and which is declared inside Ada.Text_IO. Having done that we can call Put with a single parameter, the value of type My_Float to be output, in which case a standard default format is used – or we can add further parameters controlling the format. This is best illustrated by a few examples and we will suppose that the type My_Float was declared to have 7 decimal digits as in Section 2.3.

If we do not supply any format parameters then an exponent notation is used with 7 significant digits, 1 before the point and 6 after (the 7 matches the precision given in the declaration of My_Float). There is also a leading space or minus sign. The exponent consists of the letter E followed by the exponent sign (+ or –) and then a two-digit decimal exponent.

We can override the default by providing three further parameters which give respectively, the number of characters before the point, the number of characters after the point, and the number of characters after E. However, there is still always only one digit before the point and at least one after it (if we foolishly ask for zero digits after then we get one digit). If we do not want exponent notation then we simply specify the last parameter as zero and we then get normal decimal notation.

The effect is shown by the following statements with the output given as a comment. For clarity the output is surrounded by quotes and s designates a space; in reality there are no quotes and spaces are spaces.

```
Put(12.34);                -- "s1.234000E+01"
Put(12.34, 3, 4, 2);       -- "ss1.2340E+1"
Put(12.34, 3, 4, 0);       -- "s12.3400"
```

The output of values of integer types follows a similar pattern. In this case we similarly instantiate the generic package Integer_IO inside Ada.Text_IO which applies to all integer types with the particular type such as My_Integer. We can then call Put with a single parameter, the value of type My_Integer, in which case a standard default field is used, or we can add a further parameter specifying the field. The default field is the smallest that will accommodate all values of the type My_Integer allowing for a leading minus sign. Thus for the range of My_Integer, the default field is 8. It should be noted that if we specify a field which is too small then it is expanded as necessary. So

```
Put(123);                  -- "sssss123"
Put(-123);                 -- "ssss-123"
Put(123, 4);               -- "s123"
Put(123, 0);               -- "123"
```

Simple text input is similarly performed by a call of Get with a parameter that must be a variable of the appropriate type.

A call of Get with a floating or integer parameter will expect us to type in an appropriate number at the keyboard; this may have a decimal point only if the parameter is of a floating type. It should also be noted that leading blanks (spaces) and newlines are skipped. A call of Get with a parameter of type Character will read the very next character, and this can be neatly used for controlling the flow of an interactive program, thus

```
C: Character;
...
Put("Do you want to stop?  Answer Y if so. ");
Get(C);
if C = 'Y' then
   ...
```

For simple programs that do not have to be portable, the instantiations can be avoided if we just use the predefined types Integer and Float since the predefined library contains nongeneric versions with names Ada.Integer_Text_IO and Ada.Float_Text_IO respectively. So all we need write is

```
use Ada.Integer_Text_IO, Ada.Float_Text_IO;
```

and then we can call Put and Get without more ado.

An even simpler approach which is useful for testing is to use the attribute Image described in Appendix 1, thus to output the value of N of type Integer we simply write

```
Put(N'Image);
```

This technique also works for arrays and records. To output the value of the array Pascal in Section 2.4, we can write Put(Pascal'Image); and that might output the array as follows according to how the calculation is progressing

```
[ 0 => 1,  1 =>  4,  2 => 6,  3 => 4,  4 => 1, ... ]
```

Note that the string is in square brackets. The value of the record My_Buffer at the end of Section 2.4 could be output by Put(My_Buffer'Image); which would output the record much as follows

```
( Data => [1 => 'X', 2 => 'Y', ...], Start => 1, Finish => 3 )
```

That concludes a brief introduction to input–output which has been of a rather cookbook nature. Hopefully, it has provided enough to enable the reader to drive some trial examples as well as giving some further flavour to the nature of Ada.

Exercise 4.2

1 Which of the calls of Put discussed above would produce different results if we had used the type Integer rather than My_Integer? Consider both a 16-bit and a 32-bit implementation of Integer.

4.3 Numeric library

The numeric library comprises the package Ada.Numerics plus a number of child packages. The package Ada.Numerics is as follows

```
package Ada.Numerics
    with Pure is                              -- aspect clause
  Argument_Error: exception;
  Pi: constant := 3.14159_26535_89793_23846_26433_83279_
                                      50288_41971_69399_37511;
  π: constant := Pi;
  e: constant := 2.71828_18284_59045_23536_02874_71352_
                                      66249_77572_47093_69996;
end Ada.Numerics;
```

This contains the exception Argument_Error which is raised if something is wrong with the argument of a numeric function (such as attempting to take the square root of a negative number) and the useful constants Pi and e. These constants (strictly known as named numbers as explained in Chapter 6) are given to 50 decimal places. Technically of course the literals must all be on a single line but the page width of this book (unlike the *ARM*) cannot cope. Note also that we can use the Greek letter π instead of Pi. This is an illustration of the fact that we can use other alphabets such as Greek and Cyrillic for identifiers if we wish.

Note how the pure property is indicated by the aspect clause which starts with **with**.

One child package of Ada.Numerics provides the elementary functions such as Sqrt and again illustrates the use of the generic mechanism. Its specification is

```
generic
   type Float_Type is digits <>;
package Ada.Numerics.Generic_Elementary_Functions
      with Pure, Nonblocking is
   function Sqrt(X: Float_Type'Base) return Float_Type'Base;
   ...      -- and so on
end;
```

Again there is a single generic parameter giving the floating type. In order to call the function Sqrt we must first instantiate the generic package much as we did for Float_IO, thus (assuming appropriate with and use clauses)

```
package My_Elementary_Functions is
      new Generic_Elementary_Functions(My_Float);
use My_Elementary_Functions;
```

and we can then write a call of Sqrt directly.

The rather long name My_Elementary_Functions follows the recommended practice in the *ARM*. In fact it should be noted that there is a nongeneric version for the type Float with the name Ada.Numerics.Elementary_Functions for the convenience of those using the predefined type Float.

A little point to note is that the parameter and result of Sqrt is written as Float_Type'Base; the reason for this is explained in Chapter 17 when we look at numeric types in more detail.

We emphasize that the exception Ada.Numerics.Argument_Error is raised if the parameter of a function such as Sqrt is unacceptable. This contrasts with our hypothetical function Sqrt introduced earlier which we assumed raised the predefined exception Constraint_Error when given a negative parameter. When we deal with exceptions in detail in Chapter 15 we shall see that it is often better to declare and raise our own exceptions rather than use the predefined ones.

Two other child packages are those for the generation of random numbers. Each contains a function Random. One returns a value of the type Float within the range zero to one and the other returns a random value of a discrete type (an integer type or an enumeration type). We will look briefly at the latter and refer the reader to Chapter 23 for full details of both. The specification is essentially as follows

```
generic
   type Result_Subtype is (<>);
package Ada.Numerics.Discrete_Random is
   type Generator is limited private;
   function Random(Gen: Generator) return Result_Subtype;
   ...      -- plus other facilities
end Ada.Numerics.Discrete_Random;
```

This introduces a number of new points. The most important is the form of the generic formal parameter which indicates that the actual type must be a discrete type. The pattern echoes that of an enumeration type in much the same way as that for the floating generic parameter in Generic_Elementary_Functions echoed the declaration of a floating type. Thus we see that the discrete types are another example of a class of types as discussed in Section 3.4.

A small point is that the type Generator is declared as limited. This simply means that assignment is not available for the type (or at least not for the partial view as seen by the client).

The random number generator is used as in the following fragment which illustrates the simulation of tosses of a coin

```
use Ada.Numerics;
type Coin is (Heads, Tails);
package Random_Coin is new Discrete_Random(Coin);
use Random_Coin;
G: Generator;
C: Coin;
...
loop
   C := Random(G);
   ...
end loop;
...
```

Having declared the type Coin we then instantiate the generic package. We then declare a generator and use it as the parameter of successive calls of Random. The generator technique enables us to declare several generators and thus run several independent random sequences at the same time.

This concludes a brief survey of the standard numeric packages; further details will be found in Chapter 23.

4.4 Running a program

We are now in a position to put together a complete program using the proper input–output facilities. As a first example the following program outputs the multiplication tables up to ten. Each number is printed in a column five characters wide.

```
with Ada.Text_IO, Ada.Integer_Text_IO;
use Ada.Text_IO, Ada.Integer_Text_IO;
procedure Multiplication_Tables is
begin
   for Row in 1 .. 10 loop
      for Column in 1 .. 10 loop
         Put(Row * Column, 5);
      end loop;
      New_Line;
   end loop;
end Multiplication_Tables;
```

As a further example we will rewrite the procedure Print_Roots of Section 2.2 and also use the standard mathematical library. For simplicity we will first use the predefined type Float. The program becomes

```
with Ada.Text_IO;
with Ada.Float_Text_IO;
with Ada.Numerics.Elementary_Functions;
procedure Print_Roots is
   use Ada.Text_IO;
   use Ada.Float_Text_IO;
   use Ada.Numerics.Elementary_Functions;

   X: Float;
begin
   Put("Roots of various numbers");

   ...   -- and so on as before
end Print_Roots;
```

Note that we can put the use clauses inside the procedure Print_Roots as shown or we can place them immediately after the with clauses as in the example of the multiplication tables.

If we want to write a portable version then the general approach would be

```
with Ada.Text_IO;
with Ada.Numerics.Generic_Elementary_Functions;
procedure Print_Roots is
   type My_Float is digits 7;
   package My_Float_IO is new Ada.Text_IO.Float_IO(My_Float);
   use My_Float_IO;
   package My_Elementary_Functions is
        new Ada.Numerics.Generic_Elementary_Functions(My_Float);
   use My_Elementary_Functions;

   X: My_Float;
begin
   Put("Roots of various numbers");

   ...   -- and so on as before
end Print_Roots;
```

To have to write all that introductory stuff each time is rather a burden, so we will put it in a standard package of our own and then compile it once so that it is permanently available and can be accessed without more ado. We include the type My_Integer as well and write

```
with Ada.Text_IO;
with Ada.Numerics.Generic_Elementary_Functions;
package Etc is
   type My_Float is digits 7;
   type My_Integer is range -1000_000 .. +1000_000;

   package My_Float_IO is new Ada.Text_IO.Float_IO(My_Float);
   package My_Integer_IO is new Ada.Text_IO.Integer_IO(My_Integer);

   package My_Elementary_Functions is
        new Ada.Numerics.Generic_Elementary_Functions(My_Float);
end Etc;
```

Having compiled Etc a typical program could look like

```
with Ada.Text_IO, Etc;
use Ada.Text_IO, Etc;
procedure Program is
   use My_Float_IO, My_Integer_IO, My_Elementary_Functions;
   ...
end Program;
```

where we would naturally only supply the internal use clauses for those packages we were actually using in the particular program.

(The author had great difficulty in identifying an appropriate and short name for the package Etc and hopes that he is forgiven for the pun on etcetera.)

An alternative approach, rather than declaring everything in the one package Etc, is to first compile a tiny package just containing the types My_Float and My_Integer and then to compile the various instantiations as individual library packages (remember from Section 4.1 that a library unit can be just an instantiation).

We might then have the following four library units

```
package My_Numerics is
   type My_Float is digits 7;
   type My_Integer is range –1000_000 .. +1000_000;
end My_Numerics;

with My_Numerics;  use My_Numerics;
with Ada.Text_IO;
package My_Float_IO is new Ada.Text_IO.Float_IO(My_Float);

with My_Numerics;  use My_Numerics;
with Ada.Text_IO;
package My_Integer_IO is new Ada.Text_IO.Integer_IO(My_Integer);

with My_Numerics;  use My_Numerics;
with Ada.Numerics.Generic_Elementary_Functions;
package My_Elementary_Functions is
     new Ada.Numerics.Generic_Elementary_Functions(My_Float);
```

With this approach we only need to include (via with clauses) the particular packages as required and our program is thus likely to be smaller if we do not need them all. We could even arrange the packages as an appropriate hierarchy.

The reader should now be in a position to write complete simple programs. The examples in this book have mostly been written as fragments rather than complete programs for two reasons; one is that complete programs take up a lot of space (and can be repetitive) and the other is that Ada is really all about software components anyway. On the other hand complete programs show how the components fit together and perhaps also act as templates for the reader to follow. Accordingly this book includes a small number of complete programs the first of which is immediately after this chapter.

One important matter remains to be addressed and that is how to compile and build a complete program. A major benefit of Ada is that consistency is maintained between separately compiled units so that the integrity of strong typing is preserved across compilation unit boundaries. It is therefore illegal to build a program out of inconsistent units; the means whereby this is prevented will depend upon the implementation.

A related issue is the order in which units are compiled. This is dictated by the idea of dependency. There are three main causes of dependency

- a body depends upon the corresponding specification,
- a child depends upon the specification of its parent,
- a unit depends upon the specifications of those it mentions in a with clause.

The key rule is that a unit can only be compiled if all those on which it depends are present in the library environment. (A discussion of what is meant by the program library environment will be found in Chapter 13.)

Thus in the example above, the package My_Float_IO cannot be compiled until after the package My_Numerics has been written and entered into the library. But there is no dependency between My_Float_IO and My_Integer_IO.

An important consequence of the rules is that the user A of a package B is quite independent of the body of B. The body of B can be changed and recompiled without having to recompile the user A. But of course the body of B cannot be compiled without the specification of B being present.

Many compilation systems permit several library units to be compiled together with the text all in a single file but of course their order within the file must be such that each unit follows all those on which it depends. A common special case is that a package specification and body can be compiled together provided the text of the specification precedes that of the body.

Complete simple programs might be presented in a single file. Thus if we wrote a program using the package Shapes of Exercise 3.2(1) and all of the text were contained in one file then the outline structure of the file might be

```
package Objects is ...      -- spec and body of Objects

with Objects;  use Objects;
package Shapes is ...       -- spec and body of Shapes

with Shapes;  use Shapes;
procedure Main is ...       -- the main subprogram
```

The dependency rules also need to be considered if a unit is modified and then recompiled; typically all those units which depend on it (directly or indirectly) will also have to be recompiled in order to preserve consistency of the total program (a good system will do such recompilations automatically).

Unfortunately it is not possible to explain how to manipulate the library, call the Ada compiler and then build a complete program or indeed how to call our Ada program because this depends upon the implementation and so we must leave the reader to find out how to do these last vital steps from the documentation for the implementation concerned.

This brings us to the end of our brief survey of the main features of Ada. We have in fact encountered most of the main concepts although very skimpily in some cases. We have for example said very little about the use of subtypes which impose constraints on types.

Hopefully the reader will have grasped the key structural concepts and especially the forms of abstraction discussed in Chapter 3. The remainder of this book takes us through various topics in considerable detail starting with the small-scale aspects. In looking at these aspects we should not lose sight of the big picture to which we will return in Part 3. Moreover, we will generally use the types Integer and Float for simplicity but the reader will no doubt remember that they are not portable.

Exercise 4.4

1 Write a program to output a table of square roots of numbers from 1 up to some limit specified by the user in response to a suitable question. Print the numbers as integers and the square roots to 6 decimal places in two columns. Use Set_Col to set the second column position so that the program can be easily modified. Note that a value N of type Integer can be converted to the corresponding My_Float value by writing My_Float(N). Use the package Etc.

2 Declare the packages My_Float_IO, My_Integer_IO and My_Elementary_Functions as child packages of My_Numerics. Avoid unnecessary with and use clauses.

3 Write a program to read in the lengths of the three sides of a triangle and then print out its area. Use the package Shapes of Exercise 3.2(**1**).

4 Write a program to generate 100 random days of the week and count how many of them are Sundays. Output the answer in a suitable format.

5 Write a program to print out Pascal's triangle described in Section 2.4. Read in the value of an Integer variable Size which gives the final row number. Ensure that the triangle is properly aligned and has a suitably aligned caption. Print each number in a field of width 4; this will neatly accommodate values of Size up to 12. Note that a value N of type Integer can be converted to type Count by writing Count(N).

Program 1

Magic Moments

This first example of a complete program pulls together various aspects of the type Object and its descendants which were discussed in Chapter 3. This opening description is followed by the text of the program and then some notes on specific points.

The program is in two phases. It first reads in the dimensions of a number of objects and then computes and prints out a table of their properties.

The root package Geometry contains the abstract type Object together with various abstract operations. The function MI returns the moment of inertia of the object about its centre (see Exercise 3.3(**5**)). The function Name returns a string describing the type of the object. Each concrete type is then declared in its own child package.

The child package Geometry.Circles declares the type Circle which is derived from the abstract type Object and has one additional component, Radius. It also declares concrete functions Area, MI, and Name which implement the corresponding abstract operations of the root type Object.

The child packages Geometry.Points, Geometry.Triangles, and Geometry.Squares then similarly declare types Point, Triangle, and Square with appropriate concrete functions.

There are also a number of other child packages containing related material. The package Geometry.Magic contains the class wide functions Moment and MO; Moment gives the vertical moment about the origin whereas MO gives the moment of inertia about the origin. The package Geometry.Lists contains entities for defining and manipulating lists of objects. The package Geometry.IO contains functions for reading the properties of the various types of objects; each such function returns a pointer to a newly created object.

There are then two library subprograms to do the two major phases of activities, namely Build_List and Tabulate_Properties. Finally, the main subprogram Magic_Moments essentially just calls these two other library subprograms.

```
package Geometry is
   type Object is abstract tagged
      record
         X_Coord: Float;
         Y_Coord: Float;
      end record;

   function Distance(O: Object) return Float;
   function Area(O: Object) return Float
                                        is abstract;
   function MI(O: Object) return Float is abstract;
   function Name(O: Object) return String
                                        is abstract;
end;

with Ada.Numerics.Elementary_Functions;
use Ada.Numerics.Elementary_Functions;
package body Geometry is
   function Distance(O: Object) return Float is
   begin
      return Sqrt(O.X_Coord**2 + O.Y_Coord**2);
   end Distance;
end Geometry;
-------------------------------------------

package Geometry.Magic is
   function Moment(OC: Object'Class)
                                     return Float;
   function MO(OC: Object'Class) return Float;
end;

package body Geometry.Magic is
   function Moment(OC: Object'Class)
                                     return Float is
   begin
      return OC.X_Coord * OC.Area;
   end Moment;

   function MO(OC: Object'Class) return Float is
   begin
      return OC.MI + OC.Area * OC.Distance**2;
   end MO;
end Geometry.Magic;
-------------------------------------------
```

```
package Geometry.Circles is
   type Circle is new Object with
      record
         Radius: Float;
      end record;

   function Area(C: Circle) return Float;
   function MI(C: Circle) return Float;
   function Name(C: Circle) return String;
end;

with Ada.Numerics;
package body Geometry.Circles is

   function Area(C: Circle) return Float is
   begin
      return Ada.Numerics.Pi * C.Radius**2;
   end Area;

   function MI(C: Circle) return Float is
   begin
      return 0.5 * C.Area * C.Radius**2;
   end MI;

   function Name(C: Circle) return String is
   begin
      return "Circle";
   end Name;

end Geometry.Circles;
---------------------------------------

package Geometry.Points is
   type Point is new Object with null record;

   function Area(P: Point) return Float;
   function MI(P: Point) return Float;
   function Name(P: Point) return String;
end;

package body Geometry.Points is

   function Area(P: Point) return Float is
   begin
      return 0.0;
   end Area;

   function MI(P: Point) return Float is
   begin
      return 0.0;
   end MI;

   function Name(P: Point) return String is
   begin
      return "Point";
   end Name;

end Geometry.Points;
---------------------------------------

package Geometry.Triangles is
   type Triangle is new Object with
      record
         A, B, C: Float;   -- lengths of sides
      end record;

   function Area(T: Triangle) return Float;
   function MI(T: Triangle) return Float;
   function Name(T: Triangle) return String;
end;

with Ada.Numerics.Elementary_Functions;
use Ada.Numerics.Elementary_Functions;
package body Geometry.Triangles is

   function Area(T: Triangle) return Float is
      S: constant Float := 0.5 * (T.A + T.B + T.C);
   begin
      return Sqrt(S * (S - T.A) * (S - T.B) * (S - T.C));
   end Area;

   function MI(T: Triangle) return Float is
   begin
      return T.Area * (T.A**2 + T.B**2 + T.C**2)
                                             / 36.0;
   end MI;

   function Name(T: Triangle) return String is
   begin
      return "Triangle";
   end Name;

end Geometry.Triangles;
---------------------------------------

package Geometry.Squares is
   type Square is new Object with
      record
         Side: Float;
      end record;

   function Area(S: Square) return Float;
   function MI(S: Square) return Float;
   function Name(S: Square) return String;
end;

package body Geometry.Squares is

   function Area(S: Square) return Float is
   begin
      return S.Side**2;
   end Area;

   function MI(S: Square) return Float is
   begin
      return S.Area * S.Side**2 / 6.0;
   end MI;

   function Name(S: Square) return String is
   begin
      return "Square";
   end Name;

end Geometry.Squares;
---------------------------------------
```

```
package Geometry.Lists is
   type Pointer is access Object'Class;

   type Cell is
      record
         Next: access Cell;
         Element: Pointer;
      end record;
   type List is access Cell;

   procedure Add_To_List(The_List: in out List;
                          Obj_Ptr: in Pointer);
end;

package body Geometry.Lists is

   procedure Add_To_List(The_List: in out List;
                          Obj_Ptr: in Pointer) is
   begin
      The_List := new Cell'(The_List, Obj_Ptr);
   end Add_To_List;

end Geometry.Lists;
--------------------------------------------

with Geometry.Lists;
package Geometry.IO is
   function Get_Circle return Lists.Pointer;
   function Get_Triangle return Lists.Pointer;
   function Get_Square return Lists.Pointer;
end;

with Ada.Text_IO; use Ada.Text_IO;
with Ada.Float_Text_IO; use Ada.Float_Text_IO;
with Geometry.Circles;
with Geometry.Triangles;
with Geometry.Squares;
package body Geometry.IO is

   function Get_Circle return Lists.Pointer is
      use Circles;
      X_Coord: Float;
      Y_Coord: Float;
      Radius: Float;
   begin
      Get(X_Coord);
      Get(Y_Coord);
      Get(Radius);
      return new Circle'(X_Coord, Y_Coord,
                                    Radius);
   end Get_Circle;

   function Get_Triangle return Lists.Pointer is
      use Triangles;
      X_Coord: Float;
      Y_Coord: Float;
      A, B, C: Float;
   begin
      Get(X_Coord);
      Get(Y_Coord);
      loop
         Get(A); Get(B); Get(C);
         -- check to ensure a valid triangle
         exit when A < B+C and B < C+A and
                                    C < A+B;
         Put("Sorry, not a triangle, " &
               "enter sides again please");
         New_Line;
      end loop;

      return new Triangle'(X_Coord, Y_Coord,
                                    A, B, C);
   end Get_Triangle;

   function Get_Square return Lists.Pointer is
      use Squares;
      X_Coord: Float;
      Y_Coord: Float;
      Side: Float;
   begin
      Get(X_Coord);
      Get(Y_Coord);
      Get(Side);
      return new Square'(X_Coord, Y_Coord,
                                    Side);
   end Get_Square;

end Geometry.IO;
--------------------------------------------

with Geometry.Lists; use Geometry;
with Geometry.IO; use Geometry.IO;
with Ada.Text_IO; use Ada.Text_IO;
procedure Build_List(The_List: in out Lists.List) is
   Code_Letter: Character;
   Object_Ptr: Lists.Pointer;
begin
   loop

      loop          -- loop to skip leading spaces
         Get(Code_Letter);
         exit when Code_Letter /= ' ';
      end loop;

      case Code_Letter is
         when 'C' | 'c' =>       -- expect a circle
            Object_Ptr := Get_Circle;
         when 'T' | 't' =>       -- expect a triangle
            Object_Ptr := Get_Triangle;
         when 'S' | 's' =>       -- expect a square
            Object_Ptr := Get_Square;
         when others =>
            exit;
      end case;

      Lists.Add_To_List(The_List, Object_Ptr);
   end loop;
end Build_List;
--------------------------------------------
```

```
with Geometry.Lists; use Geometry.Lists;
with Geometry.Magic; use Geometry.Magic;
with Ada.Text_IO; use Ada.Text_IO;
with Ada.Float_Text_IO; use Ada.Float_Text_IO;
procedure Tabulate_Properties(The_List: List) is
  Local: access Cell := The_List;
  This_One: Pointer;
begin
  New_Line;
  Put("          X      Y      Area    " &
      "   MI          MO          Moment");
  New_Line;
  while Local /= null loop
    This_One := Local.Element;
    New_Line;
    Put(This_One.Name); Set_Col(10);
    Put(This_One.X_Coord, 4, 2, 0); Put(' ');
    Put(This_One.Y_Coord, 4, 2, 0); Put(' ');
    Put(This_One.Area, 6, 2, 0); Put(' ');
    Put(This_One.MI, 6, 2, 0); Put(' ');
    Put(MO(This_One.all), 6, 2, 0); Put(' ');
    Put(Moment(This_One.all), 6, 2, 0);
    Local := Local.Next;
  end loop;
end Tabulate_Properties;

---------------------------------------

with Build_List;
with Tabulate_Properties;
with Geometry.Lists; use Geometry.Lists;
with Ada.Text_IO;
with Ada.Float_Text_IO;
use Ada;
procedure Magic_Moments is
  The_List: List := null;
begin
  Text_IO.Put("Welcome to Magic Moments");
  Text_IO.New_Line(2);
  Text_IO.Put("Enter C, T or S followed by " &
              "coords and dimensions");
  Text_IO.New_Line;
  Text_IO.Put("Terminate list with any other letter");
  Text_IO.New_Line(2);
  Build_List(The_List);
  Tabulate_Properties(The_List);
  Text_IO.New_Line(2);
  Text_IO.Put("Finished");
  Text_IO.New_Line;
  Text_IO.Skip_Line(2);
end Magic_Moments;
```

The overall structure is always a matter for debate. It could be argued that the main subprogram and the two other library subprograms should all be children of Geometry. Doing so would reduce the number of with clauses required.

The application of use clauses is also a matter of eternal debate. In some units they have been applied liberally whereas in others only the top level (Ada or Geometry) has been given in a use clause. This has been done largely to illustrate the possibilities. Remember that a child unit never needs a with or use clause for its parent. But a child does need a with clause in order to access a sibling.

There are a number of different approaches to reading in the details of the objects in Build_List and that shown is very straightforward. Note the use of the case statement whose structure and purpose is hopefully fairly evident. Each object is described by a letter giving its type followed by the required dimensions; any other letter terminates the list of objects. There is also a certain amount of repetition – for example we should perhaps read the common components X_Coord and Y_Coord for all the types in one place only. The function Get_Triangle checks that the three sides do form a triangle in accordance with Euclid, Book I, Proposition 20: '*Παντὸς τριγώνου αἱ δύο πλευραὶ τῆς λοιπῆς μείζονές εἰσι πάντῃ μεταλαμβανόμενει.*' – 'In any triangle two sides taken together in any manner are greater than the remaining one.' However, there are no checks that the dimensions of any of the objects are actually positive.

The procedure Tabulate_Properties outputs the various properties in a table. The while form of loop causes the loop to stop at the end of the list. The various values are output using a fixed format with a space character between them so that overlength numbers do not run together. Note also that the function Name returns a string whose length depends upon the specific type.

An interesting point is that Tabulate_Properties only interrogates those aspects that are common to all objects (that is, common components such as X_Coord or common operations such as Area). We could add another dispatching operation to the root package Geometry that printed the specific properties (such as the radius in the case of a circle) either directly or perhaps by calling subprograms in Geometry.IO.

This leads to a final point. Dispatching on a class wide value such as one of type Pointer is possible because the code knows the specific type of the object being pointed to and can choose the specific operation accordingly. This means that dispatching on output is feasible. But dispatching on input is difficult because until we have read the object we do not know its type. This is why we declared distinct subprograms Get_Circle, Get_Triangle and so on for reading each type. However, we can use an object constructor and then dispatch as explained in Section 21.7.

Part 2

Algorithmic Aspects

Chapter 5	**Lexical Style**	65
Chapter 6	**Scalar Types**	73
Chapter 7	**Control Structures**	101
Chapter 8	**Arrays and Records**	117
Chapter 9	**Expression Structures**	151
Chapter 10	**Subprograms**	165
Chapter 11	**Access Types**	193
Program 2	**Sylvan Sorter**	227

This second part covers the small-scale algorithmic features of the language in detail. These correspond to the areas covered by simple languages such as Pascal and C although Ada has much richer facilities in these areas.

Chapter 5 deals with the lexical detail which needs to be described but can perhaps be skimmed on a first reading and just referred to when required. It also briefly introduces aspect specifications which are an important feature of Ada 2012 and Ada 2022.

Chapter 6 is where the story really begins and covers the type model and illustrates that model by introducing most of the scalar types. Chapter 7 then discusses the control structures which are very straightforward. Chapter 8 covers arrays in full but only the simplest forms of records; it is also convenient to introduce characters and strings in this chapter since strings in Ada are treated as arrays of characters. Chapter 8 is quite long compared with a corresponding discussion on Pascal or C largely because of the named notation for array aggregates which is an important feature for writing readable programs. It has become even longer with the introduction of more general forms of aggregates in Ada 2022.

Chapter 9 discusses the more elaborate forms of expressions introduced in Ada 2012 and Ada 2022, namely if expressions, case expressions, quantified expressions, and declare expressions.

Subprograms are discussed in Chapter 10 and at this point we can write serious lumps of program. Again the named notation enriches the discussion and includes the mechanism for default parameters.

Chapter 11 contains a discussion on access types which correspond to pointers in Pascal and C. Although the concept of an access type is easy to understand, nevertheless the rules regarding accessibility might be found difficult on a first reading. These rules give much greater flexibility than Pascal whilst being carefully designed to prevent dangling references which can so easily cause a program to crash in C.

This second part concludes with a complete program which illustrates various algorithmic features and especially the use of access types.

Those familiar with Ada 2012 will find that most changes are in Chapters 8 and 9 which lay the foundation for improving the contract capabilities of Ada which we will meet in Chapter 16.

5 Lexical Style

5.1 Syntax notation
5.2 Lexical elements
5.3 Identifiers
5.4 Numbers
5.5 Comments
5.6 Pragmas and aspects

In the previous chapters, we illustrated the general appearance of Ada programs with some simple examples. In this chapter we get down to serious detail.

We start with some rather unexciting but essential material – the detailed construction of things such as identifiers and numbers which make up the text of a program. However, it is no good learning a human language without getting the spelling sorted out. And as far as programming languages are concerned, compilers are usually unforgiving regarding such corresponding matters.

We also take the opportunity to introduce the notation used to describe the syntax of Ada. In general we will not use this syntax notation to introduce concepts but, in some cases, it is the easiest way to do so. Moreover, if the reader wishes to consult the *ARM* then knowledge of the syntax notation is necessary. For completeness and easy reference the full syntax will be found on the website.

5.1 Syntax notation

The syntax of Ada is described using a variant of Backus–Naur Form (BNF). In this notation, syntactic categories are represented by lower case names; some of these contain embedded underlines to increase readability. A category is defined in terms of other categories by a sort of equation known as a production. Some categories are atomic and cannot be decomposed further – these are known as terminal symbols. A production consists of the name being defined followed by the special symbol ::= and its defining sequence. Other symbols used are

- [ ] square brackets enclose optional items,
- { } braces enclose optional items which may be omitted, appear once or be repeated many times,
- | a vertical bar separates alternatives.

Sometimes the name of a category is prefixed by a word in italics. In such cases the prefix aims to convey some semantic information and can be treated as a form of comment as far as the context free syntax is concerned. Where relevant, a production is presented in a form that shows the recommended layout.

5.2 Lexical elements

The international standard describes Ada 2022 in terms of the ISO 32-bit 10646:2020 set. This is very broad and covers the Latin characters including accented forms plus the Greek and Cyrillic alphabets and various ideographs as well. It is essentially equivalent to Unicode 4.0. The general intent is that Ada has a multicultural approach appropriate to the 21st century. In this book we will use the Latin and Greek alphabets.

A line of Ada text can be thought of as a sequence of groups of characters known as lexical elements. We have already met several forms of lexical elements in previous chapters. Consider for example

```
Age := 21;      -- John's age
```

This consists of five such elements

- the identifier `Age`
- the compound delimiter `:=`
- the numeric literal `21`
- the single delimiter `;`
- the comment `-- John's age`

Other classes of lexical element are strings and character literals; they are dealt with in Chapter 9.

Individual lexical elements may not be split by spaces but otherwise spaces may be inserted freely in order to improve the appearance of the program. A most important example of this is the use of indentation to reveal the overall structure. Naturally enough a lexical element must fit on one line.

The *ARM* does not prescribe how one proceeds from one line of text to the next. This need not concern us. We can just imagine that we type our Ada program as a series of lines using whatever mechanism the keyboard provides for starting a new line.

The delimiters are either the following single delimiters

```
& ' ( ) + - * / < = > , . : ; | @ [ ]
```

or the following compound delimiters

`=>`	for aggregates, cases, etc.	`..`	for ranges
`**`	exponentiation	`<>`	the 'box' for types and defaults
`:=`	assignment	`/=`	not equals
`>=`	greater than or equals	`<=`	less than or equals
`<<`	label bracket	`>>`	the other label bracket

Special care should be taken that the compound delimiters do not contain spaces.

However, spaces may occur in strings and character literals where they stand for themselves, and also in comments. Note that adjacent words and numbers must be separated from each other by spaces otherwise they would be confused. Thus we must write **end loop** rather than **endloop**.

5.3 Identifiers

We mentioned above that Ada is defined in terms of the full ISO 32-bit character set. As a consequence we can write identifiers using many different alphabets. The basic rule is that an identifier consists of a letter possibly followed by one or more letters or digits with embedded isolated underlines. Either case of letter can be used. Moreover, the meaning attributed to an identifier does not depend upon the case of the letters. So identifiers which differ only in the case of corresponding letters are considered to be the same. But, on the other hand, the underline characters are considered to be significant.

In some alphabets the correspondence between upper and lower case letters is not straightforward. Thus in Greek there are two lower case letters equivalent to s (σ and ς, the latter being used only at the end of a word) but only one corresponding upper case letter Σ. These three characters are all the same so far as Ada identifiers are concerned. Accented characters such as é and ç may be used and they are of course distinct from the unaccented forms. Similarly, ligatures such as Æ and æ may also be used.

Ada does not impose any limit on the number of characters in an identifier. Moreover all are significant. There may however be a practical limit since an identifier must fit onto a single line and an implementation is likely to impose some maximum line length. However, this maximum must be at least 200 and so identifiers of up to 200 characters will always be accepted. Programmers are encouraged to use meaningful names such as Time_Of_Day rather than cryptic meaningless names such as T. (But please not Time_of_Day.) Long names may seem tedious when first writing a program but in the course of its lifetime a program is read much more often than it is written and clarity aids subsequent understanding both by the original author and by others who may be called upon to maintain the program. Of course, in short mathematical or abstract subprograms, simple identifiers such as X and Y may be appropriate.

Identifiers are used to name all the various entities in a program. However, some words are reserved for special syntactic significance and may not be used as identifiers. We have already encountered several of these such as **if**, **procedure**, and **end**. There are 74 reserved words in Ada 2022; they are listed in Appendix 1. For readability they are printed in boldface in this book, but that is not important. In program text they could, like identifiers, be in either case or indeed in a mixture of cases – procedure, PROCEDURE and Procedure are all acceptable. Nevertheless, some discipline aids understanding, and a good convention is to use lower case for the reserved words and leading capitals for all others. But this is a matter of taste.

There are minor exceptions regarding the reserved words **access**, **delta**, **digits**, **mod**, and **range**. As will be seen later, they are also used as attributes Access, Delta, Digits, Mod, and Range. However, when so used they are always preceded by an apostrophe and so there is no confusion.

Some identifiers such as Integer and True have a predefined meaning from the package Standard. These are not reserved and can be reused although to do so is usually unwise since the program could become very confusing.

Exercise 5.3

1 Which of the following are not legal identifiers and why?

(a) Ada	(d) Ευρεκα	(g) X_
(b) fish&chips	(e) Time__Lag	(h) tax rate
(c) RATE-OF-FLOW	(f) 77E2	(i) goto

5.4 Numbers

Numbers (or numeric literals to use the proper jargon) take two forms according to whether they denote an integer (an exact whole number) or a real (an approximate and not usually whole number). This is a good opportunity to illustrate the use of the syntax.

```
numeric_literal ::= decimal_literal | based_literal
decimal_literal ::= numeral [. numeral] [exponent]
numeral ::= digit {[underline] digit}
exponent ::= E [+] numeral | E -numeral
digit ::= 0 | 1 | 2 | 3 | 4 | 5 | 6 | 7 | 8 | 9
```

A subtle point is that we can use Arabic and other forms of numerals in identifiers but not for numeric literals.

The important distinguishing feature of a real literal is that real literals always contain a decimal point whereas integer literals never do. Real literals can be used as values of any real type which covers both floating point and fixed point types.

Ada is strict on mixing up types. It is illegal to use an integer literal where the context demands a real literal and vice versa. Thus

```
Age: Integer := 21.0;
Weight: Float := 150;
```

are both illegal.

The simplest form of integer literal is just a sequence of decimal digits. If the literal is very long it should be split up into groups of digits by inserting isolated underlines thus

```
123_456_789
```

In contrast to identifiers such underlines are, of course, of no significance other than to make the literal easier to read.

The simplest form of real literal is a sequence of decimal digits containing a decimal point. Note that there must be at least one digit on either side of the decimal point. Again, isolated underlines may be inserted to improve legibility provided they are not adjacent to the decimal point; thus

```
3.14159_26536
```

Both integer and real literals can have an exponent. This consists of the letter E (either case) followed by a signed or unsigned decimal integer and indicates the power of ten by which the preceding simple literal is to be multiplied. The exponent cannot be negative in the case of an integer literal – otherwise it might not be a whole number. (An exponent of -0 is not allowed for an integer literal but it is for a real literal.) Thus the real literal 98.4 could be written with an exponent in any of the following ways

```
9.84E1     98.4e0     984.0e-1     0.984E+2
```

Note that 984e-1 would not be allowed.

Similarly, the integer literal 1900 could also be written as

```
19E2     190e+1     1900E+0
```

but not as 19000e-1 nor as 1900E-0 since these have negative exponents.

The exponent may itself contain underlines if it consists of two or more digits but it is unlikely that the exponent would be so large as to make this necessary.

A final facility is the ability to express a literal in a base other than 10. This is done by enclosing the digits between # characters following the base. Thus

```
2#111#
```

is a based integer literal of value $4 + 2 + 1 = 7$.

Any base from 2 to 16 inclusive can be used and, of course, base 10 can always be expressed explicitly. For bases above 10 the letters A to F are used to represent the extended digits 10 to 15. Thus

```
14#ABC#
```

equals $10 \times 14^2 + 11 \times 14 + 12 = 2126$.

A based literal can also have an exponent. But note carefully that the exponent gives the power of the base by which the simple literal is to be multiplied and not a power of 10 – unless, of course, the base happens to be 10. The exponent itself, like the base, is always expressed in normal decimal notation. Thus

```
16#A#E2
```

equals $10 \times 16^2 = 2560$ and

```
2#11#E11
```

equals $3 \times 2^{11} = 6144$.

A based literal can be real. The distinguishing mark is again the point. (We can hardly say 'decimal point' if we are not using a decimal base! A better term is radix point.) So we have

2#101.11# = $4 + 1 + {}^1/_2 + {}^1/_4 = 5.75$

7#3.0#e-1 = ${}^3/_7 = 0.\dot{4}2857\dot{1}$.

The reader may have felt that the possible forms of based literal are unduly elaborate. This is not really so. Based literals are useful – especially for fixed point types since they enable programmers to represent values in the form in which they think about them. Obviously bases 2, 8 and 16 will be the most useful.

Finally note that a numeric literal cannot be negative. A form such as -3 consists of a literal preceded by the unary minus operator.

Exercise 5.4

1 Which of the following are not legal literals and why? For those that are legal, state whether they are integer or real literals.

(a) 38.6	(e) 2#1011	(i) 16#FfF#
(b) .5	(f) 2.71828_18285	(j) 1_0#1_0#E1_0
(c) 32e2	(g) 12#ABC#	(k) 27.4e_2
(d) 32e-2	(h) E+6	(l) 2#11#e-1

2 What are the values of the following?

(a) 16#E#E1	(c) 16#F.FF#E+2
(b) 2#11#E11	(d) 2#1.1111_1111_111#E11

3 How many different ways can you express the following as an integer literal?

(a) the integer 41 (b) the integer 150

(Forget underlines, distinction between E and e, nonsignificant leading zeros and optional + in an exponent.)

5.5 Comments

It is important to add appropriate comments to a program to aid its understanding by someone else or yourself at a later date. A comment in Ada is written as an arbitrary piece of text following two hyphens (or minus signs). Thus

```
-- this is a comment
```

A comment can include any characters. So we might annotate a mathematical program with

```
-- the variable D_Alpha represents ∂α/∂t
```

A comment extends to the end of the line. The comment may be the only thing on the line or it may follow some other Ada text. A long comment needing several lines is merely written as successive comments.

```
-- this comment
-- is spread over
-- several lines.
```

It is important that the leading hyphens be adjacent and not separated by spaces. Note that in this book comments are written in italics. This makes it look pretty.

5.6 Pragmas and aspects

We conclude this chapter with a few remarks about pragmas even though they are hardly lexical elements in the sense of identifiers and numbers.

Pragmas originated in Ada 83 as a sort of compiler directive and as a convenient way of making a parenthetic remark to the compiler. As an example we can indicate that we wish the compiler to optimize our program with emphasis on saving space by writing

```
pragma Optimize(Space);
```

inside the region of program to which it is to apply.

However, as Ada evolved, pragmas were also used to apply extra detail to many entities. As an example, we can use the pragma Inline to indicate that all calls of a certain procedure are to be expanded inline so that the code is faster although more space might be occupied. Thus given some procedure Do_It we might write

```
pragma Inline(Do_It);
```

Properties such as Inline are known as aspects. So we say that the aspect Inline is applied to the procedure Do_It by the pragma Inline. However, there is a much neater way of applying such aspects. We can write

```
procedure Do_It( ... )
     with Inline;
```

so that the aspect is given with the specification of the procedure. The structure **with** Inline is known as an aspect specification. There are important advantages to this approach one being that the aspect specification is lexically tied to the procedure specification whereas a pragma might be somewhat later in the text. Accordingly, many pragmas which were valid in Ada 2005 are now considered obsolete.

As an example the package Ada in Section 4.1 could be written as

```
package Ada
     with Pure is
          -- still as white as driven snow
end Ada;
```

where the important comment hides the curious juxtaposition of **is** and **end** which the author thinks is somewhat ugly.

Generally a pragma can appear anywhere that a declaration or statement can appear and in some other contexts also. Sometimes there may be special rules regarding the position of a particular pragma. For fuller details on pragmas and aspect specifications see Appendix 1.

An example of the use of aspects is to generalize the concept of literals. There are three such aspects namely Integer_Literal, Real_Literal, and String_Literal.

The aspects Integer_Literal and Real_Literal are used in the package Ada.Numerics.Big_Numbers; see Section 23.6.

The general idea is that if we declare a type such as

```
type Thing is private
   with Integer_Literal => Make_Thing;
```

where Make_Thing is a function whose specification has the form

```
function Make_Thing(S: String) return Thing;
```

and converts a string S into a value of the type Thing, then rather than writing

```
My_T: Thing := Make_Thing("123_456_789");
```

we can simply write

```
My_T: Thing := 123_456_789;
```

The syntax of the string is expected to be the same as that used by Get in Integer_IO.

The aspect Real_Literal works in a similar way and the syntax is that of Get in Float_IO. However, in the case of Real_Literal we can also optionally provide a second function with two string parameters which overloads the first; this facility is used by Big_Numbers.Big_Reals as described in Section 23.6.

A more flexible approach is provided by the aspect String_Literal which expects a parameter of type Wide_Wide_String (which gives more freedom). There is an interesting example in the *ARM* concerning Roman Numbers which might appeal to Latin scholars. It permits both IC as well as the usual XCIX as a Roman Number of value 99. This flexibility is not unreasonable since there is a statue in Pisa inscribed with VC meaning 95 and not Victoria Cross.

And incidentally, Greek scholars will note that the plural of pragma should really be pragmata.

Checklist 5

The case of a letter is immaterial in all contexts except strings and character literals.

Underlines are significant in identifiers but not in numeric literals.

Spaces are not allowed in lexical elements, except in strings, character literals, and comments.

The presence of a point distinguishes real and integer literals.

An integer may not have a negative exponent.

Numeric literals cannot be signed.

Leading zeros are allowed in all parts of a numeric literal.

New in Ada 2022

Ada 2022 has one more reserved word than Ada 2012, namely **parallel**.

The following characters are also used as single delimiters @ [ ].

User-defined literals for integers, reals, and strings are added.

6 Scalar Types

6.1 Object declarations and assignments
6.2 Blocks and scopes
6.3 Types
6.4 Subtypes
6.5 Simple numeric types
6.6 Enumeration types
6.7 The type Boolean
6.8 Categories of types
6.9 Expression summary

This chapter lays the foundations for the small-scale aspects of Ada. We start by considering the declaration of objects and the assignment of values to them and briefly discuss the ideas of scope and visibility. We then introduce the important concepts of types, subtypes, and constraints. As examples of types, the remainder of the chapter discusses the numeric types Integer and Float, enumeration types in general, the type Boolean in particular, and the operations on them.

6.1 Object declarations and assignments

Values can be stored in objects which are declared to be of a specific type. Objects are either variables, in which case their value may change (or vary) as the program executes, or they may be constants, in which case they keep their same initial value throughout their life.

A variable is introduced into the program by a declaration which consists of the name (that is, the identifier) of the variable followed by a colon and then the name of the type. This can then optionally be followed by the := symbol and an expression giving the initial value. The declaration terminates with a semicolon. Thus we might write

```
J: Integer;
P: Integer := 38;
```

The first declaration introduces the variable J of type Integer but gives it no particular initial value. The second introduces the variable P and gives it the specific initial value of 38.

We can introduce several variables at the same time in one declaration by separating them by commas thus

```
I, J, K: Integer;
P, Q, R: Integer := 38;
```

In the second case all of P, Q and R are given the initial value of 38.

If a variable is declared and not given an initial value then great care must be taken not to use the undefined value of the variable until one has been properly given to it. If a program does use the undefined value in an uninitialized variable, its behaviour will be unpredictable; the program is said to have a bounded error as described in Section 2.6.

A common way to give a value to a variable is by using an assignment statement. In this, the name of the variable is followed by := and then some expression giving the new value. The statement terminates with a semicolon. Thus

```
J := 36;
```

and

```
P := Q + R;
```

are both valid assignment statements and place new values in J and P, thereby overwriting their previous values.

Note that := can be followed by any expression provided that it produces a value of the type of the variable being assigned to. We will discuss all the rules about expressions later, but it suffices to say at this point that they can consist of variables and constants with operations such as + and round brackets (parentheses) and so on just like an ordinary mathematical expression.

There is a lot of similarity between a declaration containing an initial value and an assignment statement. Both use := before the expression and the expression can be of arbitrary complexity.

An important difference, however, is that although several variables can be declared and given the same initial value together, it is not possible for an assignment statement to give the same value to several variables. This may seem odd but in practice the need to give the same value to several variables usually only arises with initial values anyway.

We should remark at this stage that strictly speaking a multiple declaration such as

```
A, B: Integer := E;
```

is really a shorthand for

```
A: Integer := E;
B: Integer := E;
```

This means that in principle the expression E is evaluated for each variable. This is a subtle point and does not usually matter, but we will encounter some examples later (in Sections 10.5 and 11.2) where the effect is important.

A constant is declared in a similar way to a variable by inserting the reserved word **constant** after the colon. Of course, a constant must be initialized in its declaration otherwise it would be useless. Why?

An example might be

```
G: constant Float := 6.7E-11;          -- gravitational constant G
```

In the case of numeric types, and only numeric types, it is possible to omit the type from the declaration of a constant thus

```
γ: constant := 0.57721_56649;          -- Euler's constant
```

It is then technically known as a number declaration and merely provides a name for the number. The distinction between integer and real named numbers is made by the form of the initial value. In this case it is real because of the presence of the decimal point. It is usually good practice to omit the type when declaring numeric constants for reasons which will appear later (this especially applies in the case of true numeric constants such as π, but is not so clear for physical constants such as c and G whose value depends upon the units chosen). But note that the type cannot be omitted in numeric variable declarations even when an initial value is provided.

There is an important distinction between the allowed forms of initial values in constant declarations (with a type) and number declarations (without a type). In the former case the initial value may be any expression and is evaluated when the declaration is encountered at run time whereas in the latter case it must be static and so evaluated at compile time. Full details are deferred until Chapter 17.

Exercise 6.1

1 Write a declaration of a floating point variable F giving it an initial value of one.

2 Write appropriate declarations of constants Zero and One of type Float.

3 What is wrong with the following declarations and statements?

(a) **var** I: Integer;
(b) g: **constant** := 981
(c) P, Q: **constant** Integer;
(d) P := Q := 7;
(e) MN: **constant** Integer := M*N;
(f) 2Pi: **constant** := 2.0*Pi;

6.2 Blocks and scopes

Ada carefully distinguishes between declarations which introduce new identifiers and statements which do not. It is clearly sensible that the declarations which introduce new identifiers should precede the statements that manipulate them. Accordingly, declarations and statements occur in distinct places in the program text. The simplest fragment of text which includes declarations and statements is a block.

A block commences with the reserved word **declare**, some declarations, **begin**, some statements and concludes with the reserved word **end** and the terminating semicolon. A trivial example is

```
declare
   J: Integer := 0;          -- declarations here
begin
   J := J + 1;               -- statements here
end;
```

A block is itself an example of a statement and so one of the statements in its body could be another block. This nesting of blocks can continue indefinitely.

Since a block is a statement it can be executed like any other statement. When this happens the declarations in its declarative part (the bit between **declare** and **begin**) are elaborated in order, and then the statements in the body (between **begin** and **end**) are executed in the usual way. Note the terminology: we elaborate declarations and execute statements. All that the elaboration of a declaration does is make the thing being declared come into existence and then evaluate and assign any initial value to it. At the **end** of the block all the things which were declared in the block automatically cease to exist.

The above simple example of a block is clearly rather foolish; it introduces J, adds 1 to it but then loses it before use is made of the resulting value.

Like other block structured languages, Ada has the concept of hiding. Consider

```
declare
   J, K: Integer;
begin
   ...                       -- here J is the outer one
   declare
      J: Integer;
   begin
      ...                    -- here J is the inner one
   end;
   ...                       -- here J is the outer one
end;
```

The variable J is declared in an outer block and then redeclared in an inner block. This redeclaration does not cause the outer J to cease to exist but merely makes it temporarily hidden. In the inner block J refers to the new J, but as soon as we leave the inner block, this new J ceases to exist and the outer one again becomes directly visible.

Another point to note is that the objects used in an initial value must, of course, already exist. They could be declared in the same declarative part but the declarations must precede their use. For example

```
declare
   J: Integer := 0;
   K: Integer := J;
begin
```

is allowed. This idea of elaborating declarations in order is important; the jargon is 'linear elaboration of declarations'. It is an example of the more general principle that the meaning of a program should be clear by reading the text in order without having to look ahead. One exception to this general rule concerns aspect specifications where looking ahead is vital, see Section 16.1.

We distinguish the terms 'scope' and 'visibility'. The scope is the region of text where an entity has some effect. In the case of a block, the scope of an entity extends from the start of its declaration until the end of the block. We say it is visible at a given point if its name can be used to refer to it at that point. A fuller discussion is deferred until Section 10.7 when we distinguish visibility and direct visibility and introduce named blocks. However, the following simple rules will suffice for the moment

- an object is never visible in its own declaration,
- an object is hidden by the declaration of a new object with the same identifier from the start of the declaration.

Thus the inner declaration of J in the example of nesting could never be

```
J: Integer := J;              -- illegal
```

because we cannot refer to the inner J in its own declaration and the outer J is already hidden by the inner J.

Exercise 6.2

1 How many errors can you see in the following?

```
declare
   I: Integer := 7;
   J, K: Integer
begin
   J := I+K;
   declare
      P: Integer = I;
      I: Integer := Q;
   begin
      K := P+I;
   end;
   Put(K);          -- output value of K
end;
```

6.3 Types

The Ada Reference Manual (section 3.2) says 'A type is characterized by a set of values and a set of primitive operations ...'.

In the case of the built-in type Integer, the set of values is represented by

..., –3, –2, –1, 0, 1, 2, 3, ...

and the primitive operations include

+, –, * and so on.

With certain exceptions to be discussed later (access types, arrays, tasks, and protected objects) every type has a name which is introduced in a type declaration. (The built-in types such as Integer are considered to be declared in the package Standard.) Moreover, every type declaration introduces a new type completely distinct from any other type.

The sets of values belonging to two distinct types are themselves quite distinct, although in some cases the actual lexical form of the values may be identical – which one is meant at any point is determined by the context. The idea of one lexical form representing two or more different things is known as overloading.

Values of one type cannot be assigned to variables of another type. This is the fundamental rule of strong typing. Strong typing, correctly used, is an enormous aid to the rapid development of *correct* programs since it ensures that many errors are detected at compile time. (Overused, it can tie one in knots; we will discuss this thought in Chapter 27.)

A type declaration uses a somewhat different syntax from an object declaration in order to emphasize the conceptual difference. It consists of the reserved word **type**, the identifier to be associated with the type, the reserved word **is** and then the definition of the type followed by the terminating semicolon.

The package Standard contains type declarations such as

```
type Integer is ... ;
```

The type definition between **is** and ; gives in some way the set of values belonging to the type. As a concrete example consider the following

```
type Colour is (Red, Amber, Green);
```

(This is an example of an enumeration type and will be dealt with in more detail in a later section in this chapter.)

This introduces a new type called Colour. Moreover, it states that there are only three values of this type and they are denoted by the identifiers Red, Amber, and Green.

Objects of this type can then be declared in the usual way

```
C: Colour;
```

An initial value can be supplied

```
C: Colour := Red;
```

or a constant can be declared

```
Default: constant Colour := Red;
```

We have stated that values of one type cannot be assigned to variables of another type. Therefore one cannot mix colours and integers and so

```
I: Integer;
C: Colour;
...
I := C;                    -- type mismatch, will not compile
```

is illegal. In some languages it is often necessary to implement concepts such as enumeration types by primitive types such as integers and give values such as 0, 1, and 2 to variables Red, Amber, and Green. Thus in Java one could write

```
static final int Red = 0, Amber = 1, Green = 2;
```

and then use Red, Amber, and Green as if they were literal values. Obviously the program would be easier to understand than if the code values 0, 1, and 2 had been used directly. But, on the other hand, the compiler could not detect the accidental assignment of a notional colour to a variable which was, in the mind of the programmer, just an ordinary integer. In Ada this is detected during compilation thus making a potentially tricky error quite trivial to discover.

6.4 Subtypes

We now introduce subtypes and constraints. A subtype characterizes a set of values which is just a subset of the values of some existing type. The subset is defined by means of a constraint or, in the case of access types, possibly also by a so-called null exclusion. Constraints take various forms according to the category of the type. Elaborate restrictions can be imposed by subtype predicates as explained in Section 16.4. As is usual with subsets, the subset may be the complete set. There is, however, no way of restricting the set of operations of the type. The subtype takes all the operations; subsetting applies only to the values.

(In Ada 83, the type of the subtype was known as the base type and we will sometimes use that term in an informal way. In later versions of Ada, the type name is strictly considered to denote a subtype of the type. Such a subtype is then called the first subtype. One advantage of this pedantic approach is that the *ARM* can use the term subtype rather than having to say type or subtype all the time.)

As an example suppose we wish to manipulate dates; we know that the day of the month must lie in the range 1 .. 31 so we declare a subtype thus

```
subtype Day_Number is Integer range 1 .. 31;
```

where the reserved word **range** followed by 1 .. 31 is known as a range constraint. We can then declare variables and constants using the subtype identifier in exactly the same way as a type identifier.

```
D: Day_Number;
```

We are then assured that the variable D can only be assigned integer values from 1 to 31 inclusive. The compiler will insert run-time checks if necessary to ensure that this is so; if a check fails then Constraint_Error is raised.

It is important to realize that a subtype declaration does not introduce a new distinct type. An object such as D is of type Integer, and so the following is perfectly legal from the syntactic point of view.

```
D: Day_Number;
I: Integer;
...
D := I;
```

On execution, the value of I may or may not lie in the range 1 .. 31. If it does, then all is well; if not then Constraint_Error will be raised. Assignment in the other direction

```
I := D;
```

will, of course, always work.

It is not always necessary to introduce a subtype explicitly in order to impose a constraint. We could equally have written

```
D: Integer range 1 .. 31;
```

Furthermore, a subtype need not impose a constraint. It is perfectly legal to write

```
subtype Some_Number is Integer;
```

although in this instance it is not of much value.

A subtype (explicit or not) may be defined in terms of a previous subtype

```
subtype Feb_Day is Day_Number range 1 .. 29;
```

Any additional constraint must of course satisfy existing constraints, thus

```
subtype Some_Day is Feb_Day range 0 .. 10;        -- wrong
```

would be incorrect and cause Constraint_Error to be raised.

The above examples have shown constraints with static bounds. This is not always the case; in general the bounds can be given by arbitrary expressions and so the set of values of a subtype need not be static, that is known at compile time. However, it is an important fact that the bounds of a type are always static.

In conclusion then, a subtype does not introduce a new type but is merely a shorthand for an existing type or subtype with an optional constraint. However, in later chapters we will encounter several contexts in which an explicit constraint is not allowed; a subtype has to be introduced for these cases. We refer to a subtype name as a subtype mark and to the form consisting of a subtype mark with an optional constraint or null exclusion (see Chapter 11) as a subtype indication as shown by the syntax

subtype_mark ::= *subtype*_name

subtype_indication ::= [null exclusion] subtype_mark [constraint]

Thus we can restate the previous remark as saying that there are situations where a subtype mark has to be used whereas, as we have seen here, the more general subtype indication (which includes a subtype mark on its own) is allowed in object declarations.

The sensible use of subtypes has two advantages. It can ensure that programming errors are detected earlier by preventing variables from being assigned inappropriate values. It can also increase the execution efficiency of a program. This particularly applies to array indexes (subscripts) as we shall see later.

We conclude this section by summarizing the assignment statement and the rules of strong typing. Assignment has the form

```
variable := expression;
```

and the two rules are

- both sides must have the same type,
- the expression must satisfy any constraints on the variable; if it does not, the assignment does not take place, and Constraint_Error is raised instead.

Note carefully the general principle that type errors (violations of the first rule) are detected during compilation whereas subtype errors (violations of the second rule) are detected during execution by the raising of Constraint_Error. (A clever compiler might give a warning during compilation.)

We have now introduced the basic concepts of types and subtypes. The remaining sections of this chapter illustrate these concepts further by considering in more detail the properties of the simple types of Ada.

It should be noted that subtype constraints restrict values to lie in a contiguous range; sometimes we wish to impose more flexible restrictions such as just even values. This can be done using subtype predicates which are discussed in Section 16.4.

Exercise 6.4

1 Given the following declarations

```
I, J: Integer range 1 .. 10;
K: Integer range 1 .. 20;
```

which of the following assignment statements could raise Constraint_Error?

(a) I := J; (b) K := J; (c) J := K;

6.5 Simple numeric types

Perhaps surprisingly, a full description of the numeric types of Ada is deferred until much later in this book. The problems of numerical analysis (error estimates and so on) are complex and Ada is correspondingly rich in this area so that it can cope in a reasonably complete way with the needs of the numerical specialist. For our immediate purposes such complexity can be ignored. Accordingly, in this section, we merely consolidate a simple understanding of the two predefined numeric types Integer and Float.

First a reminder. We recall from Section 2.3 that the predefined numeric types are not portable and should be used with caution. However, all the operations described here apply to all integer and floating point types, both the predefined types such as Integer and Float and those defined by the user such as My_Integer and My_Float.

As we have seen, a constraint may be imposed on the type Integer by using the reserved word **range**. This is then followed by two expressions separated by two dots which, of course, must produce values of integer type. These expressions need not be literal constants. One could have

```
P: Integer range 1 .. I+J;
```

A range can be null as would happen in the above case if I+J turned out to be zero. Null ranges may seem pretty useless but they often automatically occur in limiting cases, and to exclude them would mean taking special action in such cases.

The minimum value of the type Integer is given by Integer'First and the maximum value by Integer'Last. These are our first examples of attributes. Ada contains various attributes denoted by a single quote (strictly, an apostrophe) followed by an identifier. We say 'Integer tick First'.

The value of Integer'First will depend on the implementation but will always be negative. On a two's complement machine it will be –Integer'Last–1 whereas on a one's complement machine it will be –Integer'Last. So on a minimal 16-bit two's complement implementation we will have

```
Integer'First = –32768
Integer'Last = +32767
```

Of course, we should always write Integer'Last rather than +32767 if that is what we logically want. Otherwise program portability could suffer.

Two useful subtypes are

```
subtype Natural is Integer range 0 .. Integer'Last;
subtype Positive is Integer range 1 .. Integer'Last;
```

These are so useful that they are declared for us in the package Standard.

The attributes First and Last also apply to subtypes so

```
Positive'First = 1
Natural'Last = Integer'Last
```

We turn now to a brief consideration of the type Float. It is possible to apply a constraint to the type Float in order to reduce the range such as

```
subtype Chance is Float range 0.0 .. 1.0;
```

where the range of course is given using expressions of the type Float. There are also attributes Float'First and Float'Last. It is not really necessary to say any more at this point.

The other predefined operations that can be performed on the types Integer and Float are much as one would expect in any programming language. They are summarized below.

+, – These are either unary operators (that is, taking a single operand) or binary operators taking two operands.

In the case of a unary operator, the operand can be either Integer or Float; the result will be of the same type. Unary + effectively does nothing. Unary – changes the sign.

In the case of a binary operator, both operands must be of the same type; the result will be of that type. Normal addition or subtraction is performed.

***** Multiplication; both operands must be of the same type; again the result is of the same type.

/ Division; both operands must be of the same type; again the result is of the same type. Integer division truncates towards zero.

rem Remainder; in this case both operands must be Integer and the result is Integer. It is the remainder on division.

mod Modulo; again both operands must be Integer and the result is Integer. This is the mathematical modulo operation.

abs Absolute value; this is a unary operator and the single operand may be Integer or Float. The result is again of the same type and is the absolute value. That is, if the operand is positive (or zero), the result is the same but if it is negative, the result is the corresponding positive value.

****** Exponentiation; this raises the first operand to the power of the second. If the first operand is of type Integer, the second must be a positive integer or zero. If the first operand is of type Float, the second can be any integer. The result is of the same type as the first operand.

In addition, we can perform the operations =, /=, <, <=, > and >= which return a Boolean result True or False. Again both operands must be of the same type. Note the form of the not equals operator /=.

Although the above operations are mostly straightforward a few points are worth noting.

It is a general rule that mixed mode arithmetic is not allowed. One cannot, for example, add an integer value to a floating point value; both must be of the same type. A change of type from Integer to Float or vice versa can be done by using a type conversion which consists of the desired type name (or indeed subtype name) followed by the expression to be converted in parentheses.
So given

```
I: Integer := 3;
F: Float := 5.6;
```

we cannot write

```
I + F
```

but we must write

```
Float(I) + F
```

which uses floating point addition to give the floating point value 8.6, or

```
I + Integer(F)
```

which uses integer addition to give the integer value 9.

Conversion from **Float** to **Integer** always rounds rather than truncates, thus

```
1.4    becomes    1
1.6    becomes    2
```

and a value midway between two integers, such as 1.5, is always rounded away from zero so that 1.5 always becomes 2 and –1.5 becomes –2. Rounding can also be performed using various attributes as described in Section 17.4.

There is a subtle distinction between **rem** and **mod**. The **rem** operation gives the remainder corresponding to the integer division operation **/**. Integer division truncates towards zero; this means that the absolute value of the result is always the same as that obtained by dividing the absolute values of the operands. So

```
  7  /   3  =  2          7  rem   3  =  1
(–7) /   3  = –2        (–7) rem   3  = –1
  7  / (–3) = –2          7  rem (–3) =  1
(–7) / (–3) =  2        (–7) rem (–3) = –1
```

The remainder and quotient are always related by

```
(I/J) * J + I rem J = I
```

and it will also be noted that the sign of the remainder is always equal to the sign of the first operand I (the dividend).

However, **rem** is not always satisfactory. If we plot the values of I **rem** J for a fixed value of J (say 5) for both positive and negative values of I we get the pattern shown in Figure 6.1. The pattern is symmetric about zero and consequently changes its incremental behaviour as we pass through zero. The **mod** operation, on the other hand, does have uniform incremental behaviour as shown in Figure 6.2.

The **mod** operation enables us to do normal modulo arithmetic. For example

```
(A+B) mod n = (A mod n + B mod n) mod n
```

for all values of A and B both positive and negative. For positive n, A **mod** n is always in the range 0 .. n–1, whereas for negative n, A **mod** n is always in the range n+1 .. 0. Of course, modulo arithmetic is only usually performed with a positive value for n. But the **mod** operator gives consistent and sensible behaviour for negative values of n also.

We can look upon **mod** as giving the remainder corresponding to division with truncation towards minus infinity. So

```
  7  mod   3  =  1
(–7) mod   3  =  2
  7  mod (–3) = –2
(–7) mod (–3) = –1
```

In the case of **mod** the sign of the result is always equal to the sign of the second operand whereas with **rem** it is the sign of the first operand.

In summary, it is perhaps worth saying that integer division with negative operands is rare. The operators **rem** and **mod** only differ when just one operand is negative. It will be found that in such cases it is almost always **mod** that is wanted.

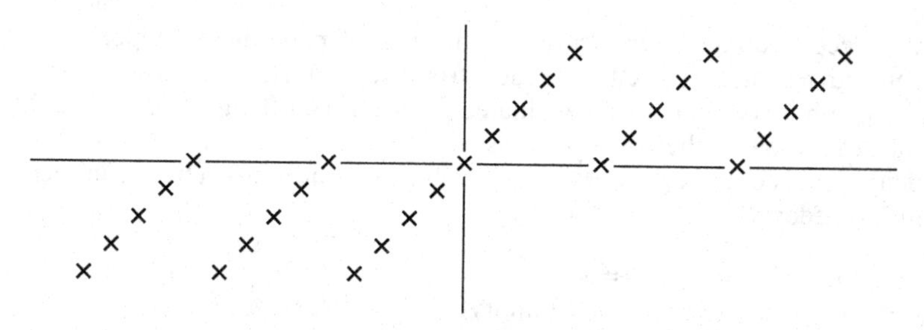

Figure 6.1 Behaviour of I **rem** 5 around zero.

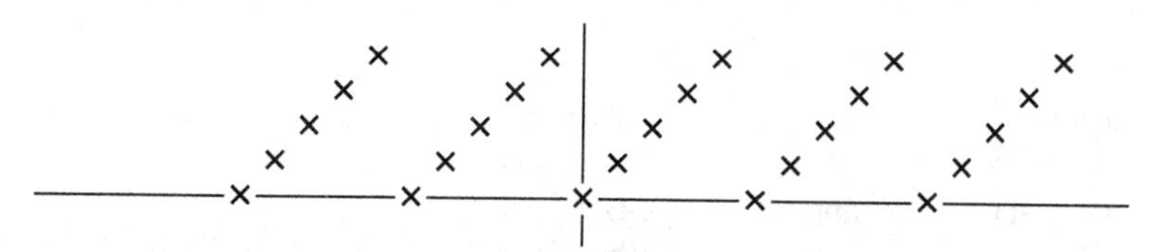

Figure 6.2 Behaviour of I **mod** 5 around zero.

Moreover, note that if we are performing modular arithmetic in general then we should perhaps be using modular types; these are unsigned integer types with automatic range wraparound and are discussed in detail in Chapter 17.

Finally, some notes on the exponentiation operator **. For a positive second operand, the operation corresponds to repeated multiplication. So

```
  3**4 = 3*3*3*3 = 81
3.0**4 = 3.0*3.0*3.0*3.0 = 81.0
```

A very subtle point is that the repeated multiplications can be performed by appropriate squarings as an optimization although as we shall see in Section 26.5 there can be slight differences in accuracy in the floating point case.

The second operand can be 0 and, of course, the result is then always the value one

```
  3**0 = 1
3.0**0 = 1.0
  0**0 = 1
0.0**0 = 1.0
```

The second operand cannot be negative if the first operand is an integer, because the result might not be a whole number. In fact, the exception Constraint_Error would be raised in such a case. But it is allowed for a floating point first operand and produces the corresponding reciprocal

```
3.0**(-4) = 1.0/81.0 = 0.0123456790123...
```

We conclude this section with a brief discussion on combining operators in an expression. As is usual, the operators have different precedence levels and the natural precedence can be overruled by the use of parentheses. Operators of the same precedence are applied in order from left to right. A subexpression in parentheses obviously has to be evaluated before it can be used. But note that the order of evaluation of the two operands of a binary operator is not specified. The precedence levels of the operators we have met so far are shown below in increasing order of precedence

```
=   /=   <   <=   >   >=
+   -                      (binary)
+   -                      (unary)
*   /   mod   rem
**  abs
```

Thus

```
A/B*C       means    (A/B)*C
A+B*C+D     means    A+(B*C)+D
A*B+C*D     means    (A*B)+(C*D)
A*B**C      means    A*(B**C)
```

In general, as stated above, several operations of the same precedence can be applied from left to right and parentheses are not necessary. However, the syntax rules forbid multiple instances of the exponentiation operator without parentheses. Thus we cannot write

```
A**B**C
```

but must explicitly write either

```
(A**B)**C     or     A**(B**C)
```

This restriction avoids the risk of accidentally writing the wrong thing. Note however that the well established forms

```
A-B-C     and     A/B/C
```

are allowed. The syntax rules similarly prevent the mixed use of **abs** and ** without parentheses.

The precedence of unary minus needs care

```
-A**B     means     -(A**B)     rather than     (-A)**B
```

as in some languages. Also

```
A**-B     and     A*-B
```

are illegal. Parentheses are necessary.

Note finally that the precedence of **abs** is, confusingly, not the same as that of unary minus. As a consequence we can write

```
- abs X     but not     abs - X
```

since the latter requires parentheses as in **abs** (-X).

Exercise 6.5

1 Evaluate the expressions below given the following

```
I: Integer := 7;
J: Integer := -5;
K: Integer := 3;
```

(a) I*J*K
(b) I/J*K
(c) I/J/K
(d) J + 2 **mod** I
(e) J + 2 **rem** I
(f) K**K**K
(g) -J **mod** 3
(h) -J **rem** 3

2 Rewrite the following mathematical expressions in Ada. Use suitable identifiers of appropriate type.

(a) Mr^2 – moment of inertia of black hole
(b) $b^2–4ac$ – discriminant of quadratic
(c) $^4/_3\pi r^3$ – volume of sphere
(d) $p\pi a^4/8l\eta$ – viscous flowrate through tube

6.6 Enumeration types

Here are some examples of declarations of enumeration types starting with Colour which we introduced when discussing types in general.

```
type Colour is (Red, Amber, Green);
type Day is (Mon, Tue, Wed, Thu, Fri, Sat, Sun);
type Stone is (Amber, Beryl, Quartz);
type Groom is (Tinker, Tailor, Soldier, Sailor,
                    Rich_Man, Poor_Man, Beggar_Man, Thief);
type Solo is (Alone);
```

This introduces an example of overloading. The enumeration literal Amber can represent a Colour or a Stone. Both meanings are visible together and the second declaration does not hide the first whether they are declared in the same declarative part or one is in an inner declarative part. We can usually tell which is meant from the context, but in cases when we cannot we can use a qualification which consists of the literal in parentheses and preceded by a subtype mark (that is its type name or a relevant subtype name) and a single quote. Thus

```
Colour'(Amber)
Stone'(Amber)
```

Examples where this is necessary will occur later.

Although we can use Amber as an enumeration literal in two distinct enumeration types, we cannot use it as an enumeration literal and the identifier of a variable at the same time. The declaration of one would hide the other and they could not both be declared in the same declarative part. Later we will see that an enumeration literal can be overloaded with a subprogram.

There is no upper limit on the number of values in an enumeration type but there must be at least one. An empty enumeration type is not allowed.

Range constraints on enumeration types and subtypes are much as for integers. The constraint has the form

```
range lower_bound_expression .. upper_bound_expression
```

and this indicates the set of values from the lower bound to the upper bound inclusive. So we can write

```
subtype Weekday is Day range Mon .. Fri;
D: Weekday;
```

or

```
D: Day range Mon .. Fri;
```

and then we know that D cannot be Sat or Sun.

If the lower bound is above the upper bound then we get a null range, thus

```
subtype Colourless is Colour range Amber .. Red;
```

Note the curious fact that we cannot have a null subtype of a type such as Solo (since it only has one value).

The attributes First and Last apply to enumeration types and subtypes, so

```
Colour'First = Red
Weekday'Last = Fri
```

There are built-in functional attributes to give the successor or predecessor of an enumeration value. These consist of Succ or Pred following the type name and a single quote. Thus

```
Colour'Succ(Amber) = Green
Stone'Succ(Amber) = Beryl
Day'Pred(Fri) = Thu
```

Of course, the thing in parentheses can be an arbitrary expression of the appropriate type. If we try to take the predecessor of the first value or the successor of the last then the exception Constraint_Error is raised. In the absence of this exception we have, for any type T and any value X,

```
T'Succ(T'Pred(X)) = X
```

and vice versa.

Another functional attribute is Pos. This gives the position number of the enumeration value, that is the position in the declaration with the first one having a position number of zero. So

```
Colour'Pos(Red) = 0
Colour'Pos(Amber) = 1
Colour'Pos(Green) = 2
```

The opposite to Pos is Val. This takes the position number and returns the corresponding enumeration value. So

```
Colour'Val(0) = Red
Day'Val(6) = Sun
```

If we give a position value outside the range, as for example

```
Solo'Val(1)
```

then Constraint_Error is raised.

Clearly we always have

```
T'Val(T'Pos(X)) = X
```

and vice versa. We also note that

```
T'Succ(X) = T'Val(T'Pos(X) + 1)
```

they either both give the same value or both raise an exception.

It should be noted that these four attributes Succ, Pred, Pos, and Val may also be applied to subtypes but are then identical to the same attributes of the corresponding base type.

It is probably rather bad practice to mess about with Pos and Val when it can be avoided. To do so encourages the programmer to think in terms of numbers rather than the enumeration values and hence destroys the abstraction.

The risk of confusion can be overcome by the use of the attributes Enum_Val and Enum_Rep introduced in Ada 2022. They are described in Section 25.1.

The operators =, /=, <, <=, > and >= also apply to enumeration types. The result is defined by the order of the values in the type declaration. So

```
Red < Green    is  True
Wed >= Thu     is  False
```

The same result would be obtained by comparing the position values. So

```
T'Pos(X) < T'Pos(Y)      and      X < Y
```

are always equivalent (except that X < Y might be ambiguous).

Exercise 6.6

1 Evaluate

(a) Day'Succ(Weekday'Last)
(b) Weekday'Succ(Weekday'Last)
(c) Stone'Pos(Quartz)

2 Write suitable declarations of enumeration types for

(a) the colours of the rainbow,
(b) typical fruits.

3 Write an expression that delivers one's predicted bridegroom after eating a portion of pie containing N stones. Use the type Groom declared at the beginning of this section.

4 If the first of the month is in D where D is of type Day, then write an assignment replacing D by the day of the week of the Nth day of the month.

5 Why might X < Y be ambiguous?

6.7 The type Boolean

The type Boolean is a predefined enumeration type whose declaration in the package Standard is

```
type Boolean is (False, True);
```

Boolean values are used in constructions such as the if statement which we briefly met in Chapter 2. Boolean values are produced by the operators =, /=, <, <=, > and >= which have their expected meaning and apply to many types. So we can write constructions such as

```
if Today = Sun then
   Tomorrow := Mon;
else
   Tomorrow := Day'Succ(Today);
end if;
```

The Boolean type (we capitalize the name in memory of the English mathematician George Boole) has all the normal properties of an enumeration type, so, for instance

```
False < True = True !!
Boolean'Pos(True) = 1
```

We could even write the curious

```
subtype Always is Boolean range True .. True;
```

The type Boolean also has other operators which are as follows

not This is a unary operator and changes True to False and vice versa. It has the same precedence as **abs**.

and This is a binary operator. The result is True if both operands are True, and False otherwise.

or This is a binary operator. The result is True if one or other or both operands are True, and False only if they are both False.

xor This is also a binary operator. The result is True if one or other operand but not both are True. (Hence the name – eXclusive OR.) Another way of looking at it is to note that the result is True if and only if the operands are different. (The operator is known as 'not equivalent' in some languages.)

The effects of **and**, **or** and **xor** are summarized in the usual operator tables shown in Figure 6.3. The precedences of **and**, **or** and **xor** are equal to each other but lower than that of any other operator. In particular they are of lower precedence than the relational operators =, /=, <, <=, > and >=. As a consequence, parentheses are not needed in expressions such as

```
P < Q and I = J
```

and	F	T
F	F	F
T	F	T

or	F	T
F	F	T
T	T	T

xor	F	T
F	F	T
T	T	F

Figure 6.3 Operator tables for **and**, **or** and **xor**.

However, although the precedences are equal, **and**, **or** and **xor** cannot be mixed up in an expression without using parentheses (unlike + and – for instance). So

B **and** C **or** D is illegal

whereas

I + J – K is legal

We have to write

B **and** (C **or** D) or (B **and** C) **or** D

in order to emphasize which meaning is required.

The reader familiar with other programming languages will remember that **and** and **or** usually have a different precedence. The problem with this is that the programmer often gets confused and writes the wrong thing. It is to prevent this that Ada makes them the same precedence and insists on parentheses. Of course, successive applications of the same operator are permitted so

B **and** C **and** D is legal

and, as usual, evaluation goes from left to right although it does not matter since **and**, **or** and **xor** are associative. Thus, even (A **xor** B) **xor** C and A **xor** (B **xor** C) are always the same.

Take care with **not**. Its precedence is higher than **and**, **or** and **xor** as in other languages and so

not A **or** B means (**not** A) **or** B rather than **not** (A **or** B)

Boolean variables and constants can be declared and manipulated in the usual way

```
Danger: Boolean;
Signal: Colour;
...
Danger := Signal = Red;
```

The variable Danger is then True if the signal is Red. We can then write

```
if Danger then
   Stop_Train;
end if;
```

Note that we do not have to write

```
if Danger = True then
```

although this is perfectly legal; it just misses the point that Danger is already of type Boolean and so can be used directly as the condition.

A worse sin is to write

```
if Signal = Red then
   Danger := True;
else
   Danger := False;
end if;
```

rather than

```
Danger := Signal = Red;
```

The literals True and False could be overloaded by declaring for example

```
type Answer is (False, Dont_Know, True);
```

but to do so might make the program rather confusing.

Finally, it should be noted that it is often clearer to introduce our own two-valued enumeration type rather than use the type Boolean. Thus instead of

```
Wheels_OK: Boolean;
...
if Wheels_OK then
```

it is much better (and safer!) to write

```
type Wheel_State is (Up, Down);
Wheel_Position: Wheel_State;
...
if Wheel_Position = Up then
```

since whether the wheels are OK or not depends upon the situation. OK for landing is different from being OK for cruising. The enumeration type removes any doubt as to which is meant.

Exercise 6.7

1 Write declarations of constants T and F having the values True and False.

2 Using T and F from the previous exercise, evaluate

(a) T **and** F **and** T

(b) **not** T **or** T

(c) F = F = F = F

(d) (F = F) = (F = F)

(e) T < T < T < T

3 Evaluate

```
(A /= B) = (A xor B)
```

for all combinations of values of Boolean variables A and B.

6.8 Categories of types

At this point we pause to consolidate the material given in this chapter so far. The types in Ada can be grouped into various categories as shown in Figure 6.4. The broad grouping is into elementary types and composite types. Ultimately everything is essentially made up of elementary types.

Elementary types are divided into access types which are dealt with in Chapter 11 and scalar types some of which we have discussed in this chapter. Arrays and simple records are dealt with in Chapter 8. Tagged record types and interfaces are dealt with in Chapter 14 when we discuss object oriented programming. The other composite types, task and protected types which concern concurrent programming, are dealt with in Chapter 20. Record, task and protected types may also be parameterized with so-called discriminants which are discussed in Chapter 18.

The scalar types themselves can be subdivided into real types and discrete types. The real types are subdivided into floating point types and fixed point types. Our sole example of a real type has been the floating point type Float – the other real types are discussed in Chapter 17. The other types discussed in this chapter, Integer, Boolean and enumeration types in general, are discrete types – the only other kinds of discrete types to be introduced are other integer types again dealt with in Chapter 17, and character types which are in fact a form of enumeration type and are dealt with in Chapter 8. Note that the integer types are subdivided into signed integer types such as Integer and the modular types.

The key distinction between the discrete types and the real types is that the former have a clear-cut set of distinct separate (that is, discrete) values. The type Float, on the other hand, should be thought of as having a continuous set of values – we know in practice that a finite digital computer must implement a real type as actually a set of distinct values but this is really an implementation detail, the underlying mathematical concept is of a continuous set of values.

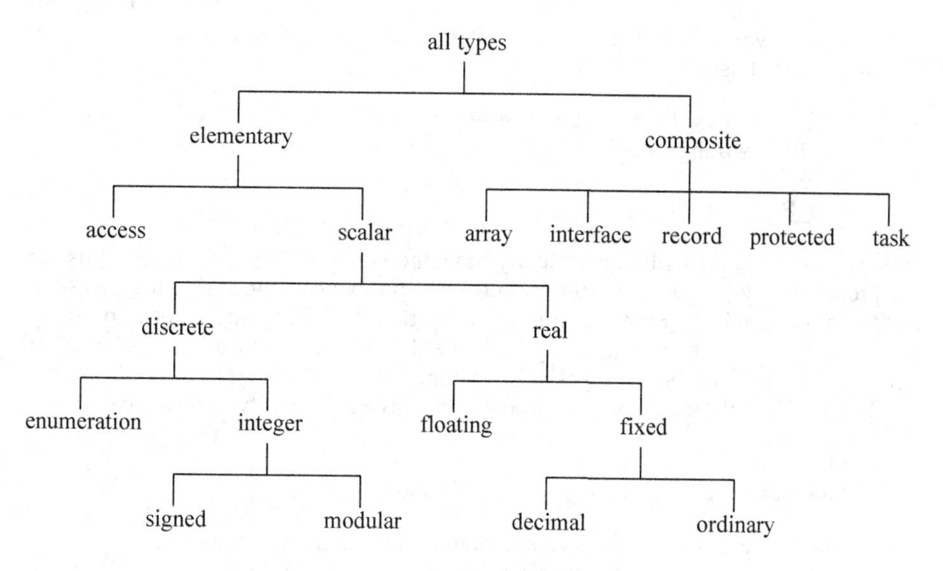

Figure 6.4 Ada type hierarchy.

The attributes Pos, Val, Succ, and Pred apply to all discrete types (and subtypes) because the operations reflect the discrete nature of the values. We explained their meaning with enumeration types in Section 6.6. In the case of the type Integer the position number is simply the number itself so

```
Integer'Pos(N) = N
Integer'Val(N) = N
Integer'Succ(N) = N+1
Integer'Pred(N) = N-1
```

The application of these attributes to integers does at first sight seem pretty futile, but when we come to the concept of generic units in Chapter 19 we will see that it is convenient to allow them to apply to all discrete types.

Furthermore, despite our previous remarks about the continuous nature of real types, the attributes Pred and Succ (but not Pos and Val) also apply to all real types. They return the adjacent implemented number.

The attributes First and Last also apply to all scalar types and subtypes including real types. The attribute Range provides a useful shorthand since S'Range is equivalent to S'First .. S'Last. The attributes First_Valid and Last_Valid apply to static scalar subtypes and are of value with static predicates, see Section 16.4.

There are two other functional attributes for all scalar types and subtypes. These are Max and Min which take two parameters and return the maximum or minimum value respectively. Thus

```
Integer'Min(5, 10) = 5
```

Again we emphasize that Max, Min, Pos, Val, Succ, and Pred for a subtype are identical to the corresponding operations on the base type, whereas in the case of First, Last, and Range this is not so.

If an object is declared without an initial value then it is undefined unless a default initial value is given for the subtype. This can be done using the aspect Default_Value thus

```
type Colour is (Red, Amber, Green)
  with Default_Value => Red;
...
The_Signal: Colour;
```

and then The_Signal will automatically have the value Red. We cannot do this for the predefined types such as Integer and Float, but we can for our own types such as My_Integer and My_Float mentioned in Section 2.3. This aspect does not apply to access types which automatically have the default value of **null** as explained in Section 11.2. For composite types see Sections 8.2 and 8.6.

Finally, note the difference between type conversion and type qualification.

```
Float(I)                -- conversion
Integer'(I)             -- qualification
```

In the case of a conversion we are changing the type, whereas in the case of a qualification we are just stating it (often to overcome an ambiguity). As a mnemonic aid *q*ualification uses a *q*uote.

If we use a subtype name then Constraint_Error could be raised. Thus

```
Positive(F)
```

converts the value of F to integer and then checks that it is positive, whereas

```
Positive'(I)
```

just checks that I is positive. In both cases the result is the checked value which is then used in an overall expression; these checks cannot just stand alone. All subtype properties including predicates (see Section 16.4) are checked (AI12-100).

Exercise 6.8

1 Evaluate

(a) Boolean'Min(True, False) (b) Weekday'Max(Tue, Sat)

6.9 Expression summary

All the operators introduced so far are shown in Table 6.1 grouped by precedence level. In all the cases of binary operators except for **, the two operands must be of the same type.

We have actually now introduced all the operators of Ada except for one (&) although as we shall see there are further possible meanings to be added.

There are also two membership tests which apply to all scalar types (among others). These are **in** and **not in**. They are technically not operators although their precedence is the same as that of the relational operators =, /= and so on. They enable us to test whether a value lies within a specified range (including the end values) or satisfies a constraint implied by a subtype. The first operand is therefore a scalar expression, the second is a range or a subtype mark and the result is, of course, of type Boolean. Simple examples are

```
I not in 1 .. 10
I in Positive
Today in Weekday
```

Note that this is one of the situations where we have to use a subtype mark rather than a subtype indication. We could not replace the last example by

```
Today in Day range Mon .. Fri
```

although we could write

```
Today in Mon .. Fri
```

Ada seems a bit curious here!

The test **not in** is equivalent to using **in** and then applying **not** to the result, but **not in** is usually more readable. So the first expression above could be written as

```
not (I in 1 .. 10)
```

where the parentheses are necessary.

Table 6.1 Simple scalar operators.

Operator	*Operation*	*Operand(s)*	*Result*
and	conjunction	Boolean	Boolean
or	inclusive or	Boolean	Boolean
xor	exclusive or	Boolean	Boolean
=	equality	any	Boolean
/=	inequality	any	Boolean
<	less than	scalar	Boolean
<=	less than or equals	scalar	Boolean
>	greater than	scalar	Boolean
>=	greater than or equals	scalar	Boolean
+	addition	numeric	same
–	subtraction	numeric	same
+	identity	numeric	same
–	negation	numeric	same
*	multiplication	Integer	Integer
		Float	Float
/	division	Integer	Integer
		Float	Float
mod	modulo	Integer	Integer
rem	remainder	Integer	Integer
**	exponentiation	Integer, Natural	Integer
		Float, Integer	Float
not	negation	Boolean	Boolean
abs	absolute value	numeric	same

The reason that **in** and **not in** are not technically operators is that they cannot be overloaded as is explained in Chapter 10 when we deal with subprograms in detail.

More elaborate forms of membership tests which were introduced in Ada 2012 are described in Section 9.1.

There are also two short circuit control forms **and then** and **or else** which like **in** and **not in** are not technically classified as operators.

The form **and then** is closely related to the operator **and**, whereas **or else** is closely related to the operator **or**. They may occur in expressions and have the same precedence as **and**, **or**, and **xor**. The difference lies in the rules regarding the evaluation of their operands.

In the case of **and** and **or**, both operands are always evaluated but the order is not specified. In the case of **and then** and **or else** the left hand operand is always evaluated first and the right hand operand is only evaluated if it is necessary in order to determine the result.

So in

X **and then** Y

X is evaluated first. If X is false, the answer is false whatever the value of Y so Y is not evaluated. If X is true, Y has to be evaluated and the value of Y is the answer.

Similarly in

```
X or else Y
```

X is evaluated first. If X is true, the answer is true whatever the value of Y so Y is not evaluated. If X is false, Y has to be evaluated and the value of Y is the answer.

The forms **and then** and **or else** should be used in cases where the order of evaluation matters. A common circumstance is where the first condition protects against the evaluation of the second condition in circumstances that could raise an exception.

Suppose we need to test

```
I/J > K
```

and we wish to avoid the risk that J is zero. In such a case we could write

```
J /= 0 and then I/J > K
```

and we would then know that if J is zero there is no risk of an attempt to divide by zero. The observant reader will realize that this is not a very good example because one could usually write I>K*J (assuming J positive) – but even here we could get overflow. Better examples occur with arrays and access types and will be mentioned in due course.

Like **and** and **or**, the forms **and then** and **or else** cannot be mixed without using parentheses.

We now summarize the primary components of an expression (that is the things upon which the operators operate) that we have met so far. They are

- identifiers such as Colour
- literals such as 4.6, 2#101#
- type conversions such as Integer(F)
- qualified expressions such as Colour'(Amber)
- attributes such as Integer'Last
- function calls such as Day'Succ(Today)

A full consideration of functions will be found in Chapter 10. However, it is worth noting at this point that a function with one parameter is called by following its name by the parameter in parentheses. The parameter can be any expression of the appropriate type and could include further function calls. We will assume for the moment that we have available a simple mathematical library containing familiar functions such as

Sqrt	square root
Log	logarithm to base 10
Ln	natural logarithm
Exp	exponential function
Sin	sine
Cos	cosine

In each case they take a parameter of type Float and return a result of type Float.

We are now in a position to write statements such as

```
Root := (–B+Sqrt(B**2–4.0*A*C)) / (2.0*A);
Sin2x := 2.0*Sin(X)*Cos(X);
```

A note on errors. The reader will have noticed that whenever anything could go wrong we have usually stated that the exception Constraint_Error will be raised. This is a general exception which applies to all sorts of violations of ranges. Constraint_Error is also raised if something goes wrong with the evaluation of an arithmetic expression itself before an attempt is made to store the result. An obvious example is an attempt to divide by zero.

As well as exceptions (which are discussed in detail in Chapter 15) there are erroneous constructs and bounded errors as mentioned in Chapter 2. The use of a variable before a value has been assigned to it (either explicitly or through the aspect Default_Value) is an important example of a bounded error. In addition there are situations where the order of evaluation is not defined and which could give rise to a nonportable program. Two cases which we have encountered so far where the order is not defined are

- The destination variable in an assignment statement may be evaluated before or after the expression to be assigned.
- The order of evaluation of the two operands of a binary operator is not defined.

(In the first case it should be realized that the destination variable could be an array component such as A(I+J) and so the expression I+J has to be evaluated as part of evaluating the destination variable; we will deal with arrays in Chapter 9.) Examples where these orders matter cannot be given until we deal with functions in Chapter 10.

This is perhaps a good moment to note that the use of Default_Value does have risks. Although it will prevent a variable from having a junk value, it can hide errors where we have inadvertently read an object before assigning a value to it. We can of course always make the default some easily recognized value such as 999 which will then show up in a debugger.

Other forms of expressions including if expressions, case expressions, and quantified expressions are described in Chapter 9 in which we also discuss further forms of membership tests.

We finish this chapter by introducing an intriging new feature of Ada 2022. It is the ability to use the symbol @ as a neat shorthand for the left hand side of an assignment. Thus rather than X := X +1; we can write X := @ + 1; That is not particularly lovely and sort of mimics C++. But there are two terrific advantages. It can be used in any context and not just for addition. But the big advantage is when the left hand side is a bit cumbersome. Thus the assignment in the Game of Life in Section 11.4 can be written as

```
C.Total_Neighbour_Count := @ + C.Neighbour_Count(I).all;
```

which is crystal clear. One does not have to pore over the text to understand just what is going on. Why was @ chosen for this role? Well it sticks out like a sore thumb and is most unlikely to be confused with anything else. Note that it is referred to as the Target Name Symbol (AI12-125-3).

Exercise 6.9

1 Rewrite the following mathematical expressions in Ada.

(a) $2\pi\sqrt{l/g}$ – period of a pendulum

(b) $\dfrac{m_0}{\sqrt{1-v^2/c^2}}$ – mass of relativistic particle

(c) $\sqrt{(2\pi n)}.n^n.e^{-n}$ – Stirling's approximation for $n!$ (integral n)

2 Rewrite **1**(c) replacing n by the real value x.

Checklist 6

Declarations and statements are terminated by a semicolon.
Initialization, like assignment, uses :=.
Any initial value is evaluated for each object in a declaration.
Elaboration of declarations is linear.
The identifier of an object may not be used in its own declaration.
Each type definition introduces a quite distinct type.
A subtype is not a new type but merely a shorthand for a type with a possible constraint.
A type is always static, a subtype need not be.
Distinguish **rem** and **mod** for negative operands.
Exponentiation with a negative exponent only applies to real types.
Take care with the precedence of the unary operators.
A scalar type cannot be empty, a subtype can.
Max, Min, Pos, Val, Succ, and Pred on subtypes are the same as on the base type.
First and Last are different for subtypes.
Qualification uses a quote.
Order of evaluation of binary operands is not defined.
Distinguish **and**, **or** and **and then**, **or else**.
Subtypes can be defined by a null exclusion as well as by a constraint.
The aspect Default_Value for scalar subtypes was added in Ada 2012.
Further forms of membership tests are described in Section 9.1.
If expressions, case expressions, and quantified expressions are described in Chapter 9.
There are also raise expressions which are described in Section 15.2.
Subtype predicates are described in Section 16.4.

New in Ada 2022

An assignment can use @ on the right hand side as an abbreviation for the left hand side. It is known as the Target Name Symbol.

Declare expressions and reduction expressions are added, see Sections 9.5 and 9.6.

Parallel forms of blocks are also added, see Section 22.8.

Attributes Enum_Val and Enum_Rep are added, see Section 25.1.

7 Control Structures

7.1 If statements
7.2 Case statements
7.3 Loop statements
7.4 Goto statements and labels
7.5 Statement classification

This chapter describes the three bracketed sequential control structures of Ada. These are the if statement, the case statement, and the loop statement. These three statements enable programs to be written with a clear flow of control without using goto statements and labels. However, for pragmatic reasons, Ada does have a goto statement and this is also described in this chapter.

The three control structures exhibit a similar bracketing style. There is an opening reserved word **if**, **case** or **loop** and this is matched at the end of the structure by the same reserved word preceded by **end**. The whole is, as usual, terminated by a semicolon. So we have

```
if              case              loop
  ...             ...               ...
end if;         end case;         end loop;
```

In the case of the loop statement the word **loop** can be preceded by an iteration scheme commencing with **for** or **while**.

Corresponding structures for expressions such as if expressions and case expressions which were introduced in Ada 2012 are described in Chapter 9.

7.1 If statements

The simplest form of if statement starts with the reserved word **if** followed by a Boolean expression and the reserved word **then**. This is then followed by a sequence of statements which will be executed if the Boolean expression turns out to be True. The end of the sequence is indicated by the closing **end if**. The Boolean expression can, of course, be of arbitrary complexity and the sequence of statements can be of arbitrary length.

A simple example is

```
if Hungry then
   Cook;
   Eat;
   Wash_Up;
end if;
```

In this, Hungry is a Boolean variable and Cook, Eat and Wash_Up are various subprograms describing the details of the activities. The statement Eat; merely calls the corresponding subprogram (subprograms are dealt with in detail in Chapter 10).

The effect of this if statement is that if variable Hungry is True then we call the subprograms Cook, Eat and Wash_Up in sequence and otherwise we do nothing. In either case we then obey the statement following the if statement.

Note how we indent the statements to show the flow structure of the program. This is most important since it enables the program to be understood so much more easily. The **end if** should be underneath the corresponding **if** and the **then** is best placed on the same line as the **if**.

Sometimes, if the whole statement is very short it can all go on one line

```
if X < 0.0 then X := -X; end if;
```

Note that **end if** will usually be preceded by a semicolon. This is because semicolons terminate statements rather than separate them as in some languages. But see Section 7.4 regarding labels at the end of a sequence of statements.

Often we will want to do alternative actions according to the value of the condition. In this case we add **else** followed by the alternative sequence to be obeyed if the condition is False. We saw an example of this in the previous chapter

```
if Today = Sun then
   Tomorrow := Mon;
else
   Tomorrow := Day'Succ(Today);
end if;
```

In Ada 2012 and Ada 2022 this can be rewritten using an if expression thus

```
Tomorrow :=
      (if Today = Sun then Mon else Day'Succ(Today));
```

as will be explained in detail in Chapter 9.

The statements in the sequences in an if statement after **then** and **else** can be quite arbitrary and so could be further nested if statements.

Suppose we have to solve the quadratic equation

$$ax^2 + bx + c = 0$$

The first thing to check is a. If $a = 0$ then the equation degenerates into a linear equation with a single root $-c/b$. (Mathematicians will understand that the other root has slipped off to infinity.) If a is not zero then we test the discriminant $b^2 - 4ac$ to see whether the roots are real or complex. We could program this as

```
if A = 0.0 then
        -- linear case
else
   if B**2 - 4.0*A*C >= 0.0 then
        -- real roots
   else
        -- complex roots
   end if;
end if;
```

Observe the repetition of **end if**. This is rather ugly and occurs sufficiently frequently to justify an additional construction. This uses the reserved word **elsif** as follows

```
if A = 0.0 then
        -- linear case
elsif B**2 - 4.0*A*C >= 0.0 then
        -- real roots
else
        -- complex roots
end if;
```

This construction emphasizes the essentially equal status of the three cases and also the sequential nature of the tests.

The **elsif** part can be repeated an arbitrary number of times and the final **else** part is optional. The behaviour is simply that each condition is evaluated in turn until one that is True is encountered; the corresponding sequence is then obeyed. If none of the conditions turns out to be True then the else part, if any, is taken; if there is no else part then none of the sequences is obeyed.

Note the spelling of **elsif**. It is the only reserved word of Ada that is not an English word (apart from operators such as **xor**). Note also the layout – we align **elsif** and **else** with the **if** and **end if** and all the sequences are indented equally.

As a further example, suppose we are drilling soldiers and they can obey four different orders described by

```
type Move is (Left, Right, Back, On);
```

and that their response to these orders is described by calling subprograms Turn_Left, Turn_Right and Turn_Back or by doing nothing at all respectively. Suppose that the variable Order of type Move contains the order to be obeyed. We could then write the following

```
if Order = Left then
   Turn_Left;
else
   if Order = Right then
      Turn_Right;
```

```
      else
         if Order = Back then
            Turn_Back;
         end if;
      end if;
   end if;
```

But it is far clearer and neater to write

```
   if Order = Left then
      Turn_Left;
   elsif Order = Right then
      Turn_Right;
   elsif Order = Back then
      Turn_Back;
   end if;
```

This illustrates a situation where there is no **else** part. However, although better than using nested if statements, this is still a bad solution because it obscures the symmetry and mutual exclusion of the four cases ('mutual exclusion' means that by their very nature only one can apply). We have been forced to impose an ordering on the tests which is quite arbitrary and not the essence of the problem. The proper solution is to use the case statement as we shall see in the next section.

There is no directly corresponding contraction for **then if**. Instead we can often use the short circuit control form **and then** which was discussed in Section 6.9.

So, rather than writing

```
   if J > 0 then
      if I/J > K then
         Action;
      end if;
   end if;
```

we can instead write

```
   if J > 0 and then I/J > K then
      Action;
   end if;
```

and this avoids the repetition of **end if**.

Exercise 7.1

1 The variables Day, Month and Year contain today's date. They are declared as

```
   Day: Integer range 1 .. 31;
   Month: Month_Name;
   Year: Integer range 1901 .. 2399;
```

where

```
   type Month_Name is (Jan, Feb, Mar, Apr, May, Jun,
                       Jul, Aug, Sep, Oct, Nov, Dec);
```

Write statements to update the variables to contain tomorrow's date. What happens if today is 31 Dec 2399?

2 X and Y are two variables of type Float. Write statements to swap their values, if necessary, to ensure that the larger value is in X. Use a block to declare a temporary variable T.

7.2 Case statements

A case statement allows us to choose one of several sequences of statements according to the value of an expression. For instance, the example of the drilling soldiers should be written as

```
case Order is
   when Left => Turn_Left;
   when Right => Turn_Right;
   when Back => Turn_Back;
   when On => null;
end case;
```

All possible values of the expression must be provided for in order to guard against accidental omissions. If, as in this example, no action is required for one or more values then the null statement has to be used. The null statement, written as just **null**; does absolutely nothing but its presence indicates that we truly want to do nothing. The sequence of statements here, as in the if statement, must contain at least one statement. (There is no empty statement as in Pascal and C.)

It often happens that the same action is desired for several values of the expression. Consider the following

```
case Today is
   when Mon | Tue | Wed | Thu => Work;
   when Fri                   => Work; Party;
   when Sat | Sun             => null;
end case;
```

This expresses the idea that on Monday to Thursday we go to work. On Friday we also go to a party. At the weekend we do nothing. The alternatives are separated by the vertical bar character. Note again the use of a null statement.

If several successive values have the same action then it is more convenient to use a range

```
when Mon .. Thu => Work;
```

We can also express the idea of a default action to be taken by all values not explicitly stated by using the reserved word **others**. So we can write

```
case Today is
   when Mon .. Thu => Work;
   when Fri        => Work; Party;
   when others     => null;
end case;
```

It is possible to have ranges as alternatives. The various possibilities can best be seen from the formal syntax (note that in this production for discrete_choice_list, the vertical bar stands for itself and is not a metasymbol).

```
case_statement ::=
    case selecting_expression is
      case_statement_alternative
     {case_statement_alternative}
    end case;

case_statement_alternative ::=
    when discrete_choice_list => sequence_of_statements

discrete_choice_list ::= discrete_choice { | discrete_choice}

discrete_choice ::= choice_expression | discrete_subtype_indication
                  | range | others

subtype_indication ::= [null exclusion] subtype_mark [constraint]

subtype_mark ::= subtype_name

range ::= range_attribute_reference
        | simple_expression .. simple_expression
```

We see that **when** is followed by one or more discrete choices separated by vertical bars and that a discrete choice may be a choice expression, a discrete subtype indication, a range or **others**. A choice expression, of course, just gives a single value – Fri being a trivial example (a choice expression is an expression excluding membership tests which could be confusing because they might include vertical bars as well). A subtype indication is a subtype mark with optionally a null exclusion (which is not possible in this context) or an appropriate constraint which in this context has to be a range constraint which is just the reserved word **range** followed by the syntactic form range. A discrete choice can also be just a range on its own which is typically two simple expressions separated by two dots – Mon .. Thu being a simple example; a range can also be given by a range attribute which we will meet in Section 8.1. Finally, a discrete choice can be just **others**.

There is considerable flexibility and we can write any of

```
Mon .. Thu
Day range Mon .. Thu
Weekday range Mon .. Thu
```

which all mean the same. Note that there is not much point in using the subtype name followed by a constraint since the constraint alone will do. However, it might be useful to use a subtype name alone when that exactly corresponds to the range required. So we could rewrite the example as

```
case Today is
   when Weekday => Work;
                   if Today = Fri then Party; end if;
   when others    => null;
end case;
```

although this solution feels untidy.

There are various other restrictions that the syntax does not tell us. One is that if we use **others** then it must appear alone and as the last alternative. As stated earlier it covers all values not explicitly covered by the previous alternatives (one can write **others** even if there are no other cases left).

Another very important restriction is that all the choices must be static so that they can be evaluated at compile time. Thus all expressions in discrete choices must be static – in practice they will usually be literals as in our examples. Similarly, if a choice is just a subtype such as Weekday then it too must be static.

At the beginning of this section we remarked that all possible values of the expression after **case** must be provided for. This usually means all values of the type of the expression. However, if the expression is of a simple form and belongs to a static subtype (that is one whose constraints are static expressions and so can be determined at compile time) then only values of that subtype need be provided for. In other words, if the compiler can tell that only a subset of values is possible then only that subset need and must be covered. The simple forms allowed for the expression are the name of an object of the static subtype (and that includes a function call whose result is of the static subtype) or a qualified or converted expression whose subtype mark is that of the static subtype.

Thus in the example, since Today is declared to be of type Day without any constraints, all values of the type Day must be provided for. However, if Today were of the static subtype Weekday then only the values Mon .. Fri would be possible and so only these could and need be covered. Even if Today is not constrained we can still write the expression as the qualified expression Weekday'(Today) and then again only Mon .. Fri are possible. So we could write

```
case Weekday'(Today) is
   when Mon .. Thu => Work;
   when Fri        => Work; Party;
end case;
```

but, of course, if Today happens to take a value not in the subtype Weekday (that is, Sat or Sun) then Constraint_Error will be raised. Mere qualification cannot prevent Today from being Sat or Sun. So this is not really a solution to our original problem.

As further examples, suppose we had variables

```
I: Integer range 1 .. 10;
J: Integer range 1 .. N;
```

where N is not static. Then we know that I belongs to a static subtype (albeit anonymous) whereas we cannot say the same about J. If I is used as an expression in a case statement then only the values 1 .. 10 have to be catered for, whereas if J is so used then the full range of values of type Integer (Integer'First .. Integer'Last) have to be catered for.

The above discussion on the case statement has no doubt given the reader the impression of considerable complexity. It therefore seems wise to summarize the key points which will in practice need to be remembered

- Every possible value of the expression after **case** must be covered once and once only.
- All values and ranges after **when** must be static.
- If **others** is used it must be last and on its own.

Finally, note that the use of **others** will be necessary if the expression after **case** is of the type *universal_integer*. This type will be discussed in Chapter 17 when we consider numeric types in detail. The main use of the type occurs with integer literals. However, certain attributes such as Pos are of the type *universal_integer* and so if the case expression is of the form Character'Pos(C) then **others** will be required even if the values it covers could never arise.

Exercise 7.2

1. Rewrite Exercise 7.1(**1**) to use a case statement to set the correct value in End_Of_Month.
2. A vegetable gardener digs in winter, sows seed in spring, tends the growing plants in summer and harvests the crop in the autumn or fall. Write a case statement to call the appropriate subprogram Dig, Sow, Tend, or Harvest according to the month M. Declare appropriate subtypes if desired.
3. An improvident man is paid on the first of each month. For the first ten days he gorges himself, for the next ten he subsists and for the remainder he starves. Call subprograms Gorge, Subsist and Starve according to the day D. Assume the variable End_Of_Month has been set and that D is declared as

```
D: Integer range 1 .. End_Of_Month;
```

7.3 Loop statements

Loop statements can take various forms including special forms for iterating over containers. The simplest form of loop statement is

```
loop
   sequence_of_statements
end loop;
```

The statements of the sequence are then repeated indefinitely unless one of them terminates the loop by some means. So immortality could be represented by

```
loop
   Work;
   Eat;
   Sleep;
end loop;
```

As a more concrete example consider the problem of computing the base e of natural logarithms from the infinite series

$$e = 1 + 1/1! + 1/2! + 1/3! + 1/4! + \ldots$$

where

$$n! = n \times (n-1) \times (n-2) \ldots 3 \times 2 \times 1$$

A possible solution is

```
declare
   E: Float := 1.0;
   I: Integer := 0;
   Term: Float := 1.0;
begin
   loop
      I := I + 1;
      Term := Term / Float(I);
      E := E + Term;
   end loop;
   ...
```

Each time around the loop a new term is computed by dividing the previous term by I. The new term is then added to the sum so far which is accumulated in E. The term number I is an integer because it is logically a counter and so we have to write Float(I) as the divisor. The series is started by setting values in E, I and Term which correspond to the first term (that for which I = 0).

The computation then goes on for ever with E becoming a closer and closer approximation to *e*. In practice, because of the finite accuracy of the computer, Term will become zero and continued computation will be pointless. But in any event we presumably want to stop at some point so that we can do something with our computed result. We can do this with the statement

```
exit;
```

If this is obeyed inside a loop then the loop terminates at once and control passes to the point immediately after **end loop**.

Suppose we decide to stop after N terms of the series – that is when I = N. We can do this by writing the loop as

```
loop
   if I = N then exit; end if;
   I := I + 1;
   Term := Term / Float(I);
   E := E + Term;
end loop;
```

The construction

```
if condition then exit; end if;
```

is so common that a special shorthand is provided

```
exit when condition;
```

So we now have

```
loop
   exit when I = N;
   I := I + 1;
   Term := Term / Float(I);
   E := E + Term;
end loop;
```

Although an exit statement can appear anywhere inside a loop – it could be in the middle or near the end – a special form of loop is provided for the frequent case where we want to test a condition at the start of each iteration. This is the while statement and uses the reserved word **while** followed by the condition for the loop to be continued. So we could write

```
while I /= N loop
   I := I + 1;
   Term := Term / Float(I);
   E := E + Term;
end loop;
```

The condition is naturally evaluated each time around the loop.

The final form of loop is the for statement which allows a specific number of iterations with a loop parameter taking in turn all the values of a range of a discrete type. Our example could be recast as

```
for I in 1 .. N loop
   Term := Term / Float(I);
   E := E + Term;
end loop;
```

where I takes the values 1, 2, 3, ..., N.

The loop parameter I is implicitly declared by its appearance after **for** and does not have to be declared outside. It takes its type from the range and within the loop behaves as a constant so that it cannot be changed except by the loop mechanism itself. When we leave the loop (by whatever means) I ceases to exist (because it was implicitly declared by the loop) and so we cannot read its final value from outside.

We could leave the loop by an exit statement – if we wanted to know the final value we could copy the value of I into a variable declared outside the loop thus

```
if condition_to_exit then
   Last_I := I;
   exit;
end if;
```

The values of the range are normally taken in ascending order. Descending order can be specified by writing

```
for I in reverse 1 .. N loop
```

but the range itself is always written in ascending order.

It is not possible to specify a numeric step size of other than 1. This should not be a problem since the vast majority of loops go up by steps of 1 and almost all the rest go down by steps of 1. The very few which do behave otherwise can be explicitly programmed using the while form of loop.

The range can be null (as for instance if N happened to be zero or negative in our example) in which case the sequence of statements will not be obeyed at all. Of course, the range itself is evaluated only once and cannot be changed inside the loop.

Thus

```
N := 4;
for I in 1 .. N loop
   ...
   N := 10;
end loop;
```

results in the loop being executed just four times despite the fact that N is changed to ten.

Our examples have all shown the lower bound of the range being 1. This, of course, need not be the case. Both bounds can be arbitrary dynamically evaluated expressions. Furthermore, the loop parameter need not be of integer type. It can be of any discrete type, as determined by the range.

We could, for instance, simulate a week's activity by

```
for Today in Mon .. Sun loop
   case Today is
      ...
   end case;
end loop;
```

This implicitly declares Today to be of type Day and obeys the loop with the values Mon, Tue, ..., Sun in turn.

The other forms of discrete range (using a type or subtype name) are of advantage here. The essence of Mon .. Sun is that it embraces all the values of the type Day. It is therefore better to write the loop using a form of discrete range that conveys the idea of completeness

```
for Today in Day loop
   ...
end loop;
```

And again since we know that we do nothing at weekends anyway we could write

```
for Today in Day range Mon .. Fri loop
```

or better

```
for Today in Weekday loop
```

It is interesting to note a difference regarding the determination of types in the case statement and for statement. In the case statement, the type of a range after **when** is determined from the type of the expression after **case**. In the for statement, the type of the loop parameter is determined from the type of the range after **in**. The dependency is the other way round.

It is therefore necessary for the type of the range to be unambiguous in the for statement. This is usually the case but if we had two enumeration types with two overloaded literals such as

```
type Planet is (Mercury, Venus, Earth, Mars, Jupiter,
                Saturn, Uranus, Neptune, Pluto);
type Roman_God is (Janus, Mars, Jupiter, Juno, Vesta,
                   Vulcan, Saturn, Mercury, Minerva);
```

then

```
for X in Mars .. Saturn loop                -- illegal
```

would be ambiguous and the compiler would not compile our program. We could resolve the problem by qualifying one of the expressions

```
for X in Planet'(Mars) .. Saturn loop
```

or (probably better) by using a form of range giving the type explicitly

```
for X in Planet range Mars .. Saturn loop
```

On the other hand

```
for X in Mars .. Vulcan loop
```

is not ambiguous because Vulcan can only be a Roman god and not a planet.

When we have dealt with numerics in more detail we will realize that the range 1 .. 10 is not necessarily of type Integer (it might be Long_Integer). A general application of the rule that the type must not be ambiguous in a for statement would lead us to have to write

```
for I in Integer range 1 .. 10 loop
```

However, this would be very tedious and so in such cases the type Integer can be omitted and is then implied by default. We can therefore conveniently write

```
for I in 1 .. 10 loop
```

More general expressions are also allowed so that we could also write

```
for I in -1 .. 10 loop
```

although we recall that -1 is not a literal as explained in Section 5.4. We will return to this topic in Section 17.1.

Finally, we reconsider the exit statement. The simple form encountered earlier always transfers control to immediately after the innermost embracing loop. However, loops may be nested and sometimes we may wish to exit from a nested construction. As an example suppose we are searching in two dimensions

```
for I in 1 .. N loop
   for J in 1 .. M loop
      -- if values of I and J satisfy
      -- some condition then leave nested loop
   end loop;
end loop;
```

A simple exit statement in the inner loop would merely take us to the end of that loop and we would have to recheck the condition and exit again. This can be avoided by naming the outer loop and using the name in the exit statement thus

```
Search:
for I in 1 .. N loop
   for J in 1 .. M loop
      if condition_OK then
         I_Value := I;
         J_Value := J;
         exit Search;
      end if;
   end loop;
end loop Search;
-- control passes here
```

A loop is named by preceding it with an identifier and colon. (It looks remarkably like a label in other languages but it is not and cannot be 'gone to'.) The identifier must be repeated between the corresponding **end loop** and the semicolon. The conditional form of exit can also refer to a loop by name; so

```
exit Search when condition;
```

transfers control to the end of the loop named Search if the condition is True.

A simple variation in Ada 2022 that can be applied to any iteration is the introduction of a filter to eliminate certain values of the iterator from consideration (AI12-250). Thus if we wanted to iterate over all values of J from 1 to 100 except those that are a multiple of 7 we could write

```
for J in 1 .. 100 when J mod 7 /= 0 loop ...
```

If we wanted to exclude Tuesday from some activity we might write

```
for Today in Day when Today /= Tuesday loop ...
```

Filters can also be used with other forms of iterator (see Section 8.1); thus given a two dimensional real array AA, writing

```
for E of AA when E < 0.0 loop
   E := 0.0;
end loop;
```

changes the value of all negative components of AA to zero.

In Ada 2012 and Ada 2022, other forms of for loop can be used with arrays (see the end of Section 8.1) and with containers (see Chapter 24).

Exercise 7.3

1 The statement Get(I); reads the next value from the input file into the integer variable I. Write statements to read and add together a series of numbers. The end of the series is indicated by a dummy negative value.

2 Write statements to determine the power of 2 in the factorization of N. Compute the result in Count but do not alter N.

3 Compute

$$g = \sum_{i=1}^{n} 1/i - \log n$$

(As $n \to \infty$, $g \to \gamma = 0.577215665...$, Euler's constant.)

7.4 Goto statements and labels

Many will be surprised that a modern programming language should contain a goto statement at all. It is now considered to be extremely bad practice to use goto statements because of the resulting difficulty in proving correctness of the program, maintenance and so on.

So why provide a goto statement? The main reason concerns automatically generated programs. If we try to transliterate (by hand or machine) a program from some other language into Ada then the goto will probably be useful. Another example might be where the program is generated automatically from some high level specification. Finally, there may be cases where the goto is the neatest way – perhaps as a way out of some deeply nested structure – but the alternative of raising an exception (see Chapter 15) could also be considered.

In order to put us off using gotos and labels (and perhaps so that our manager can spot them if we do) the notation for a label is unusual and stands out like a sore thumb. A label is an identifier enclosed in double angled brackets thus

```
<<The_Devil>>
```

and a goto statement takes the expected form of the reserved word **goto** followed by the label identifier and semicolon

```
goto The_Devil;
```

Perhaps the most common use of a goto statement is to transfer control to just before the end of a loop in the case when an iteration proves useless and we wish to try the next iteration. Thus we might write

```
for I in ... loop
   ...                                          -- various statements
   if not OK then goto End_Of_Loop; end if;
   ...                                          -- other statements
<<End_Of_Loop>>                                 -- now try another iteration
end loop;
```

Usually a label always immediately precedes a statement. But in Ada 2012 and Ada 2022 it can also be at the end of a sequence of statements as here. In earlier versions of Ada we had to have an explicit null statement at the end of the loop.

A goto statement cannot be used to transfer control into an if, case, or loop statement nor between the arms of an if or case statement. But it can transfer control out of a loop or out of a block.

7.5 Statement classification

The various statements in Ada can be classified as shown in Figure 7.1. All statements can have one or more labels. The simple statements cannot be decomposed lexically into other statements whereas the compound statements can be so decomposed and can therefore be nested. Statements are obeyed sequentially unless one of them is a control statement (or an exception is raised) or is a parallel block or parallel loop. The return statement has both simple and compound forms.

Further detail on the assignment statement is in the next chapter when we discuss composite types. Procedure calls and return statements are discussed in Chapter 10 and the raise statement which is concerned with exceptions is discussed in Chapter 15. The code statement is mentioned in Chapter 25. The remaining traditional statements (entry call, requeue, delay, abort, accept, and select) concern tasking and are dealt with in Chapter 20.

The parallel blocks and parallel loops which were introduced in Ada 2022 are dealt with in Chapter 22.

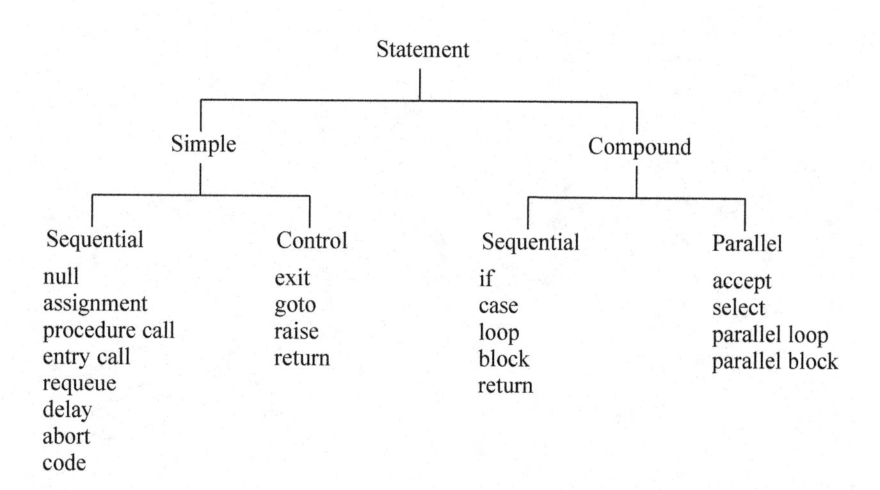

Figure 7.1 Classification of statements.

Checklist 7

Statement brackets must match correctly.

Use **elsif** where appropriate.

The choices in a case statement must be static.

All possibilities in a case statement must be catered for.

If **others** is used it must be last and on its own.

The expression after **case** can be qualified in order to reduce the alternatives.

A loop parameter behaves as a constant.

A named loop must have the name at both ends. Avoid gotos.

Use the recommended layout.

The return statement has both simple and compound forms.

Labels are permitted at the end of a sequence of statements in Ada 2012.

There are other forms of for loops for arrays and containers.

New in Ada 2022

For loops can be used in array aggregates, see Section 8.8, and as part of reduction expressions, see Section 9.6.

There are also parallel forms of blocks and for loops, see Section 22.8.

Filters can be applied to a loop parameter.

8 Arrays and Records

8.1 Arrays
8.2 Array types
8.3 Array aggregates
8.4 Characters and strings
8.5 Arrays of arrays and slices
8.6 One-dimensional array operations
8.7 Records
8.8 General aggregates

This chapter describes the main composite types which are arrays and records. It also completes our discussion of enumeration types by introducing characters and strings. At this stage we discuss arrays fairly completely but consider only the simpler forms of records. Tagged records and interfaces which permit extension and polymorphism are discussed in Chapter 14 and discriminated records which include variant records are deferred until Chapter 18. Tasks and protected types which are also classified as composite are discussed in Chapter 20.

8.1 Arrays

An array is a composite object consisting of a number of components all of the same type (strictly, subtype). An array can be of one, two or more dimensions. A typical array declaration might be

```
A: array (Integer range 1 .. 6) of Float;
```

This declares A to be a variable object which has six components, each of which is of type Float. The individual components are referred to by following the array name with an expression in parentheses giving an integer value in the discrete range 1 .. 6. If this expression, known as the index value, has a value outside the bounds of the range, then the exception Constraint_Error will be raised. We could set zero in each component of A by writing

```
for I in 1 .. 6 loop
   A(I) := 0.0;
end loop;
```

An array can be of several dimensions, in which case a separate range is given for each dimension. So

```
AA: array (Integer range 0 .. 2, Integer range 0 .. 3) of Float;
```

is an array of 12 components in total, each of which is referred to by two integer index values, the first in the range 0 .. 2 and the second in the range 0 .. 3. The components of this two-dimensional array could all be set to zero by a nested loop thus

```
for I in 0 .. 2 loop
   for J in 0 .. 3 loop
      AA(I, J) := 0.0;
   end loop;
end loop;
```

The discrete ranges do not have to be static; one could have

```
N: Integer := ... ;
B: array (Integer range 1 .. N) of Boolean;
```

and the value of N at the point when the declaration of B is elaborated would determine the number of components in B. Of course, the declaration of B might be elaborated many times during the course of a program – it might be inside a loop for example – and each elaboration will give rise to a new life of a new array and the value of N could be different each time. Like other declared objects, the array B ceases to exist once we pass the end of the block containing its declaration. Because of 'linear elaboration of declarations' both N and B could be declared in the same declarative part but the declaration of N would have to precede that of B.

The discrete range in an array index follows similar rules to that in a for statement. An important one is that a range such as 1 .. 6 implies type Integer so we could have written

```
A: array (1 .. 6) of Float;
```

However, an array index subtype could be a subtype of any discrete type. We could for example have

```
Hours_Worked: array (Day) of Float;
```

This array has seven components denoted by Hours_Worked(Mon), ..., Hours_Worked(Sun). We could set suitable values in these variables by

```
for D in Weekday loop
   Hours_Worked(D) := 8.0;
end loop;
Hours_Worked(Sat) := 0.0;
Hours_Worked(Sun) := 0.0;
```

If we only wanted to declare the array Hours_Worked to have components corresponding to Mon .. Fri then we could write

```
Hours_Worked: array (Day range Mon .. Fri) of Float;
```

or (better)

```
Hours_Worked: array (Weekday) of Float;
```

Arrays have various attributes relating to their indices. A'First and A'Last give the lower and upper bound of the first (or only) index of A. So using our last declaration of Hours_Worked

```
Hours_Worked'First = Mon
Hours_Worked'Last = Fri
```

A'Length gives the number of values of the first (or only) index. So

```
Hours_Worked'Length = 5
```

A'Range is short for A'First .. A'Last. So

```
Hours_Worked'Range   is   Mon .. Fri
```

The same attributes can be applied to the various dimensions of a multi-dimensional array by adding the dimension number in parentheses. It has to be a static expression. So, in the case of our two-dimensional array AA we have

```
AA'First(1)  = 0                 AA'First(2)  = 0
AA'Last(1)  = 2                  AA'Last(2)  = 3
AA'Length(1)  = 3                AA'Length(2)  = 4
```

and

```
AA'Range(1)   is   0 .. 2        AA'Range(2)   is   0 .. 3
```

The first dimension is assumed if (1) is omitted. It is perhaps better practice to specifically state (1) for multidimensional arrays and omit it for one-dimensional arrays.

It is always best to use attributes whenever possible in order to reflect relationships between entities because it generally means that if a program is modified, then the modifications are localized.

The Range attribute is particularly useful with loops. Our earlier examples are better written as

```
for I in A'Range loop
   A(I) := 0.0;
end loop;
for I in AA'Range(1) loop
   for J in AA'Range(2) loop
      AA(I, J) := 0.0;
   end loop;
end loop;
```

The Range attribute can also be used in a declaration. Thus

```
J: Integer range A'Range;
```

is equivalent to

```
J: Integer range 1 .. 6;
```

If a variable is to be used to index an array as in A(J) it is usually best if the variable has the same constraints as the discrete range in the array declaration. This will usually minimize the run-time checks necessary. It has been found that in such circumstances it is usually the case that the index variable J is assigned to less frequently than the array component A(J) is accessed. We will return to this topic in Section 15.3.

The array components we have seen are just variables in the ordinary way. They can therefore be assigned to and used in expressions.

Like other variable objects, arrays can be given an initial value. This will often be denoted by an aggregate which is the literal form for an array value. The simplest form of aggregate (a positional aggregate) consists of a list of expressions giving the values of the components in order, separated by commas and enclosed in parentheses. So we could initialize the array A by

```
A: array (1 .. 6) of Float := (0.0, 0.0, 0.0, 0.0, 0.0, 0.0);
```

In the case of a multidimensional array the aggregate is written in a nested form

```
AA: array (0 .. 2, 0 .. 3) of Float := ((0.0, 0.0, 0.0, 0.0),
                                        (0.0, 0.0, 0.0, 0.0),
                                        (0.0, 0.0, 0.0, 0.0));
```

and this illustrates that the first index is the 'outer' one. Or thinking in terms of rows and columns, the first index is the row number.

An aggregate must be complete. If we initialize any component of an array, we must initialize them all.

The initial values for the individual components need not be literals, they can be any expressions. These expressions are evaluated when the declaration is elaborated but the order of evaluation of the expressions in the aggregate is not specified.

An array can be declared as constant in which case an initial value is mandatory as explained in Section 6.1. Constant arrays are of frequent value as look-up tables. The following array can be used to determine whether a particular day is a working day or not

```
Work_Day: constant array (Day) of Boolean :=
                   (True, True, True, True, True, False, False);
```

An interesting example would be an array enabling tomorrow to be determined without worrying about the end of the week.

```
Tomorrow: constant array (Day) of Day :=
                   (Tue, Wed, Thu, Fri, Sat, Sun, Mon);
```

For any day D, Tomorrow (D) is the following day.

It should be noted that the array components can be of any definite type or subtype. Also the dimensions of a multidimensional array can be of different discrete types. An extreme example would be

```
Strange: array (Colour, 2 .. 7, Weekday range Tue .. Thu)
                        of Planet range Mars .. Saturn;
```

Note we said that the component type must be definite; the distinction between definite and indefinite types is explained in the next section.

Ada 2012 and Ada 2022 provide a much neater way of iterating over an array when similar actions are to be performed on each component. So rather than

```
for I in A'Range loop
   A(I) := A(I) + 1.0;
end loop;
```

we can instead write

```
for E of A loop
   E := E + 1.0;
end loop;
```

The loop parameter E thus takes the identity of each component in turn. In the case of the two-dimensional array AA, rather than the double loop shown earlier we can set each component to zero by simply writing

```
for E of AA loop
   E := 0.0;
end loop;
```

and it iterates over the array in the expected manner. The index of the last dimension varies fastest matching the behaviour in the traditional version. Note **of** rather then **in** in the loop specification.

This alternative iteration mechanism was introduced into Ada 2012 largely for improving iteration over containers as will be explained in Chapter 21 but it also applies to arrays for convenience and uniformity. See also Section 7.2 for filters.

Exercise 8.1

1 Declare an array F of integers with index running from 0 to N. Write statements to set the components of F equal to the Fibonacci numbers given by

$$F_0 = 0, \; F_1 = 1, \; F_i = F_{i-1} + F_{i-2} \qquad i > 1$$

2 Write statements to find the index values I, J of the maximum component of

```
A: array (1 .. N, 1 .. M) of Float;
```

3 Declare an array Days_In_Month giving the number of days in each month. See Exercise 7.1(**1**). Use it to rewrite that example. See also Exercise 7.2(**1**).

4 Declare an array Yesterday analogous to the example Tomorrow above.

5 Declare a constant array Bor such that

```
Bor(P, Q) = P or Q
```

6 Declare a constant unit matrix Unit of order 3. A unit matrix is one for which all components are zero except those whose indices are equal which have value one.

8.2 Array types

The arrays we introduced in the previous section did not have an explicit type name. They were in fact of various anonymous types. This is one of the few cases in Ada where an object can be declared without naming the type – the other cases are access types, tasks, and protected objects.

Reconsidering the first example in the previous section, we could write

```
type Vector_6 is array (1 .. 6) of Float;
A: Vector_6;
```

where we have declared A using the type name in the usual way.

An advantage of using a type name is that it enables us to assign whole arrays that have been declared separately. If we also have

```
B: Vector_6;
```

then we can write

```
B := A;
```

which has the effect of

```
B(1) := A(1);  B(2) := A(2);  ... B(6) := A(6);
```

although the order of assigning the components is not relevant.

On the other hand if we had written

```
C: array (1 .. 6) of Float;
D: array (1 .. 6) of Float;
```

then D := C; is illegal because C and D are not of the same type. They are of different types both of which are anonymous. The underlying rule of type equivalence is that every type definition introduces a new type and in this case the syntax tells us that an array type definition is the piece of text from **array** up to (but not including) the semicolon.

Moreover, even if we had written

```
C, D: array (1 .. 6) of Float;
```

then D := C; would still have been illegal. This is because of the rule mentioned in Section 6.1 that such a multiple declaration is only a shorthand for the two declarations above. There are therefore still two distinct type definitions even though they are not explicit.

Whether or not we introduce a type name for particular arrays depends very much on the abstract view of each situation. If we are thinking of the array as a complete object in its own right then we should use a type name. If, on the other hand, we are thinking of the array as merely an indexable conglomerate not related as a whole to other arrays then it should probably be of an anonymous type.

Arrays like Tomorrow and Work_Day of the previous section are good examples of arrays which are of the anonymous category. To be forced to introduce a type name for such arrays would introduce unnecessary clutter and a possibly false sense of abstraction.

On the other hand, if we are manipulating lots of arrays of type Float of length 6 then there is a common underlying abstract type and so it should be named. The reader might also like to reconsider the example of Pascal's triangle in Section 2.4.

The model for array types introduced so far is still not satisfactory. It does not allow us to represent an abstract view that embraces the commonality between arrays which have different bounds but are otherwise of the same type. In particular, it would not allow the writing of subprograms which could take an array of arbitrary bounds as an actual parameter. So the concept of an unconstrained array type is introduced in which the constraints for the indices are not given. Consider

```
type Vector is array (Integer range <>) of Float;
```

(The compound symbol <> is read as 'box'.)

This says that Vector is the name of a type which is a one-dimensional array of Float components with an Integer index. But the lower and upper bounds are not given; **range** <> is meant to convey the notion of information to be added later.

When we declare objects of type Vector we must supply the bounds. We can do this in various ways. We can introduce a subtype and then declare the objects.

```
subtype Vector_5 is Vector(1 .. 5);
V: Vector_5;
```

Or we can declare the objects directly

```
V: Vector(1 .. 5);
```

In either case the bounds are given by an index constraint which takes the form of a discrete range in parentheses. All the usual forms of discrete range can be used.

The index can also be given by a subtype name, thus

```
type P is array (Positive range <>) of Float;
```

in which case the actual bounds of any declared object must lie within the range implied by the index subtype Positive. An exception to this rule is that the object might have a null range such as 1 .. 0 in which case it would be a null object.

Note that the index subtype must be given by a subtype mark and not by a subtype indication; this avoids the horrid double use of **range** which could otherwise occur as in

```
type Nasty is array (Integer range 1 .. 100 range <>) of ... ;  -- illegal
```

We can now see that when we wrote

```
type Vector_6 is array (1 .. 6) of Float;
```

this was effectively a shorthand for

```
subtype index is Integer range 1 .. 6;
type anon is array (index range <>) of Float;
subtype Vector_6 is anon(1 .. 6);
```

Another useful array type declaration is

```
type Matrix is array (Integer range <>, Integer range <>) of Float;
```

And again we could introduce subtypes thus

```
subtype Matrix_3 is Matrix(1 .. 3, 1 .. 3);
M: Matrix_3;
```

or we could declare the objects directly

```
M: Matrix(1 .. 3, 1 .. 3);
```

An important point to notice is that an array subtype must give all the bounds or none at all. It would be legal to introduce an alternative name for Matrix by

```
subtype Mat is Matrix;
```

in which no bounds are given, but we could not have a subtype that just gave the bounds for one dimension but not the other.

In all of the cases we have been discussing, the ranges need not have static bounds. The bounds could be any expressions and are evaluated when the index constraint is encountered. We could have

```
M: Matrix(1 .. N, 1 .. N);
```

and then the upper bounds of M would be the value of N when M is declared. A range could even be null as would happen in the above case if N turned out to be zero. In this case the matrix M would have no components at all.

There is a further way in which the bounds of an array can be supplied; they can be taken from an initial value. Remember that all constants must have an initial value and variables may have one. The bounds can then be taken from the initial value if they are not supplied directly.

The initial value can be any expression of the appropriate type but will often be an aggregate as shown in the previous section. The form of aggregate shown there consisted of a list of expressions in parentheses. Such an aggregate is known as a positional aggregate since the values are given in position order. In the case of a positional aggregate used as an initial value and supplying the bounds, the lower bound is S'First where S is the subtype of the index. The upper bound is deduced from the number of components. (The bounds of positional aggregates in other contexts will be discussed in the next section.)

There is also a subtle difference between supplying the bounds explicitly in the object declaration or from an initial value. Thus we might have

```
V1: Vector(1 .. 5);                    -- bounds from subtype
V2: Vector := (1 .. 5 => 0.0);         -- bounds from initial value
```

The bounds of V2 are deduced from the initial value given as an aggregate (explained in detail in the next section). In general V1 and V2 will behave in the same way. But there are some situations where there are differences. Although both have the same actual subtype, the so-called nominal subtype is different, see for example the discussion on the aspect Unchecked_Union in Section 25.5.

As another example, suppose we had

```
type W is array (Weekday range <>) of Day;
Next_Work_Day: constant W := (Tue, Wed, Thu, Fri, Mon);
```

then the lower bound of the array is Weekday'First = Mon and the upper bound is Fri. It would not have mattered whether we had written Day or Weekday in the declaration of W because Day'First and Weekday'First are the same.

We can also use the box notation with anonymous array types as follows

```
Next_Work_Day: array (Weekday range <>) of Day :=
                              (Tue, Wed, Thu, Fri, Mon);
```

and there is no need to declare the intermediate type W. Again the bounds are deduced from the aggregate and we have chosen not to make the array a constant.

Using initial values to supply the bounds needs care. Consider

```
Unit_2: constant Matrix := ((1.0, 0.0), (0.0, 1.0));
```

intended to declare a 2 × 2 unit matrix with Unit_2(1, 1) = Unit_2(2, 2) = 1.0 and Unit_2(1, 2) = Unit_2(2, 1) = 0.0.

But disaster! We have actually declared an array whose lower bounds are Integer'First which might be –32768 (2^{15}) or perhaps –2147483648 (2^{31}) or some such number, but is most certainly not 1.

If we declared the type Matrix as

```
type Matrix is array (Positive range <>, Positive range <>) of Float;
```

then all would have been well since Positive'First = 1. So beware that array bounds deduced from an initial value may lead to nasty surprises.

We continue by returning to the topic of whole array assignment. In order to perform such assignment it is necessary that the array expression and the destination array variable have the same type and that the components can be matched. This does not mean that the bounds have to be equal, but merely that the number of components in corresponding dimensions is the same. In other words so that one array can be slid onto the other, giving rise to the term 'sliding semantics'. So we can write

```
V: Vector(1 .. 5);
W: Vector(0 .. 4);
...
V := W;
```

Both V and W are of type Vector and both have five components.

Sliding may occur in several dimensions so the following is also valid

```
P: Matrix(0 .. 1, 0 .. 1);
Q: Matrix(6 .. 7, N .. N+1);
...
P := Q;
```

Equality and inequality of arrays follow similar sliding rules to assignment. Two arrays may only be compared if they are of the same type. They are equal if corresponding dimensions have the same number of components and the matching components are themselves equal. Note, however, that if the dimensions of the two arrays are not of the same length then equality will return False whereas an attempt to assign one array to the other will naturally cause Constraint_Error.

Although assignment and equality can only occur if the arrays are of the same type, nevertheless an array value of one type can be converted to another type if the index types are convertible and the component subtypes are statically the same. The usual notation for type conversion is used. So if we have

```
type Vector is array (Integer range <>) of Float;
type Row is array (Integer range <>) of Float;

V: Vector(1 .. 5);
R: Row(0 .. 4);
```

then

```
R := Row(V);
```

is valid. In fact, since Row is unconstrained, the bounds of Row(V) are those of V. The normal assignment rules then apply. However, if the conversion uses a constrained type or subtype then the bounds are those of the type or subtype and the number of components in corresponding dimensions must be the same. Array type conversion is of particular value when subprograms from different libraries are used together as we shall see in Section 10.3.

Note that the component subtypes in a conversion must be statically the same – the technical term is that they must statically match. Remember that the component subtype could be constrained as in the array Strange at the end of the previous section. Static matching means that either the constraints are static and have the same bounds or else that the component subtypes come from the same elaboration. In this latter case they need not be static. So if we had

```
subtype S is Integer range 1 .. N;          -- N is some variable
type A1 is array (Index) of S;
type A2 is array (Index) of S;
type B1 is array (Index) of Integer range 1 .. N;
type B2 is array (Index) of Integer range 1 .. N;
```

then types A1 and A2 can be converted to each other but types B1 and B2 cannot. On the other hand, if we replaced N in all the declarations by a static expression such as 10 then all the subtypes would statically match and so all the array types could be converted to each other. The key point is that the matching has to be done statically (that is, at compile time) and not that the subtypes are necessarily static.

This is a good moment to explain the terms definite and indefinite. A definite subtype is one for which we can declare an object without an explicit constraint or initial value. Thus the subtype Vector_5 is definite. A type such as Vector on the other hand is indefinite since we cannot declare an object of the type without supplying the bounds either from an explicit constraint or from an initial value. Scalar types such as Integer are also definite. Other forms of indefinite type will be encountered in due course.

It will be recalled from Section 6.8 that a scalar subtype can have the aspect Default_Value so that objects of the type not explicitly initialized when declared will take that value rather than be undefined. The aspect Default_Component_Value can similarly be given for an array type whose components are of a scalar subtype.

Thus we might have

```
type Vector is array (Integer range <>) of Float
  with Default_Component_Value => 999.999;

subtype Vec is Vector
  with Default_Component_Value => 0.0;
```

The components of any object of type Vector are then 999.999 by default whereas in the case of the subtype Vec we have overridden the default to be zero. Crazy default values could be revealing when testing.

We conclude this section by observing that the attributes First, Last, Length, and Range, as well as applying to array objects, may also be applied to array types and subtypes provided they are constrained and so are definite. Hence Vector_6'Length is permitted and has the value 6 but Vector'Length is illegal.

Exercise 8.2

1. Declare an array type Bbb corresponding to the array Bor of Exercise 8.1(**5**).

2. Declare a two-dimensional array type suitable for operator tables on values of

   ```
   subtype Ring5 is Integer range 0 .. 4;
   ```

   Then declare addition and multiplication tables for modulo 5 arithmetic. Use the tables to write the expression (A + B) * C using modulo 5 arithmetic and assign the result to D where A, B, C and D have been appropriately declared. See Section 6.5.

8.3 Array aggregates

In the previous sections we introduced the idea of a positional aggregate. There is another form of aggregate known as a named aggregate in which the component values are preceded by the corresponding index value and =>. (The symbol => is akin to the 'pointing hand' sign used for indicating directions.) A simple example would be

```
(1 => 0.0, 2 => 0.0, 3 => 0.0, 4 => 0.0, 5 => 0.0, 6 => 0.0)
```

with the expected extension to several dimensions. The bounds of such an aggregate are self-evident and so our problem with the unit 2 × 2 matrix of the previous section could be overcome by writing

```
Unit_2: constant Matrix := (1 => (1 => 1.0, 2 => 0.0),
                            2 => (1 => 0.0, 2 => 1.0));
```

The rules for named aggregates are very similar to the rules for the alternatives in a case statement. Each choice can be given as a series of alternatives each of which can be a single value or a discrete range. We could therefore rewrite some previous examples as follows

```
A: array (1 .. 6) of Float := (1 .. 6 => 0.0);
```

```
Work_Day: constant array (Day) of Boolean :=
                          (Mon .. Fri => True, Sat | Sun => False);
```

In contrast to a positional aggregate, the index values need not appear in order. We could equally have written

```
(Sat | Sun => False, Mon .. Fri => True)
```

We can also use **others** but then as for the case statement it must be last and on its own (and there do not have to be any more values).

Array aggregates may not mix positional and named notation except that **others** may be used at the end of a positional aggregate.

It should also be realized that although we have been showing aggregates as initial values, they can be used generally in any place where an expression of an array type is required. They can also be the argument of type qualification.

The rules for deducing the bounds of an aggregate depend upon the form of the aggregate and its context. There are rather a lot of cases to consider and this makes the rules seem complicated although they are quite natural. We will first give the rules and then some examples of the consequences of the rules.

There are three kinds of aggregates to be considered

- Named without **others**, these have self-evident bounds.
- Positional without **others**, the number of elements is known but the actual bounds are not.
- Named or positional with **others**, neither the bounds nor the number of elements is known.

There are two main contexts to be considered according to whether the target type is constrained or unconstrained. In addition, the rules for qualification are special. So we have

- Unconstrained, gives no bounds.
- Constrained generally, gives the bounds, sliding usually permitted.
- Constrained qualification, gives the bounds, sliding never permitted.

The philosophy regarding sliding is that it is generally useful and so should be allowed and that aggregates should be no different from other array values in this respect; we saw some examples of sliding in the context of assignment in the previous section. Other contexts that behave like assignment will be met in due course when we discuss subprogram parameters and results in Chapter 9 and generic parameters in Chapter 17.

However, qualification is more in the nature of an assertion and so sliding is forbidden since it would be wrong to change the value in any way. If the bounds are not exactly the same then Constraint_Error is raised.

We will now consider the three kinds of aggregates in turn; the various combinations are summarized in Table 8.1.

A named aggregate without **others** has known bounds and will slide if permitted and necessary.

Table 8.1 Array aggregates and contexts.

Context	*Named*	*Positional*	*With* **others**
Unconstrained	OK bounds from aggregate	lower bound is S'First	illegal
Constrained like assignment	length must be same, could slide	length same, bounds from target	bounds from target
Constrained qualification	bounds must exactly match	length same, bounds from target	bounds from target

If a positional aggregate without **others** is used in a context which does not give the bounds then the lower bound is by default taken to be S'First where S is the index subtype and the upper bound is then deduced from the number of components. They never need to slide.

Aggregates with **others** are particularly awkward since we cannot deduce either the bounds or the number of elements. They can therefore only be used in a context that gives the bounds and consequently can never slide. Given the bounds, the components covered by **others** follow on from those given explicitly in the positional case and are simply those not given explicitly in the named case.

One way of supplying the bounds for an aggregate with **others** is to use qualification as we did to distinguish between overloaded enumeration literals. In order to do this we must have an appropriate (constrained) type or subtype name. So we might introduce

```
type Schedule is array (Day) of Boolean;
```

and can then write an expression such as

```
Schedule'(Mon .. Fri => True, others => False)
```

Note that when qualifying an aggregate we do not, as for an expression, need to put it in parentheses because it already has parentheses.

We have already considered the use of an aggregate as an initial value and providing the bounds in the example of Unit_2; this of course was an unconstrained context. An aggregate with **others** is thus not allowed. In the previous section we saw the effect of using a positional aggregate in such a context since it gave surprising bounds. In this section we saw how a named aggregate was more appropriate.

We will now consider the context of assignment which behaves much the same as a declaration with an initial value where the initial value is not being used to supply the bounds. These are constrained contexts and can therefore supply the bounds of an aggregate if necessary.

So both of the following are permitted

```
Work_Day: constant array (Day) of Boolean :=
                (Mon .. Fri => True, others => False);

Work_Day: constant array (Day) of Boolean :=
                (True, True, True, True, True, others => False);
```

Further insight might be obtained by another example. Consider

```
type Vector is array (Integer range <>) of Float;
V: Vector(1 .. 5) := (3 .. 5 => 1.0, 6 | 7 => 2.0);
```

which shows a named aggregate as the initial value of V. The bounds of the named aggregate are self-evident being 3 and 7 and so sliding occurs and the net result is that components V(1) .. V(3) have the value 1.0 and V(4) and V(5) have the value 2.0.

On the other hand, writing

```
V := (3 .. 5 => 1.0, others => 2.0);
```

has the rather different effect of setting V(3) .. V(5) to 1.0 and V(1) and V(2) to 2.0.

The point is that the bounds of the aggregate are taken from the context and there is no sliding. Aggregates with **others** never slide.

Similarly, no sliding occurs in

```
V := (1.0, 1.0, 1.0, others => 2.0);
```

and this results in setting V(1) .. V(3) to 1.0 and V(4) and V(5) to 2.0. It is clear that care is necessary when using **others**.

Array aggregates really are rather complicated and we still have a few points to make. The first is that in a named aggregate all the ranges and values before => must be static (as in a case statement) except for one situation. This is where there is only one alternative consisting of a single choice – it could then be a dynamic range or even a single dynamic value. An example might be

```
A: array (1 .. N) of Integer := (1 .. N => 0);
```

This is valid even if N is zero (or negative) and then gives a null array and a null aggregate. The following example illustrates a general rule that the expression after => is evaluated once for each corresponding index value; of course it usually makes no difference but consider

```
A: array (1 .. N) of Integer := (1 .. N => 1/N);
```

If N is zero then there are no values and so 1/N is not evaluated and Constraint_Error cannot occur. The reader will recall from Section 6.1 that a similar multiple evaluation also occurs when several objects are declared and initialized together.

In order to avoid awkward problems with null aggregates, a null choice is only allowed if it is the only choice. Foolish aggregates such as

```
(7 .. 6 | 1 .. 0 => 0)
```

are thus forbidden, and there is no question of the lower bound of such an aggregate.

Another point is that although we cannot mix named and positional notation within an aggregate, we can, however, use different forms for the different components and levels of a multidimensional aggregate. So the initial value of the matrix Unit_2 could also be written as

```
(1 => (1.0, 0.0),        or     ((1 => 1.0, 2 => 0.0),
 2 => (0.0, 1.0))                (1 => 0.0, 2 => 1.0))
```

or even as

```
(1 => (1 => 1.0, 2 => 0.0),
 2 => (      0.0,      1.0))
```

and so on.

Note also that the Range attribute stands for a range and therefore can be used as one of the choices in a named aggregate. However, we cannot use the range attribute of an object in its own initial value. Thus

```
A: array (1 .. N) of Integer := (A'Range => 0);   -- illegal
```

is not allowed because an object is not visible until the end of its own declaration. However, we could write

```
A: array (1 .. N) of Integer := (others => 0);
```

and this is better than repeating 1 .. N because it localizes the dependency on N.

A traditional positional aggregate cannot contain just one component because otherwise it would be ambiguous – we could not distinguish it from a scalar value that happened to be in parentheses. Such an aggregate of one component must therefore always use the named notation. So

```
A: array (1 .. 1) of Integer := (99);          -- illegal
A: array (1 .. 1) of Integer := (1 => 99);     -- legal
```

In Ada 2022, array aggregates can also be enclosed in square brackets in which case this difficulty does not arise, see Section 8.8.

Note also the following obscure case

```
A: array (N .. N) of Integer := (N => 99);
```

which shows a single choice and a single dynamic value being that choice.

The final feature is the ability to use <> as the expression with named notation in which case it means the default value. Thus we can write

```
P: array (1 .. 1000) of Integer := (1 => 2, 2 .. 1000 => <>);
```

The array P has its first component set to 2 and the others are undefined. It might be that P is to be used to hold the first 1000 prime numbers and the algorithm requires the first prime to be provided. The box cannot be used with positional notation but it is allowed with **others**. So we could write

```
(2, others => <>)
```

The box notation is not particularly useful with the types we have seen so far but it is very important with limited types which we will meet in Chapter 11.

The reader will by now have concluded that arrays in Ada are somewhat complicated. That is a fair judgement, but in practice there should be few difficulties. There is always the safeguard that if we do something wrong, the compiler will inevitably tell us. In cases of ambiguity, qualification solves the problems provided we have an appropriate type or subtype name to use.

Finally, note that the named aggregate notation can greatly increase program legibility. It is especially valuable in initializing large constant arrays and guards against the accidental misplacement of individual values. Consider

```
type Event is (Birth, Accession, Death);
type Monarch is (William_I, William_II, Henry_I, ...,
                  Victoria, Edward_VII, George_V, ... );
...
Royal_Events: constant array (Monarch, Event) of Integer :=
     (William_I   => (1027, 1066, 1087),
      William_II  => (1056, 1087, 1100),
      ...
      Victoria    => (1819, 1837, 1901),
      Edward_VII  => (1841, 1901, 1910),
      George_V    => (1865, 1910, 1936),
      ...                                );
```

The accidental interchange of two lines of the aggregate causes no problems, whereas if we had just used the positional notation then an error would have been introduced and this might have been tricky to detect.

Exercise 8.3

1 Rewrite the declaration of the array Days_In_Month in Exercise 8.1(**3**) using a named aggregate for an initial value.

2 Declare a constant Matrix whose bounds are both 1 .. N where N is dynamic and whose components are all zero.

3 Declare a constant Matrix as in **2** but make it a unit matrix.

4 Declare a constant two-dimensional array which gives the numbers of each atom in a molecule of the various aliphatic alcohols. Declare appropriate enumeration types for both the atoms and the molecules. Consider methanol CH_3OH, ethanol C_2H_5OH, propanol C_3H_7OH, and butanol C_4H_9OH.

8.4 Characters and strings

We now complete our discussion of enumeration types by introducing character types. In the enumeration types seen so far such as

```
type Colour is (Red, Amber, Green);
```

the values are represented by identifiers. It is also possible to have an enumeration type in which some or all of the values are represented by character literals.

A character literal is a further form of lexical element. It consists of a single character within a pair of single quotes. The character must be one of the graphic characters; this might be a space but it must not be a control character such as horizontal tabulate or newline.

This is a situation where there is a distinction between upper and lower case letters. Thus the character literals 'A' and 'a' are distinct.

So we could declare an enumeration type

```
type Roman_Digit is ('I', 'V', 'X', 'L', 'C', 'D', 'M');
...
Dig: Roman_Digit := 'D';
```

All the usual properties of enumeration types apply.

```
Roman_Digit'First = 'I'
Roman_Digit'Succ('X') = 'L'
Roman_Digit'Pos'('M') = 6
Dig < 'L' = False
```

There is a predefined enumeration type Character which is (naturally) a character type. We can think of its declaration as being of the form

```
type Character is (nul, ... , '0', '1', '2', ... , 'A', 'B', 'C', ... , 'a', 'b', 'c', ... , 'ÿ');
```

where the literals which are not graphic character literals (such as *nul*) are not really identifiers either (which is why they are represented here in italics). The predefined type Character represents the standard ISO 8-bit set, ISO 8859-1 commonly known as Latin-1. It describes the set of characters normally used for input and output and includes the various accented characters used in European languages, for example the last character is lower case y diaeresis; for the full declaration of type Character see Section 23.1.

It is possible to refer to the non-graphic characters as Ada.Characters.Latin_1. Nul and so on (or with a suitable use clause as Latin_1.Nul or simply Nul). We can also refer to the graphic characters (other than digits and the normal 26 upper case letters) by name; this is useful for displaying program text on output devices which do not support all the graphic characters. Finally, we can refer to those characters also in the 7-bit ASCII set as ASCII.Nul and so on but this feature is obsolescent.

There are also predefined types Wide_Character (16-bit) and Wide_Wide_Character (32-bit) corresponding to the Basic Multilingual Plane (BMP) and the full set of ISO 10646 respectively. The first 256 positions of Wide_Character correspond to those of the type Character and the first 65536 positions of Wide_Wide_Character correspond to those of the type Wide_Character.

The existence of these predefined types results in overloading of some of the literals. So an expression such as

```
'X' < 'L'
```

is ambiguous. We do not know whether it is comparing characters of the type Character, Wide_Character, or Wide_Wide_Character (or even Roman_Digit). In order to resolve the ambiguity we must qualify one or both literals thus

```
Character'('X') < 'L'                    -- False
Roman_Digit'('X') < 'L'                  -- True
```

As well as the predefined type Character there is also the predefined array type String

```
type String is array (Positive range <>) of Character;
```

This is a perfectly normal array type and obeys all the rules of the previous section. So we can write

```
S: String (1 .. 7);
```

to declare an array of range 1 .. 7. The bounds can also be deduced from the initial value thus

```
G: constant String := ('P', 'I', 'G');   -- bounds are 1 and 3
```

where the initial value takes the form of a normal positional aggregate. The lower bound of G (that is, G'First) is 1 since the index subtype of String is Positive and Positive'First is 1.

There is another notation for positional aggregates whose components are character literals. This is the string literal. So we can more conveniently write

```
G: constant String := "PIG";
```

The string literal is the last lexical element to be introduced. It consists of a sequence of printable characters and spaces enclosed in double quotes. A double quote may be represented in a string by two double quotes so that

```
('A', '"', 'B')      is the same as      "A""B"
```

The string may also have just one character or may be null. The equivalent aggregates using character literals have to be written in named notation or use square brackets as described in Section 8.8 thus

```
['A']                is the same as      "A"
(1 .. 0 => <>)       is the same as      ""
```

Another rule about a lexical string is that it must fit onto a single line. Moreover it cannot contain control characters such as *soh*. And, of course, as with character literals, the two cases of alphabet are distinct in strings.

In Section 8.6 we shall see how to overcome the limitations that a string must fit onto a single line and yet cannot contain control characters.

A major use for strings is for creating text to be output. A simple sequence of characters can be output by a call of the (overloaded) subprogram Put. Thus

```
Put("The Countess of Lovelace");
```

will output the text

```
The Countess of Lovelace
```

onto some appropriate file.

There are also types Wide_String and Wide_Wide_String defined as arrays of the corresponding character types. One consequence is that comparisons between literal strings are also ambiguous like comparisons between individual literals and so also have to be qualified as illustrated in Section 8.6.

However, the lexical string is not reserved just for use with the predefined types String, Wide_String, and Wide_Wide_String. It can be used to represent an array of any character type. We can write

```
type Roman_Number is array (Positive range <>) of Roman_Digit;
```

and then

```
Nineteen_Eighty_Four: constant Roman_Number := "MCMLXXXIV";
```

or indeed

```
Four: array (1 .. 2) of Roman_Digit := "IV";
```

Of course the compiler knows nothing of our interpretation of Roman numbers and the reason for our choice of identifiers!

Exercise 8.4

1 Declare a constant array Roman_To_Integer which can be used for table look-up to convert a Roman_Digit to its normal integer equivalent (e.g. converts 'C' to 100).

2 Given an object R of type Roman_Number write statements to compute the equivalent integer value V. It may be assumed that R obeys the normal rules of construction of Roman numbers.

8.5 Arrays of arrays and slices

The components of an array can be of any definite subtype. Remember that a definite subtype is one for which we can declare an object (without an explicit constraint or initial value). Thus we can declare arrays of any scalar type; we can also declare arrays of arrays. So we can have

```
type Matrix_3_6 is array (1 .. 3) of Vector_6;
```

where, as in Section 8.2

```
type Vector_6 is array (1 .. 6) of Float;
```

However, we cannot declare an array type whose component subtype is indefinite such as an unconstrained array (just as we cannot declare an object which is an unconstrained array). So we cannot write

```
type Matrix_3_N is array (1 .. 3) of Vector;   -- illegal
```

On the other hand, nothing prevents us declaring an unconstrained array of constrained arrays thus

```
type Matrix_N_6 is array (Integer range <>) of Vector_6;
```

It is instructive to compare the differences between declaring an array of arrays

```
AOA: Matrix_3_6;     or     AOA: Matrix_N_6(1 .. 3);
```

and the similar multidimensional array

```
MDA: Matrix(1 .. 3, 1 .. 6);
```

Aggregates for both are completely identical, for example

```
((1.0, 2.0, 3.0, 4.0, 5.0, 6.0),
 (4.0, 4.0, 4.0, 4.0, 4.0, 4.0),
 (6.0, 5.0, 4.0, 3.0, 2.0, 1.0))
```

but component access is quite different, thus

```
AOA(I)(J)
MDA(I, J)
```

where in the case of AOA the internal structure is naturally revealed. The individual rows of AOA can be manipulated as arrays in their own right, but the structure of MDA cannot be decomposed. So we could change the middle row of AOA to zero by

```
AOA(2) := (1 .. 6 => 0.0);
```

but a similar technique cannot be applied to MDA.

Arrays of arrays are not restricted to one dimension, we can have a multi-dimensional array of arrays or an array of multidimensional arrays; the notation extends in an obvious way.

Arrays of strings are revealing. Consider

```
type String_Array is array (Positive range <>,
                            Positive range <>) of Character;
```

which is an unconstrained two-dimensional array type. We can then declare

```
Farmyard: constant String_Array := ("pig", "cat", "dog",
                                    "cow", "rat", "hen");
```

where the bounds are conveniently deduced from the aggregate. But we cannot have a ragged array where the individual strings are of different lengths such as

```
Zoo: constant String_Array := ("aardvark", "baboon",
        "camel", "dolphin", "elephant", ..., "zebra");      -- illegal
```

This is a real nuisance and means we have to pad out the strings with spaces so that they are all the same length

```
Zoo: constant String_Array := ("aardvark   ",
                               "baboon     ",
                               "camel      ",
                               ...
                               "zebra      ");
```

The next problem is that we cannot select an individual one of the strings. We might want to output a particular one and so perhaps attempt

```
Put(Farmyard(5));
```

hoping to print the text

```
rat
```

but this is not allowed since we can only select an individual component of an array and, in the case of this two-dimensional array, this is just one character.

An alternative approach is to use an array of arrays. A problem here is that the component in the array type declaration has to be constrained and so we have to decide on the length of our strings right from the beginning thus

```
type String_3_Array is array (Positive range <>) of String(1 .. 3);
```

and then

```
Farmyard: constant String_3_Array := ("pig", "cat", "dog",
                                       "cow", "rat", "hen");
```

With this formulation we can indeed select an individual string as a whole and so the statement Put(Farmyard(5)); now works. However, we still cannot declare our Zoo as a ragged array; we will return to this topic in Section 11.2 when another approach will be discussed.

We thus see that arrays of arrays and multidimensional arrays each have their own advantages and disadvantages.

A special feature of one-dimensional arrays is the ability to denote a slice of an array object. A slice is written as the name of the object (variable or constant) followed by a discrete range in parentheses.

So given

```
S: String(1 .. 10);
```

then we can write S(3 .. 8) to denote the middle six characters of S. The bounds of the slice are the bounds of the range and not those of the index subtype. We could write

```
T: constant String := S(3 .. 8);
```

and then T'First = 3, T'Last = 8.

The bounds of the slice need not be static but can be any expressions. A slice would be null if the range turned out to be null.

The use of slices emphasizes the nature of array assignment. The value of the expression to be assigned is completely evaluated before any components are assigned. No problems arise with overlapping slices. So

```
S(1 .. 4) := "BARA";
S(4 .. 7) := S(1 .. 4);
```

results in S(1 .. 7) = "BARBARA". S(4) is only updated after the expression S(1 .. 4) is safely evaluated. There is no risk of setting S(4) to 'B' and then consequently making the expression "BARB" with the final result of

```
"BARBARB"
```

The ability to use slices is another consideration in deciding between arrays of arrays and multidimensional arrays. With our second Farmyard we can write

```
Pets: String_3_Array(1 .. 2) := Farmyard(2 .. 3);
```

which uses sliding assignment so that the two components of Pets are "cat" and "dog". Moreover, if we had declared the Farmyard as a variable rather than a constant then we could also write

```
Farmyard(1)(1 .. 2) := "ho";
```

which turns the "pig" into a "hog"! We can do none of these things with the old Farmyard.

Exercise 8.5

1 Write a single assignment statement to swap the first two rows of AOA.

2 Declare the second Farmyard as a variable. Then change the cow into a sow.

3 Assume that R contains a Roman number. Write statements to see if the last digit of the corresponding decimal Arabic value is a 4; if so change it to a 6.

8.6 One-dimensional array operations

Many of the operators that we met in Chapter 6 may also be applied to one-dimensional arrays.

The operators **and**, **or**, **xor** and **not** may be applied to one-dimensional Boolean arrays. For the binary operators, the two operands must have the same length and be of the same type. The underlying scalar operation is applied to matching components and the resulting array is again of the same type. The bounds of the result are the same as the bounds of the left or only operand.

Consider the following declarations and assignments

```
type Bit_Row is array (Positive range <>) of Boolean;
A, B: Bit_Row(1 .. 4);
C, D: array (1 .. 4) of Boolean;
T: constant Boolean := True;
F: constant Boolean := False;
...
A := (T, T, F, F);
B := (T, F, T, F);

A := A and B;
B := not B;
```

The result is that A now equals (T, F, F, F), and B equals (F, T, F, T). But note that C **and** D would not be allowed since C and D are of different (and anonymous) types because of the rules regarding type equivalence mentioned in Section 8.2. This is clearly a case where it is appropriate to give a name to the array type because we are manipulating the arrays as complete objects.

Note that these operators also use sliding semantics, like assignment as explained in Section 8.2, and so only demand that the types and the number of components be the same. The bounds themselves do not have to be equal. However, if the number of components are not the same then, naturally, Constraint_Error will be raised.

Boolean arrays can be used to represent sets. Consider

```
type Primary is (R, Y, B);
type Colour is array (Primary) of Boolean;
C: Colour;
```

then there are $8 = 2 \times 2 \times 2$ values that C can take. C is, of course, an array with three components and each of these has value True or False; the three components are

```
C(R),     C(Y)     and     C(B)
```

The 8 possible values of the type Colour can be represented by suitably named constants as follows

```
White:  constant Colour := (F, F, F);
Red:    constant Colour := (T, F, F);
Yellow: constant Colour := (F, T, F);
Blue:   constant Colour := (F, F, T);
Green:  constant Colour := (F, T, T);
Purple: constant Colour := (T, F, T);
Orange: constant Colour := (T, T, F);
Black:  constant Colour := (T, T, T);
```

and then we can write expressions such as

```
Red or Yellow
```

which is equal to Orange and

```
not Black
```

which is White.

So the values of our type Colour are effectively the set of colours obtained by taking all combinations of the primary colours represented by R, Y, and B. The empty set is the value of White and the full set is the value of Black. We are using the paint pot mixing colour model rather than light mixing. A value of True for a component means that the primary colour concerned is mixed in our pot. The murky mess we got at junior school from mixing too many colours together is our black!

The operations **or**, **and** and **xor** may be interpreted as set union, set intersection and symmetric difference. A test for set membership can be made by inspecting the value of the appropriate component of the set. Thus

```
C(R)
```

is True if R is in the set represented by C. We cannot use the predefined operation **in** for this. A literal value can be represented using the named aggregate notation, so we might denote Orange by

```
(R | Y => T, others => F)
```

A more elegant way of doing this will appear in the next chapter.

We now consider the relational operators. The equality operators = and /= apply to almost all types anyway and we gave the rules for arrays when we discussed assignment in Section 8.2.

The ordering operators <, <=, > and >= may be applied to one-dimensional arrays of a discrete type. (Note discrete.) The result of the comparison is based upon the lexicographic (that is, dictionary) order using the predefined order relation for the components. Remembering that the upper and lower case letters are distinct and the upper case ones occur earlier in the type Character, then for the type String the following strings are in lexicographic order

```
"" , "A" , "AZZ" , "CAT" , "CATERPILLAR" , "DOG" , "cat"
```

Strings are compared component by component until they differ in some position. The string with the lower component is then lower. If one string runs out of components as in CAT versus CATERPILLAR then the shorter one is lower. The null string is lowest of all.

Because of the existence of the types Wide_String and Wide_Wide_String we cannot actually write comparisons such as

```
"CAT" < "DOG"              -- illegal
"CCL" < "CCXC"             -- illegal
```

because they are ambiguous since we do not know whether we are comparing type String, Wide_String, Wide_Wide_String or even Roman_Number. We must qualify one or both of the strings. This is done in the usual way but a string, unlike an ordinary aggregate, has to be placed in extra parentheses otherwise we would get an ugly juxtaposition of a single and double quote. So

```
Wide_String'("CAT") < "DOG"              -- True
String'("CCL") < "CCXC"                  -- True
Roman_Number'("CCL") < "CCXC"            -- False
```

Note that the compiler is too stupid to know about the interpretation of Roman numbers in our minds and has said that 250 < 290 is false. The only thing that matters is the order relation of the characters 'L' and 'X' in the type definition. In the next chapter we will see how we can redefine < so that it works ‘properly’ for Roman numbers.

The ordering operators also apply to general expressions and not just to literal strings

```
Nineteen_Eighty_Four < "MM"                    -- True
```

The ordering operators can be applied to arrays of any discrete types. So

```
(1, 2, 3) < (2, 3)
(Jan, Jan) < (1 => Feb)
```

The predefined operators <=, > and >= are defined by analogy with <.

We finally introduce a new binary operator & which denotes concatenation of one-dimensional arrays. It has the same precedence as binary plus and minus. The two operands must be of the same type and the result is an array of the same type whose value is obtained by juxtaposing the two operands. The length of the result is thus the sum of the lengths of the operands.

The lower bound of the result depends upon whether the underlying array type is constrained or not. If it is unconstrained (considered the usual case) then the lower bound is that of the left operand as for other operators. However, if it is constrained then the lower bound is that of the array index subtype. (If the left operand is null the result is simply the right operand.)

So

```
"CAT" & "ERPILLAR" = "CATERPILLAR"
```

Concatenation can be used to construct a string which will not fit on one line

```
"This string goes " &
"on and on"
```

One or both operands of & can also be a single value of the component type. If the left operand is such a single value then the lower bound of the result is always the lower bound of the array index subtype.

```
"CAT" & 'S' = "CATS"
'S' & "CAT" = "SCAT"
'S' & 'S' = "SS"
```

This is useful for representing the control characters such as CR and LF in strings. So, using an abbreviated form rather than Ada.Characters.Latin_1.CR, we can write

```
"First line" & Latin_1.CR & Latin_1.LF & "Next line"
```

Of course, it might be neater to declare

```
CRLF: constant String := (Latin_1.CR, Latin_1.LF);
```

and then write

```
"First line" & CRLF & "Next line"
```

The operation & can be applied to any one-dimensional array type and so we can apply it to our Roman numbers. Consider

```
R: Roman_Number(1 .. 5);
S: String(1 .. 5);

R := "CCL" & "IV";
S := "CCL" & "IV";
```

This is valid. The context tells us that in the first case we apply & to two Roman numbers whereas in the second we apply it to two values of type String. There is no ambiguity as in

```
B: Boolean := "CCL" < "IV";                  -- illegal
```

where the context does not distinguish the various string types.

Exercise 8.6

1 Put the eight possible constants White ... Black of the type Colour in ascending order as determined by the operator < applied to one-dimensional arrays.

2 Evaluate

(a) Red **or** Green
(b) Black **xor** Red
(c) **not** Green

3 Show that **not** (Black **xor** C) = C is true for all values of C.

4 Why did we not write

```
(Jan, Jan) < (Feb)
```

5 Put in ascending order the following values of type String: "ABC", "123", "abc", "Abc", "abC", "aBc".

6 Given

```
C: Character;
S: String(5 .. 10);
```

What are the lower bounds of

(a) C & S (b) S & C (c) "" & S

7 Given

```
type TC is array (1 .. 10) of Integer;
type TU is array (Natural range <>) of Integer;
AC: TC;
AU: TU(1 .. 10);
```

What are the bounds of

(a) AC(6 .. 10) & AC(1 .. 5)
(b) AC(6) & AC(7 .. 10) & AC(1 .. 5)
(c) AU(6 .. 10) & AU(1 .. 5)
(d) AU(6) & AU(7 .. 10) & AU(1 .. 5)

8.7 Records

As stated at the beginning of this chapter, we are only going to consider the simplest form of record at this point. Discussions of tagged and discriminated records will be found in Chapters 14 and 18 respectively.

A record is a composite object consisting of named components which may be of different types. In contrast to arrays, we cannot have anonymous record types – they all have to be named. Consider

```
type Month_Name is (Jan, Feb, Mar, Apr, May, Jun,
                    Jul, Aug, Sep, Oct, Nov, Dec);

type Date is
   record
      Day: Integer range 1 .. 31;
      Month: Month_Name;
      Year: Integer;
   end record Date;
```

This declares the type Date to be a record containing three named components: Day, Month, and Year. Repeating the type name after **end record** is optional and allowed in Ada 2022.

We can declare variables and constants of record types in the usual way.

```
D: Date;
```

declares an object D which is a date. The individual components of D can be denoted by following D with a dot and the component name. Thus we could write

```
D.Day := 4;
D.Month := Jul;
D.Year := 1776;
```

in order to assign new values to the individual components.

Records can be manipulated as whole objects. Literal values can be written as aggregates much like arrays; both positional and named forms can be used. So we could write

```
D: Date := (4, Jul, 1776);
E: Date;
```

and then

```
E := D;
```

or

```
E := (Month => Jul, Day => 4, Year => 1776);
```

The reader will be relieved to know that much of the complexity of array aggregates does not apply to records. This is because the number of components is always known.

In a positional aggregate the components come in order. In a named aggregate they may be in any order.

A named aggregate cannot use a range because the components are not considered to be closely related. The vertical bar and **others** can be used but of course the expression must be appropriate for all the components covered.

An extra possibility for records is that the positional and named notations can be mixed in one aggregate. But if this is done then the positional components must come first and in order (without holes) as usual. So in other words, we can change to the named notation at any point in the aggregate but must then stick to it. The above date could therefore also be expressed as

```
(4, Jul, Year => 1776)
(4, Year => 1776, Month => Jul)
```

and so on.

It is possible to give default expressions for some or all of the components in the type declaration. Note that there is no need for an aspect corresponding to Default_Value for scalars and Default_Component_Value for arrays. In the case of records, we simply write

```
type Complex is
  record
    Re: Float := 0.0;
    Im: Float := 0.0;
  end record Complex;
```

or more succinctly

```
type Complex is
  record
    Re, Im: Float := 0.0;
  end record;
```

declares a record type containing two components of type Float and gives a default expression of 0.0 for each. This record type represents a complex number $x + iy$ where Re and Im are the values of x and y. The default value thus represents (0, 0), the origin of the Argand plane. We can now declare

```
C1: Complex;
C2: Complex := (1.0, 0.0);
```

The object C1 will now have the values 0.0 for its components by default. In the case of C2 we have explicitly overridden the defaults. Note that even if there are default expressions, an aggregate must always be complete in order to provide so-called full coverage analysis. This ensures that if we extend a type with additional components, then we have to modify all aggregates to match and so cannot inadvertently forget some components.

In this example both components have the same type and so the following named forms are also possible

```
(Re | Im => 1.0)
(others => 1.0)
```

We can also use the box notation with record aggregates. Thus we might declare

```
C3: Complex := (Re => 1.0, Im => <>);
```

in which case the component Re takes the explicitly given value whereas the component Im takes its default value (if any). Remember that the box can only be used with named notation although since we can mix named and positional notation in record aggregates we could also write (1.0, Im => <>).

A minor point is that we can write **others** => <> even if the components covered are of different types (or there are no more components) since no expression is involved. We can also use the vertical bar with the box even if the components are of different types for the same reason.

The only operations predefined on record types are = and /= as well as assignment (unless the type is limited as discussed in Chapter 12). Other operations must be performed at the component level or be explicitly defined by a subprogram as we shall see in the next chapter.

A record type may have any number of components. It may even have none in which case we must either write the component list as **null**; or use the abbreviated form

```
type Hole is null record;
```

The aggregate for the (null) value of such a type can be written as (**null record**) or even as (**others** => <>).

An aggregate for a record with just one component must always use named notation in order to avoid confusion with an expression in parentheses.

The components of a record type can be of any definite type; they can be other records or arrays. However, if a component is an array then it must be fully constrained and moreover it must be of a named type and not an anonymous type. And obviously a record cannot contain an instance of itself.

The components cannot be constants but the record as a whole can be. Thus

```
I: constant Complex := (0.0, 1.0);
```

is allowed and represents the square root of –1.

A more elaborate example of a record is given by

```
type Subject is (Theology, Classics, Mathematics);
type Scores is array (Subject) of Integer range 0 .. 100;
type Student is
  record
    Birth: Date;
    Finals: Scores := (others => 0);
  end record Student;
```

The record Student has two components, the first is another record, a Date, the second an array giving examination results in various subjects. The array has default values of zeros.

We can now write

```
Fred: Student;
Fred.Birth := (19, Aug, 1984);
Fred.Finals := (5, 15, 99);
```

and this would be equivalent to

```
Fred := (Birth => (19, Aug, 1984), Finals => (5, 15, 99));
```

The notation is as expected. It is better to use an aggregate rather than a series of assignments to individual components since it ensures that they are all given a value. If another component is added later and we fail to update an aggregate then the compiler will tell us.

For components of objects we proceed from left to right using the dot notation to select components of a record and indices in parentheses to select components of an array and ranges in parentheses to slice arrays. There is no limit. We could have an array of students

```
People: array (1 .. N) of Student;
```

and then statements such as

```
People(6).Birth.Day := 19;
People(8).Finals(Classics .. Mathematics) := (50, 50);
```

A final point concerns the evaluation of expressions in a record declaration. An expression in a constraint applied to a component is evaluated when the record type is elaborated. Suppose the type Student also has a component

```
Name: String(1 .. N) := (others => ' ');
```

then the length of the component Name will be the value of N when the type Student is elaborated. Of course, N need not be static and so if the type declaration is in a loop, for example, then each execution of the loop might give rise to a type with a different size component. However, for each elaboration of the record type declaration all objects of the type will have the same component size.

On the other hand, a default expression in a record type is only evaluated when an object of the type is declared and only then if no explicit initial value is provided. Of course, in simple cases, like our type Complex, it makes no difference but it could bring surprises. For example suppose we write the component Name as

```
Name: String(1 .. N) := (1 .. N => ' ');
```

then the length of the component Name is the value of N when the record type is declared whereas when a Student is subsequently declared without an initial value, the aggregate will be evaluated using the value of N which then applies. Of course, N may by then be different and so Constraint_Error will be raised. This is rather surprising; we do seem to have strayed into an odd backwater of Ada!

Exercise 8.7

1 Declare three variables C1, C2 and C3 of type Complex. Write one or more statements to assign (a) the sum, (b) the product, of C1 and C2 to C3.

2 Write statements to find the index of the first student of the array People born on or after 1 January 1980.

8.8 General aggregates

This section describes a number of new features introduced in Ada 2022. It might be argued that some of this material should be in Section 8.3. However, it would make that section rather indigestible so we treat it separately.

We take the opportunity to illustrate some changes using the syntax notation.

```
array_aggregate ::= positional_array_aggregate | null_array_aggregate
        | named_array_aggregate

positional_array_aggregate ::=
        (expression , expression {, expression})
        | (expression {, expression} , others => expression)
        | (expression {, expression} , others => <>)
        | '[' expression {, expression} [, others => expression] ']'
        | '[' expression {, expression} , others => <> ']'

null_array_aggregate ::= '[' ']'

named_array_aggregate ::=
        (array_component_association_list)
        | '[' array_component_association_list ']'
```

and so on.

An important innovation is that square brackets can be used as an alternative to parentheses (round brackets). This only applies to their use for array aggregates and the new container aggregates (see the introduction to Chapter 24) but not for record aggregates and certainly not for subprogram calls (AI12-212).

Incidently, note that in the syntax square brackets standing for themselves are enclosed in single quotes. Square brackets in the syntax otherwise denote that the enclosed item is optional. The syntax of positional_array_aggregate shows both uses of square brackets.

Using square brackets in aggregates means that we can have a positional aggregate with just one component whereas using parentheses needs at least two components as we see from the syntax.Thus

```
A: array (1 .. 1) of Integer := [99];         -- legal
A: array( 1 .. 1) of Integer := (99);         -- illegal
```

The other new item in the syntax above is the introduction of null_array_aggregate which again uses square brackets.

So we can write

```
type Vector is array (Positive range <>) of Float;
Z: Vector := [];
```

Note that Z'First is 1 and Z'Last is 0 as expected. The usual rules apply; the lower bound is as for a positional aggregate, the upper bound is its predecessor. If there is no predecessor then Constraint_Error is raised.

Multidimensional array aggregates can use a mixture of round and square brackets provided the pairs match. However the individual aggregates at inner levels could be different. Thus in the case of a two dimensional array, we might have

```
M: Matrix := ( [ P, Q ], ( R, S ) );
```

Another useful innovation is that for loops can be embedded in array aggregates in order to indicate a range of indexes (AI12-61). This naturally only applies to named aggregates and not to positional aggregates. For example, (for playing darts as in Section 16.4), we might have.

```
Doubles: array (1 .. 20) of Integer := [for J in 1 .. 20 => 2*J];
```

which is equivalent to

```
Doubles: array (1 .. 20) of Integer := (2, 4, 6, 8, 10, 12, 14, 16, 18, 20,
                                        22, 24, 26, 28, 30, 32, 34, 36, 38, 40);
```

An identity square matrix of order N might be

```
Ident_N: Matrix := [for I in 1 .. N =>
                              [for J in 1 .. N => (if I = J then 1.0 else 0.0)]];
```

Note that, for illustration, this uses square brackets for the array itself but the internal conditional expression has to be in parentheses (round brackets).

Note also that an aggregate with a for loop can also include individual items and ranges so an extreme example might be

```
(for K in 2 | 4 | 5 .. 9 => K*2, 1 | 3 => 12, others => 0)      -- strange
```

A filter can also be used in an aggregate. If B is an integer array, and we write

```
A := (for E of B when E >= 0 => E);
```

then A is simply B with all negative components removed.

Another innovation is the introduction of delta aggregates. These enable an aggregate to be described in terms of an existing value with changes.

A simple example occurs when we want to change just one component of a record but leave the others unchanged. This often occurs with postconditions (see Section 16.2). Perhaps we are updating a personnel record on someone's birthday and need to check that only the person's age has been increased by 1 and all other entries are unchanged. Thus

```
Post => Data = (Data'Old with delta Age => Age'Old + 1);
```

Two curious examples of updating a date are

```
Tomorrow := ((Yesterday with delta Day => 12) with delta Month => April));
Tomorrow := (Yesterday with delta Day => 12, Month => April);
```

The first illustrates that the mechanism can be nested; the second is more sensible.

An example with arrays might be where we are updating a list of how many times we have got various doubles at darts.

There are a number of restrictions on the use of delta aggregates. In the case of an array we cannot use the box symbol <> or **others** and the dimension of the array has to be one and the component type must not be limited. In the case of records there must not be limited components or discriminants.

Exercise 8.8

1 How many different ways can an aggregate for a unit matrix of integers be declared using positional notation?

2 Declare a constant array of integers being the first ten triangular numbers. The triangular numbers are 1, 3, 6, 10 and so on.

Checklist 8

Array types can be anonymous, but record types cannot.

Aggregates must always be complete.

Distinguish constrained array types (definite types) from unconstrained array types (those with <>).

The component subtype of an array must be definite.

Named and positional notations cannot be mixed for array aggregates – they can for records.

An array aggregate with **others** must have a context giving its bounds. It never slides.

A choice in an array aggregate can only be dynamic or null if it is the only choice.

The attributes First, Last, Length, and Range apply to array objects and constrained array types and subtypes but not to unconstrained types and subtypes.

For array assignment to be valid, the number of components must be equal for each dimension – not the bounds.

The cases of alphabet are distinct in character literals and strings.

A record component cannot be of an anonymous array type.

A default component expression is only evaluated when an uninitialized object is declared.

There are also types for wide and wide-wide characters and strings..

The aspect Default_Component_Value was introduced in Ada 2012.

New in Ada 2022

In a record type declaration, **end record** may optionally be followed by the type name.

Array aggregates can also be given using square brackets. A positional aggregate with just one component and a null array with no components are permitted using square brackets. Array aggregates can also contain loops. Both array and record aggregates can be given as delta aggregates where an aggregate is given as a variation of an existing value.

Ada 2022 also has container aggregates, see Section 24.2.

More string expressions are static. Thus S = "abc" is static if S is static, see Section 27.1.

9 Expression Structures

9.1 Membership tests
9.2 If expressions
9.3 Case expressions
9.4 Quantified expressions
9.5 Declare expressions
9.6 Reduction expressions

This chapter describes additional forms of expressions introduced in Ada 2012 and Ada 2022. These are related to the three bracketed sequential control structures of Ada described in the previous chapter. Thus, corresponding to the if statement there is the if expression and corresponding to the case statement there is the case expression. There are also quantified expressions which are related to loop statements. We also discuss new forms of membership tests added in Ada 2012.

The incentive to introduce conditional expressions into Ada was triggered by the addition of pre- and postconditions which are described in Chapter 16. Without adding conditional forms of expressions these pre- and postconditions would have been cumbersome and required lots of small functions. However, these additional forms of expressions are applicable in other situations as well and it seems appropriate to describe them here.

Other forms of expression added in Ada 2022 are declare expressions which are essentially the expression equivalent of a block and reduction expressions which enable certain statements to be grouped together and reduced to a single expression.

9.1 Membership tests

Membership tests were introduced in Section 6.9 where we showed how they could be used to check whether a value lies within a specified range or satisfies a constraint implied by a subtype.

Thus given I of type Integer we could write

```
if I in 1.. 10 then ...
```

and given the type Day describing all the days of the week and the subtype Weekday we could test whether a variable D of the type Day had a value within the subtype by writing

```
if D in Weekday then ...
```

These were the only forms of membership test allowed in Ada 2005, but more general forms were introduced in Ada 2012.

Suppose we have an enumeration type for the names of months and perhaps related subtypes thus

```
type Month_Name is (Jan, Feb, Mar, Apr, May, Jun,
                    Jul, Aug, Sep, Oct, Nov, Dec);

subtype Spring is Month_Name range March .. May;
```

and wish to check whether it is safe to eat an oyster (there has to be an R in the month). In Ada 2005 we can write

```
if M in Jan .. April or M in Sep .. Dec then
```

which means repeating M and then perhaps worrying about whether to use **or** or **or else**. Alternatively we might write

```
if M not in May .. August then
```

but this seems somewhat unnatural.

However, a membership test can have several possible membership choices separated by the vertical bar so we can more naturally write

```
if M in Jan .. April | Sep .. Dec then
```

The individual membership choices can be single expressions, subtypes, or ranges. Thus the following are all permitted

```
if N in 6 | 28 | 496 then      -- N is small and perfect

if M in Spring | June | October .. December then
                               -- combination of subtype, single value, range

if X in 0.5 .. Z | 2.0*Z .. 10.0 then         -- not discrete or static

if Obj in Triangle | Circle then              -- with tagged types

if Letter in 'A' | 'E' | 'I' | 'O' | 'U' then  -- characters
```

Membership tests are permitted for any type and values do not have to be static. However, it should be remembered that uses of the vertical bar in case statements and aggregates do require the type to be discrete and the values to be static.

Another important point about membership tests is that the membership choices (that is the items separated by the bars) are evaluated in order. As soon as one is found such that the value of the test is known to be true (or false if **not** is present) then the test as a whole is determined and the other membership choices are not evaluated. This is therefore the same as using short circuit forms such as **or else**.

It is often convenient to use a membership test before a conversion to ensure that the conversion will succeed. This avoids raising an exception which then has to be handled. Thus we might have

```
subtype Score is Integer range 1 .. 60;
Total: Integer;
S: Score;
...                          -- compute Total somehow
if Total in Score then
   S := Score(Total);        -- reliable conversion
   ...                       -- now use S knowing that it is OK
else
   ...                       -- Total was excessive
end if;
```

If we are indexing some arrays whose range is Score then it is an advantage to use S as an index since we know it will work and no checks are needed.

There are other uses for membership tests which we will encounter in due course. For example, we can use a membership test to check accessibility which is described in Chapter 11.

An important issue is that the whole purpose of a membership test X **in** S is to find out whether a condition is satisfied. We want the result of a test to be true or false and not to raise an exception. However, if the evaluation of S could itself raise an exception then we have done something silly; this is discussed in detail in Section 16.6.

Exercise 9.1

1 Assume that N is a positive integer. Assign true or false to B according to whether or not the value of N is a prime less than 20.

2 Assume that Letter is a value of the type Character. Write a test to see whether Letter is one of the first five or last five letters of the alphabet.

9.2 If expressions

A simple example of an if expression was illustrated in Section 7.1 when we noted that we could write

```
Tomorrow := (if Today = Sun then Mon else Day'Succ(Today));
```

rather than using an if statement.

Another situation where an if expression is useful is if we have alternative calls of the same subprogram but with just one parameter being different. Thus we might wish to call a procedure P thus

```
if X > 0 then
   P(A, B, D, E);
else
   P(A, C, D, E);
end if;
```

This is cumbersome but using an if expression we can simply write

```
P(A, (if X > 0 then B else C), D, E);
```

Note that there is no closing **end if**. One reason is simply that it is logically unnecessary since there can only be a single expression after **else** and also **end if** would be obtrusively heavy. However, in order to aid clarity, an if expression is always enclosed in parentheses. If the context already has parentheses then additional ones are not necessary. Thus in the case of a procedure call with a single parameter, we can just write

```
P(if X > 0 then B else C);
```

However, if the call uses named notation (see Section 10.4) then additional parentheses are needed as in

```
P(Para => (if X > 0 then B else C));
```

As expected, a series of tests can be done using **elsif** thus

```
P(if X > 0 then B elsif X < 0 then C else D);
```

and expressions can be nested

```
P(if X > 0 then (If Y > 0 then B else C) else D);
```

Without the rule requiring enclosing parentheses this could be written as

```
P(if X > 0 then If Y > 0 then B else C else D);
```

which seems more than a little confusing.

There is a special rule if the type of the expression is Boolean. In that case a final else part can be omitted and is taken to be true by default. Thus the following are equivalent

```
Q(if C1 then C2 else True);
Q(if C1 then C2);
```

Such abbreviations occur frequently in preconditions which are discussed in Chapter 16. If we write

```
Pre => (if P1 > 0 then P2 > 0)
```

then this has the obvious meaning that the precondition requires that if P1 is positive then P2 must also be positive. However, if P1 is not positive then the precondition is true and it doesn't matter about P2.

Curiously enough the two expressions

```
(if C1 then C2)
(not C1 or C2)
```

have exactly the same value and are in fact the **imp** (implies) operator which is present in some languages. The truth table for this phantom operator is shown in Figure 9.1.

imp	F	T
F	T	F
T	T	T

Figure 9.1 Operator table for the non-existent operator **imp**.

There are important rules regarding the types of the various dependent expressions in the branches of an if expression. Basically, they have to all be of the same type or convertible to the same expected type. But there are some interesting situations.

If the expression is the argument of a type conversion then effectively the conversion is considered pushed down to the dependent expressions. Thus

```
X := Float(if P then A else B);
```

is equivalent to

```
X := (if P then Float(A) else Float(B));
```

As a consequence we can write

```
X := Float(if P then 27 else 0.3);
```

and it does not matter that 27 and 0.3 are not of the same type.

Similar situations arise with other conversions. Using the examples of Section 3.3 we might declare a variable V of a class wide type such as Object'Class. See also Section 14.2. We can initialize V with a value of a type derived from Object such as

```
V: Object'Class := A_Circle;
```

where A_Circle is an object of type Circle which is derived from the type Object.

The initial value can also be given by a conditional expression such as

```
V: Object'Class := (if B then A_Circle else A_Triangle);
```

where A_Circle and A_Triangle are objects of specific types Circle and Triangle which are themselves derived from the root type Object. Thus the individual expressions do not have to be of the same specific type provided that they are all derived from the same root type. Effectively, the implicit conversion is pushed down to each dependent expression.

If the expected type is a specific tagged type then the rules for the various branches are similar to the rules for calling a subprogram with several controlling operands as described in Section 14.4. Briefly, they all have to be dynamically tagged (that is class wide) or all have to be statically tagged; they might all be tag indeterminate in which case the conditional expression as a whole is tag indeterminate.

Some obscure situations arise. Remember that the controlling expression of an if statement can be any Boolean type. Now consider

```
type My_Boolean is new Boolean;
My_Cond: My_Boolean := ...
if (if K > 10 then X = Y else My_Cond) then                -- illegal
   ...
end if;
```

The problem here is that X = Y is of type Boolean but My_Cond is of type My_Boolean. Moreover, the expected type for the condition in the if statement is any Boolean type so the poor compiler cannot make up its mind. This foolishness could be overcome by putting a type conversion around the if expression.

Similar rules regarding the various dependent parts of an if expression apply to staticness. If all parts are static then the expression as a whole is static.

The initial value of a named number has to be static (see Section 6.1). So if we wish to set the initial value of a named number Febdays to 29 or 28 and there is a static Boolean Leap indicating whether it is a leap year or not then we can write

```
Febdays: constant := (if Leap then 29 else 28);
```

Attempting to do this in earlier versions of Ada was awkward. One had to write something horrid like

```
Febdays: constant := Boolean'Pos(Leap)*29 + Boolean'Pos(not Leap)*28;
```

which is truly gruesome. Similar disgusting expressions might also be contrived for the call of the procedure where only one parameter was different. Thus we could write

```
P(A, Boolean'Pos(X>0)*B + Boolean'Pos(X<=0)*C, D, E);
```

Static expressions introduce situations where all parts of an expression need not be evaluated. Consider

```
X := (If B then P else Q);
```

If B, P, and Q are all static then the expression as a whole is static as mentioned above. If B is true then the answer is P and there is not any need to even look at Q. We say that Q is statically unevaluated and indeed it does not matter that if Q had been evaluated then it would have raised an exception.

If we write

```
Answer := (if Count = 0 then 0.0 else Total/Count);
```

then naturally, having evaluated Count and finding it to be zero, then we know that the answer is 0.0 and we do not have to evaluate Total/Count so there is no risk of dividing by zero. Similar situations arise with short circuit conditions described in Section 6.9.

Exercise 9.2

1 Assign to Days_In_Month the number of days in a month M. Assume that the year is in the range 1901 .. 2099 as in Exercise 7.1(**1**).

9.3 Case expressions

Case expressions have much in common with if expressions and the two are collectively known as conditional expressions.

Thus given a variable D of the familiar type Day, we can assign the number of hours in a working day to the variable Hours by

```
Hours := (case D is
              when Mon .. Thurs => 8.0,
              when Fri => 6.0,
              when Sat | Sun => 0);
```

A slightly more adventurous example taking full account of leap years and involving nested if expressions is

```
Days_In_Month := (case M is
                     when September | April | June | November => 30,
                     when February =>
                        (if Year mod 100 = 0 then
                           (if Year mod 400 = 0 then 29 else 28)
                        else
                           (if Year mod 4 = 0 then 29 else 28)),
                     when others => 31);
```

The reader is invited to improve this!

Note the similarity to the rules for if expressions. There is no closing **end case**. Case expressions are similarly always enclosed in parentheses but they can be omitted if the context already provides parentheses such as in a subprogram call with a single positional parameter.

The inner structure is just like that of a case statement. The individual choices must be static and an optional **others** clause may be last. The rules for the expression after **case** are the same as well and of course all values of the appropriate subtype must be covered.

If M and Year are static then the case expression as a whole is static. If M is static and equal to September, April, June, or November then the value is statically known to be 30 so that the expression for February is not evaluated even if Year is not static. (Again we say that the expression is statically unevaluated.) Note that the various choices are evaluated in order.

The rules regarding the types of the dependent expressions are exactly as for if expressions. Thus if the case expression is the argument of a type conversion then the conversion is effectively pushed down to all the dependent expressions.

It is always worth emphasizing that an important advantage of case constructions is that they give a coverage check. Thus suppose we have an enumeration type describing various animals

```
type Animal is (Bear, Cat, Dog, Horse, Wolf);
```

Note that some of these animals might be considered to be pets and thus need feeding whereas others might be considered to be pests and should not be fed (what a difference a single s makes!).

So if A is an animal and Feed_It is a Boolean then we might write

```
Feed_It := A in Cat | Dog | Horse;
```

using the more elaborate form of membership test described in Section 9.1. However, this is unwise since if we add Rabbit to the type Animal then (presuming it is a pet and not wild) then it will not be fed unless we remember to add Rabbit to the membership test.

It is better to write

```
Feed_It := (case A is
               when Cat | Dog | Horse => True,
               when Bear | Wolf => False);
```

and then if we add Rabbit to the type Animal then we must add it to one arm of the case expression otherwise it will fail to compile.

It is interesting to look for examples of case statements to see whether they can more conveniently be rewritten using case expressions. For example, the procedure Build_List of Program 1, Magic Moments, has a case statement whose purpose is to assign a value of the type Object'Class to Object_Ptr according to a character typed by the user. However, if the character is incorrect the effect is to exit the surrounding loop. One could rewrite this as

```
Object_Ptr := (case Code_Letter is
                  when 'C' | 'c' =>           -- expect a circle
                     Get_Circle
                  when 'T' | 't' =>           -- expect a triangle
                     Get_Triangle
                  when 'S' | 's' =>           -- expect a square
                     Get_Square
                  when others =>
                     null);
if Object_Ptr = null then exit; end if;
```

This certainly captures the expected behaviour that all sensible branches assign to Object_Ptr but makes the exceptional case more awkward. We will discuss this example again in Section 15.2 when we introduce raise expressions.

Note that conditional expressions can themselves be part of a larger expression. Thus we might have an if expression within a case expression and so on.

Exercise 9.3

1 Write a statement using a case expression to assign the length of the name of Today to the integer L. Thus if today is Monday the value to be assigned is 6. Do not use any attributes.

2 A sexist and ageist company provides a pension for all staff aged 60 and over when they retire. The basic pension is 500 euros per month and is incremented by 50 euros at age 70 and 80. No pension is given to those aged over 100. It is reduced by 10% for females and increased by 5% for those who are disabled. A

special terminal bonus of 100 per month is given to all in the year when they are 100. Given variables Age, Gender, and Disabled write an expression giving their monthly pension to the nearest euro and assign it to a variable Pension of type Integer. Consider whether it is best to use a case expression or if expression for each part of the calculation.

9.4 Quantified expressions

Another new form of expression is the quantified expression. Quantified expressions are closely related to for loops and might be considered to be a sort of loop expression. However, the type of a quantified expression is always a Boolean type.

As a simple example consider

```
B := (for all K in A'Range => A(K) = 0);
```

which assigns True to B if every component of the array A has value 0. If A'Range is null then the value of the quantified expression is True, see AI12-258.

A quantified expression always starts with **for** and is then followed by a quantifier which in this case is the reserved word **all**. The other possible quantifier is the reserved word **some** so we might instead have

```
B := (for some K in A'Range => A(K) = 0);
```

which assigns True to B if some component of the array A has value 0. If A'Range is null then the value of the quantified expression is False.

The expression after the => is always a Boolean expression. It is known technically as the predicate which is simply a fancy term used in logic to mean a truth value which in Ada we refer to as a Boolean expression.

Note that the loop parameter is almost inevitably used in the predicate. A quantified expression is very much like a for statement except that we evaluate the expression after => on each iteration rather than executing one or more statements. The iteration is somewhat implicit and the words **loop** and **end loop** do not appear.

The expression is evaluated for each iteration in the appropriate order (**reverse** can be inserted after **in**) and the iteration stops as soon as the value of the expression is determined. Thus in the case of **for all**, as soon as one value is found to be False, the overall expression is False whereas in the case of **for some** as soon as one value is found to be True, the overall expression is True. An iteration could raise an exception which would then be propagated in the usual way.

Like conditional expressions, a quantified expression is always enclosed in parentheses which can be omitted if the context already provides them, such as in a procedure call with a single positional parameter.

The forms **for all** and **for some** are technically known as the universal quantifier and existential quantifier respectively.

Note that in mathematics we use the symbols ∀ and ∃ to mean 'for all' and 'there exists'. Thus we might write

$$\forall\ x, \exists\ y,\ \text{s.t.}\ x + y = 0$$

which means of course that for all values x, there exists a value y such that $x+y$ equals zero. In other words, every value has a corresponding negative (although it does not say it is unique).

Readers might like to contemplate whether the symbols ∀ and ∃ are the usual letters A and E inverted and reversed respectively or perhaps simply rotated. (Casting one's mind back to the days of hot lead it is clear that they are just rotated.)

The type of a quantified expression in Ada can be any Boolean type (that is the predefined type Boolean or perhaps My_Boolean derived from Boolean). The predicate must be of the same type as the expression as a whole. Thus if the predicate is of type My_Boolean then the quantified expression is also of type My_Boolean.

Quantified expressions can be nested. So we might check that all components of a two-dimensional array AA are zero by writing

```
B := (for all I in AA'Range(1) =>
          (for all J in AA'Range(2) => AA(I, J) = 0));
```

This can be done rather more neatly using the iterator form using **of** rather than **in** as mentioned in Section 8.1. We just write

```
B := (for all E of AA => E = 0);
```

which iterates over all elements of the array AA however many dimensions it has.

Of course, we cannot always use the abbreviated form. Thus suppose we wanted to see if an array A has every component equal to its index value. The quantified expression describing this might be

```
(for all I in A'Range => A(I) = I);
```

and this cannot be rewritten in the other form because the predicate involves the index value other than for indexing the array.

Quantified expressions were introduced primarily for use in preconditions and postconditions which will be discussed in Chapter 16. However, they can be used in any context requiring an expression. Thus we might test whether an integer N is prime by

```
RN := Integer(Sqrt(Float(N)));
if (for some K in 2 .. RN => N mod K = 0) then    ...  -- N not prime
```

or we might reverse the test by

```
if (for all K in 2 .. RN => N mod K /= 0) then    ...  -- N is prime
```

Beware that this is not a recommended technique if N is at all large!

We could also use a quantified expression for checking that it is safe to eat an oyster which we did using a membership test in Section 9.1. Thus we might write

```
if (for some C of Month_Name'Image(M) => C = 'R' or C = 'r') then
```

which assumes that we have given the full names January, February, and so on for the enumeration literals in the type Month_Name.

Exercise 9.4

1 The array A of integers has been sorted into ascending order. Write a quantified expression describing this. Permit duplication.

9.5 Declare expressions

A declare expression is the expression equivalent of a block in much the same way that an if expression corresponds to an if statement. Consequently a declare expression enables us to introduce local names within an expression. An obvious use is to avoid repeating a complex subexpression which occurs several times in the overall expression. Thus if we wanted to specify a postcondition which was that $v = (x+y+z) + (x+y+z)^2 + (x+y+z)^3$ we can write

```
Post => V = (declare W: constant Integer := X + Y + Z;
                begin W + W**2 + W**3);
```

Similar rules to those for conditional expressions apply as expected. There is no **end** and the declare expression must be enclosed in parentheses but they can be omitted to avoid duplication. The expression following **begin** is known as the body expression; its type determines the type of the declare expression as a whole.

There are important restrictions on the allowed declarations. The objects have to be constant and not of a limited type. This prevents such horrors as declaring a task object which would raise all sorts of problems. Another point is that the type of the object is not allowed to be an access type and we must not involve the words **aliased**, Access, or Unchecked_Access.

As well as the declaration of constant objects, object renaming declarations are also allowed. But again there are similar restrictions on the renaming concerning limited types and access types. However, harmless renaming to avoid repetition is allowed. The above example could in fact be written as

```
Post => V = (declare W renames Integer'(X + Y + Z);
                begin W + W**2 + W**3);
```

As noted in Section 13.7, in Ada 2022 we do not have to give the subtype in a renaming declaration because it is taken from the entity being renamed (if given it was ignored in previous versions of Ada anyway). But the entity being renamed must have the syntactic form of a name so we could not write

```
W renames (X + Y + Z)
```

but adding the qualification makes it a name and so all is well. Either way we have to reveal the type Integer!

A declare expression can include other compound expressions such as conditional expressions and quantified expressions and of course internal declare expressions. Similarly, a conditional expression can include declare expressions internally as well and so on.

A final point is that a declare expression need not declare any objects at all; this matches the fact that a block need not declare anything either.

9.6 Reduction expressions

Reduction expressions are another example of how certain statements can be embedded within an expression. The general idea is that starting from a set of values, it can be transformed in two stages into a single value. The first stage enables the set to be transformed into another set and the second stage then consolidates that set into a single value using the attribute Reduce.

Suppose we wish to compute the *n*th square pyramidal number. That is the sum of the squares of the first *n* integers. It can be done by

```
Pn := 0;                    -- Pn is of type Integer
for J in 1 .. N loop
   Pn := Pn + J**2;
end loop;
```

We can consider this as starting with a set of integers. The first stage is to convert it to a set of squares and the second stage is to add them together. Using a reduction expression this can be written as

```
Pn := ([for J in 1 .. N => J**2]'Reduce("+", 0));
```

The part in square brackets is known as the value sequence. The square brackets cannot be replaced by parentheses. The value sequence consists of the usual start of a for loop and then => followed by an expression, typically involving the for loop variable. Note that the loop cannot include **reverse**.

Following the value sequence is the Reduce attribute which has two parameters. The first is the operation being used to accumulate the final result and the second is the starting value for the accumulation (this will typically be zero or one).

An example using multiplication for the accumulation is the factorial function which can be written as an expression function using a reduction expression thus

```
function Factorial(N: Natural) return Natural is
   ([for J in 1 .. N => J]'Reduce("*", 1));
```

Note that this does work correctly if N is zero, the loop is null and the reduction expression is simply its initial value which is 1.

Other forms of iterator can also be used. Thus to add together all the elements of the two dimensional array AA we write

```
Total := [for E of AA => E]'Reduce("+", 0.0);
```

In this case no transformation of the original sequence is required so it can be abbreviated to simply

```
Total := AA'Reduce("+", 0.0);
```

Moreover, the maximum of that set of numbers is given by

```
Maximum := AA'Reduce(Float'Max, 0.0);
```

Reduction expressions can also be used with parallel loops as in Section 22.8.

Exercise 9.6

1 Write a reduction expression giving the sum of the top *n* layers of a triangular pyramid of cannonballs. Remember that the layers have 1, 3, 6, 10 balls and so on.

Checklist 9

Membership choices can be expressions, ranges, and subtypes.

Membership choices need not be static.

If expressions, case expressions, and quantified expressions are always in parentheses.

Conditional expressions do not have **end if** or **end case**.

The choices in a case expression must be static.

All possibilities in a case expression must be catered for.

If **others** is used it must be last and on its own.

The expression after **case** can be qualified in order to reduce the alternatives.

This chapter does not apply to Ada 2005.

The word **some** was reserved in Ada 2012 for use in quantified expressions.

New in Ada 2022

Declare expressions and reduction expressions are new.

10 Subprograms

10.1 Functions
10.2 Operators
10.3 Procedures
10.4 Aliasing
10.5 Named & default parameters
10.6 Overloading
10.7 Declarations, scopes, and visibility

Subprograms are perhaps the oldest form of abstraction and existed long before the introduction of high level languages. Subprograms enable a unit of code to be encapsulated and thereby reused and also enable the code to be parameterized so that it can be written without knowing the actual data to which it is to be applied.

In Ada, subprograms fall into two categories: functions and procedures. Functions are called as components of expressions and return a value as part of the expression, whereas procedures are called as statements standing alone.

As we shall see, the actions to be performed when a subprogram is called are usually described by a subprogram body. Subprogram bodies are declared in the usual way in a declarative part which may for instance be in a block or indeed in another subprogram.

10.1 Functions

A function is a form of subprogram that can be called as part of an expression. In Chapter 6 we met examples of calls of functions such as Day'Succ and Sqrt.

We now consider the form of a function body which describes the statements to be executed when the function is called. For example the body of the function Sqrt might have the form

```
function Sqrt(X: Float) return Float is
   R: Float;
begin
   -- compute value of Sqrt(X) in R
   return R;
end Sqrt;
```

All function bodies start with the reserved word **function** and the designator of the function being defined. If the function has parameters the designator is followed by a list of parameter specifications in parentheses. Each gives the identifiers of one or more parameters, a colon and then its type or subtype. If there are several such specifications then they are separated by semicolons. Examples of parameter lists are

```
(I, J, K: Integer)
(Left: Integer; Right: Float)
```

The parameter list, if any, is then followed by **return** and the type or subtype of the result of the function. In the case of both parameters and result the type or subtype must be given by a subtype mark and not by a subtype indication with an explicit constraint; the reason will be mentioned in Section 10.7. (But we can give a null exclusion which concerns access types and will be discussed in Chapter 11.)

The part of the body we have described so far is called the function specification. It specifies the function to the outside world in the sense of providing all the information needed to call the function.

After the specification comes **is** and then the body proper which is just like a block – it has a declarative part, **begin**, a sequence of statements, and then **end**. As in the case of a block, the declarative part can be empty, but there must be at least one statement in the sequence of statements. Between **end** and the terminating semicolon we may repeat the designator of the function. This is optional but, if present, must correctly match the designator after **function**.

It is often necessary or just convenient to give the specification on its own but without the rest of the body. In such a case it is immediately followed by a semicolon thus

```
function Sqrt(X: Float) return Float;
```

and is then correctly known as a function declaration – although often still informally referred to as a specification. The uses of such declarations will be discussed in Section 10.7.

In general, the parameters of a subprogram can be of three modes, **in**, **in out** and **out**. The modes are described in Section 10.3 when we discuss procedures. Up until Ada 2005, functions could only have parameters of mode **in** and the discussion in this section assumes that all parameters are of mode **in**.

The formal parameters (of mode **in**) of a function act as local constants whose values are provided by the corresponding actual parameters. When the function is called the declarative part is elaborated in the usual way and then the statements are executed. A return statement is used to indicate the value of the function call and to return control back to the calling expression.

Thus considering our example suppose we had

```
S := Sqrt(T + 0.5);
```

then first T + 0.5 is evaluated and then Sqrt is called. Within the body the parameter X behaves as a constant with the initial value given by T + 0.5. It is rather as if we had

```
X: constant Float := T + 0.5;
```

The declaration of R is then elaborated. We then obey the sequence of statements which we assume compute the square root of X and assign it to R. The last statement is **return** R; this passes control back to the calling expression with the result of the function being the value of R. This value is then assigned to S.

The expression in a return statement can be of arbitrary complexity and must be of the same type as and satisfy any constraints implied by the subtype mark given in the function specification (and any null exclusion, see Section 11.3). If the constraints are violated then the exception Constraint_Error is raised. (A result of an array type can slide as mentioned later in this section.)

A function body may have several return statements. The execution of any one of them will terminate the function. Thus the function Sign which takes an integer value and returns +1, 0 or −1 according to whether the parameter is positive, zero or negative could be written as

```
function Sign(X: Integer) return Integer is
begin
   if X > 0 then
      return +1;
   elsif X < 0 then
      return -1;
   else
      return 0;
   end if;
end Sign;
```

So we see that the last lexical statement of the body need not be a return statement since there is one in each branch of the if statement. Any attempt to 'run' into the final end will raise the exception Program_Error. This is our first example of a situation giving rise to Program_Error; this exception is generally used for situations which would violate the run-time control structure.

Each call of a function produces a new instance of any objects declared within it (including parameters of course) and these disappear when we leave the function. It is therefore possible for a function to be called recursively without any problems. So the factorial function could be declared as

```
function Factorial(N: Positive) return Positive is
begin
   if N = 1 then
      return 1;
   else
      return N * Factorial(N-1);
   end if;
end Factorial;
```

If we write

```
F := Factorial(4);
```

then the function calls itself until, on the fourth call (with the other three calls all partly executed and waiting for the result of the call they did before doing the multiply) we find that N is 1 and the calls then all unwind and all the multiplications are performed.

Note that there is no need to check that the parameter N is positive since the parameter is of the subtype Positive. So calling Factorial(–2) will result in Constraint_Error. Of course, Factorial(10_000) could result in the computer running out of space in which case Storage_Error would be raised. The more moderate call Factorial(50) would undoubtedly cause overflow and thus raise Constraint_Error.

A formal parameter may be of any type but in general the type must have a name. The one exception is access parameters which are discussed in Chapter 11. So a parameter cannot be of an anonymous type such as

```
array (1 .. 6) of Float
```

In any event no actual parameter (other than an aggregate) could match such a formal parameter even if it were allowed since the actual and formal parameters must have the same type and the rules of type equivalence require that it must be named. Again access parameters are slightly different.

A formal parameter can be of an unconstrained array type such as

```
type Vector is array (Integer range <>) of Float;
```

In such a case the bounds of the formal parameter are taken from those of the actual parameter.

Consider

```
function Sum(A: Vector) return Float is
   Result: Float := 0.0;
begin
   for I in A'Range loop
      Result := Result + A(I);
   end loop;
   return Result;
end Sum;
```

then we can write

```
V: Vector(1 .. 4) := (1.0, 2.0, 3.0, 4.0);
S: Float;
...
S := Sum(V);
```

The formal parameter A then takes the bounds of the actual parameter V. So for this call we have

```
A'Range    is    1 .. 4
```

and the effect of the loop is to compute the sum of A(1), A(2), A(3), and A(4). The final value of Result which is returned and assigned to S is therefore 10.0.

The function Sum can therefore be used to sum the components of a vector with any bounds and in particular where the bounds are not known until the program executes.

A function could have a constrained array subtype as a formal parameter. However, remember that we cannot apply the constraint in the parameter list as in

```
function Sum_5(A: Vector(1 .. 5)) return Float      -- illegal
```

but must use the name of a constrained array type or subtype thus

```
subtype Vector_5 is Vector(1 .. 5);
...
function Sum_5(A: Vector_5) return Float
```

An actual parameter corresponding to such a constrained formal array must have the same number of components; sliding is allowed as for assignment. So we could have

```
W: Vector(0 .. 4);
...
S := Sum_5(W);
```

The actual parameter of a function can also be an aggregate (including a string). In fact the behaviour is exactly as for an initial value described in Section 8.3. If the formal parameter is unconstrained then the aggregate must supply its bounds and so cannot contain **others**. If the formal parameter is constrained then it provides the bounds; an aggregate without **others** could slide. But remember that an aggregate with **others** never slides.

As another example consider

```
function Inner(A, B: Vector) return Float is
   Result: Float := 0.0;
begin
   for I in A'Range loop
      Result := Result + A(I)*B(I);
   end loop;
   return Result;
end Inner;
```

This computes the inner product of the two vectors A and B by adding together the sum of the products of corresponding components. This is an example of a function with more than one parameter. Such a function is called by following the function name by a list of expressions giving the values of the actual parameters separated by commas and in parentheses. The order of evaluation of the actual parameters is not defined.

So

```
V: Vector(1 .. 3) := (1.0, 2.0, 3.0);
W: Vector(1 .. 3) := (2.0, 3.0, 4.0);
X: Float;
...
X := Inner(V, W);
```

results in X being assigned the value

```
1.0 * 2.0 + 2.0 * 3.0 + 3.0 * 4.0 = 20.0
```

Note that the function Inner is not written well since it does not check that the bounds of A and B are the same. It is not symmetric with respect to A and B since I takes (or tries to take) the values of the range A'Range irrespective of B'Range. So if the array W had bounds of 0 and 2, Constraint_Error would be raised on the third time around the loop. If the array W had bounds of 1 and 4 then an exception would not be raised but the result might not be as expected.

It would be nice to ensure the equality of the bounds by placing a constraint on B at the time of call but this cannot be done. The best we can do is simply check the bounds for equality inside the function body and perhaps explicitly raise Constraint_Error if they are not equal

```
if A'First /= B'First or A'Last /= B'Last then
   raise Constraint_Error;
end if;
```

(The use of the raise statement is described in detail in Chapter 15.)

We saw above that a formal parameter can be of an unconstrained array type. Similarly, a function result can be an array whose bounds are not known until the function is called. The result type can be an unconstrained array and the bounds are then obtained from the expression in the return statement.

As an example the following function returns a vector which has the same bounds as the parameter but whose component values are in the reverse order

```
function Rev(A: Vector) return Vector is
   R: Vector(A'Range);
begin
   for I in A'Range loop
      R(I) := A(A'First+A'Last-I);
   end loop;
   return R;
end Rev;
```

The variable R is declared to be of type Vector with the same bounds as A. Note how the loop reverses the value. The result takes the bounds of the expression R. Sadly, the function is called Rev rather than Reverse because **reverse** is a reserved word.

The matching rules for results of both constrained and unconstrained arrays are the same as for parameters. Sliding is allowed and so on.

If a function returns a record or array value then a component can be immediately selected, indexed or sliced as appropriate without assigning the value to a variable. Indeed the result is treated as a (constant) object in its own right. So

```
Rev(Y)(I)
```

denotes the component indexed by I of the array returned by the call of Rev.

It should be noted that a parameterless function call, like a parameterless procedure call, has no parentheses. There is thus a possible ambiguity between calling a function with one parameter and indexing the result of a parameterless call; such an ambiguity could be resolved by, for example, renaming the functions as will be described in Section 13.7.

We conclude this section by discussing the other form of return statement which we mention here for completeness although its importance will not be obvious until we deal with matters such as limited types in Section 12.5 and tasks in Section 22.2. This is the extended return statement. We could rewrite the last example as follows

```
function Rev(A: Vector) return Vector is
begin
   return R: Vector(A'Range) do
      for I in A'Range loop
         R(I) := A(A'First+A'Last-I);
      end loop;
   end return;
end Rev;
```

An extended return statement is bracketed between **return** and **end return**. After **return** there is the declaration of a variable which is to be the result; it can but need not have an initial value. This declaration is then followed by the reserved word **do** and then a sequence of statements and finally the closing **end return**. Running into the **end return** causes control to pass back to the function call with the value of the variable (in this case R) as the result.

An extended return statement can include any statements except other extended return statements. It can include blocks and thus internal declarations. It can also include a simple return statement which then causes control to be returned. But a simple return statement must not have an expression since the result to be returned is given by the variable of the extended return statement. So the structure might be

```
return R: T := E do
   if ... then
      ...
      return;                    -- returns R
   end if;
   ...
end return;
```

The return object R can be marked as **aliased** (see Section 11.4) and as **constant** for uniformity with other declarations.

The example of the function Rev illustrates that although the return object must be constrained, the type given in the specification need not be constrained. Thus the specification has the unconstrained array type Vector, but the extended return statement declares R to be of the constrained subtype Vector(A'Range). This necessary constraint could also be given by the initial value of R. A similar situation occurs with class wide types as discussed in Section 14.2.

A function could have several extended return statements perhaps in the branches of if or case statements. The specification might give the result as being unconstrained whereas individual branches might return results with different constraints. An example with discriminants will be found in Section 18.3 where the function Frankenstein has the return type Person but individual branches return objects of type Man or Woman. Similarly, the result type might be a class wide type such as Object'Class introduced in Section 3.3 but individual branches might return a Circle, a Triangle, a Square, and so on.

If we leave an extended return statement directly by a goto or exit statement then this does not cause a return from the function.

A variation is that the **do** ... **end return** part can be omitted so we might have

```
function Zero return Complex is
begin
  return C: Complex;
end Zero;
```

in which case the initial value of the variable C is the result. If the type Complex is as in Section 8.7 then the returned value will be (0.0, 0.0) because that is the default value of the type Complex.

The extended return statement is vital in certain situations but for the moment we can treat it as just an alternative syntax. But it makes it clear from the beginning what is to be returned and ensures that we cannot forget to return the object.

If a function is short then it can often be written as an expression function in which there is no return statement but whose result is given simply by an expression in parentheses. Expression functions are important in contracts (see Chapter 16). Their practicality is increased by the introduction of conditional, quantified, declare, and reduction expressions described in Chapter 9.

For example, the function Sign described above can be rewritten as simply

```
function Sign(X: Integer) return Integer is
  (if X > 0 then +1 elsif X < 0 then -1 else 0);
```

Note that, as in the case of if expressions, double parentheses are never needed. The same rule applies if a function returns an aggregate. So a function to return the conjugate of a complex number (as in Section 8.7) might be

```
function Conjugate(C: Complex) return Complex is
  (C.Re, -C.Im);            -- no need for double parentheses
```

But if the aggregate is given as a string, then the parentheses are necessary. Thus we might have

```
function Piggy return String is
  ("PIG");
```

In versions of Ada upto Ada 2005, functions could only have parameters of mode **in**. However, in Ada 2012 and Ada 2022, they can have parameters of any mode. The various modes are described in Section 10.3. One consequence of this is that there are various rules concerning aliasing as described in Section 10.4.

Exercise 10.1

1. Write a function Even which returns True or False according to whether its Integer parameter is even or odd.

2. Rewrite the factorial function so that the parameter may be positive or zero but not negative. Remember that the value of Factorial(0) is to be 1. Use the subtype Natural introduced in Section 6.5.

3. Write a function Outer that forms the outer product of two vectors. The outer product $\boldsymbol{C}$ of two vectors $\boldsymbol{A}$ and $\boldsymbol{B}$ is a matrix such that $C_{ij} = A_i B_j$.

4 Write a function Make_Colour which takes an array of values of type Primary and returns the corresponding value of type Colour. See Section 8.6. Check that Make_Colour((R, Y)) = Orange.

5 Rewrite the function Inner to use sliding semantics so that it works providing the arrays have the same length. Raise Constraint_Error (as outlined above) if the arrays do not match. Use an extended return statement.

6 Write a function Make_Unit that takes a single parameter N and returns a unit $N \times N$ matrix with components of type Float. Use the function to declare a constant unit $N \times N$ matrix. See Exercise 8.3(3).

7 Write a function GCD to return the greatest common divisor of two nonnegative integers. Use Euclid's algorithm that

$$\gcd(x, y) = \gcd(y, x \bmod y) \quad y \neq 0$$
$$\gcd(x, 0) = x$$

Write the function using recursion and then rewrite it using a loop statement.

10.2 Operators

In the previous section we stated that a function body commenced with the reserved word **function** followed by the designator of the function. In all the examples of that section the designator was in fact an identifier. However, it can also be a character string provided that the string is one of the language operators in double quotes. These are

abs	**and**	**mod**	**not**	**or**	**rem**	**xor**
=	/=	<	<=	>	>=	
+	–	*	/	**	&	

In such a case the function defines a new meaning of the operator concerned. As an example we can rewrite the function Inner of the previous section as an operator thus

```
function "*" (A, B: Vector) return Float is
   Result: Float := 0.0;
begin
   for I in A'Range loop
      Result := Result + A(I)*B(I);
   end loop;
   return Result;
end "*";
```

We call this new function by the normal syntax of uses of the operator "*". Thus instead of

```
X := Inner(V, W);
```

we now write

```
X := V * W;
```

This meaning of "*" is distinguished from the existing meanings of integer and floating point multiplication by the context provided by the types of the actual parameters V and W and the type of the destination X.

The giving of several meanings to an operator is another instance of overloading which we have already met with enumeration literals. The rules for the overloading of subprograms in general are discussed later in this chapter. It suffices to say at this point that any ambiguity can usually be resolved by qualification. Overloading of predefined operators is not new. It has existed in most programming languages for the past seventy years.

We can now see that the predefined meanings of all operators are as if there were a series of functions with declarations such as

```
function "+" (Left, Right: Float) return Float;
function "<" (Left, Right: Float) return Boolean;
function "<" (Left, Right: Boolean) return Boolean;
```

Moreover, every time we declare a new type, new overloadings of predefined operators such as "=" and "<" may be created.

Observe that the predefined operators always have Left and Right as formal parameter names (real mathematicians would prefer X and Y which have served the community well since the days of Newton; but Ada had to be awkward!).

Although we can add new meanings to operators we cannot change the syntax of the call. Thus the number of parameters of "*" must always be two and the precedence cannot be changed and so on. The operators "+" and "–" are unusual in that a new definition can have either one parameter or two parameters according to whether it is to be called as a unary or binary operator. Thus the function Sum could be rewritten as

```
function "+" (A: Vector) return Float is
   Result: Float := 0.0;
begin
   for I in A'Range loop
      Result := Result + A(I);
   end loop;
   return Result;
end "+";
```

and we would then write

```
S := +V;
```

rather than

```
S := Sum(V);
```

Function bodies whose designators are operators often contain interesting examples of uses of the operator being overloaded. Thus the body of "*" contains a use of "*" in A(I)*B(I). There is, of course, no ambiguity since the expressions A(I) and B(I) are of type Float whereas our new overloading is for type Vector. Sometimes there is the risk of accidental recursion. This particularly applies if we try to replace an existing meaning rather than add a new one.

Apart from the operator "/=" there are no special rules regarding the types of the operands and results of new overloadings. Thus a new overloading of "=" need not return a Boolean result. On the other hand, if it is Boolean then a corresponding new overloading of "/=" is implicitly created. Moreover, explicit new overloadings of "/=" are also allowed provided only that the result type is not Boolean.

The membership tests **in** and **not in** and the short circuit forms **and then** and **or else** cannot be given new meanings. That is why we said in Section 6.9 that they were not technically classed as operators.

In the case of operators represented by reserved words, the characters in the string can be in either case. Thus a new overloading of **or** can be declared as "or" or "OR" or even "Or".

In the previous section, it was mentioned that Ada 2012 introduced expression functions and permitted functions to have parameters of any mode. These extensions do not apply to operators.

Exercise 10.2

1. Write a function "<" that operates on two Roman numbers and compares them according to their corresponding numeric values. That is, so that "CCL" < "CCXC". See Exercise 8.4(2).
2. Write functions "+" and "*" to add and multiply two values of type Complex. See Exercise 8.7(1).
3. Write a function "<" to test whether a value of type Primary is in a set represented by a value of type Colour. See Section 8.6.
4. Write a function "<=" to test whether one value of type Colour is a subset of another.
5. Write a function "<" to compare two values of the type Date of Section 8.7.

10.3 Procedures

The other form of subprogram is a procedure; a procedure is called as a statement standing alone in contrast to a function which is always called as part of an expression. We have already encountered a number of examples of procedure calls where there are no parameters such as Work; Party; Action; and so on. We now look at procedures in general.

The body of a procedure is similar to that of a function. The differences are

- a procedure starts with **procedure**,
- its name must be an identifier,
- it does not return a result.

In Ada 2012 and Ada 2022, functions as well as procedures can have parameters of any mode. There are three different modes **in**, **out**, and **in out**. In earlier versions of Ada, functions could only have parameters of mode **in**.

The mode of a parameter is indicated by following the colon in the parameter specification by **in** or by **out** or by **in out**. If the mode is omitted then it is taken to be **in**. We usually give the mode **in** for procedures but omit it for functions. The form of parameter known as an access parameter is actually an **in** parameter. Access parameters are discussed with access types in Chapter 11.

In the case of functions; the examples earlier in this chapter omitted **in** but could have been written, for instance, as

```
function Sqrt(X: in Float) return Float;
function "*" (A, B: in Vector) return Float;
```

The general effect of the three modes can be summarized as follows.

in The formal parameter is a constant initialized by the value of the associated actual parameter.

in out The formal parameter is a variable initialized by the actual parameter; it permits both reading and updating of the value of the associated actual parameter.

out The formal parameter is an uninitialized variable; it permits updating of the value of the associated actual parameter.

Note that both **in out** and **out** parameters behave as normal variables within the subprogram but the key difference is that an **in out** parameter is always initialized by the actual parameter whereas an **out** parameter is not.

The fine detail of the behaviour depends upon whether a parameter is passed by copy or by reference. Parameters of scalar types (and access types, see Chapter 11) are always passed by copy and we will consider them first.

As a simple example of the modes **in** and **out** consider

```
procedure Add(A, B: in Integer; C: out Integer) is
begin
   C := A + B;
end Add;
```

with

```
P, Q: Integer;
...
Add(2+P, 37, Q);
```

On calling Add, the expressions 2+P and 37 are evaluated (in any order) and are the initial values of the formals A and B which behave as constants. The value of A+B is then assigned to the formal variable C. On return the value of C is assigned to the variable Q. Thus it is (more or less) as if we had written

```
declare
   A: constant Integer := 2+P;        -- in
   B: constant Integer := 37;         -- in
   C: Integer;                        -- out
begin
   C := A + B;                        -- body
   Q := C;                            -- out
end;
```

As an example of the mode **in out** consider

```
procedure Increment(X: in out Integer) is
begin
   X := X + 1;
end;
...
I: Integer;
...
Increment(I);
```

On calling Increment, the value of I is the initial value of the formal variable X. The value of X is then incremented. On return, the final value of X is assigned to the actual parameter I. So it is rather as if we had written

```
declare
   X: Integer := I;
begin
   X := X + 1;
   I := X;
end;
```

For any scalar type (such as Integer) the modes thus correspond simply to copying the value **in** at the call or **out** upon return or both in the case of **in out**.

If the mode is **in** then the actual parameter may be any expression of the appropriate type or subtype. If the mode is **out** or **in out** then the actual parameter must be a variable (it could not be a constant or an aggregate for example). The identity of such a variable is determined when the subprogram is called and cannot change during the call.

Suppose we had

```
I: Integer;
A: array (1 .. 10) of Integer;

procedure Silly(X: in out Integer) is
begin
   I := I + 1;
   X := X + 1;
end;
```

then the statements

```
A(5) := 1;
I := 5;
Silly(A(I));
```

result in A(5) becoming 2, I becoming 6, but A(6) is not affected.

If a parameter is an array or record then the mechanism of copying, described above, may generally be used but alternatively an implementation may use a reference mechanism in which the formal parameter provides direct access to the actual parameter. A program that depends on the particular mechanism because of aliasing is said to have a bounded error. See the exercises at the end of this section.

Parameters of certain types (and arrays and records with any components of those types) are always passed by reference. It so happens that we have not yet dealt with any of these types which are: task and protected types (Chapter 20), tagged record types (Chapter 14) and explicitly limited types (Chapter 12). Private types behave as the corresponding full type (Chapter 12).

Note that because a formal array parameter takes its bounds from the actual parameter, the bounds are always copied in at the start even in the case of an **out** parameter. Of course, for simplicity, an implementation could always copy in the whole array anyway. Indeed, when **out** parameters are passed by copy certain types are always copied in so that dangerous undefined values do not arise. This applies to access types (Chapter 11), discriminated record types (Chapter 18) and record types which have components with default initial values such as the type Complex in Section 8.7.

We now consider the question of constraints on parameters; these are similar to the rules for function results which were mentioned in Section 10.1.

In the case of scalar parameters the situation is as expected from the copying model. For an **in** or **in out** parameter any constraint on the formal must be satisfied by the value of the actual at the beginning of the call. Conversely for an **in out** or **out** parameter any constraint on the variable which is the actual parameter must be satisfied by the value of the formal parameter upon return from the subprogram. Any constraint imposed by the result of a function must also be satisfied.

In the case of arrays the situation is somewhat different. If the formal parameter is a constrained array type, the association is just as for assignment, the number of components in each dimension must be the same but sliding is permitted. If, on the other hand, the formal parameter is an unconstrained array type, then, as we have seen, it takes its bounds from those of the actual. The foregoing applies irrespective of the mode of the array parameter. Similar rules apply to function results; if the result is a constrained array type then the expression in the result can slide. Moreover, beware that passing results of an unconstrained array type (such as String) may be inefficient.

In the case of the simple records we have discussed so far there are no constraints and so there is nothing to say. The parameter and result mechanism for other types will be discussed in detail when they are introduced.

We noted above that an actual parameter corresponding to a formal **out** or **in out** parameter must be a variable. This allows the actual parameter in turn to be an **out** or **in out** formal parameter of some outer subprogram.

A further possibility is that an actual parameter can also be a type conversion of a variable provided, of course, that the conversion is allowed. As an example, since conversion is allowed between numeric types, we can write

```
F: Float;
...
Increment(Integer(F));
```

If F initially had the value 2.3, it would be converted to the integer value 2, incremented to give 3 and on return converted to 3.0 and finally assigned back to F.

This conversion (technically known as a view conversion) of **in out** or **out** parameters is particularly useful with arrays. Suppose we write a library of subprograms applying to our type Vector and then acquire from someone else some subprograms written to apply to the type Row of Section 8.2. The types Row and Vector are essentially the same; it just so happened that the authors used different names. Array type conversion allows us to use both sets of subprograms without having to change the type names systematically.

As a final example consider the following

```
procedure Quadratic(A, B, C: in Float;
                    Root_1, Root_2: out Float; OK: out Boolean) is
   D: constant Float := B**2 - 4.0*A*C;
begin
   if D < 0.0 or A = 0.0 then
      OK := False;
      return;
   end if;
   Root_1 := (-B+Sqrt(D)) / (2.0*A);
   Root_2 := (-B-Sqrt(D)) / (2.0*A);
   OK := True;
end Quadratic;
```

The procedure Quadratic attempts to solve the equation

$$ax^2 + bx + c = 0$$

If the roots are real they are returned via the parameters Root_1 and Root_2 and OK is set to True. If the roots are complex (D < 0.0) or the equation degenerates (A = 0.0) then OK is set to False. Note the use of the return statement. Since this is a procedure there is no result to be returned and so the word **return** is not followed by an expression. It just updates the **out** or **in out** parameters as necessary and returns control back to where the procedure was called. Note also that unlike a function we can 'run' into the **end**; this is equivalent to obeying **return**. Naturally, procedures cannot have extended return statements.

The procedure could be used in a sequence such as

```
declare
   L, M, N: Float;
   P, Q: Float;
   Status: Boolean;
begin
   ...  -- sets values into L, M and N
   Quadratic(L, M, N, P, Q, Status);
   if Status then
      -- roots are in P and Q
   else
      -- fails
   end if;
end;
```

The reader will note that if OK is set to False then no value is assigned to the **out** parameters Root_1 and Root_2. The copy rule for scalars then implies that the corresponding actual parameters become undefined. This is probably bad practice.

An alternative is to make these parameters of mode **in out** so that the initial values of the actual parameters are left unchanged in the case of no real roots.

Another approach is to use the aspect Default_Value. But remember that we cannot apply Default_Value to a predefined type. So this would encourage us to define our own real type thus

```
type Real is new Float
  with Default_Value => 0.0;
```

An important effect of Default_Value is that parameters of mode **out** are treated somewhat as if they were of mode **in out** and will remain unchanged if not updated (AI12-74). Note carefully that they will not take the default value of 0.0 and so the existing values of P and Q will remain unchanged. Of course, if we had declared a local variable R_Temp of type Real then it would take the initial value of 0.0.

This technique of copying in parameters of mode **out** has existed in Ada for access types since Ada 83. Remember that access types have a default initial value of **null**. Similarly, when a parameter is of an unconstrained array type such as String, information regarding the bounds will need to be copied in whatever the mode. In the case of a scalar type any such initial copying will be done 'in the raw' without applying any subtype checking such as range constraints.

A final point as noted in Section 6.7 is that it is often better to introduce our own two-valued enumeration type rather than use the predefined type Boolean. The above example would be much clearer if we had declared

```
type Roots is (Real_Roots, Complex_Roots);
```

with other appropriate alterations.

Exercise 10.3

1 Write a procedure Swap to interchange the values of the two parameters of type Float.

2 Rewrite the function Rev of Section 10.1 as a procedure with a single parameter. Use it to reverse an array R of type Row. Consider rewriting it with two parameters.

3 Why is the following unwise?

```
A: Vector(1 .. 1);

procedure P(V: Vector) is
begin
   A(1) := V(1) + V(1);
   A(1) := V(1) + V(1);
end;
...
A(1) := 1.0;
P(A);
```

10.4 Aliasing

Aliasing means having two names for the same object and can be a source of problems. Thus in the example in Exercise 10.3(**3**) above, within the procedure P, the arrays A and V are the same array in the case when P is called with A as a parameter. The difficulty is that the array A is accessible by two routes, directly and indirectly via the parameter P.

The big problem in this example is that the behaviour depends upon whether the parameter is passed by reference or by value. If it is passed by reference, then both assignments are doubling A(1) so that its final value is 4.0. However, if the array is passed by value, then A and V are different and both assignments assign 2.0 to A(1) and its final value is 2.0. The program has a bounded error and a good compiler will provide a warning or might raise Program_Error.

In Ada 2012 and Ada 2022, functions can have parameters of any modes. This can be useful and avoids introducing unnecessary subterfuges. For example, we might write a very simple pseudo-random number generator thus

```
function My_Random(Seed: in out Integer) return Integer is
   N: constant Integer := ... ;
   M: constant Integer := ... ;
begin
   Seed := Seed * N mod M;
   return Seed;
end My_Random;
```

where the values of N and M are suitably chosen so that the sequence iterates over a large number of the possible values in a seemingly haphazard manner. We typically initialize the sequence by giving an odd value to the variable we choose to define the sequence and then call My_Random as required

```
XXX: Integer := 12345;
...
loop
   R := My_Random(XXX);
   ...
end loop;
```

(This works surprisingly well with even quite small relatively prime values for N and M such as 5^5 and 2^{13}.)

Without parameters of mode **in out**, we have to use a procedure rather than a function (annoying), make Seed non-local (very naughty) or use a private type for Seed with perhaps a component of an access type (sly). The predefined library uses the sly approach, see Section 23.5. See also Section 16.6 regarding global state.

However, permitting functions to have parameters of all modes gives rise to further opportunities for unexpected behaviour resulting from aliasing.

The problems arise largely because (to aid optimization, it is said), Ada does not define the order of evaluation of a number of things such as the operands of a binary operator and the parameters in a subprogram call (there is a list in Section 27.4).

It is far too late to do anything about specifying these orders of evaluation so the approach taken is to prevent as much aliasing as possible. Accordingly, Ada 2012 introduced a number of rules which keep the problems to a minimum.

First, there are rules for determining when two names are *known to denote the same object*. Thus they denote the same object if

- both names statically denote the same stand-alone object or parameter; or
- both names are selected components, their prefixes are known to denote the same object, and their selector names denote the same component.

and so on with similar rules for dereferences, indexed components, and slices. There is also a rule about renaming so that if we have

```
C: Character renames S(5);
```

then C and S(5) are known to denote the same object. The index naturally has to be static. Renaming is discussed in Section 13.7.

A further step is to define when two names *are known to refer to the same object*. This covers some cases of overlapping. Thus given a record R of type T with a component C, we say that R and R.C are known to refer to the same object. Similarly with an array A we say that A and A(K) are known to refer to the same object (K does not need to be static in this example).

Given these definitions we can now state the two basic restrictions. The first concerns parameters of elementary types:

- For each name N that is passed as a parameter of mode **in out** or **out** to a call of a subprogram S, there is no other name among the other parameters of mode **in out** or **out** to that call of S that is known to denote the same object.

Roughly speaking this comes down to saying two or more parameters of mode **out** or **in out** of an elementary type cannot denote the same object. This applies to both functions and procedures.

As an example consider

```
procedure Do_It(Double, Triple: in out Integer) is
begin
   Double := Double * 2;
   Triple := Triple * 3;
end Do_It;
```

with

```
Var: Integer := 2;
...
Do_It(Var, Var);                          -- illegal from Ada 2012
```

The key problem is that parameters of elementary types are always passed by copy and the order in which the parameters are copied back is not specified. Thus Var might end up with either the value of Double or the value of Triple.

The other restriction concerns constructions which have several constituents that can be evaluated in any order and can contain function calls. Basically it says:

- If a name N is passed as a parameter with mode **out** or **in out** to a function call that occurs in one of the constituents, then no other constituent can involve a name that is known to refer to the same object.

Constructions cover many situations such as aggregates, assignments, ranges and so on as mentioned earlier.

This rule excludes the following aggregate

```
(Var, F(Var))                      -- illegal from Ada 2012
```

where F has an **in out** parameter.

The rule also excludes the assignment

```
Var := F(Var);                     -- illegal
```

if the parameter of F has mode **in out**. Remember that the destination of an assignment can be evaluated before or after the expression. So if Var were an array element such as A(I) then the behaviour could vary according to the order. To encourage good practice, it is also forbidden even when Var is a stand-alone object.

Similarly, the procedure call

```
Proc(Var, F(Var));                 -- illegal
```

is illegal if the parameter of F has mode **in out**. Examples of overlapping are also forbidden such as

```
ProcA(A, F(A(K)));                 -- illegal
ProcR(R, F(R.C));                  -- illegal
```

assuming still that F has an **in out** parameter and that ProcA and ProcR have appropriate profiles because, as explained above, A and A(K) are known to refer to the same object as are R and R.C.

On the other hand

```
Proc(A(J), F(A(K)));               -- OK
```

is permitted provided that J and K are different objects because this is only a problem if J and K happen to have the same value.

The intent is to detect situations that are clearly troublesome. Other situations that might be troublesome (such as if J and K happen to have the same value) are allowed, since to prevent them would make many programs illegal that are not actually dubious. This would cause incompatibilities and upset many users whose programs are perfectly correct.

10.5 Named and default parameters

The forms of subprogram call we have been using so far have given the actual parameters in positional order. As with aggregates we can also use the named notation in which the formal parameter name is also supplied; the parameters do not then have to be in order.

So we could write

```
Quadratic(A => L, B => M, C => N,
                            Root_1 => P, Root_2 => Q, OK => Status);
Increment(X => I);
Add(C => Q, A => 2+P, B => 37);
```

We could even write

```
Increment(X => X);
```

as we will see in Section 10.6.

This notation can also be used with functions

```
F := Factorial(N => 4);
S := Sqrt(X => T+0.5);
X := Inner(B => W, A => V);
```

The named notation cannot, however, be used with operators called with the usual infixed syntax (such as V*W) because there is clearly no convenient place to put the names of the formal parameters.

As with record aggregates, the named and positional notations can be mixed and any positional parameters must come first and in their correct order. However, unlike record aggregates, each parameter must be given individually and **others** may not be used. So we could write

```
Quadratic(L, M, N, Root_1 => P, Root_2 => Q, OK => Status);
```

The named notation leads into the topic of default parameters. It sometimes happens that one or more **in** parameters usually take the same value on each call; we can give a default expression in the subprogram specification and then omit it from the call.

Consider the problem of ordering a dry martini in the United States. One is faced with choices described by the following enumeration types

```
type Spirit is (Gin, Vodka);
type Style is (On_The_Rocks, Straight_Up);
type Trimming is (Olive, Twist);
```

The default expressions can then be given in a procedure specification thus

```
procedure Dry_Martini(Base: Spirit := Gin;
                      How: Style := On_The_Rocks;
                      Plus: Trimming := Olive);
```

Typical calls might be

```
Dry_Martini(How => Straight_Up);
Dry_Martini(Vodka, Plus => Twist);
Dry_Martini;
Dry_Martini(Gin, Straight_Up);
```

The first call uses the named notation; we get gin, straight up plus olive. The second call mixes the positional and named notations; as soon as a parameter is omitted the named notation must be used. The third call illustrates that all parameters can be omitted. The final call shows that a parameter can, of course, be supplied even if it happens to take the same value as the default expression; in this case it avoids using the named form for the second parameter.

Note that default expressions can only be given for **in** parameters. They cannot be given for operators but they can be given for functions designated by identifiers. Such a default expression (like a default expression for an initial value in a record type declaration) is only evaluated when required; that is, it is evaluated each time the subprogram is called and no corresponding actual parameter is supplied. Hence the default value need not be the same on each call although it usually will be. Default expressions are widely used in the standard input–output package to provide default formats.

Default expressions illustrate the subtle rule that a parameter specification of the form

```
P, Q: in Integer := E
```

is strictly equivalent to

```
P: in Integer := E; Q: in Integer := E
```

(The reader will recall a similar rule for object declarations; it also applies to record components.) As a consequence, the default expression is evaluated for each omitted parameter in a call. This does not usually matter but would be significant if the expression E included a function call with side effects.

Exercise 10.5

1 Write a function Add which returns the sum of the two integer parameters and takes a default value of 1 for the second parameter. How many different ways can it be called to return N+1 where N is the first actual parameter?

2 Rewrite the specification of Dry_Martini to reflect that you prefer Vodka at weekends. Hint: declare a function to return your favourite spirit according to the global variable Today.

10.6 Overloading

We saw in Section 10.2 how new meanings could be given to existing language operators. This overloading applies to subprograms in general.

A subprogram will overload an existing meaning rather than hide it provided that its specification is sufficiently different. Hiding will occur if the number, order and base types of the parameters and any result are the same; this level of matching is known as type conformance. A procedure cannot hide a function and vice versa. Note that the names of the parameters, their mode and the presence or absence of constraints or default expressions do not matter. One or more overloaded subprograms may be declared in the same declarative part.

Subprograms and enumeration literals can overload each other. In fact an enumeration literal is formally thought of as a parameterless function with a result of the enumeration type. There are two kinds of uses of identifiers – the overloadable ones and the non-overloadable ones. At any point an identifier either refers to a single entity of the non-overloadable kind or to one or many of the overloadable kind. A declaration of one kind hides the other kind and cannot occur in the same declaration list.

Each use of an overloaded identifier has to have a unique meaning as determined by the so-called overload resolution rules which use information such as the types of parameters. If the rules do not lead to a single meaning then the program is ambiguous and thus illegal.

Such ambiguities can often be resolved by qualification as we saw when the operator "<" was used with the character literals in Section 9.4. As a further example consider the British Channel Islands; the largest three are Guernsey, Jersey, and Alderney. There are styles of woollen garments named after each

```
type Garment is (Guernsey, Jersey, Alderney);
```

and breeds of cattle named after two of them (the Alderney breed became extinct as a consequence of World War II)

```
type Cow is (Guernsey, Jersey);
```

and we can perhaps imagine shops that sell both garments and cows according to

```
procedure Sell(Style: Garment);
procedure Sell(Breed: Cow);
```

The statement

```
Sell(Alderney);
```

is not ambiguous since Alderney has only one interpretation. However

```
Sell(Jersey);
```

is ambiguous since we cannot tell whether a cow or garment is being sold. One way of resolving the ambiguity is to use type qualification thus

```
Sell(Cow'(Jersey));
```

In this example the ambiguity could also be resolved by using a named parameter call thus

```
Sell(Breed => Jersey);
```

We conclude by noting that ambiguities typically arise only when there are several overloadings. In the case here both Sell and Jersey are overloaded; in the example in Section 8.4 both the operator "<" and the literals 'X' and 'L' were overloaded.

10.7 Declarations, scopes, and visibility

We said earlier that it is sometimes necessary or just convenient to give a subprogram specification on its own without the body. The specification is then followed by a semicolon and is known as a subprogram declaration. A complete subprogram, which always includes the full specification, is known as a subprogram body.

Subprogram declarations and bodies must, like other declarations, occur in a declarative part and a subprogram declaration must be followed by the corresponding body in the same declarative part. (Strictly speaking it must be in the same declarative region as we shall see when we discuss the impact of packages on visibility in Section 13.6 – but in the case of declarations in blocks and subprograms this comes to the same thing.)

An example of where it is necessary to use a subprogram declaration occurs with mutually recursive procedures. Suppose we wish to declare two procedures F and G which call each other. Because of the rule regarding linear elaboration of declarations we cannot write the call of F in the body of G until after F has been declared and vice versa. Clearly this is impossible if we just write the bodies because one must come second. However, we can write

```
procedure F( ... );          -- declaration of F

procedure G( ... ) is        -- body of G
begin
   ...
   F( ... );
   ...
end G;

procedure F( ... ) is        -- body of F repeats
begin                        -- its specification
   ...
   G( ... );
   ...
end F;
```

and then all is well.

If the specification is repeated then it must be given in full and the two must be the same. Technically we say that the two profiles in the specification must have full conformance. The profile is the formal parameter list plus result type if any. Some slight variation is allowed provided the static meaning is the same. For example: a numeric literal can be replaced by another numeric literal with the same value; an identifier can be replaced by a dotted name as described later in this section; an explicit mode **in** can be omitted; a list of parameters of the same subtype can be given distinctly; and of course the lexical spacing can be different. Thus the following two profiles are fully conformant

```
(X: in Integer := 1000; Y, Z: out Integer)
(X: Integer := 1e3; Y: out Integer; Z: out Integer)
```

It is worth noting that one reason for not allowing explicit constraints in parameter specifications is to remove any problem regarding conformance since there is then no question of evaluating constraint expressions twice and possibly having different results because of side effects. No corresponding question arises with default expressions (which are of course written out twice) since they are only evaluated when the subprogram is called.

There are other, less rigorous, levels of conformance which we will meet when we discuss access to subprogram types in Section 11.8 and renaming in Section

13.7. We have already met the weakest level of conformance which is called type conformance and controls the hiding of one subprogram declaration by another. As we saw in the previous section, hiding occurs provided just the types in the profiles are the same.

Another important situation where we have to write a subprogram declaration as well as a body, occurs in Chapter 12 when we discuss packages. Even if not always necessary, it is sometimes clearer to write subprogram declarations as well as bodies. An example might be in the case where many subprogram bodies occur together. The subprogram declarations could then be placed together at the head of the declarative part in order to act as a summary of the bodies to come.

Since subprograms occur in declarative parts and themselves contain declarative parts, they may be textually nested without limit. The normal hiding rules applicable to blocks described in Section 6.2 also apply to declarations in subprograms. (The only complication concerns overloading as discussed in the previous section.) We are also now in a position to describe the difference between visibility and direct visibility as illustrated by the nested procedures in Figure 10.1.

Just as for the example in Section 6.2, the inner J hides the outer one and so the outer J is not directly visible inside the procedure Q after the start of the declaration of the inner J.

However, we say that it is still visible even though not directly visible because we can refer to the outer J by the so-called dotted notation in which the identifier J is prefixed by the name of the unit immediately containing its declaration followed by a dot. So within Q we can refer to the outer J as P.J as illustrated by the initialization of L.

If the prefix is itself hidden then it can always be written the same way. Thus the inner J could be referred to as P.Q.J.

An object declared in a block cannot usually be referred to in this way since a block does not normally have a name. However, a block can be named in a similar way to a loop as shown in the following

```
Outer:
declare
   J: Integer := 0;
begin
   ...
   declare
      K: Integer := J;
      J: Integer := 0;
      L: Integer := Outer.J;
   begin
      ...
   end;
end Outer;
```

Here the outer block has the identifier Outer. Unlike subprograms, but like loops, the identifier has to be repeated after the matching **end**. Naming the block enables us to initialize the inner declaration of L with the value of the outer J.

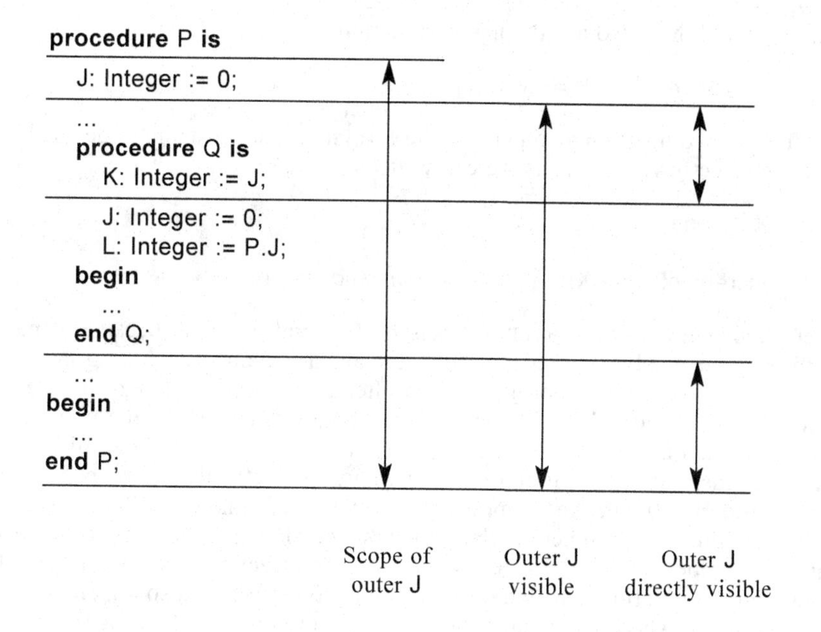

Figure 10.1 Scope and visibility.

Within a loop it is possible to refer to a hidden loop parameter in the same way. We could even rewrite the example in Section 8.1 of assigning zero to the elements of AA as

```
LL:
for I in AA'Range(1) loop
   for I in AA'Range(2) loop
      AA(LL.I, I) := 0.0;
   end loop;
end loop LL;
```

although one would be a little crazy to do so!

It should be noted that the dotted notation can always be used even if it is not necessary.

This notation can also be applied to operators. Thus the variable Result declared inside the operator "*" (see Section 10.2) can be referred to as "*".Result. And equally if "*" were declared inside the block Outer then it could be referred to as Outer."*". If it is called with this form of name then the normal function call must be used

```
X := Outer."*"(V, W);
```

Indeed, the functional form can always be used as in

```
X := "*"(V, W);
```

and we could then also use the named notation

```
X := "*"(A => V, B => W);
```

The named notation also permits the formal parameter name to be used even if it is not directly visible. Thus we can write

```
X: Integer;
...
Increment(X => X);          -- from Section 10.3
```

even though the formal parameter X is not generally visible because it has been hidden by the newly declared X used as the actual parameter.

As we have seen, subprograms can alter global variables and therefore have side effects. (A side effect is one brought about other than via the parameter and result mechanism.) It is generally considered rather undesirable to write subprograms, especially functions, that have side effects. However, some side effects are beneficial. Any subprogram which performs input–output has a side effect on the file; a function delivering successive members of a sequence of random numbers only works because of its side effects; if we need to count how many times a function is called then we use a side effect; and so on. However, care must be taken when using functions with side effects that the program is correct since there are various circumstances in which the order of evaluation is not defined. There are various rules regarding aliasing which we met in Section 10.4 which help us to avoid many problems in this area.

We conclude this section with a brief discussion of the hierarchy of **exit**, **return**, and **goto** and the scopes of block and loop identifiers and labels.

A **return** statement terminates the execution of the immediately embracing subprogram. It can occur inside an inner block or inside a loop in the subprogram and therefore also terminate the loop. An extended return statement may include inner simple return statements, blocks, and loops but not other extended return statements.

An **exit** statement terminates the named or immediately embracing loop. It can also occur inside an inner block or extended return statement but cannot occur inside a subprogram declared in the loop and thereby also terminate the subprogram. A **goto** statement can transfer control out of a loop or block or an extended return statement but not out of a subprogram.

As far as scope is concerned, identifiers of labels, blocks, and loops behave as if they are declared at the end of the declarative part of the immediately embracing subprogram or block (or package or task body). Moreover, distinct identifiers must be used for all blocks, loops, and labels inside the same subprogram (or package or task body) even if some are in inner blocks. Thus two labels in the same subprogram cannot have the same identifier even if they are inside different inner blocks. This rule reduces the risk of goto statements going to the wrong label particularly when a program is amended.

Checklist 10

Parameter and result subtypes must not be given by a subtype indication with an explicit constraint.

Formal parameter specifications are separated by semicolons not commas.

A function must return a result and should not run into its final end although a procedure can.

"/=" can only be explicitly defined if the result is not Boolean.

The order of evaluation of parameters is not defined.

The parameter and result mechanism allows an array to slide.

Scalar parameters are passed by copy. Arrays and simple records may be passed by copy or by reference.

A default parameter expression is only evaluated when the subprogram is called and the corresponding parameter is omitted.

The extended return statement was added in Ada 2005.

The abbreviated from of function known as an expression function was introduced in Ada 2012.

Functions can have parameters of any mode in Ada 2012.

New rules on aliasing were introduced Ada 2012.

New in Ada 2022

Expression functions can have the new aspect Static and so occur in static expressions, see Section 27.1.

The aspect No_Return can be applied to functions as well as to procedures, see Section 15.5.

11 Access Types

11.1 Flexibility versus integrity
11.2 Access types and allocators
11.3 Null exclusion and constraints
11.4 Aliased objects
11.5 Accessibility
11.6 Access parameters
11.7 Anonymous access types
11.8 Access to subprograms
11.9 Storage pools

This last chapter concerning the small-scale aspects of Ada is about access types. These are often known as pointer types, reference types or simply addresses in other languages and provide indirect access to other entities.

Playing with pointers is like playing with fire. Fire is perhaps the most important tool known to man. Carefully used, fire brings enormous benefits; but when fire gets out of control, disaster strikes. Pointers have similar characteristics but are well tamed in the form of access types in Ada.

This taming is done through the notion of accessibility which is discussed in some detail in Sections 11.5 and 11.6. Parts of these sections might be found hard to digest and could well be skipped at a first reading.

There are two forms of access types, those which access objects and those which access subprograms. Their type may be named or anonymous. Of particular importance are parameters of an anonymous access type. The related access discriminants are discussed in Chapter 18.

11.1 Flexibility versus integrity

The manipulation of objects by referring to them indirectly through values of other objects is a common feature of most programming languages. It is also a contentious topic since, although the technique provides considerable flexibility, it is also the cause of many programming errors and moreover, used incautiously, can make programs very hard to understand and maintain. It is worth a historical digression in order to place Ada in perspective.

Algol 68 was an early language to use indirection and used the term references. Indeed the definition of Algol 68 revolved around references to such an extent that it created the impression that the language was academically elaborate. There were also technical difficulties of dangling references, that is variables pointing to objects that no longer exist.

BCPL, from which C was derived, used references or pointers as the foundation of its storage model. Arrays were just seen as objects that could be referred to dynamically by adding an index to a base address. Thus the natural implementation model was made visible in the language itself. This almost negative abstraction was perhaps a great mistake and a step backwards in the evolution of abstraction as the driving force in language design.

C has inherited the BCPL model and it is considered quite normal practice in C to add integers to pointers (addresses) thereby creating implementation dependencies and complete freedom to do silly things.

Pascal was an example of austerity in the use of pointers. Pascal only allowed pointers to objects created in a distinct storage area commonly called the heap. The rules in Pascal were such that it was not possible to create a pointer to an object declared in the normal way on the stack. There was thus no risk of leaving the scope of the referenced object while still within the scope of the referring object and thereby leaving the latter pointing nowhere.

Ada 83 followed the Pascal model and so although secure was inflexible. This inflexibility was particularly noticeable when interfacing to programs written in other languages such as C and especially for interaction with graphical user interfaces. It only had what we now call pool specific types.

Ada 95 added more flexibility while keeping the security inherent in the strongly typed model. It introduced general access types and thus enabled pointers to items not in a storage pool (the Ada term for a heap). Security was maintained through the introduction of accessibility rules whose general objective was to prevent dangling references. Most accessibility rules were applied statically and thus had no run-time overhead. However, this was sometimes inflexible and so more dynamic techniques were also provided by access parameters which were of an anonymous type. Finally, simple access to subprogram types were introduced.

However, the Ada 95 model was somewhat inconsistent. Although it introduced general access types and access to subprogram types, it only permitted anonymous types when used as parameters, did not solve the downward closure problem (see Section 11.8) and was inconsistent in its treatment of null values and constants.

Ada 2005 swept away all these inconsistencies. Access types can be named or anonymous in (almost) all contexts. Null values and constant values are handled more uniformly and the introduction of anonymous access to subprogram types as parameters solved the downward closure problem.

Ada 2012 made a few changes in this area. There are some corrections to the rules regarding checking accessibility and some simplification to conversions between access types. Perhaps the biggest improvement is the ability to create subpools of storage pools thereby giving the user much more control over the allocation and deallocation of storage for groups of items.

Java is currently popular. It has pointers which are called references. In fact almost everything is declared using references although this is hidden from the user. This means that Java is inappropriate for high integrity applications.

Finally, note that Ada uses pointers less than many languages because it allows dynamically sized objects without using a heap.

11.2 Access types and allocators

In the case of the types we have met so far, the name of an object is bound permanently to the object itself, and the lifetime of an object extends from its declaration until control leaves the unit containing the declaration. This is far too restrictive for many applications where a more fluid control of the allocation of objects is desired. In Ada this can be done by using an access type. Objects of an access type, as the name implies, provide access to other objects and these other objects can be allocated in a storage pool independent of the block structure. One of the simplest uses of an access type is for list processing. Consider

```
type Cell;                              -- incomplete declaration
type Cell_Ptr is access Cell;

type Cell is                            -- complete declaration
   record
      Next: Cell_Ptr;
      Value: Integer;
   end record;

L: Cell_Ptr;
```

These declarations introduce the type Cell_Ptr which accesses Cell. The variable L can be thought of as a reference variable which can only point to objects of type Cell; these are records with two components, Next which is also a Cell_Ptr and can therefore access (point to or reference) other objects of type Cell, and Value of type Integer (the 'real' contents of the record). The records can therefore be formed into a linked list. Initially there are no record objects, only the single pointer L which by default takes the initial value **null** which points nowhere. We could have explicitly given L this default value thus

```
L: Cell_Ptr := null;
```

Note the circularity in the definitions of Cell_Ptr and Cell. Because of this circularity and the need to avoid forward references, it is necessary first to give an incomplete declaration of Cell. Having done this we can declare Cell_Ptr and then complete the declaration of Cell. Between the incomplete and complete declarations, we say that we have an incomplete view of the type Cell. Such an incomplete view can only be used for a few purposes the most important of which is the definition of an access type. Moreover, the incomplete and complete declarations must be in the same list of declarations except for one case concerning private types; see Section 12.5.

When we deal with anonymous access types in Section 11.7, we shall see that it is not always necessary to use incomplete types in these circumstances. However, incomplete types are also important with mutually related types as we shall see in Section 13.5 and so it is helpful to illustrate their use.

The accessed objects are created by the execution of an allocator which can (but need not) provide an initial value. An allocator consists of the reserved word **new** followed by either just the type of the new object or a qualified expression providing also the initial value of the object. The result of an allocator is an access

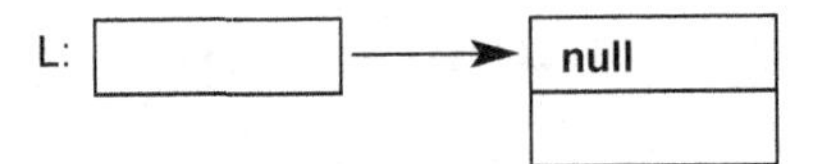

Figure 11.1 An access object.

value which can then be assigned to a variable of the access type. The new object itself will be in a storage pool as described in Section 11.9.

So the statement

```
L := new Cell;
```

creates a record of type Cell and then assigns to L a designation of (reference to or pointer to) the object. We can picture the result as in Figure 11.1.

Note that the Next component of the record takes the default value **null** whereas the Value component is undefined.

The components of the object referred to by L can be accessed using the normal dotted notation. So we could assign 37 to the Value component by

```
L.Value := 37;
```

If we attempt to do this when L has the value **null**, the exception Constraint_Error is raised. This check prevents nasty errors which can arise with other languages.

Alternatively we could have provided an initial value with the allocator

```
L := new Cell'(null, 37);    -- or new Cell'(Next => null, Value => 37);
```

The initial value here takes the form of a qualified aggregate, and as usual has to provide values for all the components irrespective of whether some have default initial expressions. The allocator could also have been used to initialize L when it was declared

```
L: Cell_Ptr := new Cell'(null, 37);
```

Distinguish carefully the types Cell_Ptr and Cell. The variable L is of type Cell_Ptr which accesses Cell whereas Cell is the accessed type which follows **new**.

Suppose we now want to create a further record and link it to the existing record. We could do this in three steps by declaring a further variable

```
N: Cell_Ptr;
```

and then executing

```
N := new Cell'(L, 10);
L := N;
```

The effect of these three steps is illustrated in Figure 11.2. Note how the assignment statement

```
L := N;
```

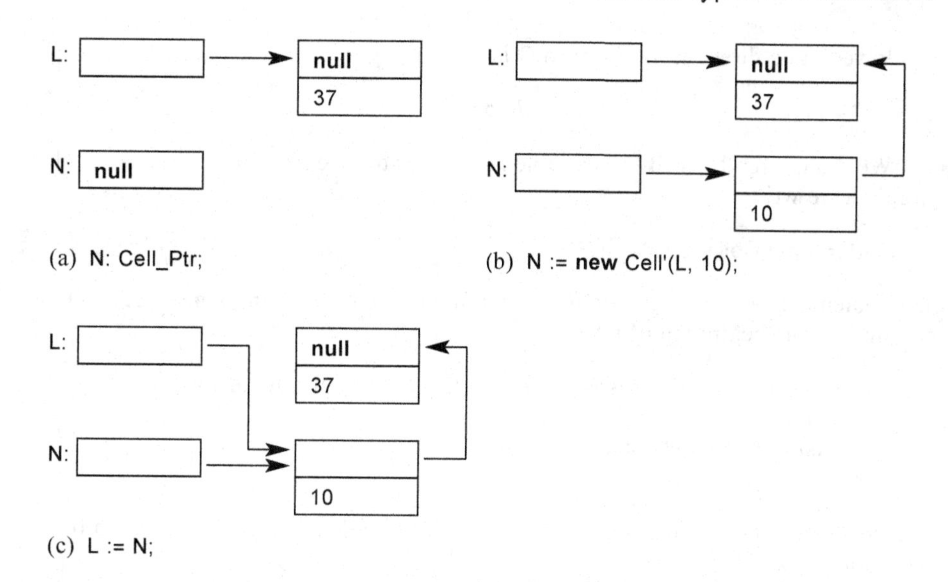

Figure 11.2 Extending a list.

copies the access values (that is, the pointers) and not the objects. If we wanted to copy the objects we could do it component by component

```
L.Next := N.Next;  L.Value := N.Value;
```

or by using **all**

```
L.all := N.all;
```

L.**all** refers to the whole object accessed by L. Using **all** is called dereferencing and is often automatic – we can think of L.Value as short for L.**all**.Value.

Similarly

```
L = N
```

will be true if L and N both refer to the same object, whereas

```
L.all = N.all
```

will be true if the objects referred to happen to have the same value.

We could declare a constant of an access type but since it is a constant we must supply an initial value

```
C: constant Cell_Ptr := new Cell'(null, 0);
```

The fact that C is constant means that it must always refer to the same object. However, the value of the object could itself be changed. So

```
C.all := L.all;
```

is allowed but a direct assignment to C is not

```
C := L;                         -- illegal
```

We did not really need the variable N in order to extend the list since we could simply have written

```
L := new Cell'(L, 10);
```

This statement can be made into a general procedure for creating a new record and adding it to the beginning of a list

```
procedure Add_To_List(List: in out Cell_Ptr; V: in Integer) is
begin
  List := new Cell'(List, V);
end Add_To_List;
```

The new record containing the value 10 can now be added to the list accessed by L by writing

```
Add_To_List(L, 10);
```

The parameter passing mechanism for access types is defined to be by copy like that for scalar types. However, in order to prevent an access value from becoming undefined an **out** parameter is always copied in at the start. Remember also that an uninitialized access object takes the specific default value **null**. These two facts prevent undefined access values which could cause a program to go berserk.

The value **null** is useful for determining when a list is empty. The following function returns the sum of the Value components of the records in a list

```
function Sum(List: Cell_Ptr) return Integer is
  Local: Cell_Ptr := List;
  S: Integer := 0;
begin
  while Local /= null loop
    S := S + Local.Value;
    Local := Local.Next;
  end loop;
  return S;
end Sum;
```

Observe that we have to make a copy of List because formal parameters of mode **in** are constants. The variable Local is then used to work down the list until we reach the end. The function works even if the list is empty.

A more elaborate data structure is the binary tree. This can be represented by nodes each of which has a value plus pointers to two subnodes (subtrees) one or both of which could be null. Appropriate declarations are

```
type Node;
type Node_Ptr is access Node;
```

```
type Node is
   record
      Left, Right: Node_Ptr;
      Value: Float;
   end record;
```

As an interesting example of the use of trees consider the following procedure Sort which sorts the values in an array into ascending order

```
procedure Sort(A: in out Vector) is
   Index: Integer;
   Tree: Node_Ptr := null;

   procedure Insert(T: in out Node_Ptr; V: in Float) is
   begin
      if T = null then
         T := new Node'(null, null, V);
      elsif V < T.Value then
         Insert(T.Left, V);
      else
         Insert(T.Right, V);
      end if;
   end Insert;

   procedure Output(T: in Node_Ptr) is
   begin
      if T /= null then
         Output(T.Left);
         A(Index) := T.Value;
         Index := Index + 1;
         Output(T.Right);
      end if;
   end Output;

begin                                -- body of Sort
   for I in A'Range loop
      Insert(Tree, A(I));
   end loop;
   Index := A'First;
   Output(Tree);
end Sort;
```

The recursive procedure Insert adds a new node containing the value V to the tree in such a way that the values in the left subtree of a node are always less than the value at the node and the values in the right subtree are always greater than (or equal to) the value at the node. The recursive procedure Output copies the values in the tree into the array A by first outputting the left subtree, then copying the value at the node and finally outputting the right subtree.

The procedure Sort simply builds up the tree by calling Insert with each of the components of the array in turn and then calls Output to copy the ordered values back into the array.

The access types we have met so far have referred to records. This will often be the case but an access type can refer to any type, even another access type. So we could have an access to an integer thus

```
type Ref_Int is access Integer;
R: Ref_Int := new Integer'(46);
```

Note that the value of the integer referred to by R is denoted by R.**all**. So we can write

```
R.all := 13;
```

to change the value from 46 to 13.

An access type can also refer to an indefinite type such as the type String. In Section 8.5 we noted, when declaring the Zoo, that the animals (or rather their names) all had to have the same length. However, with a bit of ingenuity, we can overcome this by using access types. Consider

```
type A_String is access String;
```

which enables us to declare variables which can access strings of any size. So we can write

```
A: A_String := new String'("Hello");
```

We see that although we no longer have to pad the strings to a fixed length, we now have the burden of the allocation. However, we can craftily write

```
function "+" (S: String) return A_String is
begin
  return new String'(S);
end "+";
```

and then

```
type A_String_Array is array (Positive range <>) of A_String;

Zoo: constant A_String_Array :=
          (+"aardvark", +"baboon", ..., +"very long animal... ", ..., +"zebra");
```

Remember from Section 8.5 that we can declare an array of any definite subtype; all access types are definite even if they designate objects of an indefinite type and so we can declare arrays of the type A_String. (This flexibility of access types is especially crucial for object oriented programming as we saw in the brief overview in Section 3.3.)

With this formulation there is no limit on the length of the strings and we have more or less created a ragged array. However, the length of a particular string cannot be changed. Moreover, there is the overhead of the access value which is significant if the strings are short. We will return to the topic of ragged arrays in Sections 11.4 and 18.2.

A few final points of detail. Allocators illustrate the importance of the rules regarding when and how often an expression is evaluated in certain contexts. For example, an expression in an aggregate is evaluated for each index value concerned and so

```
A: array (1 .. 10) of Cell_Ptr := (others => new Cell);
```

creates an array of ten components and initializes each of them to access a different new cell.

As a further example

```
A, B: Cell_Ptr := new Cell;
```

creates two new cells (see Section 6.1), whereas

```
A: Cell_Ptr := new Cell;
B: Cell_Ptr := A;
```

naturally creates only one. Remember also that default expressions for record components and subprogram parameters are re-evaluated each time they are required; if such an expression contains an allocator then a new object will be created each time.

If an allocator provides an initial value then this can take the form of any qualified expression. So we could have

```
L: Cell_Ptr := new Cell'(N.all);
```

in which case the object is given the same value as the object referred to by N. We could have

```
K: Integer := 46;
R: Ref_Int := new Integer'(K);
```

in which case the new object takes the value of K.

In this section we have only discussed the use of access types with allocated objects. In Section 11.4 we shall see how to use them with declared objects.

Exercise 11.2

1. Write a

   ```
   procedure Append(First: in out Cell_Ptr; Second: in Cell_Ptr);
   ```

   which appends the list Second (without copying) to the end of the list First. Take care of the special case where First is **null**.

2. Write a function Size which returns the number of nodes in a tree.

3. Write a function Copy which makes a complete copy of a tree.

4. Write the converse unary function "+" which takes an A_String as parameter and returns the corresponding String. Use this function to output the camel in the Zoo.

5. Write a function "&" to concatenate two parameters of type A_String.

11.3 Null exclusion and constraints

In the previous section we noted that if we dereference a value of an access type such as Cell_Ptr as in L.Value, and L happens to have the value **null** then Constraint_Error is raised. This means that there has to be a check on all such dereferences and this is clearly somewhat inefficient.

Accordingly, we can specify that an access type does not have the value **null** by adding a null exclusion when it is declared thus

```
type Cell_Ptr is not null access Cell;
```

All variables of the type now have to be declared with an initial value. This would not be very helpful with this list example because we would have to introduce some other technique for identifying the end of a list (the last cell could point to itself perhaps or the list could be cyclic).

Alternatively, we could leave Cell_Ptr as before and declare a subtype thus

```
type Cell_Ptr is access Cell;
subtype NN_Cell_Ptr is not null Cell_Ptr;
...
M: NN_Cell_Ptr := L;                          -- checks not null
```

There will then be a check to ensure that L does not have the value **null** before it is assigned to M. But when we write M.Value there is no need to check M. Another approach is to give the null exclusion when we declare M thus

```
M: not null Cell_Ptr := L;
```

A null exclusion has much in common with a constraint. But there are differences. An important one is that a null exclusion can appear in a parameter profile but a constraint cannot. Thus

```
procedure P(X: in out not null Ref_Int);          -- legal
procedure Q(X: in out Integer range 1 .. N);      -- illegal
function F(X: Integer) return not null Ref_Int;   -- legal
```

As explained in Section 10.7, we cannot use a constraint in a parameter specification because of possible problems of conformance and the constraint being evaluated twice. No such problem arises with a null exclusion. A null exclusion can also appear in a function result as shown for the function F above.

From time to time we say that two subtypes have to statically match. An example occurred with array conversions in Section 8.2. Where relevant, null exclusions also have to match.

We now turn to the question of constraints and access types. These can be applied to both the type being accessed and to the access type itself. This discussion may be found a bit confusing but careful attention to the detail will show that it is quite logical.

Considering first the type being accessed, we could have

```
type Ref_Pos is access Positive;
```

or equivalently

```
type Ref_Pos is access Integer range 1 .. Integer'Last;
```

The values in the objects referred to are all constrained to be positive. We could write either of

```
RP: Ref_Pos := new Positive'(10);
RP: Ref_Pos := new Integer'(10);
```

Note that if we wrote **new** Positive'(0) then Constraint_Error would be raised because 0 is not of subtype Positive. However, if we wrote **new** Integer'(0) then Constraint_Error is only raised because of the context of the allocator.

Access types can also refer to constrained and unconstrained arrays. We met the type A_String in the previous section. We could also have

```
type Ref_Matrix is access Matrix;
R: Ref_Matrix;
```

where the type Matrix is as in Section 8.2, and then obtain new matrices with an allocator where the bounds must be provided either through an explicit initial value thus

```
R := new Matrix'(1 .. 5 => (1 .. 10 => 0.0));          -- note quote
```

or by applying constraints

```
R := new Matrix(1 .. 5, 1 .. 10);                      -- no quote
```

Note the subtle distinction whereby a quote is needed in the case of the full initial value but not when we just give the constraints. This is because the first takes the form of a qualified expression whereas the second is just a subtype indication.

We could not write just

```
R := new Matrix;
```

because all array objects must have bounds (must be definite). Moreover, once allocated the bounds of a particular matrix cannot be changed. However, since R itself is unconstrained it can refer to matrices of different bounds from time to time. So we could then write

```
R := new Matrix(1 .. 10, 1 .. 5);              -- different bounds
```

We now turn to considering the case where the access type itself is constrained. We can do this by introducing a subtype

```
subtype Ref_Matrix_3x4 is Ref_Matrix(1 .. 3, 1 .. 4);
R_3x4: Ref_Matrix_3x4;
```

and R_3x4 can then only reference matrices with corresponding bounds. Alternatively we could have directly written

```
R_3x4: Ref_Matrix(1 .. 3, 1 .. 4);
```

The objects to be accessed can be allocated using an explicit constraint

```
R_3x4 := new Matrix(1 ..3, 1 .. 4);
```

or through declaring a subtype

```
subtype Matrix_3x4 is Matrix(1 .. 3, 1 .. 4);
...
R_3x4 := new Matrix_3x4;
```

This is allowed because the subtype Matrix_3x4 supplies the array bounds.

Moreover, because the subtype supplies the bounds we could use it to qualify an aggregate and so we could initialize the new object by

```
R_3x4 := new Matrix_3x4'(others => (others => 0.0));
```

which is an interesting example of an array aggregate with **others**.

It is important to appreciate that, unlike the unconstrained variable R declared above, R_3x4 cannot refer to matrices with different bounds. As a consequence the following would raise Constraint_Error

```
R_3x4 := new Matrix(1 .. 4, 1 ..3);
```

Components of an accessed array can be referred to in the usual way, so

```
R(1, 1) := 0.0;
```

sets component (1, 1) of the matrix accessed by R to zero. The whole matrix can be referred to by **all**. So

```
R_3x4.all := (1 .. 3 => (1 .. 4 => 0.0));
```

sets all components of the matrix accessed by R_3x4 to zero.

We can think of R(1, 1) as an abbreviation for R.**all**(1, 1). As with records, dereferencing is automatic. We can also write attributes such as R'First(1) or R.**all**'First(1). Similarly, slicing is permitted for one-dimensional arrays.

11.4 Aliased objects

In Section 11.2 we saw how access types provide a means of manipulating objects created by allocators. Access types can also be used to provide indirect access to declared objects. The declaration of the type has to include the word **all** and is then known as a general access type. Thus we write

```
type Int_Ptr is access all Integer;
```

and then we can assign the 'address' of any variable of type Integer to a variable of type Int_Ptr provided that the designated variable is marked as **aliased** thus

```
IP: Int_Ptr;
I: aliased Integer;
...
IP := I'Access;
```

We can then read and update the variable I indirectly through the access variable IP by, for example

```
IP.all := 25;
```

Observe how the access value (the pointer) assigned to IP is created by the use of the Access attribute. We can only apply the Access attribute to objects declared as aliased (or those considered by default to be aliased such as tagged type parameters, see Section 14.2).

There are two reasons for specifically requiring that an object be marked as **aliased**; one is as a warning to the programmer that the object might be manipulated indirectly; the other is for the compiler so that it does not allocate space for the object in a nonstandard way (such as in a register) and thereby prevent access in the normal manner (it conversely enables the compiler to perform optimizations easily if it is not marked).

As mentioned earlier, there are accessibility rules which ensure that dangling references cannot arise. They are dealt with in detail in Section 11.6 but the general principle is that we can only apply the Access attribute to an object whose lifetime is at least that of the access type.

A variation is that we can restrict the access to be read-only by replacing **all** in the type definition by **constant**. This allows read-only access to any variable and also to a constant. So we can write

```
type Const_Int_Ptr is access constant Integer;
CIP: Const_Int_Ptr;
I: aliased Integer;
C: aliased constant Integer := 1815;
...
CIP := I'Access;            -- access to a variable, or
CIP := C'Access;            -- access to a constant
```

But we could not assign C'Access to the variable IP of the general access type Int_Ptr since that would enable us to write indirectly to the constant C.

The type accessed by a general access type can of course be any type such as an array or record. We can thus build chains from records statically declared. Note that we can also use an allocator to generate general access values so that a chain could include a mixture of records from both storage mechanisms.

The components of an array can also be aliased as in

```
AI: array (1 .. 100) of aliased Integer;
...
IP := AI(I)'Access;
```

Finally, note that the accessed object could also be a component of a record type. Thus we could point directly to a component of a record (provided that the component is marked as aliased). In a fast implementation of Conway's Game of Life a cell might contain access values directly referencing the component of its eight neighbours containing the counter (having value 0 or 1) indicating whether the cell is alive or dead. So we can write

```
type Ref_Count is access constant Integer range 0 .. 1;
type Ref_Count_Array is array (Integer range <>) of Ref_Count;

type Cell is
  record
    Life_Count: aliased Integer range 0 .. 1;
    Total_Neighbour_Count: Integer range 0 .. 8;
    Neighbour_Count: Ref_Count_Array(1 .. 8);
  end record;
```

We can now link the cells together by statements such as

```
This_Cell.Neighbour_Count(1) := Cell_To_The_North.Life_Count'Access;
```

Any constraints on the accessed object (Life_Count) must statically match those on the access type (Ref_Count); in this case they both have static bounds of 0 and 1. We met static matching in Section 8.2 when discussing the rules for conversion between array types – the same rules apply here.

The heart of the computation which computes the sum of the life counts in the neighbours might be

```
C.Total_Neighbour_Count := 0;
for I in C.Neighbour_Count'Range loop
  C.Total_Neighbour_Count :=
      C.Total_Neighbour_Count + C.Neighbour_Count(I).all;
end loop;
```

We also made the type Ref_Count an **access constant** type since we only need read access to the counter; this prevents us accidentally updating it indirectly.

General access types can also refer to composite types so we might have

```
type String_Ptr is not null access all String;
A_String: aliased String := "Hello";
SP: String_Ptr := A_String'Access;
```

The string could be changed indirectly by for example

```
SP.all := "Howdy";                    -- changes A_String
```

But of course SP.**all** := "Hi"; would raise Constraint_Error because the length of the string is constrained to be 5 by the initial value of A_String.

Static matching between constraints has to be strictly observed when applying the Access attribute and in fact an array with bounds obtained from an initial value is considered different from one with an explicit constraint (they have different nominal subtypes as mentioned in Section 8.2). So

```
Another_String: aliased String(1 .. 7) := "Welcome";
...
SP := Another_String'Access;          -- illegal
```

is illegal because the explicit constraint on Another_String does not statically match the accessed type of String_Ptr (which is unconstrained). But A_String does match String_Ptr since there is no explicit constraint on A_String.

General access types can also be used to declare ragged arrays as for example a table of strings of different lengths such as the Zoo discussed in Section 11.2. We might declare an access type thus

```
type G_String is access constant String;
```

and the strings can then be declared in the normal way (with bounds obtained from their initial values). Without the use of aliasing we would have to allocate the strings dynamically using an allocator. Nevertheless this technique is cumbersome because all the individual strings have to be named objects.

Conversion between different general access types is permitted. The accessed types must be the same and any constraints must statically match (but see Section 14.2 if the accessed type is tagged and Section 18.5 if it has discriminants). But an access to constant type cannot be converted to an access to variable type since otherwise we might obtain write access to a constant. So we can write

```
CIP := Const_Int_Ptr(IP);
```

but not

```
IP := Int_Ptr(CIP);            -- illegal
```

Access to constant types are useful for providing read-only access to allocated objects. Thus if we have

```
type Const_T_Ptr is access constant T;
type T_Ptr is access all T;
My_Object: T_Ptr := new T'( ... );
Your_View: Const_T_Ptr := Const_T_Ptr(My_Object);
```

then although the variable My_Object enables the value of the allocated object to be changed, this cannot be done using Your_View. It is important to allocate the object using the type T_Ptr. If we did it with the type Const_T_Ptr then we would be stuck with an object that could never be changed and of course the type conversion in the opposite direction would not be possible.

Finally, note that **aliased** can be used generally for parameters and for the return object in an extended return statement. An example with parameters occurs in the functions such as Reference in containers (see Section 24.2).

Exercise 11.4

1 Using the type G_String declare the Zoo of Section 11.2 with declared strings (without allocators).

2 Declare a world of *N* by *M* cells for the Game of Life. Link them together in an appropriate manner. Use a dummy dead cell for the boundary so that the heart of the computation remains unchanged.

3 Reformulate the cells by making the access type refer to the cell as a whole rather than the internal component; how would this change the heart of the computation and the answer to the previous exercise?

11.5 Accessibility

The accessibility rules are designed to provide reasonable flexibility with complete security from dangling references. A simple static strategy applies in most circumstances and ensures that no run-time checks are required.

The basic rule is that the lifetime of an accessed object must be at least as long as that of the access type. Lifetime is a bit like scope except that it refers to the dynamic existence of the entity whereas scope refers to its (potential) static visibility.

As an illustration of the problem consider

```
procedure Main is
  type AI is access all Integer;
  Ref1: AI;
begin
  declare
    Ref2: AI;
    I: aliased Integer;
  begin
    Ref2 := I'Access;        -- illegal
    ...
    Ref1 := Ref2;
  end;
  ...
  declare
    ...   -- some other variables
  begin
    Ref1.all := 0;
  end;
end Main;
```

This is illegal because we have applied Access to a variable I with a lesser lifetime than the access type AI. The problem is not so much with the assignment to Ref2 which has the same lifetime as I but the fact that we can later assign Ref2 to Ref1 and Ref1 has a longer lifetime than I. The eventual assignment of 0 to Ref1.**all** could in principle overwrite almost anything because the value in Ref1 now refers to where I was and this space could now be used by some other variable of a different type (maybe another access type).

Other rules would have been possible. It might have been decreed that the assignment to Ref1 be forbidden or maybe that the final assignment to Ref1.**all** was the real culprit. For a number of reasons these alternatives are not sensible because they require extensive run-time checks or violate the idea that values can be assigned freely between variables of the same subtype.

So the actual rule is simply that the access value cannot be created in the first place if there is a risk that it might outlive the object that it refers to. In principle this is a dynamic rule but in most situations the check can be performed statically. This is illustrated by the following

```
procedure Main is
   type T is ...
   X: aliased T;
begin
   declare
      type A is access all T;
      Ptr: A;
   begin
      Ptr := X'Access;
      ...
   end;
end Main;
```

The object X outlives the access type A and so the assignment of X'Access to Ptr is permitted. In this example the dynamic lifetime follows from the simple static structure. However, because of the existence of procedure calls, we know that the dynamic structure at any time can include more active levels than just those in the static structure visible from the point concerned. It might thus be thought that static checks would not be satisfactory at all. However, because we can only write X'Access at points where both X and the access type A are in scope, it can in fact be shown that static checks based on the scope are exactly equivalent to dynamic checks based on the lifetime in examples such as this.

Of course a particular program might be safe even though it violates the static rules. If we are absolutely convinced that a program is safe then the alternative attribute Unchecked_Access can be applied by writing

```
Ref2 := I'Unchecked_Access;
```

The checks are then omitted and we are at risk. The first program above would then execute but be erroneous; anything could happen. See also Section 25.4 concerning how to minimize the risks of Unchecked_Access.

As we have seen, it is possible to convert one access type to another. In order to ensure that the lifetime of the accessed object is at least that of the target access type, conversion is only allowed if the target access type has a lifetime not greater then the source access type. Consider

```
declare
   type AI is access all Integer;
   I: aliased Integer;
   RefI: AI := I'Access;
begin
   declare
      type AJ is access all Integer;
      J: aliased Integer;
      RefJ: AJ := J'Access;
   begin
      RefI := AI(RefJ);          -- illegal
      ...
   end;
   ...
end;
```

The conversion to the outer type is not permitted since RefI would then have a dangling reference on exit from the inner block. On the other hand, a conversion from RefI to RefJ is perfectly safe and so is permitted. Note that the check can be performed statically.

11.6 Access parameters

In the simple cases considered in the previous section, static checks always suffice. Many programs use access types at the outermost level only and then everything has the same lifetime. But sometimes the static rules are too severe and a program is illegal even though nothing could go wrong. Consider

```
procedure Main is
   type T is ...
   procedure P is
      type A is access all T;
      Ptr: A := X'Access;                        -- illegal, X not in scope
   begin
      ...
   end P;
begin
   declare
      X: aliased T;
   begin
      P;
   end;
end Main;
```

This is illegal because X is out of scope at the point where we have written X'Access. However, if we could have assigned the 'address' of X to Ptr then nothing could have gone wrong because the lifetime of the type A is less than that of the variable X. This is such a common situation that the rules for parameters of an anonymous type are dynamic so that access values can be passed more flexibly. Such parameters are known as access parameters. Consider

```
procedure Main is
   type T is ...
   procedure P(Ptr: access T) is
   begin
      ...
   end P;
begin
   declare
      X: aliased T;
   begin
      P(X'Access);
   end;
end Main;
```

The parameter Ptr is an access parameter (as mentioned earlier, these are classed as **in** parameters although the word **in** never appears). It is initialized with the value of X'Access and all is well. Within the body of P we can dereference Ptr and manipulate the components of X as expected. The parameter Ptr is of course a constant and cannot be changed although the object it refers to can naturally be changed by assignment.

Type conversion between an access parameter and another access type is possible. Just allowing conversion to an inner type (this can be checked statically) would be too restrictive and so conversion to an outer type is also permitted and in this case an accessibility check is carried out at run time to ensure that the lifetime of the object referred to is not less than the target access type. The following is therefore allowed

```
procedure Main is
   type T is ...
   type A is access all T;
   Ref: A;
   procedure P(Ptr: access T) is
   begin
      ...
      Ref := A(Ptr);              -- dynamic check on conversion
   end P;
   X: aliased T;
begin
   P(X'Access);
   ...                            -- can now manipulate X via Ref
end Main;
```

Here we have converted the access parameter to the outer type A and assigned it to the variable Ref. For the call of P with X'Access this is safe because the lifetime of X is not less than the type A. But it might not be safe for all calls of P (such as in the previous example where X was declared in a local block). So the check has to be dynamic. If it fails then Program_Error is raised.

In order for this check to be possible all access parameters carry with them an indication of the accessibility of the actual parameter. Typically all that is necessary is to pass the static depth of the original object and this is then checked against the depth of the target type on the conversion.

Perhaps surprisingly this simple model with its combination of static checks in most circumstances and dynamic checks on some conversions of access parameters gives exactly the correct degree of flexibility combined with complete security.

The actual parameter corresponding to an access parameter can be (i) an access to an aliased object such as X'Access, (ii) another access parameter with the same accessed type, (iii) a value of a named or anonymous access type again with the same accessed type, or (iv) an allocator. In each case an appropriate indication of accessibility is passed. An access parameter, like other **in** parameters, can have a default expression and naturally this must be of one of these four forms.

An access parameter can be (i) used to provide access to the accessed object by dereferencing, (ii) passed as an actual parameter to another access parameter, or (iii) converted to another access type.

Dynamic accessibility checks occur on conversion to a named access type such as

```
Ref := A(Ptr);
```

or (which is really equivalent) if we write

```
Ref := Ptr.all'Access;
```

which dereferences to give the accessed object and then re-creates a reference to it.

An access parameter can be passed on to another access parameter; typically the accessibility indication is passed on unchanged but in the unusual circumstance where the called subprogram is internal to the calling subprogram, the accessibility level is replaced by that of the (statically known) formal calling parameter if less than the original actual parameter. Note that access parameters can have a null exclusion or be constant or both.

Access parameters are often an alternative to **in out** parameters. They both provide read and write access to the data concerned. And indeed, in the case of a record the components are referred to in the same way because of automatic dereferencing. However, the form of the actual parameter is different. Given

```
type T is ...
type A is access all T;
X: T;
Ref: A := new T'( ... );

procedure PIO(P: in out T);
procedure PA(Ptr: access T);
```

then calls of PIO take the form

```
PIO(X);  PIO(Ref.all);
```

whereas calls of PA require that variables be aliased so we have

```
X: aliased T;
...
PA(X'Access);  PA(Ref);
```

When we come to Chapter 14 we shall see that if the type T is a tagged type then objects such as X and Ref can call procedures using the prefixed notation – in which case the forms of the calls are exactly the same thus

```
X.PIO;  Ref.PIO;
X.PA;  Ref.PA;
```

An historical important difference between access and in out parameters was that in Ada 2005 a function can have an access parameter but not an in out parameter. This difference is eliminated in Ada 2012 and Ada 2022 because as we saw in the previous chapter, functions can now have parameters of any mode.

The full benefit of access parameters cannot be illustrated at the moment but will become clear when we discuss object oriented programming in Chapter 14 and access discriminants in Chapter 18.

The reader will probably have found the whole topic of accessibility and access parameters rather tedious. Accessibility is a bit like visibility; if we do something silly the compiler will tell us and we need not bother with the details of the rules most of the time. In any case, practical programs usually have a quite flat structure and problems will rarely arise. It is, however, very comforting to know that provided we avoid Unchecked_Access then dangling references will not arise and our program will not crash in a heap.

Exercise 11.6

1 Analyse the checks on the type conversions in the following indicating whether they are dynamic or static and whether they pass or fail.

```
procedure Main is
   type T is ...
   type A1 is access all T;
   Ref1: A1;
   procedure P(Ptr: access T) is
      type A2 is access all T;
      Ref2: A2;
   begin
      Ref1 := A1(Ptr);                    -- OK?
      Ref2 := A2(Ptr);                    -- OK?
      ...
   end P;
   X1: aliased T;
begin
   declare
      X2: aliased T;
   begin
      P(X1'Access);
      P(X2'Access);
   end;
end Main;
```

2 Similarly analyse the following. Note that the chained call of P2 from P1 is the interesting situation where the accessibility level has to be adjusted if the level of the original object is deeper than that of the formal parameter Ptr2. For convenience the level of the identifiers is indicated by their name. Note that the level of a formal parameter is one deeper than the subprogram itself.

```
procedure Main is
   type T is ...
   type A1 is access all T;
   Ref1: A1;
```

```
      procedure P1(Ptr2: access T) is
         type A2 is access all T;
         Ref2: A2;
         procedure P2(Ptr3: access T) is
            type A3 is access all T;
            Ref3: A3;
         begin
            Ref1 := A1(Ptr3);                    -- OK?
            Ref2 := A2(Ptr3);                    -- OK?
            Ref3 := A3(Ptr3);                    -- OK?
         end P2;
      begin
         P2(Ptr2);        -- chained call
      end P1;
      X1: aliased T;
   begin
      declare
         X2: aliased T;
      begin
         declare
            X3: aliased T;
         begin
            P1(X1'Access);
            P1(X2'Access);
            P1(X3'Access);
         end;
      end;
   end Main;
```

11.7 Anonymous access types

Giving a name to every access type can be a nuisance in some circumstances and so anonymous access types were introduced in Ada 2005. For example we can write

```
type Cell is
   record
      Next: access Cell;
      Value: Integer;
   end record;
```

Not only does this avoid introducing the named type Cell_Ptr but it also avoids the incomplete type declaration of Cell. An interesting point is that the name of the type Cell is being used in its own declaration. This also occurs with other types which we shall meet in due course.

In the case of components of an anonymous access type, the type is deemed to be declared at the same level as the enclosing declaration and the usual accessibility rules then apply. Thus the declaration

```
type R is
   record
      C: access Integer;
   end record;
```

is essentially equivalent to

```
type anon is access all Integer;
type R is
   record
      C: anon;
   end record;
```

However, different rules apply to stand-alone objects such as

```
V: access Integer;
```

In Ada 2005 this was treated in the same way as the component C. However, this can give rise to problems and so, in Ada 2012 and Ada 2022, the object V carries with it the accessibility of its current value as happens with access parameters described in the previous section.

Conversions from an access parameter to a local object of an anonymous type are permitted

```
procedure P(Ptr: access T) is
   Local_Ptr: access T;
begin
   Local_Ptr := Ptr;                          -- implicit conversion
   ...
end P;
```

In Ada 2012 and Ada 2022, the accessibility information contained in the access parameter is copied to the local variable (whereas in Ada 2005 it was lost).

Conversions between anonymous access types are permitted; consider

```
A_List: access Cell := ...
...
A_List := A_List.Next;                        -- implicit conversion
```

Note that there is strictly a conversion here between the anonymous access type of A_List and that of the component Next. But they are really the same type and so the conversion is permitted. Sometimes it is convenient to name a type and conversions between named and anonymous types are also allowed in many circumstances.

```
type List is access all Cell;
X: List := ...                                -- of named type
Y: access Cell := ...                         -- of anonymous type
X := List(Y);                  -- conversion to named type
Y := X;                        -- conversion to anonymous type
```

A conversion to a named access type can use the type name whereas a conversion to an anonymous type does not — it cannot because it does not have a name! In Ada 2005, conversions to a named type always had to use the type name in the conversion. However, Ada 2012 is more lenient and the name is only required if the conversion could fail a check (typically an accessibility check). In the example above the conversion cannot fail so we could have simply written

```
X := Y;                          -- OK in Ada 2012 and Ada 2022
```

The other possible check that might fail is when tags are checked when converting between tagged types, see Section 14.4.

Of course, if we attempted to convert an access to Integer to an access to Float then this is just illegal and the program will not compile.

Conversions concerning null exclusions do not need to be named. For example, suppose we declare Y to be of an anonymous type with a null exclusion and attempt to assign X to Y.

```
X: List := ...                          -- of named type
Y: not null access Cell := ...          -- of anonymous type
Y := X;                                 -- legal, but check needed
```

A run-time check is required to ensure that null is not being assigned to Y; Constraint_Error is raised if the check fails. Named conversions are never required when the conversion concerns a subtype property such as a null exclusion. This is much the same as converting between a value in an object of type Integer and assigning it to a variable of subtype Day_Number in Section 6.4. We do not use a named conversion but Constraint_Error will be raised if the check fails.

We can also use **constant** with anonymous access types. We need to distinguish

```
ACT: access constant T := X1'Access;
CAT: constant access T := X1'Access;
```

In the first case ACT is a variable and can be used to access different objects X1 and X2 of type T. But it cannot be used to change the value of those objects. In the second case CAT is a constant and can only refer to the object given in its initialization. But we can change the value of the object that CAT refers to.

We can also write

```
CACT: constant access constant T := X1'Access;
```

The object CACT is then a constant and provides read-only access to the object X1 to which it refers. It cannot be changed to refer to another object such as X2, nor can the value of X1 be changed via CACT.

Note that we never write **all** when declaring an anonymous access type in contrast to named general access types where we had

```
type A is access all T;
```

Anonymous access types are always general and so the use of **all** is not necessary (or allowed).

Anonymous access types can also be used as subprogram parameters and results. Thus we might have

```
function F(X, Y: access Cell) return not null access Cell;
procedure P(Z: access constant Integer; W: access Integer);
```

A null exclusion and **constant** can be used with both parameters and results. Parameters of anonymous access types are known as access parameters and were discussed in the previous section. Access parameters have mode **in** although this is not explicitly stated. Thus in the case of the procedure P, the parameter Z provides read-only access to the object being referenced whereas the parameter W provides both read and write access to the object being referenced. But the parameters Z and W themselves cannot be changed.

We conclude with a few words regarding a somewhat fictitious type known as *universal_access*. We cannot directly refer to this type but it is used to explain the behaviour of **null** and equality with anonymous access types. (We shall encounter other universal types when we discuss numeric types in more detail in Chapter 17.) The first point is that the literal **null** is considered to be of this type *universal_access* and appropriate implicit conversion rules allow **null** to be assigned to access objects. And the other is that there is a predefined function "=" in the package Standard with specification

```
function "=" (Left, Right: universal_access) return Boolean;
```

and it is this predefined function which is usually called when we compare anonymous access values. There is a corresponding "/=" as usual.

Now suppose we have

```
type A is access Integer;
R, S: access Integer;
...
if R = S then
```

Since we can do an implicit conversion from the anonymous access type of R and S to the type A, there is confusion as to whether the comparison uses the equality operator of the type *universal_access* or that of the type A. Accordingly, there is a preference rule that states that in the case of ambiguity there is a preference for equality of the type *universal_access*. Similar preference rules apply to *root_integer* and *root_real* (see Chapter 17).

Another example is instructive. Suppose we wish to do a deep comparison of two linked lists defined by the type Cell. We might wish to replace the predefined function by our own recursive function "=" thus

```
function "=" (L, R: access Cell) return Boolean is
begin
   if L = null or R = null then              -- universal =
      return L = R;                          -- universal =
   elsif L.Value = R.Value then
      return L.Next = R.Next;                -- recurses OK
   else
      return False;
   end if;
end "=";
```

This does work because the calls of "=" in the first two lines call the predefined "=" rather than the new version being declared because of the preference rule. This did not work in Ada 2005 which did not have this preference rule and so we had to replace the two lines after **begin** by

```
if Standard."=" (L, null) or Standard."=" (R, null) then
   return Standard."=" (L, R);
```

which was a bit ugly.

An important point regarding anonymous access types is that they cannot be used if we wish to deallocate storage using Unchecked_Deallocation as described in Section 25.2.

Exercise 11.7

1 Which of the following are legal?

```
ACT := X2'Access;
ACT.all := X2;
CAT := X2'Access;
CAT.all := X2;
```

11.8 Access to subprograms

The ability to pass subprograms as parameters of other subprograms has been a feature of most languages since Fortran and Algol 60. A notable exception was Ada 83 in which all binding of subprogram calls to the actual subprogram was determined statically. There were a number of reasons for taking such a static approach in Ada 83. There was concern for the implementation cost of dynamic binding, it was also clear that the presence of dynamic binding would reduce the provability of programs, and moreover it was felt that the introduction of generics where subprograms could be passed as parameters would cater for practical situations where formal procedure parameters were used in other languages.

Starting from Ada 2005, access to subprogram types behave in a very similar way to access to object types. They can be named or anonymous, null exclusions are permitted, and types and objects can be marked as constant. There are also similar accessibility rules that prevent a program from doing terrible things. There is one key difference and this concerns anonymous access types as parameters. In the case of objects we have seen that access parameters carry a dynamic indication of the accessibility level of the actual parameter. But anonymous access to subprogram parameters do not carry such an indication and thus behave somewhat differently. We shall start by considering simple named access to subprogram types and then look at parameters of anonymous access to subprogram types.

An access to subprogram value can be created by the Access attribute and a subprogram can be called indirectly by dereferencing such an access value. Thus we can write

```
type Math_Function is access function (F: Float) return Float;
Do_It: Math_Function;
X, Theta: Float;
```

and Do_It can then 'point to' functions such as Sin, Cos, Tan, and Sqrt which we will assume have specifications such as

```
function Sin(X: Float) return Float;
```

We can then assign an appropriate access to subprogram value to Do_It thus

```
Do_It := Sin'Access;
```

and later indirectly call the subprogram currently referred to by Do_It by writing

```
X := Do_It(Theta);
```

This is really an abbreviation for X := Do_It.**all**(Theta); but just as with many other uses of access types the **.all** is not usually required although it would be necessary if there were no parameters.

The access to subprogram mechanism can be used to program general dynamic selection and to pass subprograms as parameters. It also allows program call-back to be implemented in a natural and efficient manner.

The following procedure applies the function passed as parameter to all the elements of an array

```
procedure Iterate(Func: in not null Math_Function; V: in out Vector) is
begin
   for I in V'Range loop
      V(I) := Func(V(I));
   end loop;
end Iterate;

A_Vector: Vector := (100.0, 4.0, 0.0, 25.0);
...
Iterate(Sqrt'Access, A_Vector);       -- A_Vector is now (10.0, 2.0, 0.0, 5.0)
```

We can similarly have access to procedure types. Thus we might have algorithms for encryption of messages and apply them indirectly as follows

```
type Converter is not null access procedure (S: in out String);

procedure Encrypt(Text: in out String);
procedure Decrypt(Text: in out String);

procedure Apply(Proc: in Converter; To: in out String) is
begin
   Proc(To);                    -- indirect call
end Apply;

Sample: String := "Send reinforcements. We're going to advance.";
...
Apply(Proc => Encrypt'Access, To => Sample);
```

Tradition has it that the message becomes "Send 3/4d, we're going to a dance."! (In old British currency, 3/4d, that is three shillings and four pence, is pronounced 'three-and-fourpence'.)

Note that the last two examples used **not null** with the access to subprogram type. It will usually be the case that a null exclusion is appropriate since a null value would be unusual and the null exclusion avoids run-time checks.

Simple classic numerical codes can also be implemented in the traditional way. Thus an integration routine might have the following specification

```
type Integrand is not null access function (X: Float) return Float;

function Integrate(Fn: Integrand; Lo, Hi: Float) return Float;
```

for the evaluation of

$$\int_{lo}^{hi} f(x)dx$$

and we might then write

```
Area := Integrate(Log'Access, 1.0, 2.0);
```

which will compute the area under the curve for log(x) from 1.0 to 2.0. Within the body of the function Integrate there will be calls of the actual subprogram passed as parameter; this is a simple form of call-back.

A common paradigm within the process control industry is to implement sequencing control through successive calls of a number of interpreter actions. A sequence compiler might interactively build an array of such actions which are then obeyed. Thus we might have

```
type Action is access procedure;
Action_Sequence: array (1 .. N) of Action;

...   -- build the array
...   -- and then obey it
for I in Action_Sequence'Range loop
   Action_Sequence(I).all;
end loop;
```

where we note the need for .**all** because there are no parameters.

There are a number of rules which ensure that access to subprogram values cannot be misused. Subtype conformance matching between the profiles ensures that the subprogram always has the correct number and type of parameters and that any constraints statically match. Subtype conformance is weaker than the full conformance required between the body and specification of the same subprogram as described in Section 10.6. Subtype conformance ignores the formal parameter names and also the presence, absence or value of default initial expressions.

Accessibility rules also apply to access to subprogram types and ensure that a subprogram is not called out of context. Thus we can only apply the Access attribute if the subprogram has a lifetime at least that of the access type. This means that the simple integration routine using the named type Integrand does not work in many cases. Thus suppose we wish to integrate a function such as Exp(X**2) where Exp is some library function. We might try

```
procedure Main is
   function F(X: Float) return Float is
   begin
      return Exp(X**2);
   end F;

   Result, L, H: Float;
begin
   ...                                  -- set bounds in L and H say
   Result := Integrate(F'Access, L, H);        -- illegal
   ...
end Main;
```

but this is illegal because the subprogram F has a lifetime less than that of the access type Integrand. In this particular case the problem can be overcome by declaring F itself at the same level as Integrand. But this is not always possible if functions are nested in some way as for example if we had a double integral.

The proper solution is to change the function Integrate to use an anonymous access to subprogram type as parameter and thereby dispense with the troublesome type Integrand. The specification becomes

```
function Integrate(Fn: not null access function(X: Float) return Float;
                   Lo, Hi: Float) return Float;
```

and now we can write Integrate(F'Access, ...) as required. Note that the profile for Integrate includes the profile for the anonymous type of the parameter Fn. This nesting of profiles causes no problems. The use of an access to a local procedure in this way is often called a downward closure.

We will now look at the rules for type conversions and see how they prevent unsafe assignments to global variables. The following illustrates both access to object and access to subprogram parameters.

```
type AOT is access all Integer;
type APT is access procedure (X: in out Float);

Evil_Obj: AOT;
Evil_Proc: APT;

procedure P(Objptr: access Integer;
            Procptr: access procedure (X: in out Float)) is
begin
   Evil_Obj := AOT(Objptr);              -- may fail at run time
   Evil_Proc := APT(Procptr);            -- always fails at compile time
end P;

declare
   An_Obj: aliased Integer;
   procedure A_Proc(X: in out Float) is
   begin ... end A_Proc;
begin
   P(An_Obj'Access, A_Proc'Access);      -- legal
end;
```

```
...
Evil_Obj.all := 0;            -- would assign to nowhere
Evil_Proc.all( ... );         -- would call nowhere
```

The procedure P has an access to object parameter Objptr and an access to subprogram parameter Procptr; they are both of anonymous type. The call of P in the local block passes the addresses of a local object An_Obj and a local procedure A_Proc to P. This is permitted. We now attempt to assign the parameter values from within P to global objects Evil_Obj and Evil_Proc with the intent of assigning indirectly via Evil_Obj and calling indirectly via Evil_Proc after the object and procedure referred to no longer exist.

Both of these wicked deeds are prevented by the accessibility rules. In the case of the object parameter Objptr, the accessibility level of the actual An_Obj is greater than that of the type AOT and the conversion is prevented at run time and Program_Error is raised. But if An_Obj had been declared at the same level as AOT and not within an inner block then the conversion would have been permitted.

However, somewhat different rules apply to anonymous access to subprogram parameters. They do not carry an indication of the accessibility level of the actual parameter but simply treat it as if it were infinite (strictly – deeper than anything else). This prevents the conversion to the type APT and all is well; this is detected at compile time. But note that if the procedure A_Proc had been declared at the same level as APT then the conversion would still have failed because the accessibility level is treated as infinite.

There are a number of reasons for the different treatment of anonymous access to subprogram types. A big problem is that named access to subprogram types are implemented in the same way as C pointers to functions in most compilers. Permitting the conversion from anonymous access to subprogram types to named ones would thus have caused problems because that model does not work for many implementations. Carrying the accessibility level around would not have prevented these conversions. The key goal is just to provide a facility corresponding to that in Pascal and not to encourage too much fooling about with access to subprogram types. Recall that the attribute Unchecked_Access is permitted for access to object types but is clearly far too dangerous for access to subprogram types.

Conversion between access to subprogram types also requires that the profiles have subtype conformance and that any null exclusion is not violated.

It is of course possible for a record to contain components whose types are access to subprogram types. We will now consider a possible fragment of the system which drives the controls in the cockpit of some mythical Ada Airlines. There are a number of physical buttons on the console and we wish to associate different actions corresponding to pushing the various buttons.

```
type Button;

type Response_Ptr is access procedure (B: in out Button);

type Button is
  record
    Response: Response_Ptr;
    ...  -- other aspects of the button
  end record;
```

```
procedure Associate(B: in out Button; ... );
procedure Set_Response(B: in out Button; R: in Response_Ptr);
procedure Push(B: in out Button);
```

A button is represented as a record containing a number of components describing properties of the button (position of message on the display for example). The component Response is an access to a procedure which is the action to be executed when the button is pushed. Note carefully that the button value is passed to this procedure as a parameter so that the procedure can obtain access to the other components of the record describing the button. Incidentally, observe that the incomplete type Button is allowed to be used for a parameter or result of the access to subprogram type Response_Ptr before its full type declaration in order to break the otherwise inevitable circularity.

The procedure Set_Response assigns an appropriate access value to the component Response and the procedure Associate makes the connection between the physical button and the software button and fills in the other components. Other functions (not shown) provide access to them. The procedure Push is called when any physical button is pushed, the parameter indicating its identity. The bodies of the subprograms might be as follows

```
procedure Set_Response(B: in out Button; R: in Response_Ptr) is
begin
   B.Response := R;          -- set procedure value in record
end Set_Response;

procedure Push(B: in out Button) is
begin
   B.Response(B);            -- indirect call
end Push;
```

We can now set the specific actions we want when a button is pushed. We might want some emergency action to happen when a big red button is pushed.

```
Big_Red_Button: Button;

procedure Emergency(B: in out Button) is
begin
   Broadcast("mayday");
   ...
   Eject_Pilot;
end Emergency;
...

Associate(Big_Red_Button, ... );
Set_Response(Big_Red_Button, Emergency'Access);
...
if Disaster then
   Push(Big_Red_Button); ...
```

It is interesting to observe that we could not have used anonymous access to subprogram types instead of the named type Response_Ptr. This is because the assignment to B.Response in the procedure Set_Response would then violate the accessibility rules since the accessibility level of the anonymous parameter R would be infinite. But we could have made the component of the record of an anonymous type.

Finally, note that certain subprograms are considered intrinsic which means that they are essentially built in to the compiler. We say that their calling convention is Intrinsic whereas normal subprograms have calling convention Ada. The Access attribute cannot be applied to intrinsic subprograms.

Examples of intrinsic subprograms which we have met so far are the predefined operations in Standard such as "+", all enumeration literals (which, as mentioned in Section 10.6, are thought of as parameterless functions), an implicitly declared "/=" as a companion to "=" (see Section 10.2), and attributes that are subprograms such as Pred and Succ. Other intrinsic subprograms are those implicitly declared by derivation (Section 12.3) except in the case of tagged types (Section 14.1), and subprograms immediately declared in protected bodies (Section 20.4).

An important use of access to subprogram types is for interfacing to subprograms written in other languages; this involves the use of various aspects referring to calling conventions and is discussed in Section 25.5. See also Section 20.4 for access types referring to protected operations. We shall meet many examples of the use of anonymous access to subprogram types as downward closures when we discuss the container library in Chapter 24.

Exercise 11.8

1 Using the function Integrate show how to evaluate

$$\int_0^P e^t \sin t \, dt$$

2 Declare the specification of an appropriate function for finding a root of the equation $f(x) = 0$, and then show how you would find the root of

$$e^x + x = 7$$

3 Write a further function Integrate to evaluate

$$\int_{LX}^{HX}\int_{LY}^{HY} F(x, y) \, dy \, dx$$

in which $F(x, y)$ is an arbitrary function of the two variables x and y. This function Integrate should use the existing function Integrate for a single variable.

11.9 Storage pools

Accessed objects are allocated in a space called a storage pool associated with the access type. This pool will typically cease to exist when the scope of the access type is finally left but by then all the access variables will also have ceased to exist, so no dangling reference problems can arise.

If an allocated object becomes inaccessible because no declared objects refer to it directly or indirectly then the storage it occupies may be reclaimed so that it can be reused for other objects. An implementation may (but need not and most do not) provide a garbage collector to do this.

Alternatively, there is a mechanism whereby a program can indicate that an object is no longer required; if, mistakenly, there are still references to such allocated objects then the use of such references is erroneous. For details see Sections 25.2 and 25.4 which also describe how a user may create and control individual storage pools and subpools.

It is important to realize that each declaration of an access type introduces a new logically distinct set of accessed objects. Such sets might reside in different storage pools. Two access types can refer to objects of the same type but the access objects must not refer to objects in the wrong set. So we could have

```
type Ref_Int_A is access all Integer;
type Ref_Int_B is access all Integer;
RA: Ref_Int_A := new Integer'(10);
RB: Ref_Int_B := new Integer'(20);
```

The objects created by the two allocators are both of the same type but the access values are of different types determined by the context of the allocator and the objects might be in different pools. But we can convert between the types by using the type name because the accessed types are the same.

So we have

```
RA.all := RB.all;           -- legal, copies the values of the objects
RA := RB;                   -- illegal, different types
RA := Ref_Int_A(RB);        -- legal, has explicit conversion
```

If a named access type omits **all** then we call it a pool specific type because it can only access objects allocated in a pool. The key differences between pool specific types and general access to object types are

- pool specific types can only refer to allocated objects and never to aliased declared objects,
- pool specific types cannot be marked as constant,
- conversion between different pool specific types is forbidden.

There is an exception to the last rule. If one type is derived from another as discussed in Section 12.3, then conversion between them is permitted because they share the same storage pool.

However, although we cannot generally convert between pool specific types, we can convert from a pool specific type to a general type, but conversion in the opposite direction is not permitted. Similarly, we can convert from a pool specific type to an anonymous type but not in the opposite direction. The reason is simply that the general or anonymous type might designate an object that is not in a pool.

Checklist 11

An incomplete declaration can only be used in an access type.

The scope of an allocated object is that of the access type.

Access objects have a default initial value of **null**.

An allocator in an aggregate is evaluated for each index value.

An allocator with a complete initial value uses a quote.

A general named access type has **all** or **constant** in its definition.

An anonymous access type never has **all** but may have **constant**. Anonymous access types are general.

The attribute Access can only be applied to non-intrinsic subprograms and to aliased objects.

Beware Unchecked_Access.

Conversion to a pool specific access type is not allowed.

Anonymous access to object parameters (access parameters) carry a dynamic indication of the accessibility of the actual parameter. But anonymous access to subprogram parameters have an infinite accessibility level.

The only anonymous access types prior to Ada 2005 were access parameters (and access discriminants).

Access parameters could not be marked **constant** in Ada 95.

Null exclusions were added in Ada 2005.

Anonymous access to subprogram types were added in Ada 2005.

Stand-alone objects of an anonymous access type carry the accessibility level of their value with them in Ada 2012.

Conversions do not need to be named in Ada 2012 unless a check could fail.

There is a preference rule for predefined equality of *universal_access* in Ada 2012.

New in Ada 2022

There are a number of minor improvements to the accessibility rules. See AI12-345 on dynamic accessibility and AI12-406 on static accessibility.

Program 2

Sylvan Sorter

This program uses the treesort algorithm of Section 11.2 except that it sorts a list rather than an array.

The program first reads in a sequence of positive integers terminated by the value zero (or a negative value) and builds it into a list which is printed in a tabular format. This unsorted list is then converted into a binary tree which is printed in the usual representation. The binary tree is then converted back into an ordered list and finally the sorted values are printed. The program repeats until it encounters an empty sequence (just a zero or negative number).

The two library packages Lists and Trees contain the key data structures Cell and Node and associated operations. Although we have not yet discussed packages, private types and compilation units in depth this is much better than putting everything inside the main subprogram which is the style which would have to be adopted using just the algorithmic facilities described in Part 2. The types List and Tree are private types. Their full type reveals that they are access types referring to Cell and Node. Thus the inner structure of the types Cell and Node are not used outside their defining packages. The small package Page declares a constant defining the page width and a subtype defining a line of text.

There are then a number of library subprograms for performing the main activities, namely Read_List, Print_List, Print_Tree, Convert_List_To_Tree and Convert_Tree_To_List. Finally, the main subprogram calls these within an outer loop.

This program illustrates the use of both named and anonymous access types. The type Cell uses an anonymous type for its component whereas the type Tree uses a named type.

Using an anonymous type in Tree would result in many named conversions in Ada 2005 although, as mentioned in Section 11.7, the conversions would not need to be named in Ada 2012 or Ada 2022 because they could not fail.

```
package Lists is
   type List is private;

   function Is_Empty(L: List) return Boolean;
   procedure Clear(L: out List);
   function Make_List(V: Integer) return List;
   procedure Take_From_List(L: in out List;
                                         V: out Integer);
   procedure Append(First: in out List;
                             Second: in List);
private
   type Cell is
      record
         Next: access Cell;
         Value: Integer;
      end record;

   type List is access all Cell;
end;

package body Lists is
   function Is_Empty(L: List) return Boolean is
   begin
      return L = null;
   end Is_Empty;

   procedure Clear(L: out List) is
   begin
      L := null;
   end Clear;

   function Make_List(V: Integer) return List is
   begin
      return new Cell'(null, V);
   end Make_List;

   procedure Take_From_List(L: in out List;
                                         V: out Integer) is
   begin
      V := L.Value;
      L := List(L.Next);
   end Take_From_List;
```

```
   procedure Append(First: in out List;
                    Second: in List) is
      Local: access Cell := First;
   begin
      if First = null then
         First := Second;
      else
         while Local.Next /= null loop
            Local := Local.Next;
         end loop;
         Local.Next := Second;
      end if;
   end Append;

end Lists;
```

```
package Trees is
   type Tree is private;

   function Is_Empty(T: Tree) return Boolean;
   procedure Clear(T: out Tree);
   procedure Insert(T: in out Tree; V: in Integer);
   function Depth(T: Tree) return Integer;

   function Left_Subtree(T: Tree) return Tree;
   function Right_Subtree(T: Tree) return Tree;
   function Node_Value(T: Tree) return Integer;

private
   type Node;
   type Tree is access Node;
   type Node is
      record
         Left, Right: Tree;
         Value: Integer;
      end record;
end;

package body Trees is
   function Is_Empty(T: Tree) return Boolean is
   begin
      return T = null;
   end Is_Empty;

   procedure Clear(T: out Tree) is
   begin
      T := null;
   end Clear;

   procedure Insert(T: in out Tree; V: in Integer) is
   begin
      if T = null then
         T := new Node'(null, null, V);
      elsif V < T.Value then
         Insert(T.Left, V);
      else
         Insert(T.Right, V);
      end if;
   end Insert;

   function Depth(T: Tree) return Integer is
   begin
      if T = null then return 0; end if;
      return 1 + Integer'Max(Depth(T.Left),
                             Depth(T.Right));
   end Depth;

   function Left_Subtree(T: Tree) return Tree is
   begin
      return T.Left;
   end Left_Subtree;

   function Right_Subtree(T: Tree) return Tree is
   begin
      return T.Right;
   end Right_Subtree;

   function Node_Value(T: Tree) return Integer is
   begin
      return T.Value;
   end Node_Value;

end Trees;
```

```
package Page is
   Width: constant Integer := 40;
   subtype Line is String(1 .. Width);
end Page;
```

```
with Lists;  use Lists;
with Ada.Integer_Text_IO;  use Ada;
procedure Read_List(L: out List;
                    Max_Value: out Integer) is
   Value: Integer;
begin
   Max_Value := 0;
   Clear(L);
   loop
      Integer_Text_IO.Get(Value);
      exit when Value <= 0;
      if Value > Max_Value then
         Max_Value := Value;
      end if;
      Append(L, Make_List(Value));
   end loop;
end Read_List;
```

```
function Num_Size(N: Integer) return Integer is
begin  -- allows for leading space
   return Integer'Image(N)'Length;
end Num_Size;
```

```
with Page;  with Num_Size;
with Lists;  use Lists;
with Ada.Text_IO, Ada.Integer_Text_IO;  use Ada;
```

```
procedure Print_List(L: in List;
                     Max_Value: in Integer) is
   Temp: List := L;
   Value: Integer;
   Count: Integer := 0;
   Field: constant Integer :=
                        Num_Size(Max_Value);
begin
   Text_IO.New_Line;
   loop
      exit when Is_Empty(Temp);
      Take_From_List(Temp, Value);
      Count := Count + Field;
      if Count > Page.Width then
         Text_IO.New_Line;
         Count := Field;
      end if;
      Integer_Text_IO.Put(Value, Field);
   end loop;
   Text_IO.New_Line(2);
end Print_List;
-----------------------------------------

with Page;  with Num_Size;
with Trees;  use Trees;
with Ada.Text_IO;  use Ada;
procedure Print_Tree(T: in Tree;
                     Max_Value: in Integer) is
   Max_Width: Integer :=
         Num_Size(Max_Value) * 2**(Depth(T)-1);

   A: array (1 .. 4*Depth(T)-3) of Page.Line :=
                  (others => (others => ' '));

   procedure Put(N, Row, Col, Width: Integer) is
      Size: Integer := Num_Size(N);
      Offset: Integer := (Width - Size + 1)/2;
      Digit: Integer;
      Number: Integer := N;
   begin
      if Size > Width then
         for I in 1 .. Integer'Max(Width, 1) loop
            A(Row)(Col+I-1) := '*';
         end loop;
      else
         A(Row)(Col+Offset..Col+Offset+Size-1) :=
                           Integer'Image(Number);
      end if;
   end Put;

   procedure Do_It(T: Tree;
                   Row, Col, W: Integer) is
      Left: Tree := Left_Subtree(T);
      Right: Tree := Right_Subtree(T);
   begin
      Put(Node_Value(T), Row, Col, W);
      if not (Is_Empty(Left) and
                          Is_Empty(Right)) then
         A(Row+1)(Col+W/2) := '|';
      end if;

      if not Is_Empty(Left) then
         A(Row+2)(Col+W/4 .. Col+W/2) :=
                           (others => '-');
         A(Row+3)(Col+W/4) := '|';
         Do_It(Left, Row+4, Col, W/2);
      end if;

      if not Is_Empty(Right) then
         A(Row+2)(Col+W/2 .. Col+3*W/4) :=
                           (others => '-');
         A(Row+3)(Col+3*W/4) := '|';
         Do_It(Right, Row+4, Col+W/2, (W+1)/2);
      end if;
   end Do_It;

begin
   if Max_Width > Page.Width then
      Max_Width := Page.Width;
   end if;
   Do_It(T, 1, 1, Max_Width);
   Text_IO.New_Line;
   for I in A'Range loop
      Text_IO.New_Line
      Text_IO.Put(A(I));
   end loop;
   Text_IO.New_Line(2);
end Print_Tree;
-----------------------------------------

with Lists, Trees;  use Lists, Trees;
procedure Convert_List_To_Tree(L: in List;
                               T: out Tree) is
   Temp: List := L;
   Value: Integer;
begin
   Clear(T);
   loop
      exit when Is_Empty(Temp);
      Take_From_List(Temp, Value);
      Insert(T, Value);
   end loop;
end Convert_List_To_Tree;
-----------------------------------------

with Lists, Trees;  use Lists, Trees;
procedure Convert_Tree_To_List(T: in Tree;
                               L: out List) is
   Right_L: List;
begin
   if Is_Empty(T) then Clear(L); return; end if;
   Convert_Tree_To_List(Left_Subtree(T), L);
   Append(L, Make_List(Node_Value(T)));
   Convert_Tree_To_List(Right_Subtree(T), Right_L);
   Append(L, Right_L);
end Convert_Tree_To_List;
-----------------------------------------
```

```
with Lists, Read_List, Print_List;
with Trees, Print_Tree;
with Convert_List_To_Tree;
with Convert_Tree_To_List;
with Ada.Text_IO; use Ada.Text_IO;
procedure Sylvan_Sorter is
   The_List: Lists.List;
   The_Tree: Trees.Tree;
   Max_Value: Integer;
begin
   Put("Welcome to the Sylvan Sorter");
   New_Line(2);
   loop
      Put_Line("Enter list of positive integers" &
                                        " ending with 0");
      Read_List(The_List, Max_Value);
      exit when Lists.Is_Empty(The_List);
      Print_List(The_List, Max_Value);
      Convert_List_To_Tree(The_List, The_Tree);
      Print_Tree(The_Tree, Max_Value);
      Convert_Tree_To_List(The_Tree, The_List);
      Print_List(The_List, Max_Value);
   end loop;
   New_Line;
   Put_Line("Finished"); Skip_Line(2);
end Sylvan_Sorter;
```

This program has deliberately been written in a rather extravagant manner using recursion quite freely. The list processing is particularly inefficient because the list is very simply linked and so adding an item to the end requires traversal of the list each time (this is in procedure Append). It would clearly be better to use a structure with links to both ends of the list as illustrated in Exercise 12.5(**3**).

Most of the program is straightforward except for the procedure Print_Tree. This has internal subprograms Put and Do_It. The basic problem is that the obvious approach of printing the tree recursively by first printing the left subtree then the node value and then the right subtree requires going up and down the page. However, Text_IO always works down the page and so cannot be used directly. Accordingly, we declare an array of strings A sufficient to hold all of the tree; this array is initialized to all spaces. The number of rows is computed by using the function Depth which returns the depth of the tree. Each fragment of tree is printed in the form

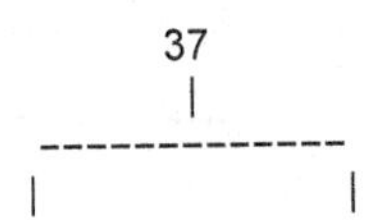

and so takes four lines. Hence the number of lines to be output is 4*Depth(T)–3.

The maximum value in the tree is remembered in Max_Value when reading the numbers in the first place. The minimum field width to print all numbers satisfactorily is then computed as Num_Size(Max_Value) which allows at least one space between adjacent numbers. This is then used to compute the overall width of the tree. If this is less than the page width then all is well. If not then it is set to the page width and we carry on regardless. The procedure Do_It is then called.

The procedure Do_It has parameters indicating the row and column of the array A from where it is to start printing and the field width available. It calls itself recursively with suitable values of these parameters for the left and right subtrees. If either subtree is empty the corresponding lines and recursive call of itself are omitted. The node value is output by a call of the procedure Put.

The procedure Put also has format parameters; it uses the attribute Image and prints the number centrally within the field. If it is unable to print the number within the given field then it outputs asterisks in the Fortran tradition (at least one); this will only happen if the width of the overall tree had to be reduced because it exceeded the page width – even then it might not happen if the numbers at the deepest part of the tree are relatively small. If the tree is very deep then the lines will eventually overlap and the whole thing will degenerate into rows of lines and asterisks; but it will keep going on to the bitter end!

This program, although curious, does illustrate many algorithmic features of Ada, such as slices, recursion, the attributes Max and Image and so on.

A stylistic point is that in some subprograms we have used the technique of first testing for the special case and, having performed the appropriate action, then leaving the subprogram via a return statement. Some readers might consider this bad practice and that we should really use an else part for the normal action and avoid the use of return in the middle of the subprogram. In defence of the approach taken one can say that mathematical proofs often dispose of the simple case in this way and moreover it does take a lot less space and means that the main algorithm is less indented.

Another stylistic point is that many of the variables declared to be of type Integer could well be of subtype Natural or Positive. We leave to the reader the task of considering this improvement.

A final point is that the program makes much use of heap storage. If the implementation has no garbage collector then it will eventually run out of space and raise Storage_Error. We will return to this issue in Chapter 25.

Part 3

The Big Picture

Chapter 12	**Packages and Private Types**	233
Chapter 13	**Overall Structure**	267
Program 3	**Rational Reckoner**	301
Chapter 14	**Object Oriented Programming**	305
Chapter 15	**Exceptions**	365
Chapter 16	**Contracts**	389
Chapter 17	**Numeric Types**	427
Chapter 18	**Parameterized Types**	449
Chapter 19	**Generics**	479
Chapter 20	**Tasking**	511
Chapter 21	**Object Oriented Techniques**	561
Chapter 22	**Tasking Techniques**	599
Program 4	**Super Sieve**	645

This third part is the core of the book and is largely about abstraction. As mentioned in Chapter 1, the evolution of programming languages is essentially about understanding various aspects of abstraction and here we look in depth at the facilities for data abstraction, object oriented programming, and programming by contract.

Chapter 12 shows how packages and private types are the keystone of Ada and can be used to control visibility by giving a client and server different views of an object. This chapter also introduces the simplest ideas of type derivation and inheritance and the very important notion of a limited type which is a type (strictly a view of a type) for which copying is not permitted. Limited types are important for modelling those real-world objects for which copying is inappropriate. Chapter 13 then discusses the hierarchical library structure and the facilities for separate compilation.

A third complete program follows Chapters 12 and 13 and illustrates the use of the hierarchical library in the construction of an abstract data type and associated operations.

The basic facilities for object oriented programming are then introduced in Chapter 14. This covers type extension and inheritance, dynamic polymorphism, dispatching, class wide types, abstract types, and interfaces. This chapter concentrates very much on the basic nuts and bolts, and a further discussion of OOP is deferred until Chapter 21 when other aspects of the language have been introduced.

Chapter 15 then covers exceptions and is followed by Chapter 16 which discusses the very important topic of contracts which was perhaps the most important new feature introduced in Ada 2012 and which is enhanced in Ada 2022 by the consideration of global state.

Chapter 17 is a detailed discussion of numeric types including modular types and the rather specialized fixed point types which were not discussed in Chapter 6.

Chapters 18 and 19 return to the theme of abstraction by considering two forms of parameterization. Chapter 18 discusses the parameterization of types by discriminants. Chapter 19 discusses genericity which enables both subprograms and packages to be parameterized in many different ways. Genericity provides static polymorphism checked at compile time in contrast to the dynamic polymorphism of type extension.

Chapter 20 completes the main discussion of core features by describing the concepts relating to tasking; these include tasks which are units which execute in parallel and protected objects which provide shared access to common data without risk of interference.

At this point the most important features of the language have been described and the main purpose of the final two chapters is to illustrate how the various features work together.

Chapter 21 discusses various aspects of OOP such as the use of interfaces and object constructors; it also outlines the subtle new iterators introduced in Ada 2012 which simplify the use of containers.

Chapter 22 extends the discussion on tasking by introducing synchronized interfaces which combine the properties of OOP and concurrent programming; it also describes selected topics from the specialized annexes and introduces the lightweight parallel features added in Ada 2022.

This part concludes with a program which illustrates various aspects of tasking and generics.

12 Packages and Private Types

12.1 Packages
12.2 Private types
12.3 Primitive operations and derived types
12.4 Equality
12.5 Limited types
12.6 Resource management

The previous chapters described the small-scale features of Ada in some detail. These features correspond to the areas addressed by the pioneering languages prior to about 1975. This chapter and the next discuss the important concepts of abstraction and programming in the large which entered into languages generally around 1980. The more recent concepts of type extension, inheritance, and polymorphism are dealt with in Chapter 14.

In this chapter we discuss packages (which is really what Ada is all about) and the important concept of a private type. The topic of library units and especially library packages is discussed in detail in the next chapter.

12.1 Packages

In this section we introduce the concept of an Abstract State Machine (ASM) and in the next section we extend this to Abstract Data Types (ADT). The general idea of abstraction is to distinguish the inner details of how something works from an external view of it. This is a common idea in everyday life. We can use a watch to look at the time without needing to know how it works. Indeed the case of the watch hides its inner workings from us.

One of the major problems with early simple languages, such as C and Pascal, is that they do not offer enough control of visibility. For example, suppose we have a stack implemented as an array plus a variable to index the current top element, thus

```
Max: constant := 100;
S: array (1 .. Max) of Integer;
Top: Integer range 0 .. Max;
```

We might then declare a procedure Push to add an item and a function Pop to remove an item.

```
procedure Push(X: in Integer) is
begin
   Top := Top + 1;
   S(Top) := X;
end Push;

function Pop return Integer is
begin
   Top := Top - 1;
   return S(Top + 1);
end Pop;
```

In a simple block structured language it is not possible to have access to the subprograms Push and Pop without also having direct access to the variables S and Top. As a result we can bypass the intended protocol of only accessing the stack through calls of Push and Pop. For example, we might make use of our knowledge of the implementation details and change the value of Top directly without corresponding changes to the array S. This could well result in a call of Pop returning a junk value which was not placed on the stack by an earlier call of Push.

The Ada package overcomes this by allowing us to place a wall around a group of declarations and only permit access to those which we intend to be visible. A package actually comes in two parts: the specification which gives the interface to the outside world, and the body which gives the hidden details.

The above example should be written as

```
package Stack is                            -- specification
   procedure Push(X: in Integer);
   function Pop return Integer;
end Stack;

package body Stack is                       -- body
   Max: constant := 100;
   S: array (1 .. Max) of Integer;
   Top: Integer range 0 .. Max;

   procedure Push(X: in Integer) is
   begin
      Top := Top + 1;
      S(Top) := X;
   end Push;

   function Pop return Integer is
   begin
      Top := Top - 1;
      return S(Top + 1);
   end Pop;

begin                                       -- initialization
   Top := 0;
end Stack;
```

The package specification (strictly declaration) starts with the reserved word **package**, the name of the package and **is**. This is followed by declarations of the entities which are to be visible. (There might then be a private part as discussed in the next section.) A package specification finishes with **end**, its name (optionally) and the terminating semicolon. In the example we just have the declarations of the two subprograms Push and Pop.

The package body also starts with **package** but this is then followed by **body**, the name and **is**. We then have a normal declarative part, **begin**, sequence of statements, **end**, optional name and terminating semicolon.

In this example the declarative part contains the variables which represent the stack and the bodies of Push and Pop. The sequence of statements between **begin** and **end** is executed when the package is declared and can be used for initialization. If there is no need for an initialization sequence, the **begin** can be omitted. Indeed, in this example we could equally have performed the initialization by writing

```
Top: Integer range 0 .. Max := 0;
```

The package is another case where we need distinct subprogram declarations and bodies. Indeed, we cannot put a body into a package specification. Moreover, if a package specification contains the specification of a subprogram, then the package body must contain the corresponding subprogram body; we say that the body completes the subprogram. We can think of the package specification and body as being just one large declarative region with only some items visible. However, a subprogram body can be declared in a package body without its specification having to be given in the package specification. Such a subprogram would be internal to the package and could only be called from within, either from other subprograms, some of which would presumably be visible, or perhaps from the initialization sequence.

There is one variation to the rule that the whole of a subprogram cannot appear in a package specification and that concerns expression functions which were introduced in Section 10.1. A very short function can be written in one lump thus

```
function Sign(X: Integer) return Integer is
  (if X > 0 then +1 elsif X < 0 then -1 else 0);
```

and such an expression function can appear in a package specification and does not need a body. Moreover, if we do indeed just give the specification of the function in a package specification thus

```
function Sign(X: Integer) return Integer;
```

then in the corresponding package body we can give the function body in full using the long form as on the third page of Section 10.1 or we can use the abbreviated form provided by an expression function as the completion of the function.

Packages are typically declared at the outermost or library level of a program and often compiled separately as described in detail in Chapter 13. But packages may be declared in any declarative part such as that in a block, subprogram or indeed another package. If a package specification is declared inside another package specification then, just as for subprograms, the body of the inner one must be declared in the body of the outer one. And again both specification and body of the inner package could be in an outer package body in which case the inner package would not be visible outside that body.

Apart from the rule that a package specification cannot contain bodies, it can contain any of the other kinds of declarations we have met.

The elaboration of a package body consists simply of the elaboration of the declarations inside it followed by the execution of the initialization sequence if there is one. The package continues to exist until the end of the scope in which it is declared (or forever in the case of a library package). Entities declared inside the package have the same lifetime as the package itself. Thus the variables S and Top retain their values between successive calls of Push and Pop.

Now to return to the use of our package. The package itself has a name and the entities in its visible part (the specification) can be thought of as components of the package in some sense. It is natural therefore that, in order to call Push, we must also mention Stack. In fact the dotted notation is used. So we could write

```
declare
   package Stack is           -- specification
      ...                     -- and
      ...                     -- body
   end Stack;
begin
   ...
   Stack.Push(M);
   ...
   N := Stack.Pop;
   ...
end;
```

Inside the package we would call Push as just Push, but we could still write Stack.Push just as in Chapter 10 we saw how we could refer to a local variable X of procedure P as P.X. Inside the package we can refer to S or Stack.S, but outside the package, Max, S, and Top are not accessible in any way.

Packages such as Stack are often referred to as abstract state machines. They contain internal state (in this case the hidden variables S and Top) and are abstract because access is through some protocol independent of how the state is implemented. The term encapsulation is also used to refer to the way in which a package encapsulates the items inside it.

It would in general be painful always to have to write Stack.Push to call Push from outside. Instead we can write **use** Stack; as a sort of declaration and we may then refer to Push and Pop directly. The use clause could follow the declaration of the specification of Stack in the same declarative part or could be in another declarative part where the package is visible. So we could write

```
declare
   use Stack;
begin
   ...
   Push(M);
   ...
   N := Pop;
   ...
end;
```

The use clause is like a declaration and similarly has a scope to the end of the block. Outside we would have to revert to the dotted notation. We could have an inner use clause referring to the same package – it would do no harm.

Two or more packages could be declared in the same declarative part. Generally, we could arrange all the specifications together and then all the bodies, or alternatively the corresponding specifications and bodies could be together. Thus we could have spec A, spec B, body A, body B, or spec A, body A, spec B, body B. The rules governing the order are simply

- linear elaboration of declarations,
- specification must precede body for same package (or subprogram).

The specification of a package may contain things other than subprograms. Indeed an important case is where it does not contain subprograms at all but merely a group of related variables, constants, and types. In such a case the package needs no body. It does not provide any hiding properties but merely gives commonality of naming. (Note that a body could be provided; its only purpose would be for initialization.)

However, a package declared at the library level is only allowed to have a body if it requires one for some reason such as providing the body for a subprogram declared in the specification. This avoids some awkward surprises which could otherwise occur as will be explained in Section 27.3.

As an example we could provide a package containing the type Day and some useful related constants.

```
package Diurnal is
   type Day is (Mon, Tue, Wed, Thu, Fri, Sat, Sun);
   subtype Weekday is Day range Mon .. Fri;
   Tomorrow: constant array (Day) of Day :=
                              (Tue, Wed, Thu, Fri, Sat, Sun, Mon);
   Next_Work_Day: constant array (Weekday) of Weekday :=
                                        (Tue, Wed, Thu, Fri, Mon);
end Diurnal;
```

Several examples of packages without bodies occur in the predefined library such as Ada.Numerics which we met in Chapter 4. We shall see other important examples when we deal with abstract tagged types and interfaces in Chapter 14.

A final point. A subprogram cannot be called successfully during the elaboration of a declarative part if its body appears later. This did not prevent the mutual recursion of the procedures F and G in Section 10.6 because in that case the call of F only occurred when we executed the sequence of statements of the body of G. But it can prevent the use of a function in an initial value. So

```
function A return Integer;
I: Integer := A;
```

is incorrect, and would result in Program_Error being raised.

A similar rule applies to subprograms in packages. If we call a subprogram from outside a package but before the package body has been elaborated, then Program_Error will be raised.

Exercise 12.1

1 The sequence defined by

$$X_{n+1} = (X_n \times 5^5) \bmod 2^{13}$$

provides a crude source of pseudorandom numbers. The initial value X_0 should be an odd integer in the range 0 to 2^{13}.

Write a package Random containing a procedure Init to initialize the sequence and a function Next to deliver the next value in the sequence.

2 Write a package Complex_Numbers which makes visible the type Complex, a constant I = $\sqrt{-1}$, and functions +, –, *, / acting on values of type Complex. See Exercise 10.2(**2**).

12.2 Private types

We have seen how packages enable us to hide internal objects from the user of a package. Private types enable us to create Abstract Data Types in which the details of the construction of a type are hidden from the user.

In Exercise 12.1(**2**) we wrote a package Complex_Numbers providing a type Complex, a constant I and some operations on the type. The specification of the package was

```
package Complex_Numbers is
   type Complex is
      record
         Re, Im: Float;
      end record;

   I: constant Complex := (0.0, 1.0);

   function "+" (X: Complex) return Complex;        -- unary +
   function "–" (X: Complex) return Complex;        -- unary –

   function "+" (X, Y: Complex) return Complex;
   function "–" (X, Y: Complex) return Complex;
   function "*" (X, Y: Complex) return Complex;
   function "/" (X, Y: Complex) return Complex;
end;
```

The trouble with this formulation is that the user can make use of the fact that the complex numbers are held in cartesian representation. Rather than always using the complex operator "+", the user could also write things like

```
C.Im := C.Im + 1.0;
```

rather than the more abstract

```
C := C + I;
```

In fact, with the above package, the user has to make use of the representation in order to construct values of the type.

We might wish to prevent use of knowledge of the representation so that we could change the representation to perhaps polar form at a later date and know that the user's program would still be correct. We can do this with a private type. Consider

```
package Complex_Numbers is
   type Complex is private;                        -- visible part
   I: constant Complex;                                 -- deferred constant
   function "+" (X: Complex) return Complex;
   function "-" (X: Complex) return Complex;
   function "+" (X, Y: Complex) return Complex;
   function "-" (X, Y: Complex) return Complex;
   function "*" (X, Y: Complex) return Complex;
   function "/" (X, Y: Complex) return Complex;
   function Cons(R, I: Float) return Complex;
   function Re_Part(X: Complex) return Float;
   function Im_Part(X: Complex) return Float;
private                                            -- private part
   type Complex is
      record
         Re, Im: Float;
      end record;
   I: constant Complex := (0.0, 1.0);                  -- constant value
end;
```

The part of the package specification before the reserved word **private** is the visible part and gives the information available externally to the package. The type Complex is declared to be private. This means that outside the package nothing is known of the details of the type. The only operations available are assignment, = and /= plus those added by the writer of the package as subprograms specified in the visible part.

We may also declare constants of a private type such as I in the visible part. The initial value cannot be given in the visible part because the details of the type are not yet known. Hence we just state that I is a constant; we call it a deferred constant.

After **private** comes the so-called private part in which we have to give the details of types declared as private and give the initial values of any deferred constants.

A private type can be implemented in any way consistent with the operations visible to the user. It can be a record as we have shown; equally it could be an array, an enumeration type and so on; it could even be declared in terms of another private type. In our case it is fairly obvious that the type Complex is naturally implemented as a record; but we could equally have used an array of two components such as

```
type Complex is array (1 .. 2) of Float;
```

After the full type declaration of a private type we have to give full declarations of any deferred constants of the type including their values.

It should be noted that as well as the functions +, –, * and / we have also provided Cons to create a complex number from its real and imaginary components and Re_Part and Im_Part to return the components. Some such functions are necessary because the user no longer has direct access to the internal structure of the type. Of course, the fact that Cons, Re_Part and Im_Part correspond to our thinking externally of the complex numbers in cartesian form does not prevent us from implementing them internally in some other form as we shall see in a moment.

The body of the package is as shown in the answer to Exercise 12.1(**2**) plus the additional functions which are trivial. It is therefore

```
package body Complex_Numbers is
   ...   -- unary + -

   function "+" (X, Y: Complex) return Complex is
   begin
      return (X.Re + Y.Re, X.Im + Y.Im);
   end "+";

   ...   -- and - * / similarly

   function Cons(R, I: Float) return Complex is
   begin
      return (R, I);
   end Cons;

   function Re_Part(X: Complex) return Float is
   begin
      return X.Re;
   end Re_Part;

   ...   -- and Im_Part similarly

end Complex_Numbers;
```

The package Complex_Numbers could be used in a fragment such as

```
declare
   use Complex_Numbers;
   C, D: Complex;
   F: Float;
begin
   C := Cons(1.5, -6.0);
   D := C + I;                  -- Complex +
   F := Re_Part(D) + 6.0;       -- Float +
   ...
end;
```

Outside the package we can declare variables and constants of type Complex in the usual way. Note the use of Cons to create the effect of a complex literal. We cannot, of course, do mixed operations between the type Complex and the type Float. Thus we cannot write

```
C := 2.0 * C;
```

but instead must write

```
C := Cons(2.0, 0.0) * C;
```

If this is felt to be tedious we could add further overloadings of the operators to allow mixed operations.

Let us suppose that for some reason we now decide to represent the complex numbers in polar form. The visible part of the package will be unchanged but the private part could now become

```
private
   π: constant := 3.14159_26536;
   type Complex is
      record
         R: Float;
         θ: Float range 0.0 .. 2.0*π;
      end record;
   I: constant Complex := (1.0, 0.5*π);
end;
```

Note how the named number π is for convenience declared in the private part; anything other than a body can be declared in a private part if it suits us – we are not restricted to just declaring the types and constants in full. Things declared in the private part are also available in the body. An alternative to declaring our own number π is to use the value in the package Ada.Numerics or at least to initialize our own number with that value; see Section 13.7.

The body of the package Complex_Numbers will now need completely rewriting. Some functions will become simpler and others will be more intricate. In particular it will be convenient to provide a function to normalize the angle θ so that it lies in the range 0 to 2π. The details are left for the reader.

However, since the visible part has not been changed the user's program will not need changing; we are assured of this since there is no way in which the user could have written anything depending on the details of the private type. Nevertheless, as we shall see in the next chapter, the user's program will need recompiling because of the general dependency rules. This may seem slightly contradictory but the compiler needs the information in the private part in order to be able to allocate storage for objects of the private type declared in the user's program. If we change the private part the size of the objects could change and then the object code of the user's program would change even though the source was the same and compiled separately.

We can therefore categorize the three distinct parts of a package as follows: the visible part which gives the logical interface to clients, the private part which gives the physical interface, and the body which gives the implementation details. When Ada was first designed some thought was given to the idea that these three parts might be written as distinct units but it was dismissed on the grounds that it would be very tedious.

It is important to appreciate that there are two quite different views of a private type such as Complex. Outside the package we know only that the type is private; inside the package and after the full type declaration we know all the properties implied by the declaration. Thus we see that we have two different views of the type according to where we are and hence which declaration we can see; these are known as the *partial view* and the *full view* respectively.

Between a private type declaration and the later full type declaration, the type is in a curiously half-defined state (technically it is not frozen which means that we do not yet know all about it). Because of this there are severe restrictions on its use, the main ones being that it cannot be used to declare variables or allocate objects. But it can be used to declare deferred constants, other types and subtypes and subprogram specifications (also entries of tasks and protected types).

Thus we could write

```
type Complex_Array is array (Integer range <>) of Complex;
```

and then

```
C: constant Complex_Array;
```

in the visible part. But until the full declaration is given we cannot declare variables of the type Complex or Complex_Array.

However, we can declare the specifications of subprograms with parameters of the types Complex and Complex_Array and can even supply default expressions. Such default expressions can use deferred constants and functions.

Deferred constants provide another example of a situation where information has to be repeated and thus be consistent. Both deferred and full declarations must have the same type; if they both supply constraints (directly or using a subtype) then they must statically match (as defined in Section 8.2). However, it is possible for the deferred declaration just to give the type and then for the full declaration to impose a constraint. Similarly if the deferred constant is marked as aliased or has a null exclusion then the full one must match. Deferred constants can also be of an anonymous access type.

So in the case of the array C above the full declaration might be

```
C: constant Complex_Array(1 .. 10) := ( ... );
```

or even

```
C: constant Complex_Array := ( ... );
```

where in the latter case the bounds are taken from the initial value.

The type Complex_Array illustrates the general distinction between the partial view and the full view that you can only use what you can see. As an example consider the operator "<". (Remember that "<" only applies to arrays if the component type is discrete, see Section 8.6.) Outside the package we cannot use "<" since we do not know whether or not the type Complex is discrete. Inside the package we find that it is not discrete and so still cannot use "<". If it had been discrete we could have used "<" after the full type declaration but of course we still could not use it outside. On the other hand, slicing is applicable to all one-dimensional arrays and so can be used both inside and outside the package.

Deferred constants are not necessarily of a private type, any constant in a package specification can be deferred to the private part. Indeed the constant C is not of a private type since Complex_Array is not itself private.

Another use for a deferred constant is to provide read-only access to a variable declared in the private part. Consider

```
    type Int_Ptr is access constant Integer;
    The_Ptr: constant Int_Ptr;                -- deferred constant
private
    The_Variable: aliased Integer;
    The_Ptr: constant Int_Ptr := The_Variable'Access;
```

The external user can read the value of The_Variable via The_Ptr but cannot change it. However, The_Variable can of course be updated by subprograms declared in the package or by any initialization sequence of the package body. This technique might be used to provide access to a 'constant' table of data which has to be computed dynamically. This is another example of having two views of something, in this case a constant view and a variable view.

Note that the above could be written using anonymous access types thus

```
    The_Ptr: constant access constant Integer;      -- deferred constant
private
    The_Variable: aliased Integer;
    The_Ptr: constant access constant Integer := The_Variable'Access;
```

which provides a neat illustration of the use of **constant access constant**.

A deferred constant can also be used to import a constant from an external system; see Section 25.5.

Note also that we could provide a parameterless function

```
function I return Complex;
```

instead of a deferred constant. One advantage is that we can change the returned value without changing the package specification and so having to recompile the user's program. Of course in the case of I we are unlikely to need to change the value anyway! Another advantage is that functions can be overloaded so that we could have several constants (of different types) with the same name such as Unit or Empty. Yet another advantage is mentioned in the next section when we consider derived types. In general, a function (with aspect Inline; see Section 25.1) is nearly always better than a deferred constant.

We conclude this section by summarizing the overall structure of a package regarding the declaration of subprograms. A package is generally in three parts: the visible part of the specification, the private part of the specification, and the body.

A procedure comes usually in two parts: its specification and its body. The specification says what it does whereas the body says how it does it. Sometimes it is not necessary to give a distinct specification since the first part of the body repeats it anyway. If there is a distinct specification then we say that the body is the completion of the procedure. Possible arrangements are

- Procedure spec in visible part of package; body in package body. Procedure can be called from outside the package and from within body.
- Procedure spec in private part of package; body in package body. Procedure can be called from places that can see the private part and from within the body.

- Procedure spec in package body; body in package body. Procedure can be called from within body.
- There is no procedure spec; body in package body. Procedure can be called from within body.

The above also applies to functions. However, the situation is more flexible because of the introduction of expression functions. An expression function can be used just as a shorthand for a body. But an expression function can also be used as a complete function without a specification or to complete a distinct specification where both specification and completion are in the package specification. An important situation is where the specification is in the visible part and is completed by an expression function in the private part and so has access to the entities in the private part; see function Check_In in Section 16.3. In recursive situations we might declare both in the visible part; some examples are in the *Ada 2012 Rationale*.

Exercise 12.2

1 Write further functions "*" to allow mixed multiplication of real and complex numbers. Consider functions declared both inside and outside the package.

2 Rewrite the fragment of user program for complex numbers omitting the use clause.

3 Complete the package Rational_Numbers whose visible part is

```
package Rational_Numbers is

   type Rational is private;
   function "+" (X: Rational) return Rational;            -- unary +
   function "-" (X: Rational) return Rational;            -- unary -
   function "+" (X, Y: Rational) return Rational;
   function "-" (X, Y: Rational) return Rational;
   function "*" (X, Y: Rational) return Rational;
   function "/" (X, Y: Rational) return Rational;          -- binary division

   function "/" (X: Integer; Y: Positive) return Rational;    -- constructor
   function Numerator(R: Rational) return Integer;
   function Denominator(R: Rational) return Positive;

private
   ...
end;
```

A rational number is a number of the form N/D where N is an integer and D is a positive integer. For predefined equality to work it is essential that rational numbers are always reduced by cancelling out common factors. This may be done using the function GCD of Exercise 10.1(7). Ensure that an object of type Rational has an appropriate default value of zero.

4 Why does

```
function "/" (X: Integer; Y: Positive) return Rational;
```

not hide the predefined integer division?

12.3 Primitive operations and derived types

This seems a good moment to discuss the question of which operations really belong to a type. In Section 6.3 we noted that 'A type is characterized by a set of values and a set of primitive operations ...'. The set of values is pretty obvious but the set of primitive operations is not intuitively obvious nor is it obvious why the definition of such a set should be important.

The key to why we need to define the set of primitive operations of a type lies in the concept of a class which we mentioned in Chapter 3. Types can be grouped into various categories with common properties. These categories play two roles in Ada, as the basis for type derivation and extension and also as the basis for formal parameter types of generics. We will introduce the ideas by first considering the simplest form of derived types; the more flexible tagged types which can be extended will be discussed in Chapter 14 and generic parameters will be discussed in Chapter 19.

Sometimes it is useful to introduce a new type which is similar in most respects to an existing type but is nevertheless a distinct type. If T is a type we can write

```
type S is new T;
```

and then S is said to be a derived type and T is the parent type of S. Remember that we are only discussing untagged types in this section.

A type plus all the types derived from it form a category of types called a derivation class or simply a class. Classes are very important and the types in a class share many properties. For example if T is a record type then S will be a record type whose components have the same names and so on.

The set of values of a derived type is a copy of the set of values of the parent. An important instance of this is that if the parent type is an access type then the derived type is also an access type and they share the same storage pool. Note that we say that the set of values is a copy; this reflects that they are truly different types and values of one type cannot be assigned to objects of the other type; however, as we shall see in a moment, conversion between the two types is possible. The notation for literals and aggregates (if any) is the same and any default initial expressions for the type or its components are the same.

The primitive operations of a type are

- predefined operations such as assignment, predefined equality, appropriate attributes and so on, and
- for a derived type, primitive operations inherited from its parent (these can be overridden), and
- for a type (immediately) declared in a package specification, subprograms with a parameter or result of the type also declared in the package specification – including any private part.

The predefined types follow the same rules, for example the primitive operations of Integer include the operators such as "+", "–" and "<" which are notionally declared in Standard along with Integer.

Enumeration literals are also primitive operations of a type since they are considered to be parameterless functions returning a value of the type. So in the case of the type Boolean, False and True are primitive operations. It is rather as if there were functions such as

```
function True return Boolean is
begin
   return Boolean'Val(1);
end;
```

So the general idea is that a type has certain predefined primitive operations, it inherits some from its parent (if any) and more can be added.

Note especially that an operation inherited from the parent type can be overridden by explicitly declaring a new subprogram with the same identifier in the same declarative region as the derived type declaration itself (a declarative region is a list of declarations such as in a block or subprogram or a package specification and body taken together). The new subprogram must have type conformance with the overridden inherited operation.

A subtle point is that inherited primitive operations are also intrinsic so that the Access attribute cannot be applied to them whereas it can be applied to explicitly declared operations. This rule does not apply to tagged types (see Section 14.1).

There are a couple of minor rules which arise from consideration of the freezing of the type (see Section 25.1). We cannot derive from a private type until after its full type declaration. Also if we derive from a type in the same package specification as that in which it is declared then the derived type will inherit all the primitive operations of the parent declared so far. Further operations might then be declared for the parent but these will not be inherited by the derived type.

Although derived types are distinct, nevertheless because of their close relationship, a value of one type can be directly converted to another type if they have a common ancestor. Consider

```
type Light is new Colour;
type Signal is new Colour;
type Flare is new Signal;
```

These types form a hierarchy rooted at Colour. We can convert from any one type to another and do not have to give the individual steps. So we can write

```
L: Light;
F: Flare;
...
F := Flare(L);
```

and we do not need to laboriously write

```
F := Flare(Signal(Colour(L)));
```

The introduction of derived types extends the possibility of conversion between array types discussed in Section 8.2. In fact a value of one array type can be converted to another array type if the component subtypes statically match and the index types are the same or convertible to each other.

The reader might wonder at the benefit of derived types and why we might wish to have one type very like another.

One use for derived types is when we want to use an existing type, but wish to avoid the accidental mixing of objects of conceptually different types. Suppose we wish to count apples and oranges. Then we could declare

```
type Apples is new Integer;
type Oranges is new Integer;

No_Of_Apples: Apples;
No_Of_Oranges: Oranges;
```

Since Apples and Oranges are derived from the type Integer they both have the operation "+". So we can write statements such as

```
No_Of_Apples := No_Of_Apples + 1;
No_Of_Oranges := No_Of_Oranges + 1;
```

but we cannot inadvertently write

```
No_Of_Apples := No_Of_Oranges;
```

If we did want to convert the oranges to apples we would have to write

```
No_Of_Apples := Apples(No_Of_Oranges);
```

The numeric types will be considered in Chapter 17 in more detail but it is worth mentioning here that strictly speaking a type such as Integer has no literals. Literals such as 1 and integer named numbers are of a type known as *universal_integer* and implicit conversion to any integer type occurs if the context so demands. Thus we can use 1 with Apples and Oranges because of this implicit conversion and not because the literal is inherited.

Now suppose that we have overloaded procedures to sell apples and oranges

```
procedure Sell(N: Apples);
procedure Sell(N: Oranges);
```

Then we can write

```
Sell(No_Of_Apples);
```

but (much as the problem we had with cows and garments in Section 10.6), writing Sell(6); is ambiguous because we do not know which fruit we are selling. We can resolve the ambiguity by qualification thus

```
Sell(Apples'(6));
```

When a subprogram is inherited a new subprogram is not actually created. A call of the inherited subprogram is really a call of the parent subprogram; **in** and **in out** parameters are implicitly converted just before the call; **in out** and **out** parameters or a function result are implicitly converted just after the call. So

```
My_Apples + Your_Apples
```

is effectively

```
Apples(Integer(My_Apples) + Integer(Your_Apples))
```

Another example of the use of derived types to avoid errors will be found in the answer to Exercise 23.4(**2**) where they are used in order to distinguish the hands of Jack and Jill when they are playing paper, stone, and scissors.

An important use of derived types is with private types when we wish to express the fact that a private type is to be implemented as an existing type such as Integer or Cell_Ptr. We shall see an example of this in Section 12.5.

Looking back at the type Complex of the previous section we can see that the primitive operations are assignment, predefined equality and inequality and the subprograms "+", ..., Im_Part declared in the package specification.

In Exercise 12.2(**2**) we noted that it was tedious to use the operators of the type Complex without a use clause since, given objects A, B, C of type Complex, we would have to write

```
C := Complex_Numbers."+"(A, B);
```

which is painful to say the least. However, there is a strong school of thought that use clauses are bad for you since they obscure the origin of entities. In order to alleviate this dilemma, Ada 95 introduced the so-called use type clause. This can be placed in a declarative part just like a use package clause and has similar scope. It makes just the primitive operators of a type directly visible. Thus we might write

```
declare
   use type Complex_Numbers.Complex;
   A, B, C: Complex_Numbers.Complex;
   ...
begin
   ...
   C := A + B;
```

Note that the full dotted notation is still required for referring to other entities in the package such as the function Cons and the type Complex itself.

A further form of use clause was introduced in Ada 2012. This is the use all type clause written thus

```
use all type Complex_Numbers.Complex;
```

This makes directly visible all the operations of a type and not just those denoted by operators. It therefore includes Cons, Re_Part and Im_Part but not the type Complex itself since that is not a primitive operation.

Another example might be instructive. Suppose we have a package Palette containing a type Colour and a function for mixing colours thus

```
package Palette is
   type Colour is (Red, Orange, Yellow, Green, Blue, ... );
   function Mix(This, That: Colour) return Colour;
end Palette;
```

In another package without any use clauses, if we want to mix Red and Green to make a mucky colour we have to write the laborious

```
Mucky: Palette.Colour := Palette.Mix(Palette.Red, Palette.Green);
```

Adding a **use type** clause makes no difference but if we provide a **use all type** clause then we can write

```
Mucky: Palette.Colour := Mix(Red, Green);
```

Remember that enumeration literals behave as functions and so are primitive operations. But the type name itself still has to refer to the package name unless we provide a use package clause.

If we wished to introduce a distinct type derived from Complex for an electromagnetic application (much as we had distinct types to count apples and oranges) then we could perhaps write

```
type Field is new Complex;
```

and we could then ensure that values of the field would not inadvertently get mixed up with other complex numbers. However, it is important to note that although the type Field inherits all the operations of Complex, it does not inherit the constant I. To maintain the inheritance abstraction it is necessary to make I into a function.

A few words on constraints and null exclusions: we can derive from a subtype using the more general subtype indication. So we could have

```
type Chance is new Float range 0.0 .. 1.0;
```

in which case the underlying derived type is derived from the underlying base type. It is as if we had written

```
type anon is new Float;
subtype Chance is anon range 0.0 .. 1.0;
```

and so Chance denotes a constrained subtype of the (anonymous) derived type. The set of values of the new derived type is actually (a copy of) the full set of values of the type Float. The operations of the type "+", ">" and so on also work on the full unconstrained set of values. So given

```
C: Chance;
```

we can legally write

```
C > 2.0
```

even though 2.0 could never be successfully assigned to C. The Boolean expression is, of course, always false (unless C was uninitialized and by chance had a silly value).

As a further example of constraints it is instructive to consider the specification of an inherited subprogram in more detail. It is obtained from that of the parent by simply replacing all instances of the parent base type in the original specification by the new derived type. Subtypes are replaced by equivalent subtypes with corresponding constraints and default initial expressions are converted by adding a type conversion. Any parameters or result of a different type are left unchanged. As an abstract example consider

```
type T is ... ;
subtype S is T range L .. R;
```

```
function F(X: T; Y: T := E; Z: Q) return S;
```

where E is an expression of type T and the type Q is quite unrelated. If we write

```
type TT is new T;
```

then it is as if we had also written

```
subtype SS is TT range TT(L) .. TT(R);
```

and the specification of the inherited function F will then be

```
function F(X: TT; Y: TT := TT(E); Z: Q) return SS;
```

in which we have replaced T by TT, S by SS, added the conversion to the expression E but left the unrelated type Q unchanged. Note that the parameter names are naturally the same.

Derived types are in some ways an alternative to private types. Derived types have the advantage of inheriting literals but they often have the disadvantage of inheriting too much. Thus, we could derive types Length and Area from Float.

```
type Length is new Float;
type Area is new Float;
```

We would then be prevented from mixing lengths and areas but we would have inherited the ability to multiply two lengths to give a length and to multiply two areas to give an area as well as hosts of irrelevant operations such as exponentiation. Such unnecessary inherited operations can be eliminated by overriding them with abstract subprograms such as

```
function "*" (Left, Right: Length) return Length is abstract;
```

An abstract subprogram has no body and can never be called; any attempt to do so is detected during compilation. Abstract subprograms are very important with tagged types and the rules are rather different as we shall see in Chapter 14. An important difference is that the related concepts of abstract types and interfaces do not apply to untagged types.

Two more topics mainly used with tagged types but also applicable to untagged types deserve mention. One is the concept of a null procedure. We can write

```
procedure Option(X: in T) is null;
```

Such a procedure declaration cannot be given a body but behaves as if it had a body consisting of

```
procedure Option(X: in T) is
begin
   null;
end Option;
```

If declared in the same specification as X then it will be inherited by a type derived from X and it could then be overridden by a procedure with a conventional body.

There is of course no such thing as a null function because a function must return a result.

Another feature is that a subprogram can be given a so-called overriding indicator. Suppose that the type Colour has a primitive procedure Option thus

```
procedure Option(C: in Colour);
```

then when we declare the type Signal derived from Colour we can replace the inherited procedure with a new version. We can write

```
overriding
procedure Option(S: in Signal);
```

The overriding indicator then confirms that we did indeed mean to override the inherited operation. This is a safeguard against a number of errors. Thus if we inadvertently wrote Operation rather than Option so that the procedure did not override then the program would fail to compile. On the other hand, if we do introduce a new operation then we can write

```
not overriding
procedure Operation(S: in Signal);
```

Overriding indicators are always optional (see Section 14.6). They can also be used on a subprogram body and in other constructions such as renaming (Section 13.7), generic instantiations (Section 19.1) and with entries (Section 22.3).

We finish this section by mentioning a curious anomaly concerning the type Boolean which really does not matter so far as the normal user is concerned. If we derive a type from Boolean then the predefined relational operators =, < and so on continue to deliver a result of the predefined type Boolean whereas the logical operators **and**, **or**, **xor**, and **not** are inherited normally and deliver a result of the derived type. We cannot go into the reason here other than to say that it relates to the fact that the relational operators for all types return a result of type Boolean (see also Section 23.1). However, it does mean that the theorem of Exercise 6.7(**3**) that **xor** and /= are equivalent only applies to the type Boolean and not to a type derived from it.

Exercise 12.3

1 Declare a package Metrics containing types Length and Area with appropriate redeclarations of the various operations "*", "/" and "**".

12.4 Equality

In the previous section we saw that predefined equality was an operation of most types; however, it is sometimes inappropriate and needs to be overridden. Remember from Section 10.2 that if we do redefine "=" and it returns the type Boolean then a corresponding function "/=" is automatically implied.

As an example, recall that in Section 12.1 we introduced the package Stack as an abstract state machine which declared a single stack. It is perhaps more useful to define a package providing an abstract data type so that we can declare several stacks. Consider

```
package Stacks is
   type Stack is private;
   procedure Push(S: in out Stack; X: in Integer);
   procedure Pop(S: in out Stack; X: out Integer);
   function "=" (S, T: Stack) return Boolean;
private
   Max: constant := 100;
   type Integer_Vector is array (Integer range <>) of Integer;

   type Stack is
      record
         S: Integer_Vector(1 .. Max);
         Top: Integer range 0 .. Max := 0;
      end record;

end;
```

Each object of type Stack is a record containing an array S and integer Top. Note that Top has a default initial value of zero. This ensures that when we declare a stack object, it is correctly initialized to be empty. Note also the introduction of the type Integer_Vector because a record component may not be of an anonymous array type.

The body of the package could be

```
package body Stacks is

   procedure Push(S: in out Stack; X: in Integer) is
   begin
      S.Top := S.Top + 1;
      S.S(S.Top) := X;
   end Push;

   procedure Pop(S: in out Stack; X: out Integer) is
   begin
      X := S.S(S.Top);
      S.Top := S.Top - 1;
   end Pop;

   function "=" (S, T: Stack) return Boolean is
   begin
      if S.Top /= T.Top then
         return False;
      end if;
      for I in 1 .. S.Top loop
         if S.S(I) /= T.S(I) then
            return False;
         end if;
      end loop;
      return True;
   end "=";

end Stacks;
```

This example illustrates many points. The parameter S of Push has mode **in out** because we need both to read from and to write to the stack. We have made Pop a procedure for uniformity with Push although in Ada 2012 and Ada 2022 it could be a function declared thus

```
function Pop(S: in out Stack) return Integer is ...
```

which might be more convenient. So in Ada 2005 we have to write

```
Pop(A_Stack, Y);          -- using a procedure Pop
```

whereas in Ada 2012 and Ada 2022 we can write

```
Y := Pop(A_Stack);        -- using a function Pop
```

Remember that the ability for a function (in 2012 and 2022) to have parameters of any mode does not extend to operators. However, "=" can be a function because we only need to read the values of the two stacks and not to update them.

The function "=" has the interpretation that two stacks are equal only if they have the same number of items and the corresponding items have the same value. It would obviously be quite wrong to compare the whole records because the unused components of the arrays would also be compared. Incidentally, the function body could be written more simply using slices – see Exercise 12.4(**3**).

The type Stack is a typical example of a data structure where the value of the whole is more than just the sum of the parts; the interpretation of the array S depends on the value of Top. Cases where there is such a relationship usually need redefinition of equality.

A minor point is that we are using the identifier S in two ways: as the name of the formal parameter denoting the stack and as the array inside the record. There is no conflict because, although the scopes overlap, the regions of visibility do not as is explained in Section 13.6. Of course, it is somewhat confusing for the reader and not good practice but it illustrates the freedom of choice of record component names.

The package could be used in a fragment such as

```
declare
   use Stacks;
   My_Stack: Stack;
   Your_Stack: Stack;
   ...
begin
   Push(My_Stack, N);
   ...
   Push(Your_Stack, M);
   ...
   if My_Stack = Your_Stack then
      ...
   end if;
   ...
end;
```

It is interesting to consider how we might check that a stack is empty. We could compare it against a constant Empty declared in the visible part of the package. But a much better technique would be to provide a function Is_Empty and possibly a corresponding function Is_Full.

Having declared our own equality for the type Stack we might decide to declare a type such as

```
type Stack_Array is array (Integer range <>) of Stack;
```

This array type has predefined equals (see Section 8.2) but it is important to note that it works in terms of the original predefined equals for the type Stack and not our redefined version. We thus probably need to redefine equality for such an array type (this will also apply to slices of that type). We might even choose to define equality to return a Boolean array with each element indicating the equality of the matching components. Remember that if we do define equality to return a type other than just Boolean then a new version of "/=" will not automatically be provided and so we might need to redefine that as well.

The principle that objects of a composite type T are equal if and only if the components are equal is often called composability of equality. We have seen that equality does not compose for arrays such as Stack_Array where equality has been redefined for the component type. But equality does compose for record types. So if a component of a record type has had equality redefined then this redefinition will apply to the record type also. (In Ada 2005, composability of record types applied only if tagged.) It is important in some applications to know that types do compose and in fact all the private types declared in the predefined library are composable.

Situations where equality needs redefinition usually only arise with composite types where the meaning of the whole is more than just the sum of the parts. Nevertheless, we can redefine equality for elementary types. If we do then the meaning of certain intrinsic structures effectively involving predefined equality is not changed. Thus the case statement always uses predefined equality in choosing the sequence to be obeyed.

Another interesting example is provided by a type such as Rational of Exercise 12.2(**3**); it would be quite reasonable to allow manipulation – including assignment – of values which were not reduced provided that equality is suitably redefined. If we wish to use predefined equality then we must reduce all values to a canonical form in which component by component equality is satisfactory. In the case of the type Stack implemented as an array, a suitable form is one in which all unused elements of the array had a standard dummy value such as zero.

Exercise 12.4

1 Rewrite the specification of Stacks to include a constant Empty in the visible part.

2 Write functions Is_Empty and Is_Full for Stacks.

3 Rewrite the **function** "=" (S, T: Stack) **return** Boolean; using slices.

4 Write a suitable body for

```
function "=" (A, B: Stack_Array) return Boolean;
```

Make it conform to the normal rules for array equality which we mentioned towards the end of Section 8.2. Also write appropriate equality functions returning an array of Boolean values.

5 Rewrite Stacks so that predefined equality is satisfactory.

6 Redefine "=" for the type Rational of Exercise 11.2(3) assuming that the package Rational_Numbers does not reduce values to a canonical form using GCD. Would this be sensible?

12.5 Limited types

The primitive operations of a record type or private type can be completely restricted to just those declared in the package specification along with the type itself. We write one of

```
type T is limited private;

type R is limited record ... end record;
```

In such a case assignment (copying) and predefined = and /= are not defined for the type. Of course we can (re)define equality for any type anyway and so the important issue in deciding whether a type should be limited is whether we wish to prevent copying. Preventing copying is often important when a type is implemented in terms of access types or has components of access types.

A limited private type may be implemented as a nonlimited type in which case assignment (and predefined equality) will be available in the private part and body of the package. We thus see that limitedness is a property of a view of a type.

The advantage of making a private type limited is that the package writer has complete control over the objects of the type – the copying of resources can be monitored and so on. It also gives more freedom of implementation for the full type as we shall see in a moment.

We now consider an alternative formulation of the type Stack and suppose that we wish to impose no maximum stack size other than that imposed by the overall size of the computer. This can be done by representing the stack as an access type referring to a list as follows

```
package Stacks is
   type Stack is limited private;
   procedure Push(S: in out Stack; X: in Integer);
   procedure Pop(S: in out Stack; X: out Integer);
   function "=" (S, T: Stack) return Boolean;
private
   type Cell is
      record
         Next: access Cell;                -- anonymous type
         Value: Integer;
      end record;
   type Stack is access all Cell;          -- named type
end;
```

```
package body Stacks is
   procedure Push(S: in out Stack; X: in Integer) is
   begin
      S := new Cell'(S, X);
   end;

   procedure Pop(S: in out Stack; X: out Integer) is
   begin
      X := S.Value;
      S := Stack(S.Next);                    -- conversion
   end;

   function "=" (S, T: Stack) return Boolean is ...

end Stacks;
```

We have avoided an incomplete type declaration by making the component Next of an anonymous type. This does however mean that a type conversion is required in Pop when assigning the component Next to S. In Ada 2005 this conversion must be named as shown but in Ada 2012 and Ada 2022 the name can be omitted since it cannot fail; see Section 11.7.

When the user declares a stack

```
S: Stack;
```

it automatically takes the default initial value **null** which denotes that the stack is empty. If we call Pop when the stack is empty then this will result in attempting to evaluate

```
null.Value
```

and this will raise Constraint_Error. The only way in which Push can fail is by running out of storage; an attempt to evaluate

```
new Cell'(S, X)
```

could raise Storage_Error. (See Section 25.4 for a discussion on how Pop could relinquish the storage that is no longer accessible.)

This formulation of stacks is one in which we have made the type limited private. Assignment would, of course, copy only the pointer to the stack rather than the stack itself and would have resulted in a complete mess.

Observe that assignment and predefined equality are available for the type Stack in the private part and body of the package. But we have to provide an appropriate definition of equality for use outside. This needs some care to avoid an infinite recursion as discussed in Section 11.7. But the following works

```
function "=" (S, T: Stack) return Boolean is
   SS: access Cell := S;
   TT: access Cell := T;
begin
   while SS /= null and TT /= null loop
      if SS.Value /= TT.Value then
         return False;
```

```
      end if;
      SS := SS.Next;
      TT := TT.Next;
    end loop;
    return SS = TT;          -- True if both null
  end "=";
```

The key point is that we have distinguished between the type Stack itself and the type of SS so that the comparison of SS against null does not recursively invoke the function being declared.

We could of course have used an incomplete declaration in defining the type Cell. This would have avoided the type conversion in Pop. We could avoid any problems in the function "=" by introducing a derived type so that we can distinguish between the type Stack and its representation. We would then have

```
type Cell;
type Cell_Ptr is access Cell;

type Cell is
  record
    Next: Cell_Ptr;
    Value: Integer;
  end record;

type Stack is new Cell_Ptr;
```

The type Stack now has all the operations of the type Cell_Ptr but we can replace the inherited equality by our own definition without difficulty.

In Section 11.2 we stated that if we had to write an incomplete declaration first because of circularity (as in this type Cell) then there was an exception to the rule that the complete declaration had to occur in the same list of declarations. The exception is that a private part and the corresponding package body are treated as a single list of declarations as far as this rule is concerned.

So, the complete declaration of the type Cell could be moved from the private part to the body of the package Stacks. This might be an advantage since (as we shall see in Section 13.1), it then follows from the dependency rules that a user program would not need recompiling just because the details of the type Cell are changed. In implementation terms this is possible because it is assumed that values of all access types occupy the same space – typically a single word. However, if the complete declaration is in the body then the type cannot be used as an access parameter of a primitive operation or as a parameter or result of an access to subprogram type.

We can write subprograms with limited private types as parameters outside the defining package. As a simple example, the following procedure enables us to determine the top value on the stack without permanently removing it.

```
procedure Top_Of(S: in out Stack; X: out Integer) is
begin
  Pop(S, X);
  Push(S, X);
end Top_Of;
```

It is worth emphasizing that a private type such as Stack presents two views known as the partial view and the full view. Remember that within the private part (after the full type declaration) and within the body, any private type (limited or not) is treated in terms of how it is represented using the full view. On the other hand, outside the defining package and in the visible part we can only see the partial view as defined by the private declaration and the visible operations. The important principle that the properties of a type available at a given point depend entirely upon the type declaration visible from that point and not what is otherwise visible from that point is discussed further in Section 14.6 when we consider private extensions.

The parameter mechanism for a private type is that corresponding to how the type is represented. This applies both inside and outside the package. Of course, outside the package, we know nothing of how the type is represented and therefore should make no assumption about the mechanism used.

In the case of the limited type Stack, it is actually implemented as an access type and these are always passed by copy. However if the type (or in the case of a private type, the full type) is explicitly limited then parameters are always passed by reference.

Note also that the primitive operations of the type are those declared in the package specification including the private part. Thus some may be private and only known to the full view. We will return to the topic of primitive operation and views of a type when we consider tagged types in Chapter 14. Some rules for tagged types and derivation are somewhat different (such as conversion) and it is important to remember that in this chapter we have only been discussing the properties of untagged types.

The key property of limited types is that copying is not permitted. This prevents assignment statements but does not prevent the giving of an initial value providing that this does not involve copying. There are two ways in which such an initial value can be created, one is by using an aggregate and the other is by a constructor function.

Suppose for example that we wish to have a constructor function that makes up a stack with one integer item already in place. Consider

```
function Make_One(X: Integer) return Stack is
begin
   return new Cell'(null, X);
end Make_One;
```

This function (which has to be declared in the package Stacks because it uses the full view of the type) could then be used in a declaration outside the package thus

```
S: Stack := Make_One(7);                    -- initialization legal
```

with the result that the initial value of S is a stack containing the single item 7. However we could not write

```
S := Make_One(8);                           -- assignment illegal
```

outside the package because this is an assignment statement and copying is not allowed.

The mechanisms involved are easier to understand if we use a somewhat artificial type which is explicitly limited and not private. Thus consider the type

```
type T is limited
  record
    A: Integer;
    B: Boolean;
  end record;
```

The components of the record are not limited but the record as a whole is limited. We can initialize an object of the type T using an aggregate

```
V: T := (A => 10, B => True);
```

We should think of the individual components of the variable V as being initialized individually *in situ* – an actual aggregated value is not created and then assigned. Limited aggregates enable a constant of a limited type to be declared. They can also be used in a number of other contexts as well, such as

- a component of an enclosing record or array aggregate,
- the default expression for a record component,
- the expression in an initialized allocator,
- an actual parameter or default parameter expression of mode **in**,
- the result in a return statement,

as well as for generic parameters of limited types (see Chapter 19).

Limited constructor functions can be used in exactly the same places as limited aggregates. Moreover, if the type is private as well as limited then aggregates cannot be used but constructor functions can. A trivial function would be

```
function Init_T(X: Integer; Y: Boolean) return T is
begin
  return (A => X, B => Y);
end Init_T;
```

and we could then write

```
V: T := Init_T(10, True);
```

which has exactly the same effect as the declaration of V directly using an aggregate.

Moreover, the function builds the aggregate in the return statement directly in the variable V so that no copying occurs. So the address of V has to be passed as a hidden parameter to the function but this is not visible to the user. Also, the only allowed forms of expression in the return statement are aggregates and calls of other functions.

The function is more flexible than the aggregate since it can do arbitrary calculations with its parameters before constructing the return aggregate. Even more flexibility can be obtained by using an extended return statement as mentioned in Section 10.1. We can write

```
function Init_T( ... ) return T is
begin
   return X: T := ... do
      ...                         -- do arbitrary computation on X
   end return;
end Init_T;
```

This is particularly useful when the type is indefinite such as a discriminated record containing an array and we do not know the bounds of the array statically so that we cannot write an aggregate. Other applications occur with tasks and protected records; see Section 22.2.

We have seen that there are occasions when it is an advantage to make a type limited since this prevents the user from making copies. This is typically a sensible thing to do when the implementation might involve access types or the object represents a resource as illustrated in the next section. Some types are inherently limited as we shall see when we discuss task and protected types in Section 20.7.

A composite type containing a limited component is itself limited. A simple example of a limited composite type is given by an array of a limited type and we might reconsider the type Stack_Array of the previous section. In the case where the component is limited the array is also limited and so does not have predefined equality; but it could be defined and indeed the definition given as the answer to Exercise 12.4(**4**) would be satisfactory.

Limited types are very important in object oriented programming since there are many situations where an object represents a real-world entity and where making a copy would be quite inappropriate. As mentioned above explicitly limited types are always passed by reference. There are therefore good reasons for always making a type explicitly limited if assignment is not appropriate. A good example of a limited type is the type Person in Section 18.5.

Exercise 12.5

1 Write functions Is_Empty and Is_Full for the type Stack using the formulation where Stack is derived from Cell_Ptr.

2 Write the function "=" for the type Stack also using the formulation where Stack is derived from Cell_Ptr.

3 Complete the package whose visible part is

```
package Queues is
   Empty: exception;
   type Queue is limited private;
   procedure Join(Q: in out Queue; X: in Item);
   procedure Remove(Q: in out Queue; X: out Item);
   function Length(Q: Queue) return Integer;
private
```

Items join a queue at one end and are removed from the other so that a normal first-come–first-served protocol is enforced. Trying to remove an item from an empty queue raises the exception Empty. Implement the queue as a singly linked list; maintain pointers to both ends of the list so that scanning of the list is avoided. The function Length returns the number of items in the queue.

12.6 Resource management

An important example of the use of a limited private type is in providing controlled resource management. Consider the simple human model where each resource has a corresponding unique key. This key is then issued to the user when the resource is allocated and then has to be shown whenever the resource is accessed. So long as there is only one key and copying and stealing are prevented we know that the system is foolproof. A mechanism for handing in keys and reissuing them is usually necessary if resources are not to be permanently locked up. Typical human examples are the use of metal keys with safe deposit boxes, credit cards and so on.

Now consider the following

```
package Key_Manager is
   type Key is limited private;
   procedure Get_Key(K: in out Key);
   procedure Return_Key(K: in out Key);
   function Valid(K: Key) return Boolean;
   procedure Action(K: in Key; ... );
   ...
private
   Max: constant := 100;                        -- number of keys
   type Key_Code is new Integer range 0 .. Max;
   subtype Key_Range is Key_Code range 1 .. Key_Code'Last;
   type Key is limited                          -- explicitly limited
      record
         Code: Key_Code := 0;
      end record;
end;

package body Key_Manager is
   Free: array (Key_Range) of Boolean := (others => True);

   function Valid(K: Key) return Boolean is
   begin
      return K.Code /= 0;
   end Valid;

   procedure Get_Key(K: in out Key) is
   begin
      if K.Code = 0 then
         for I in Free'Range loop
            if Free(I) then
               Free(I) := False;
               K.Code := I;
               return;
            end if;
         end loop;              -- all keys in use if end of loop reached
      end if;
   end Get_Key;
```

```
        procedure Return_Key(K: in out Key) is
        begin
           if K.Code /= 0 then
              Free(K.Code) := True;
              K.Code := 0;
           end if;
        end Return_Key;

        procedure Action(K: in Key; ... ) is
        begin
           if Valid(K) then
              ...
           end if;
        end Action;

     end Key_Manager;
```

The type Key is represented by an explicitly limited record with a single component Code of the type Key_Code. This type is derived from Integer in order to prevent confusion between codes and any other integers in the program. The component Code has a default value of 0 which represents an unused key. Values in the range 1 .. Max represent the allocation of the corresponding resource.

When we declare a variable of type Key it automatically takes an internal code value of zero. In order to use the key we must first call the procedure Get_Key; this allocates the first free key number to the variable. The key may then be used with various procedures such as Action which represents a typical request for some access to the resource guarded by the key.

Finally, the key may be relinquished by calling Return_Key. So a typical fragment of user program might be

```
     declare
        use Key_Manager;
        My_Key: Key;
     begin
        ...
        Get_Key(My_Key);
        ...
        Action(My_Key, ... );
        ...
        Return_Key(My_Key);
        ...
     end;
```

A variable of type Key can be thought of as a container for a key. When initially declared the default value indicates that the container is empty. The type Key is a record partly because only record components could take default initial values in early versions of Ada when this example was first written. Note how various possible misuses of keys are overcome.

- The user is unable to make a copy of a key because the type Key is limited and assignment is thus not available. Moreover, it is explicitly limited and so always passed by reference and never by copy.

- If the user calls Get_Key with a variable already containing a valid key then no new key is allocated. It is important not to overwrite an old valid key otherwise that key would be lost.
- A call of Return_Key resets the variable to the default state so that the variable cannot be used as a key until a new one is issued by a call of Get_Key. Moreover, the user cannot make a copy and retain it because the type is limited.

The function Valid is provided so that the user can see whether a key variable contains the default value or an allocated value. It is obviously useful to call Valid after Get_Key to ensure that the key manager was able to provide a new key value; note that once all keys are issued, a call of Get_Key does nothing.

One apparent flaw is that there is no compulsion to call Return_Key before the scope containing the declaration of My_Key is left. The key would then be lost for ever and ever. This corresponds to the real life situation of losing a key (although in our model it can never be found again – it is as if it had been thrown into a black hole). We can overcome this using a controlled type as described in Section 14.7.

Exercise 12.6

1 Complete the package whose visible part is

```
package Bank is
   type Money is new Natural;
   type Key is limited private;
   procedure Open_Account(K: in out Key; M: in Money);
      -- open account with initial deposit M
   procedure Close_Account(K: in out Key; M: out Money);
      -- close account and return balance
   procedure Deposit(K: in Key; M: in Money);
      -- deposit amount M
   procedure Withdraw(K: in out Key; M in out Money);
      -- withdraw amount M; if account does not contain M
      -- then return what is there and close account
   function Statement(K: Key) return Money;
      -- returns a statement of current balance
   function Valid(K: Key) return Boolean;
      -- checks the key is valid
private
   ...
```

2 Assuming that your solution to the previous question allowed the bank the use of the deposited money, reformulate the private type to represent a home savings box or safe deposit box where the money is in a box kept by the user.

3 A thief writes the following

```
declare
   use Key_Manager;
   My_Key: Key;
```

```
      procedure Cheat(Copy: in out Key) is
      begin
         Return_Key(My_Key);
         Action(Copy, ... );
         ...
      end;
   begin
      Get_Key(My_Key);
      Cheat(My_Key);
      ...
   end;
```

He attempts to return his key and then use the copy. Why is he thwarted?

4 A vandal writes the following

```
   declare
      use Key_Manager;
      My_Key: Key;
      procedure Destroy(K: out Key) is null;   -- a null procedure
   begin
      Get_Key(My_Key);
      Destroy(My_Key);
      ...
   end;
```

He attempts to destroy the value in his key by calling a procedure which does not update the **out** parameter; he anticipates that this will result in a junk value being assigned to the key. Why is he thwarted?

Checklist 12

Variables inside a package exist between calls of subprograms of the package.

For predefined equality to be sensible, the values should be in a canonical form.

Predefined equality for an array type uses predefined equality of its components.

A nonlimited private type can be implemented in terms of another private type provided it is also nonlimited.

A limited private type can be implemented in terms of any private type limited or not.

Explicitly limited types are always passed by reference.

Objects of limited types can be initialized *in situ* by aggregates and function calls in Ada 2005.

The concepts of limited aggregates and limited constructor functions were new in Ada 2005.

Overriding indicators and null procedures were new in Ada 2005.

Completion of an access type cannot be deferred to a body in some circumstances in Ada 2005.

Expression functions were new in Ada 2012 and can be used as a complete function in a package specification or body.

Expression functions can also be used as a completion in all parts of a package. An important case is where the specification is given in the visible part and the completion occurs in the private part.

The use all type clause was new in Ada 2012.

All record types compose for equality in Ada 2012.

New in Ada 2022

The new aspect Default_Initial_Condition can be given for a private type, see Section 16.2; this aspect is used in containers, see Section 24.2.

13 Overall Structure

13.1 Library units
13.2 Subunits
13.3 Child library units
13.4 Private child units
13.5 Mutually dependent units
13.6 Scope, visibility, and accessibility
13.7 Renaming
13.8 Programs, partitions, and elaboration

In this chapter we discuss the hierarchical library structure and other mechanisms for separate compilation which were outlined in Chapter 4.

Many languages ignore the simple fact that programs are written in pieces, compiled separately and then joined together. Indeed large programs should be thought of as being composed out of a number of subsystems which themselves have internal structure. There are clearly two distinct requirements.

There is a requirement for decomposing a large coherent program into a number of internal subcomponents. Such a decomposition has particular advantages when a large program is being developed by a team.

There is also a requirement for the creation of a program library where subsystems are written for general use and consequently are written before the programs that use them. Within this structure it is convenient to decompose the interface presented to future clients so that they may select only those parts of a system that are required.

This chapter also contains a further discussion on scope and visibility and summarizes the curious topic of renaming.

13.1 Library units

We will start by considering which units of Ada can be compiled separately and the general idea of dependency.

The most common units of compilation are package specifications and bodies and subprograms. Such units may be compiled individually or for convenience several could be submitted to the compiler together. Thus we could compile the specification and body of a package together but, as we shall see, it may be more convenient to compile them individually. As usual a subprogram body alone is sufficient to define the subprogram fully.

A library unit can also be a generic package or subprogram or an instantiation of one as discussed further in Section 27.3.

Compilation units are kept in a program library in some form. We will describe the behaviour in general terms but an implementation may use any model that satisfies two major requirements: a program cannot be built out of units that are not consistent, and a unit cannot be compiled until all units upon which it depends are present in the library.

There are two obvious compilation models according to the form in which the compiler needs the information regarding dependency. In one, the source model, the compiler only needs access to the source text of the other units; in the other, the object model, the compiler requires the other units to be compiled as well. (The object model was the one presumed in Ada 83.) In order to cover such variation we use phrases such as 'entered into the library'.

So, once entered into the library, a unit can be used by any subsequently compiled unit but the using unit must indicate the dependency by a with clause.

As a simple example suppose we compile the package Stack of Section 12.1. This package depends on no other unit and so it needs no with clause. We will compile both specification and body together so the text submitted will be

```
package Stack is
   ...
end Stack;
package body Stack is
   ...
end Stack;
```

As well as producing the object code corresponding to the package, the compiler has to ensure that the program library has all the information required in order to compile subsequent dependent units. In the case of the source model nothing extra will be needed, whereas for the object model some encoded form of the information in the specification will need to be inserted.

We now suppose that we write a procedure Main which will use the package Stack. The procedure Main is going to be the main subprogram in the usual sense. It will have no parameters and we can imagine that it is called by some magic outside the language itself. The Ada language does not prescribe that the main subprogram should have the identifier Main; it is merely a convention which we are adopting here because it has to be called something.

The text we submit to the compiler could be

```
with Stack;
procedure Main is
   use Stack;
   M, N: Integer;
begin
   ...
   Push(M);
   ...
```

```
        N := Pop;
        ...
    end Main;
```

The with clause goes before the unit so that the dependency of the unit on other units is clear at a glance. A with clause may not be embedded in an inner scope.

On encountering a with clause the compiler retrieves from the program library the information describing the interface presented by the unit mentioned (in source or encoded form) so that it can check that the unit being compiled uses the interface correctly. Thus if the procedure Main tries to call Push with the wrong number or type of parameters then this will be detected during compilation. This thorough checking between separately compiled units is a key feature of Ada.

If a unit is dependent on several other units then they can be mentioned in the one with clause, or it might be more convenient to use distinct with clauses. Thus we could write

```
with Stack, Diurnal;
procedure Main is ...
```

or equally

```
with Stack; with Diurnal;
procedure Main is ...
```

For convenience we can place a use clause after a with clause. Thus

```
with Stack;  use Stack;
procedure Main is ...
```

and then Push and Pop are directly visible without more ado. The use and with clauses preceding a compilation unit are collectively known as the context clause of the unit. A use clause in a context clause can only refer to packages mentioned earlier in with clauses in the context clause.

A very minor point is that we can even gratuitously repeat the same with clause and use clause in a context clause.

Only direct dependencies need be given in a with clause. Thus if package P uses the facilities of package Q which in turn uses the facilities of package R, then, unless P also directly uses R, the with clause for P should mention only Q. The user of P does not care about R and should not need to know since otherwise the hierarchy of development would be made more complicated.

Note carefully that a context clause (both with clauses and use clauses) in front of a package or subprogram declaration will also apply to the corresponding body. It can but need not be repeated. Of course, the body may have additional dependencies which will need indicating with its own context clause anyway. Dependencies which apply only to the body should not be given with the specification since otherwise the independence of the body and the specification would be reduced.

If a package specification and body are compiled separately then the body must be compiled after the specification has been entered into the library. We say that the body is dependent on the specification. However, any unit using the package is dependent only on the specification and not the body. If the body is changed in a

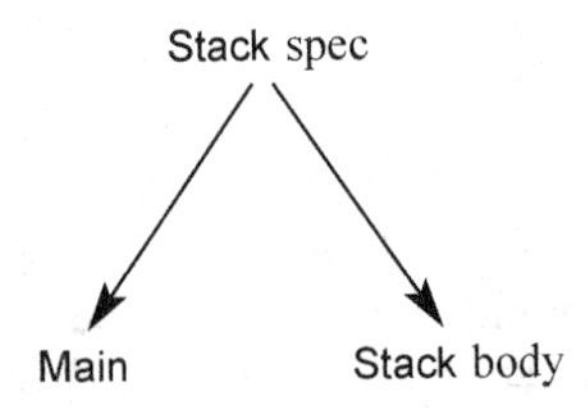

Figure 13.1 Dependencies between units.

manner consistent with not changing the specification, any unit using the package will not need recompiling. The ability to compile specification and body separately simplifies program maintenance and clearly distinguishes the logical interface from the physical implementation.

The dependencies between the specification and body of the package Stack and the procedure Main are illustrated by the graph in Figure 13.1.

The general rule regarding the order of compilation is simply that a unit must be compiled after all units on which it depends are entered into the library. Consequently, if a unit is changed and so has to be recompiled then all dependent units must also be recompiled before they can be linked into a total program.

There is one package that need not (and indeed cannot) be mentioned in a with clause. This is the package Standard which effectively contains the declarations of all the predefined types such as Integer and Boolean and their predefined operations. The package Standard is described in more detail in Section 23.1.

Finally, there are two important rules regarding library units. They must have distinct names; they cannot be overloaded. Moreover they cannot be operators. These rules simplify the implementation of an Ada program library on top of a conventional file system or database.

Exercise 13.1

1 The package D and subprograms P and Q and Main have with clauses as follows

specification of D	no with clause
body of D	**with** P, Q;
subprogram P	no with clause
subprogram Q	no with clause
subprogram Main	**with** D;

Draw a graph showing the dependencies between the units. How many different orders of compilation are possible (a) if the library uses the source model, (b) if it uses the object model?

13.2 Subunits

In this section we introduce a further form of compilation unit known as a subunit. The body of a package, subprogram (or task or protected object, see Chapter 20) can be 'taken out' of an immediately embracing library unit and itself compiled separately. The body in the embracing unit is then replaced by a body stub. As an example suppose we remove the bodies of the subprograms Push and Pop from the package Stack. The body of Stack would then become

```
package body Stack is
   Max: constant := 100;
   S: array (1 .. Max) of Integer;
   Top: Integer range 0 .. Max;
   procedure Push(X: Integer) is separate;        -- stub
   function Pop return Integer is separate;        -- stub
begin
   Top := 0;
end Stack;
```

The removed units are termed subunits; they may then be compiled separately. They have to be preceded by **separate** followed by the name of the parent unit in parentheses. Thus the subunit Push becomes

```
separate (Stack)
procedure Push(X: Integer) is
begin
   Top := Top + 1;
   S(Top) := X;
end Push;
```

and similarly for Pop. Like distinct specifications and bodies, the specification in the subunit must have full conformance with that in the stub.

In the above example the parent unit is (the body of) a library unit. The parent body could itself be a subunit; in such a case its name must be given in full using the dotted notation starting with the ancestor library unit. Thus if R is a subunit of Q which is a subunit of P which is a library unit, then the text of R must start

```
separate (P.Q)
```

As with library units and for similar reasons, the subunits of a library unit must have distinct identifiers and not be overloaded. But we could have a subunit P.S.R as well as P.Q.R and indeed P.R. Also, subunits cannot be operators.

A subunit is dependent on its parent body (and any library units explicitly mentioned) and so must be compiled after they are entered into the library.

Visibility within a subunit is as at the corresponding body stub – it is exactly as if the subunit were plucked out with its environment intact and full type checking is maintained. As a consequence any context clause applying to the parent body need not be repeated just because the subunit is compiled separately. Moreover, it is possible to give the subunit access to additional library units by preceding it with its own context clause (possibly including use clauses). Any such context clause precedes **separate**. So the text of R might commence

```
with X;  use X;
separate (P.Q)
...
```

A possible reason for doing this might be if we can then remove any reference to library unit X from the parent P.Q and so reduce the dependencies. This would give us greater freedom with recompilation; if X were recompiled for some reason then only R would need recompiling as a consequence and not also Q.

Note that a with clause only refers to library units and never to subunits. Finally observe that several subunits or a mixture of library units, library unit bodies, and subunits can be compiled together.

Exercise 13.2

1 Suppose that the package Stack is written with separate subunits Push and Pop. Draw a graph showing the dependencies between the five units: procedure Main, procedure Push, function Pop, package specification Stack, package body Stack. How many different orders of compilation are possible (a) if the library uses the source model, (b) if it uses the object model?

13.3 Child library units

One of the great strengths of Ada is the library package where the distinct specification and body decouple the user interface to a package (the specification) from its implementation (the body). This enables the details of the implementation and the clients to be recompiled separately without interference provided the specification remains stable.

However, although the simple structure we have seen so far works well for smallish programs it is not satisfactory when programs become large or complex. There are two aspects of the problem: the coarse control of visibility of private types and the inability to extend without recompilation.

There are occasions when we wish to write two distinct packages which nevertheless share a private type. We cannot do this with unrelated packages. We either have to make the type not private so that both packages can see it with the unfortunate consequence that all the client packages can also see the type; this breaks the abstraction. Or, on the other hand, if we wish to keep the abstraction, then we have to merge the two packages together and this results in a large monolithic package with increased recompilation costs.

The other aspect of the difficulty arises when we wish to extend an existing system by adding more facilities to it. If we add to a package specification then naturally we have to recompile it but moreover we also have to recompile all existing clients even if the additions have no impact upon them.

Another similar problem with a simple flat structure is the potential clash of names between library units in different parts of the system.

These problems are solved by the introduction of a hierarchical library structure containing child packages and child subprograms. There are two kinds of children: public children and private children. We will consider public children in this section and then private children in the next section.

Consider the familiar example of a package for the manipulation of complex numbers as described in Section 12.2. It contains the private type itself plus the arithmetic operations and also subprograms to construct and decompose a complex number taking a cartesian view.

```
package Complex_Numbers is
   type Complex is private;
   ...
   function "+" (X, Y: Complex) return Complex;
   ...
   function Cons(R, I: Float) return Complex;
   function Re_Part(X: Complex) return Float;
   function Im_Part(X: Complex) return Float;
private
   ...
end Complex_Numbers;
```

We have deliberately not shown the completion of the private type since it is largely immaterial how it is implemented. Although this package gives the user a cartesian view of the type, nevertheless it certainly does not have to be implemented that way as we saw in Chapter 12.

Some time later we might need to additionally provide a polar view by the provision of subprograms which construct and decompose a complex number from and to its polar coordinates. We can do this without disturbing the existing package and its clients by adding a child package as follows

```
package Complex_Numbers.Polar is
   function Cons_Polar(R, θ: Float) return Complex;
   function "abs" (X: Complex) return Float;
   function Arg(X: Complex) return Float;
end Complex_Numbers.Polar;
```

and within the body of this package we can access the full details of the private type Complex. Note the use of the Greek letter θ for the angle.

(If this example seems all too mathematical then consider the equivalent problem of representing the position of a ship in terms of either latitude and longitude or range and bearing with respect to some fixed location.)

Note the notation, a package having the name P.Q is a child package of its parent package P. We can think of the child package as being declared inside the declarative region of its parent but after the end of the specification of its parent; most of the visibility rules stem from this model. In other words the declarative region defined by the parent (which is primarily the specification and body of the parent, see Section 13.6) also includes the space occupied by the text of the children; but it is important to realize that the children are inside that region and do not just extend it. The rules are worded this way to make it clear that a child subprogram is not a primitive operation of a type declared in its parent's specification because the child is not declared in the specification but after it. (Primitive operations were introduced in Section 12.3.)

In just the same way, root library packages can be thought of as being declared in the declarative region of the package Standard and after the end of its specification. So library units are children of Standard.

An important special visibility rule is that the private part (if any) and the body of a child have visibility of the private part of the parent. In other words they have visibility of the whole of the specification of the parent and not just the visible part.

However, the visible part of a (public) child package does not have visibility of the private part of its parent; if it did it would allow renaming (discussed later in this chapter) and hence the export of the hidden private details to any client; this would break the abstraction of the private type (this rule does not apply to private children as we shall see in the next section).

The body of the child package for the complex number example can now be written using the full view of the type Complex. Assuming that it is indeed implemented as a record with components Re and Im of type Float, then access to some elementary functions will be required. So the body might be

```
with Ada.Elementary_Functions;
use Ada.Elementary_Functions;
package body Complex_Numbers.Polar is
   function Cons_Polar(R, θ: Float) return Complex is
   begin
      return (R*Cos(θ), R*Sin(θ));
   end Cons_Polar;
   ...
end Complex_Numbers.Polar;
```

In order to access the subprograms of the child package the client must, of course, have a with clause for the child. This also implicitly provides a with clause for the parent as well so we need not write both. Thus we might have

```
with Complex_Numbers.Polar;
package Client is ...
```

and then within Client we can access the various subprograms in the usual way by writing Complex_Numbers.Re_Part or Complex_Numbers.Polar.Arg and so on.

Direct visibility can be obtained by use clauses as expected. However, a use clause for the child does not imply one for the parent; but, because of the model that the child is in the declarative region of the parent, a use clause for the parent makes the child name itself directly visible. So writing

```
with Complex_Numbers.Polar;  use Complex_Numbers;
```

now allows us to refer to the subprograms as Re_Part and Polar.Arg respectively.

We could of course have added

```
use Complex_Numbers.Polar;
```

and we would then be able to refer to the subprogram in Polar just as Arg.

There is a rule that a use clause does not take effect until the end of the context clause. So we could not abbreviate this last use clause to just **use** Polar; on the grounds that we already have direct visibility of the parent.

Finally, note that any context clause on the parent also applies to its children (in the same way that it applies to its body and any subunits) and transitively to their bodies and children and so on. Moreover, a child does not need a with clause or use clause for its parent.

Child packages neatly solve both the problem of sharing a private type over several compilation units and the problem of extending a package without having to recompile the clients. They thus provide a form of programming by extension.

A package may of course have several children. In fact with hindsight it might have been more logical to have developed our complex number package as three packages: a parent containing the private type and the four arithmetic operations and then two child packages, one giving the cartesian view and the other giving the polar view of the type. At a later date we could add yet another package providing perhaps the trigonometric functions on complex numbers and again this can be done without modifying what has already been written and thus without the risk of introducing errors.

There are a number of other examples which we have already met where child packages might be useful. We could add mixed operations between types Complex and Float in a child package (see Exercise 12.2(**1**)). We could belatedly add the functions Is_Empty and Is_Full or declare a deferred constant Empty for the package Stacks of Section 12.4 in a child package.

Declaring such a deferred constant in a child package is only possible because the private part of a child has access to the private part of its parent and so can see the full view of the type. So we might have

```
package Stacks.More is
   Empty: constant Stack:
private
   Empty: constant Stack := ((others => 0), 0);
end Stacks.More;
```

We could use this technique to declare other constants of type Complex and give them appropriate initial values.

In general the private part and body of a child package can use types declared in the private part of its parent. However, a quirk can arise in the case of access types. Recall from Section 12.5 that an accessed type such as Cell in a private part can be completed in the body. In such a case the incomplete type cannot be used by a child package. This is not really surprising but is simply a consequence of the fact that a child cannot see the full type declaration.

Finally, it is very important to realize that the child mechanism is hierarchical. Children may have children to any level so we can build a complete tree providing decomposition of facilities in a natural manner. A child may have a private part and this is then visible from its children but not from its parent.

Siblings have much the same relationship to each other as two unrelated packages at the top level. Thus a child can obviously only have visibility of a sibling previously entered into the library; a child can only see the visible part of its siblings; a child needs a with clause in order to access a sibling; and so on. Moreover, since siblings both have direct visibility of their parent, they can refer to each other without using the parent name as a prefix.

As already mentioned, a child (specification and body) automatically depends upon its parent (and grandparent) and needs no with clause for them. But other dependencies must be explicitly mentioned. Thus a parent body may depend upon its children and grandchildren but their specification would have to be entered into the library first and the parent body would need with clauses for them.

One very important use of the hierarchical structure was discussed in Chapter 4 where we saw that the predefined library is structured as packages System, Interfaces, and Ada each of which has numerous child packages.

We have already superficially met the package Ada.Characters.Latin_1 which is a grandchild of Ada when discussing the type Character in Section 8.4. This package contains constants for the various control characters and so we now see why we can refer to them as Ada.Characters.Latin_1.Nul and so on. Writing

```
use Ada.Characters.Latin_1;
```

allows this to be abbreviated to simply Nul.

Note the distinction between child packages and lexically nested packages such as the (obsolescent) package ASCII in Standard. Lexically nested packages cannot be separately compiled (although their bodies could be subunits) and just have single identifiers as their name in their declaration. But externally they are both referred to using the same notation and as a consequence it is illegal to have a nested package and a child package with the same identifier because it would be ambiguous. Another difference is that the private part of a nested package is never visible to a child. We shall see further similarities and differences between nested and child packages in the next section.

Although we have concentrated on child packages in this section, child subprograms are also very useful; there are some examples in the next section.

Exercise 13.3

1 Rewrite the package Complex_Numbers as a parent and two children.

2 Draw the dependency graph for the hierarchy of the previous exercise.

3 If we derive the type Field (see Section 12.3) from Complex as now structured, then which operations are inherited?

13.4 Private child units

In the previous section we introduced the concept of hierarchical child units and showed how these allowed extension and continued privacy of private types without recompilation. However, the whole idea was based around the provision of additional facilities for the client. The specifications of the additional units were all visible to the client.

In the development of large subsystems it often happens that we would like to decompose the system for implementation reasons but without giving any additional visibility to clients.

In Section 13.2 we saw how a body could be separately compiled as a subunit. However, although a subunit can be recompiled without affecting other subunits at the same level, any change to its specification requires its parent body and hence all sibling subunits to be recompiled.

Greater flexibility is provided by a form of child unit that is totally private to its parent. In order to illustrate this idea consider the following outline of an operating system.

```
package OS is
   -- parent package defines types used throughout the system
   type File_Descriptor is private;
   ...
private
   type File_Descriptor is new Integer;
end OS;

package OS.Exceptions is
   -- exceptions used throughout the system
   File_Descriptor_Error,
   File_Name_Error,
   Permission_Error: exception;
end OS.Exceptions;

private package OS.Internals is
   ...
end OS.Internals;

procedure OS.Interpret(Command: in String);

with OS.Exceptions;
package OS.File_Manager is
   type File_Mode is (Read_Only, Write_Only, Read_Write);
   function Open(File_Name: String; Mode: File_Mode)
                                             return File_Descriptor;
   procedure Close(File: in File_Descriptor);
   ...
end OS.File_Manager;

private package OS.Internals.Debug is
   ...
end OS.Internals.Debug;
```

In this example the parent package contains the types used throughout the system. There are three public child units, the package OS.Exceptions containing various exceptions, the package OS.File_Manager which provides file open/close routines (note the explicit with clause for its sibling OS.Exceptions) and a procedure OS.Interpret which interprets a command line passed as a parameter. (Incidentally this illustrates that a child unit can be a subprogram as well as a package. It can actually be any library unit and that includes a generic declaration and a generic instantiation.) There is also a private child package called OS.Internals and a private grandchild called OS.Internals.Debug.

A private child (distinguished by starting with the word **private**) can be declared at any point in the child hierarchy. The general idea is that the private children can be used by the parent and the public children as part of their implementation but cannot be used by the clients.

Before describing the rules for private children in detail we have to introduce a new form of with clause, the so-called private with clause. As mentioned in Section 12.2, when Ada was being designed some thought was given to the idea that the visible part, the private part, and the body might be written as three distinct entities, perhaps even as

```
with A;
package P is
   ...                                   -- visible part
end;

with B;
package private P is                     -- just dreaming
   ...                                   -- private part
end;

with C;
package body P is
   ...                                   -- body
end;
```

The idea would have been that a with clause on the private part would have applied just to the private part and the body. However, it was clear that this would have been an administrative nightmare in many situations and so the two-part specification and body emerged with the private part lurking at the end of the visible part of the specification (and sharing its context clause). This was undoubtedly the right decision in general. The division into just two parts supports separate compilation well and although the private part is not part of the logical interface to the user it does provide information about the physical interface and that is needed by the compiler.

However, for some purposes this is a nuisance and a variation of the with clause can be used. If we write

```
private with Q;
package P is ...
```

then Q is visible to the private part (and body) of P but not to the visible part. In other words it behaves as a with clause on the private part which then transitively applies to the body as well. Such a private with clause can also be placed on a body in which case it is simply treated as a normal with clause.

We are not allowed to have a use clause in the same context clause as the private with clause but we can always put a use clause in the private part thus

```
private with Q;
package P is
   ...                      -- Q not visible here
private
   use Q;
   ...                      -- entities in Q directly visible here
end P;
```

We can now return to the visibility rules for private children. These are similar to those for public children but there are two extra rules.

The first extra rule is that a private child is only ever visible within the subtree of the hierarchy whose root is its parent. And moreover, within that tree, it is never visible to the visible parts of any public siblings although it can be visible to their private parts. In order for it to be visible to the private part of a public child, the specification of the public child must have a private with clause for the private sibling. A normal with clause is not permitted.

In the example as written above, the private package OS.Internals is visible to the bodies of OS, of OS.File_Manager, and of OS.Interpret (OS.Exceptions has no body anyway) and it is also visible to both the specification and the body of OS.Internals_Debug. But it is not visible outside OS and a client package certainly cannot access OS.Internals at all.

However, if the private part of OS.File_Manager also needed to access the package OS.Internals then it must have a private with clause as well, thus

```
with OS.Exceptions; private with OS.Internals;
package OS.File_Manager is ...
```

The other extra rule is that the visible part of a private child can access the private part of its parent. This is quite safe because it cannot export information about a private type to a client because it is not itself visible. Nor can it export information indirectly via its public siblings because, as we have seen, it is never visible to their visible parts but only to their private parts and bodies.

We can now safely implement our system in the package OS.Internals and we can create a subtree for the convenience of development and extensibility. We might then have a third level in the hierarchy containing packages such as OS.Internals.Devices, OS.Internals.Access_Rights and so on.

The introduction of child units extends the categorization into logical, physical, and implementation parts mentioned in Section 12.2. The visible part of a package plus the visible parts of its public children form the logical interface to external clients; the private parts of the parent and public children and visible and private parts of private children form the physical interface; and then the various bodies form the implementation part.

The various relationships are illustrated in Figure 13.2 which shows a package plus two public children and two private children. The solid lines with arrows indicate direct visibility which does not require with clauses; these lines can be followed transitively. Thus a private part can see the corresponding visible part and a body can see the corresponding private part and thus the visible part. Similarly, the private part of a public child can see the private part and thus the visible part of its parent but the visible part of a public child can only see the visible part of its parent.

The effect of typical with clauses is shown by broken lines. Of course a with clause only ever gives visibility of a visible part. Moreover, a line representing a with clause (normal or limited) can never go to the right; thus the specification of a public child cannot have a normal with clause for a private child, but its body can. Having followed a broken line we can then transitively follow any solid line. Thus if a client has a with clause for a child then it automatically gains access to the parent as well. One example of a private with clause is shown by a dotted line.

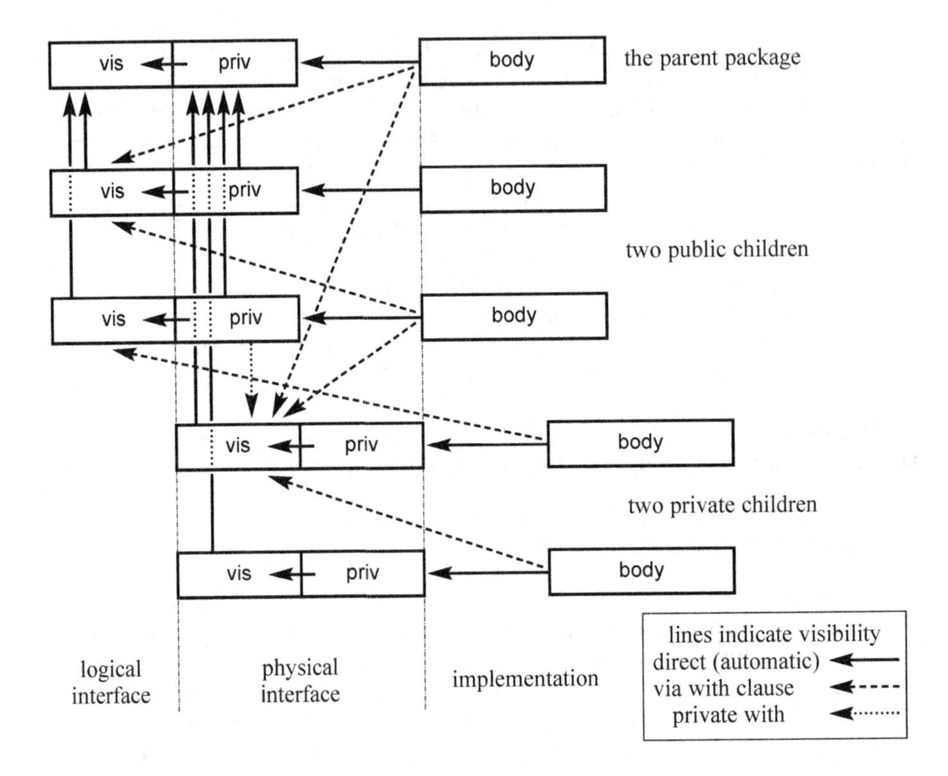

Figure 13.2 Visibility throughout a hierarchy.

Further decomposition to grandchildren is as expected. Thus a public child of a private child is effectively an extension of the specification of the private child and is visible to the bodies of its public aunts and to all parts of its private aunts. But a private child of a private child is not visible to its aunts at all.

There are a number of similarities between child and nested packages. Nested packages in a package specification behave rather like public children, whereas nested packages in a package body behave like private children. But there are differences: nested packages in a body can access earlier declarations in the body whereas a child (public or private) cannot access declarations in its parent body at all. Similar remarks apply to nested subprograms.

The rules also apply to private and public child subprograms. Thus the specification of a public child subprogram cannot have a with clause for a private sibling, but it can have a private with clause for a private sibling. A distinct body can have a with clause for a private sibling.

As a curious example, if we decided to replace the subunits of Section 13.2 by children, then the visible subprograms would become public children and a private child could be used for the declarations of the data. The structure might be as follows

```
package Stack is
end;

procedure Stack.Push(X: Integer);

function Stack.Pop return Integer;

private package Stack.Data is
   Max: constant := 100;
   S: array (1 .. Max) of Integer;
   Top: Integer range 0 .. Max := 0;
end Stack.Data;

with Stack.Data; use Stack.Data;
procedure Stack.Push(X: Integer) is
begin
   Top := Top + 1;
   S(Top) := X;
end Stack.Push;

...   -- Pop similarly
```

The packages Stack and Stack.Data have no bodies; note in particular that the initialization of Top is done in its declaration; it could be done in a body but this requires the aspect Elaborate_Body as discussed in Section 13.8. The package Stack has an empty specification as well and so is a null package. It only exists in order to provide commonality of naming so that the private package can be declared.

An interesting point is that the use clause for Stack.Data in the context clause cannot be abbreviated to just **use** Data; but it could be so written in the declarative part of Stack.Push. The reason is that within the context clause the principle of linear elaboration means we do not know that we are in fact about to encounter a child unit of Stack. However, once inside Stack.Push we have direct visibility of Stack. Another point is that if Stack.Push did not have a distinct specification then the with clause for Stack.Data would have to be a private with clause. This structure occurs in Program 3, the Rational Reckoner.

Other arrangements are possible. For example the data could be in the private part of the package Stack and the subprograms would no longer need with clauses.

Although we have introduced private with clauses largely in the context of private children, nevertheless they have other uses. If we write

```
private with P;
package Q is ...
```

then we are assured that the package Q cannot inadvertently access P in the visible part. Thus writing **private with** provides additional documentation information which can be useful to both human reviewers and program analysis tools. So if we have a situation where a private with clause is all that is needed then we should use it rather than a normal with clause.

We conclude this section by summarizing the various visibility rules.

- A specification never needs a with clause for its parent; it may have one for a sibling except that a public child specification may only have a private with clause for a private sibling; it may not have a with clause for its own child. A public child subprogram without a separate specification cannot have a with clause but can have a private with clause for a private sibling.

- A body never needs a with clause for its parent; it may have one for a sibling (public or private); it may have one for its own child.
- A context clause on a specification applies to the body and any children (and transitively to subunits and grandchildren ...).
- The entities of the parent are directly visible within a child; a use clause is not required.
- A private child is never visible outside the tree rooted at its parent. And within that tree it is not visible to the visible parts of public siblings.
- The private part and body of any child can access the private part of its parent (and grandparent ...).
- In addition the visible part of a private child can also access the private part of its parent (and grandparent ...).
- A with clause for a child automatically implies with clauses for all its ancestors.
- A use clause for a unit makes the child units accessible by simple name (this only applies to child units for which there is also a with clause).

These rules may seem a bit complex but actually stem from just a few considerations of consistency. Questions regarding access to children of sibling units and other remote relatives follow by analogy with an external client viewing the appropriate subtree.

Some consequences of the rules for child units are worth noting. We can use the same identifier for an entity in both parent and child. So given an entity X in a package P we could declare another entity X in a child P.Q thus

```
package P is
   X: Integer;
end P;

package P.Q is
   Y: Integer := X;
   X: Integer := P.X;
end P.Q;
```

On the other hand we cannot use the same name for an entity in P and for the child itself. Thus the child could not be called P.X.

Exercise 13.4

1. Rewrite the body of the package Rational_Numbers of Exercise 12.2(**3**) so that the functions Normal and GCD are in a private child package Rational_Numbers.Slave.

2. Declare (in outline) an additional child package Complex_Numbers.Trig for computing the trigonometric functions Sin, Cos, etc. of complex numbers. Use a private child function Sin_Cos for the common part of the calculation. Ensure that with and use clauses are correct.

13.5 Mutually dependent units

The library structures described so far are not entirely adequate when two types depend upon each other in some way. For example, suppose we have two types Line and Point describing a geometrical configuration and that each is defined in terms of the other. For simplicity we assume that each line goes through three points and that each point is on three lines. We can do this with incomplete types thus

```
type Point;
type Line;                              -- incomplete types

type Point is
   record
      L, M, N: access Line;
   end record;

type Line is
   record
      P, Q, R: access Point;
   end record;
```

For simple examples this is satisfactory but for more elaborate situations we might wish to declare the types and their associated operations in distinct packages. Suppose the types are declared in two child packages Geometry.Points and Geometry.Lines. We would like to write

```
with Geometry.Points;
package Geometry.Lines is
   type Line is
      record
         P, Q, R: access Points.Point;
      end record;
   ...
end Geometry.Lines;
```

with a similar declaration for Geometry.Points. But this is not possible because each package would have a with clause for the other and a with clause can only refer to a package already in the library. Clearly, both packages cannot come first.

The solution is to use a limited with clause for one of them thus

```
limited with Geometry.Lines;
package Geometry.Points is
   type Point is
      record
         L, M, N: access Lines.Line;
      end record;
   ...
end Geometry.Points;
```

and this package can be placed in the library first.

By writing **limited with** Geometry.Lines; we get access to all the types visible in the specification of Geometry.Lines but as if they were declared as incomplete.

In other words we get incomplete views of the types. We can then do all the things we can normally do with incomplete types such as use them to declare access types. (Of course the implementation checks later that Geometry.Lines does actually have a type Line.) We also get visibility of the names of internal packages.

Incidentally, there really is a finite geometry with only three points on each line and three lines through each point. This is the Fano plane named after the Italian mathematician Gino Fano (1871–1952). There are just seven points and seven lines in the plane. See for example *Gems of Geometry* by the author.

Although it is only necessary to use a limited with clause on one package in order to break the circularity, it is likely that for symmetry we would place limited with clauses on the specifications of both packages. Assuming that both bodies follow both specifications then they can have normal with clauses as usual.

Incomplete types can be completed by any type other than another incomplete type in Ada 2012 and Ada 2022. They can be completed by a private type thus

```
type T1;                                   -- incomplete type
type T2(X: access T1) is private;
type T1(X: access T2) is private;          -- completion
```

This example uses access discriminants which are explained in Section 18.7.

Here is another example which is perhaps more realistic. This concerns employees and the departments of the organization in which they work. The information about employees needs to refer to the departments and the departments need to refer to the employees. We assume that the material regarding employees and departments is quite large so that we naturally wish to declare the two types in distinct packages Employees and Departments. So we would like to say

```
with Departments;  use Departments;
package Employees is
   type Employee is private;
   procedure Assign_Employee(E: in out Employee;
                             D: in out Department);
   type Dept_Ptr is access all Department;
   function Current_Department(E: Employee) return Dept_Ptr;
   ...
end Employees;

with Employees;  use Employees;
package Departments is
   type Department is private;
   procedure Choose_Manager(D: in out Department;
                            M: in out Employee);
   ...
end Departments;
```

We cannot write this because each package has a with clause for the other and they cannot both be declared (or entered into the library) first.

We assume of course that the type Employee includes information about the Department for whom the Employee works and the type Department contains information regarding the manager of the department and presumably a list of the other employees as well – note that the manager is naturally also an Employee.

Without limited with clauses we have to put everything into one package thus

```
package Workplace is
   type Employee is private;
   type Department is private;
   procedure Assign_Employee(E: in out Employee;
                             D: in out Department);
   type Dept_Ptr is access all Department;
   function Current_Department(E: Employee) return Dept_Ptr;
   procedure Choose_Manager(D: in out Department;
                            M: in out Employee);
private
   ...
end Workplace;
```

Not only does this give rise to huge cumbersome packages but it also prevents us from using the proper abstractions. Thus the types Employee and Department have to be declared in the same private part and so are not protected from each other's operations.

But using limited with clauses we can write

```
limited with Departments;
package Employees is
   type Employee is private;
   procedure Assign_Employee(E: in out Employee;
                             D: in out Departments.Department);
   type Dept_Ptr is access all Departments.Department;
   function Current_Department(E: Employee) return Dept_Ptr;
   ...
end Employees;

limited with Employees;
package Departments is
   type Department is private;
   procedure Choose_Manager(D: in out Department;
                            M: in out Employees.Employee);
   ...
end Departments;
```

It is important to understand that a limited with clause does not impose a dependency. Thus if a package A has a limited with clause for B, then A does not depend on B as it would with a normal with clause, and so B does not have to be compiled before A or placed into the library before A.

Note the terminology: we say that we have a limited view of a package if the view is provided through a limited with clause. So a limited view of a package provides an incomplete view of its visible types. And by an incomplete view we mean as if they were incomplete types.

Although we do not generally know whether an incomplete type is implemented by copy or by reference (unless it is tagged, see Section 14.2), nevertheless we can use an incomplete type as parameter or result provided that we do know by a point where the subprogram is actually called or the body itself is encountered since it is only at those points that the compiler has to grind out the code for the call or the body. Earlier versions of Ada required the mechanism to be known when the specification was reached and so in Ada 2005, the parameter D of Assign_Employee had to be an access parameter.

Interesting examples of mutual dependency can occur within a hierarchy. Suppose we have a parent package App, a public child App.User_View, and a private child App.Secret_Details.

In the previous section we saw that a public child could have a private with clause for a private sibling. A public and private child might also have a mutual dependency and if the public child is entered into the library first then it has to have a form of with clause for the private child that combines the restrictions of a limited with clause and a private with clause. This is called a limited private with clause and is written as follows.

```
limited private with App.Secret_Details;
package App.User_View is ...
```

A parent package can never have a normal with clause for a child since the parent must be placed in the library first. But the parent can have a limited with clause for a public child and a limited private with clause for a private child. The overall structure might be

```
limited with App.User_View;
limited private with App.Secret_Details;
package App is
   ...                      -- limited view of User_View
private
   ...                      -- limited view of Secret_Details
end App;

limited private with App.Secret_Details;
package App.User_View is
   ...                      -- full view of App, no view of Secret_Details
private
   ...                      -- limited view of Secret_Details
end App.User_View;

with App.User_View;
private package App.Secret_Details is
   ...                      -- full view of App and User_View
end App.Secret_Details;
```

There are various restrictions on the use of limited with and use clauses. The most important are that a limited with clause can only appear on a package specification (and not on a body for example) and that a use clause cannot apply to a package for which we only have a limited view.

The second restriction is perhaps surprising and means that we cannot write

```
limited with P;  use P;     -- illegal
```

The reason is to avoid various difficulties that can arise in obscure circumstances – these not concern the general user.

The fact that a limited with clause cannot appear on a body is no hardship because, as mentioned earlier, we would expect that all the specifications would come first and so we would be able to place a normal with clause on a body.

Remember from the previous section that a with clause on a parent also applies to all its children. The same applies to a limited with clause. Moreover, we can have a limited with clause on a parent unit and a corresponding normal with clause on a child. Thus the package App has a limited with clause for User_View and the child App.Secret_Details has a normal with clause for User_View. But the reverse is not permitted. Note that a child package cannot have a limited with clause for its parent (but it can have a normal with clause for its parent although it never needs one).

We cannot have both a limited with and normal with clause for the same unit in the same context clause because they imply very different views of the unit.

```
with P; limited with P;    -- illegal
```

But we can have both a private with and normal with clause in the same context clause thus

```
with P; private with P;    -- legal
```

the private with is just ignored.

13.6 Scope, visibility, and accessibility

Having just introduced the concepts of library and child units it seems appropriate to summarize the major points regarding scope and visibility which will be relevant to the everyday use of Ada. For some of the fine detail, the reader is referred to the *ARM*. We have already mentioned the term declarative region. Blocks and subprograms are examples of declarative regions and the scope rules associated with them were described in Sections 6.2 and 10.7. We now have to consider the broad effect of the introduction of packages and library units.

A package specification and body together constitute a single declarative region. Thus if we declare a variable X in the specification then we cannot redeclare X in the body (except of course in an inner region such as a local subprogram).

In the case of a declaration in the visible part of a package, its scope extends from the declaration to the end of the scope of the package itself. If the package is inside the visible part of another package then this means that, applying the rule again, the scope extends to the end of that of the outer package and so on.

In the case of a declaration in a package body or in the private part of a package, its scope extends to the end of the package body.

If a unit is a library unit then its scope includes just those units which depend upon it; these are its children, body, and subunits as appropriate plus those mentioning the unit in a with clause.

The rules for child units follow largely from the model of the child being in the declarative region of the parent and just after the specification of the parent. So the scope of a declaration in the specification of the parent extends through all child units. There is one important variation in the case of a declaration in the private part of a library unit: its scope does not include the visible part of any public children.

We recall from Section 10.7 that a declaration is directly visible if we can refer to it by just its direct name such as X. If not directly visible then it may still be visible and can be referred to by the dotted notation such as P.X.

Moreover, there can be places within its scope where a declaration is not visible at all. For example, a scalar, array, or record object is not visible in its own declaration anywhere and a package is not visible until the reserved word **is** of its declaration.

In the case of the simple nesting of blocks and subprograms a declaration is otherwise directly visible throughout its scope and so can be referred to by its direct name except where hidden by another declaration. Where not directly visible it can be referred to by using the dotted notation where the prefixed name is that of the unit embracing the declaration.

In the case of a declaration in a package the same rules apply inside the package. Outside the package, a declaration in the visible part is visible but not directly visible unless we write a use clause.

The declarations directly visible at a given point are those directly visible before considering any use clauses plus those made directly visible by use clauses.

The basic rule is that an identifier declared in a package is made directly visible by a use clause provided the same identifier is not also in another package with a use clause and also provided that the identifier is not already directly visible anyway. If these conditions are not met then the identifier is not made directly visible and we have to continue to use the dotted notation.

A slightly different rule applies if all the identifiers are subprograms or enumeration literals. In this case they all overload each other and all become directly visible. If the specifications have type conformance (and so would normally hide each other) then the identifiers are still directly visible although things like formal parameter names may be needed to identify a particular call.

The general purpose of these rules is to ensure that adding a use clause cannot silently change the meaning of an existing piece of text. (This is the so-called Beaujolais effect. A bottle (maybe it was a case) of Beaujolais was once offered to anyone finding an example illustrating this phenomenon. Ada 83 did have some Beaujolais effects but Ada is now thought to be free from them.)

We have only given a brief sketch of the main visibility rules here and the reader is probably confused. In practice there should be no problems since the Ada compiler will, we hope, indicate any ambiguities or other difficulties and things can always be put right by using a dotted name or a qualification.

There are other rules regarding record component names, subprogram parameters and so on which are as expected. For example, there is no conflict between an identifier of a record component and another use of the identifier outside the type definition itself. Consider

```
declare
   type R is
      record
         I: Integer;
      end record;

   type S is
      record
```

```
      I: Integer;
   end record;
   AR: R;
   AS: S;
   I: Integer;
begin
   ...
   I := AR.I + AS.I;          -- legal
   ...
end;
```

The visibility of the I in the type R extends from its declaration until the end of the block and so it can be referred to using the dotted notation. However, its direct visibility is confined to within the declaration of R. Another example is the use of S both as the stack and as the internal array in Section 12.4.

Similar considerations prevent conflict in named aggregates and in named parameters in subprogram calls where the name before => is not directly visible.

Note that a use clause can mention several packages. However, a use clause does not take effect until the semicolon. Suppose we have nested packages

```
package P1 is
   package P2 is
      ...
   end P2;
   ...
end P1;
```

then outside P1 we could write

```
use P1; use P2;
```

or

```
use P1, P1.P2;
```

but not

```
use P1, P2;                            -- illegal
```

We could even write **use** P1.P2; to gain direct visibility of the entities in P2 but not those in P1.

Similar rules apply in the case of child packages. So assuming

```
package P1 is
   ...
end P1;
package P1.P2 is
   ...
end P1.P2;
```

then we can again write exactly the same use clauses as in the case of the nested packages.

Remember that a use clause in a context clause can only refer to packages mentioned in a with clause earlier in the context clause (and it must be a normal with clause and not a limited with or private with clause); moreover a use clause in a context clause does not take effect until the end of the context clause. So we cannot write something like

```
with P1.P2;  use P1;  use P2;                -- illegal
```

but have to spell it out in detail as

```
with P1.P2;  use P1;  use P1.P2;
```

Note that writing **with** P1.P2; implies **with** P1; but **use** P1.P2; does not imply **use** P1; as well. This is because dependency on the child implies dependency on the parent but there is no reason why we should not choose to have direct visibility of internal declarations as we wish.

We can repeat a with or use clause as well, thus we can even write

```
with P1, P1;  use P1;  with P1;  use P1, P1;
```

The main reason for permitting this is that it might be helpful if the text of a program is automatically generated from various components.

As mentioned in Section 12.3, there is also the use type clause which makes the primitive operators of a type directly visible. This avoids making all declarations directly visible but allows the convenience of infixed notation.

Although a use clause is much like a declaration nevertheless the range of its effect is not always the same. Thus a use clause in the visible part of a package has no effect outside the package whereas, of course, the declaration of an entity such as a variable in the visible part of a package is visible outside the package. But a use clause in a block or subprogram has a similar scope to normal declarations in a block or subprogram.

Finally, there is the package Standard. This contains all the predefined entities and moreover every root library unit should be thought of as being declared as a child of Standard. This explains why an explicit with clause for Standard is not required. Another important consequence is that, provided we do not hide the name Standard by redefining it, a library unit P can always be referred to as Standard.P. Hence, in the absence of anonymous blocks, loops, and overloading, every identifier in the program has a unique name commencing with Standard. We could even refer to the predefined operators in this way

```
Four: Integer := Standard."+" (2, 2);
```

It is probably good advice not to redefine Standard; indeed declaring a package Standard simply results in Standard.Standard which is very confusing.

We conclude this section by remarking that the existence of packages plays no part in the accessibility rules. Packages are static scope walls whereas accessibility concerns dynamic lifetimes; packages can therefore be ignored in considering levels of accessibility.

13.7 Renaming

Renaming enables another name to be given to an existing entity. This is often convenient for providing a local abbreviation. We start by considering objects.

Consider the case of a large program with many library units and suppose that the unit we are in has with clauses for several packages This, That, and The_Other. If we have use clauses for all these packages then it is not clear from which package an arbitrary identifier such as X has been imported; it might be This.X, That.X, or The_Other.X. In the absence of use clauses we have to use the full dotted notation and the origin of everything is then obvious. However, it is also accepted that long meaningful identifiers should be generally used. A long meaningful package name followed by the long meaningful name of an entity in the package is often too much. However, we can introduce an abbreviation by renaming an object thus

```
V: Float renames Aircraft_Data.Current_Velocity;
```

and then compactly use V in local computation and yet still have the full identification available in the text of the current unit.

Up to Ada 2012, object renaming had to supply a subtype name. However, any constraints implied by the subtype name are ignored. Consider

```
Someday: Weekday := Mon;                    -- as in Section 6.6
Anyday: Day renames Sunday;
```

and then attempt to assign Sun to Anyday, we will get Constraint_Error because the subtype of Anyday is still Weekday. In Ada 2022 we can simply write

```
Anyday renames Someday;
```

Of course, if we attempt to assign Sun to Anyday then we still get Constraint_Error but we are no longer telling a fib.

Subprograms can also be renamed. Thus suppose we wish to use both the function Inner and the equivalent operator "*" of Chapter 10 without declaring two distinct subprograms. We can write one of

```
function "*" (X, Y: Vector) return Float renames Inner;
```

or the reverse

```
function Inner(X, Y: Vector) return Float renames "*";
```

according to which we declare first. Overriding indicators can also be given if desired.

Renaming of operators can be used to avoid the use of prefixed notation in the absence of a use clause; the alternative of a use type clause should be considered as described in Section 12.3.

Renaming is also useful in the case of library units. Thus we might wish to have two or more overloaded subprograms and yet compile them separately. This cannot be done directly since library units must have distinct names. However, differently named library units could be renamed so that the user sees the required effect. The restriction that a library unit cannot be an operator can similarly be overcome. The same tricks can be done with subunits.

Note also that a library unit can be renamed as another library unit. Ada 83 did not have child units and the text input–output package was simply Text_IO. In Ada 95, Ada.Text_IO was renamed as Text_IO for backward compatibility.

If a subprogram is renamed, the number, types, and modes of the parameters (and result if a function) must be the same. This is called mode conformance.

However, any constraints on the parameters or result in the new subprogram are ignored; those on the original still apply. This is because calls using the renaming are simply compiled as calls of the original subprogram. Thus we might have

```
procedure P(X: in Positive);
procedure Q(Y: in Natural) renames P;
...
Q(0);                        -- raises Constraint_Error
```

The call of Q raises Constraint_Error because zero is not an allowed value of Positive. The constraint implied by Natural on the renaming is completely ignored.

But the same does not apply to null exclusion in the case of parameters of access types. There is a general philosophy that 'null exclusions never lie'. In other words if we give a null exclusion then the entity must exclude null; however, if no null exclusion is given then the entity might nevertheless exclude null. (A good example of this occurs in Section 14.4 when we discuss access parameters and tagged types.) In the case of renaming we might have

```
procedure P(X: not null access T);
procedure Q(Y: access T) renames P;          -- legal
...
Q(null);                                      -- raises Constraint_Error
```

The call of Q raises Constraint_Error because the parameter excludes null even though there is no explicit null exclusion on the renaming. On the other hand

```
procedure P(X: access T);
procedure Q(Y: not null access T) renames P;    -- illegal
```

is illegal because the null exclusion in the renaming is a lie.

Although the number of parameters must match, the presence, absence, or value of default parameters do not have to match. So renaming can be used to introduce, change or delete default expressions; the default parameters associated with the new name are those shown in the renaming declaration. Hence renaming cannot be used as a trick to give an operator default values. Similarly, parameter names do not have to match but naturally the new names must be used for named parameters of calls using the new subprogram name.

We can also provide the body of a subprogram as simply a renaming of another subprogram. The text looks exactly the same as when renaming a subprogram specification. Thus in the body of the package Simple_IO of Section 2.2 we might implement Put for the type String as simply

```
procedure Put(S: in String) renames Ada.Text_IO.Put;
```

There are, moreover, two extra rules when renaming is used for subprogram bodies. All profile subtypes must statically match and the calling conventions must

Table 13.1 Conformance matching of profiles.

Level	*Matches*	*Used for*
type	types	hiding (see Section 10.6) overriding untagged primitive ops (12.3)
mode	+ modes	renaming as spec generic subprograms (19.2, 19.3)
subtype	+ subtypes statically + convention	renaming as body access to subprograms (11.8) overriding tagged primitive ops (14.1) requeue (20.9)
full	+ names + defaults + null exclusions	distinct specs and bodies (8.6) stubs and subunits (13.2) entry specs and bodies/accept (20.2, 20.4) repeated discriminants (18.1)

be the same; this is called subtype conformance and is the same as that required for access to subprogram types. The reason for requiring subtype conformance is so that a simple jump to the old subprogram can be compiled as the call to the new one.

A subprogram renaming is treated as a renaming as body rather than as a renaming as specification only if a body is needed to complete an existing specification. Note that a renaming as body can occur in a private part; this is discussed further in the answer to Exercise 19.2(4).

Renamings as bodies may require an elaboration check and so if misused could raise Program_Error. More interestingly, it is possible to create circularities using renaming as bodies; some are illegal because of freezing rules (see Section 25.1), but some result in infinite loops.

The reader may be baffled by all the various levels of conformance we have mentioned from time to time. For convenience they are summarized in Table 13.1 which gives them in increasing order of strength.

The unification of subprograms and enumeration literals is further illustrated by the fact that an enumeration literal can be renamed as a parameterless function with the appropriate result. For example

```
function Ten return Roman_Digit renames 'X';
```

As mentioned above, object renaming simply provides a new name for an existing object and if a subtype name is supplied then any constraints implied are ignored. (We cannot give an explicit constraint in an object renaming.)

Somewhat stricter rules apply to access types. In the case of anonymous access types we might write

```
Local_Ptr: access T renames Ptr;
Local_Const: access constant T renames ACT;
Local_Sub: access procedure (X: Integer) renames P;
```

In the case of access to object types, the designated subtypes must statically match and they must both be constant or not. In the case of access to subprogram types the profiles must have subtype conformance.

It will be recalled from Section 11.7 that if we copy an access parameter to a local variable then the accessibility information was lost in Ada 2005 although preserved in Ada 2012 and Ada 2022. Similarly, renaming does not lose accessibility information, so we can write

```
procedure P(Ptr: access T) is
   Local_Ptr: access T renames Ptr;
...
```

and then Local_Ptr retains the accessibility information of the parameter Ptr.

For all access to object type renamings (named or anonymous) any null exclusion given on the renaming must not lie.

Renaming can also be used to partially evaluate the name of an object. Suppose we have an array of records such as the array People in Section 8.7 and that we wish to scan the array and print out the dates of birth in numerical form. We could write

```
for S in People'Range loop
   Put(People(S).Birth.Day);  Put(" : ");
   Put(Month_Name'Pos(People(S).Birth.Month)+1);  Put(" : ");
   Put(People(S).Birth.Year);
end loop;
```

It is clearly painful to repeat People(S).Birth each time. We could declare a variable D of type Date and copy People(S).Birth into it, but this would be very wasteful if the record were at all large. We could also use an access variable to refer to it but this would mean making the variable aliased as well as introducing an access type which might otherwise not be necessary. A better technique is to use renaming thus

```
for S in People'Range loop
   declare
      D: Date renames People(S).Birth;
   begin
      Put(D.Day);  Put(" : ");
      Put(Month_Name'Pos(D.Month)+1);  Put(" : ");
      Put(D.Year);
   end;
end loop;
```

Beware that renaming does not correspond to text substitution – the identity of the object is determined when the renaming occurs. If any variable in the name subsequently changes then the identity of the object does not change. Remember that any constraints implied by the subtype mark in the renaming declaration are totally ignored; those on the original object still apply.

Package renaming takes a very simple form and is particularly useful with long child names thus

```
package Rights renames OS.Internals.Access_Rights;
```

The abbreviated name could for example be used in a with clause and maybe as an alternative to a use clause. Sadly, the abbreviated name cannot be used as part of the name for declaring a further child. So we could not write

```
package Rights.Data is                    -- illegal
```

but have to spell it out in pitiless detail

```
package OS.Internals.Access_Rights.Data is
```

In summary, renaming can be applied to objects (variables and constants), components of composite objects (including slices of arrays), exceptions (see Chapter 15), subprograms, and packages.

Although renaming does not directly apply to types an almost identical effect can be achieved by the use of a subtype

```
subtype S is T;
```

It can also be used with incomplete types as a convenient abbreviation

```
subtype Dept is Departments.Department;        -- see Section 13.5
```

but constraints and null exclusions cannot be used with incomplete types.

Note also that renaming cannot be applied to a named number. This is partly because renaming requires a type name and named numbers do not have an explicit type name (we will see in Chapter 17 that they are of so-called universal types which cannot be explicitly named). Thus the constant e (the base of natural logarithms) in the package Ada.Numerics cannot be given a local renaming for brevity (not that it is exactly long anyway!). However, there is no need, we can just declare another named number

```
E: constant := Ada.Numerics.e;
```

which is no disadvantage because the named numbers are not run-time objects anyway and so there is no duplication.

Finally, note that renaming does not automatically hide the old name nor does it ever introduce a new entity; it just provides another way of referring to an existing entity or, in other words, another view of it (that is why new constraints in a renaming declaration are ignored). Renaming can be very useful at times but the indiscriminate use of renaming should be avoided since the aliases introduced can make understanding the program more difficult.

We conclude with an interesting change in Ada 2022 which is that we can also rename a value denoted by a name but not by a general expression. However, if an expression is qualified then it is classed as a name and so can also be renamed. Thus we have

```
Thirteen: constant := 13;                 -- OK
Tough_Luck renames 13;                    -- illegal
Unlucky renames Thirteen;                 -- OK
Lucky renames Integer'(1 + 2 + 4);        -- OK - a secret 7
```

Exercise 13.7

1. Declare a renaming of the literal Mon of type Day from the package Diurnal of Section 12.1.
2. Declare a renaming of Diurnal.Next_Work_Day.
3. Declare Pets as a renaming of part of the second Farmyard of Section 8.5.
4. Rename the operator "+" from the package Complex_Numbers of Section 12.2 so that it can be used in infix notation without a use or use type clause.
5. Declare a renaming to avoid the repeated evaluation of World(I, J) in the answer to Exercise 11.4(**2**).

13.8 Programs, partitions, and elaboration

We mentioned in Section 13.1 the idea that a program starts execution by some external magic calling the main subprogram. In fact, in the general case, a complete program comprises a number of partitions each of which can have its own main subprogram. Partitions typically have their own address space and communication between partitions is restricted although the typing rules are enforced across the whole program. The details of this topic are outside the core language and are covered by the Distributed Systems annex discussed very briefly in Section 26.3. We will restrict ourselves here to programs consisting of just one partition.

The first thing that happens when a program starts is that the library units have to be elaborated. This has to be done before the main subprogram is called because it might depend upon the library units. (Incidentally, there need not be a main subprogram at all, see Section 27.3.)

Elaboration of some library units may be trivial, but in the case of a package the initial values of top level objects must be evaluated and if the body has an initialization part then that must be executed.

The order of these elaborations is not precisely specified but it must be consistent with the dependencies between the units. In addition, the pragma Elaborate (or Elaborate_All) can be used to specify that a body is to be elaborated before the current unit. This may be necessary to prevent ambiguities or even to prevent Program_Error. Consider the situation mentioned at the end of Section 12.1 thus

```
package P is
   function A return Integer;
end P;

package body P is
   function A return Integer is
   begin
      return ... ;
   end A;
end P;
```

```
with P;
package Q is
   I: Integer := P.A;
end Q;
```

The three units can be compiled separately and the dependency requirements are that the body of P and the specification of Q must both be compiled after the specification of P is entered into the library. But there is no need for the body of P to be compiled before the specification of Q. However, when we come to elaborate the three units it is important that the body of P be elaborated before the specification of Q otherwise Program_Error will be raised by the attempt to call the function A in the declaration of the integer I.

The key point is that the dependency rules only partially constrain the order in which the units are elaborated and so we need some way to impose order at the library level much as the linear text imposes elaboration order within a unit. This can be done by various pragmas. Thus writing

```
with P; pragma Elaborate(P);                -- context clause of Q
package Q is
   I: Integer := P.A;
end Q;
```

specifies that the body of P is to be elaborated before the specification of Q. The pragma goes in the context clause of Q and can refer to one or more of the library units mentioned earlier in the context clause. The pragma can be used even if the with clause is a private with clause but not a limited with clause.

The related pragma Elaborate_All specifies that all library units needed by the named units are elaborated before the current unit (and is thus transitive).

We now turn to consider various aspects that traditionally were indicated using pragmas but now are typically enforced by aspect specifications.

The aspect Elaborate_Body specifies that a body is to be elaborated immediately after the corresponding specification. We mentioned in Section 12.1 that a library package can only have a body if required by some rule. If we want the package Stack.Data of Section 13.4 to have a body in order to initialize Top then we can use the aspect Elaborate_Body to force a body to be required, thus

```
private package Stack.Data
      with Elaborate_Body is
   ...
end Stack.Data;
```

If the dependencies and any elaboration aspects and pragmas are such that no consistent order of elaboration exists then the program is illegal; if there are several possible orders and the behaviour of the program depends on the particular order then it will not be portable.

There are three other aspects relating to elaboration. They are Preelaborate, Pure, and Preelaborable_Initialization.

The aspect Preelaborate essentially states that the unit can be elaborated before the program executes; generally this means that it is all static and has no code. It doesn't mean that it necessarily will be elaborated before the program runs but simply that it could be if the implementation were up to it; but certainly all such preelaborable units will be elaborated before other units which are not marked as preelaborable. This concept is important for certain real-time and distributed systems.

The aspect Pure indicates that the unit is not only preelaborable but also has no state. As noted in Chapter 4, this applies to the package Ada which traditionally was written as

```
package Ada is
  pragma Pure(Ada);
end Ada;
```

Pure units can only depend upon other pure units; preelaborable units can only depend upon other preelaborable units including pure units.

Many aspects can be given by using a pragma or by an aspect specification as outlined in Section 5.6. Thus we can also write

```
package Ada
    with Pure is
end Ada;
```

although maybe that seems a bit odd because of the juxtaposition of **is** and **end**.

However, using an aspect specification might be useful for development purposes since we could change the state of a whole group of units in one blow by a structure such as

```
package State_Control is
  Purity: constant Boolean := True;
end State_Control;

...

with State_Control;
package P1
    with Pure => State_Control.Purity is
  ...
end P1;
-- and so on for packages P2, P3 etc
```

We could then set Purity false for adding some debugging material during development and then set it to true for final production.

The final aspect is Preelaborable_Initialization. It is used in situations like the following

```
package Q
    with Preelaborate is
  type T is private
    with Preelaborable_Initialization;
private
```

```
    type T is
      record
        C: Integer := 7;
      end record;
  end Q;

  with Q;
  package P
      with Preelaborate is
    Obj: Q.T;
  end P;
```

The package Q is preelaborable because it does not declare any objects. But the package P does declare an object and is only permitted to be preelaborable because the object has static initialization. But the object is of a private type and so this fact would not normally be visible to P. However, the package Q includes the aspect Preelaborable_Initialization for the type T and this promises that T will indeed be preelaborable. Without the aspect Preelaborable_Initialization on T, the package P would not be preelaborable and the aspect Preelaborate on P would be illegal.

Many examples of the use of these three aspects, Pure, Preelaborate, and Preelaborable_Initialization, will be found in the predefined library discussed in Chapter 23. A minor advantage of using aspect specifications rather than pragmas is that several aspects can be given together whereas a bunch of pragmas looks ugly and takes more space.

A somewhat related pragma is Restrictions. This states that certain aspects of the language are not used. Most restrictions concern the Real-Time Systems or High Integrity annexes but a few relate to the core language.

The restriction No_Dependence asserts that the partition does not use a particular library unit. Thus we might write

```
pragma Restrictions(No_Dependence => Ada.Text_IO);
```

The parameter need not be a predefined unit such as Ada.Text_IO but could be any unit. Care is needed to spell the name correctly. Thus writing No_Dependence => Supperstring will not guard against using the package Superstring.

We can also write

```
pragma Restrictions(No_Implementation_Pragmas,
                                        No_Implementation_Attributes);
```

These do not apply to the whole partition but only to the compilation or library environment concerned. This helps us to ensure that implementation parts of a program are identified. There is also a similar restriction No_Obsolescent_Features which ensures that we do not use features that are now deemed obsolescent such as the package ASCII.

A full list of all the predefined restrictions will be found in Appendix 1.

Exercise 13.8

1 The library package P has a Boolean variable B in its visible part. Library packages Q and R both have initialization parts; that in Q sets P.B to True whereas that in R sets P.B to False. Prevent any ambiguities by the use of elaborate pragmas.

Checklist 13

A library unit cannot be compiled until other library units mentioned in its with clause are entered into the library.

A body cannot be compiled until the corresponding specification is entered into the library.

A subunit cannot be compiled until its parent body is entered into the library.

A child cannot be compiled until the specification of its parent is entered into the library.

A package specification and body form a single declarative region.

A context clause on a specification also applies to the body and any children. Similarly a context clause on a body also applies to any subunits.

A library package can only have a body if it needs one to satisfy other language rules.

Do not attempt to redefine Standard.

Renaming is not text substitution.

Limited with and private with clauses were added in Ada 2005.

The pragma Preelaborable_Initialization was added in Ada 2005.

The Restrictions identifiers No_Dependence, No_Implementation_Attributes, No_Implementation_Pragmas, and No_Obsolescent_Features were added in Ada 2005.

An incomplete type can be completed by a private type in Ada 2012.

The parameter mechanism for subprogram parameters does not need to be known when the subprogram specification is declared, but only when a call is made or the body declared.

New in Ada 2022

The type can be omitted when renaming an object; all entities denoted by a name can also be renamed.

A preelaborable unit can call static functions and a few other system functions.

Program 3

Rational Reckoner

This program illustrates the use of a hierarchical library to create an abstract data type and its associated operations. It is based on the type Rational of Exercise 12.2(3).

The root package Rational_Numbers declares the type Rational and the usual arithmetic operations. The private child package Rational_Numbers.Slave contains the functions Normal and GCD as in Exercise 13.4(1). Thus Normal cancels any common factors in the numerator and denominator and returns the normalized form. It is called by various operations in the parent package. It would be possible to restructure Normal so that it was a private child function.

Appropriate input–output subprograms are declared in the public child Rational_Numbers.IO.

In order to exercise the rational subsystem a simple interpretive calculator is added as the main subprogram. This enables the user to read in rational numbers, perform the usual arithmetic operations upon them, and print out the result.

The numbers are held in a stack which is operated upon in the reverse Polish form (Polish after the logician Jan Łucasiewicz). The root package Rat_Stack contains subprograms Clear, Push, and Pop. The private child package Rat_Stack.Data contains the actual data, and the public child procedure Rat_Stack.Print_Top prints the top item. This structure is somewhat similar to that described in Section 13.4 and enables the stack size to be changed with minimal effect on the rest of the system.

The main subprogram is decomposed into subunits Process and Get_Rational in order to isolate the dependencies and clarify the structure.

```
package Rational_Numbers is
   type Rational is private;

   -- unary operators
   function "+" (X: Rational) return Rational;
   function "-" (X: Rational) return Rational;

   -- binary operators
   function "+" (X, Y: Rational) return Rational;
   function "-" (X, Y: Rational) return Rational;
   function "*" (X, Y: Rational) return Rational;
   function "/" (X, Y: Rational) return Rational;

   -- constructor and selector functions
   function "/" (X: Integer; Y: Positive) return Rational;
   function Numerator(R: Rational) return Integer;
   function Denominator(R: Rational) return Positive;
private
   type Rational is
      record
         Num: Integer := 0;        -- numerator
         Den: Positive := 1;       -- denominator
      end record;
end;

----------------------------------------

private package Rational_Numbers.Slave is
   function Normal(R: Rational) return Rational;
end;

package body Rational_Numbers.Slave is

   function GCD(X, Y: Natural) return Natural is
   begin
      if Y = 0 then
         return X;
      else
         return GCD(Y, X mod Y);
      end if;
   end GCD;

   function Normal(R: Rational) return Rational is
      G: Positive := GCD(abs R.Num, R.Den);
   begin
      return (R.Num/G, R.Den/G);
   end Normal;

end Rational_Numbers.Slave;

----------------------------------------
```

```
with Rational_Numbers.Slave;
package body Rational_Numbers is
   use Slave;

   function "+" (X: Rational) return Rational is
   begin
      return X;
   end "+";

   function "-" (X: Rational) return Rational is
   begin
      return (-X.Num, X.Den);
   end "-";

   function "+" (X, Y: Rational) return Rational is
   begin
      return Normal((X.Num*Y.Den + Y.Num*X.Den,
                                          X.Den*Y.Den));
   end "+";

   function "-" (X, Y: Rational) return Rational is
   begin
      return Normal((X.Num*Y.Den - Y.Num*X.Den,
                                          X.Den*Y.Den));
   end "-";

   function "*" (X, Y: Rational) return Rational is
   begin
      return Normal((X.Num*Y.Num, X.Den*Y.Den));
   end "*";

   function "/" (X, Y: Rational) return Rational is
      N: Integer := X.Num*Y.Den;
      D: Integer := X.Den*Y.Num;
   begin
      if D < 0 then D := -D;  N := -N; end if;
      return Normal((Num => N, Den => D));
   end "/";

   function "/" (X: Integer; Y: Positive)
                                   return Rational is
   begin
      return Normal((Num => X, Den => Y));
   end "/";

   function Numerator(R: Rational)
                                   return Integer is
   begin
      return R.Num;
   end Numerator;

   function Denominator(R: Rational)
                                   return Positive is
   begin
      return R.Den;
   end Denominator;

end Rational_Numbers;
```

```
package Rational_Numbers.IO is
   procedure Get(X: out Rational);
   procedure Put(X: in Rational);
end;
```

```
with Ada.Text_IO, Ada.Integer_Text_IO;
use Ada;
with Rational_Numbers.Slave;
package body Rational_Numbers.IO is

   procedure Get(X: out Rational) is
      N: Integer;              -- numerator
      D: Integer;              -- denominator
      C: Character;
      EOL: Boolean;            -- end of line
   begin
      -- read the (possibly) signed numerator
      -- this also skips spaces and newlines
      Integer_Text_IO.Get(N);
      Text_IO.Look_Ahead(C, EOL);
      if EOL or else C /= '/' then
         raise Text_IO.Data_Error;
      end if;
      Text_IO.Get(C);     -- remove the / character
      Text_IO.Look_Ahead(C, EOL);
      if EOL or else C not in '0' .. '9' then
         raise Text_IO.Data_Error;
      end if;
      -- read the unsigned denominator
      Integer_Text_IO.Get(D);
      if D = 0 then
         raise Text_IO.Data_Error;
      end if;
      X := Slave.Normal((N, D));
   end Get;

   procedure Put(X: in Rational) is
   begin
      Integer_Text_IO.Put(X.Num, 0);
      Text_IO.Put('/');
      Integer_Text_IO.Put(X.Den, 0);
   end Put;

end Rational_Numbers.IO;
```

```
with Rational_Numbers;
use Rational_Numbers;
package Rat_Stack is
   Error: exception;
   procedure Clear;
   procedure Push(R: in Rational);
   function Pop return Rational;
end;
```

```
private package Rat_Stack.Data is
   Max: constant := 4;          -- stack size
   Top: Integer := 0;
   Stack: array (1 .. Max) of Rational;
end Rat_Stack.Data;
```

```
with Rat_Stack.Data;
package body Rat_Stack is
   use Data;

   procedure Clear is
   begin
      Top := 0;
   end Clear;

   procedure Push(R: in Rational) is
   begin
      if Top = Max then
         raise Error;
      end if;
      Top := Top + 1;
      Stack(Top) := R;
   end Push;

   function Pop return Rational is
   begin
      if Top = 0 then
         raise Error;
      end if;
      Top := Top - 1;
      return Stack(Top + 1);
   end Pop;

end Rat_Stack;
----------------------------------------

with Rational_Numbers.IO;
with Ada.Text_IO;
private with Rat_Stack.Data;
procedure Rat_Stack.Print_Top is
   use Data;
begin
   if Top = 0 then
      Ada.Text_IO.Put("Nothing on stack");
   else
      Rational_Numbers.IO.Put(Stack(Top));
   end if;
   Ada.Text_IO.New_Line;
end Rat_Stack.Print_Top;
----------------------------------------

with Rat_Stack;
with Ada.Text_IO;  use Ada.Text_IO;
procedure Rational_Reckoner is
   C: Character;
   Control_Error, Done: exception;
   procedure Process(C: Character) is separate;

begin
   Put("Welcome to the Rational Reckoner");
   New_Line(2);
   Put_Line("Operations are + - * / ? ! plus eXit");
   Put_Line("Input rational by #[sign]digits/digits");
   Rat_Stack.Clear;
   loop
      begin
         Get(C);
         Process(C);
      exception
         when Rat_Stack.Error =>
            New_Line;
            Put_Line("Stack overflow/underflow, " &
                       "stack reset");
            Rat_Stack.Clear;
         when Control_Error =>
            New_Line;
            Put_Line("Unexpected character, " &
                       "not # + - * / ? ! or X");
         when Done =>
            exit;
      end;
   end loop;
   New_Line;
   Put_Line("Finished");
   Skip_Line(2);
end Rational_Reckoner;
----------------------------------------

with Rat_Stack.Print_Top;  use Rat_Stack;
with Rational_Numbers;  use Rational_Numbers;
separate(Rational_Reckoner)
procedure Process(C: Character) is
   R: Rational;
   procedure Get_Rational(R: out Rational)
                                       is separate;
begin
   case C is
      when '#' =>
         Get_Rational(R);
         Push(R);
      when '+' =>
         Push(Pop + Pop);
      when '-' =>
         R := Pop;  Push(Pop - R);
      when '*' =>
         Push(Pop * Pop);
      when '/' =>
         R := Pop;  Push(Pop / R);
      when '?' =>
         Print_Top;
      when '!' =>
         Print_Top;  R := Pop;
      when ' ' =>
         null;
      when 'X' | 'x' =>
         raise Done;
      when others =>
         raise Control_Error;
   end case;
end Process;
----------------------------------------
```

```
with Rational_Numbers.IO;
separate(Rational_Reckoner.Process)
procedure Get_Rational(R: out Rational) is
begin
  loop
    begin
      IO.Get(R);
      exit;
    exception
      when Data_Error =>
        Skip_Line; New_Line;
        Put_Line("Not a rational, try again ");
        Put('#');
    end;
  end loop;
end Get_Rational;
```

The procedures for input and output of rational numbers are in a distinct package since it is likely that alternative formats might be tried and this avoids changing the root package.

For simplicity the procedure Get uses the predefined procedure Get for reading integers in Ada.Integer_Text_IO. The expected format for a rational number consists of an optional sign, a sequence of digits, a solidus (/), and then another sequence of digits; for example -34/67. Using the integer Get for the denominator automatically causes any leading spaces and newlines to be skipped and any sign to be processed. It then uses the procedure Look_Ahead to look at the next characters (see Section 23.6) and raises the predefined exception Ada.IO_Exceptions.Data_Error (via the renaming in Ada.Text_IO, see Sections 23.5 and 23.6) if the numerator is not immediately followed by a solidus and a digit. It then reads the (unsigned) denominator and checks that it is not zero. The behaviour with regard to skipping spaces and newlines follows the same general pattern as for the predefined subprograms Get for the predefined types. Note, however, that by using the predefined Get for integers, we do actually allow the numerator and denominator to be in based notation and to have an exponent!

By contrast, Put is almost trivial although more elaborate formats could be devised.

The functions Normal and GCD are also in a distinct package. Again this enables alternative algorithms to be used without recompiling the entire system. Note that since the denominator is never zero, the test Y = 0 in the top level call of GCD never succeeds and so at least one recursion always occurs. The parameters could be interchanged to avoid this. An alternative approach is to use the iterative form as in the answer to Exercise 10.1(7).

A further discussion on the structure of these packages will be found on the web.

The stack subsystem has been divided into several units mainly so that Print_Top is distinct and alternative formats can be tried without disturbing the whole of the stack system. Thus Print_Top has a private with clause giving access to the data in the private package Rat_Stack.Data. Another approach might have been to put the data in the private part of the root package Rat_Stack but this would mean recompiling the whole system if the stack depth (currently 4) were changed.

The main subprogram has been decomposed into subunits largely to clarify the exception handling and avoid deep indentation; it also reduces the dependencies somewhat. Exception handling is discussed in detail in Chapter 15 but it is hoped that the simple structure shown here is quite clear. The calculator is driven by the procedure Process which manipulates the system according to a control character. The character # signifies that a rational number is to follow; the characters +, –, * and / cause the appropriate operations on the top two items of the stack (care has to be taken with - and / that the items are used in the correct order); the characters ? and ! cause the top item to be printed and ! also causes the top item to be deleted. Spaces are ignored. Thus the sequence

```
#3/4 #2/3 + !
```

results in 17/12 being output. The character X or x causes the system to terminate.

If the stack is misused then Rat_Stack.Error is raised and the stack reset to empty. An unexpected control character raises Control_Error and the system continues after a suitable warning. Note also that termination is indicated by the raising of the exception Done; this might be considered bad practice since termination in this case is not an exceptional situation. On the other hand it is quite convenient to bundle it in with the other exceptions.

The reading of a rational number is carried out by the subunit Get_Rational. Spaces and newlines are allowed between the # and the number but not within the number itself. If Data_Error is raised because the sequence following # is not recognized then the user is invited to resubmit a sequence and a further # is output as a prompt. Note also how a call of Skip_Line removes any unused characters.

The system could clearly be extended in various ways. For example, a prompt might be output whenever input is expected. Another improvement would be to handle Constraint_Error which is raised if any of the numeric operations overflow.

14 Object Oriented Programming

14.1 Type extension
14.2 Polymorphism
14.3 Abstract types and interfaces
14.4 Primitive operations and tags
14.5 Views and redispatching
14.6 Private types and extensions
14.7 Controlled types
14.8 Multiple inheritance
14.9 Multiple implementations

We now come to a discussion of the basic features of Ada which support what is generally known as object oriented programming or OOP. Ingredients of OOP include the ability to extend a type with new components and operations, to identify a specific type at run time and to select a particular operation depending on the specific type. A major goal is the reuse of existing reliable software without the need for recompilation.

This chapter concentrates on the fundamental ideas of type extension and polymorphism. Related topics are type parameterization (using discriminants) and genericity; these are discussed in Chapters 18 and 19 respectively. Finally, Chapter 21 considers how these various aspects of OOP all fit together especially with regard to multiple inheritance.

The reader might find it helpful to reread Sections 3.2 and 3.3 before considering this chapter in detail.

14.1 Type extension

In Section 12.3 we saw how it was possible to declare a new type as derived from an existing type and how this enabled us to use strong typing to prevent inadvertent mixing of different uses of similar types. We also saw that primitive operations were inherited by the derived type and could be overridden and, moreover, that further primitive operations could be added if the derivation were in a package specification.

We now introduce a more flexible form of derivation where it is possible to add additional components to a record type as well as additional operations. This gives rise to a possible tree of types where each type contains the components of its parent plus other components as well. Since the types are clearly different, although with common properties, it is convenient to be able to deal with an object of any type in the tree and to determine the type of the object at run time. It is clear therefore that each object has to have some additional information indicating its type. This additional information is provided by a hidden component called the tag.

Accordingly, record types can be extended on derivation provided they are marked as tagged. Private types implemented as records can also be marked as tagged.

We saw a simple example of a hierarchy of tagged types plus associated primitive operations in Chapter 3 where we declared the types Object, Circle and so on. These could be declared in one or several packages as illustrated in the answer to Exercise 3.2(**1**), thus

```
package Objects is

  type Object is tagged
    record
      X_Coord: Float;
      Y_Coord: Float;
    end record;

  function Distance(O: Object) return Float;
  function Area(O: Object) return Float;

end Objects;

with Objects; use Objects;
package Shapes is

  type Circle is new Object with
    record
      Radius: Float;
    end record;

  function Area(C: Circle) return Float;

  type Point is new Object with null record;

  type Triangle is new Object with
    record
      A, B, C: Float;
    end record;

  function Area(T: Triangle) return Float;

end Shapes;
```

In this example the type Object is the root of a tree of types and Circle, Point, and Triangle are derived from it. Note how the extra components are indicated and that in the case of the type Point where no extra components are added we have to indicate this explicitly by **with null record**; this makes it clear to the reader that it is a tagged type since every tagged type has **tagged** or **with** (or **interface** as discussed in Section 14.3) in its declaration.

The primitive operations of Object are Area and Distance and these are inherited by Circle, Point, and Triangle. Although Distance is appropriate for all the types, the function Area is redefined for Circle and Triangle. (However, as we saw in Section 3.3, the function Area for Object should really be abstract so that it cannot be accidentally inherited; see Section 14.3. Moreover, Distance would be better with a class wide parameter as noted in the next section.)

Type conversion is always allowed towards the root of the tree, but an extension aggregate is required in the opposite direction in order to give values for any additional components. So we can write

```
O: Object := (1.0, 0.5);
C: Circle := (0.0, 0.0, 34.7);
T: Triangle;
P: Point;
...
O := Object(C);
...
C := (O with 41.2);
T := (O with A => 3.0, B => 4.0, C => 5.0);
P := (O with null record);
```

An extension aggregate can use positional or named notation and the familiar **with null record** is required when there are no extra components.

The expression before **with** can be an expression of any appropriate ancestor type, it does not have to be of the immediate parent type. So moving into another dimension we might have

```
type Cylinder is new Circle with
   record
      Height: Float;
   end record;

Cyl: Cylinder;
```

and then we could write any of

```
Cyl := (O with Radius => 41.2, Height => 231.6);
Cyl := (C with Height => 231.6);
Cyl := (Object(T) with 41.2, 231.6);
```

In the last case we first convert the triangle T to an object and then extend it to give a cylinder. However, note that the type Cylinder is not sensible since a cylinder is not a form of circle at all. We will return to this topic in Section 21.1.

When we come to Section 14.7 we will find that it is sometimes not possible to give an expression of an ancestor type; as an alternative we can simply provide a subtype mark instead. In fact we can always do this and then the components corresponding to the ancestor type are initialized by default (if at all) as for any object of the type. So writing

```
C := (Object with Radius => 41.2);
```

will result in the circle having components as if we had written

```
Obj: Object;                    -- no initial value
C: Circle := (Obj with Radius => 41.2);
```

and so the coordinates of C are not defined.

This is clearly not very sensible in this case but we might have chosen to declare the type Object so that it was by default at the origin thus

```
type Object is tagged
  record
    X_Coord: Float := 0.0;
    Y_Coord: Float := 0.0;
  end record;
```

and then the circle would also by default be at the origin.

At this point we pause to consider the main commonalities and differences between type derivation with tagged types which we have just been discussing and type derivation with other types as discussed in Section 12.3.

The common points are

- existing components are inherited,
- inheritance, overriding, and addition of primitive operations are allowed in the same places; additional operations are only allowed if the derivation occurs in a package specification,
- derivation can occur in the same package specification as the parent and inherits all its primitive operations declared so far.

On the other hand, the differences are

- derivation from a tagged type freezes it so that no further operations can be added,
- a tagged record type can have additional components,
- type conversion is only allowed towards the ancestor for a tagged type; both ways for untagged types,
- inherited operations of tagged types have the same convention as the parent type whereas inherited operations of untagged types are always intrinsic; remember that the Access attribute cannot be applied to intrinsic operations,
- if an inherited operation is overridden then the conformance requirements are different; in the case of a tagged type it must have subtype conformance, whereas for an untagged type it only has to have type conformance.

The reason for the last difference will become apparent later.

We now consider a more extensive example which illustrates the use of tagged types to build a system as a hierarchy of types. We will see how this allows the system to be extended without recompilation of its central part.

Our system concerns the processing of reservation requests for a mythical Ada Airlines (not low cost but naturally very reliable). There are presumably various aspects to this: the creation of a request; its processing by some central system; and then reporting back indicating success or failure. There are three categories of travel, Basic, Nice, and Posh. The better categories have options which can be requested when making the reservation. Nice passengers are given a choice of seat (Aisle or Window) and a choice of meal which can be Green (vegetarian), White (fish or fowl) or Red (for the carnivores). Posh passengers are also given onward personal ground transport (or Personal Onward Surface Help).

We concentrate on the part of the system that processes the requests and consider the hierarchy of types required. The package specification might be

```
package Reservation_System is

   type Position is (Aisle, Window);
   type Meal_Type is (Green, White, Red);

   type Reservation is tagged
      record
         Flight_Number: Integer;
         Date_Of_Travel: Date;
         Seat_Number: String(1 .. 3) := "   ";
      end record;

   procedure Make(R: in out Reservation);
   procedure Select_Seat(R: in out Reservation);

   type Basic_Reservation is new Reservation with null record;

   type Nice_Reservation is new Reservation with
      record
         Seat_Sort: Position;
         Food: Meal_Type;
      end record;

   overriding
   procedure Make(NR: in out Nice_Reservation);              -- overrides
   procedure Order_Meal(NR: in Nice_Reservation);

   type Posh_Reservation is new Nice_Reservation with
      record
         Destination: Address;
      end record;

   overriding
   procedure Make(PR: in out Posh_Reservation);              -- overrides
   procedure Arrange_Limo(PR: in Posh_Reservation);

end Reservation_System;
```

We start with the root type Reservation which contains the components common to all reservations. It also has a procedure Make which performs all those actions common to making a reservation of any category. The type Basic_ Reservation is simply a copy of Reservation (note **with null record**;) and could be dispensed with; Basic_Reservation inherits the procedure Make from Reservation. The type Nice_Reservation extends Reservation and provides its own procedure Make thus overriding the inherited version. The type Posh_Reservation further extends Nice_Reservation and similarly provides its own procedure Make. The procedures Select_Seat, Order_Meal, and Arrange_Limo are called by the various procedures Make as required. The main purpose of Select_Seat is to fill in the seat number with a string such as "56A"; hence the parameter has in out mode.

Note the use of the reserved word **overriding** which precedes the overriding procedures Make. This is optional as explained in Section 12.3 and is an indication to the compiler (and anyone maintaining the program later) that this truly is an overriding of an existing inherited operation. This is a safeguard and the program will fail to compile if we accidentally spell Make incorrectly or give it the wrong number of parameters. We can also write **not overriding** when declaring a new operation such as Select_Seat or the first time we declare Make.

The package body might be as follows

```
package body Reservation_System is

  procedure Make(R: in out Reservation) is
  begin
    Select_Seat(R);
  end Make;

  procedure Make(NR: in out Nice_Reservation) is
  begin
    Make(Reservation(NR));            -- make as plain reservation
    Order_Meal(NR);
  end Make;

  procedure Make(PR: in out Posh_Reservation) is
  begin
    Make(Nice_Reservation(PR));       -- make as nice reservation
    Arrange_Limo(PR);
  end Make;

  procedure Select_Seat(R: in out Reservation) is separate;
  procedure Order_Meal(NR: in Nice_Reservation) is separate;
  procedure Arrange_Limo(PR: in Posh_Reservation) is separate;

end Reservation_System;
```

Each distinct body for Make contains just the code immediately relevant to the type and delegates other processing back to its parent using an explicit type conversion. This avoids repetition of code and simplifies maintenance. Note carefully that all type checking is static; the choice of Make is done with simple overload resolution based on the known type of the parameter. (The reader may feel concerned that the procedure Select_Seat does not have enough information to work for nice and posh passengers; do not worry – all will be revealed in Section 14.5.)

If at a later date the growing Ada Airlines purchases some secondhand Concordes (alas no more) then a new reservation type such as Supersonic_Reservation will be required. This can be added without recompiling (and perhaps more importantly, without retesting) the existing code.

```
with Reservation_System;
package Supersonic_Reservation_System is
```

```
   type Supersonic_Reservation is
                    new Reservation_System.Reservation with
     record
        Champagne: Vintage;
        ...   -- other supersonic components
     end record;

   overriding
   procedure Make(SR: in out Supersonic_Reservation);
   ...
end Supersonic_Reservation_System;
```

We could have made this package a child of Reservation_System. This would have avoided the need for a with clause and emphasized that it was all really part of the same system. The entities in the parent package would also be immediately visible and use clauses or dotted names would be avoided. Furthermore, as we shall see in Section 14.6, the fact that the child package can see the private part of its parent can be very important.

Exercise 14.1

1. Declare a point P with the same coordinates as a given circle C.
2. Declare an object R of type Reservation and assign to it an appropriate flight number and date. Then declare an object NR of the type Nice_Reservation with common components the same as R and a window seat and vegetarian meal.
3. Rewrite the package Supersonic_Reservation_System as a child package of Reservation_System.

14.2 Polymorphism

The facilities we have seen so far have allowed us to define a new type as an extension of an existing one. We have introduced the different categories of Reservation as distinct but related types. What we also need is a means to manipulate any kind of Reservation and to process it accordingly. We do this through the introduction of the notion of class wide types which provide dynamic polymorphism.

Each tagged type T has an associated type denoted by T'Class. This type comprises the union of all the types derived from T. The values of T'Class are thus the values of T and all its derived types; the type T is known as the root type of the class. Moreover, a value of any type derived from T can always be converted to the type T'Class (implicitly in certain contexts).

So, for example, in the case of the type Reservation the types can be pictured as in Figure 14.1. A value of any of the reservation types can be converted to Reservation'Class. Note carefully that Nice_Reservation'Class is not the same as Reservation'Class; the former consists just of Nice_Reservation and Posh_Reservation.

Each value of a class wide type has a tag which identifies its particular type at run time. Thus the tag acts as a hidden component as mentioned earlier.

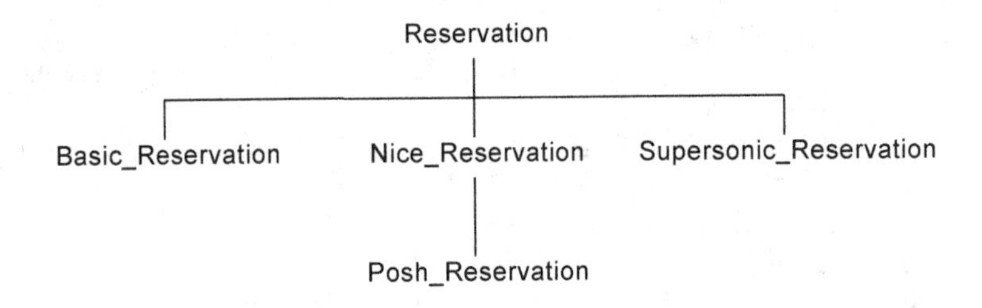

Figure 14.1 A tree of types.

The type T'Class is treated as an indefinite type (like an unconstrained array type); this is because we cannot possibly know how much space could be required by any value of a class wide type because the type might be extended. As a consequence, although we can declare an object of a class wide type, it must be constrained; however, the constraint is not given explicitly but by initializing it with a value of a specific type and it is then constrained by the tag of that type. So we could write

```
NR: Nice_Reservation;
...
RC: Reservation'Class := NR;
```

although this is not very helpful. Of more importance is the fact that a formal parameter can be of a class wide type and the actual parameter can then be of any specific type in the class. (Note again the analogy with arrays; a formal can be of an unconstrained array type and the actual can then be constrained.)

We now continue our example by considering how we might queue a series of reservation requests and process them in sequence by some central routine. The essence of the problem is that such a routine cannot assume knowledge of the individual types since we want it to work (without recompilation) even if we extend the system by adding a new reservation type to it.

The central routine could thus take a class wide value as its parameter so we might have

```
procedure Process_Reservation(RC: in out Reservation'Class) is
   ...
begin
   ...
   Make(RC);                -- dispatch according to tag
   ...
end Process_Reservation;
```

In this case we do not know which procedure Make to call until run time because we do not know which specific type the reservation belongs to. However, RC is of a class wide type and so its value includes a tag indicating the specific type of the value. The choice of Make is then determined by the value of this tag; the parameter is then implicitly converted to the appropriate specific reservation type before being passed to the appropriate procedure Make.

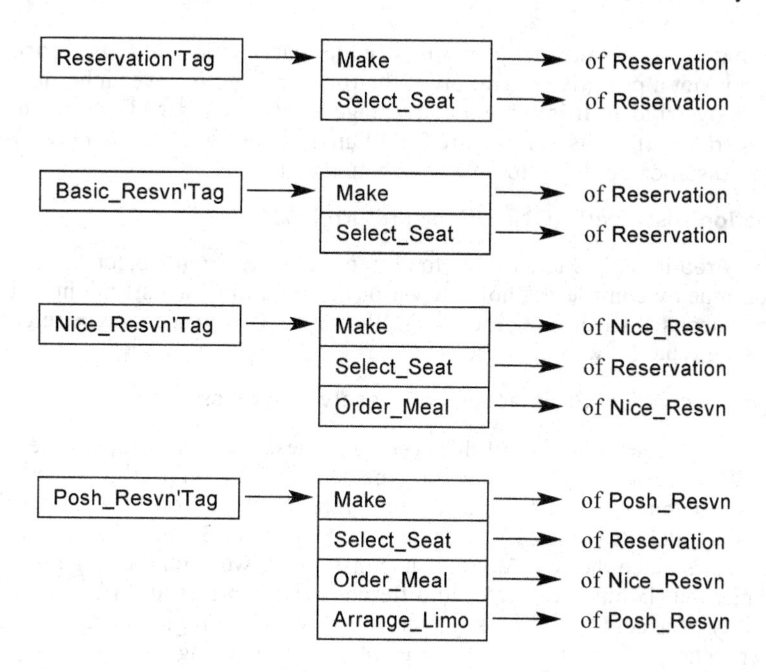

Figure 14.2 Tags and dispatch tables.

This run-time choice of procedure is called dispatching and is key to the flexibility of class wide programming. It is important to realize that dispatching can be implemented very efficiently. A possible implementation in this case is to make the tag point to a dispatch table each entry of which in turn points to the code of the body of a primitive operation. Each value of a tagged type will have the tag at a standard place such as at the beginning of the value. This model is illustrated in Figure 14.2 which shows how the various operations are inherited, replaced or added. (For simplicity, we have omitted the predefined primitive operations such as equality.)

The reason that dispatching is efficient is that there is never any need to check anything at run time. The dispatch table is arranged so that the displacements are the same for all types in the class. Also we know that every operation in the class is present because operations cannot be removed on derivation; only replaced or added. Moreover, as mentioned in the previous section, they always have the same convention and have subtype conformance so that the call always works dynamically.

(It should be added that the simple model described here needs to be extended when we come to include the effect of multiple inheritance to be discussed in Section 14.8. But it illustrates the general ideas.)

We now see the importance of the distinction between Reservation'Class and Nice_Reservation'Class. We can only dispatch to Order_Meal from the latter class wide type since only the latter has Order_Meal as a primitive operation of every type in the class.

Note that a procedure with a class wide parameter such as Process_Reservation is not a primitive operation of the root type and is never inherited and so cannot be overridden. It is often an advantage to use a class wide rather than a primitive operation if by its very nature it will apply to all types in the class. Thus the function Distance applying to objects is better written as

```
function Distance(O: Object'Class) return Float;
```

since unlike Area it applies unchanged to all types derived from Object.

We continue by considering how the various reservation requests might be held on a heterogeneous list awaiting processing. We can declare an access type referring to a class wide type. So we can write

```
type Reservation_Ptr is access all Reservation'Class;
```

in which case an access variable of this type could designate any value of the class wide type. We cannot change the specific type of the object referred to at any time into another type but we can from time to time refer to objects of different specific types. (This is much as the access variable R of Section 11.3 can refer to matrices of different sizes since the type Matrix is unconstrained; we cannot change the size of a particular matrix but R can refer to different sized matrices at different times.) The flexibility of access types is a key factor in class wide programming.

A heterogeneous list can be made in the obvious way using

```
type Cell is
  record
    Next: access Cell;
    Element: Reservation_Ptr;
  end record;
```

and the central routine can then manipulate the reservations using an access value as parameter

```
procedure Process_Reservation(RP: in Reservation_Ptr) is
  ...
begin
  ...
  Make(RP.all);               -- dispatch to appropriate Make
  ...
end Process_Reservation;
...
List: access Cell;            -- list of reservations
...
while List /= null loop       -- process the list
  Process_Reservation(List.Element);
  List := List.Next;
end loop;
```

In this case, the value of the object referred to by RP is of a class wide type and so includes a tag indicating the specific type. The parameter RP.**all** is thus dereferenced, the value of the tag gives the choice of Make and the parameter is then implicitly converted before being passed to the chosen procedure Make as before.

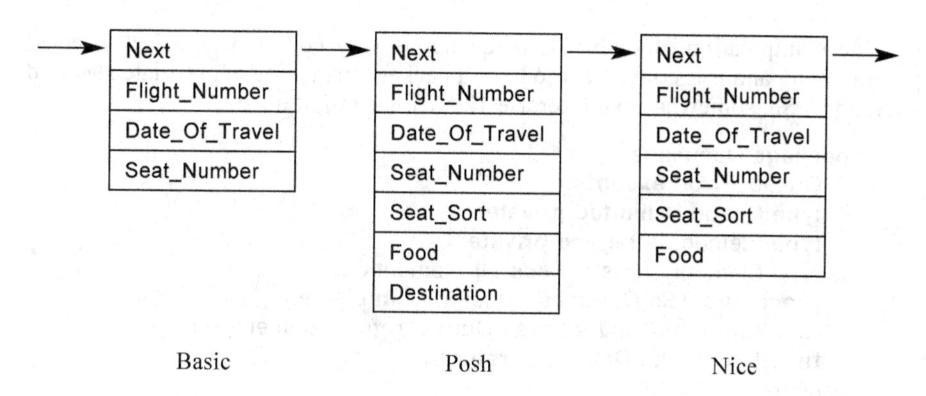

Figure 14.3 A heterogeneous list.

It is fundamental in class wide programming to manipulate objects via references; this is largely because the objects may be of different sizes. As a consequence, references to the objects will typically be on a list of some kind and will move from list to list as they are processed within the system. (The objects themselves will not move and so, as mentioned in Section 12.5, can be of a limited type.) If, as is likely, each object is only on one list at a time, then we can adopt a rather neater approach to chaining the objects together.

The general idea is that we first declare some root type which contains the pointer to the next item in the list and then the types to be placed on the list can be extended from the root type. Consider

```
type Element is tagged;                  -- tagged incomplete
type Element_Ptr is access all Element'Class;

type Element is tagged
   record
      Next: Element_Ptr;
   end record;
```

We have used a form of incomplete type which states that the full type will be tagged and so permits the use of the Class attribute. Objects of any type in the class Element'Class can be linked together through the one common element. We can now modify the reservation system so that the type Reservation is

```
type Reservation is new Element with
   record
      Flight_Number: Integer;
      Date_Of_Travel: Date;
      Seat_Number: String(1 .. 3) := "   ";
   end record;
```

with the rest of the system as before. The various reservations can now be joined together to form a list as illustrated in Figure 14.3.

The manipulation of such lists or queues is very common in object oriented programming and it is convenient to have a package providing appropriate standard operations. So, somewhat like Exercise 12.5(**3**), we might have

```
package Queues is
   Queue_Error: exception;
   type Queue is limited private;
   type Element is tagged private;
   type Element_Ptr is access all Element'Class;
   procedure Join(Q: access Queue; E: in Element_Ptr);
   function Remove(Q: access Queue) return Element_Ptr;
   function Length(Q: Queue) return Integer;
private
   type Element is tagged
      record
         Next: Element_Ptr;
      end record;
   type Queue is limited
      record
         First, Last: Element_Ptr;
         Count: Integer := 0;
      end record;
end Queues;
```

This package illustrates many points. The inner workings are hidden from the user by making both the type Queue and the type Element private. The type Queue is also limited since assignment must not be allowed (it would mess up the internal pointers); we have also made the full type explicitly limited just to ensure that the implementation of the body does not inadvertently attempt to assign a Queue either. (This is a useful safeguard because remember that we might later write a child package and that could see the private part also.) The type Element is given as tagged private. This means that the full type must also be tagged and permits the user to extend from the type without knowing its details.

The subprograms Join and Remove take an access parameter rather than an in out parameter. One advantage in Ada 2005 was that Remove could then be a function which is perhaps more convenient to use (but in Ada 2012 and Ada 2022 a function can have an in out parameter anyway). A possible disadvantage is that we have to specifically create a reference to the queue and this means marking it as aliased or creating the queue with an allocator.

Using this package we can now declare the root type Reservation and so on

```
with Queues;
package Reservation_System is
   ...
   type Reservation is new Queues.Element with
      record
         ...
      end record;
   ...
end Reservation_System;
```

and then create and place reservations on a queue by statements such as

```
The_Queue: access Queue := new Queue;
...
New_Resvn: Reservation_Ptr := new Nice_Reservation;
...
Join(The_Queue, Element_Ptr(New_Resvn));
...
```

Removing a reservation from the queue can later be done by

```
Next_Resvn: Reservation_Ptr;
...
Next_Resvn := Reservation_Ptr(Remove(The_Queue));
Process_Reservation(Next_Resvn);
```

Note very carefully that we have to explicitly convert the result of Remove to the type Reservation_Ptr. This is because Remove returns a result of the type Element_Ptr. Nothing would have prevented us from putting any type derived from Element on the queue and so there is no guarantee that it is indeed a reservation. The conversion performs a check that the type referred to is indeed a member of Reservation'Class and Constraint_Error is raised if the check fails.

In Section 21.4 we shall see how we can ensure that only reservations are placed on such a queue.

In Ada 2012 and Ada 2022, the parameter Q of Join and Remove can be of mode in out since a function can have in out parameters. The_Queue can then be of type Queue rather than an access type.

Having introduced class wide types and access types to tagged types this seems a good moment to summarize the rules concerning type conversions and tagged types. The general principle is that we can only convert towards an ancestor, and run-time checks may be needed to ensure this if the source is a class wide type; if the check fails then Constraint_Error is raised. In detail

- Conversion between two specific types is only permitted towards the ancestor.
- Conversion from a specific type to a class wide type of any ancestor type is allowed. Conversion to a class wide type that is not of an ancestor is not allowed.
- Conversion from a class wide type to a specific type is allowed provided the type of the actual value is a descendant of the specific type. A dynamic check may be required.
- Conversion between two class wide types is allowed provided the actual value is in the target class. A dynamic check may be required.

(The terms ancestor type and descendant type include the type itself.)

Note that conversion from a specific type to a class wide type is allowed implicitly on initialization and parameter passing (which are similar contexts) but that an explicit named conversion must be given on assignment. This rule is relaxed in Ada 2012 and Ada 2022 if no check is needed on the conversion.

Conversion between access types is allowed provided the designated types can be converted in the same direction. If the conversion requires a check at run time that could fail then the conversion must be to a named access type. This is in contrast to checks concerning null exclusion where the destination type can always be an anonymous access type. See Section 11.7. Of course, all that happens when we convert between access types is that we get a different view of the same object and so the new view must be an allowed interpretation. We will come back to the topic of views and conversions in more detail in Section 14.5.

The fact that a specific type is essentially a subtype of a class wide type (we say the class wide type covers the specific type) is also relevant in extended return statements which were introduced in Section 10.1. We might have a function

```
function Get_Next( ... ) return Reservation'Class is
begin
   ...
   return NR: Nice_Reservation do
      ...
   end return;
   ...
end Get_Next;
```

This is permitted since the type of NR is covered by the return type of the function.

We conclude by noting that tagged type parameters are always passed by reference and considered aliased. The similarities and differences between in out and access parameters were discussed in Section 11.6. Note also that dispatching occurs with all modes of formal parameters including access parameters (and access results) but does not occur with parameters and results of a named access type.

Finally, note that the model of tagged types that we have introduced so far only permits strict trees of types and thus single inheritance. In the next section we shall introduce interfaces which permit multiple inheritance as described in Section 14.8. The very simple model of a dispatch table given in Figure 14.2 then has to be extended but the details need not concern the user.

Exercise 14.2

1 Declare a procedure that will print the area of any geometrical object of a type derived from the type Object.

2 In a traditional world women do not have beards and men do not bear children. However, all persons have a date of birth. Declare a type Person with the common component Birth of type Date (as in Section 8.7) and then derived types Man and Woman that have additional components as appropriate indicating whether they have a beard or not and how many children they have borne respectively.

3 Declare procedures Print_Details for Person, Man, and Woman which output information regarding the current values of their components. Then declare a procedure Analyse_Person which takes a parameter of the class wide type Person'Class and calls the appropriate procedure Print_Details.

4 Write the body of the package Queues.

14.3 Abstract types and interfaces

It is sometimes convenient to declare a type solely to act as the foundation upon which other types can be built by derivation and not in order to declare objects of the type itself. We can do this by marking a tagged type as abstract. An abstract type can have abstract primitive subprograms; these have no body and cannot be called but simply act as placeholders for operations to be added later. We thus might write

```
package P is
   type T is abstract tagged
      record ... end record;
   procedure Op(X: T) is abstract;
   ...
end P;
```

It is a very important rule that it is not possible to create an object of an abstract type. Hence no object can ever have a tag corresponding to an abstract type and so it is never possible to attempt to dispatch to a primitive subprogram of an abstract type. However, an abstract type can have components and can also have concrete operations.

An interface is much like an abstract type but is more restricted. An interface cannot have components and it cannot have concrete operations (operations that are not abstract) except for null procedures which, it will be recalled from Section 12.3, behave as if they have a null body. Like an abstract type, we cannot declare objects of an interface. So we might write

```
package Q is
   type Int is interface;
   procedure Op(X: Int) is abstract;
   procedure N(X: Int) is null;
   procedure Action(X: Int'Class);          -- class wide operation
   ...
end Q;
```

Although an interface cannot have any concrete operations other than null procedures, we can declare a concrete procedure with a class wide parameter such as Action. Remember that such operations are not primitive and so not inherited.

Generally speaking, interfaces are more useful than abstract types because they permit multiple inheritance as explained in Section 14.8.

A concrete type (one that is not abstract) can be derived from an interface or from an abstract type provided that all inherited abstract subprograms are replaced by concrete subprograms which therefore have bodies (it is in this sense that the abstract subprograms are placeholders). We often speak of the concrete type as implementing the interface or abstract type.

We can now reformulate the example of processing reservations so that the root type Reservation is an interface and then build the specific types upon it. This enables us to program and compile all the infrastructure routines, such as Process_Reservation in the previous section, that deal with reservations in general without any concern at all for the individual reservation types and indeed before deciding what they should contain.

The baseline package can then simply become

```
package Reservation_System is
   type Reservation is interface;
   type Reservation_Ptr is access all Reservation'Class;
   procedure Make(R: in out Reservation) is abstract;
end Reservation_System;
```

in which we have declared the type Reservation as an interface with just the procedure Make as an abstract subprogram; remember that it does not have a body and hence the package also has no body.

We can now develop the reservation infrastructure and then later add the normal reservation system containing the three types of reservations. We introduce a child package as follows

```
package Reservation_System.Subsonic is

   type Position is (Aisle, Window);
   type Meal_Type is (Green, White, Red);

   type Basic_Reservation is new Reservation with
      record
         Flight_Number: Integer;
         Date_Of_Travel: Date;
         Seat_Number: String(1 .. 3) := "   ";
      end record;

   -- now provide concrete subprogram for abstract Make
   -- and add other subprograms as necessary

   overriding
   procedure Make(BR: in out Basic_Reservation);
   procedure Select_Seat(BR: in out Basic_Reservation);

   type Nice_Reservation is new Basic_Reservation with
      record
         Seat_Sort: Position;
         Food: Meal_Type;
      end record;

   overriding
   procedure Make(NR: in out Nice_Reservation);
   procedure Order_Meal(NR: in Nice_Reservation);

   type Posh_Reservation is new Nice_Reservation with
      record
         Destination: Address;
      end record;

   overriding
   procedure Make(PR: in out Posh_Reservation);
   procedure Arrange_Limo(PR: in Posh_Reservation);

end Reservation_System.Subsonic;
```

In this revised formulation we must provide a concrete procedure Make for the concrete type Basic_Reservation in order to implement the interface Reservation. The procedure Select_Seat now takes a parameter of type Basic_Reservation and the type Nice_Reservation is more naturally derived from Basic_Reservation.

Note carefully that we did not make Select_Seat an abstract subprogram in the package Reservation_System. There was no need; it is only Make that is required by the general infrastructure such as the procedure Process_Reservation and to add anything else would weaken the abstraction and clutter the interface with unnecessary operations.

We have chosen to include the overriding indicator **overriding** whenever we declare a further Make. This ensures that if we make a simple typographical error such as misspelling Make or getting the parameters wrong then the compiler will detect the error and not just add a bogus operation. We could of course add **not overriding** to the declarations of Select_Seat and so on.

We also have to make corresponding changes to the package body. This is left as an exercise for the reader.

When we now add the Supersonic_Reservation we can choose to derive this from the baseline Reservation as before or perhaps from some other point in the tree picking up the existing facilities of one of the other levels.

As an example of the use of an abstract type rather than an interface, consider the types Person, Man, and Woman of Exercise 14.2(**2**). We really do not want to be able to declare objects of the type Person because they are incomplete; we want all real persons to be either of the type Man or Woman and the type Person is merely a convenience for the common properties such as the component Birth of type Date. We can prevent the declaration of objects of the type Person by making it abstract

```
type Person is abstract tagged
  record
    Birth: Date;
  end record;
```

and then the types Man and Woman can be derived as before.

Despite the type Person being abstract there is no reason why we should not continue to declare the concrete procedure Print_Details of Exercise 14.2(**3**)

```
procedure Print_Details(P: in Person) is
begin
  Print_Date(P.Birth);
end Print_Details;
```

Although we will never dispatch to this procedure because there can never be an object of the type Person itself, nevertheless it can be used to print the information common to the types derived from Person. We will return to this topic in Section 14.5.

It is possible to derive an abstract type from a concrete type or from another abstract type. In either case we can replace inherited concrete subprograms by abstract ones or concrete ones, add additional abstract or concrete subprograms and so on. The overall rule is simply that a concrete type cannot have abstract subprograms. It is possible to derive an interface from one or more other interfaces but not from types whether abstract or concrete. See Section 14.8.

Although we can declare a concrete subprogram with parameters of an abstract type such as Print_Details above, a function returning an abstract type (or an access to an abstract type) must always be abstract.

An important related situation occurs in the case of a primitive function that returns a concrete type when that type is extended. Clearly the function cannot return a value of the extended type (since it does not know how to provide values for the new components) and so the function requires overriding with a new definition providing an appropriate result (an example is the type Text_Map in Section 24.10). If the derived type is declared as abstract then we do not need to provide a new definition – the inherited one just becomes abstract. These rules do not apply if the extension is in fact null since the function can be inherited in the normal way.

We conclude with a final remark on the syntax. A tagged type declaration always has just one of **interface**, **tagged**, and **with** (but none if not a tagged type).

Exercise 14.3

1 Declare a function Further that takes two parameters of type Object from Section 14.1 and returns that further from the origin by comparing their distances. Can this be inherited and applied to objects of types Point and Circle?

2 Reformulate the type Object as an abstract type. Make the function Area abstract and make Distance take a class wide parameter. (Remember that it is best to use a class wide parameter rather than inheritance if by its very nature the operation will apply to all types in the class without change.) Then declare types Point and Circle etc. from the type Object.

3 Declare a function Further that takes two parameters of type Object'Class and returns the further one by comparing their distances.

4 Repeat the previous exercise for a function Bigger that bases the comparison on the areas of the objects.

5 Write the body of the package Reservation_System.Subsonic.

14.4 Primitive operations and tags

In this section we consider in a little more detail some of the fundamental properties of tagged types and their operations.

It will be remembered from Section 12.3 that the primitive operations of a type are

- various predefined operations such as assignment and equality,
- those inherited from its ancestors and,
- those declared in the same package specification as the type itself.

We have said ancestors here rather than parent because of the existence of multiple inheritance to be discussed in Section 14.8.

For the sake of discussion we assume that we have the following declarations of packages Root and Shapes much as in the answer to Exercise 14.3(**2**)

```
package Root is
   type Object is abstract tagged
      record
         X_Coord, Y_Coord: Float;
      end record;

   function Area(O: Object) return Float is abstract;
   function MI(O: Object) return Float is abstract;
   function Distance(O: Object'Class) return Float;

end Root;

package body Root is
   function Distance(O: Object'Class) return Float is
   begin
      return Sqrt(O.X_Coord**2 + O.Y_Coord**2);
   end Distance;
end Root;

with Root;
package Shapes is
   type Circle is new Root.Object with
      record
         Radius: Float;
      end record;

   function Area(C: Circle) return Float;
   function MI(C: Circle) return Float;

   type Triangle is new Root.Object with
      record
         A, B, C: Float;        -- lengths of sides
      end record;

   function Area(T: Triangle) return Float;
   function MI(T: Triangle) return Float;

   ...   -- and so on for other types such as Square
end Shapes;
```

Then (apart from the predefined operations) the primitive operations of Object are Area and MI. The function Distance is not a primitive operation because it is class wide. Similarly, the primitive operations of Circle and Triangle are the overridden versions of Area and MI.

We can call these operations by the usual technique of giving the subprogram name and then the parameters in parentheses (only one parameter in these cases). Thus we might write

```
A := Shapes.Area(A_Circle);
D := Root.Distance(A_Triangle);
M := Shapes.MI(A_Square);
```

These calls all mention the name of the package containing the operation which means that Distance has to be called differently from the others. Moreover, we might later decide to restructure the hierarchy into a set of packages thus

```
package Geometry is
  type Object is abstract ...
  ...   -- functions Area, Mi, Distance
end Geometry;

package Geometry.Circles is
  type Circle is new Object with...
  ...   -- functions Area, MI
end Geometry.Circles;

package Geometry.Triangles is
  type Triangle is new Object with...
  ...   -- functions Area, MI
end Geometry.Triangles;
```

This is a much more elegant structure and avoids having to write Root.Object when doing the extensions. But the assignments now become

```
A := Geometry.Circles.Area(A_Circle);
D := Geometry.Distance(A_Triangle);
M := Geometry.Squares.MI(A_Square);
```

This is really all rather a nuisance and unnecessarily emphasizes the location of the operations. In object oriented programming it is the object that should be dominant and not the location of the operations. Accordingly, an alternative notation is allowed for calling operations of tagged types.

The rule is that if an operation Op on a type T is declared in a package P and X is of type T, then a call

```
P.Op(X, other paras)        -- package P mentioned
```

can be replaced by

```
X.Op(other paras)           -- package P not mentioned
```

provided that

- T is a tagged type,
- Op is a primitive (dispatching) or class wide operation of T,
- X is the first parameter of Op.

The reason there is never any need to mention the package is that, by starting from the object, we can identify its type and thus the primitive operations of the type. Note that a class wide operation can be called in this way only if it is declared at the same place as the primitive operations of T (or one of its ancestors).

Using this notation the various assignments now become

```
A := A_Circle.Area;
D := A_Triangle.Distance;
M := A_Square.MI;
```

and of course since there are no extra parameters in these cases, there are no parentheses either. It is important to note that these calls are independent of the package structure which helps with program maintenance.

This prefixed notation also has some other advantages. One is that it unifies the notation for calling a function with a single parameter and directly reading a component of the object. Thus we can write uniformly

```
X := A_Circle.X_Coord;
A := A_Circle.Area;
```

(Of course if we were foolish and had a visible component Area as well as a function Area then we could not call the function in this way.)

Another advantage is that explicit dereferencing is often not necessary when using access types. Suppose we have

```
type Pointer is access all Geometry.Object'Class;
...
This_One: Pointer := A_Circle'Access;
```

Then using the package notation we have to write

```
Put(This_One.X_Coord); ...
Put(This_One.Y_Coord); ...
Put(Geometry.Area(This_One.all));
```

whereas using the prefixed notation we can uniformly write

```
Put(This_One.X_Coord); ...
Put(This_One.Y_Coord); ...
Put(This_One.Area);
```

It is important to note that the first parameter of an operation plays a special role since it has to be of the tagged type concerned. Treating the first parameter especially can seem odd in some circumstances such as when there is symmetry among the parameters. Accordingly, we shall only use the prefixed notation in examples when it seems appropriate.

We will now turn to the rules for dispatching. A parameter of a primitive operation which is of the type concerned is known as a controlling parameter and a corresponding actual parameter is a controlling operand. The basic principle is that dispatching only occurs when calling a primitive operation and a controlling operand is class wide. Thus the call

```
Make(RC);                -- RC of type Reservation'Class
```

in the procedure Process_Reservation in Section 14.2 is a dispatching call. The value of the tag of RC is used to determine which procedure Make to call and this is determined at run time. We say that it is dynamically tagged.

On the other hand a call such as

```
Make(Reservation(NR));    -- NR of type Nice_Reservation
```

in the package body of Reservation_System in Section 14.1 is not a dispatching call because the type of the operand is the specific type Reservation as a result of the explicit type conversion.

The subprograms Make, Select_Seat and so on have just a single controlling parameter. However, a primitive operation can have several controlling parameters but they must all be of the same type.

Thus it would not be possible to declare

```
procedure Something(C: Circle; T: Triangle);
```

in the specification of the package Shapes above containing the declarations of the types Circle and Triangle. Such a procedure could of course be declared in a different package but it would then not be a primitive operation of either Circle or Triangle. It could also be declared as a child subprogram because child subprograms are not primitive operations either. Another possibility is to make one parameter class wide so that it then becomes a primitive operation of the other type. (Incidentally, this restriction only applies to tagged types – a subprogram can be a primitive operation of more than one type provided at most one is tagged.)

A controlling parameter may be of any mode and it can also be an access parameter of the type. A function can also have a controlling result in which case the type must be the same as that of any controlling parameters.

The principle that all controlling operands and results of a call must be of the same type is very important. If they are statically determined then this is checked at compile time. If they are dynamically determined (for example, variables of a class wide type) then again the actual values must all be of the same specific type and of course this check has to be made at run time (the tags are compared) and Constraint_Error is raised if the check fails. In order to avoid confusion a mixed situation whereby some operands are static and some are dynamic is not allowed.

As an example consider the function

```
function Is_Further(X, Y: Object) return Boolean;
```

which we assume is primitive together with

```
T1, T2: Triangle;
C1, C2: Circle;
Obj_1: Object'Class := ... ;        -- must be initialized
Obj_2: Object'Class := ... ;        -- because a class wide type
```

Then we can write calls such as

```
Is_Further(T1, T2)            -- non-dispatching, type Triangle
Is_Further(C1, C2)            -- non-dispatching, type Circle
Is_Further(Obj_1, Obj_2)      -- dispatching
```

and in the last case a check is made before the call that Obj_1'Tag equals Obj_2'Tag. On the other hand the following are illegal for the reasons stated

```
Is_Further(T1, C2)              -- illegal - mixed specific types
Is_Further(Obj_1, T2)           -- illegal - mixed static and dynamic
```

and both these situations are detected at compile time. A type conversion could be used to overcome the restriction in the second case, thus

```
Is_Further(Obj_1, Object'Class(T2))       -- legal
```

It is also possible to dispatch on the result of a function when the context of the call determines the specific type. Suppose we have the following further primitive operations

```
function Unit return Object;
function Double(X: Object) return Object;
function Is_Bigger(X: Object; Y: Object := Unit) return Boolean;
```

(We assume now that Object is not abstract.) These rather unlikely functions might behave as follows. Unit returns an object of unit size such as a circle of unit diameter or a triangle of unit side. Double returns an object with twice the linear size as that of the parameter. Is_Bigger compares the area of two objects and by default takes a unit object as the second parameter.

The functions Unit and Double have controlling results. Consider

```
Is_Bigger(T1, Unit)             -- non-dispatching, type Triangle
Is_Bigger(Obj_1, Unit)          -- dispatching
```

In the first case, the controlling operand T1 is static and determines that the call of Unit is also static; the call of Unit is thus chosen at compile time to be the Unit with result of type Triangle. In the second case, the controlling operand Obj_1 is dynamic and determines the type at run time; in this case the call of Unit dispatches to the particular Unit with the same tag as Obj_1; there is no run-time check because only one controlling operand is used to determine the type. The call of Unit is thus like a chameleon and adapts to the circumstances; we say that it is *tag indeterminate*.

The situation can be nested, for example

```
Is_Bigger(Obj_1, Double(Obj_2))          -- dispatching
```

in which case the tags of Obj_1 and Obj_2 are checked to ensure that they are the same; Constraint_Error is raised if they are not. The more elaborate expression Double(Unit) is also tag indeterminate so we can have

```
Is_Bigger(C1, Double(Unit))          -- non-dispatching
Is_Bigger(Obj_1, Double(Unit))       -- dispatching
```

In the second case the call of Double is then determined by the specific type of Obj_1 and this in turn determines the call of Unit.

We can also use a call of a function such as Unit to determine a default value. Thus we can have

```
Is_Bigger(T1)                   -- non-dispatching
Is_Bigger(Obj_1)                -- dispatching
```

and in the first case the default call of Unit is statically determined to be that of type Triangle whereas in the second case the call of Unit is dynamically determined by the specific type of the value of Obj_1.

It is interesting to note that a default expression for a controlling operand has to be tag indeterminate and so has to be a call of a function such as Unit or an expression such as Double(Unit). The reason is that we need to be able to use the default expression in both dispatching and non-dispatching contexts.

Note that an overridden operation does not have to have the same pattern of defaults as its parent; defaults are in the eye of the caller. Overload resolution identifies the declaration relevant to a call whether dispatching or not and it is the text of that declaration which gives any default expressions.

The use of overload resolution explains why the following call with two different class wide parameters fails even though the tags might be the same

```
Is_Further(Obj_1, CC)        -- illegal if CC is of type Circle'Class
```

This fails because neither the declaration of Is_Further for type Object nor that for type Circle is applicable. Such a mixed call would only be acceptable if there was a function Is_Further with at least one class wide formal parameter.

Similarly, if all controlling operands and any result are tag indeterminate then the situation is ambiguous and thus illegal. For example

```
Is_Further(Unit, Unit)        -- illegal
```

We are assuming that both Triangle and Circle are visible and so we do not know which Is_Further and Unit to apply. Again the overload resolution fails.

On the other hand, if only the type Circle is visible then such examples are not ambiguous and the type of Unit is *statically* determined to be Circle. An interesting example is that we could use a tag indeterminate expression as the initial value for a class wide object thus

```
CC: Circle'Class := Unit;     -- non-dispatching
```

The above discussion may have seemed a bit tedious but the underlying ideas are really quite simple and are aimed to make things as explicit as possible so that surprises are minimized or at least show up at compile time. A further example of the use of the various rules will be found in Section 14.9.

Assignment is a dispatching operation in the general case as we shall see in Section 14.7 when we consider controlled types. All the above rules apply and so both sides must have the same type and be statically or dynamically determined together.

So we can write

```
Obj_1 := Obj_2;
```

and there is a run-time check to ensure that the tags are the same (remember that a class wide object is constrained by its initial value). But, we cannot write

```
T1 := Obj_1;                  -- illegal
Obj_2 := C2;                  -- illegal
```

because in both cases we are mixing static and dynamic tags. We therefore have to provide an appropriate conversion such as

```
Obj_2 := Object'Class(C2);          -- run-time check
```

On the other hand, *implicit* conversion to a class wide type is permitted when providing an initial value in a declaration or as an actual parameter.

Equality is also a primitive operation but has somewhat different rules. If we compare two class wide values and they have different tags then the result False is returned rather than raising Constraint_Error which would occur with assignment or any other operation with two controlling operands. There is a strong analogy with arrays. We can only assign one array to another if the lengths are the same but equality simply returns False if the lengths are different and never raises Constraint_Error. But operations such as **and** and **or** on one-dimensional arrays raise Constraint_Error if the lengths are different.

Equality is also different with regard to inheritance; this is because it is normally predefined and so is always expected to be available and have sensible properties. On the other hand we might wish to redefine equality; directly inheriting a redefined version could bring surprises. Suppose we decide that two values of the type Object are to be considered equal if they are located within some small distance, epsilon, of the same point. We might declare

```
function "=" (A, B: Object) return Boolean is
begin
   return (A.X_Coord - B.X_Coord)**2
          + (A.Y_Coord - B.Y_Coord)**2 < Epsilon**2;
end "=";
```

The normal rules for inheritance would mean that this would be inherited unchanged by Point and Circle. This would be very surprising for the type Circle since it would ignore the Radius component completely. On the other hand, without redefinition of "=" for the type Object, the predefined equality for Circle would have applied predefined equality to all its components including the Radius. Because this drastic change of behaviour would be so surprising, what actually happens is that the equality for Object is not inherited for Circle but simply incorporated into the predefined equality for Circle. So two circles would be equal if their centres were within epsilon and their radii exactly equal.

The full rule for predefined equality of a type extension is that the primitive operation (possibly redefined) is used for the parent part and for any tagged components in the extension whereas predefined equality is always used for untagged components in the extension.

Recalling the discussion in Section 12.4, we see that tagged types (and indeed all record types in Ada 2012 and Ada 2022) always compose for equality. We are guaranteed that all private types in the predefined library that export equality compose properly – many will be implemented as tagged types.

Having redefined equality for Object, if we now wanted to redefine equality for Circle so that two circles are equal if they are at exactly the same location but their radii are within epsilon then we can either use the underlying coordinates directly or we can retrieve the original equality for the type Object if we had had the foresight to rename it before the redefinition (a two pass algorithm!) by writing

```
function Old_Equality(A, B: Object) return Boolean renames "=";
```

(Such a renaming is sometimes called a squirrelling renaming.) Of course a renaming after the redefinition will refer to the new operation. All renamings in a package specification are genuine primitive operations and have their own slots in the dispatch table which are initialized with the operation that is current at the point of the renaming. These new names thus denote primitive operations in their own right and can themselves be overridden on later inheritance. This might seem peculiar because the principle of renaming is that no new entity is ever created. But this is still true, the new slots just give further ways of referring to existing entities.

We mentioned earlier that an access parameter can also be a controlling parameter. However, it is an important rule that a controlling operand can never be null since null has no tag and the tag is needed for dispatching or checking. So we should write

```
procedure Whatever(Ptr: not null access Object);
```

if Whatever is a primitive operation. If we attempt to call Whatever with a null parameter then Constraint_Error will be raised.

We can actually omit **not null** for compatibility with Ada 95 in the case of an access parameter that is a controlling parameter but the behaviour remains as if it were given. We recall the rule when discussing renaming in Section 13.7 that 'null exclusions never lie'. So if present it must be heeded, but if absent a null exclusion might still apply for other reasons as in this case.

Another example of this rule concerns renaming, null exclusions, and primitive operations (this example is strictly for the collector of Ada curios). Consider

```
package P is
   type T is tagged ...
   procedure One(X: access T);                      -- excludes null

   package Inner is
      procedure Deux(X: access T);                  -- includes null
      procedure Trois(X: not null access T);        -- excludes null
   end Inner;

   use Inner;

   procedure Two(X: access T) renames Deux;         -- illegal
   procedure Three(X: access T) renames Trois;      -- legal
   ...
```

The procedure One is a primitive operation of T and its parameter X is therefore a controlling parameter and so excludes null even though this is not explicitly stated. However, the declaration of Two is illegal. It is trying to be a primitive operation of T and therefore its controlling parameter X has to exclude null. But Two is a renaming of Deux whose corresponding parameter does not exclude null and so the renaming is illegal. (Remember that primitive operations have to be declared immediately inside the package concerned and so subprograms declared in an inner package are not primitive.) On the other hand the declaration of Three is permitted because the parameter of Trois does exclude null.

We have mentioned the existence of the tag of an object as being effectively a hidden component and we have seen how the value of the tag of a class wide object is used for dispatching. An object thus has the property of being self-describing; it carries an indication of the identity of its type with it.

In the previous section we considered the formulation of the type Person as

```
type Person is abstract tagged
  record
    Birth: Date;
  end record;
type Man is new Person with
  record
    Bearded: Boolean;
  end record;

type Woman is new Person with
  record
    Children: Integer;
  end record;
```

Note that, perhaps worryingly, there is no explicit component indicating the sex of a person; it might appear as if there was no way to find out one's sex! But luckily this information is not lost since it is implicit in the tag. The tag can be implicitly tested by membership tests such as

```
if P in Woman then
  ...                          -- special processing for Women
end if;
```

where P is of the class wide type Person'Class.

Indeed, it is also possible to test the tag explicitly using the attribute Tag which can be applied to a value of a class wide type and to a tagged type itself. So we could alternatively have written

```
if P'Tag = Woman'Tag then
```

The value of the attribute Tag is of the private (but nonlimited) type Tag declared in the package Ada.Tags whose visible part has the form

```
package Ada.Tags
    with Preelaborate, Nonblocking, Global => in out synchronized is
  type Tag is private
    with Preelaborable_Initialization;
  No_Tag: constant Tag;
  function Expanded_Name(T: Tag) return String;
  function Parent_Tag(T: Tag) return Tag;
  ...   -- other functions
  function Is_Abstract(T: Tag) return Boolean;
  Tag_Error: exception;
private
```

We can declare variables of the type Tag in the usual way. The function Expanded_Name returns the full dotted name of the type of the tag as a string (in upper case and starting with a root library unit). So if Obj is an object of type Object'Class whose value happens to have the specific type Circle where Circle is declared in the library package Objects then

```
Expanded_Name(Obj'Tag)
```

would return the string "OBJECTS.CIRCLE". There are also versions of Expanded_Name that return the result as a Wide_String or Wide_Wide_String.

The function Is_Abstract tells us whether the tag is that of an abstract type. It was added in Ada 2012.

It is important to remember that the attribute Tag cannot be applied to a value of a specific type but only to a value of a class wide type or to the *name* of a specific type. (The reason is explained in the next section.)

We could also write

```
if P in Woman'Class then
```

and this would then cover any types derived from Woman as well. It is possible to do the corresponding test using tags by using the function Parent_Tag also declared in Ada.Tags. But it is perhaps a little more elegant to use membership tests wherever possible.

The package Ada.Tags has a number of other functions which will be explained in Sections 21.7 and 23.9.

Exercise 14.4

1. Define "=" for the type Circle so that two circles are equal if they are within epsilon of the same location and their radii are within epsilon of each other. Assume that "=" for Object has been appropriately redefined.

2. Now define "=" for the type Circle so that they have to be at the same location but their radii are within epsilon of each other. Assume that Old_Equality was declared and renames the original equality on the type Object.

14.5 Views and redispatching

We now consider the rules for type conversion once more. Recall that the basic rule is that type conversion is only allowed towards the root of a tree of tagged types and so we can convert a Nice_Reservation into a Reservation. On the other hand we cannot convert a specific type away from the root; we have to use an extension aggregate even if there are no extra components.

We can however convert a value of a class wide type to a specific type thus

```
NR := Nice_Reservation(RC);
```

where RC is of the type Reservation'Class. In such a case there is a run-time check that the current value of the class wide object RC is of a specific type for which the conversion is possible. Hence it must be of the type Nice_Reservation or derived from it so that the conversion is not away from the root of the tree. In other words there is a check that the value of RC is actually in Nice_Reservation'Class. Constraint_Error is raised if the check fails.

Some conversions are what is known as view conversions. This means that the underlying object is not changed but we merely get a different view of it. (Much as the private view and full view of a type are just different views; the type is still the same.) We met an example of a view conversion in Section 10.3 when we called the procedure Increment with the view conversion Integer(F) as parameter.

Most conversions of tagged types are view conversions. For example the conversion in

```
Make(Reservation(NR));
```

is a view conversion. The value passed to the call of Make (with parameter of type Reservation) is in fact the same value as held in NR (tagged types are *always* passed by reference) but we can no longer see the components relating to the type Nice_Reservation. And in fact the tag still relates to the underlying value and this might even be the tag for Posh_Reservation because it could have been view converted all the way towards the root of the tree.

Moreover, if we did an assignment as in

```
NR := Nice_Reservation(PR);
```

then the tag of NR is of course not changed. All that happens is that the components appropriate to the type of NR are copied from the object PR. We can also have a view conversion as the destination of an assignment

```
Nice_Reservation(PR) := NR;
```

and the components copied are then just those appropriate to the type of the view. We could even have a conversion on both sides such as

```
Object(A_Circle) := Object(A_Triangle);
```

This makes the X_Coord and Y_Coord of the circle the same as those of the triangle. Other components, and in particular the tag, are not changed.

It is indeed an important principle that the tag of an object (both specific and class wide) is never changed. In particular, the fact that a view conversion does not change the tag is very important for what is called redispatching.

It often happens that after one dispatching operation we apply a further common (and inherited) operation and so need to dispatch once more to an operation of the original type. If the original tag were lost then this would not be possible. An example of the seeds of this difficulty already lies in our reservation system for Ada Airlines.

Consider again

```
procedure Make(NR: in out Nice_Reservation) is
begin
   Make(Reservation(NR));        -- make as plain reservation
   Order_Meal(NR);
end Make;
```

in which there is a call of the procedure Order_Meal. This call is not a dispatching call because the parameter is of a specific type and indeed there is only one procedure Order_Meal. Inside the body of Order_Meal we would expect to deal just with an order from a nice reservation and would not anticipate having to take account of the fact that the order might have originated from a posh passenger. (Although one would hope that posh meals are indeed better than nice meals.)

Actually we *could* write

```
procedure Order_Meal(NR: Nice_Reservation) is
   NRC: Nice_Reservation'Class := NR;
begin
   if NRC in Posh_Reservation then
      ...                       -- order a posh meal
   else
      ...                       -- order a nice meal
   end if;
end Order_Meal;
```

where we have regained the original type by converting to the class wide type Nice_Reservation'Class.

We could avoid the burden of the assignment by putting the conversion in the test itself

```
if Nice_Reservation'Class(NR) in Posh_Reservation then
```

or by introducing a renaming

```
NRC: Nice_Reservation'Class renames Nice_Reservation'Class(NR);
```

although this is rather a mouthful.

Note also that we could alternatively have written the test as

```
if NRC'Tag = Posh_Reservation'Tag then
```

but remember that we cannot apply the attribute Tag to an object of a specific type. So we could not have avoided the introduction of the class wide variable by writing

```
if NR'Tag = Posh_Reservation'Tag then      -- illegal
```

This is disallowed because it would be very confusing to allow NR'Tag since we would naturally expect this always to be Nice_Reservation'Tag and to find that it had some other value would be strange.

However, it is of course against the spirit of the game to mess about inside Order_Meal to see if it was actually ordered by a posh passenger. The whole idea of programming by extension is that one should be able to write the body for Order_Meal without considering how the type system might be extended later. Indeed, the type Posh_Reservation (like Supersonic_Reservation) might be in a later package in which case we could not here refer to the type Posh_Reservation at all.

The proper approach is to use redispatching. We write a distinct procedure

```
procedure Order_Meal(PR: in Posh_Reservation);
```

for posh passengers and redispatch in the body of Make as follows

```
procedure Make(NR: in out Nice_Reservation) is
begin
   Make(Reservation(NR));      -- make as plain reservation
   Order_Meal(Nice_Reservation'Class(NR));      -- redispatch
end Make;
```

Redispatching occurs because we have converted the parameter to the class wide type Nice_Reservation'Class (remember that dispatching occurs if the actual parameter is class wide and the formal is specific). So it works properly and our posh passenger will now get a posh meal instead of just a nice one.

We will now leave our reservation system noting that there is an analogous possible difficulty with Select_Seat; it seems likely that the requests of nice and posh passengers for an aisle or window seat might be overlooked. We leave the consideration of this as an exercise.

The fact that view conversion does not change the tag explains why we can safely and usefully declare a concrete procedure for an abstract type such as the procedure Print_Details for the type Person. We can never declare an object of the type Person and so a dispatching call of this procedure is not possible. A static call is however possible with a view conversion as in

```
procedure Print_Details(W: in Woman) is
begin
   Print_Details(Person(W));                    -- view conversion
   Print_Integer(W.Children);
end;
```

but since the tag never changes no harm can arise.

As another example consider the innocuous looking

```
procedure Swap(X, Y: in out Object) is
   T: Object := X;
begin
   X := Y;  Y := T;
end Swap;
```

and consider its inheritance by the type Circle.

This does not work as we might have hoped because it only swaps the Object part of the Circle. Remember that inheritance only applies to the subprogram specification and the code of the body is completely unchanged unless overridden. All references to the type Object in the body remain as references to the Object view of the Circle. So the inherited version is effectively

```
procedure Swap(X, Y: in out Circle) is
   T: Object := Object(X);
begin
   Object(X) := Object(Y);  Object(Y) := T;
end Swap;
```

One solution is to write

```
procedure Swap(X, Y: in out Object'Class) is
  T: Object'Class := X;
begin
  X := Y;  Y := T;
end Swap;
```

where the assignments are effectively done by dispatching. Being a class wide operation it cannot be overridden. Alternatively

```
procedure Swap(X, Y: in out Object) is
  T: Object'Class := Object'Class(X);
begin
  Object'Class(X) := Object'Class(Y);  Object'Class(Y) := T;
end Swap;
```

can be overridden if desired and again does the assignments according to the specific type. Both solutions work if Object is abstract whereas the original would not even compile in that case.

The above examples can be summarized by considering the case of a type T with primitive operations A and B. Suppose furthermore that the implementation of A is conveniently performed by calling B so that the body of A has the form

```
procedure A(X: T) is
begin
  ...
  B(X);
  ...
end A;
```

Now suppose that a new type TT is declared as an extension of T and that the inherited operation A appears satisfactory but the inherited operation B needs to be overridden by a new version

```
procedure B(X: TT) is ...
```

This raises the problem of whether the inherited version of A applying to the type TT should internally call B of the type T or the overridden B for the type TT. The answer will clearly depend upon the particular application. If we want it to continue to call B for the type T then the code as above is correct. However, if we want it to call the overridden B applicable to TT then we must change the call of B in the body of A to

```
B(T'Class(X));
```

so that it then dispatches to the relevant B. Of course the behaviour of A for the type T remains unchanged since it just dispatches to the same specific operation as before. The danger in all this is that the (potential) error lies in the body of A for T and only shows up when the type is extended.

Looking back at the previous examples, the operations Make and Order_Meal play the roles of A and B in the first example whereas Swap and := play these roles in the second example.

Care should always be taken when inheriting any operation with a controlling operand which is an **out** parameter to ensure that the inherited version does provide appropriate values for any additional components. There is no corresponding risk with a function with a controlling result since such a function requires overriding when the type is extended as we saw in Section 14.3.

It is hoped that the discussion in this and the previous section has not seemed overly complex and detailed. Object oriented programming may be very flexible but it has its pitfalls and it is important that the reader be aware of these. Ada strives for clarity. The basic rule is that dispatching is only used if the actual parameter is of a class wide type and this is always clear at the point of the call. This simple rule coupled with the fact that the original tag is never lost should cause fewer surprises than the more obscure rules of some languages.

We conclude this section with a brief summary of the main points regarding tagged types.

- Record (and private) types can be tagged. Values of tagged types carry a tag with them. The tag indicates the specific type. A tagged type can be extended on derivation with additional components. The tag of an object can never be changed.
- The primitive operations of a type are those implicitly declared, plus, in the case of a type declared in a package specification, all subprograms with a parameter or result of that type also declared in the package specification.
- Primitive operations are inherited on derivation and can be overridden. If the derivation occurs in a package specification then further primitive operations can be added.
- Types and subprograms can be declared as abstract. An abstract subprogram does not have a body but one can be provided on derivation. An interface is a form of abstract type that has no components and no concrete operations other than null procedures. Only interfaces and abstract tagged types can have abstract primitive subprograms.
- T'Class denotes the class wide type rooted at T. It is an indefinite type. An appropriate access type can designate any value of T'Class.
- Objects of a class wide type may be declared but they must be initialized and are then constrained by the tag of the initial value. Formal parameters of a class wide type are similarly constrained by the actual parameter.
- Type conversion must always be towards an ancestor. Implicit conversion from a specific type to an ancestor class wide type is allowed when passing parameters and on initialization.
- Parameters of tagged types are always passed by reference. They are also considered aliased so that the Access attribute can be applied.
- Calling a primitive operation with an actual parameter of a class wide type results in dispatching: that is the run-time selection of the operation according to the tag.

Finally, do remember that the tag is an intrinsic part of an object of a tagged type and can be thought of as a hidden component. However, unlike other components, the tag of an object can never be changed.

Exercise 14.5

1 In the revised procedure Make for Nice_Reservation why could we not write

```
Order_Meal(Reservation'Class(NR));      -- redispatch
```

2 Add a procedure Select_Seat appropriate for nice passengers (and better) and rewrite the procedure Make for the type Reservation to dispatch on Select_Seat.

14.6 Private types and extensions

We now come to a detailed consideration of the interaction between type extension, private types, and child packages.

In Section 14.2 we saw that the type Element in the package Queue was declared as

```
type Element is tagged private;
```

The full type then also has to be tagged. The partial view could in fact be declared as abstract thus

```
type Element is abstract tagged private;
```

and this would be an advantage in this case since there is no point in the user being able to create objects of the type Element.

The rules regarding matching of full and partial views are as expected from the general principle that the full view must deliver the properties promised by the partial view.

If the partial (external) view is tagged then the external client can do type extension and so the full view must also be tagged but the reverse is not true. The partial view can be untagged and the full view can be tagged. Of course the external client cannot use any of the properties of type extension in such a situation. A similar pattern applies to the property of being abstract. If the partial view is abstract then the full view need not be but, on the other hand, if the partial view is not abstract then the full view cannot be abstract either.

Note that there is no such thing as a private interface. This is largely because interfaces have no components and so nothing to hide.

Limited types are somewhat different if both views are tagged; both partial and full view must be limited or not together. Another rule is that a tagged record can only have a limited component if the record is explicitly limited, or in the case of a type extension, if the ultimate ancestor is explicitly limited. These rules prevent potential difficulties with extension and dispatching.

Abstract and private types pose small problems with aggregates. We can only give a normal aggregate if all the components are visible. However, we can always use an extension aggregate if the ancestor part is private

```
Some_Element: Element;
...
(Some_Element with ... )
```

or (and this is especially useful if the type is abstract so that the object Some_Element cannot be declared), we can simply use the subtype name as explained in Section 14.1 thus

```
(Element with ... )
```

and components corresponding to the type Element will be default initialized. An alternative technique (perhaps a dirty trick) is to use a view conversion of some existing object of a concrete type derived from Element. Thus we might have

```
(Element(An_Object) with ... )
```

The type Element illustrates that we can extend from a private tagged type with additional components visible to the user even though the original components are hidden from the user. The reverse is also possible as we saw in Section 3.2; we can extend an existing type with additional components which are hidden from the external user. We could declare a type Shape using a private extension declaration and make visible the fact that the type Shape is derived from Object and yet keep the additional components hidden. Thus

```
package Hidden_Shape is
   type Shape is new Object with private;  -- private extension
   ...
private
   type Shape is new Object with            -- full type declaration
      record
         ...                   -- the private components
      end record;
   function Area(S: Shape) return Float;
end Hidden_Shape;
```

It is not necessary for the full declaration of Shape to be derived directly from the type Object. There might be one or more intermediate types such as Circle; all that matters is that Shape is ultimately derived from Object. If there are no extra components then we write **with null record**; as expected.

Of course, the type Shape could never be an existing type. Writing

```
type Shape is new Circle with null record;
```

in the private part does not make Shape the same as Circle but derived from it.

We can declare and override primitive operations (such as Area) in the private part or in the visible part. New and overridden operations in the visible part behave as expected; they are visible to both client and server. But operations in the private part bring interesting possibilities.

A new operation in the private part will have a new slot in the dispatch table even though it is not visible for all views of the type. It is an important principle that there is only one dispatch table for a type and it may be that some operations are not visible for some views.

A minor point is that an abstract type is not allowed to have private abstract operations because it would not be possible to override them and so it would be impossible to extend the type. This would be a serious violation of the abstraction because the poor external user should be totally unaware of any private operations. A similar restriction is that a function with a controlling result cannot be a private operation because again it could not be overridden.

The final case is where we override an inherited visible operation such as Area for Shape in the private part. Despite not being directly visible, nevertheless the overridden operation will be used whether called directly or indirectly through dispatching. It is an important principle that the same operation is called both directly and indirectly for the same tag – this allows a fragment to be tested with static binding and then we know we will still get the same effect if the final program does dispatching.

As an example of the use of private extensions we will rewrite the basic geometry system as a collection of abstract data types. We declare each derived type in its own child package and make these packages into a hierarchy matching the type hierarchy. This ensures that each body can see all the components even though they are hidden from the external client. Remember that the private part and body of a child package can see the private part of the parent package. We might write

```
package Geometry is
   type Object is abstract tagged private;
   function Area(O: Object) return Float is abstract;
   function Distance(O: Object'Class) return Float;

   procedure Move(O: in out Object'Class; X, Y: in Float);
   function X_Coord(O: Object'Class) return Float;
   function Y_Coord(O: Object'Class) return Float;
private
   type Object is abstract tagged
      record
         X_Coord: Float := 0.0;
         Y_Coord: Float := 0.0;
      end record;
end Geometry;

package Geometry.Circles is
   type Circle is new Object with private;
   function Area(C: Circle) return Float;
   function Make_Circle(Radius: Float) return Circle;
   function Get_Radius(C: Circle) return Float;
private
   type Circle is new Object with
      record
         Radius: Float;
      end record;
end Geometry.Circles;

package Geometry.Triangles is
   type Triangle is new Object with private;
   function Area(T: Triangle) return Float;
   function Make_Triangle(A, B, C: Float) return Triangle;
   procedure Get_Sides(T: in Triangle; A, B, C: out Float);
```

```
private
   ...
end Geometry.Triangles;
```

and so on. Since the types are now private we have provided class wide selector functions for accessing the coordinates (these can have the same identifier as the hidden components) and specific constructor and selector subprograms such as Make_Circle, Get_Radius and so on.

We have not provided individual procedures for setting the two coordinates but a single procedure Move. This ensures that both coordinates get changed consistently. Thus if we wish to move an object from its default location at the origin (0.0, 0.0) to say (3.0, 4.0) then we write

```
A_Triangle.Move(3.0, 4.0);                -- using prefixed notation
```

whereas if we had used two procedure calls such as

```
A_Triangle.Set_X_Coord(3.0);
A_Triangle.Set_Y_Coord(4.0);
```

then it might seem as if the triangle was transitorily at the point (3.0, 0.0). There are other risks as well – we might forget to set one component or accidentally set the same component twice. Other more serious risks will become apparent when we deal with tasking in Chapter 20.

This example also reveals another advantage of the prefixed notation. We can still write

```
X := A_Triangle.X_Coord;
```

irrespective of whether the component is read directly or via a function.

Constructor functions pose a problem. Suppose we decide to introduce a distinct type for equilateral triangles by declaring a type Equilateral_Triangle in a further child package Geometry.Triangles.Equilateral. The first interesting point is that this introduces a different form of specialization in which we do not need any additional components but rather a constraint between existing components, namely that the sides are equal. We need a new constructor function that takes just a single parameter.

But the real problem is that, as currently written, the new type will inherit the function Make_Triangle from its parent type (moreover, it does not require overriding because there are no additional components). But it will have three parameters and any overridden function will also need three parameters contrary to our requirements. We just do not want Make_Triangle for the new type at all, we want Make_Equilateral with a single parameter.

In order to solve this problem we have to arrange that Make_Triangle is not a primitive operation. This can be done in a number of ways. It could return the class wide type Triangle'Class – this works but is clumsy because a type conversion would be required after each call in order to obtain the specific type Triangle. We could put the function in a child package – it could then still see the private part but would no longer be primitive.

But perhaps the best solution is to make it a child function of the package Geometry.Triangles thus

```
function Geometry.Triangles.Make_Triangle(A, B, C: Float) return Triangle;
```

An advantage of this approach is that a child function can be accessed without any extra use clause just as if it were a primitive operation. Moreover, we do not have to think of a name for a containing package.

The same approach using child subprograms can be taken with selector subprograms but there is often no need. In conclusion the package and function for equilateral triangles might be just

```
package Geometry.Triangles.Equilateral is
   type Equilateral_Triangle is new Triangle with private;
private
   type Equilateral_Triangle is new Triangle with null record;
end;

with Geometry.Triangles.Make_Triangle;
function Geometry.Triangles.Equilateral.Make_Equilateral(Side: Float)
                                        return Equilateral_Triangle is
begin
   return (Make_Triangle(Side, Side, Side) with null record);
end;
```

The extension aggregate in the return statement converts the type Triangle returned by Make_Triangle to Equilateral_Triangle using a null extension as described in Section 14.1. Note also that the package has no body.

Clearly this is a trivial example but it illustrates important principles. A constructor function such as Make_Triangle will often have to check that its parameters are consistent and perhaps raise an exception declared in the root package if they are not. The reader might care to consider the introduction of a type Isosceles and how we might convert between different forms of triangle. We will encounter a further example of this kind in Section 18.4.

We will now reconsider the reservation system using private extensions of the interface Reservation but by contrast will continue to declare all the types in a single package. Consider

```
package Reservation_System is
   type Reservation is interface;
   procedure Make(R: in out Reservation) is abstract;
   type Basic_Reservation is new Reservation with private;
   type Nice_Reservation is new Reservation with private;
   type Posh_Reservation is new Reservation with private;
private
   type Basic_Reservation is new Reservation with
      record
         Flight_Number: Integer;
         Date_Of_Travel: Date;
         Seat_Number: String(1 .. 3) := "   ";
      end record;
```

```
   overriding
   procedure Make(BR: in out Basic_Reservation);          -- overrides
   not overriding
   procedure Select_Seat(BR: in out Basic_Reservation);        -- new

   type Position is (Aisle, Window);
   type Meal_Type is (Green, White, Red);

   type Nice_Reservation is new Basic_Reservation with
     record
        Seat_Sort: Position;
        Food: Meal_Type;
     end record;

   procedure Make(NR: in out Nice_Reservation);            -- overrides
   procedure Select_Seat(NR: in out Nice_Reservation);        -- ditto
   procedure Order_Meal(NR: in Nice_Reservation);             -- new

   type Posh_Reservation is new Nice_Reservation with
     record
        Destination: Address;
     end record;

   procedure Make(PR: in out Posh_Reservation);            -- overrides
   procedure Order_Meal(PR: in Posh_Reservation);             -- ditto
   procedure Arrange_Limo(PR: in Posh_Reservation);           -- new
end Reservation_System;
```

Externally all that is visible is that there are the various types and there is a procedure Make – this is all that is necessary in order to write a procedure such as Process_Reservation. The relationships between the types are not visible since they are just shown as deriving from Reservation; this is a good illustration of the fact that the full declaration of a private type need not be directly derived from the ancestor type given in the private extension.

Note also that Select_Seat and Order_Meal are private primitive operations (we have properly provided the various versions so that passengers get their appropriate choices of seat and meal). Moreover, although the redeclarations of Make are also in the private part nevertheless they are externally callable because the original operation Make is visible.

In Section 12.3 we mentioned that we cannot add further primitive operations to a type after a type has been derived from it. This is a consequence of the freezing rules which are discussed in Chapter 25; these rules concern when the representation of a type is determined or frozen and the effect of it being frozen. The two rules that concern us here are that first, we cannot add more primitive operations to a type once its representation is frozen (since this determines the dispatch table) and secondly, declaring a derived type freezes its parent (if not already frozen). Luckily this second rule applies only to the full type declaration and not to a private extension. Otherwise we could not declare the private dispatching operations such as Select_Seat. The full declaration of Basic_Reservation then freezes the type Reservation and prevents us from adding further primitive operations to it. Another point is that we cannot give the full type declaration corresponding to a type extension until after the full declaration of its parent. It is therefore important that the various types and operations are declared in the proper order especially if they are all in the same package.

We can now add the supersonic package

```
package Reservation_System.Supersonic is
   type Supersonic_Reservation is new Reservation with private;
private
   type Supersonic_Reservation is new ... with
      record
         ...
      end record;

   procedure Make(SR: in out Supersonic_Reservation);
   procedure Select_Seat(SR: in out Supersonic_Reservation);
   ...
end Reservation_System.Supersonic;
```

The type Supersonic_Reservation can now be an extension of any member of the tree as appropriate and, being in a child package, it can see the full details of the various types. Note that Select_Seat overrides properly despite being a private primitive operation of Reservation.

We did not bother to give overriding indicators in all of the above but clearly there would have been no difficulty in doing so. But sometimes it is not known whether a subprogram is going to override and that is why they are optional (apart from considerations of compatibility with Ada 95). Consider

```
package P is
   type NT is new T with private;
   procedure Op(X: NT);
private
```

Now suppose the type T does not have an operation Op. Then clearly it would be wrong to write

```
package P is
   type NT is new T with private;          -- T has no Op
   overriding                              -- illegal
   procedure Op(X: NT);
private
```

because that would violate the information known in the partial view.

But suppose that in fact it turns out that in the private part the type NT is actually derived from TT (itself derived from T) and that TT does have an operation Op.

```
private
   type NT is new TT with ...              -- TT has Op
end P;
```

In such a case it turns out in the end that Op is in fact overriding after all. We can then put an overriding indicator on the body of Op since at that point we do know that it is overriding.

Equally of course we should not specify **not overriding** for Op in the visible part because that might not be true either (since it might be that TT does have Op). However, if we did put **not overriding** on the partial view then that would not in itself be an error but would simply constrain the full view not to be overriding and thus ensure that TT does not have Op.

Of course if T itself has Op then we could and indeed should put an overriding indicator in the visible part since we know that to be the truth at that point.

The general rule is not to lie. But the rules are slightly different for overriding and not overriding. For overriding it must not lie at the point concerned. For not overriding it must not lie anywhere.

This asymmetry is a bit like presuming the prisoner is innocent until proved guilty. We sometimes start with a view in which an operation appears not to be overriding and then later on we find that it is overriding after all. But the reverse never happens – we never start with a view in which it is overriding and then later discover that it was not. So the asymmetry is real and justified.

We cannot expect to find all the bugs in a program through syntax and static semantics; the key goal of overriding indicators is to provide a simple way of finding most of them.

We conclude this section by expanding on the remark made in Section 12.5 that the properties of a type visible from a given point depend upon the declaration of that type which is visible from that point. Consider

```
package P is
   type T is tagged private;
private
   type T is tagged
      record
         X: Integer;
      end record;
end;

with P; use P;
package Q is
   type DT is new T with
      record
         Y: Integer;
      end record;
end;
```

Now suppose that the body of P has a with clause for Q and consider visibility of an object A of the type DT from within the body of P. We might write

```
with Q; use Q;
package body P is
   A: DT;
   ...
begin
   A.X := 5;                  -- illegal
   T(A).X := 5;               -- legal
end P;
```

Although DT is extended from T and the full declaration of T is visible from within the body of P, nevertheless the component X of A of type DT is not visible. The visibility is controlled by what we can see at the point where DT is declared and at that point we cannot see the component X. However, within the body of P, we could carry out a view conversion of A to the type T and then access the component X.

One consequence is that the additional component of DT could also have the identifier X. However, this would not be possible if DT were declared in a child package of P because then the private part and the body of the child would have direct visibility of both components and a clash of names would arise.

Exercise 14.6

1 Could we declare the functions Further of Exercises 14.3(**2**) and 14.3(**3**) for the type Shape in the private part of the package Hidden_Shape?

14.7 Controlled types

A very interesting example of the use of type extension is provided by considering the facilities for controlled types. These allow a user complete control over the initialization and finalization of objects and also provide the capability for user-defined assignment.

The general principle is that there are three distinct primitive activities concerning the control of objects

- initialization after creation,
- finalization before destruction,
- adjustment after assignment,

and the user is given the ability to provide appropriate procedures which are called to perform whatever is necessary at various points in the life of an object. These procedures are Initialize, Finalize, and Adjust and they take the object as a parameter of mode in out.

To see how this works, consider

```
declare
   A: T;                    -- create A, Initialize(A)
begin
   ...
   A := E;                  -- Finalize(A), copy value, Adjust(A)
   ...
end;                        -- Finalize(A)
```

After A is declared and any normal default initialization carried out, the Initialize procedure is called. On an assignment, Finalize is first called to tidy up the old object about to be overwritten and thus destroyed, the physical copy is then made and finally Adjust is called to do whatever might be required for the new copy. At the end of the block Finalize is called once more before the object is destroyed.

Note, of course, that the user does not physically write the calls of the three control procedures, they are called automatically by the compiled code.

In the case of a declaration with an initial value

```
A: T := E;                    -- create A, copy value, Adjust(A)
```

the calls of Initialize and Finalize that would occur with distinct declaration and assignment are effectively cancelled out (but see the note below for the case when the expression E is an aggregate).

There are many other situations where the control procedures are invoked such as when calling allocators, evaluating aggregates and so on; the details are omitted but the principles will be clear.

In the case of a nested structure where inner components might themselves be controlled, such inner components are initialized and adjusted before the object as a whole and on finalization everything is done in the reverse order.

In order for a type to be controlled it has to be extended from one of two tagged types declared in the library package Ada.Finalization whose specification is as follows

```
package Ada.Finalization
     with Pure, Nonblocking => False is

   type Controlled is abstract tagged private
     with Preelaborable_Initialization;

   procedure Initialize(Object: in out Controlled) is null;
   procedure Adjust(Object: in out Controlled) is null;
   procedure Finalize(Object: in out Controlled) is null;

   type Limited_Controlled is abstract tagged limited private
     with Preelaborable_Initialization;

   procedure Initialize(Object: in out Limited_Controlled) is null;
   procedure Finalize(Object: in out Limited_Controlled) is null;

private
   ...
end Ada.Finalization;
```

We see that there are distinct abstract types for nonlimited and limited types. Naturally enough the Adjust procedure does not exist in the case of limited types because they cannot be copied.

This package also provides an example of null procedures. Remember that if a procedure specification has **is null** then the procedure behaves as if it has a null body. That is, if it is called, nothing happens.

As a simple example, suppose we wish to declare a type and keep track of how many objects (values) of the type are in existence and also record the identity number of each object in the object itself. We could declare

```
with Ada.Finalization; use Ada.Finalization;
package Tracked_Things is

   type Thing is new Controlled with
      record
         Identity_Number: Integer;
         ...      -- other data
      end record;

   overriding
   procedure Initialize(Object: in out Thing);

   overriding
   procedure Adjust(Object: in out Thing);

   overriding
   procedure Finalize(Object: in out Thing);

end Tracked_Things;

package body Tracked_Things is

   The_Count: Integer := 0;
   Next_One: Integer := 1;

   procedure Initialize(Object: in out Thing) is
   begin
      The_Count := The_Count + 1;
      Object.Identity_Number := Next_One;
      Next_One := Next_One + 1;
   end Initialize;

   procedure Adjust(Object: in out Thing)
         renames Initialize;

   procedure Finalize(Object: in out Thing) is
   begin
      The_Count := The_Count - 1;
   end Finalize;

end Tracked_Things;
```

In this example we have considered each value of a thing to be a new one and so Adjust is the same as Initialize and we can conveniently use a renaming declaration to provide the body as was mentioned in Section 13.7. An alternative approach might be to consider new things to be created only when an object is first declared (or allocated). This variation is left as an exercise.

Another point is that we have used overriding indicators for the three overridden operations. This is good practice in this important area.

The observant reader will note that the identity number is visible to users of the package and thus liable to abuse. We can overcome this by using a child package in which we extend the type. This enables us to provide different views of a type and effectively allows us to create a type with some components visible to the user and some components hidden.

Consider

```
package Tracked_Things is
   type Identity_Controlled is abstract tagged private;
private
   type Identity_Controlled is abstract new Controlled with
     record
        Identity_Number: Integer;
     end record;

   overriding
   procedure Initialize ...
   ...   -- etc.
end Tracked Things;

package Tracked_Things.User_View is

   type Thing is new Identity_Controlled with
     record
        ...   -- visible data
     end record;

end Tracked_Things.User_View;
```

In this arrangement we first declare a private type Identity_Controlled just containing the component Identity_Number (this component being hidden from the user) and then in the child package we further extend the type with the visible data which we wish the user to see. Note carefully that the type Identity_Controlled is abstract so objects of this type cannot be declared and moreover the user cannot even see that it is actually a controlled type. We also declare Initialize, Adjust, and Finalize in the private part and so they are also hidden from the user.

Using a child package allows any subprograms declared in it to access the hidden identity number. Note also that several different types could be derived from Identity_Controlled all sharing the same control mechanism. Other arrangements are also possible.

We finish this section with a few observations on the package Ada.Finalization. The procedures Initialize, Adjust, and Finalize are null and this will often be an appropriate default. The types Controlled and Limited_Controlled are of course abstract.

A key reason for making the default procedures null is because it will often be the case that the Finalize procedure for a type will naturally include a call of the Finalize procedure for the parent type. We might write

```
type T is new Parent with
  record
     ...   -- additional components
  end record;
...
procedure Finalize(Object: in out T) is
begin
  ...        -- operations to finalize additional components
  Finalize(Parent(Object));
end Finalize;
```

This is particularly relevant if the parent is a generic formal parameter (see Exercise 21.3). So all we might know is that the parent is some controlled type; since the default Finalize is null it can always be called with impunity.

In some obscure circumstances Finalize can be called twice for the same object unless it is of a limited type. (It can happen with the abort statement; see Chapter 20.) It is therefore advisable to write Finalize so that any second call does not matter. Another thing to beware is that Finalize must not raise an exception – this would cause a bounded error.

Controlled types provide a good example of the use of the form of extension aggregate where the ancestor part is just given by a subtype mark. We can typically write

```
X: T := (Controlled with ... );
```

Note that we cannot easily give an expression for the ancestor part since its type is abstract. We could however, do a dirty trick with a view conversion by writing Controlled(A_Thing) for the ancestor part as explained in the previous section.

When an aggregate provides an initial value as in this example, the value is created directly in X and Adjust is not called for the object as a whole. (This is similar to the use of aggregates to initialize limited types as described in Section 12.5.) This behaviour is vital for the case of declaring an initialized object in the same package as the type T since calling any new Adjust for T at this stage would otherwise raise Program_Error because the body of the new Adjust will not yet have been elaborated.

We conclude with an important warning. Remember to spell Finalize with -ize and not with -ise. Declaring a procedure Finalise will simply add a new primitive operation leaving the inherited Finalize still applicable. This is a good example of the dangers inherent in the flexibility of dynamic binding. Misspelling when overriding a primitive operation can cause an error which can only be detected by debugging whereas misspelling a statically called subprogram usually causes a compile-time error. But, if we consistently use overriding indicators as recommended then writing

```
overriding
procedure Finalise( ... );                    -- illegal
```

will indeed be detected at compile time because the procedure Finalise is not an overriding but a new operation.

(The author is delighted that those who spell incorrectly are in danger. Those who recall their classical Greek will appreciate that -ize is etymologically correct. It is correct British English too as stated clearly by the *Oxford English Dictionary*, but sadly many British publishers as well as even *The Times* now use -ise inherited from our Gallic friends across the Channel. Maybe this illustrates that it is always best to inherit from the ultimate ancestor!)

Exercise 14.7

1 Rewrite Initialize, Adjust, and Finalize as necessary for the situation where we only consider new things to be created when an object is declared or allocated.

2 Show how to modify the key manager of Section 12.6 so that Return_Key is automatically called on scope exit.

14.8 Multiple inheritance

We now come to a very important facility introduced in Ada 2005. In Ada 95, the inheritance model was strictly single inheritance. In the examples so far we have always had a strict tree of types descended from a single root type or interface. Some other languages such as C++ have embraced full multiple inheritance and got into a bit of a mess. However, Java introduced the notion of interfaces which work well and interfaces in Ada are the key to multiple inheritance.

We have already seen some examples of interfaces when discussing the reservation system. In this section we give a more detailed description of the rules regarding interfaces and one simple example of multiple inheritance but more elaborate examples must wait until Chapter 21 since it is helpful to have explained the generic mechanism as well.

General multiple inheritance has problems. Suppose that we have a type T with some components and operations. Perhaps

```
type T is tagged
   record
      A: Integer;
      B: Boolean;
   end record;

procedure Op1(X: T);
procedure Op2(X: T);
```

We then derive two new types from T thus

```
type T1 is new T with
   record
      C: Character;
   end record;

procedure Op3(X: T1);
                          -- Op1 and Op2 inherited, Op3 added

type T2 is new T with
   record
      C: Colour;
   end record;

procedure Op1(X: T2);
procedure Op4(X: T2);
                          -- Op1 overridden, Op2 inherited, Op4 added
```

Now suppose that we were able to derive a further type from both T1 and T2 by perhaps writing

```
type TT is new T1 and T2 with null record;        -- illegal
```

This is about the simplest example one could imagine. We have added no further components or operations. But what would TT have inherited from its two parents?

There is a general rule that a record cannot have two components with the same identifier so presumably it has just one component A and one component B. But what about C? Does it inherit the character or the colour? Or is it illegal because of the clash? Suppose T2 had a component D instead of C. Would that be allowed? Would TT then have four components?

And then consider the operations. Presumably it has both Op1 and Op2. But which implementation of Op1? Is it the original Op1 inherited from T via T1 or the overridden version inherited from T2? Clearly it cannot have both. But there is no reason why it cannot have both Op3 and Op4, one inherited from each parent.

The problems arise when inheriting *components* from more than one parent and inheriting different *implementations* of the same operation from more than one parent. There is no problem with inheriting the same specification of an operation from two parents.

These observations provide the essence of the solution. At most one parent can have components and at most one parent can have concrete operations – for simplicity they have to be the same parent. But abstract operations can be inherited from several parents. This can be phrased as saying that this kind of multiple inheritance is about merging contracts to be satisfied rather than merging algorithms or state.

As we have already seen, an interface is an abstract tagged type with no components and no concrete operations except for null procedures. And hence there is no problem with inheriting from several interfaces plus one normal tagged type.

So the main type derivation rule in Ada is that a tagged type can be derived from zero or one conventional tagged types plus zero or more interface types. Thus

```
type NT is new T and Int1 and Int2 with ... ;
```

where Int1 and Int2 are interface types. The normal tagged type if any has to be given first in the declaration. The first type is known as the parent so the parent could be a normal tagged type or an interface. The other types are known as progenitors. Additional components and operations are allowed in the usual way.

The term progenitors may seem strange but the term ancestors in this context is confusing and so a new term is introduced. Progenitors comes from the Latin *progignere*, to beget, and so is very appropriate.

Suppose that the interface Int1 and type T are in packages with operations thus

```
package P1 is
   type Int1 is interface;
   procedure Op1(X: Int1) is abstract;
   procedure N1(X: Int1) is null;
end P1;

package P is
   type T is tagged record ... end record;
   procedure Opt1( ... );
   procedure Opt2( ... );
end P;
```

We could now write

```
with P1, P;
package PNT is
   type NT is new P.T and P1.Int1 with ... ;
   procedure Op1(X: NT);                    -- concrete procedure
   ...  -- possibly other ops of NT
end PNT;
```

We must of course provide a concrete procedure for Op1 inherited from the interface Int1 since we have declared NT as a concrete type. We could also provide an overriding for N1 but if we do not then we simply inherit the null procedure of Int1. We could also override the inherited operations Opt1 and Opt2 from T in the usual way and possibly add some quite new ones.

Interfaces can be composed from other interfaces thus

```
type Int2 is interface;
...
type Int3 is interface and Int1;
...
type Int4 is interface and Int1 and Int2;
```

When we compose interfaces in this way we can add new operations so that the new interface such as Int4 will have all the operations of both Int1 and Int2 plus possibly some others declared specifically as operations of Int4. All these operations must be abstract or null and there are fairly obvious rules regarding what happens if two or more of the progenitors have the same operation. Thus a null procedure overrides an abstract one with the same (that is conformant) profile and repeated operations with the same profile must have the same convention.

We refer to all the interfaces in an interface list as progenitors. So Int1 and Int2 are the progenitors of Int4. The first one is not a parent – that term is only used when deriving a type as opposed to composing an interface. Note that the term ancestor covers all generations whereas parent and progenitors are first generation only.

Similar rules apply when a tagged type is derived from another type plus one or more interfaces as in the case of the type NT which was

```
type NT is new T and Int1 and Int2 with ... ;
```

In this case it might be that T already has some of the operations of Int1 and/or Int2. If so then the operations of T must match those of Int1 or Int2.

We informally speak of a specific tagged type as implementing an interface from which it is derived (directly or indirectly). Thus in the above example the tagged type NT must implement all the operations of the interfaces Int1 and Int2. If the type T already implements some of the operations then the type NT will automatically implement them because it will inherit the implementations from T. It could of course override such inherited operations in the usual way.

The normal rules apply in the case of functions. Suppose one operation is a function F thus

```
package P2 is
  type Int2 is interface;
  function F(Y: Int2) return Int2 is abstract;
end P2;
```

then the type NT must provide a concrete function to override F (unless the type extension of NT is null) even though T might already have such a conforming operation as in

```
package P is
  type T is tagged record ...
  function F(X: T) return T;
  ...
end P;
```

The new function F will then override the functions F of both T and Int2.

Class wide types also apply to interface types and an interface can have class wide operations such as

```
procedure Action(X: Int1'Class);
```

In the body of Action we would typically have calls of the primitive operations of the interface and these would be dispatching calls thus

```
procedure Action(X: Int1'Class) is
begin
  ...
  Op1(X);                  -- dispatching call
  ...
  N1(X);                   -- dispatching call
end Action;
```

An example of this important structure will be found in Section 21.2.

Interfaces can also be used in private extensions. Thus

```
  type PT is new T and Int2 and Int3 with private;
  ...
private
  type PT is new T and Int2 and Int3 with ... ;
```

An important rule regarding private extensions is that the full view and the partial view must agree with respect to the set of interfaces they implement. Thus although the parent in the full view need not be T but can be any type derived from T, the same is not true of the interfaces which must be such that they both implement the same set exactly. This rule is important in order to prevent a client type from overriding private operations of the parent if the client implements an interface added in the private part. This is sometimes known as the rule that there are 'no hidden interfaces'.

We can compose mixtures of limited and nonlimited interfaces but if any one of them is nonlimited then the resulting interface must not be specified as limited. This is because it must implement the equality and assignment operations implied by the nonlimited interface. Similar rules apply to types which implement one or more interfaces.

If an interface is limited then it must be stated explicitly

```
type LI is limited interface;                 -- limited
type NLI is interface;                        -- nonlimited
type I is limited interface and LI;           -- limited
```

This contrasts with the situation regarding normal types

```
type LT is tagged limited ...                 -- limited
type NLT is tagged ..                         -- nonlimited
type T is new LT with ...                     -- limited
```

In the case of types, limitedness is inherited from the parent type and does not have to be stated (this is for compatibility with Ada 83 and Ada 95) whereas in the case of interfaces it always has to be stated. However, **limited** can optionally be given on a normal type derivation and is always required if we derive a limited type from a limited interface thus

```
type T is limited new LT with ...             -- limited optional
type T is limited new LI with ...             -- limited mandatory
```

These rules all really come down to the same thing. If any parent or progenitor is nonlimited then the descendant must be nonlimited. In other words if a type (including an interface) is limited then all its ancestors must also be limited.

We conclude with an amusing example. Suppose we wish to inherit from both the type Geometry.Object and a type People.Person. This might be in order to simulate the Flatlanders of the book *Flatland* written by Edwin Abbott in 1884. This book is a satire on class structure and concerns a world in which people are flat geometrical objects. The working classes are triangles, the middle classes are other polygons and the aristocracy are circles. Sadly, all females are two-sided and thus simply a line segment. We would therefore like to write something like

```
type Flatlander is new People.Person and Geometry.Object with ...
```

where the type Person might be in a package People thus

```
package People is
   type Person is abstract tagged
      record
         Birth: Date;
         ...
      end record;
   ...                              -- various operations on Person
end People;
```

The earlier type Object had components. We cannot inherit from two types with components and so one of them must be turned into an interface. We can turn the type Object into an interface by writing

```
package Geometry is
   type Object is interface;

   procedure Move(O: in out Object; X, Y: in Float) is abstract;
   function X_Coord(O: Object) return Float is abstract;
   function Y_Coord(O: Object) return Float is abstract;
   function Area(O: Object) return Float is abstract;
   ...
end Geometry;
```

We can now declare an abstract type Flatlander. Remember that the first type is the only one that can have components and so the type Person has to be first in the list. In other words the type Person is the parent type and the type Object is a progenitor type. We write

```
package Flatland is
   type Flatlander is abstract new Person and Object with private;

   procedure Move(F: in out Flatlander; X, Y: Float);
   function X_Coord(F: Flatlander) return Float;
   function Y_Coord(F: Flatlander) return Float;

private
   type Flatlander is abstract new Person and Object with
      record
         X_Coord, Y_Coord: Float := 0.0;
      end record;
end;
```

The type Flatlander will now inherit the components such as Birth from the type Person, any operations of the type Person and the abstract operations of the type Object. Moreover, it is convenient to declare the coordinates as components since we need to do that eventually and we can then override the inherited abstract operations Move, X_Coord and Y_Coord with concrete ones. The type Flatlander is abstract since we are not yet in a position to provide a concrete version of a function such as Area. The package body is straightforward.

We can now declare a type Square suitable for Flatland (when originally written the book was published anonymously under the pseudonym A_Square) as follows

```
package Flatland.Squares is

   type Square is new Flatlander with
      record
         Side: Float;
      end record;

   function Area(S: Square) return Float;

end;

package body Flatland.Squares is

   function Area(S: Square) is
```

```
    begin
      return S.Side**2;
    end Area;

  end Flatland.Squares;
```

and all the operations are thereby implemented. By way of illustration we have made the the extra component Side of the type Square directly visible. We can now declare

```
  A_Square: Square := (Flatlander with Side => 4.00);
```

and he will have all the properties of a square and a person and will be located by default at the origin. Incidentally, Dr Abbott had a proper education at St John's College, Cambridge. He read theology, classics, and mathematics just like the type Student of Section 8.7.

This concludes a basic description of the use of interfaces for multiple inheritance. Another example will be found in the next section and more elaborate examples will be found in Chapter 21. Note also that there are other forms of interfaces, namely synchronized interfaces, task interfaces, and protected interfaces. They will be described in Chapter 22.

14.9 Multiple implementations

We conclude this chapter on the fundamentals of OOP in Ada with a discussion on an important characteristic of object oriented programming – the ability to provide different implementations of a single abstraction. Of course one can do this statically by writing a package with alternative bodies but only one body can appear in one program.

Interfaces and inheritance enable different types to be treated as different realizations of a common abstraction. The tag of an object indicates its implementation and allows a dynamic binding between the client and the appropriate implementation.

We can thus develop different implementations of a single abstraction, such as a family of set types, as in the next example. We naturally start with an interface type

```
  package Abstract_Sets is

    type Set is interface;

    function Empty return Set is abstract;
    function Unit(E: Element) return Set is abstract;
    function Union(S, T: Set) return Set is abstract;
    function Intersection(S, T: Set) return Set is abstract;
    procedure Take(From: in out Set; E: out Element) is abstract;
  end Abstract_Sets;
```

This package provides an abstract specification of sets where we assume that Element is some discrete subtype such as Integer or Colour. The type Set is an interface. The package also defines a set of abstract primitive operations for the interface Set. An implementation of the abstraction can then be created by deriving a concrete type from the interface Set and providing concrete operations for that implementation.

The primitive operations are Empty which returns the null set, Unit which builds a set of one element, Union and Intersection as expected, and Take which removes one (arbitrary) element from the set.

One possible implementation would be to use a Boolean array as in Section 8.6 where each element represents the presence or absence of a member in the set as follows.

```
with Abstract_Sets;
package Bit_Vector_Sets is

   type Bit_Set is new Abstract_Sets.Set with private;

   function Empty return Bit_Set;
   function Unit(E: Element) return Bit_Set;
   function Union(S, T: Bit_Set) return Bit_Set;
   function Intersection(S, T: Bit_Set) return Bit_Set;
   procedure Take(From: in out Bit_Set; E: out Element);

private

   type Bit_Vector is array (Element) of Boolean;

   type Bit_Set is new Abstract_Sets.Set with
      record
         Data: Bit_Vector;
      end record;
end Bit_Vector_Sets;

package body Bit_Vector_Sets is

   function Empty return Bit_Set is
   begin
      return (Data => (others => False));
   end;

   function Unit(E: Element) return Bit_Set is
      S: Bit_Set := Empty;
   begin
      S.Data(E) := True;
      return S;
   end;

   function Union(S, T: Bit_Set) return Bit_Set is
   begin
      return (Data => S.Data or T.Data);
   end;
      ...                      -- Intersection
      ...                      -- and Take
end Bit_Vector_Sets;
```

Such an implementation is only appropriate if the number of values in the subtype Element is not too large (note that we would typically pack the array anyway using the aspect Pack described in Chapter 25).

An alternative implementation more appropriate to sparse sets might be based on using a linked list containing the elements present in a set. For such an implementation we would have to redefine equality and assignment as well as the abstract operations; we will return to this in a moment.

But the really interesting thing is that we could then write a program which contained both forms of sets; we could convert from one representation to any other by using

```
procedure Convert(From: in Set'Class; To: out Set'Class) is
   Temp: Set'Class := From;
   E: Element;
begin
   -- build target set, element by element
   To := Empty;
   while Temp /= Empty loop
      Take(Temp, E);
      To := Union(To, Unit(E));
   end loop;
end Convert;
```

This works by extracting the elements one at a time from the source set From and then building them up into the target set To. It is instructive to consider the fine details of how it dispatches onto the appropriate operations according to the specific type of its parameters using the rules described in Section 14.4.

The first action is to copy the original set into the class wide variable Temp. This avoids damaging the original set. Remember that all variables of class wide types such as Temp have to be initialized since class wide types are indefinite.

We then have

```
To := Empty;
```

which illustrates the case of a tag being indeterminate. The function Empty has a controlling result but no controlling operands to determine the tag; the choice of function to call has to be determined by the tag of the class wide parameter To which is the destination of the assignment.

We then come to

```
while Temp /= Empty loop
```

where the dispatching equality operator has two controlling operands. Since Empty is tag indeterminate the tag of Temp is used to determine which Empty and then which equality operator to call.

The statement

```
Take(Temp, E);
```

has the single controlling operand Temp and so dispatches according to the tag of Temp which is that of From.

Finally the statement

```
To := Union(To, Unit(E));
```

also causes dispatching to Unit and Union according to the tag of To.

An interesting property of the procedure Convert is that nothing can go wrong; it is not possible for two controlling operands to have different tags and thus raise Constraint_Error. Indeed there are no checks but just enough information to do the dispatching correctly. The algorithm used by Convert also works if the *formal* parameters are specific types in which case no dispatching occurs since all the types are known statically – the variable Temp would also have to be of the appropriate specific type as well.

We mentioned above that assignment is also a dispatching operation although this is not often apparent. In this example, however, if the type of From were a linked list then a deep copy would be required otherwise the original value could be damaged when the copy is decomposed. Such a deep copy can be performed by using a controlled type as described in Section 14.7.

The general idea is that the set is implemented as a record containing a single component of the controlled type Inner which itself contains a single component of the type Cell_Ptr which is an access to the usual linked list containing the actual elements. Assigning the record also assigns the inner component and this invokes the procedure Adjust for the controlled component which then copies the whole list. The implementation might be as follows

```
with Abstract_Sets;
with Ada.Finalization; use Ada.Finalization;
package Linked_Sets is

   type Linked_Set is new Abstract_Sets.Set with private;
   ...   -- the various operations on Linked_Set

private
   type Cell;
   type Cell_Ptr is access Cell;

   type Cell is
      record
         Next: Cell_Ptr;
         E: Element;
      end record;

   type Inner is new Controlled with
      record
         The_Set: Cell_Ptr;
      end record;

   procedure Adjust(Object: in out Inner);

   type Linked_Set is new Abstract_Sets.Set with
      record
         Component: Inner;
      end record;

end Linked_Sets;
```

```
package body Linked_Sets is
   function Copy(P: Cell_Ptr) return Cell_Ptr is      -- deep copy
   begin
      if P = null then
         return null;
      else
         return new Cell'(Copy(P.Next), P.E);
      end if;
   end Copy;

   procedure Adjust(Object: in out Inner) is
   begin
      Object.The_Set := Copy(Object.The_Set);
   end Adjust;
   ...
end Linked_Sets;
```

The procedure Adjust for Inner performs a deep copy by calling the function Copy with its single component The_Set as parameter. Performing an assignment on the type Linked_Set (such as Temp := From;) causes Adjust to be called on its inner component thereby making the deep copy; Figure 14.4 shows the situation (a) just after the bitwise copy is made but before Adjust is called, and (b) after Adjust is called.

Note that the type Linked_Set as a whole is not controlled because it is simply derived from Abstract_Sets.Set which is not. Nevertheless it behaves rather as if it were controlled because of its controlled component and so exhibits a hidden form of multiple inheritance; see Section 21.1. But of course none of the mechanism is visible to the user. Observe that we do not need to provide a procedure Initialize and

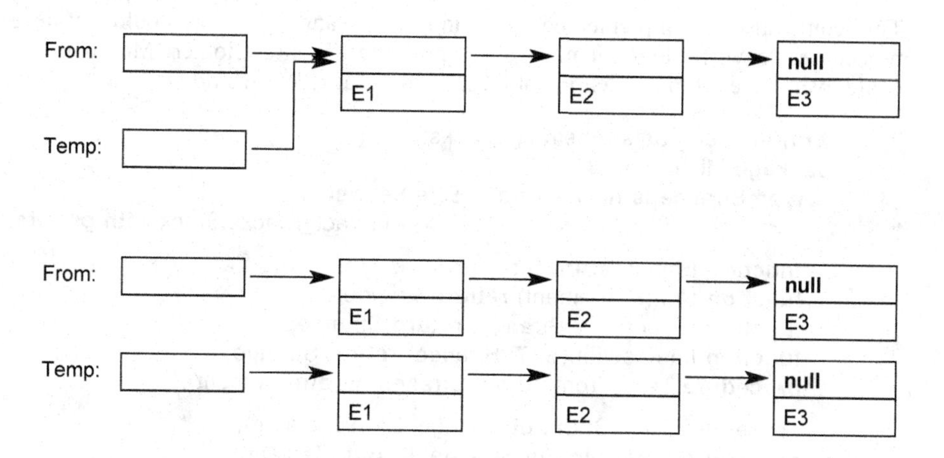

Figure 14.4 Two stages in the deep copy assignment.

that we have not bothered to provide Finalize although it would be wise to do so in order to discard unused space; see Section 25.4.

We could alternatively have used multiple inheritance so that the full type is

```
type Linked_Set is new Controlled and Abstract_Sets.Set with
  record
    The_Set: Cell_Ptr;
  end record;
```

and this avoids the introduction of the type Inner. Again the type Linked_Set is not visibly controlled. The procedure Adjust then becomes

```
procedure Adjust(Object: in out Linked_Set) is
begin
  Object.The_Set := Copy(Object.The_Set);
end Adjust;
```

The details of the remaining subprograms are left to the imagination of the reader. Some thought is necessary in order to avoid excessive manipulation. It is probably best to arrange that elements are not duplicated. It might also be advisable to keep the elements on the list in some canonical order otherwise the equality operation will be tedious.

As another example we might have an interface defining a stack thus

```
package Abstract_Stacks is

  type Stack is interface;

  function Empty return Stack is abstract;
  procedure Push(S: in out Stack; E: in Element) is abstract;
  procedure Pop(S: in out Stack; E: out Element) is abstract;
end Abstract_Stacks;
```

This could also be implemented in various ways and again we could define a procedure Convert which changed one representation into another. Moreover we could even have a type which is both a set and a stack. We would write

```
with Abstract_Sets, Abstract_Stacks;
package Mixture is
  type Strange is new Abstract_Sets.Set and
                              Abstract_Stacks.Stack with private;

  function Empty return Strange;
  function Unit(E: Element) return Strange;
  function Union(S, T: Strange) return Strange;
  function Intersection(S, T: Strange) return Strange;
  procedure Take(From: in out Strange; E: out Element);

  procedure Push(S: in out Strange; E: in Element);
  procedure Pop(S: in out Strange; E: out Element);

private
  ...
end Mixture;
```

An interesting feature here is that the function Empty is an abstract operation of both progenitors. Since they have the same profile the one operation of the type Strange can implement both. The reader might wonder how the type Strange can be both a set and a stack. That is entirely up to the writer of the package Mixture. It could be essentially a stack each of whose elements is a set or indeed vice versa. All that matters from the point of view of the language is that the various operations are implemented in a manner consistent with their profiles.

Exercise 14.9

1 Given

```
B1, B2: Bit_Set;
L1: Linked_Set;
C1: Set'Class := ... ;
C2: Set'Class := ... ;
```

then which of the following are legal and which involve a run-time check?

(a) Union(B1, B2)
(b) Union(B1, L1)
(c) Union(B1, C1)
(d) Union(C1, C2)

2 The procedure Convert will convert between any two sets but is somewhat inefficient if the two sets happen to have the same representation. Modify it to overcome this inefficiency.

3 Declare a procedure Convert that will convert between stacks of different representations. Sketch two representations of a stack using an array and linked lists much as in Sections 12.4 and 12.5.

Checklist 14

Additional primitive operations can only be added if derivation is in a package specification.

Additional primitive operations cannot be added to the parent type after a full type is derived from it.

Conversion of tagged types can only be towards the root.

Objects of a class wide type must be initialized.

Dispatching only occurs with a class wide actual parameter and a specific formal parameter.

It is not possible to derive from a class wide type.

Tagged type parameters are always passed by reference.

Tagged formal parameters are considered aliased.

Objects of an abstract type or interface are not allowed.

The operations of an interface must be abstract or null procedures.

Functions returning an abstract type must themselves be abstract.

Functions returning a tagged type require overriding on extension unless the extension is null.

The attribute Tag cannot be applied to an object of a specific type.

Equality has special rules on extension.

The tag of an object can never be changed.

An abstract type cannot have private abstract operations.

Controlling operands are never **null**.

Extension at an inner level is permitted in Ada 2005.

Incomplete types can be marked as tagged in Ada 2005.

Functions do not require overriding in Ada 2005 if the extension is null.

The prefixed notation was added in Ada 2005.

Interfaces were added in Ada 2005.

Synchronized interfaces were added in Ada 2012, see Section 22.3.

New in Ada 2022

The concept of stable properties of a type is introduced, see Section 16.5.

15 Exceptions

15.1 Handling exceptions
15.2 Declaring and raising exceptions
15.3 Checking and exceptions
15.4 Exception occurrences
15.5 Exception pragmas and aspects
15.6 Scope of exceptions

At various times in the preceding chapters we have said that if something goes wrong when the program is executed, then an exception, often Constraint_Error, will be raised. In this chapter we describe the exception mechanism and show how remedial action can be taken when an exception occurs. We also show how we may define and use our own exceptions. Exceptions concerned with interacting tasks are dealt with in Chapter 20.

15.1 Handling exceptions

We have seen that if we break various language rules then an exception may be raised when we execute the program. There are four predefined exceptions (declared in the package Standard) of which we have met three so far

Constraint_Error — This generally corresponds to something going out of range; this includes when something goes wrong with arithmetic such as an attempt to divide by zero.

Program_Error — This will occur if we attempt to violate the control structure in some way such as running into the **end** of a function, breaking the accessibility rules or calling a subprogram whose body has not yet been elaborated – see Sections 10.1, 11.6, and 12.1.

Storage_Error — This will occur if we run out of storage space, as for example if we called the recursive function Factorial with a large parameter – see Section 10.1.

The other predefined exception is Tasking_Error. This is concerned with tasking and so is dealt with in Chapter 20.

Note that for historical reasons the exception Constraint_Error is also renamed as Numeric_Error in the package Standard. Moreover, Numeric_Error is considered obsolescent and so should be avoided.

If we anticipate that an exception may occur in a part of a program then we can write an exception handler to deal with it. For example, suppose we write

```
begin
  ...  -- sequence of statements
exception
  when Constraint_Error =>
    ...  -- do something
end;
```

If Constraint_Error is raised while we are executing the sequence of statements between **begin** and **exception** then the flow of control is interrupted and immediately transferred to the sequence of statements following the =>. The clause starting **when** is known as an exception handler.

As an illustration we could compute Tomorrow from Today by writing

```
begin
  Tomorrow := Day'Succ(Today);
exception
  when Constraint_Error =>
    Tomorrow := Day'First;
end;
```

If Today is Day'Last (that is, Sun) then the attempt to evaluate Day'Succ(Today) causes the exception Constraint_Error to be raised. Control is then transferred to the handler for Constraint_Error and the statement Tomorrow := Day'First; is executed. Control then passes to the end of the block.

This is really a bad example. Exceptions should be used for rarely occurring cases or those which are inconvenient to test for at their point of occurrence. By no stretch of the imagination is Sunday a rare day. Over 14% of all days are Sundays. Nor is it difficult to test for the condition at the point of occurrence. So we should really have written

```
if Today = Day'Last then
  Tomorrow := Day'First;
else
  Tomorrow := Day'Succ(Today);
end if;
```

However, it is a simple example with which to illustrate the mechanism.

Several handlers can be written between **exception** and **end**. Thus we might have

```
begin
  ...  -- sequence of statements
exception
```

```
      when Constraint_Error | Program_Error =>
         Put("Constraint Error or Program Error occurred");
         ...
      when Storage_Error =>
         Put("Ran out of space");
         ...
      when others =>
         Put("Something else went wrong");
         ...
   end;
```

In this example a message is output according to the exception. Note the similarity to the case statement. Each **when** is followed by one or more exception names separated by vertical bars. As usual we can write **others** but it must be last and on its own; it handles any exception not listed in the previous handlers.

Note that we can mention the same exception twice in the same handler; this is useful when exceptions are renamed such as in the case of historic programs which might have a common handler for Constraint_Error and its renaming Numeric_Error.

Exception handlers can appear at the end of a block, extended return statement, subprogram body, package body (also task body and accept statement, see Chapter 20) and have access to all entities declared in that unit. The examples have shown a degenerate block in which there is no **declare** and declarative part; the block was introduced just to provide somewhere to hang the handlers. We could rewrite the bad example to determine tomorrow as a function thus

```
function Tomorrow(Today: Day) return Day is
begin
   return Day'Succ(Today);
exception
   when Constraint_Error =>
      return Day'First;
end Tomorrow;
```

It is important to realize that control can never be returned directly to the unit where the exception was raised. The sequence of statements following => replaces the remainder of the unit containing the handler and thereby completes execution of the unit. Hence a handler for a function must generally contain a return statement in order to provide the 'emergency' result.

Note that a goto statement cannot transfer control from a unit into one of its handlers or vice versa or from one handler to another. However, the statements of a handler can otherwise be of arbitrary complexity. They can include blocks, calls of subprograms and so on. A handler of a block could contain a goto statement which transferred control to a label outside the block and it could contain an exit statement if the block were inside a loop.

Observe that a handler at the end of a package body applies only to the initialization sequence of the package and not to any subprograms in the package. Such subprograms must have individual handlers if they are to deal with exceptions.

We now consider the question of what happens if a unit does not provide a handler for a particular exception. The answer is that the exception is propagated dynamically. This simply means that the unit is terminated and the exception is raised at the point where the unit was executed.

In the case of a block we therefore look for a handler in the unit containing the block and similarly for an extended return. In the case of a subprogram, the call is terminated and we look for a handler in the unit which called the subprogram. In the case of a package body we consider the unit containing the package declaration

This unwinding process is repeated until either we reach a unit containing a handler for the particular exception or come to the top level. If we find a unit containing a relevant handler then the exception is handled at that point. However, if we reach the main subprogram and have still found no handler then the main subprogram is abandoned and we can expect the run-time environment to provide us with a suitable diagnostic message. (Unhandled exceptions in tasks are dealt with in Sections 20.8 and 22.5.) Another possibility is that an unhandled exception might occur during the elaboration of a library unit in which case the program is abandoned before the main subprogram is called.

It is important to understand that exceptions are propagated dynamically and not statically. That is, an exception not handled by a subprogram is propagated to the unit calling the subprogram and not to the unit containing the declaration of the subprogram – these may or may not be the same. Moreover, if the statements in a handler themselves raise an exception then the unit is terminated and the exception propagated to the calling unit; the handler does not loop.

Finally, note that if an exception is raised in a declaration in a subprogram then the exception is propagated as explained in Section 15.6.

Exercise 15.1

Note: these are exercises to check your understanding of exceptions. They do not necessarily reflect good Ada programming techniques.

1. Assuming that calling Sqrt with a negative parameter and attempting to divide by zero both raise Constraint_Error, rewrite the procedure Quadratic of Section 10.3 without explicitly testing D and A.

2. Rewrite the function Factorial of Section 10.1 so that if it is called with a negative parameter (which would normally raise Constraint_Error) or a large parameter (which would normally raise Storage_Error or Constraint_Error) then a standard result of say -1 is returned. Hint: declare an inner function Slave which actually does the work.

15.2 Declaring and raising exceptions

Relying on the predefined exceptions to detect unusual but anticipated situations is usually bad practice because they do not provide a guarantee that the exception has in fact been raised because of the anticipated situation. Something else may have gone wrong instead.

For example, consider the package Stack of Section 12.1. If we call Push when the stack is full then the statement Top := Top + 1; will raise Constraint_Error and if we call Pop when the stack is empty then Top := Top - 1; will also raise Constraint_Error. Since Push and Pop do not have exception handlers, the exception will be propagated to the unit calling them. So we could write

```
declare
   use Stack;
begin
   ...
   Push(M);
   ...
   N := Pop;
   ...
exception
   when Constraint_Error =>
     ...    -- stack manipulation incorrect?
end;
```

Misuse of the stack now results in control being transferred to the handler for Constraint_Error. However, there is no guarantee that the exception has arisen because of misuse of the stack; something else could have gone wrong instead.

A better solution is to raise an exception specifically declared to indicate misuse of the stack. Thus the package could be rewritten as

```
package Stack is
   Error: exception;
   procedure Push(X: Integer);
   function Pop return Integer;
end Stack;

package body Stack is
   Max: constant := 100;
   S: array (1 .. Max) of Integer;
   Top: Integer range 0 .. Max;

   procedure Push(X: Integer) is
   begin
      if Top = Max then
         raise Error;
      end if;
      Top := Top + 1;
      S(Top) := X;
   end Push;

   function Pop return Integer is
   begin
      if Top = 0 then
         raise Error;
      end if;
      Top := Top - 1;
      return S(Top + 1);
   end Pop;

begin
   Top := 0;
end Stack;
```

The exception Error is declared in a similar way to a variable. Several exceptions could be declared in the same declaration.

An exception can be raised by an explicit raise statement naming the exception. The handling and propagation rules for user-defined exceptions are just as for the predefined exceptions. We can now write

```
declare
   use Stack;
begin
   ...
   Push(M);
   ...
   N := Pop;
   ...
exception
   when Error =>
      ...   -- stack manipulation incorrect
   when others =>
      ...   -- something else went wrong
end;
```

We have now successfully separated the handler for misusing the stack from the handler for other exceptions.

Note that if we had not provided a use clause then we would have had to refer to the exception in the handler as Stack.Error; the usual dotted notation applies.

What could we expect to do in the handler in the above case? Apart from reporting that the stack manipulation has gone wrong, we might also expect to reset the stack to an acceptable state although we have not provided a convenient means of doing so. A procedure Reset in the package Stack would be useful. A further thing we might do is to relinquish any resources that were acquired in the block and might otherwise be inadvertently retained. Suppose for instance that we had also been using the package Key_Manager of Section 12.6. We might then call Return_Key to ensure that a key declared and acquired in the block had been returned. Remember that Return_Key does no harm if called unnecessarily.

In the case of controlled types discussed in Section 14.7, we do not have to call Finalize since that is done automatically when we leave the block anyway.

We would probably also want to reset the stack and return the key in the case of any other exception as well; so it would be sensible to declare a procedure Clean_Up to do all the actions required. So our block might look like

```
declare
   use Stack, Key_Manager;
   My_Key: Key;

   procedure Clean_Up is
   begin
      Reset;
      Return_Key(My_Key);
   end;
```

```
begin
   Get_Key(My_Key);
   ...
   Push(M);
   ...
   Action(My_Key, ... );
   ...
   N := Pop;
   ...
   Return_Key(My_Key);
exception
   when Error =>
      Put("Stack used incorrectly");
      Clean_Up;
   when others =>
      Put("Something else went wrong");
      Clean_Up;
end;
```

Note that we could make the return of the key automatic by using a controlled type for the key as in Exercise 14.7(**2**). Leaving a unit via a handler will invoke any Finalize procedures properly.

We have rather assumed that Reset is a further procedure declared in the package Stack but note that we could write our own curious procedure externally as follows

```
procedure Reset is
   Junk: Integer;
   use Stack;
begin
   loop
      Junk := Pop;
   end loop;
exception
   when Error =>
      null;
end Reset;
```

This works by repeatedly calling Pop until Error is raised. We then know that the stack is empty. The handler needs to do nothing other than prevent the exception from being propagated; so we merely write **null**. This procedure seems a bit like trickery; it would be far better to have a reset procedure in the overall package Stack.

Sometimes the actions that require to be taken as a consequence of an exception need to be performed on a layered basis. In the above example we returned the key and then reset the stack but it is probably the case that the block as a whole cannot be assumed to have done its job correctly. We can indicate this by raising an exception as the last action of the handler, thus

```
exception
  when Error =>
    Put("Stack used incorrectly");
    Clean_Up;
    raise Another_Error;
  when others =>
    ...
end;
```

The exception Another_Error will then be propagated to the unit containing the block. We could put the statement

```
raise Another_Error;
```

in the procedure Clean_Up.

Sometimes it is convenient to handle an exception and then propagate the same exception. This can be done by just writing

```
raise;
```

This is particularly useful when we handle several exceptions with the one handler. So we might have

```
  when others =>
    Put("Something else went wrong");
    Clean_Up;
    raise;
end;
```

The current exception will be remembered even if the action of the handler raises and handles its own exceptions such as occurred in our trick procedure Reset. However, note that there is a rule that we can only write **raise**; directly in a handler and not for instance in a procedure called by the handler such as Clean_Up.

The stack example illustrates a legitimate use of exceptions. The exception Error should rarely occur and it would be inconvenient to test for the condition at each possible point of occurrence. To do that we would presumably have to provide an additional parameter to Push of type Boolean and mode **out** to indicate that all was not well, and then test it after each call. We could do the same for the function Pop, remember that functions can have an **out** parameter in Ada 2012 and Ada 2022. But maybe it would be best to use two procedures. The package specification would then become

```
package Stack is
  procedure Push(X: in Integer; B: out Boolean);
  procedure Pop(X: out Integer; B: out Boolean);
end;
```

and we would have to write

```
declare
  use Stack;
  OK: Boolean;
```

```
begin
   ...
   Push(M, OK);
   if not OK then ...     end if;
   ...
   Pop(N, OK);
   if not OK then ...     end if;
end;
```

It is clear that the use of an exception provides a better structured program.

Note that nothing prevents us from explicitly raising one of the predefined exceptions. We recall that in Section 10.1 when discussing the function Inner we stated that probably the best way of coping with parameters whose bounds were unequal was to explicitly raise Constraint_Error.

Ada 2012 introduced conditional expressions as described in Chapter 9. Another feature added to Ada 2012 after it became a standard (see AI12-22) is the ability to explicitly raise an exception within an expression. So just as we can write

```
if X > Y then
   Z := +1;
elsif X < Y then
   Z := -1;
else
   raise Some_Error;
end if;
```

we can equally write

```
Z := (if X < Y then +1 elsif X > Y then -1 else raise Some_Error);
```

An exception can similarly be explicitly raised as part of a case expression or indeed as part of any expression. But a raise expression cannot be just **raise** on its own.

In Section 9.3 we wondered whether the case statement assigning values to Object_Ptr in Program 1 might be better done as a case expression. Indeed, we could have a raise expression to handle the exceptional case of an incorrect character. But we would then need a handler to exit the loop which would not really be an improvement.

Exercise 15.2

1. Rewrite the package Random of Exercise 12.1(**1**) so that it declares and raises an exception Bad if the initial value is not odd.

2. Rewrite the answer to Exercise 15.1(**2**) so that the function Factorial always raises Constraint_Error if the parameter is negative or too large.

3. Declare a function "+" which takes two parameters of type Vector and returns their sum using sliding semantics by analogy with the predefined one-dimensional array operations described in Section 8.6. Use type Vector from Section 8.2. Raise Constraint_Error if the arrays do not match.

4 Are we completely justified in asserting that Stack.Error could only be raised by the stack going wrong?

5 Assuming that the exception Error is declared in the specification of Stacks, rewrite procedures Push and Pop of Section 12.5 so that they raise Error rather than Storage_Error and Constraint_Error.

15.3 Checking and exceptions

In the previous section we came to the conclusion that it was logically better to check for the stack overflow condition ourselves rather than rely upon the built-in check associated with the violation of the subtype of Top. At first sight the reader may well feel that this would reduce the execution efficiency of the program. However, this is not necessarily so, assuming a reasonably intelligent compiler; this stack example can be used to illustrate the advantages of the use of appropriate subtypes.

We will concentrate on the procedure Push; similar arguments apply to the function Pop.

First, consider the original package Stack of Section 12.1. In that we had

```
S: array (1 .. Max) of Integer;
Top: Integer range 0 .. Max;

procedure Push(X: Integer) is
begin
   Top := Top + 1;
   S(Top) := X;
end Push;
```

If the stack is full (that is Top = Max) and we call Push then it is the assignment to Top that raises Constraint_Error. This is because Top has a range constraint. However, the only run-time check that needs to be compiled is that associated with checking the upper bound of Top. There is no need to check for violation of the lower bound since the expression Top + 1 could not be less than 1 (assuming that the value in Top is always in range). Note also that no checks need be compiled with respect to the assignment to S(Top). This is because the value of Top at this stage must lie in the range 1 .. Max (which is the index range of S) – it cannot exceed Max because this has just been checked by the previous assignment and it cannot be less than 1 since 1 has just been added to its previous value which could not have been less than 0. So just one check needs to be compiled in the procedure Push.

The type Integer is itself considered to be constrained as we shall see in Chapter 17. So if the variable Top was simply declared as

```
Top: Integer;
```

then a check would still be applied to the assignment to Top (it might be done by the hardware), and checks would also have to be compiled for the assignment to S(Top) in order to ensure that Top lay within the index range of S. So three checks would be necessary in total, one to prevent overflow of Top and two for the index range.

So applying the range constraint to Top actually reduces the number of checks required from three to one. This is typical behaviour given a compiler with a moderate degree of flow analysis. The more you tell the compiler about the properties of the variables (and assuming the constraints on the variables match their usage), the better the object code.

Now consider what happens when we add our own test as in the previous section (and we assume that Top now has its range constraint)

```
procedure Push(X: Integer) is
begin
   if Top = Max then
      raise Error;
   end if;
   Top := Top + 1;
   S(Top) := X;
end Push;
```

Clearly, we have added a check of our own. As a consequence, there is now no need for the compiler to insert the check on the upper bound of Top in the assignment statement

```
Top := Top + 1;
```

because our own check will have caused control to be transferred away via the raising of the Error exception for the one original value of Top that would have caused trouble. So the net effect of adding our own check is simply to replace the one compiler check by one of our own; the net result is the same and the object code is not less efficient.

There are two morals to this tale. The first is that we should tell the compiler the whole truth about our program; the more it knows about the properties of our variables, the more likely it is to be able to keep the checks to the appropriate minimum. In fact this is just an extension of the advantage of strong typing discussed in Section 6.3 where we saw how arbitrary run-time errors can be replaced by easily understood compile-time errors.

The second moral is that introducing our own exceptions rather than relying upon the predefined ones need not reduce the efficiency of our program. In fact it is generally considered bad practice to rely upon the predefined exceptions for steering our program and especially bad to raise the predefined exceptions explicitly ourselves. It is all too easy to mask an unexpected genuine error that needs fixing.

It should also be noted that we can always ask the compiler to omit the run-time checks by using the pragma Suppress. This is described in more detail in Section 15.5.

Finally, an important warning. Our analysis of when checks can be omitted depends upon all variables satisfying their constraints at all times. Provided checks are not suppressed we can be reasonably assured of this apart from one nasty loophole. This is that we are not obliged to supply initial values in the declarations of variables in the first place. So they can start with a junk value which does not satisfy any constraints and may not even be a value of the base type. If such a variable is read before being updated then our program has a bounded error and all our analysis is worthless. We should therefore ensure that all variables are initialized or updated before they are read. See Section 26.6. Beware however, that giving variables gratuitous initial values can obscure flow errors and so is not generally a good idea.

Exercise 15.3

1 Consider the case of the procedure Push with explicit raising of Error but suppose that there is no range constraint on Top. How many checks would then be required?

15.4 Exception occurrences

There are a number of auxiliary facilities which enable a more detailed analysis of exceptions and their cause. In the clean-up clause in Section 15.2 we wrote

```
    when others =>
       Put("Something else went wrong");
       Clean_Up;
    raise;
 end;
```

This is not very helpful since we would really like to record what actually happened. We can do this using an exception occurrence which identifies both the exception and the instance of its being raised (that is, the circumstances associated with the particular error condition).

The limited type Exception_Occurrence is declared in the child package Ada.Exceptions together with various subprograms including functions Exception_Name, Exception_Message, and Exception_Information. These functions take an exception occurrence as their single parameter and return a string (which has lower bound of 1).

As their names suggest, Exception_Name returns the name of the exception (the full dotted name in upper case) and Exception_Message and Exception_Information return two levels of more detailed information which identify the cause and location. The result of Exception_Message should be a one-line message and not contain the exception name whereas Exception_Information will include both the name and the message and might provide full details of a trace back as well. Although the details are dependent upon the implementation the general intent is that the messages are suitable for output and analysis on the system concerned.

To get hold of the occurrence we write a 'choice parameter' in the handler and this behaves as a constant of the type Exception_Occurrence. We can now write

```
    when Event: others =>
       Put("Unexpected exception: ");
       Put_Line(Exception_Name(Event));
       Put(Exception_Message(Event));
       Clean_Up;
       raise;
 end;
```

The object Event of the type Exception_Occurrence acts as a sort of marker which enables us to identify the current occurrence; its scope is the handler. Such a choice parameter can be placed in any handler.

We can also supply our own message. One way to do this is by an extended form of raise statement in which a string is supplied. Consider

```
declare
   Trouble: exception;
begin
   ...
   raise Trouble with "Doom";
   ...
   raise Trouble with "Gloom";
   ...
exception
   when Event: Trouble =>
      Put(Exception_Message(Event));
end;
```

This extended raise statement raises the exception Trouble with the string attached as the message. The call of Put in the handler will output "Doom" or "Gloom" according to which occurrence of Trouble was raised. Note that our own messages work consistently with the predefined messages; for example Exception_Information will include our message if we have supplied one. A message can also be attached to a raise expression.

Note the difference between

```
raise An_Error;             -- message is implementation-defined
raise An_Error with "";     -- message is null
```

The package Ada.Exceptions contains other types and subprograms which provide further facilities. Its specification is

```
package Ada.Exceptions
      with Preelaborate, Nonblocking, Global => in out synchronized is
   type Exception_Id is private
      with Preelaborable_Initialization;

   Null_Id: constant Exception_Id;

   function Exception_Name(Id: Exception_Id) return String;
   ...   -- also Wide_ and Wide_Wide_Exception_Name

   type Exception_Occurrence is limited private
      with Preelaborable_Initialization;
   type Exception_Occurrence_Access is
                                        access all Exception_Occurrence;
   Null_Occurrence: constant Exception_Occurrence;

   procedure Raise_Exception(E: in Exception_Id;
                             Message: in String := "")
      with No_Return;

   function Exception_Message(X: Exception_Occurrence) return String;
```

```
   procedure Reraise_Occurrence(X: in Exception_Occurrence);

   function Exception_Identity(X: Exception_Occurrence)
                                             return Exception_Id;
   function Exception_Name(X: Exception_Occurrence) return String;
   ...  -- also Wide_ and Wide_Wide_Exception_Name
   function Exception_Information(X: Exception_Occurrence) return String;

   procedure Save_Occurrence(Target: out Exception_Occurrence;
                             Source: in Exception_Occurrence);
   function Save_Occurrence(Source: Exception_Occurrence)
                                return Exception_Occurrence_Access;
   ... -- also procedures for streaming occurrences, see Section 23.9
private
   ...
end Ada.Exceptions;
```

The two subprograms Save_Occurrence enable exception occurrences to be saved for detailed later analysis. Note that the type Exception_Occurrence is limited; using subprograms rather than allowing the user to save values through assignment gives better control over the use of storage for saved exception occurrences (which could be large since they may contain extensive trace back information).

The procedure Save_Occurrence copies the occurrence from the Source to the Target. It may truncate the message associated with the occurrence to 200 characters; this corresponds to the minimum size of line length required to be supported as mentioned in Section 5.3. On the other hand, the function Save_Occurrence copies the occurrence to a newly created object and returns an access value to the new object. It is not permitted to truncate the message.

So we might have some debugging package containing an array in which perhaps up to 100 exception occurrences might be stored. A fragment of the body of a crude implementation might be

```
Dump: array (1 .. 100) of Exception_Occurrence;
Dump_Index: Integer := 0;
...
procedure Dump_Ex(E: Exception_Occurrence) is
begin
   Dump_Index := Dump_Index + 1;
   Save_Occurrence(Dump(Dump_Index), E);
end Dump_Ex;

procedure Analyse_Ex is
begin
   Put_Line("Analysis of saved occurrences:");
   for I in 1 .. Dump_Index loop
      Put_Line(Exception_Information(Dump(I)));
   end loop;
   Dump_Index := 0;          -- reset dump
end Analyse_Ex;
```

and then a handler could save an occurrence for later analysis by writing

```
    when Event: others =>
      Dump_Ex(Event);
      Clean_Up;
      raise;
  end;
```

Care is clearly needed to ensure that the dumping package does not itself raise exceptions and thereby get the system into a mess.

An occurrence may be reraised by calling the procedure Reraise_Occurrence. This is precisely equivalent to reraising an exception by a raise statement without an exception name and does not create a new occurrence (this ensures that the original cause is not lost). An important advantage of Reraise_Occurrence is that it can be used to reraise an occurrence that was stored by one of the subprograms Save_Occurrence.

Another way of raising an exception with a message is to call the somewhat redundant procedure Raise_Exception. In order to do this we need some way of referring to the exception. However, exceptions are not proper types and cannot be passed as parameters in the normal way. So instead we pass a value of the type Exception_Id which can be thought of as a global enumeration type whose literals represent the individual exceptions. The identity of an exception is provided by the attribute Identity.

So the statement

```
Raise_Exception(Trouble'Identity, "Doom");
```

is identical to

```
raise Trouble with "Doom";
```

The reason for the existence of Raise_Exception is that the form of raise statement with a message did not exist in Ada 95 and so Raise_Exception had to be used instead. The aspect No_Return applies to Raise_Exception; this is explained in the next section.

We mentioned above that the exceptions can be thought of as representing the literals of some type Exception_Id and that the Identity attribute enabled conversion from an exception to its identity. We can sort of go in the reverse direction by calling the function Exception_Name applied to an exception identity; this returns the full dotted name of the exception as a string. So

```
Exception_Name(Error'Identity)
```

might return the string "STACK.ERROR". (This is the same form as the string returned by Expanded_Name applied to tags; see Section 14.4.)

The identity of an exception associated with an occurrence is obtained by calling the function Exception_Identity. If the function Exception_Identity is applied to Null_Occurrence then it returns Null_Id and thereby provides a useful test for Null_Occurrence.

Applying Exception_Name, Exception_Message, or Exception_Information to Null_Occurrence raises Constraint_Error. Similarly, applying Raise_Exception to Null_Id also raises Constraint_Error.

Note finally that the following are equivalent

```
Exception_Name(Exception_Identity(Event))
Exception_Name(Event)
```

since there are two overloadings of the function Exception_Name.

Exercise 15.4

1 Revise the package Stack of Section 15.2 so that appropriate messages are attached to the raising of the exception Error. Change the handler to output the specific message.

15.5 Exception pragmas and aspects

There are a number of pragmas and aspects associated with exceptions. (Remember from Section 5.6 that certain properties of entities are known as aspects. Sometimes they are set by pragmas and sometimes by aspect specifications.)

First, consider the pragma Assert and the associated pragma Assertion_Policy. These enable a program to check that some condition holds and to raise an exception with a message if it does not. The exception raised is Assertion_Error in the package Ada.Assertions. Thus we might write

```
pragma Assert(Status = Green, "Help");
```

The first parameter of Assert is thus a Boolean expression and the second (and optional) parameter is a string. The parameter of Assertion_Policy is an identifier which controls the behaviour of the pragma Assert. Two policies are defined by the language, namely, Check and Ignore. Further policies may be defined by the implementation. It is also implementation defined as to which policy applies by default so we should always set the policy if we are using assertions.

The specification of the package Ada.Assertions is

```
package Ada.Assertions
    with Pure is
  Assertion_Error: exception;
  procedure Assert(Check: in Boolean);
  procedure Assert(Check: in Boolean; Message: in String);
end Ada.Assertions;
```

The pragma Assert can be used wherever a declaration or statement is allowed. Thus it might occur in a list of declarations such as

```
N: constant Integer := ... ;
pragma Assert(N > 1);
```

```
A: Real_Matrix(1 .. N, 1 .. N);
EV: Real_Vector(1 .. N);
```

and in a sequence of statements such as

```
pragma Assert(Transpose(A) = A, "A not symmetric");
EV := Eigenvalues(A);
```

(These examples use types and subprograms from the Numerics annex, see Section 26.5.) If the policy set by Assertion_Policy is Check then the above pragmas are equivalent to

```
if not N > 1 then
   raise Assertion_Error;
end if;
```

and

```
if not Transpose(A) = A then
   raise Assertion_Error with "A not symmetric";
end if;
```

Remember from the previous section that a raise statement without any explicit message is not the same as one with an explicit null message. In the former case a subsequent call of Exception_Message returns implementation defined information whereas in the latter case it returns a null string. This same behaviour thus occurs with the Assert pragma as well – providing no message is not the same as providing a null message.

If the policy set by Assertion_Policy is Ignore then the Assert pragma is ignored at run time – but of course the syntax of the parameters is checked during compilation.

The two procedures Assert in the package Ada.Assertions have an identical effect to the corresponding Assert pragmas except that their behaviour does not depend upon the assertion policy. Thus the call

```
Assert(Some_Test);
```

is always equivalent to

```
if not Some_Test then
   raise Assertion_Error;
end if;
```

In other words we could define the behaviour of

```
pragma Assert(Some_Test);
```

as equivalent to

```
if policy_identifier = Check then
   Assert(Some_Test);                -- call of procedure Assert
end if;
```

Note again that there are two procedures Assert, one with and one without the message parameter. These correspond to raise statements with and without an explicit message.

The pragma Assertion_Policy is a so-called configuration pragma and controls the behaviour of Assert throughout the units to which it applies. It is thus possible for different policies to be in effect in different parts of a partition. As well as controlling the behaviour of Assert, the pragma Assertion_Policy also controls the contract facilities such as pre- and postconditions described in the next chapter.

Another feature introduced in Ada 2005 with a related flavour is the ability to assert that a subprogram never returns by applying the aspect No_Return. This only applied to procedures in Ada 2005 and Ada 2012 but can also be applied to functions in Ada 2022. We will look at procedures first.

In the case of a procedure it asserts that the procedure never returns in the normal sense. Control can leave the procedure only by the propagation of an exception or it might loop forever (which is common among certain real-time programs). Thus we might have a procedure Fatal_Error which outputs some message and then propagates an exception which can be handled in the main subprogram. We can apply the aspect to the procedure specification using an aspect specification thus

```
procedure Fatal_Error(Message: in String)
  with No_Return;
```

We might then have

```
procedure Fatal_Error(Message: in String) is
begin
  Put_Line(Message);
  ...                              -- other last wishes
  raise Death;
end Fatal_Error;
...
procedure Main is
  ...
  Put_Line("Program terminated successfully");
exception
  when Death =>
    Put_Line("Program terminated: known error");
  when others =>
    Put_Line("Program terminated: unknown error");
end Main;
```

There are two consequences of supplying the aspect No_Return to a procedure

- The implementation checks at compile time that the procedure concerned has no explicit return statements. There is also a check at run time that it does not attempt to run into the final end – Program_Error is raised if it does as in the case of running into the end of a function.
- The implementation is able to assume that calls of the procedure do not return and so various optimizations can be made.

We might then have a call of Fatal_Error as in

```
function Pop return Integer is
begin
   if Top = 0 then
      Fatal_Error("Stack empty");            -- never returns
   elsif
      Top := Top - 1;
      return S(Top + 1);
   end if;
end Pop;
```

If No_Return applies to Fatal_Error then the compiler should not compile a jump after the call of Fatal_Error and should not produce a warning that control might run into the final end of Pop.

We could restructure the procedure Fatal_Error to raise the exception Death with the message thus

```
procedure Fatal_Error(Message: in String)
   with No_Return is
begin
   ...                              -- other last wishes
   raise Death with Message;
end Fatal_Error;
```

The exception handler for Death in the main subprogram can now use Exception_Message to print out the message.

If a subprogram has no distinct specification then the aspect specification is placed inside the body (as shown above). Note carefully that it goes before the reserved word **is**.

If we apply the aspect No_Return to a function then the function must still have a return statement but the statement is only allowed to raise an exception and thus not to actually return in the normal way. Thus we might have

```
function Foolish return Integer is
   with No_Return is
begin
   ...
   return raise Help with Another_Message;
end Foolish;
```

If several subprograms have the aspect No_Return, then using aspect specifications each must have its own. However, a pragma No_Return could apply to several subprograms declared in the same package specification. For further details of the rules regarding aspect specifications see Section 16.1.

It is vital that dispatching works correctly with procedures that do not return. A non-returning dispatching procedure can only be overridden by a non-returning procedure and so the overriding one must also have the aspect No_Return thus

```
type T is tagged ...
procedure P(X: T; ... )
   with No_Return;
...
type TT is new T with ...
overriding
procedure P(X: TT; ... )
   with No_Return;
```

The reverse is not true, a procedure that does return can be overridden by one that does not. It is possible to give the aspect No_Return for an abstract procedure, but obviously not for a null procedure. (The aspect No_Return can also be given for a generic procedure. It then applies to all instances. See Chapter 19.)

See also the somewhat related aspect Allows_Exit described in Section 21.6 regarding the curious topic of procedural iterators introduced in Ada 2022.

There are also two pragmas Suppress and Unsuppress for controlling checks. Essentially, Suppress says that specific run-time checks which give rise to exceptions such as Constraint_Error can be turned off whereas Unsuppress says they must not be turned off. Thus Suppress is merely a recommendation whereas Unsuppress is an order. One reason why Suppress is merely a recommendation is that some checks may be done by hardware. Moreover, there is no guarantee that a suppressed exception will not be raised. Indeed it could be propagated from another unit compiled with checks.

The core language checks corresponding to the exception Constraint_Error are Access_Check (checking that an access value is not null), Discriminant_Check (checking that a discriminant value is consistent with the component being accessed or a constraint), Division_Check (checking the second operand of **/**, **rem** and **mod**), Index_Check (checking that an index is in range), Length_Check (checking that the number of components of an array match), Overflow_Check (checking for numeric overflow), Range_Check (checking that various constraints are satisfied) and Tag_Check (checking that tags are equal when dispatching). There are other checks relating to the specialized annexes.

The core language checks corresponding to Program_Error are Allocation_Check (checking an obscure situation regarding allocation), Elaboration_Check (checking that the body of a unit has been elaborated), Accessibility_Check (checking an accessibility level), and Program_Error_Check (checking that the number of chunks is not negative and other curious situations).

The check corresponding to Storage_Error is Storage_Check (checking that space for a storage pool or task has not been exceeded).

The check corresponding to Tasking_Error is Tasking_Check (checking that a task has activated successfully and that a called task has not terminated).

There are also checks that raise Assertion_Error associated with the use of the Pre, Static_Predicate, and Dynamic_Predicate aspects (see Chapter 16). See the *ARM* for details.

The pragma Suppress takes the form

```
pragma Suppress(Range_Check);
```

We can also write **pragma** Suppress(All_Checks); which does the obvious thing.

The purpose of Unsuppress is to ensure that checks are performed so that we can rely upon an exception being raised. Suppose we have a type Sat_Int that we wish to behave as a saturated type. This means it never overflows but just takes the maximum or minimum value when appropriate. We could write the multiplication operator for such a type as

```
function "*" (Left, Right: Sat_Int) return Sat_Int is
  pragma Unsuppress(Overflow_Check);
begin
  return Integer(Left) * Integer(Right);
exception
  when Constraint_Error =>
    if (Left > 0 and Right > 0) or (Left < 0 and Right < 0) then
      return Sat_Int'Last;
    else
      return Sat_Int'First;
    end if;
end "*";
```

The pragma Unsuppress ensures that the code always works as intended even if checks are suppressed in the program as a whole.

We conclude with an important warning regarding Suppress. If we suppress a check and the situation nevertheless arises then the program is erroneous.

Exercise 15.5

1 Write a pragma Assert to check that an array A is sorted into ascending order. Use a quantified expression.

15.6 Scope of exceptions

To a large extent exceptions follow the same scope rules as other entities. An exception can hide and be hidden by another declaration; it can be referred to by the dotted notation and so on. An exception can be renamed

```
Help: exception renames Bank.Alarm;
```

Exceptions are different in many ways. We cannot declare arrays of exceptions, and they cannot be components of records, parameters of subprograms. In short, exceptions are not objects and so cannot be manipulated. They behave like literals of a globally defined enumeration type Exception_Id as mentioned in Section 15.4.

A very important characteristic of exceptions is that they are not created dynamically as a program executes but should be thought of as existing throughout the life of the program. Indeed the full set of exceptions in a program is only known when the program is bound and so the representation of the type Exception_Id can be thought of as occurring at bind time. This aspect of exceptions relates to the way in which they are propagated dynamically up the chain of execution rather than statically up the chain of scope. An exception can be propagated outside its scope although of course it can then only be handled anonymously by **others**. This is illustrated by the following

```
procedure Main is
  procedure P is
    X: exception;
  begin
    raise X;
  end P;
begin
  P;
exception
  when others =>
     ...                  -- X handled here
end Main;
```

The procedure P declares and raises the exception X but does not handle it. When we call P, the exception X is propagated to the block calling P where it is handled anonymously.

We could of course rediscover the textual form of the exception by writing

```
when Event: others =>
```

and then Exception_Name(Event) would return the string "MAIN.P.X".

It is even possible to propagate an exception out of its scope, where it becomes anonymous, and then back in again where it can once more be handled by its proper name. See Section 15.6a, Exception Scope, on the website.

A further illustration of the nature of exceptions is afforded by a recursive procedure containing an exception declaration. Unlike variables declared in a procedure, we do not get a new exception for each recursive call. Each recursive activation refers to the same exception – they have the same Exception_Id. Consider the following example

```
procedure F(N: Integer) is
  X: exception;
begin
  if N = 0 then
    raise X;
  else
    F(N - 1);
  end if;
exception
  when X =>
    Put("Got it!");
    raise;
  when others =>
    null;
end F;
```

Suppose we execute F(4); we get recursive calls F(3), F(2), F(1), and finally F(0). When F is called with parameter zero, it raises the exception X, handles it, prints out a confirmatory message and then reraises it. The calling instance of F (which itself had N = 1) receives the exception and again handles it as X and so on. The message is therefore printed out five times in all and the exception is finally propagated anonymously. Observe that if each recursive activation had created a different exception then the message would only be printed out once.

In most of the examples we have seen so far exceptions have been raised in statements. An exception can however also be raised in a declaration. Thus

```
N: Positive := 0;
```

would raise Constraint_Error because the initial value of N does not satisfy the range constraint 1 .. Integer'Last of the subtype Positive. An exception raised in a declaration is not handled by a handler (if any) of the unit containing the declaration but is immediately propagated up a level. This means that in any handler we are assured that all declarations of the unit were successfully elaborated and so there is no risk of referring to something that does not exist.

Finally, a warning regarding parameters of mode **out** or **in out**. If a subprogram is terminated by an exception then any actual parameter of an elementary type will not have been updated since such updating occurs on a normal return. On the other hand, a parameter of a tagged record type or an explicitly limited type is always passed by reference and so will always have been updated. For an array or other record type the parameter mechanism is not so closely specified and the actual parameter may or may not have its original value. A program assuming a particular mechanism may have a bounded error. As an example consider the procedure Withdraw of the package Bank in Exercise 12.6(**1**) and suppose that the type key were not explicitly limited. It would be incorrect to attempt to take the key away and raise an alarm as in

```
procedure Withdraw(K: in out Key; M: in out Money) is
begin
  if Valid(K) then
    if M > Amount_Remaining then
      M := Amount_Remaining;
      Free(K.Code) := True;              -- close account
      K.Code := 0;
      raise Alarm;
    else
      ...
    end if;
  end if;
end Withdraw;
```

If the parameter mechanism were implemented by copy then the bank would think that the key was now free but would have left the greedy customer with a copy. However, if the key were explicitly limited or a tagged type as in Exercise 14.7(**2**) then this problem would not arise since such types are always passed by reference.

Exercise 15.6

1 Rewrite the package Bank of Exercise 12.6(**1**) to declare an exception Alarm and raise it when any illegal banking activity is attempted.

2 Consider the following pathological procedure

```
procedure P is
begin
  P;
exception
  when Storage_Error =>
    P;
end P;
```

What happens when P is called? To be explicit suppose that there is enough stack space for only N simultaneous recursive calls of P but that on the N+1th call the exception Storage_Error is raised. How many times will P be called in all and what eventually happens?

Checklist 15

Do not use exceptions unnecessarily.

Use specific user declared exceptions rather than predefined exceptions where relevant.

Ensure that handlers return resources correctly.

Match the constraints on index variables to the arrays concerned.

Beware of uninitialized variables.

Out and in out parameters may not be updated correctly if a subprogram is terminated by an exception.

Numeric_Error is obsolescent.

The ability to add a message to raise statements was added in Ada 2005.

The pragmas Assert, Assertion_Policy, No_Return, and Unsuppress were added in Ada 2005.

Raise expressions were added in Ada 2012.

The pragma No_Return is obsolete in Ada 2012 and replaced by the corresponding aspect specification.

New in Ada 2022

A number of other checks are defined to give better control of their suppression. For example Tasking_Check for Tasking_Error (checking that a task has been activated etc).

The aspect No_Return can be applied to functions as well as to procedures.

16 Contracts

16.1 Aspect specifications
16.2 Preconditions and postconditions
16.3 Type invariants
16.4 Subtype predicates
16.5 Global state
16.6 Messages

This chapter describes the mechanisms for defining and enforcing contracts first introduced in Ada 2012. The general idea is that we add aspects defining certain requirements on the behaviour of an entity such as a subprogram and then, when the subprogram is executed, if any of these requirements are not met, an exception is raised. Ada 2022 also introduces the ability to describe the impact of a subprogram on global state.

Moreover, as mentioned in Chapter 9, the introduction of pre- and postconditions in turn triggered the introduction of conditional and quantified expressions and (in Ada 2022) declare and reduction expressions.

As well as specifying pre- and postconditions on subprograms and the impact on global state we can also specify invariant properties of private types, and additional restrictions on the values of subtypes. These properties are all given by the use of aspect specifications briefly mentioned in Section 5.6.

We start by giving further details of the rules regarding aspect specifications.

16.1 Aspect specifications

As we saw in Section 15.5 we can apply an aspect such as No_Return to a procedure using an aspect specification. We can supply several aspects by a series of aspect specifications as follows

```
procedure Do_It( ... )
  with Inline, No_Return;
```

In the general case an aspect specification supplies an expression for the aspect following =>. In the case of both Inline and No_Return, these aspects take a Boolean value so we could pedantically write

```
procedure Do_It ( ... )
  with No_Return => True,
       Inline => True;
```

There is a general rule that if an aspect requires a Boolean value then it can be omitted and by default is taken to be true. But this does not apply to the aspects Default_Value and Default_Component_Value, see Sections 6.8 and 8.2.

As we see, aspect specifications are introduced by the reserved word **with** and if there are several they are separated by commas. The word **with** is used for a number of purposes (we have already met with clauses and type extensions) and this might be considered confusing. However, with a sensible layout there should be no problems.

In the case of a subprogram without a distinct specification, the aspect specification goes in the subprogram body before **is** thus

```
procedure Do_It( ... )
  with Inline is
  ...
begin
  ...
end Do_It;
```

This arrangement is because the aspect specification is very much part of the specification of the subprogram. If a subprogram has a distinct specification then we cannot give a language-defined aspect specification on the body; this avoids problems of conformance.

If there is a stub but no specification then any aspect specification goes on the stub but not the body. Thus aspect specifications go on the first of specification, stub, and body but are never repeated. Note also that we can give aspect specifications on other forms of stubs and bodies such as package bodies (also task and entry bodies, see Chapter 20).

In the case of a stub, abstract subprogram, or null subprogram which never have bodies, the aspect specification goes after **is separate**, **is abstract**, or **is null** thus

```
procedure Action(D: in Data) is separate
  with Convention => C;

procedure Enqueue( ... ) is abstract
  with Synchronization => By_Entry;

procedure Nothing is null
  with Something;
```

The aspect Convention is described in Section 25.5; the example of the use of Synchronization is from the package Synchronized_Queue_Interfaces, a child of Ada.Containers, see Section 24.8.

The same style is followed by expression functions described in Section 10.1 thus

```
function Inc (A: Integer) return Integer is (A + 1)
  with Inline;
```

As we have seen, many aspects (such as Inline) can be set by aspect specifications and corresponding pragmas are obsolete. Others such as Pure, Preelaborate, and Preelaborable_Initialization (see Section 13.8) were preferred to be set by pragmas but now aspect specifications are typically used.

Many aspects such as Inline apply to a particular entity. However, there are some pragmas that do not relate to any particular entity and so for which an alternative aspect specification would be impossible. These include Assert and Assertion_Policy, Suppress and Unsuppress, and Restrictions.

Other properties of entities such as details of their representation can often be given by aspect specifications; these are described in Section 25.1.

In deciding between pragmas and aspect specifications, remember that an aspect specification such as Inline has to be given with the subprogram specification to which it relates; this is an advantage since all the properties of an entity can be given together and this avoids scattering them around as can happen with pragmas.

Moreover, since a corresponding pragma is split from the entity, not only might it occur much later in the text but it has to mention the entity to which it relates. A curious problem with pragmas is that because of overloading, we are not able to distinguish between two procedures Do_It with different profiles and so writing

```
pragma Inline(Do_It);
```

will apply to all preceding procedures Do_It which might be a nuisance.

There is, however, one consequence of the introduction of aspect specifications and this is that the rule of linear elaboration had to be revised. So, within an aspect specification for A we can refer forward to an entity B which has not yet been declared provided B has been declared by the time A has to be frozen. The concept of freezing is outlined in Section 25.1.

In the remainder of this chapter we will meet many aspects, such as

```
Pre => Boolean expression          Pre'Class =>
Post =>                            Post'Class =>
Type_Invariant =>                  Type_Invariant'Class =>
Static_Predicate =>                Dynamic_Predicate =>
```

In these cases, the expected type of the aspect is Boolean (that is of the predefined type Boolean or derived from it). As mentioned in Section 9.4 when discussing quantified expressions, we often refer to such expressions as predicates which is simply a mathematical term for a truth value. (And nothing to do with human linguistics where we say that in the sentence 'I love Ada', the subject is 'I' and the predicate is 'love Ada'.)

Othe aspects introduced in Ada 2022 are Global, Global'Class, Stable_Properties, and Nonblocking. These are discussed in Section 16.5.

Exercise 16.1

1 Declare a type My_Boolean derived from the predefined type Boolean with a default value of True.

16.2 Preconditions and postconditions

Preconditions and postconditions were certainly the most important of the contract features introduced in Ada 2012 and perhaps are the most important new features in Ada 2012 overall.

A precondition is given by the aspect Pre. Thus to give a precondition for the procedure Pop in a package Stacks we might augment its specification thus

```
procedure Pop(S: in out Stack; X: out Integer)
   with Pre => not Is_Empty(S);
```

where Is_Empty is a function also in the package Stacks.

In a similar way we might give a postcondition as well which might be that the stack is not full. So altogether the specification of the package Stacks in Section 12.4 might become

```
package Stacks is
   type Stack is private;

   function Is_Empty(S: Stack) return Boolean;
   function Is_Full(S: Stack) return Boolean;

   procedure Push(S: in out Stack; X: in Integer)
      with
         Pre => not Is_Full(S),
         Post => not Is_Empty(S);

   procedure Pop(S: in out Stack; X: out Integer)
      with
         Pre => not Is_Empty(S),
         Post => not Is_Full(S);

   function "=" (S, T; Stack) return Boolean;

private
   ...
end Stacks;
```

Note how the individual aspects Pre and Post are separated by commas. The final semicolon is of course the semicolon at the end of the subprogram declaration as a whole. Note also that the value of the parameter S of the functions Is_Empty or Is_Full in the pre- and postcondition is the value of the parameter S of the procedures Push and Pop when they are called. This is important regarding the postcondition; in the case of Push, the value of S when Is_Empty is called is the value after the Push operation has been performed.

Pre- and postconditions are controlled by the same mechanism as assertions using the pragma Assert. It will be recalled from Section 15.5 that these can be switched on and off by the pragma Assertion_Policy. Thus if we write

```
pragma Assertion_Policy(Check);
```

then assertions are enabled whereas if the parameter of the pragma is Ignore then all assertions are ignored.

In the case of a precondition, whenever a subprogram with a precondition is called, if the policy is Check then the precondition is evaluated and if it is false then Assertion_Error is raised and the subprogram is not entered. Similarly, on return from a subprogram with a postcondition, if the policy is Check then the postcondition is evaluated and if it is false then Assertion_Error is raised.

So if the policy is Check and Pop is called when the stack is empty then Assertion_Error is raised whereas if the policy is Ignore then the predefined exception Constraint_Error would probably be raised (depending upon how the stack had been implemented).

Note that it is not permitted to give the aspects Pre or Post for a null procedure; this is because all null procedures are meant to be interchangeable.

There are also aspects Pre'Class and Post'Class for use with tagged types (and they can be given with null procedures). The subtle topic of multiple inheritance of pre- and postconditions will be discussed in a moment.

Fine control of these aspects can be given by a variation of the pragma Assertion_Policy which allows different policies for different aspects. Thus if we want checks on for preconditions but off for postconditions we can write

```
pragma Assertion_Policy(Pre => Check, Post => Ignore);
```

Now suppose we apply a specific precondition Before and/or a specific postcondition After to a procedure P by writing

```
procedure P(P1: in T1; P2: in out T2; P3: out T3)
   with Pre => Before,
        Post => After;
```

where Before and After are Boolean expressions.

The precondition Before and the postcondition After can involve the parameters P1 and P2 and P3 and any visible entities such as other variables, constants, and functions. Note that Before can involve an **out** parameter such as P3 (inasmuch as one can always read any constraints on an **out** parameter such as the bounds if it were an array).

The attribute X'Old is useful in postconditions; it denotes the value of X on entry to P and has the same type as X (even if anonymous, AI12-32). It is typically applied to parameters of mode **in out** such as P2 but it can be applied to any visible entity such as a global variable. This can be useful for monitoring global variables which are updated by the call of P. Note that 'Old can only be used in postconditions and not in arbitrary text and it cannot be applied to objects of a limited type.

Perhaps surprisingly, 'Old can also be applied to parameters of mode **out**. For example, in the case of a parameter of a record type that is updated as a whole, nevertheless we might want to check that a particular component has not changed. Thus in updating some personal details, such as address and occupation, we might ensure that the person's date of birth and sex are not tampered with by writing

```
Post => P.Sex = P.Sex'Old and P.Dob = P.Dob'Old
```

In the case of an array, we can write A(J)'Old which means the original value of A(J). But A(J'Old) is different since it is the component of the final value of A but indexed by the old value of J. If J is not modified then it is better to say A(J)'Old rather than A'Old(J) which would imply remembering the whole array.

Remember that the result of a function is an object and so 'Old can be applied to it. Note carefully the difference between F(X)'Old and F(X'Old). The former applies F to X on entry to the subprogram and saves it. The latter saves X and applies F to it when the postcondition is evaluated. These could be different because the function F might also involve global variables which have changed.

Generally, 'Old can be applied to anything but there are restrictions on its use in certain conditional structures in which it can only be applied to statically determined objects. This is illustrated by the following

```
Table: array (1 .. 10) of Integer := ... ;
procedure P(K: in out Natural)
  with Post => K > 0 and then Table(K)'Old = 1;    -- illegal
```

The programmer's intent is that the postcondition uses a short circuit form to avoid evaluating Table(K) if K is not positive on exit from the procedure. But, 'Old is evaluated and stored on entry and this could raise Constraint_Error because K might for example be zero. This is a conundrum since the compiler cannot know whether the value of Table(K) will be needed and also K can change so it cannot know which K anyway. So such structures are forbidden.

In the case of a postcondition applying to a function F, the result of the function is denoted by the attribute F'Result. Again this attribute can only be used in postconditions.

So if we have a type My_Integer (as in Section 17.1) we might declare a procedure Pinc or a function Finc to perform an increment with postconditions thus

```
procedure Pinc(X: in out My_Integer)
  with Post => X = X'Old+1;

function Finc(X: My_Integer) return My_Integer
  with Post => Finc'Result = X'Old+1;
```

In summary, the overall effect of calling the procedure P with checks enabled is that, after evaluating any parameters at the point of call, it as if the body were

```
if not Before then                    -- check precondition
  raise Assertion_Error;
end if;

... evaluate and store any 'Old stuff;

... call actual body of P;

if not After then                     -- check postcondition
  raise Assertion_Error;
end if;

... copy back any by-copy parameters;

... return to point of call;
```

Occurrences of Assertion_Error are propagated and so raised at the point of call; they cannot be handled inside P. Of course, if the evaluation of Before or After themselves raise some exception then that will be propagated to the point of call.

The conditions Pre and Post can also be applied to entries; see Section 20.2.

Table 16.1 Obligations and assumptions.

	Pre	Post
Call writer	obligation	assumption
Body writer	assumption	obligation

Before progressing to the problems of inheritance it is worth reconsidering the purpose of preconditions and postconditions.

- A precondition Before is an obligation on the caller to ensure that it is true before the subprogram is called and so is an assumption that the implementer of the body can rely upon on entry to the body.
- A postcondition After is an obligation on the implementer of the body to ensure that it is true on return from the subprogram and so is an assumption that the caller can rely upon on return from the body.

The symmetry is neatly illustrated by Table 16.1 above.

The simplest form of inheritance occurs with derived types that are not tagged. Suppose we declare the procedure Pinc as above with the postcondition shown and supply a body

```
procedure Pinc(X: in out My_Integer) is
begin
   X := X+1;
end Pinc;
```

and then declare a type

```
type Apples is new My_Integer;
```

then the procedure Pinc is inherited by the type Apples. So if we then write

```
No_Of_Apples: Apples;
...
Pinc(No_Of_Apples);
```

what actually happens is that the code of the procedure Pinc originally written for My_Integer is called and so the postcondition is inherited automatically.

If the user now wants to add a precondition to Pinc that the number of apples is not negative then a completely new subprogram has to be declared which overrides the old one thus

```
procedure Pinc(X: in out Apples)
   with Pre => X >= 0,
        Post => X = X'Old+1;
```

and a new body has to be supplied (which will of course in this curious case be essentially the same as the old one). Note carefully that this example used My_Integer and not the predefined type Integer. This was to illustrate inheritance which only occurs with primitive operations. Remember that the primitive operations of Integer are just those declared with the type Integer in Standard and so Pinc could not be a primitive operation of Integer.

We now turn to tagged types and first continue to consider the specific conditions Pre and Post. As a familiar example, consider the hierarchy consisting of a type Object and the direct descendants Circle, Square, and Triangle.

Suppose the type Object is

```
type Object is tagged
  record
    X_Coord, Y_Coord: Float;
  end record;
```

and we declare a function Area thus

```
function Area(O: Object) return Float
  with Pre => O.X_Coord > 0.0,
       Post => Area'Result = 0.0;
```

This imposes a requirement on the caller that the function is called only with objects with positive *x*-coordinate (for some obscure reason), and a requirement on the implementer of the body that the area is zero (raw objects are just points and have no area).

If we now declare a type Circle as

```
type Circle is new Object with
  record
    Radius: Float;
  end record;
```

and override the inherited function Area then the Pre and Post conditions on Area for Object are not inherited and we have to supply new ones, perhaps

```
function Area(C: Circle)
  with Pre => C.X_Coord - C.Radius > 0.0,
       Post => Area'Result > 3.1 * C.Radius**2 and
               Area'Result < 3.2 * C.Radius**2;
```

The conditions ensure that all of the circle is in the right half-plane and that the area is about right!

So the rules so far are exactly as for the untagged case. If an operation is not overridden then it inherits the conditions from its ancestor but if it is overridden then those conditions are lost and new ones have to be supplied. And if no new ones are supplied then the pre- and postconditions are by default taken to be True.

In conclusion, the conditions Pre and Post are always given on a specification but in a sense are very much part of the actual body. One consequence of this is that an abstract subprogram cannot have Pre and Post conditions because an abstract subprogram has no body. Pre- and postconditions can also be given on an expression function but not if it is just acting as a completion.

We now consider the class wide conditions Pre'Class and Post'Class which are subtly different. The first point is that the class wide ones apply also to all descendants even if the operations are overridden. In the case of Post'Class if an overridden operation has no postcondition given then it is taken to be True (as in the case of Post). But in the case of Pre'Class, if an overridden operation has no precondition given then it is only taken to be True if no other Pre'Class applies (no other is inherited). We will now look at the consequences of these rules.

It might be that we want certain conditions to hold throughout the hierarchy, perhaps that all objects concerned have a positive *x*-coordinate and nonnegative area. In that case we can use class wide conditions.

```
function Area(O: Object) return Float
   with Pre'Class => O.X_Coord > 0.0,
        Post'Class => Area'Result >= 0.0;
```

Now when we declare Area for Circle, the aspects Pre'Class and Post'Class from Object will be inherited by the function Area for Circle. Note that within a class wide condition a formal parameter of type T is interpreted as of T'Class. Thus O is of type Object'Class and thus applies to Circle. The inherited postcondition is simply that the area is not negative and uses the attribute 'Result.

If we do not supply conditions for the overriding Area for Circle and simply write

```
overriding
function Area(C: Circle) return Float;
```

then the precondition inherited from Object still applies. For the postcondition, not only is the postcondition from Object inherited but there is also an implicit postcondition of True. So the applicable conditions for Area for Circle are

```
Pre'Class for Object

Post'Class for Object
True
```

Suppose on the other hand that we give explicit Pre'Class and Post'Class for Area for Circle thus

```
overriding
function Area(C: Circle) return Float
   with Pre'Class => ... ,
        Post'Class => ... ;
```

We then find that the applicable conditions for Area for Circle are

```
Pre'Class for Object
Pre'Class for Circle

Post'Class for Object
Post'Class for Circle
```

Incidentally, it makes a lot of sense to declare the type Object as abstract so that we cannot declare pointless objects. In that case Area might as well be abstract as well. Although we cannot give conditions Pre and Post for an abstract operation we can still give the class wide conditions Pre'Class and Post'Class.

If the hierarchy extends further, perhaps Equilateral_Triangle is derived from Triangle which itself is derived from Object, then we could add class wide conditions to Area for Triangle and these would also apply to Area for Equilateral_Triangle. And we might add specific conditions for Equilateral_Triangle as well. So we would then find that the following apply to Area for Equilateral_Triangle

```
Pre'Class for Object
Pre'Class for Triangle
Pre for Equilateral Triangle

Post'Class for Object
Post'Class for Triangle
Post for Equilateral_Triangle
```

The postconditions are quite straightforward, all apply and all must be True on return from the function Area. The compiler can see all these postconditions when the code for Area is compiled and so they are all checked in the body. Note that any default True makes no difference because B **and** True is the same as B.

However, the rules regarding preconditions are perhaps surprising. The specific precondition Pre for Equilateral_Triangle must be True (checked in the body) but so long as just one of the class wide preconditions Pre'Class for Object and Triangle is true then all is well. Note that class wide preconditions are checked at the point of call. Do not get confused over the use of the word apply. They all apply but only the ones seen at the point of call are actually checked.

The reason for this state of affairs concerns dispatching and especially redispatching. Consider the case of Ada airlines in Section 14.1 which has Basic, Nice, and Posh passengers. Remember that Basic passengers just get a seat. Nice passengers also get a meal and Posh passengers also get a limo. The types Reservation, Nice_Reservation, and Posh_Reservation form a hierarchy with Nice_Reservation being extended from Reservation and so on. The facilities are assigned when a reservation is made by calling an appropriate procedure Make thus

```
procedure Make(R: in out Reservation) is
begin
   Select_Seat(R);
end Make;

procedure Make(NR: in out Nice_Reservation) is
begin
   Make(Reservation(NR));
   Order_Meal(NR);
end Make;

procedure Make(PR: in out Posh_Reservation) is
   Make(Nice_Reservation(PR));
   Arrange_Limo(PR);
end Make;
```

Each Make calls its ancestor in order to avoid duplication of code and to ease maintenance.

A variation involving redispatching (see Section 14.5) introduces two different procedures Order_Meal, one for Nice passengers and one for Posh passengers. We then need to ensure that Posh passengers get a posh meal rather than a nice meal. We write

```
procedure Make(NR: in out Nice_Reservation) is
begin
   Make(Reservation(NR));
                              -- now redispatch to appropriate Order_Meal
   Order_Meal(Nice_Reservation'Class(NR));
end Make;
```

Now suppose we have a precondition Pre'Class on Order_Meal for Nice passengers and one on Order_Meal for Posh passengers. The call of Order_Meal sees that it is for Nice_Reservation'Class and so the code includes a test of Pre'Class on Nice_Reservation. It does not necessarily know of the existence of the type Posh_Reservation and cannot check Pre'Class on that Order_Meal. At a later date we might add Supersonic passengers and this can be done without recompiling the rest of the system so it certainly cannot do anything about checking Pre'Class on Order_Meal for Supersonic_Reservation which does not exist when the call is compiled. So when we eventually get to the body of one of the procedures Order_Meal all we know is that some Pre'Class on Order_Meal has been checked somewhere. And that is all that the writer of the code of Order_Meal can rely upon. Note that nowhere does the compiled code actually 'or' a lot of preconditions together.

In summary, class wide preconditions are checked at the point of call. Class wide postconditions and both specific preconditions and postconditions are checked in the actual body.

A small point to remember is that a class wide operation such as

```
procedure Do_It(X: in out T'Class);
```

is not a primitive operation of T and so although we can specify Pre and Post for Do_It we cannot specify Pre'Class and Post'Class for Do_It.

We noted above that the aspects Pre and Post cannot be specified for an abstract subprogram because it doesn't have a body. They cannot be given for a null procedure either, since we want all null procedures to be identical and do nothing and that includes no conditions.

An innovation in Ada 2022 is that pre- and postconditions can also be applied to access to subprogram types. As a very simple example consider the type Math_Function in Section 11.8 which is

```
type Math_Function is access function (F: Float) return Float;
```

If we wanted to ensure that any call would only accept positive parameters then we could write

```
type Math_Function is access function (F: Float) return Float
   with Pre => F >= 0.0;
```

and similarly for postconditions.

This is perhaps a good moment to mention the related topic of default initial conditions. In Section 6.8, we mentioned that a default value can be given for objects of a type when it is declared, thus

```
type Colour is (Red, Amber, Green)
  with Default_Value => Red;
```

In the case of arrays we can use the aspect Default_Component_Value as in Section 8.2. And in the case of records we can give a default value to individual components in the record type declaration itself. Ada 2022 also enables us to declare how a private object can be default initialized. The aspect Default_Initial_Condition can be applied to a private type or to a private extension (and to the corresponding generic parameters).

Consider the type Bit_Set from Section 14.9. In that example as written, when a Bit_Set is declared it is undefined and it is assumed that the user will assign something sensible to it. But it is far better to ensure that objects do have a sensible initial value by for example writing

```
type Bit_Set is new Abstract_Sets.Set with private
  with Default_Initial_Condition => Bit_Set = Empty;
```

where the function Empty returns an empty set. A better approach might be to declare a function Is_Empty which returns an empty set giving

```
  with Default_Initial_Condition => Is_Empty(Bit_Set);
```

When an object of type Bit_Set is declared (and after calling the appropriate Initialize if it is a controlled type) a check is performed to ensure that the default initial condition is satisfied. If it fails then Assertion_Error is raised much as if it were a check on a postcondition.

Indeed, we could control the checks by writing

```
pragma Assertion_Policy(Pre => Ignore, Post => Check,
                        Default_Initial_Condition => Check);
```

If a type with a Default_Initial_Condition is extended then the policy applies to the descendant type as well. Thus several such policies can apply to a type.

We now turn to the question of multiple inheritance and progenitors and the so-called Liskov Substitution Principle (LSP). The usual consequence of LSP is that in the case of preconditions they are combined with 'or' (thus weakening) and the rule for postconditions is that they are combined with 'and' (thus strengthening). But the important thing is that a relevant concrete operation can be substituted for the corresponding operations of all its relevant ancestors.

In Ada, a type T can have one parent and several progenitors as described in Section 14.8. Thus we might have

```
type T is new P and G1 and G2 with ...
```

where P is the parent and G1 and G2 are progenitors. Remember that a progenitor cannot have components and cannot have concrete operations (apart possibly for null procedures). So the operations of the progenitors have to be abstract or null and cannot have Pre and Post conditions. However, they can have Pre'Class and Post'Class conditions. It is possible that the same operation Op is primitive for more than one of these.

Thus the progenitors G1 and G2 might both have an operation Op thus

```
procedure Op(X: in G1) is abstract;
procedure Op(X: in G2) is abstract;
```

If they are conforming (as they are in this case) then the one concrete operation Op of the type T derived from both G1 and G2 will implement both of these. (If they don't conform then they are simply overloadings and two operations of T are required.) Hence the one Op for T can be substituted for the Op of both G1 and G2 and LSP is satisfied.

Now suppose both abstract operations have pre- and postconditions. Take postconditions first, we might have

```
procedure Op(X: in G1) is abstract
   with Post'Class => After1;

procedure Op(X: in G2) is abstract
   with Post'Class => After2;
```

Users of the Op of G1 will expect the postcondition After1 to be satisfied by any implementation of that Op. So if using the Op of T which implements the abstract Op of G1, it follows that Op of T must satisfy the postcondition After1. By a similar argument regarding G2, it must also satisfy the postcondition After2.

It thus follows that the effective postcondition on the concrete Op of T is as if we had written

```
procedure Op(X: in T)
   with Post'Class => After1 and After2;
```

But of course we don't actually have to write that since we simply write

```
overriding
procedure OP(X: in T);
```

and it automatically inherits both postconditions and the compiler inserts the appropriate code in the body. Remember that if we don't give a condition then it is True by default but 'anding' in True makes no difference.

If we do provide another postcondition thus

```
overriding
procedure OP(X: in T)
   with Post'Class => After_T;
```

then the overall class wide postcondition to be checked before returning will be After1 **and** After2 **and** After_T.

Now consider preconditions. Suppose the two versions of Op are

```
procedure Op(X: in G1) is abstract
  with Pre'Class => Before1;

procedure Op(X: in G2) is abstract
  with Pre'Class => Before2;
```

Assuming that there is no corresponding Op for P, we must provide a concrete operation for T thus

```
overriding
procedure Op(X: in T)
  with Pre'Class => Before_T;
```

This means that at a point of call of Op the precondition to be checked is Before_T **or** Before1 **or** Before2. As long as this is satisfied it does not matter that Before1 and Before2 might have been different.

If we do not provide an explicit Pre'Class then the condition to be checked at the point of call is Before1 **or** Before2.

An interesting case arises if a progenitor (say G1) and the parent have a conforming operation. Thus suppose P itself has the operation

```
procedure Op(X: in P);
```

and moreover that the operation is not abstract. Then (ignoring preconditions for the moment) this Op for P is inherited by T and thus provides a satisfactory implementation of Op for G1 and all is well.

Now suppose that Op for P has a precondition thus

```
procedure Op(X: in P)
  with Pre'Class => Before_P;
```

and that Before_P and Before1 are not the same. If we do not provide an explicit overriding for Op, it would be possible to call the body of Op for P when the precondition it knows about, Before_P, is False (since Before1 being True would be sufficient to allow the call to proceed). This would effectively mean that no class wide preconditions could be trusted within the subprogram body and that would be totally unacceptable. So in this case there is a special rule that an explicit overriding is required for Op for T.

If Op for P is abstract then a concrete Op for T must be provided and the situation is just as in the case for the Op for G1 and G2.

If T itself is declared as abstract (and P is not abstract and Op for P is concrete) then the inherited Op for T is abstract.

(These rules are similar to those for functions returning a tagged type, see Section 14.3, when the type is extended; it has to be overridden unless the type is abstract in which case the inherited operation is abstract.)

We finish this somewhat mechanical discussion of the rules by pointing out that if silly inappropriate preconditions are given then we will get a silly program.

So programmers should not write preconditions that are not sensible and sensibly related to each other. Because of the generality, the compiler cannot tell so stupid things are hard to prohibit. There is no defence against stupid programmers.

A concrete example using simple numbers might help. Suppose we have a tagged type T1 and an operation Solve which takes a parameter of type T1 and perhaps finds the solution to an equation defined by the components of T1. Solve delivers the answer in an **out** parameter A with a parameter D giving the number of significant digits required in the answer. Also we impose a precondition on the number of digits D thus

```
type T1 is tagged record ...

procedure Solve(X: in T1; A: out Float; D: in Integer)
   with Pre'Class => D < 5;
```

The intent here is that the version of Solve for the type T1 always works if the number of significant digits asked for is less than 5.

Now suppose we declare a type T2 derived from T1 and that we override the inherited Solve with a new (better) version that works if the number of significant digits asked for is less than 10 thus

```
type T2 is new T1 with ...

overriding
procedure Solve(X: in T2; A: out Float; D: in Integer)
   with Pre'Class => D < 10;
```

And so on with a type T3

```
type T3 is new T2 with ...

overriding
procedure Solve(X: in T3; A: out Float; D: in Integer)
   with Pre'Class => D < 15;
```

Thus we have a hierarchy of algorithms Solve with increasing capability. Now suppose we have a dispatching call

```
An_X: T1'Class := ... ;
Solve(An_X, Answer, Digs);
```

This will dispatch to one of the Solve procedures but we do not know which one. The only precondition that applies is that on the Solve for T1 which is D < 5. That is fine because D < 5 implies D < 10 and D < 15 and so on. Thus the preconditions work because the hierarchy weakens them.

Similarly, if we have

```
An_X: T2'Class := ... ;
Solve(An_X, Answer, Digs);
```

then it will dispatch to a Solve for one of T2, T3, ..., but not to the Solve for T1. The applicable preconditions are D < 5 and D < 10 and these are notionally 'ored' together which means D < 10 is actually required. To see this suppose we supply D = Digs = 7. Then D < 5 is False but D < 10 is True so by 'oring' False and True we get True, so the call works.

On the other hand if we write

```
An_X: T2 := ... ;
Solve(An_X, Answer, Digs);
```

then no dispatching is involved and the Solve for T2 is called. But both class wide preconditions D < 5 and D < 10 apply and so again the resulting 'ored' precondition that is required is D < 10.

Now it should be clear that if the preconditions do not form a weakening hierarchy then we will be in trouble. Thus if the preconditions were D < 15 for T1, D < 10 for T2, and D < 5 for T3, then dispatching from the root will only check D < 15. However, we could end up calling the Solve for T2 which expects the precondition D < 10 and this might not be satisfied.

Care is thus needed with preconditions that they are sensibly related.

Exercise 16.2

1 The type Square is derived from Object. Declare a function Area with some appropriate pre- and postconditions.

16.3 Type invariants

There are two other facilities of a contractual nature concerning types and subtypes. One is known as type invariants and describe properties of a type that remain true and can be relied upon; this topic is described in this section. The other is known as subtype predicates which extend the idea of constraints and are described in the next section. The distinction can be confusing at first sight and the following thoughts might be helpful.

Type invariants are not like constraints since invariants apply to all values of a type, whereas constraints are used to identify just a subset of the values of a type. Invariants are only meaningful on private types; and only apply to the external view of the type as seen by the user; within the package (that is using the full view) the invariant need not always be satisfied as we shall in the examples below.

In some ways, an invariant is more like the range of values specified when declaring a new integer type, as opposed to the constraint specified when defining an integer subtype. The specified range of an integer type can be violated in the middle of an arithmetic computation, but must be satisfied by the time the value is stored back into an object of the type.

Type invariants are useful if we want to ensure that some relationship between the components of a private type always holds. Thus suppose we have a stack and wish to ensure that no value is placed on the stack equal to an existing value on the stack. We can modify the earlier example to

```
package Stacks is

   type Stack is private
     with
        Type_Invariant => Is_Unduplicated(Stack);
```

```
function Is_Empty(S: Stack) return Boolean;
function Is_Full(S: Stack) return Boolean;
function Is_Unduplicated(S: Stack) return Boolean;

procedure Push(S: in out Stack; X: in Integer)
  with
    Pre => not Is_Full(S),
    Post => not Is_Empty(S);

-- and so on
```

The function Is_Unduplicated then has to be written to check that all values of the stack are different.

Note that we have mentioned Is_Unduplicated in the type invariant before its specification. This violates the usual 'linear order of elaboration'. However, as mentioned in Section 16.1, there is a general rule that all aspect specifications are only elaborated when the entity they refer to is frozen. Recall that one of the reasons for the introduction of aspect specifications was to overcome this problem with the existing mechanisms which caused information to become separated from the entities to which it relates.

The invariant on a private type T is checked when the value can be changed from the point of view of the outside user. In the case of the stack, the invariant Is_Unduplicated will be checked when we declare a new object of type Stack and each time we call Push and Pop.

Note that any subprograms internal to the package and not visible to the user can do what they like. It is only when a value of the type Stack emerges into the outside world that the invariant is checked.

The type invariant could be given on the full type in the private part rather than on the visible declaration of the private type (but not on both). Thus the user need not know that an invariant applies to the type.

Type invariants, like pre- and postconditions, are controlled by the pragma Assertion_Policy and only checked if the policy is Check. If an invariant fails to be true then Assertion_Error is raised at the appropriate point.

Like pre- and postconditions there are both specific invariants that can be applied to any type and class wide invariants that can only be applied to tagged types.

The invariant Is_Unduplicated is a curious example because it cannot be violated by Pop anyway since if there were no duplicates then removing the top item cannot make one appear.

Moreover, Push needs to ensure that the item to be added is not a duplicate of one on the stack already and so essentially much of the checking is repeated. Indeed, when writing Push we should be able to assume that no items are already duplicated and hence all we need to do is check that the new item to be added is not equal to one of the existing items (so n comparisons). However, a general function Is_Unduplicated will typically compare all pairs and thus require a double loop (hence $n(n-1)/2$ comparisons).

The reader is invited to meditate over this conundrum. One's first reaction might be that this is a bad example. However, one way to ensure reliability is to introduce redundancy. Thus if the encoding of Is_Unduplicated and Push are done independently then there is an increased probability that any error will be detected.

The aspect Type_Invariant requires an expression of a Boolean type. The mad programmer could therefore also write

```
type Stack is private
  with Type_Invariant;
```

which would thus be True by default and so useless! Actually it might not be entirely useless since it might act as a placeholder for an invariant to be defined later and meanwhile the program will compile and execute.

Type invariants are useful whenever a type is more than just the sum of its components. Note carefully that the invariant may not hold when an object is being manipulated by a subprogram having access to the full type. In the case of Push and Pop and the invariant Is_Unduplicated this will not happen but consider the following simple example.

Suppose we have a type Point which describes the position of an object in a plane. It might simply be

```
type Point is
  record
    X, Y: Float;
  end record;
```

Now suppose we want to ensure that all points are within a unit circle. We could ensure that a point lies within a square by means of range constraints by writing

```
type Point is
  record
    X, Y: Float range -1.0 .. +1.0;
  end record;
```

but we need to ensure that X**2 + Y**2 is not greater than 1.0, and that cannot be done by individual constraints. So we might declare a type Disc_Pt with an invariant as follows

```
package Places is

  type Disc_Pt is private
    with Type_Invariant => Check_In(Disc_Pt);

  function Check_In(D: Disc_Pt) return Boolean
    with Inline;

  ...                        -- various operations on disc points

private

  type Disc_Pt is
    record
      X, Y: Float range -1.0 .. +1.0;
    end record;

  function Check_In(D: Disc_Pt) return Boolean is
    (D.X**2 + D.Y**2 <= 1.0);              -- completion

end Places;
```

Note that we have used an expression function in the private part as the completion for Check_In. They are very useful for small functions in situations like this and typically will be given the aspect Inline on the specification as shown.

Now suppose that we wish to make available to the user a procedure Flip that reflects a Disc_Pt in the line $x = y$, or in other words interchanges its X and Y components. The body might be

```
procedure Flip(D: in out Disc_Pt) is
   T: Float;                  -- temporary
begin
   T := D.X;  D.X := D.Y;  D.Y := T;
end Flip;
```

This works just fine but note that just before the assignment to D.Y, it is quite likely that the invariant does not hold. If the original value of D was (0.1, 0.8) then at the intermediate stage it will be (0.8, 0.8) and so well outside the unit circle.

So there is a general principle that an intermediate value not visible externally need not satisfy the invariant. There is an analogy with numeric types. The intermediate value of an expression can fall outside the range of the type but will be within range when the final value is assigned to the object. For example, suppose type Integer is 16 bits (a small machine) but the registers perform arithmetic in 32 bits, then a statement such as

```
J := K * L / M;
```

could easily produce an intermediate result K * L outside the range of Integer but the final value could be in range.

In many cases it will not be necessary for the user to know that a type invariant applies to the type; it is after all merely a detail of the implementation. So perhaps the above should be rewritten as

```
package Places is

   type Disc_Pt is private;
   ...                        -- various operations on disc points
private

   type Disc_Pt is
      record
         X, Y: Float range -1.0 .. +1.0;
      end record
      with Type_Invariant => Disc_Pt.X**2 + Disc_Pt.Y**2 <= 1.0;

end Places;
```

In this case we do not need to declare a function Check_In at all. Note the use of the type name Disc_Pt in the invariant expression. This is an example of the use of a type name to denote a current instance.

We now turn to consider the places where a type invariant on a private type T is checked. These are basically when it can be changed from the point of view of the outside user. They are

- after default initialization of an object of type T,
- after a conversion to type T,
- after assigning to a view conversion involving descendants and ancestors of type T,
- after a call of T'Read or T'Input,
- after a call of a subprogram declared in the immediate scope of T and visible outside that has a parameter (of any mode including an access parameter) with a part of type T or returns a result with a part of type T.

Note that by saying a part of type T, the checks not only apply to subprograms with parameters and results of type T but they also apply to parameters and results whose components are of the type T or are view conversions involving the type T.

Observe that parameters of mode **in** are also checked because, as is well known, there are accepted techniques for changing such parameters. However, checks on **in** parameters only apply to procedures and not to functions (this avoids possible problems of infinite recursion).

Beware, however, that the checks do not extend to deeply nested situations, such as components with components that are access values to objects that themselves involve type T or worse. Thus there are holes in the protection offered by type invariants. However, if the types are straightforward and the writer does not do foolish things like surreptitiously exporting access types referring to T then all will be well.

The checks on type invariants regarding parameters and results can be conveniently implemented in the body of the subprogram in much the same way as for postconditions. This saves duplicating the code of the tests at each point of call.

If a subprogram such as Flip which is visible outside is called from inside then the checks still apply. This is not strictly necessary of course, but fits the simple model of the checks being in the body and so simplifies the implementation.

Similarly, checks are applied on the initialization of deferred constants in the private part and generally to default initialization of objects in the body.

If an untagged type is derived then any existing specific invariant is inherited for inherited operations. However, a further invariant can be given as well and both will apply to the inherited operations. This fits in with the model of view conversions used to describe how an inherited subprogram works on derivation. The parameters of the derived type are view converted to the parent type before the body is called and back again afterwards. As mentioned above, view conversions are one of the places where invariants are checked.

However, if we add new operations then the old invariant does not apply to them. In truth, the specific invariant is not really inherited at all; it just comes along for free with the inherited operations that are not overridden. So if we do add new operations then we need to state the total invariant required.

Note that this is not quite the same model as specific postconditions. We cannot add postconditions to an inherited operation but have to override it and then any specific postconditions on the parent are lost. In any event, in both cases, if we want to use inheritance then we should really use tagged types and class wide aspects.

So there is also an aspect Type_Invariant'Class for use with private tagged types. The distinction between Type_Invariant and Type_Invariant'Class has similarities to that between Post and Post'Class.

The specific aspect Type_Invariant can be applied to any type but Type_Invariant'Class can only be applied to tagged types. A tagged type can have both an aspect Type_Invariant and Type_Invariant'Class.

Type_Invariant cannot be applied to an abstract type. Type_Invariant'Class is inherited by all derived types; it can also be applied to an abstract type or interface. This can ensure that all types derived from an interface have certain properties.

Note the subtle difference between Type_Invariant and Type_Invariant'Class. Type_Invariant'Class is inherited for all operations of the type but as noted above Type_Invariant is only incidentally inherited by the operations that are inherited.

An interesting rule is that Type_Invariant'Class cannot be applied to a full type declaration which completes a private type such as Disc_Pt in the example above. This is because the writer of an extension will need to see the applicable invariants and this would not be possible if they were in the private part.

So if we have a type T with a class wide invariant thus

```
type T is tagged private
   with Type_Invariant'Class => F(T);
procedure Op1(X: in out T);
procedure Op2(X: in out T);
```

and then write

```
type NT is new T with private
   with Type_Invariant'Class => FN(NT);
overriding
procedure Op2(X: in out NT);
not overriding
procedure Op3(X: in out NT);
```

then both invariants F and FN will apply to NT.

Note that the procedure Op1 is inherited unchanged by NT, procedure Op2 is overridden for NT and procedure Op3 is added.

Now consider various calls. The calls of Op1 will involve view conversions as mentioned earlier and these will apply the checks for FN and the inherited body will apply the checks for F. The body of Op2 will directly include checks for F and FN as will the body of Op3. So the invariant F is properly inherited and all is well.

Remember that if the invariants were specific and not class wide then although Op1 will have checks for F and FN, Op2 and Op3 will only check FN.

In the case of the type Disc_Pt we might decide to derive a type which requires that all values are not only inside the unit circle but outside an inner circle – that is in an annulus or ring. We use the class wide invariants so that the parent package is

```
package Places is

   type Disc_Pt is tagged private
      with Type_Invariant'Class => Check_In(Disc_Pt);

   function Check_In(D: Disc_Pt) return Boolean
      with Inline;

   ...                          -- various operations on disc points
private
```

```
    type Disc_Pt is tagged
      record
        X, Y: Float range -1.0 .. +1.0;
      end record;

    function Check_In(D: Disc_Pt) return Boolean is
      (D.X**2 + D.Y**2 <= 1.0);

  end Places;
```

And then we might write

```
  package Places.Inner is

    type Ring_Pt is new Disc_Pt with private
      with Type_Invariant'Class => Check_Out(Ring_Pt);

    function Check_Out(R: Ring_Pt) return Boolean
      with Inline;

  private
    type Ring_Pt is new Disc_Pt with null record;

    function Check_Out(R: Ring_Pt) return Boolean is
      (R.X**2 + R.Y**2 >= 0.25);

  end Places.Inner;
```

And now the type Ring_Pt has both its own type invariant but also that inherited from Disc_Pt thereby ensuring that points are within the ring or annulus.

Finally, it is worth emphasizing that it is good advice not to use inheritance with specific invariants but they are invaluable for checking internal and private properties of types.

Exercise 16.3

1. Write the function body for Is_Unduplicated assuming the general case requiring the need to make all comparisons. The implementation can be assumed to be as in Section 12.4.

2. Rewrite the function Is_Unduplicated as an expression function.

16.4 Subtype predicates

The subtype feature of Ada is very valuable and enables the early detection of errors that linger in many programs in other languages and cause disaster later. However, although valuable, the subtype mechanism is rather limited. We can only specify a contiguous range of values in the case of integer and enumeration types.

Accordingly, Ada 2012 introduced subtype predicates as an aspect that can be applied to type and subtype declarations. The requirements proved awkward to satisfy with a single feature so in fact there are two aspects: Static_Predicate and Dynamic_Predicate. They both take a Boolean expression and the key difference is that the static predicate is restricted to certain types of expressions so that it can be used in more contexts.

Suppose we are concerned with seasons and that we have a type Month thus

```
type Month is (Jan, Feb, Mar, Apr, May, ..., Nov, Dec);
```

Now suppose we wish to declare subtypes for the seasons. For most people winter is December, January, February. (From the point of view of solstices and equinoxes, winter is from December 21 until March 21 or thereabouts, but March seems to me generally more like spring rather than winter and December feels more like winter than autumn.) So we would like to declare a subtype embracing Dec, Jan, and Feb. We cannot do this with a constraint but we can use a static predicate by writing

```
subtype Winter is Month
  with Static_Predicate => Winter in Dec | Jan | Feb;
```

and then we are assured that objects of subtype Winter can only be Dec, Jan, or Feb (provided once more that the Assertion_Policy pragma has set the Policy to Check). Note the use of the subtype name (Winter) in the expression where it stands for the current instance of the subtype.

The aspect is checked whenever an object is default initialized, on assignments, on conversions, on parameter passing and so on. If a check fails then Assertion_Error is raised.

As another example, suppose we wish to specify that an integer is even. We might expect to be able to write

```
subtype Even is Integer
  with Static_Predicate => Even mod 2 = 0;         -- illegal
```

Sadly, this is illegal because the expression in a static predicate is restricted and cannot use some operations such as **mod**. In detail, the expression can only be

- a static membership test where the choice is selected by the current instance,
- a case expression whose dependent expressions are static and selected by the current instance,
- a call of the predefined operations =, /=, <, <=, >, >= where one operand is the current instance,
- an ordinary static expression,

and, in addition, a call of a Boolean logical operator **and**, **or**, **xor**, **not** whose operands are such static predicate expressions, and, a static predicate expression in parentheses.

So we see that the predicate in the subtype Even cannot be a static predicate because the operator **mod** is not permitted with the current instance. But **mod** could be used in an inner static expression.

As a consequence, we have to use a dynamic predicate thus

```
subtype Even is Integer
  with Dynamic_Predicate => Even mod 2 = 0;        -- OK
```

Note that a subtype with predicates cannot be used in some contexts such as index constraints. This is to avoid having arrays with holes and similar nasty things.

However, static predicates are allowed in a for loop meaning to try every value. So we could write

```
for M in Winter loop ...
```

Beware that the loop uses values for M in the order, Jan, Feb, Dec and not Dec, Jan, Feb as the user might have wanted.

On the other hand, dynamic predicates are not allowed in a loop in this way so we cannot write

```
for X in Even loop ...
```

but have to spell it out in detail such as

```
for X in Integer loop
  if X mod 2 = 0 then                         -- or if X in Even then
    ...                        -- body of loop
  end if;
end loop;
```

Looking at the detailed rules we see that the predicate in the subtype Winter can be a static predicate because it takes the form of a membership test where the choice is selected by the current instance and whose individual items are themselves all static. Another useful example of this kind is

```
subtype Letter is Character
  with Static_Predicate => Letter in 'A' .. 'Z' | 'a' .. 'z';
```

Static case expressions are valuable because they provide the comfort of covering all values of the current instance. Suppose we have a type Animal

```
type Animal is (Bear, Cat, Dog, Horse, Wolf);
```

We could then declare a subtype of friendly animals

```
subtype Pet is Animal
  with Static_Predicate => Pet in Cat | Dog | Horse;
```

and perhaps

```
subtype Predator is Animal
  with Static_Predicate => not (Predator in Pet);
```

or equivalently

```
subtype Predator is Animal
  with Static_Predicate => Predator not in Pet;
```

Now suppose we add Rabbit to the type Animal. Assuming that we consider that rabbits are pets and not food, we should change Pet to correspond but we might forget with awkward results. Maybe we have a procedure Hunt which aims to eliminate predators

```
procedure Hunt(P: in out Predator);
```

and we will find that our poor rabbit is hunted rather than petted!

What we should have done is use a case expression controlled by the current instance thus

```
subtype Pet is Animal
   with Static_Predicate =>
      (case Pet is
         when Cat | Dog | Horse => True,
         when Bear | Wolf => False);
```

and now if we add Rabbit to Animal and forget to update Pet to correspond then the program will fail to compile.

Note that a similar form of if expression where the current instance has to be of a Boolean type would not be useful and so is excluded.

Static subtypes with static predicates can also be used in case statements. Thus elsewhere in the program we might have

```
case Animal is
   when Pet =>          ...          -- feed it
   when Predator =>     ...          -- feed on it
end case;
```

Observe that we do not have to list all the individual animals and naturally there is no others clause. If other animals are added to Pet or Predator then this case statement will not need changing. Thus not only do we get the benefit of full coverage checking, but the code is also maintenance free. Of course if we add an animal that is neither a Pet nor Predator (Sloth perhaps?) then the case statement will need updating.

Subtype predicates, like pre- and postconditions and type invariants are similarly monitored by the pragma Assertion_Policy. If a predicate fails (that is, has value False) then Assertion_Error is raised.

Subtype predicates are checked in much the same sort of places as type invariants. Thus

- on a subtype conversion,
- on parameter passing (which covers expressions in general),
- on default initialization of an object.

Note an important difference from type invariants. If a type invariant is violated then the damage has been done. But subtype predicates are checked before any damage is done. This difference essentially arises because type invariants apply to private types and can become temporarily false inside the defining package as we saw with the procedure Flip applying to the type Disc_Pt.

If an object is declared without initialization and no default applies then any subtype predicate might be false in the same way that a subtype constraint might be violated.

Beware that subtype predicates like type invariants are not foolproof. Thus in the case of a record type they apply to the record as a whole but they are not checked if an individual component is modified.

Subtype predicates can be given for all types in principle. Thus we might have

```
type Date is
  record
    D: Integer range 1 .. 31;
    M: Month;
    Y: Integer;
  end record;
```

and then

```
subtype Winter_Date is Date
  with Dynamic_Predicate => Winter_Date.M in Winter;
```

Note how this uses the subtype Winter which was itself defined by a subtype predicate. However, Winter_Date has to have a Dynamic_Predicate because the selector is not simply the current instance but a component of it.

We can now declare and manipulate a Winter_Date

```
WD: Winter_Date := (25, Dec, 2011);
...
Do_Date(WD);
```

and the subtype predicate will be checked on the call of Do_Date. However, beware that if we write

```
WD.Month := Jun;                          -- dodgy
```

then the subtype predicate is not checked because we are modifying an individual component and not the record as a whole.

Subtype predicates can be given with type declarations as well as with subtype declarations. Consider for example declaring a type whose only allowed values are the possible scores for an individual throw when playing darts. These are 1 to 20 and doubles and trebles plus 50 and 25 for an inner and outer bull's eye.

We could write these all out explicitly

```
type Score is new Integer
  with Static_Predicate =>
    Score in 1 | 2 | 3 | 4 | 5 | 6 | 7 | 8 | 9 | 10 | 11 | 12 | 13 | 14 | 15 | 16
          | 17 | 18 | 19 | 20 | 21 | 22 | 24 | 25 | 26 | 27 | 28 | 30 | 32 | 33
          | 34 | 36 | 38 | 39 | 40 | 42 | 45 | 48 | 50 | 51 | 54 | 57 | 60;
```

But that is rather boring and obscures the nature of the predicate. We can split it down by first defining individual subtypes for singles, doubles, and trebles as follows (note the use of for loops as described in Section 8.8 for doubles and trebles)

```
subtype Single is Integer range 1 .. 20;

subtype Double is Integer
  with Static_Predicate =>
    Double in [for J in 1 .. 20 =>2*J];
```

```
subtype Treble is Integer
   with Static_Predicate =>
      Treble in [for J in 1.. 20 => 3*J];
subtype Score is Integer
   with Static_Predicate =>
      Score in Single or Score in Double or Score in Treble or
      Score in 25 | 50;
```

Note that it would be neater to write

```
subtype Score is Integer
   with Static_Predicate =>
      Score in Single | Double | Treble | 25 | 50;
```

Observe that it does not matter that the individual predicates overlap. That is a score such as 12 is a Single, a Double, and also a Treble.

If we do not mind the predicates being dynamic then we can write

```
subtype Double is Integer
   with Dynamic_Predicate =>
      Double mod 2 = 0 and Double / 2 in Single;
```

and so on. Or we could even use a quantified expression

```
subtype Double is Integer
   with Dynamic_Predicate =>
      (for some K in Single => Double = 2*K);
```

or go all the way in one lump

```
type Dyn_Score is new Integer
   with Dynamic_Predicate =>
      (for some K in 1 .. 20 => Score = K or Score = 2*K or Score = 3*K)
       or Score in 25 | 50;
```

There are some restrictions on the use of subtypes with predicates. As mentioned above, if a subtype has a static or dynamic predicate then it cannot be used as an array index subtype. This is to avoid arrays with holes. So we cannot write

```
type Winter_Hours is array (Winter) of Hours;          -- illegal
type Hits is array (Score range <>) of Integer;        -- illegal
```

Similarly, we cannot use a subtype with a predicate to declare the range of an array object or to select a slice. So if we have

```
type Month_Days is array (Month range <>) of Integer;
The_Days: Month_Days := (31, 28, 31, 30, ... );
```

then we cannot write

```
Winter_Days: Month_Days(Winter);                      -- illegal array
The_Days(Winter) := (Jan | Dec => 31, Feb => 29);     -- illegal slice
```

However, a subtype with a static predicate can be used in a for loop thus

```
for W in Winter loop ...
```

and in a named aggregate such as

```
(Winter => 10.0, others => 14.0);                    -- OK
```

but a subtype with a dynamic predicate cannot be used in these ways. Actually the restriction is slightly more complicated. If the original subtype is not static such as

```
subtype To_N is Integer range 1 .. N;
```

then even if To_N has a static predicate it still cannot be used in a for loop or named aggregate.

These rules can also be illustrated by considering the dartboard. We might like to accumulate a count of the number of times each particular score has been achieved. So we might like to declare

```
type Hit_Count is array (Score) of Integer;          -- illegal
```

but sadly this would result in an array with holes and so is forbidden. However, we could declare an array from 1 to 60 and then initialize it with 0 for those components used for hits and −1 for the unused components. Of course, we ought not to repeat literals such as 1 and 60 because of potential maintenance problems. But, we can use new attributes First_Valid and Last_Valid thus

```
type Hit_Count is array (Score'First_Valid .. Score'Last_Valid) of Integer
                                    := (Score => 0, others => -1);
```

which uses Score to indicate the used components. The attributes First_Valid and Last_Valid can be applied to any static subtype but are particularly useful with static predicates.

In detail, First_Valid returns the smallest valid value of the subtype. It takes any range and/or predicate into account whereas First only takes the range into account. Similarly Last_Valid returns the largest value. Incidentally, they are illegal on null subtypes (because null subtypes have no valid values at all).

The Hit_Count array can then be updated by the value of each hit as expected

```
A_Hit: Score := ... ;                          -- next dart
Hit_Count(A_Hit) := Hit_Count(A_Hit) + 1;
```

If we attempt to assign a value of type Integer which is not in the subtype Score to A_Hit then Assertion_Error is raised (unless a different exception is specified as explained in the next section).

After the game, we can now loop through the subtype Score and print out the number of times each hit has been achieved and perhaps accumulate the total at the same time thus

```
for K in Score loop
   New_Line;  Put(Hit);  Put(Hit_Count(K));
   Total := Total + K * Hit_Count(K);
end loop;
```

The reason for the distinction between static and dynamic predicates is that the static form can be implemented as small sets with static operations on the small sets. Hence the loop

```
for K in Score loop ...
```

can be implemented simply as a sequence of 43 iterations. However, a loop such as

```
for X in Even loop ...
```

which might look innocuous requires iterating over the whole set of integers. Thus we insist on having to write

```
for X in Integer loop
   if X in Even then ...
```

which makes the situation quite clear.

Another restriction on the use of subtypes with predicates is that the attributes First, Last, and Range cannot be applied. But Pred and Succ are permitted because they apply to the underlying type. As a consequence, if a generic body (see Section 19.1) uses First, Last, or Range on a formal type and the actual type has a subtype predicate then Program_Error is raised.

Subtype predicates can be applied to abstract types but not to incomplete types. Subtype predicates are inherited as expected on derivation. Thus if we have

```
type T is ...
   with Static_Predicate => PredS;
```

and then

```
type NT is new T
   with Dynamic_Predicate => PredD;
```

the result is that both predicates apply to NT rather as if we had written the predicate as PredS **and then** PredD. Note that they are applied in order. If any one is dynamic then restrictions on the use of subtypes with a dynamic predicate apply.

There is no need for special predicates for class wide types in the way that we have both Type_Invariant and Type_Invariant'Class. So in the general case where a tagged type is derived from a parent and several progenitors

```
type T is new P and G1 and G2 with ...
```

where P is the parent and G1 and G2 are progenitors, the subtype predicate applicable to T is simply those for P, G1, G2 all anded together.

Exercise 16.4

1 The type Rainbow is declared as

```
type Rainbow is (Red, Orange, Yellow, Green, Blue, Indigo, Violet);
```

declare a subtype Primary whose only values are Red, Yellow, and Blue.

2 Declare a subtype of Integer whose only values are positive integers less than 1000 and have remainder 1 when divided by 37. Consider both static and dynamic predicates.

16.5 Global state

The ability to optionally specify the impact of a subprogram on global variables is added in Ada 2022. This will be familiar to users of SPARK where it is mandatory (see Section 27.6 for a brief introduction to SPARK). However, Ada is rather more complex than SPARK and a full and complete description of the impact of a subprogram on global state in all circumstances would be complex to define and daunting to use. So the approach taken has a broad brush flavour (AI12-79(3), AI12-302, AI12-380).

The main feature is the new aspect Global that can be used to indicate the global objects that a unit manipulates. It is optional, but if given, it must be complete.

As a very simple example we will consider a package for manipulating random numbers and use the algorithm given in the answer to Exercise 12.1(1). In that algorithm, the numbers in the sequence are obtained by multiplying the previous number by 5^5 and then taking the result modulo 2^{13}. This gives a simple but reasonable sequence largely because of the nature of the prime numbers 2, 5, and 13. We generalize by making both the multiplier (5^5) and the modular base (2^{13}) parameters of the system.

For the sake of illustration the data being used is in a global package called Data. The subprograms for manipulating the numbers are in a package Random. The procedure Set_Up sets the values of the modular base and the multiplier, the procedure Start initiates the sequence, and the function Next gets the next random number and moves the sequence on. The packages are as follows

```
package Data is
   Modulus: Integer;
   Multiplier: Integer;
   State: Integer;
end Data;

with Data;
package Random is
   procedure Set_Up(Mo, Mu: in Integer)
      with Global => (out Data.Modulus, Data.Multiplier);
   procedure Start(Seed: in Integer)
      with Global => out Data.State;
   function Next return Integer
      with Global => (in Data.Modulus, Data.Multiplier; in out Data.State);
end;
```

```
use Data;
package body Random is

   procedure Set_Up(Mo, Mu: in Integer) is
   begin
      Modulus := Mo;
      Multiplier := Mu;
   end Set_Up;

   procedure Start(Seed: in Integer) is
   begin
      State := Seed;
   end Start;

   function Next return Integer is
   begin
      State := State * Multiplier mod Modulus;
      return State;
   end Next;

end Random;
```

The Global aspect indicates the mode in a similar manner to subprogram parameters. Several items can be grouped together with commas and semicolons separating items in an obvious manner. Lists of items are enclosed within parentheses but a single item (as in the procedure Start) can appear on its own.

This example was contrived just to illustrate the forms of the aspects. Now consider a more realistic example by putting the data within the package Random. We can put the Multiplier and Modulus in the package body but for illustration we put the State in the package specification. The result is

```
package Random is
   State: Integer;
   procedure Set_Up(Mo, Mu: in Integer);
   procedure Start(Seed: in Integer)
      with Global => out State;
   function Next return Integer
      with Global => (in out State; in Random);
end;

package body Random is
   Modulus: Integer;
   Multiplier: Integer;

   procedure Set_Up(Mo, Mu: in Integer) is
   begin
      Modulus := Mo;
      Multiplier := Mu;
   end Set_Up;

   procedure Start(Seed: in Integer) is
   begin
      State := Seed;
   end Start;
```

```
   function Next return Integer is
   begin
      State := State * Multiplier mod Modulus;
      return State;
   end Next;

end Random;
```

The only example of a global variable is now State which is manipulated by Start and Next. Of course, Modulus and Multiplier are global to the bodies of all three subprograms but are not visible to their specifications which is where the aspects are given. However, if a global aspect is given, it has to be complete and so the specification of Next has to acknowledge that it manipulates entities global to its body; this is done by using the package name Random as standing for the entities in the package body resulting in **with** Global => (**in out** State; **in** Random). Although global annotations are optional, if we give one for Set_Up then it has to be **with** Global => **out** Random;

The global aspect can also be given to a package in which case it applies to all subprograms in the package; individual subprograms could also have global aspect clauses but they would overwrite that inherited from the package.

As well as mentioning individual objects there are a number of general possibilities which use one of **null**, **all**, **synchronized**. Thus **global** => **null** means that no access to any global objects is permitted. In the case of **all** and **synchronized** the mode **in**, **out**, or **in out** has to be given. Thus **global** => **in all** means that anything can be read but nothing can be written to. A very frequent aspect is **global** => **in out synchronized** which means that access is only permitted to read and write entities that can be accessed by several tasks or threads without interference.

The global aspect can also be given explicitly as just Unspecified. In the case of a library level package with Pure, the global aspect defaults to **null** otherwise it defaults to Unspecified.

Some examples of the aspect clauses of predefined packages are as follows

```
package Ada.Tags                              -- see Section 14.4
   with Preelaborate, Nonblocking, Global => in out synchronized is

package Ada.Text_IO
   with Global => in out synchronized is

package Ada.Numerics.Big_Numbers
   with Pure, Nonblocking, Global => null is
```

For the aspect Nonblocking see Section 20.2.

Some examples of global aspects on subprograms in the predefined library are

```
procedure Flush                               -- in various I/O packages
   with Global => in out all;

function Line_Length return Count             -- in Text_IO
   with Global => in all;
```

Note that **out all** is rather unlikely since it would mean that the subprogram is free to overwrite any global variable but must not read any!

The global aspect can also be given for tasks and entries and to protected types and objects in which case it applies to all the protected operations of the type. It can also be given for access to subprogram types.

As one might expect there is also the aspect Global'Class which can be used to describe the upper bound of the global impact of a dispatching operation.

A number of aspects are defined in the High Integrity Annex. These include Use_Formal and Dispatching. The aspect Use_Formal => **null** occurs on many subprograms in the container library; see for example Section 24.2.

Another feature defined in the High Integrity Annex is the use of **overriding**. This can be used to override the mode of a parameter in a global annotation.

An important example occurs with the predefined random number generators (see Section 23.5) where the next number is obtained by calling the function Random whose specification in the floating point case since Ada 95 has been

```
function Random(Gen: Generator) return Uniformly_Distributed;
```

Now calling Random updates some hidden state which is identified by Gen. This is reflected in the global annotation by adding

```
with Global => overriding in out Gen;
```

The reader will have concluded that this is a tricky topic. Being optional it can be ignored for most purposes but it is useful in describing aspects of the behaviour of the predefined library. For more details consult the *ARM*.

Note also that there are new restriction identifiers No_Unspecified_Globals and No_Hidden_Indirect_Globals.

Users of SPARK will be pleased to see that the annotations in Ada 2022 match those of original SPARK so that the awkward annotations necessary in SPARK 2014 are no longer necessary.

Remember that the use of **overriding** and the application to generic formal parameters are defined in the High Integrity Annex and so might not be available in all implementations.

A related topic regarding global state is the concept of stable properties. It is often the case that certain properties of an entity are not changed by most operations on it. A good example is given by a file. An important property is whether it is open or closed. If it is open then most operations such as writing to it do not change that property. So openness is typically unchanged and is an example of a stable property.

In Ada 2022, the new aspects Stable_Property and Stable_Property'Class are introduced (AI12-187). These are used in the container library in order to simplify the description of various operations. Thus in the case of the type Vector we find that its definition includes

```
Stable_Properties => (Length, Capacity,...
```

This implies that the value of Length and Capacity will not (generally) change. Note carefully that Length and Capacity are functions whose only parameter is of the type Vector. The stability is enforced by the implicit addition of postconditions to all the primitive subprograms of the type Vector of the form

```
Length(V) = Length(V)'Old and Capacity(V) = Capacity(V)'Old
```

Note that the checks are only applied if the Assertion_Policy has Post => Check.

If we wish to change a stable property, such as happens if we close a file or empty a container, then we write for example

```
procedure Clear(V: in out Vector)
  with Post => Length(V) = 0;
```

and in this case, the stable property Length is not enforced while the stable property Capacity is still enforced.

The full details are rather elaborate and the interested reader is invited to study the description of the aspects in the *ARM* and in particular its application to the container library.

16.6 Messages

A number of improvements were made in this area by AI12-54-2 These concern the aspect Predicate_Failure which enables specific messages to be associated with a failure and rules regarding the order of evaluation of predicates if several apply.

The expected type of the expression defined by the aspect Predicate_Failure is String and gives the message to be associated with a failure. This can be illustrated using the type File_Type and associated operations for input–output as described in Section 23.5. We can write

```
subtype Open_File_Type is File_Type
  with
    Dynamic_Predicate => Is_Open(Open_File_Type),
    Predicate_Failure => "File not open";
```

If the predicate fails then Assertion_Error is raised with the message "File not open".

We can also use a raise expression and thereby ensure that a more appropriate exception is raised. If we write

```
Predicate_Failure => raise Status_Error with "File not open";
```

then Status_Error is raised rather than Assertion_Error with the given message. We could of course explicitly mention Assertion_Error by writing

```
Predicate_Failure => raise Assertion_Error with "A message";
```

Finally, we could omit any message and just write

```
Predicate_Failure => raise Status_Error;
```

in which case the message is null.

A related issue is discussed in AI12-71. If several predicates apply to a subtype which has been declared by a refined sequence then the predicates are evaluated in the order in which they occur. This is especially important if different exceptions are specified by the use of Predicate_Failure since without this rule the wrong exception might be raised. The same applies to a combination of predicates, null exclusions, and old-fashioned subtypes.

This can be illustrated by an extension of the above example. Suppose we have

```
subtype Open_File_Type is File_Type
  with
    Dynamic_Predicate => Is_Open(Open_File_Type),
    Predicate_Failure => raise Status_Error;
subtype Read_File_Type is Open_File_Type
  with
    Dynamic_Predicate => Mode(Read_File_Type) = In_File,
    Predicate_Failure => raise Mode_Error with "Can't read file: " &
                                               Name(Read_File_Type);
```

The subtype Read_File_Type refines Open_File_Type. If the predicate for it were evaluated first and the file was not open then the call of Mode would raise Status_Error which we would not want to happen if we wrote

```
if F in Read_File_Type then ...
```

Care is needed with membership tests which were discussed in Section 9.1. The whole purpose of a membership test (and similarly the Valid attribute) is to find out whether a condition is satisfied. So if we write

```
if X in S then
   ...                        -- do this
else
   ...                        -- do that
end if;
```

we expect the membership test to be true or false. However, if the evaluation of S itself raises some exception then the purpose of the test is violated.

It is important to understand these related topics. Consider once more a very simple predicate such as

```
subtype Winter is Month
  with Static_Predicate => Winter in Dec | Jan | Feb;
```

and suppose we declare a variable W thus

```
W: Winter := Jan;
```

If we now do

```
W := Mar;
```

then Assertion_Error will be raised because the value Mar is not within the subtype Winter (assuming that the assertion policy is Check). If, however, we would rather have Constraint_Error raised then we can modify the declaration of Winter to

```
subtype Winter is Month
  with Static_Predicate => Winter in Dec | Jan | Feb,
       Predicate_Failure => raise Constraint_Error;
```

and then obeying

```
W := Mar;
```

will raise Constraint_Error.

On the other hand suppose we declare a variable M thus

```
M: Month := Mar;
```

and then do a membership test

```
if M in Winter then
   ...                          -- do this if M is a winter month
else
   ...                          -- do this if M is not a winter month
end if;
```

then of course no exception is raised since this is a membership test and not a predicate check.

Note however, that we could write something odd such as

```
subtype Winter2 is Month
   with Dynamic_Predicate =>
                  (if Winter2 in Dec | Jan | Feb then True else raise E);
```

then the very evaluation of the predicate might raise the exception E so that

```
M in Winter2
```

will either be True or raise the exception E but will never be False. Note that in this contrived example the predicate has to be a dynamic one because a static predicate cannot include a raise expression.

So this should clarify the reasons for introducing Predicate_Failure. It enables us to give a different behaviour for when the predicate is used in a membership test as opposed to when it is used in a check and it also allows us to add a message.

Finally, it should be noted that the predicate expression might involve the evaluation of some subexpression perhaps through the call of some function. We might have a predicate describing those months that have 30 days thus

```
subtype Month30 is Month
   with Static_Predicate => Month30 in Sep | Apr | Jun | Nov;
```

which mimics the order in the nursery rhyme. However, suppose we decide to declare a function Days30 to do the check so that the subtype becomes

```
subtype Month30 is Month
   with Dynamic_Predicate => Days30(Month30);
```

and for some silly reason we code the function incorrectly so that it raises an exception (perhaps it accidentally runs into its **end** and always raises Program_Error). In this situation if we write

```
M in Month30
```

then we will indeed get Program_Error and not false.

Perhaps this whole topic can be summarized by simply saying that a membership test is not a check. Indeed a membership test is often useful in ensuring that a subsequent check will not fail as was discussed in Section 9.1.

Exercise 16.6

1 Rewrite the declaration of the subtype Even in Section 16.4 so that the exception Constraint_Error is raised with an appropriate message if the predicate is violated.

Checklist 16

Preconditions and postconditions are given by Boolean aspects; they are checked on the call of a subprogram and on its return respectively.

Remember LSP for class wide pre- and postconditions.

Type invariants are checked when the external view could change; however, they are not foolproof.

Subtype predicates are more like constraints.

These facilities are controlled by the pragma Assertion_Policy.

Violation of a predicate raises Assertion_Error unless we use the aspect Predicate_Failure to indicate a different exception.

This chapter was added in Ada 2012.

New in Ada 2022

The new aspects Global and Global'Class enable the impact of a subprogram on global state to be described.

Pre- and postconditions can also be applied to access to subprogram types.

17 Numeric Types

17.1 Signed integer types
17.2 Modular types
17.3 Real types
17.4 Floating point types
17.5 Fixed point types
17.6 Decimal types

We now come at last to a more detailed discussion of numeric types. There are two categories of numeric types in Ada: integer types and real types. The integer types are subdivided into signed integer types and modular types (these are unsigned). The real types are subdivided into floating point types and fixed point types; the fixed point types are further subdivided into ordinary fixed point types and decimal types.

There are two problems concerning the representation of numeric types in a computer. First, the range will inevitably be restricted and indeed many machines have hardware operations for various ranges so that we can choose our own compromise between range of values and space occupied by values. Secondly, it may not be possible to represent accurately all the possible values of a type. These difficulties cause problems with program portability because the constraints vary from machine to machine. Getting the right balance between portability and performance is not easy. The best performance is achieved by using types that correspond exactly to the hardware. Perfect portability requires using types with precisely identical range and accuracy and identical operations.

Ada recognizes this situation and provides numeric types in ways that allow a programmer to choose the correct balance for the application. High performance can thus be achieved while keeping portability problems to a minimum.

We start by discussing integer types because these are somewhat simpler since they are parameterized only by their range whereas the real types are parameterized by both range and accuracy.

17.1 Signed integer types

All implementations of Ada have the predefined type Integer. In addition there may be other predefined types such as Long_Integer, Short_Integer and so on with a respectively longer or shorter range than Integer (could actually be the same). The range of values of these predefined types will be symmetric about zero except for an extra negative value in two's complement machines (which now seem to dominate over one's complement machines). All predefined integer types have the same predefined operations that were described in Chapter 6 as applicable to the type Integer (except that the second operand of "**" is always of type Integer).

Thus we might find that on machine A we have types Integer and Long_Integer with

```
range of Integer:
  -32_768 .. +32_767                                    (i.e. 16 bits)
range of Long_Integer:
  -2147_483_648 .. +2147_483_647                        (i.e. 32 bits)
```

whereas on machine B we might have types Short_Integer, Integer, and Long_Integer with

```
range of Short_Integer:
  -2048 .. +2047                                        (i.e. 12 bits)
range of Integer:
  -8388_608 .. +8388_607                                (i.e. 24 bits)
range of Long_Integer:
  -140_737_488_355_328 .. +140_737_488_355_327          (i.e. 48 bits)
```

For most purposes the type Integer will suffice on either machine and that is why we have simply used Integer in examples in this book so far. However, suppose we have an application where we need to manipulate signed values up to a million. The type Integer is inadequate on machine A and to use Long_Integer on machine B would be extravagant. We *could* overcome this problem by using derived types and writing (for machine A)

```
type My_Integer is new Long_Integer;
```

and then using My_Integer throughout the program. To move the program to machine B we would just replace this one declaration by

```
type My_Integer is new Integer;
```

However, Ada enables the choice between the underlying types to be made automatically. If we write

```
type My_Integer is range -1E6 .. 1E6;
```

then the implementation will implicitly choose the smallest appropriate hardware type and it will be somewhat as if we had written

```
type My_Integer is new hardware_type range -1E6 .. 1E6;
```

where the anonymous *hardware_type* is chosen appropriately. So in effect My_Integer will be a subtype of an anonymous type based on one of the predefined types and so objects of type My_Integer will be constrained to take only the values in the range –1E6 .. 1E6 and not the full range of the anonymous type. Note that the range must have static bounds since the choice of machine type is made at compile time.

To understand what is really going on we need to distinguish between the *range* of a type and the *base range* of a type. The range of a type is the requested range whereas the base range is the actual implemented range. So in the case of My_Integer, the range is –1E6 .. 1E6 whereas the base range is the full range of the underlying anonymous machine type.

The base type can be indicated by applying the attribute Base to the type. Thus My_Integer'Base is the base type (strictly the base subtype) and My_Integer'Base 'Range is the base range; we could declare variables of this subtype by writing

```
Var: My_Integer'Base;
```

Note that the attribute Base always denotes an unconstrained subtype whereas My_Integer is constrained.

It is an important rule that range checks are applied to constrained subtypes but not to unconstrained subtypes. If a range check fails then Constraint_Error is raised. Although range checks do not apply to unconstrained subtypes nevertheless overflow checks always apply and so we either get the correct mathematical result or Constraint_Error is raised. Using a constrained subtype is more portable whereas an unconstrained subtype is likely to be more efficient. But note that overflow checks in particular may well be automatically performed by the hardware with no time penalty.

Another point regarding unconstrained types is that the compiler is permitted to optimize intermediate expressions and storage for variables by using a larger range if helpful. For example, all registers might be of 32 bits and all arithmetic be performed using 32 bits on machine A even though the type Integer itself was only 16 bits. As a consequence, if an unconstrained variable of type Integer were held in such a register then it could have values outside its 16-bit base range. This is naturally only possible for an unconstrained variable.

We can illustrate these points by considering

```
type Index is range 0 .. 20_000;
I, J, K: Index;
IB, JB: Index'Base;
```

on machine A. The range of Index is then 0 .. 20_000 as given, whereas the base range is –32_768 .. +32_767. Suppose furthermore that IB is held in a 16-bit store whereas JB is optimized and held in a 32-bit register. Now consider

```
I := 17_000;
J := 16_000;
...
K := I + J;                -- range check fails, Constraint_Error
IB := I + J;               -- overflows, Constraint_Error
JB := I + J;               -- OK
```

The addition of I and J is successfully performed in a 32-bit register. The assignment to the constrained variable K fails because the result is outside the range of Index. The assignment to IB fails because the value is outside the base range and IB is implemented using the base range. The assignment to JB succeeds because JB is in a 32-bit register.

The fact that the addition can be performed using registers with a larger range than the base range is reflected by the fact that the specification of "+" can be thought of as

```
function "+" (Left, Right: Index'Base) return Index'Base;
```

so that there are no range checks on evaluating the parameters or returning the results.

Similar considerations apply to the predefined named types such as Integer. The subtype Integer is constrained whereas Integer'Base is unconstrained. The range and base range happen to be the same. The predefined operations take the form

```
function "+" (Left, Right: Integer'Base) return Integer'Base;
```

as shown in Section 23.1. Note the strange formal parameter names Left and Right; these are the usual names for parameters of the predefined operators.

Not all the implemented base ranges need correspond to a predefined type such as Integer or Long_Integer declared in Standard. There could be others. Indeed the *ARM* recommends that only Integer and Long_Integer be given such names. The type Integer will always exist and have at least a 16-bit range; if Long_Integer does exist then it will have at least a 32-bit range.

In addition to the types Integer and Long_Integer an implementation should provide types directly corresponding to the hardware in the package Interfaces. These will have names such as Integer_8, Integer_16, ..., Integer_128 according to the supported ranges.

We can convert between one integer type and another by using the normal notation for type conversion. Given

```
type My_Integer is range –1E6 .. 1E6;
type Index is range 0 .. 20_000;
M: My_Integer;
I: Index;
```

then we can write

```
M := My_Integer(I);
I := Index(M);
```

On machine A a genuine hardware conversion is necessary but on machine B both types will have the same representation as Integer and the conversion will be null (we assume that both machines have no other hardware types than those corresponding to the predefined types mentioned at the start of this section).

All the integer types can be thought of as being ultimately derived from an anonymous type known as *root_integer*. The range of this type *root_integer* is System.Min_Int .. System.Max_Int where Min_Int and Max_Int are constants declared in the package System. These are the minimum and maximum signed integer values supported by the executing program. The type *root_integer* has all the usual integer operations. (The derivation model should not be taken too literally as we shall see when we discuss generic parameters in Section 19.2.)

The integer literals are considered to belong to another anonymous type known as *universal_integer*. The range of this type is essentially infinite. Integer numbers declared in a number declaration (see Section 6.1) such as

```
Ten: constant := 10;
```

are also of type *universal_integer*. However, there are no *universal_integer* variables and no *universal_integer* operations.

Conversion from *universal_integer* to all other integer types is implicit and does not require an explicit type conversion. In fact the type *universal_integer* behaves a bit like the class wide type for integers; that is like *root_integer*'Class. (But of course there is no tag and so the analogy cannot be taken too far.)

It should be noted that some attributes such as the function Pos in fact deliver a *universal_integer* value and since Pos can take a dynamic argument it follows that certain *universal_integer* expressions may actually be dynamic (but they immediately get converted to some other integer type – possibly *root_integer* if the context does not impose a specific type).

The initial value in a number declaration has to be static; it can be of any integer type. So

```
M: constant := 10;
MM: constant := M * M;
N: constant Integer := 10;
NN: constant := N * N;
```

are all allowed.

Sometimes an expression involving literals is ambiguous. In such a case there is a preference for the type *root_integer* if that overcomes the ambiguity. For example

```
MM: constant := M * M;
```

is apparently ambiguous since the operator "*" is defined for all the specific integer types (including *root_integer* but not *universal_integer*). It is then taken to be *root_integer* by this preference rule.

Of course, since these expressions are all static and thus evaluated at compile time the distinction is somewhat theoretical especially since static expressions are always evaluated exactly. However, it is an important rule that a static expression may not exceed the base range of the expected type (such as Integer for N above).

A fuller description of static expressions will be found in Section 27.1 but the general rule for scalar types is that if it looks static then it is static. Thus 2 + 3 is always static and not just in a context demanding a static expression.

Recall that a range such as 1 .. 10 occurring in a for statement or in an array type definition or in a quantified expression is considered to be of type Integer. This is a consequence of the preference rule and a special rule which says that a range of type *root_integer* in such contexts is taken to be Integer. Note incidentally that specifying that the range is type Integer rather than any other integer type means that the predefined type Integer does have rather special properties.

We now see why we could write

```
for I in -1 .. 10 loop
```

in Section 7.3. This is ambiguous but by preference taken to be of type *root_integer* and then the special rule considers it to be of type Integer.

The use of integer type declarations which reflect the need of the program rather than the arbitrary use of Integer and Long_Integer is good practice because it not only encourages portability but also enables us to distinguish between different kinds of objects as, for example, when we were counting apples and oranges in Section 12.3.

Consideration of separation of concerns is the key to deciding whether to use numeric constants (of a specific type) or named numbers (of type *universal_integer*). If a literal value is a pure mathematical number then it should be declared as a named number. If, however, it is a value related naturally to just one of the program types then it should be declared as a constant of that type. Thus if we are counting oranges and are checking against a limit of 100 it is better to write

```
Max_Oranges: constant Oranges := 100;
```

rather than

```
Max_Oranges: constant := 100;
```

so that the accidental mixing of types as in

```
if No_Of_Apples = Max_Oranges then
```

will be detected by the compiler.

Returning to the question of portability it should be realized that complete portability is not easily obtained. We have seen that although we can specify the ranges for our types and thus ensure that variables are constrained and can never have values outside their range, nevertheless intermediate expressions are often computed with a wider range that is not prescribed exactly. For example, assume

```
type My_Integer is range -1E6 .. +1E6;
I, J: My_Integer;
```

and consider

```
I := 10_000;
J := (I * I) / 5_000;
```

We saw that the operations actually applied to My_Integer'Base – and even then might return values outside the base range. Ignoring this last possibility for the moment, on machine A the intermediate product and final result are computed with no problem and the final result lies within the range of J. But on machine B we get Constraint_Error because the intermediate product 1E8 is outside the base range. So the program is not fully portable.

We could declare

```
function Old_Multiply(Left, Right: My_Integer) return My_Integer
                                                    renames "*";
```

```
function "*"(Left, Right: My_Integer) return My_Integer is
begin
  return Old_Multiply(Left, Right);
end "*";
```

and then this new function "*" will hide the old one; in the new one the operands and result have range checks and so our program will be portable although slow (and this example will fail on both machines!). Note that the hiding and renaming work because the constraints do not matter for the conformance rules involved (see Section 13.7). Remember that My_Integer and My_Integer'Base are both really subtypes of some anonymous type.

In conclusion, portability can be achieved at various levels. Using Integer is acceptable if we know that the values are always within 16 bits. Using Long_Integer is less wise. It is far better to use our own types such as My_Integer. If we are confident about the hardware we are using then we can use types such as Integer_32 in the package Interfaces.

Exercise 17.1

1 What types on machines A and B are used to represent

```
type P is range 1 .. 1000;
type Q is range 0 .. +32768;
type R is range -1E14 .. +1E14;
```

2 Would it make any difference if A and B were one's complement machines with the same number of bits in the representations?

3 Given

```
N: Integer := 6;
P: constant := 3;
R: My_Integer := 4;
```

what are the types of

(a) N + P (c) P + R (e) P * P
(b) N + R (d) N * N (f) R * R

4 Declare a type Longest_Integer which is the maximum supported by the implementation.

17.2 Modular types

The modular types are unsigned integer types and exhibit cyclic arithmetic. Suppose for example that we wish to perform unsigned 8-bit arithmetic (that is byte arithmetic). We can declare

```
type Unsigned_Byte is mod 256;
```

and then the range of values supported by Unsigned_Byte is 0 .. 255. The normal arithmetic operations apply but all arithmetic is performed modulo 256 and overflow cannot occur.

The modulus of a modular type has to be static. It need not be a power of two although it often will be. It might, however, be convenient to use some obscure prime number as the modulus in the implementation of hash tables. Other uses might be to represent naturally cyclic concepts such as compass bearings or the time of day expressed in appropriate units.

The logical operations **and**, **or**, **xor**, and **not** are also available on modular types; the binary operations treat the values as bit patterns; the **not** operation subtracts the value from the maximum for the type. These operations are of most use when the modulus is a power of two – in which case **not** has the expected behaviour; they are well defined for other moduli but do have some curious properties. Note that no problems arise with mixing logical operations with arithmetic operations because negative values are not involved.

In the previous section we noted that the package Interfaces contains declarations of types such as Integer_16. Corresponding to each of these there is a modular type such as Unsigned_16. For these modular types (whose modulus will inevitably be a power of two) a number of shift and rotate operations are defined. They are Shift_Left, Shift_Right, Shift_Right_Arithmetic, Rotate_Left, and Rotate_Right. They all have an identical profile such as

```
function Shift_Left(Value: Unsigned_16; Amount: Natural)
                                              return Unsigned_16;
```

These subprograms have the expected behaviour. They have calling convention Intrinsic which means that the Access attribute cannot be applied to them in order to produce an access to subprogram value.

Conversion between modular types and signed integer types is possible provided the value is not out of range of the destination. If it is then Constraint_Error is naturally raised.

Thus suppose we had

```
type Signed_Byte is range -128 .. +127;
S: Signed_Byte := -106;
U: Unsigned_Byte := Unsigned_Byte(S);
```

then the type conversion will raise Constraint_Error. However, we can use the attribute Mod which applies to any modular type and converts a *universal_integer* argument to the modular type with appropriate cyclic arithmetic. So we can write

```
U := Unsigned_Byte'Mod(S);
```

and this gives the value 150. Conversion in the opposite direction can be done in various ways. We might simply write

```
if U >= 128 then
   S := Signed_Byte(U - 128) - 128;        -- S is negative
else
   S := Signed_Byte(U);                    -- S is positive
end if;
```

We could of course do the conversions as bit patterns using unchecked conversion as described in Section 25.2. Note that literals and static expressions are always checked at compile time to ensure they are within the base range of the expected type, and so

```
U := 256;
```

is illegal. On the other hand, a computation such as

```
U := 128;
U := U + U;
```

which is performed with arithmetic modulo 256 results in zero being assigned to U.

Note that since modular types are integer types they are also discrete types and so have all the common properties of the class of discrete types. Subtypes can be declared with a reduced range (this would be unusual). Modular types can be sensibly used as array index types and for loop parameters. An example of an array index dealing with a circular buffer will be found in Section 20.4.

The attribute Modulus applies to modular subtypes and naturally returns the modulus; it is of type *universal_integer*. The attributes Pred and Succ wrap around and so never raise Constraint_Error.

Exercise 17.2

1 Given

```
X: Unsigned_Byte := 16#AB#;
Y: Unsigned_Byte := 16#CD#;
```

what are the values of

(a) X **or** Y
(b) X + Y
(c) X – Y
(d) X * Y

2 Reconsider Exercise 8.2(2) using an appropriate modular type.

3 Given

```
type Ring5 is mod 5;
A: Ring5 := 3;
B: Ring5 := 4;
```

what are the values of

(a) **not** (A **and** B)
(b) **not** A **or not** B

4 Declare a modular type suitable for compass bearings accurate to one minute of arc, then declare a constant Degree having the value of one degree. Finally declare constants representing the directions SE, NNW and NE by E. Remember that North is zero on the compass and that East is 90°. (This is a different convention from that used for complex numbers.

17.3 Real types

Integer types are exact types. But real types are approximate and introduce problems of accuracy which have subtle effects. This book is not a specialized treatise on errors in numerical analysis and so we do not intend to give all the details of how the features of Ada can be used to minimize errors and maximize portability but concentrate instead on outlining the basic principles.

Real types are subdivided into floating point types and fixed point types. Apart from the details of representation, the key abstract difference is that floating point values have a relative error whereas fixed point values have an absolute error. Concepts common to both floating and fixed point types are dealt with in this section and further details of the individual types are in subsequent sections.

There are types *universal_real* and *root_real* having similar properties to *universal_integer* and *root_integer*. Again the preference rule chooses *root_real* in the case of an ambiguity. Static operations on the type *root_real* are notionally carried out with infinite accuracy during compilation. The real literals (see Section 5.4) are of type *universal_real*. Real numbers declared in a number declaration such as

```
Pi: constant := 3.14159_26536;
```

are also of type *universal_real*. (Remember that the difference between an integer literal and a real literal is that a real literal always has a point in it.)

As well as the usual operations on a numeric type, some mixing of *root_real* and *root_integer* operands is also allowed. Specifically, a *root_real* can be multiplied by a *root_integer* and vice versa and division is allowed with the first operand being *root_real* and the second operand being *root_integer* – in all cases the result is *root_real*. So we can write either of

```
Two_Pi: constant := 2 * Pi;            -- legal
Two_Pi: constant := 2.0 * Pi;          -- legal
```

but not

```
Pi_Plus_Two: constant := Pi + 2;       -- illegal
```

because mixed addition is not defined. Note that we cannot do an explicit type conversion between *root_integer* and *root_real* although we can always convert the former into the latter by multiplying by 1.0.

Exercise 17.3

1 Given

```
Two: Integer := 2;
E: constant := 2.71828_18285;
Max: constant := 100;
```

what are the types of

(a) Two * E (b) Two * Max (c) E * Max (d) Two * Two (e) E * E (f) Max * Max

2 Given

```
N: constant := 100;
```

declare a real number R having the same value as N.

17.4 Floating point types

All implementations have a predefined type Float and may also have further predefined types Long_Float, Short_Float and so on with respectively more and less precision (could be the same). These types all have the predefined operations that were described in Chapter 6 as applicable to the type Float.

The type Float will have at least 6 digits of precision (provided the hardware can cope) and the type Long_Float (if available) will have at least 11 digits of precision. Other named types in Standard are not recommended in the interests of portability. In a similar way to the integer types, named types corresponding to the hardware supported floating point types may be declared in the package Interfaces with appropriate names such as IEEE_Float_64. (Note the contrast between Integer and Float. Integer always has at least 16 bits but, for awkward hardware, Float is permitted not to have 6 digits.)

However, as mentioned in Section 2.3, it is good practice not to use the predefined types because of potential portability problems. If we write

```
type Real is digits 7;
```

then we are asking the implementation to base Real on a predefined floating point type with at least 7 decimal digits of precision.

The number of decimal digits requested, D, must be static. The actual precision and range supported will of course depend upon the hardware. The actual precision is known as the base decimal precision and the implemented range is the base range. For some awkward hardware, the base decimal precision might not hold over the whole of the base range and so that part of the range for which it does hold is called the safe range. We are guaranteed that the safe range includes -10.0**(4*D) .. +10.0**(4*D). For most applications that is all we need to know.

For specialized applications the Numerics annex contains a description of the minimum guaranteed properties in terms of what are called model numbers. See Section 26.5.

The types Float, Long_Float, and Real are unconstrained and so no range checks are imposed on assignments to variables of these types. But overflow checks always apply and raise Constraint_Error if they fail (provided the attribute Machine_Overflows for the type is true).

It is possible to declare a constrained type by for example

```
type Risk is digits 7 range 0.0 .. 1.0;
```

in which case the range constraint must be static. We can impose a (possibly dynamic) range constraint on a floating point subtype or object by

```
R: Real range 0.0 .. 100.0;
```

or

```
subtype Positive_Real is Real range 0.0 .. Real'Last;
```

and so on. If a range is violated then Constraint_Error is raised.

If we do declare a type with a range such as Risk then the safe range will always include the range given and the 4*D rule does not apply. This might force the use of a higher precision than without the range (not in the case of Risk though which has a relatively small range compared with the precision!)

The attribute Base can be applied to all real types and so Risk'Base gives the corresponding (anonymous) unconstrained subtype. Note that in contrast to Integer, the type Float is unconstrained and consequently Float and Float'Base are the same.

As well as a portable type Real, we might declare a more accurate type Long_Real perhaps for the more sensitive parts of a calculation

```
type Long_Real is digits 12;
```

and then declare variables of the two types

```
R: Real;
LR: Long_Real;
```

as required. Conversion between these follows similar rules to integer types

```
R := Real(LR);
LR := Long_Real(R);
```

and we need not concern ourselves with whether Real and Long_Real are based on the same or different predefined types.

Again in a similar manner to integer types, conversion of real literals, real numbers and *universal_real* attributes to floating types is automatic.

It is possible to apply a digits constraint to a floating point type thus

```
subtype Rough_Real is Real digits 5;
```

Such a constraint has little effect – objects of the subtype will be held in the same way as those of the type Real. The only difference is that the digits attribute of the subtype will be 5 and this will affect the default output format; see Section 23.6. This is considered an obsolescent feature of Ada.

The detailed workings of floating point types are defined in terms of various machine attributes based on a canonical model. In this model, nonzero values are represented as

$$sign.mantissa.radix^{exponent}$$

where

sign is +1 or −1,
radix is the hardware radix such as 2 or 16,
mantissa is a fraction in the base *radix* with nonzero leading digit,
exponent is an integer.

For any subtype S of a type T, the attribute S'Machine_Radix gives the *radix* and the three attributes S'Machine_Mantissa, S'Machine_Emin, and S'Machine_Emax collectively give the maximum number of digits in the *mantissa* and the range of values of the *exponent* for which every number in the canonical form is exactly represented in the machine. They all have type universal integer.

For example, one possible representation in a 32-bit word might be in binary with 1 sign bit, 8 bits for the exponent and 23 bits for the mantissa. Since the leading mantissa bit is always 1, it need not be stored. The exponent can be held in a biased form with logical values ranging from –128 to +127. For such a type the attributes would be 2, 24, –128, and 127 respectively. This format was used on the old DEC PDP-11.

By contrast, the 32-bit IBM 370 format has a hexadecimal radix, with 1 sign bit, 7 exponent bits and 24 mantissa bits. The attributes are 16, 6, –64, and 63. Note carefully that the 24 bits only allow for 6 hexadecimal digits.

There are also the closely related model attributes Model_Mantissa, Model_Emin, Model_Epsilon, Model_Small, and the function Model. Although they have to be provided by all implementations and are thus in the core language, their full description relates to implementations supporting the Numerics annex; see Section 26.5. Nevertheless, the following relationships always hold where D is the requested decimal precision for the subtype S

$$g + D\log_{radix}10 \leq \text{Model\_Mantissa} \leq \text{Machine\_Mantissa}$$

$$\text{Model\_Emin} \geq \text{Machine\_Emin}$$

$$\text{Model\_Epsilon} = radix^{(1-\text{Model\_Mantissa})}$$

$$\text{Model\_Small} = radix^{(\text{Model\_Emin}-1)}$$

where g is 1 unless *radix* is a power of 10 in which case g is 0.

The requested precision D determines the choice of underlying type. The model attributes relate to the abstract properties of an ideal hardware type whereas the machine attributes relate to what the hardware actually does (which in many cases will be the same, but sometimes has marginal variations). Model_Epsilon is a measure of the accuracy and is typically the difference between the number one and the next number above one. Model_Small is typically the smallest positive number. These two attributes are of type *universal_real*.

Various attributes enable the specialized programmer to manipulate the internals of floating point numbers. Thus S'Exponent returns the exponent and S'Fraction returns the fraction part. These and other attributes (Adjacent, Ceiling, Compose, Floor, Leading_Part, Machine_Rounding, Remainder, Rounding, Scaling, Truncation, and Unbiased_Rounding) are briefly described in Appendix 1.

The attributes for rounding are interesting. Different applications have different requirements. The function Rounding returns the nearest integral value and for midway values rounds away from zero. For the statistically minded Unbiased_Rounding rounds to the even value if midway whereas for those who just want speed Machine_Rounding is ideal since it does whatever the hardware does.

An interesting possibility is that the hardware may distinguish between positive and negative zero. The attribute S'Signed_Zeros is then true. A negative zero would arise for example if a small negative number were divided by a large positive one such that the result underflowed to zero. The distinction can be useful in certain boundary situations; for example the function Arctan behaves differently as mentioned in Section 23.4. Note that negative zero can also be distinguished by copying its sign to a nonzero number, using S'Copy_Sign and then testing the result. But otherwise a negative zero behaves as positive zero and for example X = 0.0 is true if X is a negative zero.

The attribute S'Digits gives the number of decimal digits requested whereas the attribute S'Base'Digits gives the number of decimal digits provided.

There are also the usual attributes S'First and S'Last which are the bounds of the subtype. The base range is S'Base'First .. S'Base'Last and the safe range is S'Safe_First .. S'Safe_Last.

Digits and Base'Digits are of type *universal_integer*; Safe_First and Safe_Last are of type *universal_real*; First and Last are of the base type T, (strictly the base subtype T).

We do not intend to say more about floating point here but hope that the reader will have understood the principles involved. In general one can simply specify the precision required and all will work. But care is sometimes needed and the advice of a professional numerical analyst should be sought when in doubt. For further details the reader should consult Section 26.5 and the *ARM*.

Exercise 17.4

1 Rewrite the function Inner of Section 10.1 using the type Real for parameters and result and a local type Long_Real with 14 digits accuracy to perform the calculation in the loop.

17.5 Fixed point types

The fixed point types come in two forms: the ordinary fixed point which is based on a binary representation, and decimal fixed point which is based on a decimal representation. Ordinary fixed point is typically used only in specialized applications or for doing approximate arithmetic on machines without floating point hardware. Decimal fixed point is typically used for accountancy and is discussed in the next section.

An ordinary fixed point type declaration specifies an absolute error and also a mandatory range. It takes the form

```
type F is delta D range L .. R;
```

In effect this is asking for values in the range L to R with an accuracy of D which must be positive. D, L and R must be real and static.

The implemented values of a fixed point type are the multiples of a positive real number *small*. The actual value of *small* chosen by the implementation can be any power of 2 less than or equal to D. The base range includes all multiples of *small* within the requested range.

As an example, if we have

```
type T is delta 0.1 range -1.0 .. +1.0;
```

then *small* could be $^1/_{16}$. So the implemented values of T will always include

$$-^{15}/_{16}, -^{14}/_{16}, \ldots, -^{1}/_{16}, 0, +^{1}/_{16}, \ldots, +^{14}/_{16}, +^{15}/_{16}$$

Note carefully that L and R might actually be outside the base range by as much as *small*; the definition just allows this.

If we have a typical 16-bit implementation then there will be a wide choice of values of *small*. At one extreme *small* could be 2^{-15} which would give us a much greater accuracy but just the required range (the base range would include –1.0 but not +1.0 on a two's complement machine). At the other extreme *small* could be $^1/_{16}$ in which case we would have a much greater base range but just the required accuracy.

The type T is constrained and range checks will apply; the corresponding unconstrained base subtype is given by T'Base. So the base range as usual is T'Base'Range.

Of course, a different implementation might use just 8 bits for T. Moreover, using attribute definition clauses (which will be discussed in more detail in Chapter 25) it is possible to give the compiler more precise instructions. We can say

```
for T'Size use 5;
```

which will force the implementation to use the minimum 5 bits. Of course, it might be ridiculous for a particular architecture and in fact the compiler is allowed to refuse unreasonable requests.

We can also override the rule that *small* is a power of 2 by using an attribute definition clause. Writing

```
for T'Small use 0.1;
```

will result in the implemented numbers being multiples of 0.1. On an 8-bit implementation the base range would then be from –12.8 to +12.7. Note that if we explicitly specify *small* then any spare bits give extra range and not extra accuracy. This is yet another win for the accountants at the expense of the engineers!

In Ada 2012 and Ada 2022, both Small and Size can alternatively be set using aspect specifications when the type T is declared. Thus we might write

```
type T is delta 0.1 range –1.0 .. +1.0
  with Size => 5,
       Small => 0.1;
```

The advantage of using the default standard whereby *small* is a power of 2 is that conversion between fixed point and other numeric types can be based on shifting. The use of other values of *small* will in general mean that conversion requires implicit multiplication and division.

A standard simple example is given by

```
Del: constant := 2.0**(–15);
type Frac is delta Del range –1.0 .. 1.0;
```

which will be represented as a pure fraction on a 16-bit two's complement machine. Note that it does not really matter whether the upper bound is written as 1.0 or 1.0–Del; the largest implemented number will be 1.0–Del in either case.

A good example of the use of a specified value for *small* is given by a type representing an angle and which uses the whole of a 16-bit word

```
type Angle is delta 0.1 range –π .. π
  with Small => π * 2.0**(–15);
```

Note that the value given for the aspect Small must not exceed the value for delta. A subtle point is that it is still delta that controls the default output format through the attribute Aft, see Section 23.6.

The predefined arithmetic operations +, –, *, / and **abs** can be applied to fixed point values. Addition and subtraction can only be applied to values of the same type and, of course, return that type. Addition and subtraction are always exact.

Predefined multiplication and division are explained in terms of the anonymous type *universal_fixed*; the parameters and result are of this type which is matched by any fixed point type. As a consequence, multiplication and division are allowed between different fixed point types and the result is then of the type expected by the context. However, if the result is a parameter of a further multiplication or division then the result must first be explicitly converted to a particular type; this ensures that the scale of the intermediate value is properly determined. Multiplication and division by type Integer are also allowed and these return a value of the fixed point type. If the result of a multiplication or division is not an exact multiple of *small* then either adjacent multiple of *small* is permitted.

Conversion of real literals, real numbers, and *universal_real* attributes to fixed point types (including *universal_fixed*) is again automatic.

So given

```
F, G, H: Frac;
```

we can write

```
F := F + G;
F := F * G;
F := 0.5 * F;
F := F + 0.5;
```

However, we cannot write

```
F := F * G * H;              -- illegal
```

but must explicitly state the intermediate type such as

```
F := Frac(F * G) * H;
```

As expected, we cannot write

```
F := 2 + F;                  -- illegal
```

but we can write

```
F := 2 * F;
```

because multiplication is defined between fixed point types and Integer.

It is often necessary to define our own fixed point operations especially for multiplication and division. We might for example wish the type to be saturated as for the type Sat_Int in Section 15.5. Thus we might write

```
function "*" (Left, Right: Frac) return Frac is
   pragma Unsuppress(Overflow_Check);
begin
   return Standard."*" (Left, Right);
exception
```

```
    when Constraint_Error =>
      if (Left > 0.0 and Right > 0.0) or (Left < 0.0 and Right < 0.0) then
        return Frac'Last;
      else
        return Frac'First;
      end if;
  end "*";
```

and similar functions for addition, subtraction, and division (taking due care over division by zero and so on). This raises an interesting issue since it does not hide the predefined operations (as it would for an integer type) because the result type is Frac whereas the predefined operation has a result type of *universal_fixed*.

This could cause awkward ambiguities if it were not for a rule which says that the predefined operation for multiplication is not used if there is a user-defined operation for either operand type unless there is an explicit conversion of the result or we write Standard."*". A similar rule applies to division.

So we can write

```
H := F * G;                               -- user-defined *
H := Standard."*" (F, G);                 -- predefined *
```

A good example of the use of fixed point types might be with measurements of different but related units. Suppose we declare three types TL, TA, TV representing lengths, areas, and volumes. We use centimetres as the basic unit with an accuracy of 0.1 cm together with corresponding consistent units and accuracies for areas and volumes. We might declare

```
type TL is delta 0.1 range -100.0 .. 100.0;
type TA is delta 0.01 range -10_000.0 .. 10_000.0;
type TV is delta 0.001 range -1000_000.0 .. 1000_000.0;
for TL'Small use TL'Delta;
for TA'Small use TA'Delta;
for TV'Small use TV'Delta;

function "*" (Left: TL; Right: TL) return TA;
function "*" (Left: TL; Right: TA) return TV;
function "*" (Left: TA Right: TL) return TV;

function "/" (Left: TV; Right: TL) return TA;
function "/" (Left: TV; Right: TA) return TL;
function "/" (Left: TA; Right: TL) return TL;

XL, YL: TL;
XA, YA: TA;
XV, YV: TV;
```

These types have an explicit *small* equal to their delta and are such that no scaling is required to implement the appropriate multiplication and division operations.

Note that all three types have primitive user-defined multiplication and division operations even though, in the case of multiplication, TV only appears as a result type. Thus the predefined multiplication or division with any of these types as operands can only be considered if the result has a type conversion (or we explicitly mention Standard).

As a consequence the following are legal

```
XV := XL * XA;              -- legal, volume = length × area
XL := XV / XA;              -- legal, length = volume ÷ area
```

but the following are not because they do not match the user-defined operations

```
XV := XL * XL;              -- illegal, volume ≠ length × length
XV := XL / XA;              -- illegal, volume ≠ length ÷ area
XL := XL * XL;              -- illegal, length ≠ length × length
```

But if we insist on multiplying two lengths together then we can use an explicit conversion thus

```
XL := TL(XL * XL);          -- legal, predefined operation
```

and this uses the predefined operation.

If we need to multiply three lengths to get a volume without storing an intermediate area then we can write

```
XV := XL * XL * XL;
```

and this is unambiguous since there are no explicit conversions and so the only relevant operations are those we have declared.

As another example we return to the package Complex_Numbers of Section 12.2 and consider how we might implement the package body using a polar representation. Any reader who gave thought to the problem will have realized that writing a function to normalize an angle expressed in radians to lie in the range 0 to 2π using floating point raises problems of accuracy since 2π is not a representable number.

An alternative approach is to use a fixed point type. We can then arrange for π to be exactly represented. Another natural advantage is that fixed point types have uniform absolute error which matches the physical behaviour. The type Angle declared earlier is not quite appropriate because, as we shall see, it will be convenient to allow for angles of up to 4π.

The private part of the package could be

```
private
   π: constant := Ada.Numerics.π;
   type Angle is delta 0.1 range -4*π .. 4*π;
   for Angle'Small use π * 2.0**(-13);
   type Complex is
      record
         R: Real;
         θ: Angle range -π .. π;
      end record;
   I: constant Complex := (1.0, 0.5*π);
end;
```

where we have taken the opportunity to use the portable floating point type Real which seems more appropriate for this example.

We now need a function to normalize an angle to lie within the range of the component θ (which is neater if symmetric about zero). This is a nuisance but cannot be avoided if we use fixed point types. The function might be

```
function Normal(φ: Angle) return Angle is
begin
   if φ >= π then
      return φ – 2*π;
   elsif φ < –π then
      return φ + 2*π;
   else
      return φ;
   end if;
end Normal;
```

Another interesting point is that the values for θ that we are using do not include the upper bound of the range +π; the function Normal converts this into the equivalent –π. Unfortunately we cannot express the idea of an open bound in Ada although it would be perfectly straightforward to implement the corresponding checks.

The range for the type Angle has to be such that it will accommodate the sum of any two values of θ. However, making the lower bound of the range for Angle equal to –2*π is not adequate since there is no guarantee that the lower bound will be a represented number – it will not be in a one's complement implementation. So for safety's sake we squander a bit on doubling the range.

The various functions in the package body can now be written assuming that we have access to appropriate trigonometric functions applying to the fixed point type Angle and returning results of type Real. So we might have

```
package body Complex_Numbers is
   function Normal ...    -- as above
   ...
   function "*" (X, Y: Complex) return Complex is
   begin
     return (X.R * Y.R, Normal(X.θ + Y.θ));
   end "*";
   ...
   function RI_Part(X: Complex) return Real is
   begin
     return X.R * Cos(X.θ);
   end RI_Part;
   ...
end Complex_Numbers;
```

where we have left the more complicated functions for the enthusiastic reader.

An alternative would of course be to use a modular type much like the type of Exercise 17.2(**4**) but in that case we would have to look after the scaling ourselves.

One might think that the ideal solution would be for Ada to have a form of fixed point types with cyclic properties. However, hardware typically works in binary and so if a full cycle (that is 360°) is represented exactly, then even the angle of an equilateral triangle cannot be represented exactly. Clearly, Babylonian arithmetic using base 60 is what we want!

There are delta constraints much like the digits constraints for floating point types; again they are obsolescent and only affect output formats.

The attributes of a fixed point subtype S of a type T include S'Delta which is the requested delta, D, and S'Small which is the smallest positive represented number, *small*. Machine_Radix, Machine_Rounds, and Machine_Overflows also apply to all fixed point types. There are also the usual attributes S'First and S'Last which give the actual upper and lower bounds of the subtype.

Delta and Small are of type *universal_real*; First and Last are of type T.

Exercise 17.5

1. Given F of type Frac explain why we could write

```
F := 0.5 * F;
```

2. Write the following further function for the package Complex_Numbers implemented as in this section

```
function "**" (X: Complex; N: Integer) return Complex;
```

Remember that if a complex number z is represented in polar form (r, θ), then

$z^n \equiv (r, \theta)^n = (r^n, n\theta)$.

3. An alternative approach to the representation of angles in fixed point would be to hold the values in degrees. Rewrite the private part and the function Normal using a canonical range of 0.0 .. 360.0 for θ. Make the most of a 16-bit word but use a power of 2 for *small*.

17.6 Decimal types

Decimal types are used in specialized commercial applications and are dealt with in depth in the Information Systems annex which is outside the scope of this book. However, the basic syntax of decimal types is in the core language and it is therefore appropriate to give a very brief overview.

A decimal type is a form of fixed point type. The declaration provides a value of delta as for an ordinary fixed point type (except that in this case it must be a power of 10) and also prescribes the number of significant decimal digits. So we can write

```
type Money is delta 0.01 digits 14;
```

which will cope with values of some currency such as euros up to one trillion (billion for older Europeans) in units of one cent. This allows 2 digits for the cents and 12 for the euros so that the maximum allowed value is

€999,999,999,999.99

The usual operations apply to decimal types as to other fixed point types; they are more exactly prescribed than for ordinary fixed point types since accountants abhor rounding errors. For example, type conversion always truncates towards zero and if the result of a multiplication or division lies between two multiples of *small* then again it truncates towards zero. Furthermore, the Information Systems annex describes a number of special packages for decimal types including conversion to external format using picture strings. See Section 26.4.

The attributes Digits, Scale, and Round apply to decimal types; see Appendix 1. The attributes which apply to ordinary fixed point types also apply to decimal types. For most machines, Machine_Radix will always be 2 but some machines support packed decimal types and for them Machine_Radix will be 10.

Checklist 17

Declare explicit types for increased portability.

Beware of overflow in intermediate expressions.

Use named numbers or typed constants as appropriate.

The type Integer is constrained but Float is not.

If in doubt consult a numerical analyst.

The attribute Mod was added in Ada 2005.

The attribute Machine_Rounding was added in Ada 2005.

The rules excluding the use of predefined fixed point multiplication and division in the presence of user-defined operations did not exist in Ada 95 (and Ada 2005 reverted to the situation in Ada 83).

Aspects such as Small can be given by an aspect specification in Ada 2012.

New in Ada 2022

Floating point attributes such as Floor and Ceiling may optionally be provided for fixed point types (AI12-362-2).

The shift and rotate operations are permitted in static expressions, see Section 27.1.

18 Parameterized Types

18.1 Discriminated record types
18.2 Default discriminants
18.3 Variant parts
18.4 Discriminants and derived types
18.5 Access types and discriminants
18.6 Private types and discriminants
18.7 Access discriminants

In this chapter we describe the parameterization of types by what are known as discriminants. Discriminants are components which have special properties. All composite types other than arrays can have discriminants. In this chapter we deal with the properties of discriminants in general and their use with record types (both tagged and untagged); their use with task and protected types is discussed in Chapter 20 and Chapter 22.

Discriminants can be of a discrete type or an access type. In the latter case the access type can be a named access type or it can be anonymous. A discriminant of an anonymous access type is called an access discriminant by analogy with an access parameter.

We start by dealing with discrete discriminants of untagged types.

18.1 Discriminated record types

In the record types we have seen so far there was no formal language dependency between the components. Any dependency was purely in the mind of the programmer as for example in the case of the private type Stack in Section 12.4 where the interpretation of the array S depended on the value of the integer Top.

In the case of a discriminated record type, some of the components are known as discriminants and the remaining components can depend upon these. The discriminants can be thought of as parameterizing the type and the syntax reveals this analogy.

As a simple example, suppose we wish to write a package providing various operations on square matrices and that in particular we wish to write a function

Trace which sums the diagonal elements of a square matrix. We could contemplate using a type Matrix as in Section 8.2

```
type Matrix is array (Integer range <>, Integer range <>) of Float;
```

but the function would then have to check that the matrix passed as an actual parameter was indeed square. We would have to write something like

```
function Trace(M: Matrix) return Float is
   Sum: Float := 0.0;
begin
   if M'First(1) /= M'First(2) or M'Last(1) /= M'Last(2) then
      raise Non_Square;
   end if;
   for I in M'Range loop
      Sum := Sum + M(I, I);
   end loop;
   return Sum;
end Trace;
```

This is somewhat unsatisfactory; we would prefer to use a formulation which ensured that the matrix was always square and had a lower bound of 1. We can do this using a discriminated type. Consider

```
type Square(Order: Positive) is
   record
      Mat: Matrix(1 .. Order, 1 .. Order);
   end record;
```

This is a record type having two components: the first, Order, is a discriminant of the discrete subtype Positive and the second, Mat, is an array whose bounds depend upon the value of Order.

Variables and constants of type Square are declared in the usual way but (like array bounds) a value of the discriminant must be given either explicitly as a constraint or from an initial value. Thus the following are permitted

```
M: Square(3);
M: Square(Order => 3);
M: Square := (3, (1 .. 3 => (1 .. 3 => 0.0)));
```

The value provided for the discriminant could be any dynamic expression but once the variable is declared its constraint cannot be changed. The initial value for M could be provided by an aggregate as shown; note that the number of components in the subaggregate depends upon the first component of the aggregate. This can also be dynamic so that we could declare a Square of order N and initialize it to zero by

```
M: Square := (N, (1 .. N => (1 .. N => 0.0)));
```

We could avoid repeating N by giving the initial value separately (remember that we cannot refer to an object in its own declaration). So we could also write

```
M: Square(N);
...
M := (M.Order, (M.Mat'Range(1) => (M.Mat'Range(2) => 0.0)));
```

If we attempt to assign a value to M which does not have the correct discriminant value then Constraint_Error will be raised.

We can now rewrite the function Trace as follows

```
function Trace(M: Square) return Float is
   Sum: Float := 0.0;
begin
   for I in M.Mat'Range loop
      Sum := Sum + M.Mat(I, I);
   end loop;
   return Sum;
end Trace;
```

There is now no way in which a call of Trace can be supplied with a non-square matrix. Note that the discriminant of the formal parameter is taken from that of the actual parameter in a similar way to the bounds of an array.

Discriminants generally have much in common with array bounds. Thus we can introduce subtypes and a formal parameter can be constrained as in

```
subtype Square_3 is Square(3);
...
function Trace_3(M: Square_3) return Float;
```

and then the actual parameter of a call of Trace_3 would have to have a discriminant value of 3; otherwise Constraint_Error would be raised.

The result of a function could be of a discriminated type and, like arrays, the result could be a value whose discriminant is not known until the function is called. So we can write a function to return the transpose of a square matrix

```
function Transpose(M: Square) return Square is
begin
   return R: Square(M.Order) do                  -- extended return for fun
      for I in 1 .. M.Order loop
         for J in 1 .. M.Order loop
            R.Mat(I, J) := M.Mat(J, I);
         end loop;
      end loop;
   end return;
end Transpose;
```

A private type can have discriminants in the partial view and it must then be implemented as a type with corresponding discriminants (usually a record but it could be a task or protected type, see Section 20.10).

As an example, reconsider the type Stack in Section 12.4. We can overcome the problem that all the stacks had the same maximum length by making Max a discriminant.

```
package Stacks is
   type Stack(Max: Natural) is private;
   procedure Push(S: in out Stack; X: in Integer);
   procedure Pop(S: in out Stack; X out Integer);
   function "=" (S, T: Stack) return Boolean;
private
   type Integer_Vector is array (Integer range <>) of Integer;
   type Stack(Max: Natural) is
      record
         S: Integer_Vector(1 .. Max);
         Top: Integer := 0;
      end record;
end;
```

Each variable of type Stack now includes a discriminant component giving the maximum stack size. When we declare a stack we must supply the value thus

```
ST: Stack(Max => 100);
```

and as in the case of the type Square the value of the discriminant cannot later be changed. Note that the discriminant is visible and can be referred to as ST.Max although the remaining components are private.

The body of the package Stacks remains as before (see Section 12.4). Note that the function "=" can be used to compare stacks with different values of Max since it only compares those components of the internal array which are in use.

Discriminants bear a resemblance to subprogram parameters in several respects. For example, the subtype of a discriminant must not have an explicit constraint. This is so that the same full conformance rules can be used when a discriminant specification has to be repeated in the case of a private type with discriminants, as illustrated by the type Stack above.

We can also have deferred constants of a discriminated type and they follow similar rules to array types as discussed in Section 12.2. The deferred constant need not provide a discriminant but if it does then it must statically match that in the full constant declaration.

Suppose we wish to declare a constant Stack with a discriminant of 3. We can omit the discriminant in the visible part and merely write

```
C: constant Stack;
```

and then give the discriminant in the private part either as a constraint or through the mandatory initial value (or both)

```
C: constant Stack(3) := (3, (1, 2, 3), 3);
```

It is possible to declare a type with several discriminants. We may for instance wish to manipulate matrices which although not constrained to be square nevertheless have both lower bounds of 1. This could be done by

```
type Rectangle(Rows, Columns: Positive) is
   record
      Mat: Matrix(1 .. Rows, 1 .. Columns);
   end record;
```

and we could then declare either of

```
R: Rectangle(2, 3);
R: Rectangle(Rows => 2, Columns => 3);
```

The usual rules apply: positional values must be in order, named ones may be in any order, mixed notation can be used but positional ones must come first.

Similarly to multidimensional arrays, a subtype must supply all the constraints or none at all. We could not declare

```
subtype Row_3 is Rectangle(Rows => 3);  -- illegal
```

in order to get the equivalent of

```
type Row_3(Columns: Positive) is
  record
    Mat: Matrix(1 .. 3, 1 .. Columns);
  end record;
```

although a similar effect can be obtained with derived types (see Section 18.4).

The above examples have used discriminants as the upper bounds of arrays; they can also be used as lower bounds. In Section 18.3 we will see that they can also be used to introduce a variant part. In all these cases a discriminant must be used directly and not as part of a larger expression. So we could not declare

```
type Symmetric_Array(N: Positive) is
  record
    A: Vector(-N .. N);                    -- illegal
  end record;
```

where the discriminant N is part of the expression -N. But a discriminant can also be used in a default initial expression in which case it need not stand alone.

Exercise 18.1

1. Suppose that M is an object of the type Matrix. Write a call of the function Trace whose parameter is an aggregate of type Square in order to determine the trace of M. What would happen if the two dimensions of M were unequal?
2. Rewrite the specification of Stacks to include a constant Empty in the visible part. See also Exercise 12.4(**1**).
3. Write a function Is_Full for Stacks. See also Exercise 12.4(**2**).
4. Declare a constant Square of order N and initialize it to a unit matrix. Use the function Make_Unit of Exercise 10.1(**6**).

18.2 Default discriminants

The discriminant types we have encountered so far have been such that once a variable is declared, its discriminant cannot be changed just as the bound of an array cannot be changed. It is possible, however, to provide a default expression for a discriminant and the situation is then quite different. A variable can then be declared with or without a discriminant constraint. If one is supplied then that value overrides the default and as before the discriminant cannot be changed. If, on the other hand, a variable is declared without a value for the discriminant, then the value of the default expression is taken but it can then be changed by a complete record assignment. A type with default discriminants is said to be mutable because unconstrained variables can change shape or mutate.

An important variation occurs in the case of tagged types. Defaults for discriminants are only permitted in the case of tagged types if they are limited. This is because we do not want tagged types to be mutable but on the other hand limited types are not mutable anyway so they are allowed to have default discriminants.

A record type with default discriminants is an example of a definite type whereas one without defaults is an indefinite type. Remember that we can declare (uninitialized) objects and arrays of a definite type but not of an indefinite type, see Section 8.5.

Suppose we wish to manipulate polynomials of the form

$$P(x) = a_0 + a_1x + a_2x^2 + \dots a_nx^n$$

where $a_n \neq 0$ if $n \neq 0$.

Such a polynomial could be represented by

```
type Poly(N: Index) is
   record
      A: Integer_Vector(0 .. N);
   end record;
```

where

```
subtype Index is Integer range 0 .. Max;
```

but then a variable of type Poly would have to be declared with a constraint and would thereafter be a polynomial of that fixed size. This would be most inconvenient because the sizes of the polynomials may be determined as the consequences of elaborate calculations. For example, if we subtract two polynomials which have $n = 3$, then the result will only have $n = 3$ if the coefficients of x^3 are different.

However, if we declare

```
type Polynomial(N: Index := 0) is
   record
      A: Integer_Vector(0 .. N);
   end record;
```

then we can declare variables

```
P, Q: Polynomial;
```

which do not have constraints. The initial value of their discriminants would be zero because the default value of N is zero but the discriminants could later be changed by assignment. Note however that a discriminant can only be changed by a complete record assignment. So

```
P.N := 6;
```

would be illegal. This is quite natural since we cannot expect the array P.A to adjust its bounds by magic.

Variables of the type Polynomial could be declared with constraints

```
R: Polynomial(5);
```

but R would thereafter be constrained for ever to be a polynomial with $n = 5$.

Initial values can be given in declarations in the usual way

```
P: Polynomial := (3, (5, 0, 4, 2));
```

which represents $5 + 4x^2 + 2x^3$. Note that despite the initial value, P is not constrained.

In practice we might make the type Polynomial a private type so that we could enforce the rule that $a_n \neq 0$. Observe that predefined equality is satisfactory. We can give the discriminant in the partial view and then both the private type declaration and the full type declaration must give the default expression for N.

Note once more the similarity to subprogram parameters; the default expression is only evaluated when required and so need not produce the same value each time. Moreover, the same conformance rules apply when it has to be written out again in the case of a private type.

We can alternatively choose that the partial view does not show the discriminant at all. We might write

```
package Polynomials is
   type Polynomial is private;
   ...
private
   type Polynomial(N: Index := 0) is ...
   ...
end;
```

and then of course we cannot declare constrained polynomials outside the package because the discriminant is not visible. Of course it is vital that the full type has defaults so that the type is definite.

If we declare functions such as

```
function "-" (P, Q: Polynomial) return Polynomial;
```

then it will be necessary to ensure that the result is normalized so that a_n is not zero. This could be done by the following function

```
function Normal(P: Polynomial) return Polynomial is
   Size: Integer := P.N;
begin
   while Size > 0 and P.A(Size) = 0 loop
      Size := Size - 1;
   end loop;
   return (Size, P.A(0 .. Size));
end Normal;
```

This is a further illustration of a function returning a value whose discriminant is not known until it is called. Note the use of the array slice.

If default expressions are supplied then they must be supplied for all discriminants of the type. Moreover an object must be fully constrained or not at all; we cannot supply constraints for some discriminants and use the defaults for others.

The attribute Constrained can be applied to an object of a discriminated type and gives a Boolean value indicating whether the object is constrained or not. For any object of types such as Square and Stack which do not have default values for the discriminants this attribute will, of course, be True. But in the case of objects of a type such as Polynomial which does have a default value, the attribute may be True or False. So, using P, Q, and R declared above

```
P'Constrained = False
Q'Constrained = False
R'Constrained = True
```

We mentioned above that an unconstrained formal parameter will take the value of the discriminant of the actual parameter. In the case of an **out** or **in out** parameter, the formal parameter will be constrained if the actual parameter is constrained (an **in** parameter is constant anyway). Suppose we declare a procedure to truncate a polynomial by removing its highest order term

```
procedure Truncate(P: in out Polynomial) is
begin
   P := (P.N-1, P.A(0 .. P.N-1));
end Truncate;
```

Then Truncate(Q); will be successful, but Truncate(R); will result in Constraint_Error being raised. (We will also get Constraint_Error if we try to remove the only term of an unconstrained polynomial.)

We have seen that a discriminant can be used as the bound of an array. It can also be used as the discriminant constraint of an inner component. We could declare a type representing rational polynomials (that is one polynomial divided by another) by

```
type Rational_Polynomial(N, D: Index := 0) is
   record
      Num: Polynomial(N);
      Den: Polynomial(D);
   end record;
```

The relationship between constraints on the rational polynomial as a whole and its component polynomials is interesting. If we declare

```
R: Rational_Polynomial(2, 3);
```

then R is constrained for ever and the components R.Num and R.Den are also permanently constrained with constraints 2 and 3 respectively. However

```
P: Rational_Polynomial := (2, 3, Num => (2, (-1, 0, 1)),
                                 Den => (3, (-1, 0, 0, 1)));
```

is not constrained. This means that we can assign complete new values to P with different values of N and D. The fact that the components Num and Den are declared as constrained does not mean that P.Num and P.Den must always have a fixed length but simply that for given N and D they are constrained to have the appropriate length. So we could not write

```
P.Num := (1, (1, 1));
```

because this would violate the constraint on P.Num. However, we can write

```
P := (1, 2, Num => (1, (1, 1)), Den => (2, (1, 1, 1)));
```

because this changes everything together. Of course we can always make a direct assignment to P.Num that does not change the current value of its own discriminant.

The original value of P represented $(x^2 - 1)/(x^3 - 1)$ and the final value represents $(x + 1)/(x^2 + x + 1)$ which happens to be the same with the common factor of $(x - 1)$ cancelled. The reader will note the strong analogy between the type Rational_Polynomial and the type Rational of Exercise 12.2(**3**). We could write an equivalent function Normal to cancel common factors of the rational polynomials and the whole package of operations would then follow.

Another possible use of a discriminant is as part of the expression giving a default initial value for one of the other record components (but not another discriminant). Although the discriminant value may not be known until an object is declared, this is not a problem since the default initial expression is only evaluated when the object is declared and no other initial value is supplied. A discriminant can also be used to introduce a variant part as described in the next section.

However, we cannot use a discriminant for any other purpose. This unfortunately meant that when we declared the type Stack in the previous section we could not continue to apply the constraint to Top by writing

```
type Stack(Max: Natural) is
  record
    S: Integer_Vector(1 .. Max);
    Top: Integer range 0 .. Max := 0;      -- illegal
  end record;
```

since the use of Max in the range constraint is not allowed.

We conclude this section by reconsidering the problem of variable length strings and ragged arrays previously discussed in Sections 8.5, 11.2, and 11.4. Yet another approach is to use discriminated records. There are a number of possibilities such as

```
subtype String_Size is Integer range 0 .. 80;
type V_String(N: String_Size := 0) is
  record
    S: String(1 .. N);
  end record;
```

The type V_String is very similar to the type Polynomial (the lower bound is different). We have chosen a maximum string size corresponding to a typical page width (or historic punched card).

We can now declare fixed or varying v-strings such as

```
V: V_String := (5, "Hello");
```

We can overcome the burden of having to specify the length by writing

```
function "+" (S: String) return V_String is
begin
   return (S'Length, S);
end "+";
```

and then

```
type V_String_Array is array (Positive range <>) of V_String;

Zoo: constant V_String_Array := (+"aardvark", +"baboon",
                        +"camel", +"dolphin", +"elephant", ..., +"zebra");
```

Since v-strings have default discriminants they are definite and so we can declare unconstrained v-strings and also arrays of them. However, there is a limit of 80 on v-strings and, moreover, the storage space for the maximum size string is likely to be allocated irrespective of the actual string.

The reader might feel bemused by the number of different ways in which variable length strings can be handled. There are indeed many conflicting requirements and in order to promote portability a number of predefined packages for manipulating strings are provided as we shall see in Section 23.3.

Exercise 18.2

1. Declare a Polynomial representing zero (that is, $0x^0$).
2. Write a function "*" to multiply two polynomials.
3. Write a function "–" to subtract one polynomial from another. Use the function Normal.
4. Rewrite the procedure Truncate to raise Truncate_Error if we attempt to truncate a constrained polynomial.
5. What would be the effect of replacing the discriminant of the type Polynomial by (N: Integer := 0)?
6. Rewrite the declaration of the type Polynomial so that the default initial value of a polynomial of degree n represents x^n. Hint: use & in the initial value.
7. Write the specification of a package Rational_Polynomials. Make the type Rational_Polynomial private with visible discriminants. The functions should correspond to those of the package Rational_Numbers of Exercise 12.2(**3**).
8. Write a function "&" to concatenate two v-strings.

18.3 Variant parts

It is sometimes convenient to have a record type in which part of the structure is fixed for all objects of the type but the remainder can take one of several different forms. This can be done using a variant part and the choice between the alternatives is governed by the value of a discriminant.

In many ways the use of a variant is an alternative to a tagged type where the hidden tag plays the role of the discriminant. For example in Section 14.4 we had an abstract type Person and derived types Man and Woman; we can reformulate this using a variant part as follows

```
type Gender is (Male, Female);

type Person(Sex: Gender) is
   record
      Birth: Date;
      case Sex is
         when Male =>
            Bearded: Boolean;
         when Female =>
            Children: Integer;
      end case;
   end record;
```

This declares a record type Person with a discriminant Sex. The component Birth of type Date (see Section 8.7) is common to all objects of the type. However, the remaining components depend upon Sex and are declared as a variant part. If the value of Sex is Male then there is a further component Bearded whereas if Sex is Female then there is a component Children.

Since no default expression is given for the discriminant, all objects of the type must be constrained either explicitly or from an initial value. So we can therefore declare

```
John: Person(Male);
Barbara: Person(Female);
```

or we can introduce subtypes and so write

```
subtype Man is Person(Sex => Male);
subtype Woman is Person(Sex => Female);
John: Man;
Barbara: Woman;
```

Aggregates take the usual form but, of course, give only the components for the corresponding alternative in the variant. The value for a discriminant governing a variant must usually be static so that the compiler can check the consistency of the aggregate. We can therefore write

```
John := (Male, (19, Aug, 1937), False);
Barbara := (Female, (13, May, 1943), 2);
```

but not

```
S: Gender := Female;
...
Barbara := (S, (13, May, 1943), 2);
```

because S is not static but a variable.

There are two exceptions to the rule that the discriminant in an aggregate must be static. One occurs when there are two discriminants but the cases are nested, this is discussed in AI05-220 and thus applied to Ada 2012 as well. The other is if the discriminant can select only one variant anyway, see AI12-86.

The components of a variant can be accessed in the usual way. We could write

```
John.Bearded := True;
Barbara.Children := Barbara.Children + 1;
```

but an attempt to access a component of the wrong alternative such as John.Children would raise Constraint_Error.

Note that although the sex of an object of type Person cannot be changed, it need not be known at compile time. We could have

```
S: Gender := ...
...
Chris: Person(S);
```

where the sex of Chris is not determined until he or she is declared. The rule that a discriminant must be static applies only to aggregates.

The variables of type Person are necessarily constrained because the type has no default expression for the discriminant. It is therefore not possible to assign a value which would change the sex; an attempt to do so would raise Constraint_Error. However, as with the type Polynomial, we could declare a default initial expression for the discriminant and consequently declare unconstrained variables. Such unconstrained variables could then be assigned values with different discriminants but only by a complete record assignment.

We could therefore have

```
type Gender is (Male, Female, Neuter);

type Mutant(Sex: Gender := Neuter) is
  record
    Birth: Date;
    case Sex is
      when Male =>
        Bearded: Boolean;
      when Female =>
        Children: Integer;
      when Neuter =>
        null;
    end case;
  end record;
```

Note that we have to write **null**; as the alternative in the case of Neuter where we do not want any components. Like a null statement in a case statement this indicates that we really meant to have no components and did not omit them by accident.

We can now declare

```
The_Thing: Mutant;
```

The sex of this unconstrained mutant is neuter by default but can be changed by a whole record assignment.

Note the difference between the following

```
The_Thing: Mutant := (Neuter, (1, Jan, 1984));
It: Mutant(Neuter) := (Neuter, (1, Jan, 1984));
```

In the first case the mutant The_Thing is not constrained but just happens to be initially neuter. In the second case the object It is permanently neuter. These examples also illustrate the form of the aggregate when there are no components in the alternative; there are none so we write none – we do not write **null**.

The rules regarding the alternatives closely follow those regarding the case statement described in Section 7.2. Each **when** is followed by one or more choices separated by vertical bars and each choice is either a simple expression or a discrete range. The choice **others** can also be used but must be last and on its own. All values and ranges must be static and all possible values of the discriminant must be covered once and once only. The possible values of the discriminant are those of its static subtype (if there is one) or type. Each alternative can contain several component declarations or can be null as we have seen.

A record can only contain one variant part and it must follow other components. However, variants can be nested; a component list in a variant part could itself contain a variant part but again it must follow other components in the alternative.

Also observe that it is unfortunately not possible to use the same identifier for components in different alternatives of a variant – all components of a record must have distinct identifiers.

It is worth emphasizing the rules regarding the changing of discriminants. If an object is declared with a discriminant constraint then the constraint cannot be changed – after all it is a constraint just like a range constraint and so the discriminant must always satisfy the constraint. Because the constraint allows only a single value this naturally means that the discriminant can only take that single value and so cannot be changed.

The other basic consideration is that, for implementation reasons, all objects must have values for discriminant components. Hence if the type does not provide a default initial expression, the object declaration must and since it is expressed as a constraint the object is then consequently constrained.

There is a restriction on renaming components of a variable of a discriminated type. If the existence of the component depends upon the value of a discriminant then it cannot be renamed if the variable is unconstrained. (This only applies to variables and not to constants.) Consider

```
Hairy: Boolean renames The_Thing.Bearded;         -- illegal
Offspring: Integer renames Barbara.Children;      -- OK
```

The first is illegal because there is no guarantee that the component Bearded of the mutant The_Thing will continue to exist after the renaming even if it does exist at the moment of renaming. However, the second is valid because Barbara is a person and cannot change sex.

A similar restriction applies to the use of the Access attribute. Assuming the component Children is marked as aliased then Barbara.Children'Access is permitted whereas The_Thing.Children'Access is not.

Note, amazingly, that we can write

```
Bobby: Man renames Barbara;
```

because the constraint in the renaming declaration is ignored (see Section 13.7). Barbara has not had a sex change – she is merely in disguise!

Observe that Person is indefinite but Man, Woman, and Mutant are definite. So we can declare arrays of Man, Woman, and Mutant, but not of Person.

Although Person is indefinite we can declare a function with return type of Person and actual persons returned will be of a specific type such as Man or Woman. We can use extended return statements and write (discounting the mutant)

```
function Frankenstein(G: Gender; ... ) return Person is
begin
   if G = Male then
      return M: Person(Sex => Male) do
         ...                                    -- make a man
      end return;
   else
      return F: Person(Sex => Female) do
         ...                                    -- make a woman
      end return;
   end if;
end Frankenstein;
```

It is very instructive to consider how the types of this section might be rewritten using tagged types. The differences stem from the fact that tagged type derivation gives rise to distinct types whereas variants are simply different subtypes of the same type. The key differences using tagged types are

- An aggregate for a tagged type would not give the sex since it is inherent in the type.
- Attempting to access the wrong component John.Children is a compile-time error with tagged types whereas it raises Constraint_Error with variants.
- The secretive Chris of unknown sex could be of a class wide type and then the sex would be dynamically determined by the mandatory initial value; this is not quite so flexible because we need a whole initial value rather than just the constraint. So we have to copy another person or we have to write an extension aggregate such as (Person **with** Children => 0).
- The Mutant has no corresponding formulation using tagged types because a mutation would correspond to changing the tag and this can never happen.
- The restriction on renaming does not apply to tagged types because the tag of an object never changes and it is the tag which determines which components are present.

In deciding whether it is appropriate to use a variant or a tagged type, a key consideration is mutability. If an object can change its shape at run time then a variant must be used. If an individual object can never change its shape then a tagged type is possible. A tagged type is better if the root operations are fixed but the categories might be extended. On the other hand, if the number of categories is fixed but more operations might be added then a variant formulation might be better. Thus adding a new operation such as Name to the root package Geometry of Program 1 requires everything to be recompiled.

Exercise 18.3

1 Write a procedure Shave which takes an object of type Person and removes any beard if the object is male and raises Shaving_Error if the object is female.

2 Write a procedure Sterilize which takes an object of type Mutant and ensures that its sex is Neuter by changing it if necessary and possible and otherwise raises an appropriate exception.

3 Declare a discriminated type Object which describes geometrical objects which are either a circle, a triangle or just a point. Use the same properties as for the corresponding tagged types of Section 14.1. Then declare a function Area which returns the area of an Object.

4 Declare a discriminated type Reservation corresponding to the types of the subsonic reservation system of Section 14.3.

18.4 Discriminants and derived types

We now consider the interaction of discriminants with derived types. This is rather different according to whether the type is tagged or not and so to avoid confusion we consider the two cases separately starting with untagged types.

We can of course derive from a discriminated record such as Rectangle in Section 18.1 in the usual way by

```
type Another_Rectangle is new Rectangle;
```

and then the discriminants become inherited so that Another_Rectangle also has the two discriminants Rows and Columns.

We could also derive from a constrained subtype

```
type Square_4 is new Rectangle(Rows => 4, Columns => 4);
```

in which case Square_4 is a constrained subtype of some anonymous type with the two discriminants.

Of more interest are situations where the derived type has its own discriminants. In such a case the parent must be constrained and the new discriminants replace the old ones which are therefore not inherited. Moreover, each new discriminant must be used to constrain one or more discriminants of the parent. So the new type may have fewer discriminants than its parent but cannot have more. Remember that extension cannot occur with untagged types and so the implementation model is that each apparent new discriminant has to use the space of one of the discriminants of the parent. There are a number of interesting possibilities

```
type Row_3(Columns: Positive) is
        new Rectangle(Rows => 3, Columns => Columns);
type Square(N: Positive) is
        new Rectangle(Rows => N, Columns => N);
type Transpose(Rows, Columns: Positive) is
        new Rectangle(Rows => Columns, Columns => Rows);
```

where we have used named notation for clarity. In all these cases the names of the old discriminants are no longer available in the new view.

The type Row_3 has lost one degree of freedom since it no longer has a visible discriminant called Rows; we could of course convert to type Rectangle and then look at the Rows and find it was 3.

In the case of the type Square (we are almost back to the type Square of Section 18.1), the one new discriminant is used to constrain both old ones. Hence a conversion of a value of type Rectangle to type Square will check that the two discriminants of the rectangle are equal and raise Constraint_Error if they are not.

The type Transpose has the discriminants mapped in the reverse order. So if R is a Rectangle then R.Rows = Transpose(R).Columns; we can use the result of the conversion as a name and then select the discriminant. But of course the internal array is still the same and so this is not a very interesting example.

We now turn to a consideration of tagged types. Tagged record types can also have discriminants but they can only have defaults if limited. Thus tagged objects are never mutable. Again there is the basic rule that if the derived type has its own discriminants then these replace any old ones and so the parent type must be constrained. The new discriminants may be used to constrain the parent, but they need not. So in the tagged case, extension is allowed and the derived type may have more discriminants than its parent.

Thus an alternative implementation for the type Person might be based on

```
type Gender is (Male, Female);
type Person(Sex: Gender) is tagged
  record
    Birth: Date;
  end record;
```

and we can then extend with, for example

```
type Man is new Person(Male) with
  record
    Bearded: Boolean;
  end record;
```

The type Man naturally inherits the discriminant from Person in the sense that a man still has a component called Sex although of course it is constrained to be Male.

We could also extend from the unconstrained type and then the new type would inherit the old discriminant just as any other component

```
type Old_Person is new Person with
  record
    Pension: Money;
  end record;
```

and the Old_Person also has a component Sex (although it is not visible in the declaration but then neither is the Birth component). We cannot declare an object of type Old_Person without a constraint.

As an example of providing new discriminants (in which case the parent type must be constrained) we might declare a type Boxer who must be male (females don't box in this model of the world) and then add a discriminant giving his weight

```
type Weight is (Light, Middle, Heavy);

type Boxer(W: Weight) is new Person(Male) with
  record
    ...  -- information according to weight perhaps as variant
  end record;
```

In this case the Boxer does not have a sex component at all since the discriminant has been replaced.

Other examples are provided by the type Object and its descendants. A useful guide is that discriminants make sense for controlling structure but not for other aspects of a type. We might have

```
type Regular_Polygon(No_Of_Sides: Natural) is new Object with
  record
    Side: Float;
  end record;
...
type Pentagon is new Regular_Polygon(5) with null record;
```

There is a limit to the amount of specialization that can be performed through syntactic structures. Thus, one might think it would be nice to write

```
type Polygon(No_Of_Sides: Natural) is new Object with
  record
    Sides: Float_Array(1 .. No_Of_Sides);
  end record;
```

but of course this is no good because we also need the values of the angles. But it is unwise to make such details directly visible to the user since direct assignment to the components could result in inconsistent values which did not represent a closed polygon. So it is probably better to write

```
package Geometry.Polygons is
  type Polygon(No_Of_Sides: Natural) is new Object with private;
  ...
private
  type Polygon(No_Of_Sides: Natural) is new Object with
    record
      Sides: Float_Array(1 .. No_Of_Sides);
      Angles: Float_Array(1 .. No_Of_Sides);
    end record;
end Geometry.Polygons;
```

As discussed in Section 14.6 we now need a constructor function which is best declared as a child function Geometry.Polygons.Make_Polygon. This should check that the angles and sides given as parameters do form a closed polygon.

We might then declare

```
package Geometry.Polygons.Four_Sided is
   type Quadrilateral is new Polygon(4) with private;
   type Parallelogram is new Quadrilateral with private;
   type Square is new Parallelogram with private;
   ...
private
```

which reveals the hierarchy of the different kinds of quadrilaterals. The full type declarations all have null extensions of course. This structure ensures that operations that apply to the class of all parallelograms also apply to squares.

Exercise 18.4

1 Declare the type Boxer so that it can be of either sex.

2 Declare a child function to make a quadrilateral. Assume the existence of

```
function Geometry.Polygons.Make_Polygon(Sides, Angles: Float_Array)
                                                       return Polygon;
```

3 Declare a package containing conversion functions between the various types of quadrilateral.

18.5 Access types and discriminants

Access types can refer to discriminated record types in much the same way that they can refer to array types. In both cases they can be constrained or not. Consider the problem of representing a family tree. We could declare

```
type Person;
type Person_Name is access all Person;

type Person is limited
   record
      Sex: Gender;
      Birth: Date;
      Spouse: Person_Name;
      Mother: Person_Name;
      First_Child: Person_Name;
      Next_Sibling: Person_Name;
   end record;
```

This model assumes a monogamous and legitimate system. The children are linked together through the component Next_Sibling and a person's father is identified as the spouse of their mother. (The reader could consider how the model might be altered to accommodate more flexible social systems such as those in which marriage is permitted between persons of the same gender.)

Although the above type Person is adequate for the model, it is more interesting to use a discriminated type so that different components can exist for the different sexes and more particularly so that appropriate constraints can be applied. Consider

```
type Person(Sex: Gender);
type Person_Name is access all Person;

type Person(Sex: Gender) is limited
  record
    Birth: Date;
    Mother: Person_Name(Female);
    Next_Sibling: Person_Name;
    case Sex is
      when Male =>
        Wife: Person_Name(Female);
      when Female =>
        Husband: Person_Name(Male);
        First_Child: Person_Name;
    end case;
  end record;
```

The incomplete declaration of Person need not give the discriminants but if it does (as here) then it must also give any default initial expressions and the discriminants must conform to those in the subsequent complete declaration.

An important point is that we have made the type Person a limited type. This is because it would be quite inappropriate to copy a person although it is of course quite reasonable to copy the name of a person. We have also made the access type Person_Name general (using **all**) so that we can declare persons as ordinary variables if required as well as using the storage pool.

The component Mother is now constrained always to access a person whose sex is female (could be null). Similarly the components Wife and Husband are constrained; note that these had to have distinct identifiers and so could not both be Spouse. However, the components First_Child and Next_Sibling are not constrained and so could access a person of either sex. We have also taken the opportunity to save on storage by making the children belong to the mother only.

When an object of type Person is created by an allocator, a value must be provided for the discriminant either through an explicit initial value as in

```
Janet: Person_Name;
...
Janet := new Person'(Female, (22, Feb, 1967), Barbara, others => null);
```

or by supplying a discriminant constraint thus

```
Janet := new Person(Female);
```

Note that as in the case of arrays (see Section 11.3) a quote is needed in the case of the full initial value but not when we just give the constraint. Note also the use of **others** in the aggregate; this is allowed because the last three components all have the same base type. We could also write **others** => <>.

Naturally enough, the constraint cannot be omitted since the type Person does not have a default discriminant. For convenience we could introduce subtypes and so declare

```
subtype Man is Person(Male);
subtype Woman is Person(Female);
...
Janet := new Woman;
```

An allocated object cannot later have its discriminants changed except in one circumstance explained in Section 18.6. This general rule applies even if the discriminants have default initial expressions; objects created by an allocator are in this respect different from objects created by a normal declaration where having defaults for the discriminants always allows unconstrained objects to be declared and later to have their discriminants changed.

On the other hand, we see that despite the absence of a default initial expression for the discriminant, we can nevertheless declare unconstrained objects of type Person_Name; such objects, of course, take the default initial value **null** and so no problem arises. Thus although an allocated object cannot generally have its discriminants changed, nevertheless an unconstrained access variable could refer from time to time to objects with different discriminants.

The reason for not normally allowing an allocated object to have its discriminants changed is that it could be accessed from several constrained objects such as the components Father and it would be difficult to ensure that such constraints were not violated. Another similar rule is that we cannot have general access subtypes with constraints (as opposed to pool specific types) referring to types that have defaulted discriminants. So we cannot write

```
type Mutant_Name is access all Mutant;
subtype Things_Name is Mutant_Name(Sex => Neuter);      -- illegal
```

where the type Mutant is as in Section 18.3. For examples of the problems that this and other related rules solve see the *Ada 2005 Rationale*.

But there are no problems with Person_Name since the type Person does not have a default for its discriminant. So we can write

```
subtype Mans_Name is Person_Name(Male);
subtype Womans_Name is Person_Name(Female);
```

We can now write a procedure to marry two people.

```
procedure Marry(Bride: not null Womans_Name;
                Groom: not null Mans_Name) is
begin
   if Bride.Husband /= null or Groom.Wife /= null then
      raise Bigamy;
   end if;
   Bride.Husband := Groom;
   Groom.Wife := Bride;
end Marry;
```

The constraints on the parameter subtypes ensure that an attempt to marry people of the same sex will raise Constraint_Error at the point of call. Similarly, the explicit null exclusions on the parameters ensure that an attempt to marry a nonexistent person will also result in Constraint_Error at the point of call. We should perhaps have included the null exclusions in the definitions of the subtypes Mans_Name and Womans_Name.

Remember that although **in** parameters are constants we can change the components of the accessed objects – we are not changing the values of Bride and Groom to access different objects. So this section should really have started with a sequence such as

```
John, Barbara: Person_Name;
...
John := new Person'(Male, (19, Aug, 1937), Edith, others => null);
Barbara := new Person'(Female, (13, May, 1943), Ena, others => null);
...
Marry(Barbara, John);
```

A function could return an access value as for example

```
function Spouse(P: Person_Name) return Person_Name is
begin
   case P.Sex is
      when Male =>
         return P.Wife;
      when Female =>
         return P.Husband;
   end case;
end Spouse;
```

The result of such a function call is treated as a constant object (see the function Rev in Section 10.1) and can be directly used as part of a name so we can write

```
Spouse(P).Birth
```

to give the birthday of the spouse of P. We could even write

```
Spouse(P).Birth := Newdate;
```

but this is only possible because the function delivers an access value. It could not be done if the function actually delivered a value of type Person rather than Person_Name. Moreover, we cannot write Spouse(P) := Q; in an attempt to replace our spouse by someone else. Note also that

```
Spouse(P).all := Q.all;                     -- illegal
```

is also illegal because the type Person is limited and cannot be copied. We can copy the components individually since they are not limited. However, writing .**all** gives a dereference and is not the same as writing out the individual assignments. Of course we cannot change the sex of our spouse anyway.

The following function gives birth to a new child. We need the mother, the sex of the child and the date as parameters.

```
function New_Child(Mother: not null Womans_Name;
                   Boy_Or_Girl: Gender;
                   Birthday: Date) return Person_Name is
  Child: Person_Name;
begin
  if Mother.Husband = null then
    raise Out_Of_Wedlock;
  end if;
  Child := new Person(Boy_Or_Girl);
  Child.Birth := Birthday;
  Child.Father := Mother.Spouse;
  declare
    Last: Person_Name := Mother.First_Child;
  begin
    if Last = null then
      Mother.First_Child := Child;
    else
      while Last.Next_Sibling /= null loop
        Last := Last.Next_Sibling;
      end loop;
      Last.Next_Sibling := Child;
    end if;
  end;
  return Child;
end New_Child;
```

Observe that a discriminant constraint need not be static – the value of Boy_Or_Girl is not known until the function is called. As a consequence we cannot give the complete initial value with the allocator because we do not know which components to provide. Hence we allocate the child with just the value of the discriminant and then separately assign the date of birth and the father. The remaining components take the default value **null**. We finally link the new child onto the end of the chain of siblings which starts from the component First_Child of Mother. Note that a special case arises if the new child is the first born.

We can now write

```
Helen: Person_Name := New_Child(Barbara, Female, (28, Sep, 1969));
```

It is interesting to consider how the family saga could be rewritten using type extension and inheritance rather than discriminants. Clearly we can use an abstract root type Person and derived types Man and Woman. We also need three distinct access types. So

```
type Person is tagged;
type Man is tagged;
type Woman is tagged;
```

```
type Person_Name is access all Person'Class;
type Mans_Name is access all Man;
type Womans_Name is access all Woman;

type Person is abstract tagged limited
   record
      Birth: Date;
      Mother: Womans_Name;
      Next_Sibling: Person_Name;
   end record;

type Man is limited new Person with
   record
      Wife: Womans_Name;
   end record;

type Woman is limited new Person with
   record
      Husband: Mans_Name;
      First_Child: Person_Name;
   end record;
```

In the declaration of Man and Woman, we can omit **limited** since limitedness is always derived from a parent type but it should be stated for clarity (it has to be stated if the parent is an interface – see Section 14.8).

In practice we would undoubtedly declare the various types (in either formulation) as private although keeping the relationships between the types visible. In the tagged case the visible part might become

```
package People is

   type Person is abstract tagged limited private;
   type Man is limited new Person with private;
   type Woman is limited new Person with private;

   type Person_Name is access all Person'Class;
   type Mans_Name is access all Man;
   type Womans_Name is access all Woman;

private
```

We could then declare all the various subprograms Marry, Spouse and so on inside this package or maybe in a child package which of course would have full visibility of the details of the types. We need to take care in starting the system because we cannot assign to the various components externally. The function New_Child cannot be used initially because it needs married parents. However, a sequence such as

```
Adam := new Man;
Eve := new Woman;
Marry(Eve, Adam);
Cain := New_Child(Eve, Male, Long_Ago);
```

seems to work although it leaves some components of Adam and Eve not set properly.

We might consider declaring the component Father in the type Person with a null exclusion on the grounds that everyone has a father and so save some run-time checks. But there is then the problem of declaring Adam and Eve since they must be given a non-null Father otherwise Constraint_Error will be raised.

This is the usual chicken and egg problem. A common programming technique is to make such a first object refer to itself. But this is almost impossible if we insist on using a null exclusion. For example we could not write

```
First_Man: Person_Name := new Person'
                     (Male, Some_Date, First_Man, others => <>);
```

because we cannot refer to an object in its own declaration. One way around it is to introduce a third variant which does not have a component Father and then the first person can be of this kind but that seems very artificial. Another way is to make every person by default have their father as themselves. This can be done by a dirty trick (that is, an interesting technique) as explained in Section 18.7.

We leave to the reader the task of writing the various subprograms using the tagged formulation. One advantage is that more checking is done at compile time because Man and Woman are distinct types rather than just subtypes of Person.

Note that we did not use anonymous access types in the above. This is because there would have been a forward reference between the type Person and the component Father. We could have used anonymous access types in the discriminant model but for ease of comparison they have both been treated the same way. Moreover, for either model, it seems very appropriate for this application to talk explicitly about names as well as persons. Observe also that we made the access types general (with **all**); this is so that we can convert between them – remember that conversion is not allowed between pool specific access types.

This brings us finally to the rules regarding type conversion between different general access types referring to the same discriminated record type. Conversion is permitted provided the accessed subtypes statically match (the usual rule) but might raise Constraint_Error if the target subtype is constrained.

Exercise 18.5

1 Write a function to return a person's heir (use the variant formulation). Follow the historical rules of primogeniture applicable to monarchies – the heir is the eldest son if there is one and otherwise is the eldest daughter. Return **null** if there is no heir.

2 Write a procedure to enable a woman to get a divorce. Divorce is only permitted if there are no children.

3 Modify the procedure Marry in order to prevent incest. A person may not marry their sibling, parent, or child.

4 Rewrite Marry using the tagged type formulation; this is trivial.

5 Rewrite Spouse using the tagged type formulation. Hint: consider dispatching.

6 Rewrite New_Child using the tagged type formulation.

18.6 Private types and discriminants

In this section we bring together a number of matters relating to the control of resources through private types and discriminants.

In earlier sections we have seen how private types provide general control by ensuring that all operations on a type are through defined subprograms. Making a type also limited prevents assignment.

The introduction of discriminants adds other possibilities. We have seen that a partial view might show the discriminants as in the type Stack of Section 18.1

```
package Stacks is
   type Stack(Max: Natural) is private;
   ...
private
   type Stack(Max: Natural) is ...
```

and also that the partial view might hide them as with

```
package Polynomials is
   type Polynomial is private;
   ...
private
   type Polynomial(N: Index := 0) is ...
```

of Section 18.2 even though the full type has a discriminant; we noted that the full type has to be definite and so the discriminant has to have a default in this case.

Another possibility is that the partial view might have discriminants and then the full type could be implemented in terms of an existing type with discriminants thus

```
package Squares is
   type Square(N: Positive) is private;
   ...
private
   type Square(N: Positive) is new Rectangle(N, N);
```

using the examples of Section 18.4.

The final possibility is that the partial view might have unknown discriminants written

```
type T(<>) is private;
```

This view is considered indefinite and prevents the user from declaring uninitialized objects of the type; the partial view might also be tagged. If the writer of the package does not provide any means of initializing objects (through a function or deferred constant) then the user cannot declare objects at all.

Making the partial view limited (again it might also be tagged) by writing

```
type T(<>) is limited private;
```

prevents the user from copying objects as usual.

These forms enable the provider of the package to have complete control over all objects of the type; typically the user would be given access to objects of the type through an access type which is then often called a handle. We will consider some examples of controlling the privacy of abstractions in this way in Section 21.7.

If the partial view has unknown discriminants then the full view may or may not have discriminants; none were promised and so none need be provided. The full type can also be another type with unknown discriminants (such as being derived from one).

A class wide type is also treated as having unknown discriminants (the tag is a hidden discriminant). Remember that a class wide type is another example of an indefinite type so that uninitialized objects are not allowed.

An interesting situation occurs when the full type is mutable but the partial view does not reveal the discriminants. Consider

```
package Beings is
   type Mutant is private;
   type Mutant_Name is access Mutant;
   F, M: constant Mutant;
private
   type Mutant(Sex: Gender := Neuter) is
      record
         ...                          -- as in Section 18.3
      end record;

   F: constant Mutant := (Female, ... );
   M: constant Mutant := (Male, ... );
end Beings;
```

The user can now write

```
Chris: Mutant_Name := new Mutant'(F);     -- it's a girl, a copy of F
...
Chris.all := M;                           -- OK?  Yes! it's a boy
```

Note that we have now changed the sex of the allocated object referred to by Chris. This has to be allowed because the external view does not show that the constants M and F are different internally. This is the one situation referred to in Section 18.5 where an allocated object can have its discriminant changed.

We conclude this section with a curious example of discriminants by reconsidering the type Key of Section 12.6. We could change this to

```
type Key(Code: Natural := 0) is limited private;
```

with full type

```
type Key(Code: Natural := 0) is null record;
```

With this formulation the user can read the code number of the key, but cannot change it. There is, however, a small flaw whose detection and cure is left as an exercise. Note also that we have declared Code as of subtype Natural rather than of type Key_Code; this is because Key_Code is not visible to the user. Of course we could make Key_Code visible but this would make Max visible as well and we might not want the user to know how many keys there are.

However, as observed in Exercise 14.7(**2**), it is far better to make the type tagged and controlled so that it can be initialized and finalized properly. The inquisitive user can always be given read access to the value of the code through a function.

Exercise 18.6

1 What is the flaw in the suggested new formulation for the type Key? Hint: remember that the user declares keys explicitly. Show how it can be overcome.

18.7 Access discriminants

A discriminant may also be of an access type. This enables a record (or task or protected object) to be parameterized with some other structure with which it is associated. The access type can be a named access type or it can be anonymous in which case the discriminant is known as an access discriminant. In both cases a default value might be provided.

Discriminants of a named access type are not particularly interesting; they behave much as other components of a record except that they can be visible even though the rest of the record might be private; but they have no controlling function in the way that discrete discriminants can control array bounds and variants.

Access discriminants have similar accessibility properties to access parameters – for example, they can refer to local variables. This gives extra flexibility and so most discriminants of access types are in fact access discriminants.

A typical structure might be

```
type Data is ...

type R(D: access Data) is
   record
      ...
   end record;
```

and then a declaration of an object of type R must include an access value to an associated object of type Data. Thus

```
The_Data: aliased Data := ... ;
The_Record: R(The_Data'Access, ... );
```

An important point is that the two objects are bound together permanently. On the one hand, if the discriminant does not have a default value then it cannot be changed anyway. On the other hand, if it does have a default value then the type has to be explicitly marked as **limited**. This means that since a discriminant can only be changed by a whole record assignment and assignment is forbidden for limited types, once more it cannot be changed. Thus either way the two objects are bound together permanently. See also Section 22.4 for coextensions.

Within the type R, the discriminant D could be used as a constraint for an inner component or as the default initial value of an inner component.

An interesting special case is where the object having the discriminant is actually a component of the type the discriminant refers to! This can be used to enable a component of a record to obtain the identity of the record in which it is embedded. Consider

```
type Inner(Ptr: access Outer) is limited ...

type Outer is limited
  record
    ...
    Component: Inner(Outer'Access);
    ...
  end record;
```

The Component of type Inner has an access discriminant Ptr which refers back to the instance of the record Outer. This is because the attribute Access applied to the name of a record type inside its declaration refers to the current instance. (When we deal with tasks we will see that a similar situation arises when the name of a task type is used inside its own body; see Section 22.1.) If we now declare an object of the type Outer

```
Obj: Outer;
```

then the structure created is as shown in Figure 18.1. We call it a self-referential structure for obvious reasons. All objects of the type Outer will refer to themselves.

An important example of the use of this feature will be found in Section 22.2 where a component of a record is a task with a discriminant.

Incidentally, we can also write

```
type Cell is tagged limited
  record
    Next: access Cell := Cell'Unchecked_Access;
    ...
  end record;
```

In this case whenever an object of the type Cell is created without an explicit initial value (whether by declaration or by an allocator) the component Next refers to the new object itself. Such a mechanism might be useful in automatically providing a dummy header cell in some linked list applications as for example when we were trying to make Father have a null exclusion in the family saga of Section 18.5. Sadly, the accessibility rules require us to use Unchecked_Access in this case.

Access discriminants work together with access parameters and thereby minimize accessibility problems. Suppose we had some general procedure to manipulate the type Data

```
procedure P(A: access Data);
```

then we can make calls such as in

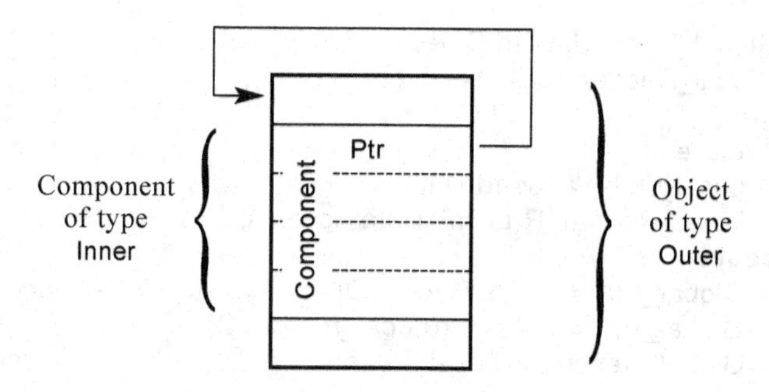

Figure 18.1 A self-referential structure.

```
declare
   My_Data: aliased Data := ...
   My_Record: R(My_Data'Access);
begin
   ...
   P(My_Record.D);                    -- legal
   ...
```

and the accessibility of the data is passed via the access parameter of P. On the other hand, using a named access type as the parameter of P thus

```
type Data_Ptr is access all Data;
procedure P(A: Data_Ptr);
```

fails because the type conversion required in the corresponding call

```
P(Data_Ptr(My_Record.D));             -- illegal
```

breaks the static accessibility rules of Section 11.5.

Examples of using access parameters with access discriminants will be found in Sections 21.5 and 22.4 and also in Section 21.4a, Multiple Views, on the website.

Exercise 18.7

1 Analyse the following and explain why the various attempts to assign the reference to the local data to the more global variable or component are thwarted.

```
procedure Main is
   type Data is ...
   type Data_Ptr is access all Data;
   type R(D: access Data) is limited
      record
         ...
      end record;
   Global_Ptr: Data_Ptr;
```

```
      Global_Data: aliased Data;
      Global_Record: R(Global_Data'Access);
   begin
      declare
         Local_Data: aliased Data;
         Local_Record: R(Local_Data'Access);
      begin
         Global_Ptr := Local_Record.D;                  -- incorrect
         Global_Ptr := Data_Ptr(Local_Record.D);        -- incorrect
         Global_Record := Local_Record;                 -- incorrect
      end;
   end Main;
```

Checklist 18

If a discriminant does not have a default expression then all objects must be constrained.

The discriminant of an unconstrained object can only be changed by a complete record assignment.

Discriminants can only be used as array bounds or to govern variants or as nested discriminants or in default initial expressions for components.

A discriminant in an aggregate and governing a variant must be static.

Any variant must appear last in a component list.

If an accessed object has a discriminant then it is always constrained.

Discriminants of tagged types can only have defaults if limited.

On derivation, discriminants are all inherited or completely replaced with a new set.

A record type with access discriminants must be explicitly limited or not have defaults for the discriminants.

Access discriminants have dynamic accessibility.

General access types to a type with default discriminants were permitted in Ada 95 (and caused awkward problems).

Aliased objects are not automatically constrained in Ada 2005. Allocated objects are also not constrained in Ada 2005 if there is a partial view without discriminants.

Tagged record types can have default discriminants if the record type is limited.

New in Ada 2022

The discriminant in an aggregate need not be static in some circumstances, see AI12-86 and also AI05-220.

Discriminants can have aspects but none are defined by the language.

19 Generics

19.1 Declarations and instantiations
19.2 Type parameters
19.3 Subprogram parameters
19.4 Package parameters
19.5 Generic library units

In this chapter we describe the generic mechanism which allows a special form of parameterization at compile time which can be applied to subprograms and packages. The generic parameters can be types of various categories (access types, limited types, tagged types and so on), subprograms and packages as well as values and objects.

Genericity is important for reuse. It provides static polymorphism as opposed to the dynamic polymorphism provided by type extension and class wide types. Being static it is intrinsically more reliable but usually less flexible. However, in the case of subprogram parameters it does not have the convention restrictions imposed by access to subprogram types and thus is particularly useful where the parameter is an operation.

Package parameters enable the composition of generic packages while ensuring that their instantiations are compatible. They can also be used to group together the parameters of other generic units to provide signatures.

19.1 Declarations and instantiations

We often get the situation that the logic of a piece of program is independent of the types involved and it therefore seems unnecessary to repeat it for all the different types to which we might wish it to apply. A simple example is provided by the procedure Swap of Exercise 10.3(1)

```
procedure Swap(X, Y: in out Float) is
   T: Float;
begin
   T := X;  X := Y;  Y := T;
end;
```

It is clear that the logic is independent of the type Float. If we also wanted to swap integers or dates we could write other similar procedures but this would be tedious. The generic mechanism allows us to avoid this. We declare

```
generic
   type Item is private;
procedure Exchange(X, Y: in out Item);

procedure Exchange(X, Y: in out Item) is
   T: Item;
begin
   T := X;  X := Y;  Y := T;
end;
```

The subprogram Exchange is a generic subprogram and acts as a kind of template. The subprogram specification is preceded by the generic formal part consisting of the reserved word **generic** followed by a (possibly empty) list of generic formal parameters. The subprogram body is written exactly as normal but note that, in the case of a generic subprogram, we always have to give both the specification and the body separately.

A generic procedure is not a real procedure and so cannot be called; but from a generic procedure we can create an actual procedure by a mechanism known as generic instantiation. For example, we may write

```
procedure Swap is new Exchange(Float);
```

This is a declaration and states that Swap is obtained from the template given by Exchange. Actual generic parameters are provided in a parameter list in the usual way. The actual parameter in this case is the type Float corresponding to the formal parameter Item. We could also use the named notation

```
procedure Swap is new Exchange(Item => Float);
```

So we have now created the procedure Swap acting on type Float and can henceforth call it in the usual way. We can make further instantiations

```
procedure Swap is new Exchange(Integer);
procedure Swap is new Exchange(Date);
```

and so on. We are here creating further overloadings of Swap which can be distinguished by their parameter types just as if we had laboriously written them out in detail.

Superficially, it may look as if the generic mechanism is merely one of text substitution and indeed in simple cases the behaviour would be the same. However, an important difference relates to the meaning of identifiers in the generic body but which are neither parameters nor local to the body. Such nonlocal identifiers have meanings appropriate to where the generic body is declared and not to where it is instantiated. If text substitution were used then nonlocal identifiers would of course take their meaning at the point of instantiation and this could give very surprising results.

We may also have generic packages. A simple example is provided by the package Stack in Section 12.1. The trouble with that package is that it only works on type Integer although of course the same logic applies irrespective of the type of the values manipulated. We can also make Max a parameter and give it a default value of 100 for convenience. We write

```
generic
   Max: Positive := 100;                    -- default 100
   type Item is private;
package Stack is
   procedure Push(X: in Item);
   function Pop return Item;
end Stack;

package body Stack is
   S: array (1 .. Max) of Item;
   Top: Integer range 0 .. Max;
   ...   -- etc. as before but with Integer replaced by Item
end Stack;
```

We can now create and use a stack of a particular size and type by instantiating the generic package as in the following

```
declare
   package My_Stack is new Stack(50, Float);
   use My_Stack;
begin
   ...
   Push(X);
   ...
   Y := Pop;
   ...
end;
```

The package My_Stack which results from the instantiation behaves just as a normal directly written out package. The use clause allows us to refer to Push and Pop directly. If we did a further instantiation perhaps using the default for Max (and remembering to used named notation because the first parameter is by default)

```
package Another_Stack is new Stack(Item => Integer);
use Another_Stack;
```

then Push and Pop are further overloadings but can be distinguished by the parameter and result type. Of course, if Another_Stack were also declared with the actual generic parameter being Float, then we would have to use the dotted notation to distinguish the instances of Push and Pop despite the use clauses.

Both generic units and generic instantiations may be library units. Thus, having compiled the generic package Stack, an instantiation could itself be separately compiled just on its own thus

```
with Stack;
package Boolean_Stack is new Stack(200, Boolean);
```

If we added an exception Error to the package as in Section 15.2 so that the generic package declaration was

```
generic
   Max: Positive := 100;
   type Item is private;
package Stack is
   Error: exception;
   procedure Push(X: Item);
   function Pop return Item;
end Stack;
```

then each instantiation would give rise to a distinct exception and because exceptions cannot be overloaded we would naturally have to use the dotted notation to distinguish them.

We could, of course, make the exception Error common to all instantiations by making it global to the generic package. It and the generic package could perhaps be declared inside a further package

```
package All_Stacks is
   Error: exception;
   generic
      Max: Positive := 100;
      type Item is private;
   package Stack is
      procedure Push(X: Item);
      function Pop return Item;
   end Stack;
end All_Stacks;

package body All_Stacks is
   package body Stack is
      ...
   end Stack;
end All_Stacks;
```

This illustrates the binding of identifiers global to generic units. The meaning of Error is determined at the point of the generic declaration irrespective of the meaning at the point of instantiation.

The above examples have illustrated formal parameters which were types and also integers. In fact generic formal parameters can be values and objects much as the parameters applicable to subprograms; they can also be types, subprograms, and packages. As we shall see in the next sections, we can express the formal types, subprograms, and packages so that we can assume in the generic body that the actual parameters have the properties we require.

Object parameters can be of mode **in** or **in out** but not **out**. As in the case of parameters of subprograms, **in** is taken by default as illustrated by Max in the example above. Explicit constraints are not permitted but null exclusions are.

An **in** generic parameter acts as a constant whose value is provided by the corresponding actual parameter. A default expression is allowed as in the case of parameters of subprograms; such a default expression is evaluated at instantiation if no actual parameter is supplied in the same way that a default expression for a subprogram parameter is evaluated when the subprogram is called if no actual parameter is supplied. Observe that an **in** generic parameter can be of a limited type; the actual parameter might then be an aggregate or constructor function call as explained in Section 12.5.

An **in out** parameter, however, acts as a variable renaming the corresponding actual parameter. The actual parameter must therefore be the name of a variable and its identification occurs at the point of instantiation using the same rules as for renaming described in Section 13.7. One such rule is that constraints on the actual parameter apply to the formal parameter and any constraints implied by the formal subtype mark are, perhaps surprisingly, completely ignored. Another rule is that if the value of any identifier which is part of the name subsequently changes then the identity of the object referred to by the generic formal parameter does not change. Because of this there is a restriction, like that on renaming, that the actual parameter cannot be a component of an unconstrained discriminated record if the very existence of the component depends on the value of the discriminant. Thus if The_Thing is a Mutant as in Section 18.3, The_Thing.Bearded could not be an actual generic parameter because The_Thing could have its Sex changed. However, The_Thing.Birth would be valid.

It will now be realized that although the notation **in** and **in out** is identical to subprogram parameters the meaning is somewhat different. Thus there is no question of copying in and out and indeed no such thing as **out** parameters.

Anonymous access types (both access to object and access to subprogram types) are also permitted for the type of the formal parameter so we can have parameters such as

```
generic
   A: access T := null;
   AN: in out not null access constant T;
   F: access function (X: Float) return Float;
   FN: not null access function (X: Float) return Float;
```

As usual, explicit null exclusions are permitted whereas explicit constraints are not. If the subtype of the formal object excludes null (as in AN and FN) then the actual must also exclude null but the reverse does not hold. Unlike access parameters of subprograms the mode can be **in** or **in out** as for any generic object parameter. Default expressions are permitted as illustrated for A and **constant** is permitted as illustrated for AN. The obvious matching rules apply such as subtype conformance in the case of F and FN.

Inside the generic body, the formal generic parameters can generally be used quite freely – but there is one important restriction. This arises because generic parameters (and their attributes) are not generally considered to be static. There are various places where an expression has to be static such as in the alternatives in a case statement or variant, and in the range in an integer type definition, or the number of digits in a floating point type definition and so on. In all these situations a generic formal parameter cannot be used because the expression would not then be static.

However, the type of the expression in a case statement and similarly the type of the discriminant in a variant may be a generic formal type provided there is an others clause – this ensures that all values are covered.

But a generic parameter is treated as static in both the visible and private part of a generic package if the actual parameter is static.

Our final example in this section illustrates the nesting of generics. The following generic procedure performs a cyclic interchange of three values and for amusement is written in terms of the generic procedure Exchange

```
generic
   type Thing is private;
procedure CAB(A, B, C: in out Thing);

procedure CAB(A, B, C: in out Thing) is
   procedure Swap is new Exchange(Item => Thing);
begin
   Swap(A, B);
   Swap(A, C);
end CAB;
```

Although nesting is allowed, it must not be recursive.

It is important to appreciate that generic subprograms are not subprograms and cannot be overloaded with subprograms or with each other. One consequence is that generic subprogram renaming does not require the parameter list to be repeated. So we can simply write

```
generic procedure Taxi renames CAB;
```

although perhaps confusing for this particular example!

But instantiations of generic subprograms are real subprograms and in the appropriate circumstances can be primitive operations. Thus if we wrote

```
package Dates is
   type Date is record ... end record;
   procedure Swap is new Exchange(Date);
```

then Swap would be a primitive operation of Date. We might even have

```
overriding
procedure Swap is new Exchange(Thing);
```

which means that Swap must be an overriding operation for the type Thing.

Another point is that generic subprograms can have preconditions and postconditions. But instantiations of generic subprograms cannot have preconditions and postconditions.

Finally, remember that the same rules for mixing named and positional notation apply to generic instantiation as to subprogram calls. Hence if a parameter is omitted, subsequent parameters must be given using named notation. Of course, a generic unit need have no parameters in which case the instantiation takes the same form as for a subprogram call – the parentheses are omitted.

Exercise 19.1

1 Write a generic package based on the package Stacks in Section 18.1 so that stacks of any type may be declared. Then declare a stack S of length 30 and type Boolean. Use named notation.

2 Write a generic package containing both Swap and CAB.

19.2 Type parameters

In the previous section we introduced types as generic parameters. The examples showed the formal parameter taking the form

```
type T is private;
```

In this case, inside the generic subprogram or package, we may assume that assignment and equality are defined for T and that T is definite so that we may declare uninitialized objects of type T. We can assume nothing else unless we specifically provide other parameters as we shall see in a moment. Hence T behaves in the generic unit much as a private type outside the package defining it; this analogy explains the notation for the formal parameter. The corresponding actual parameter must, of course, provide assignment and equality and so it can be any definite type except one that is limited. Note carefully that it cannot be an indefinite type such as String.

A formal generic type parameter can take many other forms. It can be

```
type T is limited private;
```

and in this case assignment and predefined equality are not available. The corresponding actual parameter can be any definite type, limited or nonlimited, tagged or untagged.

Either of the above forms could have unknown discriminants such as

```
type T(<>) is private;
```

in which case the type is considered indefinite within the generic unit and so uninitialized objects cannot be declared. The actual type could then be any nonlimited indefinite type such as String as well as any nonlimited definite type. It could also be a class wide type since these are considered to be indefinite.

Another possibility is that the formal type could have known discriminants such as

```
type T(X: U; Y: V; ... ) is private;
```

and the actual type must then have discriminants with statically matching subtypes. The formal type must not have default expressions for the discriminants but the actual type can. The formal type is therefore indefinite.

We can also require that the actual type be tagged by one of

```
type T is tagged private;
type T is abstract tagged private;
```

In the first case the actual type must be tagged but not abstract. The second case allows the actual type to be abstract as well and this includes the possibility of the actual type being an interface.

All the above forms can be put together in the obvious way, so an extreme example might be

```
type T(I: Index) is abstract tagged limited private;
```

in which case the actual type must have one discriminant statically matching the subtype Index, it must be tagged, it might be abstract, it might be limited.

Another possibility concerns incomplete types. Thus we might have the forms

```
type T;
type T is tagged;
```

and the actual type can then be an incomplete type. If the formal specifies **tagged** then the actual type must be tagged; if the formal does not specify **tagged** then the actual type might or might not be tagged. Discriminants can also be given and must match. Note moreover that an incomplete formal type can also be matched by an appropriate complete type.

The matching rules are expressed in terms of categories as outlined in Section 6.8. A category of types is just a set of types with common properties. The formal parameter defines the category and the actual can then be any type in that category.

Derivation classes are important categories and the forms

```
type T is new S;
type T is new S with private;
```

require that the actual type must be derived from S. The first form applies if S is not tagged and the second if S is tagged. We can also insert **abstract** after **is** in the second case. Within the generic unit all properties of the derivation class may be assumed. Unknown discriminants and limited can be added to both forms. (But known discriminants are not permitted with either of these forms.)

Most of the matching rules follow automatically from the derivation model but there are a number of additional rules. For example if S is an access type without a null exclusion then the actual type must not have a null exclusion.

A very important form of tagged type is an interface and these can also be used as generic parameters. Thus we might have

```
type F is interface;
```

The actual type could then be any interface. This is perhaps unlikely.

If we wanted to ensure that a formal interface had certain operations then we might first declare an interface A with the required operations

```
type A is interface;
procedure Op1(X: A; ... ) is abstract;
procedure N1(X: A; ... ) is null;
```

and then the generic formal parameter would be

```
type F is interface and A;
```

and then the actual interface must be descended from A and so have operations which match Op1 and N1.

A formal interface might specify several ancestors

```
type FAB is interface and A and B;
```

where A and B are themselves interfaces. And A and B or just some of them might themselves be further formal parameters as in

```
generic
   type A is interface;
   type FAB is interface and A and B;
```

This means that FAB must have both A and B as ancestors; it could of course have other ancestors as well. Note that this illustrates that one generic formal parameter can depend upon a previous formal parameter that is a type.

Formal tagged types might also be descended from interfaces. Thus we might have

```
generic
   type NT is new T and A and B with private;
```

in which case the actual type must be descended from the tagged type T and also from the interfaces A and B. The parent type T itself might be an interface or a normal tagged type. Again some or all of T, A, and B might be earlier formal parameters. Also we can explicitly state **limited** in which case all of the ancestor types must also be limited.

As we shall see in Chapter 22, interfaces can also be restricted to synchronized types such as task types and protected types and special forms of generic formal parameter apply in these cases.

The formal parameter could also be one of

```
type T is (<>);
type T is range <>;
type T is mod <>;
type T is digits <>;
type T is delta <>;
type T is delta <> digits <>;
```

In the first case the actual parameter must be a discrete type – an enumeration type or integer type. In the other cases the actual parameter must be a signed integer type, modular type, floating point type, ordinary fixed point type, or decimal type respectively. Within the generic unit the appropriate predefined operations and attributes are available.

As a simple example consider

```
generic
   type T is (<>);
function Next(X: T) return T;
```

```
function Next(X: T) return T is
begin
   if X = T'Last then
      return T'First;
   else
      return T'Succ(X);
   end if;
end Next;
```

The formal parameter T requires that the actual parameter must be a discrete type. Since all discrete types have attributes First, Last, and Succ we can use these attributes in the body in the knowledge that the actual parameter will supply them. However, if a generic body uses First, Last, or Range on a formal type and the actual type has a subtype predicate then Program_Error is raised. On the other hand the attributes First_Valid and Last_Valid can always be used; if the actual subtype does not have a subtype predicate then these attributes are equivalent to First and Last.

We could now write

```
function Tomorrow is new Next(Day);
```

so that Tomorrow(Sun) = Mon.

An actual generic parameter can also be a subtype but an explicit constraint or null exclusion is not allowed. The formal generic parameter then denotes the subtype. Thus we can have

```
function Next_Work_Day is new Next(Weekday);
```

so that Next_Work_Day(Fri) = Mon. Note how the behaviour depends on the fact that the Last attribute applies to the subtype and not to the base type so that Day'Last is Sun and Weekday'Last is Fri.

The actual parameter could also be an integer type so we could have

```
subtype Digit is Integer range 0 .. 9;
function Next_Digit is new Next(Digit);
```

and then Next_Digit(9) = 0.

Now consider the package Complex_Numbers of Section 12.2; this could be made generic so that the particular floating point type upon which the type Complex is based can be a parameter. It would then take the form

```
generic
   type Floating is digits <>;
package Generic_Complex_Numbers is
   type Complex is private;
   ...   -- as before with Float replaced by Floating
   I: constant Complex := (0.0, 1.0);
end;
```

Note that we can use the literals 0.0 and 1.0 because they are of the type *universal_real* which can be converted to whatever type is passed as actual parameter. The package could then be instantiated by for instance

```
package My_Complex_Numbers is
        new Generic_Complex_Numbers(My_Float);
```

A formal generic parameter can also be an array type. The actual parameter must then also be an array type with the same number of dimensions and statically matching index subtypes and component subtypes. The subcomponents could be marked as aliased but this must apply to either both the formal and actual type or neither of them. Similarly, either both must be unconstrained arrays or both must be constrained arrays. If constrained then the index ranges must statically match. Note that unlike ordinary array type declarations, generic formal arrays cannot have explicit constraints or null exclusions in the index subtypes and component subtypes in all cases.

As noted above it is possible for one generic formal parameter to depend upon a previous formal parameter which is a type. This will often be the case with arrays. As an example consider the function Sum in Section 10.1. This added together the components of an array of component type Float and index type Integer. We can generalize this to add together the components of any floating point array with any index type

```
generic
   type Index is (<>);
   type Floating is digits <>;
   type Vec is array (Index range <>) of Floating;
function Sum(A: Vec) return Floating;

function Sum(A: Vec) return Floating is
   Result: Floating := 0.0;
begin
   for I in A'Range loop
      Result := Result + A(I);
   end loop;
   return Result;
end Sum;
```

Note that although Index is a formal parameter it does not explicitly appear in the generic body; nevertheless it is implicitly used since the loop parameter I is of type Index.

We could instantiate this by

```
function Sum_Vector is new Sum(Integer, Float, Vector);
```

and this will give the function Sum of Section 10.1.

The matching of actual and formal arrays takes place after any formal types have been replaced in the formal array by the corresponding actual types. As an example of matching index subtypes note that if we had

```
type Vector is array (Positive range <>) of Float;
```

then we would have to use Positive (or an equivalent subtype) as the actual parameter for the Index.

The final possibility for formal type parameters is the case of an access type. The formal can be

```
type A is access T;
type A is access constant T;
type A is access all T;
```

where T may but need not be a previous formal parameter. The actual parameter corresponding to A must then be an access type with accessed type T. In the first case the actual type can be any such access type other than an access to constant type. In the second case it must be an access to constant type and in the third it must be a general access to variable type. Constraints on the accessed type must statically match. The formal type can also have a null exclusion thus

```
type A is not null access T;
```

in which case the actual type must also have a null exclusion and vice versa.

A formal access to subprogram type takes one of the forms

```
type P is access procedure ...
type F is access function ...
```

and the profiles of actual and formal subprograms must have mode conformance. This is the same conformance as applies to the renaming of subprogram specifications and so follows the model that generic parameter matching is like renaming. Null exclusions are again allowed.

Observe that there is no concept of a formal record type; a similar effect can be achieved by the use of a formal derived type.

Incidentally, we have noted that a numeric type such as

```
type My_Integer is range -1E6 .. 1E6;
```

is implemented as a hardware type that might correspond to one of the predefined types such as Integer or Long_Integer. However, it is not actually derived from one of these although ultimately all are conceptually derived from *root_integer*. So if we had

```
generic
   type T is new Integer;
```

then this could never be matched by My_Integer. Another thought is that **type** T **is range** <>; can be considered as equivalent to **type** T **is new** *root_integer*; which we cannot write.

Another typical example is provided by considering operations on sets. We saw in Section 8.6 how a Boolean array could be used to represent a set. Exercises 10.1(**4**), 10.2(**3**) and 10.2(**4**) also showed how we could write suitable functions to operate upon sets of the type Colour. The generic mechanism allows us to write a package for manipulating sets of an arbitrary type.

Consider

```
generic
   type Element is (<>);
package Set_Of is
   type Set is private;
   type List is array (Positive range <>) of Element;

   Empty, Full: constant Set;

   function Make_Set(L: List) return Set;
   function Make_Set(E: Element) return Set;
   function Decompose(S: Set) return List;

   function "+" (S, T: Set) return Set;         -- union
   function "*" (S, T: Set) return Set;         -- intersection
   function "-" (S, T: Set) return Set;         -- symmetric difference

   function "<" (E: Element; S: Set) return Boolean;    -- inclusion
   function "<=" (S, T: Set) return Boolean;            -- contains

   function Size(S: Set) return Natural;        -- no of elements
private
   type Set is array (Element) of Boolean;

   Empty: constant Set := (Set'Range => False);
   Full: constant Set := (Set'Range => True);
end;
```

The single generic parameter is the element type which must be discrete. The type Set is private so that the Boolean operations cannot be applied directly (inadvertently or malevolently). Aggregates of the type List are used to represent literal sets. The constants Empty and Full denote the empty and full set respectively. The functions Make_Set enable the creation of a set from a list of the element values or a single element value. Decompose turns a set back into a list of elements.

The operators +, * and – represent union, intersection and symmetric difference; they are chosen as more natural than the underlying **or**, **and** and **xor** which can be used to implement them. The operator < tests to see whether an element value is in a set. The operator <= tests to see whether one set is a subset of another. Finally, the function Size returns the number of element values present in a particular set.

In the private part the type Set is declared as a Boolean array indexed by the element type (which is why the element type had to be discrete). The constants Empty and Full are declared as arrays whose elements are all False and all True respectively. The body of the package is left as an exercise.

We can instantiate the package to work on the type Primary of Section 8.6 by

```
package Primary_Sets is new Set_Of(Primary);
use Primary_Sets;
```

For comparison we could then write

```
subtype Colour is Set;
White: Colour renames Empty;
Black: Colour renames Full;
```

and so on.

We can use this example to illustrate the difference between the rules for the template (the generic text as written) and an instance (the effective text after instantiation).

The first point is that the generic package is not a genuine package and in particular does not export anything. So no meaning can be attached to Set_Of.List outside the generic package and nor can Set_Of appear in a use clause. Of course, inside the generic package we could indeed write Set_Of.List if we wished to be pedantic or had hidden List by an inner redeclaration.

If we now instantiate the generic package thus

```
package Character_Set is new Set_Of(Character);
```

then Character_Set is a genuine package and so we can refer to Character_Set.List outside the package and Character_Set can appear in a use clause. In this case there is no question of writing Character_Set.List *inside* the package because the inside text is quite ethereal.

Another very important point concerns the properties of an identifier such as List. Inside the generic template we can only use the properties common to all possible actual parameters as expressed by the formal parameter notation. Outside we can additionally use the properties of the particular instantiation. So, inside we cannot write

```
S: List := "String";
```

because we do not know that the actual type is going to be a character type – it could be an integer type. However, outside we can indeed write

```
S: Character_Set.List := "String";
```

because we know full well that the actual type is, in this instance, a character type.

We can also use this example to explore the creation and composition of types. Our attempt to give the type Set the name Colour through a subtype is not ideal because the old name could still be used. We would really like to pass the name Colour in some way to the generic package as the name to be used for the type. We cannot do this and retain the private nature of the type. But we can use the derived type mechanism to create a proper type Colour from the type Set

```
type Colour is new Set;
```

Recalling the rules for inheriting primitive subprograms from Section 12.3, we note that the new type Colour automatically inherits all the functions in the specification of Set_Of (strictly the instantiation Primary_Sets) because they all have the type Set as a parameter or result type.

However, this is a bit untidy; the constants Empty and Full will not have been inherited and the type List will still be as before.

One improvement therefore is to replace the constants Empty and Full by equivalent parameterless functions so that they will also be inherited. A better approach to the type List is to make it and its index type into further generic parameters.

The visible part of the package will then just consist of the type Set and its subprograms

```
generic
   type Element is (<>);
   type Index is (<>);
   type List is array (Index range <>) of Element;
package Nice_Set_Of is
   type Set is private;
   function Empty return Set;
   function Full return Set;
   ...
private
```

We can now write

```
type Primary_List is array (Positive range <>) of Primary;
package Primary_Sets is new Nice_Set_Of(Element => Primary,
                                        Index => Positive,
                                        List => Primary_List);
type Colour is new Primary_Sets.Set;
```

The type Colour now has all the functions we want and the array type has a name of our choosing. We might still want to rename Empty and Full thus

```
function White return Colour renames Empty;
```

or we can still declare White as a constant by

```
White: constant Colour := Empty;
```

As a general rule it is better to use derived types rather than subtypes because of the greater type checking provided during compilation; sometimes, however, derived types introduce a need for lots of explicit type conversions which clutter the program, in which case the formal distinction is probably a mistake and one might as well use subtypes.

As a final example in this section, consider the following generic form of a package to manipulate stacks (see Section 12.5)

```
generic
   type Item (<>) is private;
package Stacks is
   type Stack is limited private;
   procedure Push(S: in out Stack; X: in Item);
   procedure Pop(S: in out Stack; X: out Item);
   function "=" (S,T: Stack) return Boolean;
private ...
```

The key point to note is that we have chosen to use the form of generic parameter with unknown discriminants. The actual type can then be indefinite such as a class wide type or the unconstrained array type String. This enables us to manipulate a stack of strings of different sizes such as the names of the animals. However, since the actual type can be indefinite this means that the generic body cannot declare objects of the type Item (without an initial value) or arrays of the type Item. The implementation therefore has to use allocated objects to hold the values of the type Item. The access values could themselves be held in an array in the traditional manner as in Section 12.4, thus

```
type Access_Item_Array is array (Integer range <>) of access Item;

type Stack is
  record
    S: Access_Item_Array(1 .. Max);
    Top: Integer range 0 .. Max := 0;
  end record;
```

where Max could well be another generic parameter. Alternatively the individual access values could be linked together as in Section 12.5. Either way the type has to be limited private and care is needed in deallocating unneeded objects. See also the discussion in Section 21.4.

Note that the attribute Definite can be applied to an indefinite formal type such as Item and gives a Boolean value indicating whether the actual type is definite or not. We could therefore use alternative algorithms according to the nature of the actual parameter.

Default type parameters are introduced in Ada 2022. For example

```
type T is (<>) or use Weekday;
```

requires the actual to be a discrete type with the default being the enumeration subtype Weekday. Note the **or use** in this context. Other forms of type parameter are treated similarly with obvious rules regarding compatibility.

We conclude by summarizing the general principle regarding the matching of actual to formal generic types which should now be clear. The formal type represents a category of types which have certain common properties and these properties can be assumed in the generic unit. The corresponding actual type must then supply these properties. The matching rules are designed so that this is assured by reference to the parameters only and without considering the details of the generic body. As a consequence the user of the generic unit need not see the body for debugging purposes. This notion of matching guaranteed by the parameters is termed the contract model.

Some properties cannot be passed via the parameters and have to be checked upon instantiation. The general principle is to assume the best in the specification and then check it at instantiation but to assume the worst in the body so that it cannot go wrong. The situations where this happens are mostly obscure and in some cases the offending stuff can be moved to the private part. For example if the formal type is limited then the actual need not be. The body will assume the worst (that it is limited), but in the private part it will be viewed as limited or not according to the actual parameter. Similarly, remember that the use of generic formal parameters is restricted by the rule that they are not static in the body although they will be static in the specification if the actual parameter is static.

Another point is that accessibility checks in a generic body are always dynamic; this is because a unit might be instantiated at any level. Apart from these and a few other minor cases which need not concern the normal user, the general principle that any unit can be made generic holds true.

Exercise 19.2

1 Instantiate Next to give a function behaving like **not**.

2 Rewrite the specification of the package Rational_Numbers so that it is a generic package taking the integer type as a parameter. See Exercise 12.2(**3**).

3 Rewrite the function Outer of Exercise 10.1(**3**) so that it is a generic function with appropriate parameters. Instantiate it to give the original function.

4 Write the body of the package Set_Of.

5 Rewrite the private part of Set_Of so that an object of the type Set is by default given the initial value Empty when declared.

19.3 Subprogram parameters

As mentioned earlier a generic parameter can also be a subprogram. There are a number of characteristic applications of this facility and we introduce the topic by considering the classical problem of sorting.

Suppose we wish to sort an array into ascending order. There are a number of general algorithms that can be used which do not depend on the type of the values being sorted. All we need is some comparison operation such as "<" which is defined for the type.

We might start by considering the specification

```
generic
   type Index is (<>);
   type Item is (<>);
   type Collection is array (Index range <>) of Item;
procedure Sort(C: in out Collection);
```

Although the body is largely irrelevant it might help to illustrate the problem to consider the following crude possibility

```
procedure Sort(C: in out Collection) is
   Min: Index;
   Temp: Item;
begin
   for I in C'First .. Index'Pred(C'Last) loop
      Min := I;
      for J in Index'Succ(I) .. C'Last loop
         if C(J) < C(Min) then Min := J; end if;        -- use of <
      end loop;
      Temp := C(I);  C(I) := C(Min);  C(Min) := Temp;
   end loop;
end Sort;
```

This trivial algorithm repeatedly scans the part of the array not sorted, finds the least component (which because of the previous scans will be not less than any component of the already sorted part) and then swaps it so that it is then the last element of the now sorted part. Note that because of the generality we have imposed upon ourselves, we cannot write

```
for I in C'First .. C'Last-1 loop
```

because we cannot rely upon the array index being an integer type. We only know that it is a discrete type and therefore have to use the attributes Index'Pred and Index'Succ which we know to be available since they are common to all discrete types.

However, the main point to note is the call of "<" in the body of Sort. This calls the predefined function corresponding to the type Item. We know that there is such a function because we have specified Item to be discrete and all discrete types have such a function. Unfortunately the net result is that our generic sort can only sort arrays of discrete types. It cannot sort arrays of floating types. Of course we could write a version for floating types by replacing the generic parameter for Item by

```
type Item is digits <>;
```

but then it would not work for discrete types. What we really need to do is specify the comparison function to be used in a general manner. We can do this by adding a fourth parameter which is a formal subprogram so that the specification becomes

```
generic
   type Index is (<>);
   type Item is private;
   type Collection is array (Index range <>) of Item;
   with function "<" (X, Y: Item) return Boolean;
procedure Sort(C: in out Collection);
```

The formal subprogram parameter is like a subprogram declaration preceded by **with**. (The leading **with** is necessary to avoid a syntactic ambiguity and has no other subtle purpose.)

We have also made the type Item private since the only common property now required (other than supplied through the parameters) is that the type Item can be assigned. The body remains as before.

We can now sort an array of any (nonlimited) type provided that we have an appropriate comparison to supply as parameter. So in order to sort an array of the type Vector, we first instantiate thus

```
procedure Sort_Vector is
     new Sort(Integer, Float, Vector, "<");
```

and we can then apply the procedure to the array concerned

```
An_Array: Vector( ... );
...
Sort_Vector(An_Array);
```

Note that the call of "<" inside Sort is actually a call of the function passed as actual parameter; in this case it is indeed the predefined function "<" anyway.

Passing the comparison rule gives our generic sort procedure amazing flexibility. We can, for example, sort in the reverse direction by

```
procedure Reverse_Sort_Vector is
     new Sort(Integer, Float, Vector, ">");
...
Reverse_Sort_Vector(An_Array);
```

This may come as a slight surprise but it is a natural consequence of the call of the formal "<" in

```
if C(J) < C(Min) then ...
```

being, after instantiation, a call of the actual ">". No confusion should arise because the internal call is hidden but the use of the named notation for instantiation would look curious

```
procedure Reverse_Sort_Vector is
     new Sort( ..., "<" => ">");
```

We could also sort our second Farmyard of Section 8.5 assuming it to be a variable so that the animals are in alphabetical order

```
subtype String_3 is String(1 .. 3);

procedure Sort_String_3_Array is
     new Sort(Positive, String_3, String_3_Array, "<");
...
Sort_String_3_Array(Farmyard);
```

The "<" operator passed as parameter is the predefined operation applicable to one-dimensional arrays described in Section 8.6.

The correspondence between formal and actual subprograms is such that the formal subprogram just renames the actual subprogram. Thus the matching rules regarding parameters, results and so on are as described in Section 13.7. In particular the constraints on the parameters are those of the actual subprogram and any implied by the formal subprogram are ignored. A parameterless formal function can also be matched by an enumeration literal of the result type just as for renaming.

Generic subprogram parameters (like generic object parameters) can have default values. These are given in the generic formal part and take three forms. In the above example we could write

```
with function "<" (X, Y: Item) return Boolean is <>;
```

This means that we can omit the corresponding actual parameter if there is visible at the point of *instantiation* a unique subprogram with the same designator and matching specification. With this alteration to Sort we could have omitted the last parameter in the instantiation giving Sort_Vector.

Another form of default value is where we give an explicit name for the default parameter. The usual rules for defaults apply; the default name is only evaluated if required by the instantiation but the binding of identifiers in the expression which is the name occurs at the point of *declaration* of the generic unit. In our example

```
with function "<" (X, Y: Item) return Boolean is Less_Than;
```

could never be valid because the specification of Less_Than must match that of "<" and yet the parameter Item is not known until instantiation. Valid possibilities are where the formal subprogram has no parameters depending on formal types or the default subprogram is an operation or attribute of a formal type or is itself another formal parameter. Thus we might have

```
with function Next(X: T) return T is T'Succ;
```

The final form of default parameter can only be used with formal procedures and not functions and is simply a null procedure. Thus we might have

```
with procedure P( ... ) is null;
```

As a final example of the use of our generic Sort (which we will assume now has a default parameter <> for "<"), we show how any type can be sorted provided we supply an appropriate rule.

Thus consider sorting an array of the type Date from Section 8.7. We write

```
type Date_Array is array (Positive range <>) of Date;

function "<" (X, Y: Date) return Boolean is
begin
   if X.Year /= Y.Year then
      return X.Year < Y.Year;
   elsif X.Month /= Y.Month then
      return X.Month < Y.Month;
   else
      return X.Day < Y.Day;
   end if;
end "<";

procedure Sort_Date_Array is new Sort(Positive, Date, Date_Array);
```

where the function "<" is passed through the default mechanism.

We could give the comparison rule a more appropriate name such as

```
function Earlier(X, Y: Date) return Boolean;
```

but we would then have to pass it as an explicit parameter.

Formal subprograms can be used to supply further properties of type parameters in a quite general way. Consider the generic function Sum of the previous section. We can generalize this even further by passing the adding operator itself as a generic parameter

```
generic
   type Index is (<>);
   type Item is private;
```

```
      type Vec is array (Index range <>) of Item;
      with function "+" (X, Y: Item) return Item;
   function Apply(A: Vec) return Item;

   function Apply(A: Vec) return Item is
      Result: Item := A(A'First);
   begin
      for I in Index'Succ(A'First) .. A'Last loop
         Result := Result + A(I);
      end loop;
      return Result;
   end Apply;
```

The operator "+" has been added as a parameter and Item is now just private and no longer floating. This means that we can apply the generic function to any binary operation on any type. However, we no longer have a zero value and so have to initialize Result with the first component of the array A and then iterate through the remainder. In doing this, remember that we cannot write

```
   for I in A'First+1 .. A'Last loop
```

because the type Index may not be an integer type.

Our original function Sum of Section 10.1 is now given by

```
   function Sum is new Apply(Integer, Float, Vector, "+");
```

We could equally have

```
   function Prod is new Apply(Integer, Float, Vector, "*");
```

Another possible use of formal subprograms is in mathematical applications such as integration. As we saw in Section 11.8, values of access to subprogram types can be passed as parameters to other subprograms and this is usually perfectly satisfactory. However, we might need to make the unit generic so that any floating type can be used. We could then pass the function to be integrated as a further generic parameter.

We could have a generic function

```
   generic
      type Floating is digits <>;
      with function F(X: Floating) return Floating;
   function Integrate(A, B: Floating) return Floating;
```

and then in order to integrate a particular function we must instantiate Integrate with that function as actual generic parameter. Thus suppose we needed to evaluate

$$\int_0^P e^t \sin t \, dt$$

using the type Long_Real. We would write

```
function G(T: Long_Real) return Long_Real is
begin
   return Exp(T) * Sin(T);
end;

function Integrate_G is new Integrate(Long_Real, G);
```

and then the result is given by the expression

```
Integrate_G(0.0, P)
```

In practice the function Integrate would have other parameters indicating the accuracy required and so on.

Examples such as this are often found confusing at first sight. The key point to remember is that there are two distinct levels of parameterization. First we fix the function to be integrated at instantiation and then we fix the bounds when we call the integration function thus declared. The sorting examples were similar; first we fixed the parameters defining the type of array to be sorted and the rule to be used at instantiation, and then we fixed the actual array to be sorted when we called the procedure.

Whether to use a generic or an access to subprogram parameter is to some extent a matter of taste. The generic mechanism enables the type to be parameterized as well and also permits the actual subprogram to have convention Intrinsic whereas this cannot be done with an access to subprogram parameter since the Access attribute cannot be applied to intrinsic operations. As we shall see in Chapter 24, the container library uses access to subprogram parameters for iteration but generics for sorting.

There is one other form of generic subprogram parameter which we will briefly outline here and explain its use in Section 21.8. This takes the form

```
with procedure Do_This( ... ) is abstract;
```

The formal subprogram Do_This must have controlling parameters or result of just one tagged type which may be a formal type or some other type global to the generic. The following would be illegal

```
with procedure Do_That(X1: TT1; X2: TT2) is abstract;      -- illegal
with function Fn(X: Float) return Float is abstract;        -- illegal
```

The first is illegal because it has two controlling types TT1 and TT2 and the second is illegal because it does not have any.

The actual parameter can be abstract or concrete. Remember that the overriding rules ensure that the specific operation for any concrete type will always have a concrete operation. Note also that since the operation is abstract it can only be called through dispatching.

Formal abstract subprograms can have defaults in much the same way that formal concrete subprograms can have defaults. We write

```
with procedure P(X: in out T) is abstract <>;
with function F return T is abstract Unit;
```

The first means of course that the default has to have identifier P and the second means that the default is some function Unit. It is not possible to give null as the default for an abstract parameter for various reasons. Defaults will probably be rarely used for abstract parameters.

We conclude with an important remark concerning the identification of primitive operations. A primitive operation such as "<" which applies to all discrete types (as in the first example of sorting) and occurring inside the generic unit always refers to the predefined operation even if it has been redefined for the actual type concerned. The general principle is that since the generic unit applies to the category of discrete types as a whole then the primitive operations ought to refer to those guaranteed for all members of the category. Of course if the operation is passed as a distinct parameter (explicitly or by default) then the redefined operation will apply.

On the other hand, if the formal parameter is tagged, so that the actual type is an extension of the formal, then primitive operations refer to overriding ones. The reason for the different approach is that the whole essence of tagged types is to override operations and so within the generic it is natural to refer to the operation of the actual type. For example if we wrote

```
generic
   type T is new Object with private;
package P ...
```

so that the actual type must be derived from Object then inside P we expect the function Area to refer to that of the actual type supplied such as Circle and not to that of Object. Indeed in the formulation of Exercise 14.3(**2**) it would be a disaster if it referred to the Area of Object since it is abstract.

Exercise 19.3

1 Instantiate Sort to apply to

```
type Poly_Array is array (Integer range <>) of Polynomial;
```

See Section 18.2. Define a sensible ordering for polynomials.

2 Instantiate Sort to apply to an array of the type Mutant of Section 18.3. Put neuter things first, then females, then males and within each class the younger first. Could we sort an array of the type Person from the same section?

3 Sort the array People of Section 8.7.

4 What happens if we attempt to sort an array of less than two components?

5 Describe how to make a generic sort procedure based on the procedure Sort of Section 11.2. It should have an identical specification to the procedure Sort of this section.

6 Write a generic function to search an array and return the index of the first component satisfying some criterion. Make one parameter the identity of an exception to be raised if there is no such component. Note that although an exception cannot be passed as a parameter, nevertheless a value of the type

Exception_Id can be passed; see Section 15.4. Make the exception Constraint_Error by default.

7 Write a generic function Equals to define the equality of one-dimensional arrays of a private type. See Exercise 12.4(**4**). Instantiate it to give the function "=" applying to the type Stack_Array.

8 Reconsider Exercise 11.8(**2**) using generics.

19.4 Package parameters

The last kind of formal generic parameter is the formal package. This greatly simplifies the composition of generic packages by allowing one package to be used as a parameter to another so that a hierarchy of consistently related packages can be created.

There are two basic forms of formal package parameters. The simplest is

```
with package P is new Q(<>);
```

which indicates that the actual parameter corresponding to P must be a package which has been obtained by instantiating Q which must itself be a generic package. We can also explicitly indicate the actual parameters required by the instantiation of Q thus

```
with package R is new Q(P1, P2, P3);
```

and then the actual package corresponding to R must be an instantiation of Q with matching parameters – **in** parameters must be static with the same value, subtypes must statically match and any others must denote the same entity.

It is also possible to specify just some of the parameters by using the <> notation as in aggregates. The named notation must be used for such parameters. We could write either of

```
with package S is new Q(P1, F2 => <>, F3 => <>);
with package S is new Q(P1, others => <>);
```

and in this case the actual package corresponding to S can be any package which is an instantiation of Q where the first actual parameter is P1 but the other two parameters are left unspecified. Incidentally the form Q(<>); can be seen as an abbreviation for Q(**others** => <>);

As a simple example suppose we wish to develop a package for the manipulation of complex vectors; we want to make it generic with respect to the underlying floating type. Naturally enough we will build on our simple package for complex numbers in its generic form outlined in Section 19.2

```
generic
   type Floating is digits <>;
package Generic_Complex_Numbers is
   type Complex is private;
   ...
```

```
      function "+" (X, Y: Complex) return Complex;
      ...
   end;
```

Our new generic package will need to use the various arithmetic operations exported from some instantiation of Generic_Complex_Numbers. These could all be imported as individual subprogram parameters but this would give a very long formal parameter list. Instead we can import the instantiated package as a whole. All we have to write is

```
   generic
      type Index is (<>);
      with package Complex_Numbers is
            new Generic_Complex_Numbers (<>);
   package Generic_Complex_Vectors is
      use Complex_Numbers;
      type Vector is array (Index range <>) of Complex;
      ...   -- other types and operations on vectors
   end;
```

and then we can instantiate the two packages by a sequence such as

```
   package Long_Complex is
         new Generic_Complex_Numbers(Long_Float);
   package Long_Complex_Vectors is
         new Generic_Complex_Vectors(Integer, Long_Complex);
```

Note the use clause in the specification of Generic_Complex_Vectors. Without this the component subtype of the type Vector would have to be written as Complex_Numbers.Complex. Note also that within Generic_Complex_Vectors not only do we have visibility of all the operations exported by the instantiation of Generic_Complex_Numbers but we can also refer to the formal parameter Floating. This is a special rule for formal parameters with the default form <>; we shall see its use in a moment.

We now consider a more elaborate example where the new package builds on the properties of two other packages. One is the toy package Generic_Complex_ Numbers and the other is the predefined library package for computing elementary functions. This is described in detail in Section 23.5 but for this chapter all we need is the following outline sketch

```
   generic
      type Float_Type is digits <>;
   package Ada.Numerics.Generic_Elementary_Functions is
      ...
      function Sqrt(X: Float_Type'Base) return Float_Type'Base;
      ...   -- similarly other functions such as
      ...   -- Log, Exp, Sin, Cos, Sinh, Cosh
      ...
   end;
```

(The parameters and results have subtype Float_Type'Base in order to avoid unnecessary constraint checks in intermediate expressions.)

Although the following example is of a rather mathematical nature it is hoped that the general principles will be appreciated. It follows on from the above elementary functions package and concerns the provision of similar functions but working on complex arguments. Suppose we want to provide the ability to compute Sqrt, Log, Exp, Sin, and Cos with functions such as

```
function Sqrt(X: Complex) return Complex;
```

Many readers will have forgotten that this can be done or perhaps never knew. It is not necessary to dwell on the details of how such calculations are performed or their use; the main point is to concentrate on the principles involved. These computations use various operations on the real numbers out of which the complex numbers are formed. Our goal is to write a generic package which works however the complex numbers are implemented (cartesian or polar) and also allows any floating point type as the basis for the underlying real numbers.

Here are the formulae which we will need to compute

Taking $z \equiv x + iy \equiv r(\cos\theta + i\sin\theta)$ as the argument:

$$\text{sqrt } z = r^{1/2}(\cos\theta/2 + i\sin\theta/2)$$
$$\log z = \log r + i\,\theta$$
$$\exp z = e^x(\cos y + i\sin y)$$
$$\sin z = \sin x\cosh y + i\cos x\sinh y$$
$$\cos z = \cos x\cosh y - i\sin x\sinh y$$

We thus see that we will need the functions Sqrt, Cos, Sin, Log, Exp, Cosh, and Sinh applying to the underlying floating type. We also need to be able to decompose and reconstruct a complex number using both cartesian and polar forms; we will assume that the package Generic_Complex_Numbers has been extended to do this (we will consider the question of child packages in the next section).

We can now write

```
with Ada.Numerics.Generic_Elementary_Functions;
use Ada.Numerics;
with Generic_Complex_Numbers;
generic
   with package Elementary_Functions is
         new Generic_Elementary_Functions(<>);
   with package Complex_Numbers is
         new Generic_Complex_Numbers
                                    (Elementary_Functions.Float_Type);
package Generic_Complex_Functions is
   use Complex_Numbers;

   function Sqrt(X: Complex) return Complex;
   ...

end Generic_Complex_Functions;
```

where the actual packages must be instantiations of Generic_Elementary_Functions and Generic_Complex_Numbers. Note the forms of the formal packages. Any instantiation of Generic_Elementary_Functions is allowed but the instantiation of Generic_Complex_Numbers must have Elementary_Functions.Float_Type as its actual parameter. This ensures that both packages are instantiated with the same floating type.

Note carefully that we are using the formal parameter exported from the first instantiation as the required parameter for the second instantiation. As noted above, a formal parameter is only accessible in this way when the default form <> is used. In order to reduce verbosity it is permitted to have a use clause in the generic formal list, so we could have written

```
with package Elementary_Functions is
     new Generic_Elementary_Functions(<>);
use Elementary_Functions;
with package Complex_Numbers is
     new Generic_Complex_Numbers(Float_Type);
```

although this is perhaps not so clear.

Finally, the instantiations are

```
type My_Float is digits 9;

package My_Elementary_Functions is
     new Generic_Elementary_Functions(My_Float);

package My_Complex_Numbers is
     new Generic_Complex_Numbers(My_Float);

package My_Complex_Functions is
     new Generic_Complex_Functions
                    (My_Elementary_Functions, My_Complex_Numbers);
```

and *Hey presto!* it all works. Note that, irritatingly, the complex type is just Complex and not My_Complex. However, we could use a subtype or derived type as we did for the type Colour and the package Set_Of in Section 19.2.

The reader might wonder why we did not simply instantiate the various packages inside the body of Generic_Complex_Functions and thereby avoid the package parameters. This works but could result in wasteful and unnecessary multiple instantiations since we may well need them at the user level anyway.

It is hoped that the general principles have been understood and that the mathematics has not clouded the issues. The principles are important but not easily illustrated with short examples. It should also be noted that although the package Generic_Elementary_Functions described above is exactly as in the predefined library, the complex number packages are just an illustration of how generics can be used. The Numerics annex does indeed provide packages for complex numbers and complex functions but they use a rather different approach as outlined in Section 26.5. A better example (also in the Numerics annex) is the package Generic_Complex_Arrays which takes instantiations of the two packages Generic_Real_Arrays and Generic_Complex_Types.

Another application of formal packages is where we wish to bundle together a number of related types and operations and treat them as a whole. We make them parameters of an otherwise null generic package. In essence, the successful instantiation of the generic package is an assertion that the entities have the required relationship and the group can then be referred to using the instantiated package (sometimes known as a signature).

As a very trivial example we will have noticed that the group of parameters

```
type Index is (<>);
type Item is private;
type Vec is array (Index range <>) of Item;
```

has occurred from time to time; see Sort and Apply in the previous section. For a successful instantiation of a generic unit having these as parameters it is necessary that the actual types have the required relationship. We could write

```
generic
   type Index is (<>);
   type Item is private;
   type Vec is array (Index range <>) of Item;
package General_Vector is end;
```

(note the curious juxtaposition of **is end** – we can even write **is private end**) and then rewrite Sort and Apply as

```
generic
   with package P is new General_Vector(<>);
   with function "<" (X, Y: P.Item) return Boolean is <>;
procedure Sort (V: in out P.Vec);

generic
   with package P is new General_Vector(<>);
   use P;
   with function "+" (X, Y: Item) return Item;
function Apply(A: Vec) return Item;
```

Note that the use clause in the formal list for Apply simplifies the text with some risk of obscurity. We can now perform instantiations as follows. First we might write

```
package Float_Vector is new General_Vector(Integer, Float, Vector);
```

which ensures that Integer, Float, and Vector have the required relationship. Then we can write

```
procedure Sort_Vector is new Sort(Float_Vector, "<");

function Sum is new Apply(Float_Vector, "+");
```

which should be compared with the corresponding instantiations in the previous section. Clearly this example is rather trivial and the gain obtained by simplifying the parameter lists for Sort and Apply is barely worth the effort of the extra layer of abstraction.

Another example might be the signature of a package for manipulating sets. We can write (note that the formal types are incomplete)

```
generic
   type Element;                -- incomplete
   type Set;                    -- incomplete
   with function Empty return Set is <>;
   with function Unit(E: Element) return Set is <>;
   with function Union(S, T: Set) return Set is <>;
   with function Intersection(S, T: Set) return Set is <>;
   ...
package Set_Signature is end;
```

We might then have some other generic package which takes an instantiation of this set signature. However, it is likely that we would need to specify the type of the elements but possibly not the set type and certainly not all the operations. So typically we would have

```
generic
   type My_Element is private;
   with package Sets is
                 new Set_Signature(Element => My_Element, others => <>);
```

and this illustrates a situation where one parameter is given but the others are arbitrary.

For a further discussion on this topic see Section 4.3 of the *Ada 2012 Rationale*.

Exercise 19.4

1. Write a body for the package Generic_Complex_Functions using the formulae defined above. Ignore exceptions.
2. Reconsider Exercise 19.3(**1**) using the package General_Vector.
3. A mathematical group is defined by an operation over a set of elements thus

   ```
   generic
      type Element is private;
      Identity: in Element;
      with function Op(X, Y: Element) return Element;
      with function Inverse(X: Element) return Element;
   package Group is end;
   ```

   Declare a generic function Power which takes an element E and a signed integer N as parameters and delivers the *N*th power of the element using the group operation. Then define the addition group over the type Integer and finally instantiate the power function.
4. Any finite group can be represented as a discrete type. Adapt the definition of Group of the previous exercise accordingly and then define a (generic) function Is_Group that checks whether a signature conforms to the semantic requirements of a group.

19.5 Generic library units

We now consider the interaction between generics and hierarchical libraries. Both are important tools in the construction of subsystems and it is essential that genericity be usable with the child concept.

It is an important rule that if a parent unit is generic then all its children must also be generic. The reverse does not hold; a nongeneric parent can have both generic children and nongeneric children.

If the parent unit is not generic then a generic child may be instantiated in the usual way at any point where it is visible. On the other hand, if the parent unit is itself generic, then the rules regarding the instantiation of a child are somewhat different according to whether the instantiation is inside or outside the generic hierarchy. If inside then the instantiation is as normal (see Program 4 for an example) but if outside then the parent must be instantiated first.

In effect the instantiation of the parent creates an actual unit with a generic child inside it; we can then instantiate this child provided we have a with clause for the original generic child. Note that these instantiations do not have to be as library units although they might be.

This is best illustrated with a symbolic example consisting of a package Parent and a child Parent.Child. The parent will typically have formal generic parameters but often the child will have none since it will be parameterized by those of its parent which will be visible to it anyway. So we might have

```
generic
   type T is private;
package Parent is
   ...
end Parent;

generic
package Parent.Child is
   ...
end Parent.Child;
```

We can then instantiate this hierarchy inside some other package P by writing

```
with Parent.Child;
package P is
   package Parent_Instance is new Parent(T => Some_Type);
   package Child_Instance is new Parent_Instance.Child;
   ...
end P;
```

Note that the names of the new units are quite unrelated. Moreover, the instantiation for the child refers to Parent_Instance.Child and not Parent.Child. Nevertheless we need a with clause for Parent.Child (we do not need a with clause for the parent because that for the child implies one anyway).

If we instantiate at library level then again they might have unrelated names thus

```
with Parent;
package Parent_Instance is new Parent(T => Some_Type);

with Parent.Child;
with Parent_Instance;
package Child_Instance is new Parent_Instance.Child;
```

Note how in the second case we need a with clause for both the instance of the parent and the original child.

We could also form the instances into a hierarchy with a similar structure to the original

```
with Parent;
package Parent_Instance is new Parent(T => Some_Type);

with Parent.Child;
package Parent_Instance.Child_Instance is new Parent_Instance.Child;
```

In the second case we only have one with clause; remember that a child never needs one for its parent. Note moreover, that in this hierarchical case, the child names have to be different from those in the original generic hierarchy otherwise there would be a clash of names.

It is a general principle that we have to instantiate a generic hierarchy (or as much of it as we want) unit by unit. The main reason for requiring all children of a generic unit to be generic is to provide a handle to do this; as a consequence it is often the case that the child units have no formal generic parameters.

Of course, we need not instantiate the whole of a hierarchy for a particular application; indeed, one of the reasons for the unit by unit approach is to eliminate problems concerning the impact of the addition of new children to the generic hierarchy on existing instantiations.

Exercise 19.5

1 Sketch the structure of a hierarchy of three complex number packages Generic_Complex_Numbers, Generic_Complex_Numbers.Cartesian and Generic_Complex_Numbers.Polar; see Exercise 13.3(**1**). Then rewrite the specification of Generic_Complex_Functions of the previous section using this hierarchical form.

Checklist 19

The generic mechanism is not text replacement; nonlocal name binding would be different.

Object **in out** parameters are bound by renaming.

Subprogram generic parameters are bound by renaming.

Generic subprograms may not overload – only the instantiations can.

Generic subprograms always have a separate specification and body.

Formal parameters (and defaults) may depend upon preceding parameters.

Generic formal parameters and their attributes are not static.

Ada 95 did not permit parameters to be objects of limited types of mode **in**.

The parameter forms for objects of anonymous access types were added in Ada 2005.

The forms for interfaces were added in Ada 2005.

The default **null** for formal subprogram parameters was added in Ada 2005.

The form **is abstract** for formal subprogram parameters was added in Ada 2005.

The ability to specify just some parameters of a formal package was added in Ada 2005.

Formal parameters for incomplete types were added in Ada 2012.

New in Ada 2022

Various aspects are allowed on generic formal parameters including pre- and postconditions, Atomic and Atomic_Components, Volatile and Volatile_Components, Independent and Independent_Components. New aspects such as Nonblocking are also allowed, see Section 20.2.

Specific defaults for generic type parameters can be indicated by **or use** followed by the default item.

20 Tasking

20.1 Parallelism
20.2 The rendezvous
20.3 Timing and scheduling
20.4 Protected objects
20.5 Simple select statements
20.6 Timed and conditional calls
20.7 Concurrent types and activation
20.8 Termination, exceptions, and ATC
20.9 Signalling and scheduling
20.10 Summary of structure

The final major topic to be introduced is tasking. Ada is fairly unusual in having tasking constructions built in to the language. Some might argue that such matters are the concern of the operating system and are better done by calls from an otherwise sequential program. However, built-in constructions provide greater reliability; general operating systems do not provide the control and timing needed by many applications; and every operating system is different. Building tasking into the language also makes it easier to benefit from any parallelism in the machine and increases the potential for optimization.

This chapter concentrates on the key ideas. Further examples and details especially concerning the interaction between tasking and OOP are dealt with in Chapter 22 which also covers important facilities concerning timing which are actually defined in the Real-Time Systems annex.

20.1 Parallelism

So far we have only considered sequential programs in which statements are obeyed in order. In many applications it is convenient to write a program as several parallel activities which cooperate as necessary. This is particularly true of programs which interact in real time with physical processes in the real world, simulation programs (which mimic parallel activities in the real world), and programs which wish to exploit multiprocessor architectures directly.

In Ada, parallel activities are defined by means of tasks. A task is lexically described by a form very similar to a package. It consists of a task specification describing the interface presented to other tasks and a task body describing the dynamic behaviour of the task.

```
task T is                    -- specification
   ...
end T;

task body T is               -- body
   ...
end T;
```

In some cases a task presents no interface to other tasks in which case the specification reduces to just

```
task T;
```

As a simple example of parallelism, consider a family going shopping to buy ingredients for a meal. Suppose they need meat, salad, and wine and that the purchase of these items can be done by calling procedures Buy_Meat, Buy_Salad, and Buy_Wine respectively. The whole expedition could be represented by

```
procedure Shopping is
begin
   Buy_Meat;
   Buy_Salad;
   Buy_Wine;
end;
```

However, this solution corresponds to the family buying each item in sequence. It would be far more efficient for them to split up so that, for example, mother buys the meat, the children buy the salad, and father buys the wine. They agree to meet again perhaps back at the car. This parallel solution is represented by

```
procedure Shopping is
   task Get_Salad;

   task body Get_Salad is
   begin
      Buy_Salad;
   end Get_Salad;

   task Get_Wine;

   task body Get_Wine is
   begin
      Buy_Wine;
   end Get_Wine;

begin
   Buy_Meat;
end Shopping;
```

In this formulation, mother is represented as the main processor and calls Buy_Meat directly from the procedure Shopping. The children and father are considered as subservient processors and perform the locally declared tasks Get_Salad and Get_Wine which respectively call the procedures Buy_Salad and Buy_Wine.

The example illustrates the declaration, activation, and termination of tasks. A task is a program component like a package and is declared in a similar way inside a subprogram, block, package or indeed another task body. A task specification can also be declared in a package specification in which case the task body must be declared in the corresponding package body. However, a task specification cannot be declared in the specification of another task but only in the body.

The activation of a task is automatic. In the above example the local tasks become active when the parent unit reaches the **begin** following the task declaration. Such a task will terminate when it reaches its final **end**. Thus the task Get_Salad calls the procedure Buy_Salad and then promptly terminates.

A task declared in the declarative part of a subprogram, block or task body is said to depend on that unit. It is an important rule that a unit cannot be left until all dependent tasks have terminated. This termination rule ensures that objects declared in the unit and therefore potentially visible to local tasks cannot disappear while there exists a task which could access them. (Note that a task cannot depend on a package – we will return to this later.)

It is important to realize that the main subprogram is itself considered to be called by a hypothetical main task. We can now trace the sequence of actions when this main task calls the procedure Shopping. First the tasks Get_Salad and Get_Wine are declared and then when the main task reaches the **begin** these dependent tasks are set active in parallel with the main task. The dependent tasks call their respective procedures and terminate. Meanwhile the main task calls Buy_Meat and then reaches the **end** of Shopping. The main task then waits until the dependent tasks have terminated if they have not already done so. This corresponds to mother waiting for father and children to return with their purchases.

In the general case termination therefore occurs in two stages. We say that a unit is completed when it reaches its final **end**. It will subsequently become terminated only when all dependent tasks, if any, are also terminated. Of course, if a unit has no dependent tasks then it effectively becomes completed and terminated at the same time (but see Section 20.8).

Exercise 20.1

1 Rewrite procedure Shopping to contain three local tasks so that the symmetry of the situation is revealed.

20.2 The rendezvous

In the Shopping example the various tasks did not interact with each other once they had been set active except that their parent unit had to wait for them to terminate. Generally, however, tasks will interact with each other during their lifetime.

There are two main ways in which tasks can interact: directly, by sending messages to each other, and indirectly, by common access to shared data. We consider direct communication using messages in this section and indirect communication through shared data in Section 20.4.

Messages are directly passed between tasks in Ada by a mechanism known as the rendezvous. This is similar to the human situation where two people meet, perform a transaction and then go on independently.

A rendezvous between two tasks occurs as a consequence of one task calling an entry declared in another. An entry is declared in a task specification in a similar way to a procedure in a package specification

```
task T is
   entry E( ... );
end;
```

An entry can have **in**, **out**, and **in out** parameters in the same way as a procedure and parameters of mode **in** may have defaults. However, a task entry cannot have access parameters nor can it have a result like a function. An entry is called in a similar way to a procedure

```
T.E( ... );
```

A task name cannot appear in a use clause and so the dotted notation is necessary to call the entry from outside the task. Of course, a local task could call an entry of its parent directly – the usual scope and visibility rules apply.

A task can also have a private part containing the declaration of entries which are not visible to the external user; they could be called by local tasks.

The statements to be obeyed during a rendezvous are described by corresponding accept statements in the body of the task containing the declaration of the entry. An accept statement usually takes the form

```
accept E( ... ) do
   ...        -- sequence of statements
end E;
```

The formal parameters of the entry E are repeated in the same way that a procedure body repeats the formal parameters of a corresponding procedure declaration. The **end** is optionally followed by the name of the entry. A significant difference is that the body of the accept statement is just a sequence of statements plus optional exception handlers. Any local declarations must be provided by writing a local block. The syntax is thus very similar to that of the extended return statement.

The most important difference between an entry call and a procedure call is that in the case of a procedure, the task that calls the procedure also immediately executes the procedure body whereas in the case of an entry, one task calls the entry but the corresponding accept statement is executed by the task owning the entry. Moreover, the accept statement cannot be executed until a task actually calls the entry and the task owning the entry reaches the accept statement. Naturally one of these will occur first and the task concerned will then be suspended until the other reaches its corresponding statement. When this occurs the sequence of statements of the accept statement is executed by the called task while the calling task remains suspended. This interaction is called a rendezvous. When the end of the accept statement is reached the rendezvous is completed and both tasks then proceed independently. The parameter mechanism is exactly as for a subprogram call; note that expressions in the actual parameter list are evaluated before the call is issued.

We can elaborate the shopping example by giving the task Get_Salad two entries, one for mother to hand the children the money for the salad and one to collect the salad from them afterwards. We do the same for Get_Wine.

We can also replace the procedures Buy_Salad, Buy_Wine, and Buy_Meat by functions which take money as a parameter and return the appropriate ingredient. The shopping procedure might now become

```
procedure Shopping is
   task Get_Salad is
      entry Pay(M: in Money);
      entry Collect(S: out Salad);
   end Get_Salad;

   task body Get_Salad is
      Cash: Money;
      Food: Salad;
   begin
      accept Pay(M: in Money) do
         Cash := M;
      end Pay;

      Food := Buy_Salad(Cash);

      accept Collect(S: out Salad) do
         S := Food;
      end Collect;
   end Get_Salad;

   ...      -- task Get_Wine similarly

begin
   Get_Salad.Pay(50);
   Get_Wine.Pay(100);
   The_Meat := Buy_Meat(200);
   Get_Salad.Collect(The_Salad);
   Get_Wine.Collect(The_Wine);
end Shopping;
```

The final outcome is that the various ingredients end up in the variables The_Meat, The_Salad, and The_Wine whose declarations are left to the imagination.

The logical behaviour should be noted. As soon as the tasks Get_Salad and Get_Wine become active they encounter accept statements and wait until the main task calls the entries Pay in each of them. After calling the function Buy_Meat, the main task calls the Collect entries. Curiously, mother is unable to collect the wine until after she has collected the salad from the children.

As a more abstract example consider the problem of providing a task to act as a single buffer between one or more tasks producing items and one or more tasks consuming them. The intermediate task can hold just one item.

```
task Buffer is
   entry Put(X: in Item);
   entry Get(X: out Item);
end;
```

```
task body Buffer is
  V: Item;
begin
  loop
    accept Put(X: in Item) do
      V := X;
    end Put;
    accept Get(X: out Item) do
      X := V;
    end Get;
  end loop;
end Buffer;
```

Other tasks may then dispose of or acquire items by calling

```
Buffer.Put( ... );
Buffer.Get( ... );
```

Intermediate storage for the item is the variable V. The body of the task is an endless loop which contains an accept statement for Put followed by one for Get. Thus the task alternately accepts calls of Put and Get which fill and empty the variable V.

Several different tasks may call Put and Get and consequently may have to be queued. Every entry has a queue of tasks waiting to call the entry – this queue is normally processed in a first-in–first-out manner (but see Section 26.2). It may, of course, be empty at a particular moment. The number of tasks on the queue of entry E is given by E'Count but this attribute may only be used inside the body of the task owning the entry.

This example was for illustration only; as we shall see in a moment such data manager tasks which effectively decouple the producer and consumer are usually not necessary – this is really a shared data application and is more efficiently programmed using protected objects as described in Section 20.4.

An entry may have several corresponding accept statements (usually only one and possibly even none). Each execution of an accept statement removes one task from the queue. Note the asymmetric naming in a rendezvous. The calling task must name the called task but not vice versa. Moreover, several tasks may call an entry and be queued but a task can only be on one queue at a time.

Entries may be overloaded both with each other and with subprograms and obey the same rules. An entry may be renamed as a procedure

```
procedure Write(X: in Item) renames Buffer.Put;
```

This mechanism may be useful in avoiding excessive use of the dotted notation. An entry, renamed or not, may be an actual or default generic parameter corresponding to a formal subprogram.

An entry may have no parameters, such as

```
entry Signal;
```

and it could then be called by

```
T.Signal;
```

An accept statement need have no body as in

```
accept Signal;
```

In such a case the purpose of the call is merely to effect a synchronization and not to pass information. However, an entry without parameters can have an accept statement with a body and vice versa. So we could write

```
accept Signal do
   Fire;
end;
```

in which case the task calling Signal is only allowed to continue after the call of Fire is completed. We could also have

```
accept Put(X: in Item);
```

although clearly the parameter value is not used.

There are few constraints on the statements in an accept statement. They may include entry calls, subprogram calls, blocks, and further accept statements. But an accept statement may not contain an asynchronous select (see Section 20.8), nor an accept statement for the same entry or one of the same family (see Section 20.9). Moreover, an accept statement may not itself appear in a subprogram body but must be in the sequence of statements of the task although it could be in a block or other accept statement. The execution of a **return** statement in an accept statement corresponds to reaching the final end and so terminates the rendezvous. Similarly to a subprogram body, a **goto** or **exit** statement cannot transfer control out of an accept statement.

A task may call one of its own entries but, of course, will promptly deadlock. This may seem foolish but programming languages allow lots of silly things such as endless loops and so on. We could expect a good compiler to warn us of obvious potential deadlocks.

A key property of multitasking is the ability to do more than one thing at a time. A very important consideration is that the several things being done at the same time do not interfere with each other. Ada 2022 has a number of new features that prevent or detect interference.

There are two considerations. On the one hand we need to be prevented from doing actions that interfere with each other, this is typically done by a task being blocked. On the other hand we might wish to know that a task will not be blocked when executing a lump of code.

Many actions are defined to be potentially blocking such as calling an entry or executing an accept, select, delay, or abort statement.

The new aspect Nonblocking (of type Boolean) can be applied to many entities including packages, subprograms, and entries as well as to various generic parameters. When discussing global state in Section 16.5, we noted that the packages Ada.Tags and Ada.Numerics.Big_Numbers have the aspect Nonblocking. If Nonblocking is given for a package then it is inherited by subprograms in that package as well.

If Nonblocking is given for a protected type (see Section 20.4) then we are guarded against the horror of accidentally executing a potentially blocking operation inside a protected operation since it is caught at compilation time.

If a package is Pure then it is automatically Nonblocking. Predefined operations of an elementary type such as "+" for the type Integer are also Nonblocking. Entries by their very nature are usually blocking. Some packages explicitly have Nonblocking => False. Examples are the somewhat magic packages such as Ada.Iterator_Interfaces and Ada.Finalization. For details of features that enable conflicts to be checked see Section 22.9.

The new aspect Max_Entry_Queue_Length can be applied to an entry. In the case of the task Buffer we could impose limits on the entries Put and Get by writing

```
task Buffer is
   entry Put(X: in Item)
      with Max_Entry_Queue_Length => 3;
   entry Get(X: in Item)
      with Max_Entry_Queue_Length => 7;
end;
```

The maximum length is of type Integer. Program_Error is raised in the caller if an entry is called when its queue is full. The aspect can also be applied to a task in which case it applies to all entries in the task.

Finally, note that entries can also have pre- and postconditions in a similar manner to procedures.

Exercise 20.2

Note: this is simply an exercise on using the rendezvous. A better solution is to use a protected object as described in Section 20.4.

1 Write the body of a task whose specification is

```
task Char_To_Line is
   entry Put(C: in Character);
   entry Get(L: out Line);
end;
```

where

```
type Line is array (1 .. 80) of Character;
```

The task acts as a buffer which alternately builds up a line by accepting successive calls of Put and then delivers a complete line on a call of Get.

20.3 Timing and scheduling

As we have seen, an Ada program may contain several tasks. Conceptually, it is often best to think of these tasks as each having its own personal processor so that, provided a task is not waiting for something to happen, it will actually be executing.

In practice, of course, most implementations will not be able to allocate a unique processor to each task and indeed, in many cases, there will be only one physical processor. It will then be necessary to allocate the processor(s) to the tasks that are logically able to execute by some scheduling algorithm. This can be done in many ways.

One of the simplest mechanisms is to use time slicing. This means giving the processor to each task in turn for some fixed time interval such as 10 milliseconds. If a task cannot use its turn (perhaps because it is held up awaiting a partner in a rendezvous), then a sensible scheduler would allocate its turn to the next task. Similarly, if a task cannot use all of its turn then the remaining time could be allocated to another task.

Time slicing is somewhat rudimentary since it treats all tasks equally. It is often the case that some tasks are more urgent than others and in the face of a shortage of processing power this equality is wasteful. The idea of a task having a priority is therefore introduced. A simple scheduling system would be one where each task had a distinct priority and the processor would then be given to the highest priority task which could actually run. Combinations of time slicing and priority scheduling are also possible. A system might permit several tasks to have the same priority and time slice between them.

The core of the Ada language remains silent about scheduling and leaves the details of techniques supported to the Real-Time Systems annex. In this chapter we therefore generally take the abstract view that each task has its own processor. A further discussion of priorities and scheduling algorithms is in Section 26.2.

A task may be held up for various reasons; it might be waiting for a partner in a rendezvous or for a dependent task to terminate. It can also be held up by executing a delay statement such as

```
delay 3.0;
```

This suspends the task (or main subprogram) executing the statement for three seconds. The expression after the reserved word **delay** is of a predefined fixed point type Duration and gives the period in seconds. At the expiry of the delay the task will be ready again. (Note, however, that there might not be a processor immediately available to execute the task since in the meantime a higher priority task might have obtained control.)

The type Duration is a fixed point type so that the addition of durations can be done without systematic loss of accuracy. If we add together two fixed point numbers of the same type then we always get the exact mathematical answer; this does not apply to floating point. On the other hand, we need to express fractions of a second in a convenient way and so the use of a real type rather than an integer type is much more satisfactory.

Delays can be neatly expressed by using suitable constant declarations, thus

```
Seconds: constant Duration := 1.0;
Minutes: constant Duration := 60.0;
Hours: constant Duration := 3600.0;
```

We can then write for example

```
delay 2*Hours+40*Minutes;
```

in which the expression uses the rule that a fixed point value can be multiplied by an integer giving a result of the same fixed point type.

A delay statement with a zero or negative argument has no effect (other than possibly causing rescheduling). See Section 26.2.

Although the type Duration is implementation-defined, we are guaranteed that it will allow durations (both positive and negative) of up to at least one day (86,400 seconds). Delays of more than a day (which are unusual) would have to be programmed with a loop. At the other end of the scale, the smallest value of Duration, that is Duration'Small, is guaranteed to be not greater than 20 milliseconds.

More general timing operations can be performed by using the predefined package Ada.Calendar whose specification is

```
package Ada.Calendar
    with Nonblocking, Global => in out synchronized is
  type Time is private;

  subtype Year_Number is Integer range 1901 .. 2399;
  subtype Month_Number is Integer range 1 .. 12;
  subtype Day_Number is Integer range 1 .. 31;
  subtype Day_Duration is Duration range 0.0 .. 86_400.0;

  function Clock return Time;

  function Year(Date: Time) return Year_Number;
  function Month(Date: Time) return Month_Number;
  function Day(Date: Time) return Day_Number;
  function Seconds(Date: Time) return Day_Duration;

  procedure Split(Date: in Time;
                  Year: out Year_Number;
                  Month: out Month_Number;
                  Day: out Day_Number;
                  Seconds: out Day_Duration);

  function Time_Of(Year: Year_Number;
                   Month: Month_Number;
                   Day: Day_Number;
                   Seconds: Day_Duration := 0.0) return Time;

  function "+" (Left: Time; Right: Duration) return Time;
  function "+" (Left: Duration; Right: Time) return Time;
  function "-" (Left: Time; Right: Duration) return Time;
  function "-" (Left: Time; Right: Time) return Duration;
  function "<" (Left, Right: Time) return Boolean;
  function "<=" (Left, Right: Time) return Boolean;
  function ">" (Left, Right: Time) return Boolean;
  function ">=" (Left, Right: Time) return Boolean;

  Time_Error: exception;

private
  ...      -- implementation dependent
end Ada.Calendar;
```

A value of the private type Time is a combined time and date; it can be decomposed into the year, month, day and the duration since midnight of the day concerned by the procedure Split. Alternatively, the functions Year, Month, Day, and Seconds may be used to obtain the individual values. On the other hand, the function Time_Of can be used to build a value of Time from the four constituents; the seconds parameter has a default of zero. Note the subtypes Year_Number, Month_Number, and Day_Number.

The current Time is returned by a call of the function Clock. The result is, of course, returned in an indivisible way and there is no risk of getting the time of day and the date inconsistent around midnight as there would be if there were separate functions delivering the individual components of the current time and date.

The various overloadings of "+", "–" and the ordering operators allow us to add, subtract and compare times and durations as appropriate. A careful distinction must be made between Time and Duration. Time is absolute but Duration is relative.

The exception Time_Error is raised if the parameters of Time_Of do not form a proper date. Time_Error is also raised by calls of "+" and "–" which attempt to create a time or duration outside the implemented ranges. Applying the procedure Split or the functions Year, Month, Day, or Seconds to a date whose year is out of range will also raise Time_Error.

As well as the relative delay statement mentioned above there is also an absolute delay statement; this takes a time rather than a duration as parameter. Thus to delay until a given time we write

```
delay until Some_Time;
```

where Some_Time is of type Time.

As an example of the use of the package Calendar suppose we wish a task to call a procedure Action at regular intervals, every five minutes perhaps. Our first attempt might be to write

```
loop
   delay 5*Minutes;          -- beware of drift
   Action;
end loop;
```

However, this is unsatisfactory for various reasons. First, we have not taken account of the time of execution of the procedure Action and the overhead of the loop itself, and secondly, we have seen that a delay statement sets a minimum delay only (since a higher priority task might retain the processor on the expiry of the delay). Furthermore, we might get preempted by a higher priority task at any time anyway. So we will inevitably get a cumulative timing drift. This can be overcome by writing for example

```
declare
   use Calendar;
   Interval: constant Duration := 5*Minutes;
   Next_Time: Time := First_Time;
begin
   loop
      delay until Next_Time;
      Action;
      Next_Time := Next_Time + Interval;
   end loop;
end;
```

In this formulation Next_Time contains the time when Action is next to be called; its initial value is in First_Time and it is updated exactly on each iteration by adding Interval. This solution will have no cumulative drift provided the mean duration of Action plus the overheads of the loop and updating Next_Time and so on do not exceed Interval. Of course, there may be a local drift if a particular call of Action takes a long time or other tasks temporarily use the processors. Finally, there is one other condition that must be satisfied for the required timing to be obtained: the interval has to be a multiple of *small* for the type Duration.

The type Calendar.Time is system dependent and will usually be the local civil time and thus subject to daylight saving and time zone changes. It can therefore go backwards. For many applications this is unhelpful and the alternative monotonic type Time defined in the package Ada.Real_Time should be used (see Section 22.6).

There are three child packages of Ada.Calendar providing other useful but somewhat specialized facilities. We will outline the more useful and leave the reader to consult the *ARM* for more details. The first is

```
package Ada.Calendar.Time_Zones
    with Nonblocking, Global => in out synchronized is
  type Time_Offset is range -28*60 .. 28*60;
  Unknown_Zone_Error: exception;
  function UTC_Time_Offset(Date: Time := Clock) return Time_Offset;
end Ada.Calendar.Time_Zones;
```

Time zones are described in terms of the number of minutes different from UTC (Coordinated Universal Time); this is close to but not quite the same as Greenwich Mean Time (GMT) and similarly does not suffer from leaping about in spring and falling about in the autumn.

NOTE: see AI12-336 for an explanation that in Ada 2022 we can write Local_Time_Offset as well as UTC_Time_Offset (one is a renaming).

So the function UTC_Time_Offset applied in an Ada program in Paris to a value of type Time in summer will return a time offset of 120 (one hour for European Central Time plus one hour for daylight saving); and if applied in New York in winter will return an offset of -300. Remember that the type Calendar.Time incorporates the date. To find the offset now (that is, at the time of the function call) we simply write

```
Offset := UTC_Time_Offset;
```

and then Clock is called by default.

To find what the offset was on Christmas Day 2000 we write

```
Offset := UTC_Time_Offset(Time_Of(2000, 12, 25));
```

and this should return 60 in Paris. So the poor function has to remember the whole history of local time changes since 1901 and predict them forward to 2399 – these Ada systems are pretty smart! In reality the intent is to use whatever the underlying operating system provides. If the information is not known then it can raise Unknown_Zone_Error.

A useful fact is that Clock - Duration(UTC_Time_Offset*60) gives UTC time – provided we don't do this just as daylight saving comes into effect in which case the call of Clock and that of UTC_Time_Offset might not be compatible.

More generally the type Time_Offset can be used to represent the difference between two time zones. If we want to work with the difference between New York and Paris then we could say

```
NY_Paris: Time_Offset := –360;
```

The time offset between two different places can be greater than 24 hours for two reasons. One is that the International Date Line weaves about somewhat and the other is that daylight saving time can extend the difference as well. Differences of 26 hours can easily occur and 27 hours is possible. Accordingly the range of the type Time_Offset allows for a generous 28 hours.

Another problem with time is the introduction of leap seconds to take account of the fact that the rotation of the earth is gradually slowing down. The package Ada.Calendar.Arithmetic is

```
package Ada.Calendar.Arithmetic
    with Nonblocking, Global => in out synchronized is
  type Day_Count is range
    –366*(1+Year_Number'Last – Year_Number'First) ..
    +366*(1+Year_Number'Last – Year_Number'First);
  subtype Leap_Seconds_Count is Integer range –2047 .. 2047;

  procedure Difference(Left, Right: in Time;
                       Days: out Day_Count;
                       Seconds: out Duration;
                       Leap_Seconds: out Leap_Seconds_Count);
  function "+" (Left: Time; Right: Day_Count) return Time;
  function "+" (Left: Day_Count; Right: Time) return Time;
  function "–" (Left: Time; Right: Day_Count) return Time;
  function "–" (Left, Right: Time) return Day_Count;
end Ada.Calendar.Arithmetic;
```

The procedure Difference gives the difference between two times and includes a count of the leap seconds in the interval. These versions of functions "+" and "–" apply to values of type Time and Day_Count whereas those in the parent Calendar apply only to Time and Duration and thus only work for intervals of a day or so. Moreover, these functions ignore leap seconds whereas the functions in Calendar take account of leap seconds.

The package Ada.Calendar.Formatting contains versions of Split and Time_Of which take account of time zone offsets and leap seconds. Of more practical use are the type Day_Name, the function Day_Of_Week and functions Image and Value

```
type Day_Name is (Monday, Tuesday, Wednesday, Thursday, Friday,
                                                Saturday, Sunday);
function Day_Of_Week(Date: Time) return Day_Name;
function Image(Date: Time; Include_Time_Fraction: Boolean := False;
                    Time_Zone: Time_Offset := 0) return String;
function Value(Date: String; Time_Zone: Time_Offset := 0) return Time;
function Image (Elapsed_Time: Duration;
                Include_Time_Fraction: Boolean := False) return String;
function Value(Elapsed_Time: String) return Duration;
```

The call Image(Now) might return the string "2022-02-22 14:50:00". whereas the call Image (10_000.0) would return "02:46:40". The functions Value work in reverse as expected.

Exercise 20.3

1 Write a generic procedure to call a procedure regularly. The generic parameters will be the procedure to be called, the time of the first call, the interval and the number of calls. If the time given for the first call is in the past use the current time instead.

2 Is there any difference between

```
delay until Next_Time;
delay Next_Time - Clock;
```

3 Assign the time of the next noon to the variable High_Noon of type Time.

20.4 Protected objects

Consider the problem of protecting a variable V from uncontrolled access. We might consider using a package and two procedures Read and Write

```
package Protected_Variable is
   procedure Read(X: out Item);
   procedure Write(X: in Item);
end;

package body Protected_Variable is
   V: Item := initial_value;

   procedure Read(X: out Item) is
   begin
      X := V;
   end Read;

   procedure Write(X: in Item) is
   begin
      V := X;
   end Write;

end Protected_Variable;
```

However this is very unsatisfactory. Nothing prevents different tasks in our system from calling Read and Write simultaneously and thereby causing interference. As a more specific example, suppose that the type Item is a record giving the coordinates of an aircraft or ship

```
type Item is
   record
      X_Coord: Float;
      Y_Coord: Float;
   end record;
```

Suppose that a task A acquires pairs of values and uses a call of Write to store them into V and that another task B calls Read whenever it needs the latest position. Now assume that A is halfway through executing Write when it is interrupted by task B which promptly calls Read. It is clear that B could get a value consisting of the new *x*-coordinate and the old *y*-coordinate which would no doubt represent a location where the vessel had not been. The use of such data for calculating the heading of the vessel from regularly read pairs of readings could lead to disaster.

Even more subtle problems arise if a task loses control part way through the execution of a statement. Remember that a single Ada statement will typically correspond to several machine statements and an interrupt could occur at any point.

The reader may wonder how the task A could be interrupted by task B anyway. In a single processor system with time slicing it might merely have been that B's turn came just at an unfortunate moment. Another possibility is that B might have a higher priority than A; if B had been waiting for time to elapse before taking the next reading by obeying a delay statement, then A might be allowed to execute and B's delay might expire just at the wrong moment. In practical real-time situations things are always happening at the wrong moment!

The proper solution is to use a protected object rather than a package. A protected object has a distinct specification and body in a similar style to a package or task. The specification provides the access protocol and the body provides the implementation details.

The specification of a protected object is also split into a visible part and a private part. The visible part contains the specifications of subprograms and entries providing the protocol. The private part contains the hidden shared data and also the specifications of any other subprograms and entries which are private to the object. The subprograms and entries declared in the specification (both visible and private parts) are known as protected operations.

Unlike packages and tasks, the body of a protected object cannot declare any data. It can only contain the bodies of its protected operations plus any locally declared subprograms; all data must be in the private part.

Now consider the following

```
protected Variable is
   procedure Read(X: out Item);
   procedure Write(X: in Item);
private
   V: Item := initial_value;
end Variable;

protected body Variable is

   procedure Read(X: out Item) is
   begin
      X := V;
   end Read;

   procedure Write(X: in Item) is
   begin
      V := X;
   end Write;

end Variable;
```

The protected object Variable provides controlled access to the private variable V. The procedure Read enables us to read the current value whereas the procedure Write enables us to update the value. Calls are written in the usual way using the familiar dotted notation

```
Variable.Read(Current_Value);
...
Variable.Write(New_Value);
```

Within a protected body we can have a number of subprograms and the implementation is such that (like a monitor) calls of the subprograms are mutually exclusive and thus cannot interfere with each other. This exclusivity is an intrinsic part of the implementation typically performed using a lock. A lock can be likened to a token such as used on 19th century railways to ensure unique access to a section of track; only the train with the token could proceed. Here only the task with the lock can proceed.

We could alternatively have used a function for the reading

```
function Read return Item is
begin
   return V;
end Read;
```

An important difference between a procedure and function in the protected body is that a procedure can access the private data in an arbitrary manner whereas a function is only allowed read access to the private data. The implementation is consequently permitted to perform the useful optimization of allowing multiple calls of functions at the same time thus automatically solving the basic classic readers and writers problem. However, if multiple calls are not desired, they can be prohibited by the aspect Exclusive_Functions applied to the protected type or to a singleton protected object (this aspect was introduced by AI12-129).

In the above example, the subprograms can always be executed as soon as the lock is available. Sometimes however, a system is such that certain calls cannot always be processed but must await some condition. A good example is provided by the classic bounded buffer. This allows up to N items to be buffered between the producer and the consumer. Just for fun we use a modular type as promised in Section 17.2. Consider the following

```
N: constant := 8;              -- for instance

type Index is mod N;
type Item_Array is array (Index) of Item;

protected type Buffering is
   entry Put(X: in Item);
   entry Get(X: out Item);
private
   A: Item_Array;
   In_Ptr, Out_Ptr: Index := 0;
   Count: Integer range 0 .. N := 0;
end Buffering;
```

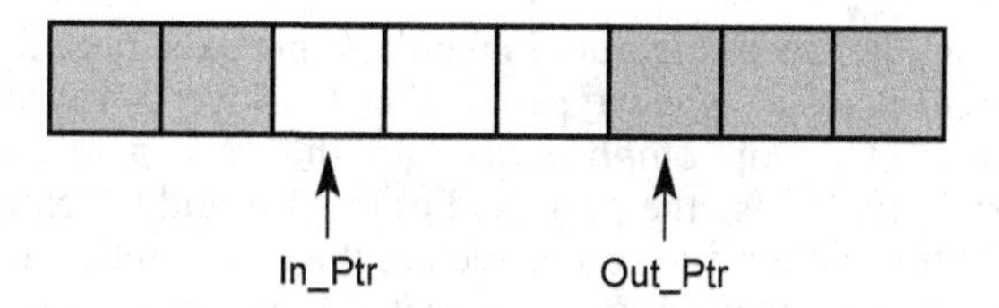

Figure 20.1 The bounded buffer.

```
protected body Buffering is
   entry Put(X: in Item) when Count < N is
   begin
      A(In_Ptr) := X;
      In_Ptr := In_Ptr + 1;  Count := Count + 1;
   end Put;

   entry Get(X: out Item) when Count > 0 is
   begin
      X := A(Out_Ptr);
      Out_Ptr := Out_Ptr + 1;  Count := Count - 1;
   end Get;
end Buffering;
```

The buffer is the array A of length N which is a number set to 8 in this example. Note that the index type of the array is the modular type Index. The variables In_Ptr and Out_Ptr index the next available and the oldest used locations of the buffer respectively and Count is the number of locations of the buffer which are full. The buffer is used cyclically so In_Ptr need not be greater than Out_Ptr. The situation in Figure 20.1 shows a partly filled buffer with Count = 5, In_Ptr = 2, and Out_Ptr = 5. The portion of the buffer in use is shaded. The variables In_Ptr, Out_Ptr, and Count are all initialized to 0 so that the buffer is initially empty.

The objective is to allow items to be added to and removed from the buffer in a first-in–first-out manner but to prevent the buffer from being overfilled or under-emptied. This is done by the introduction of protected entries which always have barrier conditions.

The syntax of an entry body is like that of a procedure body except that it always has a barrier consisting of **when** followed by a condition. An entry body can have a declarative part in contrast to an accept statement which cannot. It can also have access parameters whereas an accept statement cannot.

The example also shows how the notion of a protected object is generalized into a protected type by the addition of the reserved word **type** after **protected**. The type acts as a template for protected objects and individual objects are then declared in the usual way. The subprograms and entries of a particular protected object are called using the dotted notation. We thus write

```
My_Buffer: Buffering;
...
My_Buffer.Put(X);
```

Note the analogy between this notation and the prefixed notation for calling an operation of an object of a tagged type. In that case we can write Obj.Op(*other paras*) rather than P.Op(Obj, *other paras*). In the case here we always write Prot_Obj.Op(*paras*). In a sense the protected object is a hidden extra parameter.

As an aside, note that we have had to declare the constant N, the type Index, and the array type Item_Array external to the protected type; this is because we are only allowed components inside the protected type. There is a general rule that we cannot declare a type inside a type; a similar problem occurred with the type Stack in Section 12.4. These restrictions do not really matter since in practice the protected type is likely to be declared inside a (possibly generic) package.

The behaviour of the protected object is controlled by the barriers. When an entry is called its barrier is evaluated; if the barrier is false then the calling task is queued until circumstances are such that the barrier is true and the lock is released so that some other task can call the object meanwhile. When My_Buffer is declared, the buffer is empty, and so the barrier for Put (the condition Count < N) is true whereas the barrier for Get is false. So initially only a call of Put can be executed and a task issuing a call of Get will be queued.

The statements of the entry bodies copy the item to or from the buffer and then update In_Ptr or Out_Ptr and Count to reflect the new state. Since In_Ptr and Out_Ptr are of the modular type Index, the addition wraps around automatically as required.

At the end of the execution of an entry body (or a procedure body) of the protected object, all barriers which have queued tasks are re-evaluated thus possibly permitting the processing of an entry call which had been queued on a false barrier. So at the end of the first call of Put, if a call of Get had been queued, then the barrier is re-evaluated thus permitting a waiting call of Get to be serviced at once.

It is important to realize that there is no task associated with the buffer itself; the evaluation of barriers is effectively performed by the run-time system. Barriers are evaluated when an entry is first called and when something happens which could sensibly change the state of a barrier with a waiting task.

Thus barriers are only re-evaluated at the end of an entry or procedure body and not at the end of a protected function call because a function call cannot change the state of the protected object and so is not expected to change the values of barriers. These rules ensure that a protected object can be implemented efficiently.

It may happen that two or more entries with queued tasks have true barriers as a result of re-evaluation. In such a case the core language does not specify which queue is serviced. Moreover, within a queue, the core language by default specifies that the calls are serviced in order of arrival. The Real-Time Systems annex enables other policies (such as those based on priorities) to be specified; see Section 26.2.

Note that a barrier *could* refer to a global variable; such a variable might get changed other than through a call of a protected procedure or entry – it could be changed by another task or even by a call of a protected function; such changes will thus not be acted upon promptly. The programmer needs to be aware of this and should not use global variables in barriers without due consideration.

The barrier protection mechanism is superimposed upon the natural mutual exclusion of the protected construct thus giving two distinct levels of protection. At the end of a protected call, already queued entries (whose barriers have now become true) take precedence over other calls contending for the protected object. On the other hand, a new entry call cannot even evaluate its barrier if the protected object is busy with another call until that call (and any similar queued calls) have finished.

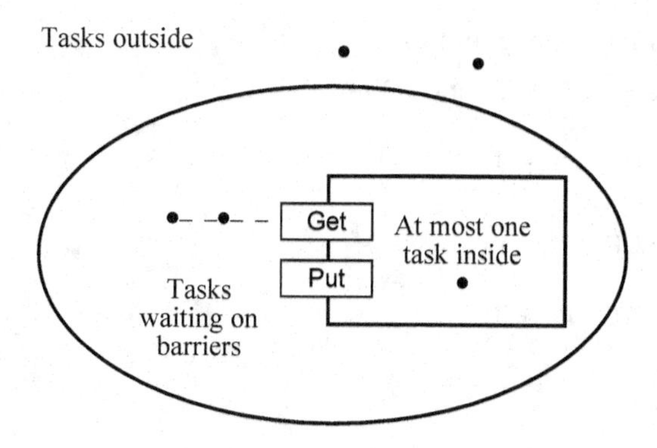

Figure 20.2 The eggshell model.

This has the following important consequence: if the state of a protected resource changes and there is a task waiting for the new state, then this task will gain access to the resource and be guaranteed that the state of the resource when it gets it is the same as when the decision to release the task was made. Unsatisfactory polling and race conditions are completely avoided.

The two-level model has been termed the eggshell model. We can envisage the protected object with its entry queues as surrounded by a shell as illustrated in Figure 20.2. The shell can only be penetrated by a new task trying to call a subprogram or entry when the protected object is quiescent. Tasks can thus be waiting at two levels, outside the shell where they are just milling around in an unstructured way contending for access to the implementation lock which guards the protected object as a whole, and inside the shell in an orderly manner on entry queues. The internal waiting tasks always take priority over the external tasks.

Figure 20.2 illustrates the state when the first two calls were of Get (and thus were queued), the third call was of Put and is being obeyed and two subsequent calls of Put are contending for the lock.

Protected objects are somewhat similar to monitors; they are both passive constructions with synchronization provided by the language run-time system. However, protected objects have the great advantage over monitors in that the protocols are described by barrier conditions (which are fairly easy to prove correct) rather than the low level and unstructured signals internal to monitors as found in Modula.

Protected types enable very efficient implementations of various semaphore and similar paradigms. For example a general semaphore might be implemented as follows

```
protected type Semaphore(Start_Count: Integer := 1) is
   entry Secure;
   procedure Release;
private
   Count: Integer := Start_Count;
end Semaphore;
```

```
protected body Semaphore is
   entry Secure when Count > 0 is
   begin
      Count := Count - 1;
   end Secure;
   procedure Release is
   begin
      Count := Count + 1;
   end Release;
end Semaphore;
```

The entry Secure and the procedure Release correspond to Dijkstra's classic P and V operations (from the Dutch *Passeren* and *Vrijmaken*). This example also illustrates that a protected type can have a discriminant which is here used to provide the initial value of the semaphore or in other words the number of items of the resource being guarded by the semaphore. The discriminant has a default value of one which corresponds to the usual binary semaphore. So we can write

```
S: Semaphore;
...
S.Secure;
...        -- protected statements
S.Release;
```

Observe that we have used the default value for the discriminant. However, this does not mean that S is mutable in the sense that we can change its discriminant as we did with the Mutant in Section 18.3. Protected objects are inherently limited and cannot be changed in any way.

The number of tasks on an entry queue is given by the attribute Count. This is often useful as part of a barrier condition as we shall see later.

Access types can refer to protected operations but the access type has to include the word **protected**. We could write

```
type Read_Ptr is access protected procedure (X: out Item);
Reader: Read_Ptr;
...
Reader := Variable.Read'Access;
```

and then a call of Reader will call the protected procedure Read of the protected object Variable as expected. The usual conformance and accessibility rules apply. An access type can similarly refer to a protected function but there are no access to entry types. Normal access to subprogram types cannot refer to protected operations and vice versa. In fact, protected operations have the special convention Protected. This includes protected operations declared in the private part. For an example see the final version of the protected object Egg in Section 22.6 where the operation Is_Done is in the private part. However, subprograms that are declared locally to the body of a protected object have convention Intrinsic and so the Access attribute cannot be applied to them at all. This permits optimized calling conventions to be used for calls of such internal subprograms.

We conclude by observing that the general principle of a protected operation is that it should be of a short duration. Moreover, certain so-called potentially blocking operations are considered bounded errors if invoked during a protected action; these include: attempting a rendezvous, an entry call, a delay, an abort and creating a task. Note that although we cannot call an entry from within a protected operation, we can call a protected subprogram of the same or another protected object; however, if we call a subprogram of the same protected object then we must do it directly (and not by calling an external subprogram that then calls back in). Finally, we can call an entry using the requeue statement described in Section 20.9.

Exercise 20.4

1. Modify the protected object Variable so that a call of Write is obeyed first.
2. Encapsulate the protected type Buffering in a generic package with the type Item as a parameter and make the size of the buffer a discriminant of the type.
3. Declare a protected object with similar semantics to the task Buffer of Section 20.2.
4. Reconsider Exercise 20.2(1) using a protected object.

20.5 Simple select statements

A protected object passively provides a number of possible services. Sometimes, the passive nature of a protected object is somewhat restrictive and we can then use a task to dynamically provide a number of alternative services. This can be done by the select statement which allows a task to select from one of many rendezvous. In order to illustrate the ideas we will now show alternative solutions to the problem of the protected variable using a task to manage the protected data. Consider

```
task Protected_Variable is
   entry Read(X: out Item);
   entry Write(X: in Item);
end;

task body Protected_Variable is
   V: Item := initial_value;
begin
   loop
      select
         accept Read(X: out Item) do
            X := V;
         end;
      or
         accept Write(X: in Item) do
            V := X;
         end;
      end select;
   end loop;
end Protected_Variable;
```

The task body consists of an endless loop containing a single select statement. A select statement starts with the reserved word **select** and finishes with **end select**; it contains two or more alternatives separated by **or**. In this example each alternative consists of an accept statement – one for Read and one for Write.

The behaviour of each execution of the select statement depends upon whether calls of Read or Write or both or neither have been made

- If neither Read nor Write has been called then the task is suspended until one or the other is called and then the corresponding accept statement is obeyed.
- If calls of Read are queued but none of Write are queued then a call of Read is accepted and vice versa with Read and Write reversed.
- If calls of both Read and Write are queued then an arbitrary choice is made.

Thus each execution of the select statement results in one of its branches being obeyed and one call of Read or Write being dealt with. We can think of the task as corresponding to a person serving two queues of customers waiting for two different services. If only one queue has customers then the server deals with it; if there are no customers then the server waits for the first irrespective of the service required; if both queues exist, the server just makes an arbitrary choice each time.

As with protected entry queues, the core language specifies that task entry queues are serviced in order of arrival by default but remains silent about which queue is serviced when more than one is possible. Again the Real-Time Systems annex allows other policies to be specified; see Section 26.2.

So each time around the loop the task Protected_Variable accepts a call of Read or Write. This prevents multiple access to the variable V since it can only deal with one call at a time but does not impose order upon the calls. Compare this with the task Buffer in Section 20.2 which imposed an order upon the calls of Put and Get.

The following variation on the protected variable introduces a guarding condition that ensures that a value cannot be read until after a first value has been written and then only if no tasks are waiting to write a further value

```
task body Protected_Variable is
   V: Item;
   Initialized: Boolean := False;
begin
   loop
      select
         when Initialized and Write'Count = 0 =>
         accept Read(X: out Item) do
            X := V;
         end;
      or
         accept Write(X: in Item) do
            V := X;
         end;
         Initialized := True;
      end select;
   end loop;
end Protected_Variable;
```

A branch of a select statement can commence with **when** condition => followed by an accept statement and then possibly some further statements. Each time the select statement is encountered any guarding conditions are evaluated. The behaviour is then as for a select statement without guards but excluding those branches for which the conditions were false. After an accept statement a branch may contain further statements. These are executed by the server task as part of the select statement but outside the rendezvous.

So the guards are conditions which have to be true before a service can be offered. The accept statement gives the service to the customer. Any statements after the accept statement are bookkeeping actions performed as a consequence of giving the service but which can be done after the customer has left.

A few points need emphasis. The guards are re-evaluated at the beginning of each execution of the select statement (but their order of evaluation is not defined). An absent guard is taken as true. If all guards turn out to be false then the exception Program_Error is raised. It should be realized that a guard need not still be true when the corresponding rendezvous is performed because it might use global variables and therefore be changed by another task.

The attribute Write'Count is the number of tasks currently on the queue for the entry Write. The use of the Count attribute in guards needs care. It gives the value when the guard is evaluated and can well change before a rendezvous is accepted. It could increase because another task joins the queue – that would not matter in this example. But it could also decrease unexpectedly and this might indeed give problems. Such difficulties do not arise with protected objects.

The simple example of the protected object Variable of Section 20.4 solved the classic reader–writer problem of allowing multiple readers and a single writer. However, as observed in the answer to Exercise 20.4(**1**), the multiple reading optimization is destroyed if barrier conditions are involved. We can overcome this by using a control task encapsulated within a package. Consider

```
package Reader_Writer is
   procedure Read(X: out Item);
   procedure Write(X: in Item);
end;

package body Reader_Writer is
   V: Item;

   task Control is
      entry Start;
      entry Stop;
      entry Write(X: in Item);
   end;

   task body Control is
      Readers: Integer := 0;
   begin
      accept Write(X: in Item) do
         V := X;
      end;
```

```
         loop
            select
               when Write'Count = 0 =>
               accept Start;
               Readers := Readers + 1;
            or
               accept Stop;
               Readers := Readers - 1;
            or
               when Readers = 0 =>
               accept Write(X: in Item) do
                  V := X;
               end;
            end select;
         end loop;
      end Control;

      procedure Read(X: out Item) is
      begin
         Control.Start;  X := V;  Control.Stop;
      end Read;

      procedure Write(X: in Item) is
      begin
         Control.Write(X);
      end Write;

   end Reader_Writer;
```

The task Control has three entries: Write to do the writing and Start and Stop associated with reading. A call of Start indicates a wish to start reading and a call of Stop indicates that reading has finished. The task is wrapped up in a package in order to provide multiple reading access through the procedure Read which can be called reentrantly; it also enforces the protocol of calling Start and then Stop around the statement X := V; the procedure Write merely calls the entry Write.

The task Control declares a variable Readers which indicates how many readers are present. It begins with an accept statement for Write to ensure that the variable is initialized and then enters a loop containing a select statement. This has three branches, one for each entry.

On a call of Start or Stop the count of number of readers is incremented or decremented. A call of Write can only be accepted if the condition Readers = 0 is true. Hence writing when readers are present is forbidden. Of course, since the task Control actually does the writing, multiple writing is prevented and, moreover, it cannot at the same time accept calls of Start and so reading is not possible when writing is in progress. However, multiple reading is allowed as we have seen. Note finally that further reading is not permitted if one or more writers are waiting.

In Section 22.3 we shall see how the various implementations of readers and writers can be brought together as different versions of a synchronized interface.

20.6 Timed and conditional calls

There are also various other forms of select statement. It is possible for one or more of the branches to start with a delay statement rather than an accept statement. Consider

```
select
   accept This( ... ) do
      ...
   end;
or
   accept That( ... ) do
      ...
   end;
or
   delay 10*Minutes;
   ...         -- time-out statements
end select;
```

If neither a call of This nor That is received within ten minutes, then the third branch is taken and the statements following the delay are executed. The task might decide that since its services are no longer apparently required it can do something else or maybe it can be interpreted as an emergency.

As an example consider a process control system where we might expect an acknowledgement from the operator that some action has been taken in response to a request. After a suitable interval we might take our own emergency action if no acknowledgement has been received. This could be written as follows

```
Operator.Call("Put out fire");

select
   accept Acknowledge;
or
   delay 1*Minutes;
   Fire_Brigade.Call;
end select;
```

A delay alternative can be guarded and indeed there could be several in a select statement although clearly only the shortest one with a true guard can be taken. It should be realized that if one of the accept statements is obeyed then any delay is cancelled – we can think of a delay alternative as waiting for a rendezvous with the clock. A delay is, of course, set from the start of the select statement and reset each time the select statement is encountered. Finally, note that it is the start of the rendezvous that matters rather than its completion as far as the time-out is concerned.

Another form of select statement is one with an else part. Consider

```
select
   accept This( ... ) do
      ...
   end;
```

```
   or
      accept That( ... ) do
         ...
      end;
   else
      ...      -- alternative statements
   end select;
```

In this case the final branch is preceded by **else** rather than **or** and consists of just a sequence of statements. The else branch is taken at once if none of the other branches can be immediately accepted. A select statement with an else part is rather like one with a branch starting **delay** 0.0; it times out at once if there are no customers to be dealt with. A select statement cannot have both an else part and delay alternatives.

There is a subtle distinction between an accept statement starting a branch of a select and an accept statement anywhere else. In the first case the accept statement is bound up with the workings of the select statement and is to some extent conditional. In the second case, once encountered, it will be obeyed come what may. The same distinction applies to a delay statement starting a branch of a select statement and one elsewhere. Thus if we change the **or** to **else** in our emergency action to give

```
   select
      accept Acknowledge;
   else
      delay 1*Minutes;
      Fire_Brigade.Call;
   end select;
```

then the status of the delay is quite different. It just happens to be one of a sequence of statements and is obeyed in the usual way. So if we cannot accept a call of Acknowledge at once, we immediately take the else part. The fact that the first statement is a delay is just fortuitous – we delay for one minute and then call the fire brigade. There is no time-out. We see therefore that the simple change from **or** to **else** causes a dramatic difference in meaning – so take care!

If a select statement has an else part then Program_Error can never be raised. The else part cannot be guarded and so will always be taken if all branches have guards and they all turn out to be false.

There are two other forms of select statement which are rather different; they concern a single entry call rather than one or more accept statements. The timed entry call allows a sequence of statements to be taken as an alternative to an entry call if it is not accepted within the specified duration. Thus

```
   select
      Operator.Call("Put out fire");
   or
      delay 1*Minutes;
      Fire_Brigade.Call;
   end select;
```

will call the fire brigade if the operator does not accept the call within one minute. The entry call can be to a task or to a protected object and it is the acceptance of the call that matters rather than its completion.

Finally there is the conditional entry call. Thus

```
select
   Operator.Call("Put out fire");
else
   Fire_Brigade.Call;
end select;
```

will call the fire brigade if the operator cannot immediately accept the call.

Timed and conditional entry calls are quite different from the general select statement. They concern only a single unguarded call and so always have exactly two branches – one with the entry call and the other with the alternative sequence of statements. Timed and conditional calls also apply to entries renamed as procedures. But they never apply to static procedure calls – see Section 22.3.

Timed and conditional calls are useful if a client task does not want to be unduly delayed when a server task or protected object is busy. They allow a client to abandon a request after waiting for a time or, in the conditional case, an impatient client can abandon it at once if not immediately served.

Timed calls on task entries need care particularly if the Count attribute is used since a decision based on its value may be invalidated because of a timed call removing a task from an entry queue. The analogous problem cannot arise with protected objects; the removal of a task from the queue of a protected entry is itself a protected action and causes the re-evaluation of barriers.

The final form of select statement is the asynchronous select statement; this is discussed in Section 20.8.

20.7 Concurrent types and activation

We observed in Section 20.4 how a protected type declaration acted as a template from which a number of similar protected objects could be declared. In the same way, it is sometimes useful to have several similar but distinct tasks. Moreover, it is often not possible to predict the number of such tasks required. For example, we might wish to create distinct tasks to follow each aircraft within the control zone of an air traffic control system. Clearly, such tasks need to be created and disposed of in a dynamic way not related to the static structure of the program. Protected types and task types (which we collectively refer to as concurrent types) follow similar rules which we now look at in more detail. We will concentrate on task types because they have additional properties resulting from their active nature.

A template for similar tasks is provided by a task type declaration. Consider

```
task type T is
   entry E( ... );
end T;

task body T is
   ...
end T;
```

This is identical to the simple task declarations we have seen so far except that the reserved word **type** follows **task** in the specification. The task body follows the same rules as before.

To create an actual task we use the normal form of object declaration thus

```
X: T;
```

and this declares a task X of type T.

Task objects and protected objects can be used in structures in the usual way. Thus we can declare records and arrays containing tasks and protected objects

```
type Rec is
  record
    CT: T;
    Count: Integer;
    Flag: Boolean;
  end record;
R: Rec;

AOT: array (1 .. 10) of T;
```

The entries of such tasks and entries and subprograms of protected objects are called using the object name; thus we write

```
X.E( ... );
AOT(I).E( ... );
R.CT.E( ... );
```

A most important consideration is that concurrent objects are not variables but behave as constants. Thus a task object declaration creates a task which is permanently represented by the object. Hence assignment is not allowed for task and protected types; the equality operations are not predefined either but they could be defined by the user. Task types and protected types are in fact forms of limited types and so could be used as the actual type corresponding to a formal generic parameter specified as limited private and as the full type in a private part corresponding to a limited private type.

In Section 22.3 we shall see that we can declare synchronized interfaces of various kinds which can then be implemented by task types and protected types. Such task types and protected types are tagged types whereas the task types and protected types in this chapter are not. This is not so strange since we have both tagged and untagged record types.

Individual tasks and protected objects do not need any initialization in the usual sense. However, a record of a type containing a task or protected object, such as one of the type Rec above, can be initialized for the sake of the other components. We then have to use the <> notation as a dummy value for the task component thus

```
A_Rec: Rec := (CT => <>, Count => 77, Flag => True);
```

Remember from Section 12.5 that a composite type such as a record with a component of a limited type is itself limited. So we cannot make an assignment to A_Rec but we can use an aggregate as an initial value when it is declared.

Concurrent types may have discriminants including access discriminants. Such discriminants have all the usual properties and parameterize objects of the type. Singleton tasks and protected objects cannot have discriminants.

Subprogram parameters may be of concurrent types; they are always passed by reference and so formal and actual parameters always refer to the same object. All modes have the same effect for task types. For a protected type, the mode **in** allows calls of protected functions only whereas modes **in out** and **out** also allow calls of protected entries and procedures.

Functions cannot have results of concurrent types but they can have results of access types referring to concurrent types. A function can also return a composite type containing a concurrent type and since such a type is limited can only be called as a constructor as described in Section 12.5. We might have

```
function Make_Rec( ... ) return Rec;
...
A_Rec: Rec := Make_Rec( ... );
```

The extended form of return statement will be found useful with composite types containing tasks – an example is the function Make_Process in Section 22.2.

In Section 20.1 we briefly introduced the idea of dependency. Each task is dependent on some unit and there is a general rule that a unit cannot be left until all tasks dependent upon it have terminated.

A task declared as a task object (or using the abbreviated singleton form) is dependent upon the enclosing block, subprogram or task body in which it is declared. Inner packages do not count in this rule – this is because a package is merely a passive scope wall and has no dynamic life. If a task is declared in a package (or nested packages) then the task is dependent upon the block, subprogram or task body in which the package or packages are themselves declared. For completeness, a task declared in a library package is said to depend on that package and we refer to it as a library task. After termination of the main subprogram, the main environment task (which calls the main subprogram) must wait for all library tasks to terminate. Only then does the program as a whole terminate.

We saw earlier that a task becomes active only when the declaring unit reaches the **begin** following the declaration. The execution of a task can be thought of as a two-stage process. The first stage, known as activation, consists of the elaboration of the declarations of the task body whereas the second stage consists, of course, of the execution of its statements. During the activation stage the parent unit is not allowed to proceed. If several tasks are declared in a unit then their activations and the subsequent execution of their statements occur independently and in parallel. But it is only when the activation of all the tasks is complete that the parent unit can continue with the execution of the statements following the **begin** in parallel with the new tasks. As an example consider the behaviour of a block containing the declarations of two tasks A and B

```
declare
   A: T;
   B: T;
begin
   ...
end;
```

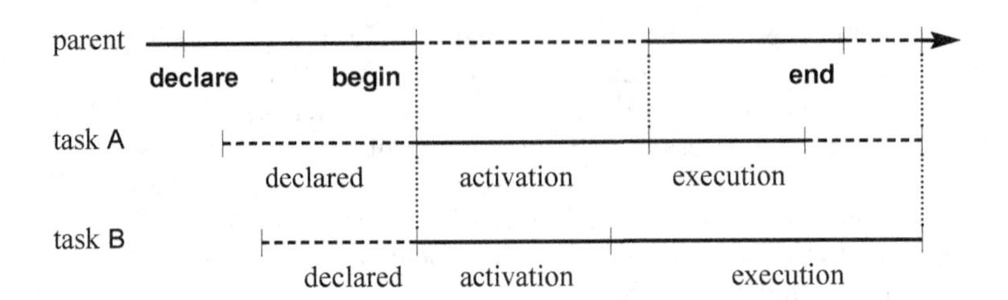

Figure 20.3 Task activation.

The activation process is depicted in Figure 20.3. Time flows from left to right and a solid line indicates that a unit is actively doing something whereas a dashed line indicates that it exists but is suspended. Points of synchronization are indicated by vertical dots between the lines. For the sake of illustration, we show task A finishing its activation after task B so that the parent resumes execution when task A enters its execution stage. We also show task A finishing its execution and therefore becoming completed and terminated before task B. The parent is shown reaching its **end** and therefore completing execution of the block after task A has terminated but before task B has terminated. The parent is therefore suspended until task B is terminated when it can then resume execution with the statements following the block.

One reason for treating task activation in this way concerns exceptions. Recall that an exception raised during the elaboration of declarations is not handled at that level but immediately propagated. So an exception raised in the declarations in a new task could not be handled by that task at all. However, since it is clearly desirable that some means be provided for detecting such an exception, it is obvious that an exception has to be raised in the parent unit. In fact the predefined exception Tasking_Error is raised irrespective of the original exception. It would make life rather difficult if this exception were raised in the parent unit after it had moved on in parallel and so it is held up until all the new tasks have been activated. The exception Tasking_Error is then raised in the parent unit as soon as it attempts to move on from the **begin**. If several of the new tasks raise exceptions during activation then Tasking_Error is only raised once. Such tasks become completed (not terminated) and do not affect sibling tasks being simultaneously activated.

The other thing that can go wrong is that an exception can occur in the declarations of the parent unit itself. In this case any new tasks which have been declared (but of course will not have been activated because the parent unit has not yet reached its **begin**) will automatically become terminated and are never activated at all.

Task objects can be declared in a package and although not dependent upon the package are nevertheless set active at the **begin** of the package body. If the package body has no initialization statements and therefore no **begin**, then a null initialization statement is assumed; if a package has no body, then a body with just a null initialization statement is assumed. So the task Control in the package Reader_Writer of Section 20.5 is set active at the end of the declaration of the package body.

Special rules apply to library task activation if the pragma Partition_ Elaboration_Policy is specified with parameter Sequential. This pragma is defined in the High Integrity Systems annex, see Section 26.6.

Tasks and protected objects can also be created through access types. We write

```
type Ref_T is access T;
```

and then we can create a task using an allocator in the usual way

```
RX: Ref_T := new T;
```

The type Ref_T is a normal access type and so assignment and equality operations on objects of the type are allowed. The entry E of the task accessed by RX can be called as expected by

```
RX.E( ... );
```

Tasks created through access types obey slightly different rules for activation and dependency. They commence activation immediately upon evaluation of the allocator whether it occurs in a sequence of statements or in an initial value – we do not wait until the ensuing **begin**. Moreover, activation is completed before the allocator returns with the access value. Furthermore, such tasks are not dependent upon the unit where they are created but are dependent upon the block, subprogram body or task body containing the declaration of the access type itself.

If tasks are components of a composite object and such an object is created using an allocator, then all the tasks in the object are created and activated before the allocator returns. There are rules that parallel those for declaring several tasks described above. Elaboration and initialization of the components occurs before activation of the tasks and the allocator only returns when all activations are completed. Similarly Tasking_Error can only be raised once and so on.

The reader will probably feel that the activation mechanism is somewhat elaborate. However, in practice, the details will rarely need to be considered. They are mentioned in order to show that the mechanism is well defined rather than because of their everyday importance.

Note that entries in a task can be called as soon as it is declared and even before activation commences – the call will just be queued. However, situations in which this is sensibly possible are rare.

As an example of the use of task types we can rewrite the procedure Iterate of Section 11.8 to use an array of local tasks, one to perform the operation Func for each array element (this might be appropriate if the operation were very time consuming and the machine had multiple processors). We have

```
procedure Iterate(Func: in Math_Function; V: in out Vector) is
   task type Dæmon is
      entry Send(Value: in Float);
      entry Get(Value: out Float);
   end;

   task body Dæmon is
      X: Float;
   begin
```

```
      accept Send(Value: in Float) do
         X := Value;
      end Send;
      X := Func(X);
      accept Get(Value: out Float) do
         Value := X;
      end Get;
   end Dæmon;

   Dæmons: array (V'Range) of Dæmon;
begin
   for I in Dæmons'Range loop
      Dæmons(I).Send(V(I));
   end loop;
   for I in Dæmons'Range loop
      Dæmons(I).Get(V(I));
   end loop;
end Iterate;
```

Another use of task types is for the creation of agents. An agent is a task that does something on behalf of another task. As an example suppose a task Server provides some service through an entry Request. Suppose also that it may take Server some time to provide the service so that it is reasonable for the calling task User to do something else while waiting for the answer to be prepared. There are various ways in which the User could collect the answer. It could call another entry Enquire; the Server task would need some means of recognizing the caller – it could do this by issuing a key on the call of Request and insisting that it be presented again when calling Enquire (or, if the Systems Programming annex is supported, the mechanism of task identification could be used; see Section 26.1). This protocol corresponds to taking something to be repaired, being given a ticket and then having to exchange it when the repaired item is collected later. An alternative approach which avoids the issue of keys, is to create an agent. This corresponds to leaving your address and having the repaired item mailed back to you. We will now illustrate this approach.

First of all we declare a task type as follows

```
task type Mailbox is
   entry Deposit(X: in Item);
   entry Collect(X: out Item);
end;

task body Mailbox is
   Local: Item;
begin
   accept Deposit(X: in Item) do
      Local := X;
   end;
   accept Collect(X: out Item) do
      X := Local;
   end;
end Mailbox;
```

A task of this type acts as a simple mailbox. An item can be deposited and collected later. The general idea is that the user gives the identity of their mailbox to the server so that the server can deposit the item in the mailbox from which the user can collect it later. We need an access type

```
type Address is access Mailbox;
```

The tasks Server and User now take the following form

```
task Server is
   entry Request(A: in Address; X: in Item);
end;

task body Server is
   Reply: Address;
   Job: Item;
begin
   loop
      accept Request(A: in Address; X: in Item) do
         Reply := A;
         Job := X;
      end;
      ...      -- work on job
      Reply.Deposit(Job);
   end loop;
end Server;

task User;

task body User is
   My_Box: Address := new Mailbox;
   My_Item: Item;
begin
   Server.Request(My_Box, My_Item);
   ...      -- do something while waiting
   My_Box.Collect(My_Item);
end User;
```

In practice the user might poll the mailbox from time to time to see if the item is ready. This is easily done using a conditional entry call.

```
select
   My_Box.Collect(My_Item);
   -- item collected successfully
else
   -- not ready yet
end select;
```

It is important to realize that the agent serves several purposes. It enables the deposit and collect to be decoupled so that the server can get on with the next job. Moreover, it means that the server need know nothing about the user; calling the user directly would require the user to be of a particular task type and this would be most unreasonable. The agent enables us to factor off the only property required of the user, namely the existence of the entry Deposit.

If the decoupling property were not required then the agent could be

```
task body Mailbox is
begin
   accept Deposit(X: in Item) do
      accept Collect(X: out Item) do
         Collect.X := Deposit.X;
      end;
   end;
end Mailbox;
```

The agent does not need a local variable in this case since it now only exists in order to aid closely coupled communication. Note the use of the dotted notation in the nested accept statements in order to distinguish the two uses of X; we could equally have written X := Deposit.X; but the use of Collect is more symmetric.

Exercise 20.7

1 Rewrite the agent task as a protected object. What are the advantages and disadvantages of doing this?

20.8 Termination, exceptions, and ATC

A task can become completed and then terminate in various ways as well as running into its final end. It will have been noticed that in some of our earlier examples, the body of a task was an endless loop and clearly never terminated. This means that it would never be possible to leave the unit on which the task was dependent. Suppose, for example, that we chose to use a task implementation of the protected variable type. If we then declared

```
PV: Protected_Variable;
```

we could not leave the unit on which PV depends without terminating the task in some way. We could, of course, add a special entry Stop and call it just before leaving the unit, but this might be inconvenient. Instead it is possible to make a task automatically terminate itself when it is of no further use by a special form of select alternative. We thus rewrite the main loop as

```
loop
   select
      accept Read(X: out Item) do
         X := V;
      end;
   or
      accept Write(X: in Item) do
         V := X;
      end;
   or
```

```
            terminate;
        end select;
    end loop;
```

The terminate alternative is taken if the unit on which the task depends has reached its end and so is completed and all sibling tasks and dependent tasks are terminated or are similarly able to select a terminate alternative. In such circumstances all the tasks are of no use since they are the only tasks that could call their entries and they are all dormant. Thus the whole set automatically terminates. In practice, this merely means that all service tasks should have a terminate alternative and will then terminate themselves without more ado.

A terminate alternative may be guarded. However, it cannot appear in a select statement with a delay alternative or an else part.

Selection of a terminate alternative is classified as normal termination – the task is under control of the situation and terminates voluntarily.

However, the terminate alternative is rarely needed since most applications of servers will use protected objects and they cause no dependency problems because they are passive. In other cases it is possible to use finalization to cause some Stop entry to be called. This has other advantages, since it can cause any cleanup actions to be performed whereas terminate cannot be followed by any other statements. (Terminate was important in Ada 83 but is almost obsolescent now.)

At the other extreme, the abort statement unconditionally terminates one or more tasks. It consists of **abort** followed by a list of task names such as

```
    abort X, AOT(3), RX.all;
```

If a task is aborted then all tasks dependent upon it or a subprogram or block currently called by it are also aborted. If the task is suspended for some reason, then it immediately becomes completed; any delay is cancelled; if the task is on an entry queue, it is removed; other possibilities are that it has not yet even commenced activation or it is at an accept or select statement awaiting a partner. If the task is not suspended, then completion will occur as soon as convenient and certainly no new communication with the task will be possible.

Complications arise if the task is engaged in a rendezvous when it is aborted, since we then also have to consider the effect on the partner. If the called task is aborted, then the calling task receives the exception Tasking_Error. On the other hand, if the calling task is aborted, then the called task is not affected; the rendezvous carries on to completion with the caller in a somewhat abnormal state and it is only when the rendezvous is complete that the caller becomes properly completed. The rationale is simple; if a task asks for a service and the server dies so that it cannot be provided then the customer should be told. On the other hand, if the customer dies, too bad – but we must avoid upsetting the server who might have the database in a critical state.

The rendezvous is thus an example of an *abort deferred region* for the caller. Similarly the operations of a protected object are abort deferred regions since it would be far too disruptive to abort a task while inside such a region. Thus we are assured that a task will not be aborted while 'inside' a protected object. Other abort deferred regions concern controlled types: the Initialize and Finalize operations and assignment of a controlled type are all abort deferred.

Note that the rules for abort are formulated in terms of completing the tasks rather than terminating them. This is because a parent task cannot be terminated until its dependent tasks are terminated and if one of those is the caller in a rendezvous with a third party then its termination will be delayed. Thus completion of the tasks is the best that can be individually enforced and their termination will then automatically occur in the usual way.

A possible use for the abort statement might be in an exception handler. Remember that we cannot leave a unit until all dependent tasks are terminated. Hence, if an exception is raised in a unit, then we cannot tidy up that unit and propagate the exception on a layered basis while dependent tasks are still alive and so one of the actions of tidying up might be to abort all such tasks. Thus the procedure Clean_Up of Section 15.2 might do just this.

The abort statement is very disruptive and should only be used in extreme situations. It might be appropriate for a command task to abort a complete subsystem in response to an operator command. Even so, it is probably always best to attempt a controlled shutdown and only resort to the abort statement as a desperate measure. Statements in the command task might be as follows

```
select
   T.Closedown;
or
   delay 60*Seconds;
   abort T;
end select;
```

If the slave task does not accept the Closedown call within a minute, then it is ruthlessly aborted. We are assuming that the slave task polls the Closedown entry at least every minute using a conditional accept statement such as

```
select
   accept Closedown;
   -- tidy up and die
else
   -- carry on normally
end select;
```

If we cannot trust the slave to close down properly even after accepting the entry call, then the command task can always issue an abort after a due interval just in case. Aborting a task which has already terminated has no effect. So the command task might read

```
select
   T.Closedown;
   delay 10*Seconds;
or
   delay 60*Seconds;
end select;
abort T;
```

Of course, even this is not foolproof since the malevolent slave might continue for ever in the rendezvous itself

```
accept Closedown do
  loop
    Put_Line("Can't catch me");
  end loop;
end Closedown;
```

Some minimal degree of cooperation is obviously needed!

The status of a task T can be ascertained by the use of two attributes. Thus T'Terminated is true if a task is terminated. The other attribute, T'Callable, is true unless the task is completed or terminated or in the abnormal state pending final abortion. The use of these attributes needs care. For example, between discovering that a task has not terminated and taking some action based on that information, the task could become terminated. However, the reverse is not possible since a task cannot be restarted and so it is quite safe to take an action based on the information that a task has terminated.

In Section 20.1 we mentioned that if a task has no dependents then completion and termination effectively occur together. However, completion and termination might not be exactly simultaneous and so it is possible for both T'Terminated and T'Callable to be false even though T has no dependents.

We continue by discussing a few remaining points on exceptions. The exception Tasking_Error is concerned with general communication failure. As we have seen, it is raised if a failure occurs during task activation and it is also raised in the caller of a rendezvous if the server is aborted. In addition, no matter how a task is completed, all tasks still queued on its entries receive Tasking_Error. Similarly calling an entry of a task that is already completed also raises Tasking_Error in the caller.

If an exception is raised during a rendezvous (as a consequence of an action by the called task) and is not handled by the accept statement, then it is propagated into both tasks as the same exception on the grounds that both need to know. Of course, if the accept statement handles the exception internally, then that is the end of the matter anyway.

It might be convenient for the called task, the server, to inform the calling task, the user, of some event by the explicit raising of an exception. In such a case it is likely that the server task will not wish to take any action and so a null handler will be required. So in outline we might write

```
begin
  select
    accept E( ... ) do
      ...
      raise Error;          -- tell user
      ...
    end E;
    ...
  end select;
exception
  when Error =>
    null;                   -- server forgets
end;
```

If an exception is not handled by a task at all, then, like the main subprogram, the task is abandoned and the exception is lost; it is not propagated to the parent unit because it would be too disruptive to do so. If the Systems Programming annex is implemented then the lost exception can be detected by using the facilities of the package Ada.Task_Termination, see Section 22.5. In any event, it might be good practice for all significant tasks to have a general handler at the outermost level in order to guard against the loss of exceptions and consequential silent death of the task.

Finally, in the case of a protected object, an exception raised during a protected operation and not handled locally is propagated to the calling task.

We conclude this section by considering the final form of select statement which is used for asynchronous transfer of control (or ATC for short). ATC enables an activity to be abandoned if some condition arises (such as running out of time) so that an alternative sequence of statements can be executed instead. This gives the capability of performing mode changes.

The general effect can of course be programmed by the introduction of an agent task and the use of the abort statement but this is a heavy solution not at all appropriate for most applications needing a mode change.

Asynchronous transfer of control is achieved by a form of select statement which comprises two parts: an abortable part and a triggering alternative. As a simple example consider

```
select
   delay 5.0;                           -- triggering alternative
   Put_Line("Calculation did not complete");
then abort
   Invert_Giant_Matrix(M);              -- abortable part
end select;
```

The general idea is that if the statements between **then abort** and **end select** do not complete before the expiry of the delay then they are abandoned and the statements following the delay are executed instead. Thus if we cannot invert our giant matrix in five seconds we give up and print a message. However, this only works if there are so-called abort completion points in the abortable part. We can ensure this if the matrix inverter does **delay** 0.0; from time to time. Or more elegantly calls Yield as mentioned in Section 26.2. (See AI12-98.)

The statement that triggers the abandonment can alternatively be an entry call instead of a delay statement. If the call returns before the computation is complete then again the computation is abandoned and any statements following the entry call are executed instead. On the other hand, if the computation completes before the entry call, then the entry call is itself abandoned. The entry call can, of course, be to a task or to a protected object.

One possible scenario is where an iterative calculation is allowed to proceed until the answer is required. Two tasks are involved, a slave task to do the calculation and a controlling task which uses the answer and moreover decides when it wants the answer. The best estimate is to be used irrespective of whether the computation has completely converged.

The best estimate to date can be kept in a protected object such as

```
protected Result is
   procedure Set_Estimate(X: in Data);
   function Get_Estimate return Data;
private
   The_Estimate: Data;
end;
```

The slave task calls the protected procedure Set_Estimate each time it produces a better estimate. The current best estimate can then be retrieved at any time by the controlling task by calling the protected function Get_Estimate.

The slave task performs its iterative calculation in the abortable part of an asynchronous select statement. The triggering event of this asynchronous select statement is used to indicate that the slave task is to stop; it can be a call on an entry Wait of yet another protected object which also has a protected procedure Signal. Thus

```
protected Trigger is
   entry Wait;
   procedure Signal;
private
   Flag: Boolean := False;
end;
```

The use of such protected objects for signalling is discussed in more detail in the next section.

The controlling task might then execute

```
Trigger.Signal;
Final_Answer := Result.Get_Estimate;
```

where the call Trigger.Signal causes the entry Wait to be accepted and this in turn stops the slave task.

An important advantage of this approach is that the slave task is decoupled from the controlling task and is able to perform its computation without having to poll some shared variable to see whether it should stop.

Exercise 20.8

1 Rewrite the protected type Buffering of Section 20.4 as a task type and then modify it so that it has the following specification

```
task type Buffering is
   entry Put(X: in Item);
   entry Finish;
   entry Get(X: out Item);
end;
```

The writing task calls Put and finally calls Finish. The reading task calls Get; a call of Get when there are no further items raises the global exception Done.

2 What happens in the following bizarre situation

```
task body Server is
begin
   accept E do
      abort Caller;
      raise Havoc;
   end E;
end Server;

task body Caller is
begin
   Server.E;
exception
   when Havoc =>
      Put_Line("What a mess");
end Caller;
```

3 Sketch the asynchronous select statement containing the iterative calculation performed by the slave task.

20.9 Signalling and scheduling

In this section we consider a few standard paradigms concerning the signalling of events between tasks and the scheduling of tasks.

The simplest form of signal is the persistent signal where the signalling of the event allows one waiting task to proceed. This can done by the following

```
protected Event is
   entry Wait;
   procedure Signal;
private
   Occurred: Boolean := False;
end Event;

protected body Event is

   entry Wait when Occurred is
   begin
      Occurred := False;
   end Wait;

   procedure Signal is
   begin
      Occurred := True;
   end Signal;

end Event;
```

Events of this kind are called persistent since, even if no task is waiting when the event occurs, it is remembered until some task issues a wait. Such events are very similar to binary semaphores with the wait operation corresponding to P and the signal to V. See Section 20.4.

Another form of signal is the broadcast signal; this allows all the currently waiting tasks to proceed and is then forgotten. So if there are no waiting tasks then the signal has no effect. This can be programmed in many ways and can be used to illustrate the requeue statement.

It sometimes happens that a service needs to be provided in two parts and that the calling task has to be suspended after the first part until conditions are such that the second part can be done.

In the example of the broadcast signal, the difficulty is to prevent tasks that call the wait operation after the event has occurred, but before the signal can be reset, from getting through. In other words, we must reset the signal in preference to letting new tasks through. The requeue statement allows us to program such preference control. An implementation is

```
protected Event is
   entry Wait;
   entry Signal;
private
   entry Reset;
   Occurred: Boolean := False;
end Event;

protected body Event is

   entry Wait when Occurred is
   begin
      null;                        -- note null body
   end Wait;

   entry Signal when True is       -- barrier is always true
   begin
      if Wait'Count > 0 then
         Occurred := True;
         requeue Reset;            -- requeue
      end if;
   end Signal;

   entry Reset when Wait'Count = 0 is
   begin
      Occurred := False;
   end Reset;

end Event;
```

In contrast to the persistent signal, the Boolean variable Occurred is normally false and is only true while tasks are being released. The entry Wait has a null body and just exists so that calling tasks can suspend themselves on its queue while waiting for Occurred to become true.

The entry Signal has a permanently true barrier and so is always processed. If there are no tasks on the queue of Wait (that is no tasks are waiting), then there is nothing to do and so it exits. On the other hand, if there are tasks waiting then it must release them in such a way that no further tasks can get on the queue and, moreover, it must then regain control so that it can reset the flag. It does this by requeuing itself on the entry Reset after setting Occurred to true to indicate that the event has occurred.

The semantics of requeue are such that this completes the action of Signal. However, remember that at the end of the body of a protected entry or procedure the barriers are re-evaluated for those entries which have tasks queued. In this case there are indeed tasks on the queue for Wait and there is also a task on the queue for Reset (the task that called Signal in the first place); the barrier for Wait is now true but of course the barrier for Reset is false since there are still tasks on the queue for Wait. A waiting task is thus allowed to execute the body of Wait (being null this does nothing) and the task thus proceeds and then the barrier evaluation repeats. The process continues until all the waiting tasks have gone when finally the barrier of Reset also becomes true. The original task which called signal now executes the body of Reset thus resetting Occurred to false so that the system is once more in its initial state. The protected object as a whole is now finally left since there are no waiting tasks on any of the barriers.

Note carefully that if any tasks had tried to call Wait or Signal while the whole process was in progress then they would not have been able to do so because the protected object as a whole was busy. This illustrates the two levels of protection and is the underlying reason why a race condition does not arise.

Another consequence of the two levels is that it still all works properly even in the face of such difficulties as timed and conditional calls and aborts. The reader may recall, for example, that by contrast, the Count attribute for entries in tasks cannot be relied upon in the face of timed entry calls.

In the case of a protected object a queued entry call can still disappear from the queue as a consequence of abort or a timed call but such removal is treated as a protected operation of the protected object and can only be performed when the object is quiescent. (Remember that protected operations are abort deferred, as mentioned in Section 20.8.) After such removal any barrier using the Count attribute for that queue will be immediately re-evaluated so that consistency is maintained. Removing a task from a queue might thus allow a task queued on a different queue to proceed.

A minor point to note is that the entry Reset is declared in the private part of the protected type and thus cannot be called from outside.

The above example has been used for illustration only. The astute reader will have observed that the condition is not strictly needed inside Signal; without it the caller will simply always requeue and then immediately be processed if there are no waiting tasks. But the condition clarifies the description. Indeed, we can actually program this example without using requeue at all; this is left as an exercise.

Note that we can requeue on an entry renamed as a procedure in much the same way as we can do timed calls on an entry renamed as a procedure.

We now illustrate a quite general technique which effectively allows the requests in a single entry queue to be handled in an arbitrary order. Consider the problem of allocating a group of resources from a set. We do not wish to hold up a later request that can be satisfied just because an earlier request must wait for the release of some of the resources it wants. One essence of the problem is that requests typically have to be handled in stages; the first stage is to say what is wanted and indeed it might be possible to satisfy the request at that stage but typically the task will then have to wait until conditions are appropriate. A practical difficulty is that guards and barriers cannot use the parameters of the call. This inevitably leads to several entry calls and requires the requeue statement.

We suppose that the resources are represented by a discrete type Resource. We can conveniently use the generic package Set_Of from Section 19.2

```
package Resource_Sets is new Set_Of(Resource);
use Resource_Sets;
...
protected Resource_Allocator is
   entry Request(S: in Set);
   procedure Release(S: in Set);
private
   entry Again(S: in Set);
   Free: Set := Full;
   Waiters: Integer := 0;
end Resource_Allocator;

protected body Resource_Allocator is

   procedure Release(S: in Set) is
   begin
      Free := Free + S;       -- return resources
      Waiters := Again'Count;
   end Release;

   entry Request(S: in Set) when True is
   begin
      if S <= Free then
         Free := Free - S;   -- allocation successful
      else
         requeue Again;      -- no good, try later
      end if;
   end Request;

   entry Again(S: in Set) when Waiters > 0 is
   begin
      Waiters := Waiters - 1;
      if S <= Free then
         Free := Free - S;   -- allocation successful
      else
         requeue Again;      -- no good, try later
      end if;
   end Again;

end Resource_Allocator;
```

This protected object has an entry Request and a procedure Release which have as parameters the set S of resources to be acquired or returned; the type Set is from the instantiation of Set_Of.

There is also a private entry Again which is very similar to the entry Request. They both check the set S against the set Free of available resources using the inclusion operator "<=" from (the instantiation of) Set_Of. If all the resources are available, Free is altered using the symmetric difference operator "–" from Set_Of. The procedure Release returns the resources passed as the parameter S by updating Free using the union operator "+" from Set_Of. Note that the declaration of Free gives it the initial value Full also from Set_Of.

A first attempt at acquiring the resources is made by calling the entry Request. If it fails then it requeues on the entry Again. The guard of Request is just True since a new call is always allowed an immediate attempt but it has to be an entry because only an entry can do a requeue. A call of Release allows all those tasks waiting on Again to have another try. The variable Waiters is set to the number of tasks waiting on the entry Again and is decremented by Again thus allowing each waiting task just one further attempt. The entry Again is similar to Request and requeues on itself if the request still fails.

This solution works perfectly; the Count attribute is always correct and no race conditions arise (a task is requeued at once and cannot get out of order because of the first-in–first-out entry queueing policy (but see Section 26.2) and new tasks cannot even enter the system). It is also proof against aborting tasks (except that resources might get lost but that can be remedied by using a controlled type for the resources); timed and conditional calls cause no problems and so on. This is all because the two-level nature of the protected object provides preference control (existing queued or requeued calls take preference over new ones) and so prevents race conditions.

Our examples of requeue have shown an entry in a protected object requeuing on another or the same entry of the same protected object. In fact requeue can be from any entry E1 to any other entry E2 including to and from and between entries of tasks. A requeue to E2 can either pass on all the parameters of the original call to E1 (implicitly) or none. Thus the destination entry E2 must either have a parameter profile with subtype conformance to the original call of E1 (in which case all the parameters are passed on) or no parameters at all. In either case the requeue statement has no explicit parameters.

Remember that entries can also have pre- and postconditions. If entry E1 does a requeue on E2 then any precondition on E2 is checked. Moreover, any postcondition on E2 must imply any postcondition on E1 (see AI12-90).

Abort also raises issues with regard to requeue. Because a requeue is normally seen as continuing the same service as was asked for by the original call, a requeued call is normally treated specially and is not allowed to be aborted since this might mess up the internal structures of the protected object. If this special treatment is not required (as here) then we can requeue with abort thus

```
requeue Again with abort;
```

We conclude this section by introducing the concept of families of entries. There are occasions when we want to compute in some way which of various entries to call. Of course we could write a case statement but it is simpler to use an entry family which is rather like a one-dimensional array of entries.

For example, requests with priorities can be handled by a family of entries. Suppose we have three levels of priority given by

```
type Priority is (Urgent, Normal, Low);
```

and that we have a task Controller providing access to some action on a type Data but with requests for the action on three queues according to their priority. This could be done with three distinct entries but it is neater to use an entry family thus:

```
task Controller is
   entry Request(Priority) (D: Data);
end;

task body Controller is
begin
   loop
      select
         accept Request(Urgent) (D: Data) do
            Action(D);
         end;
      or
         when Request(Urgent)'Count = 0 =>
         accept Request(Normal) (D: Data) do
            Action(D);
         end;
      or
         when Request(Urgent)'Count = 0 and
                      Request(Normal)'Count = 0 =>
         accept Request(Low) (D: Data) do
            Action(D);
         end;
      end select;
   end loop;
end Controller;
```

Request is a family of entries, indexed by a discrete range which in this case is the type Priority.

A protected object can also have entry families but in this case there is just one body for the whole family which takes the form

```
entry Request (for P in Priority) (D: Data) when Barrier_Condition(P) is
begin
   Action(D);
end Request;
```

where the barrier condition can depend upon the entry index. It might even be

```
function Barrier_Condition(P: Priority) return Boolean is
   Result: Boolean;
begin
   case P is
      when Urgent =>
         Result := True;
```

```
            when Normal =>
               Result := Request(Urgent)'Count = 0;
            when Low =>
               Result := Request(Urgent)'Count = 0 and
                              Request(Normal)'Count = 0;
         end case;
         return Result;
      end Barrier_Condition;
```

which would be declared within the protected object.

Clearly the approach shown here is only feasible if the number of priority values is small. In the case of the protected object, the case statement can easily be replaced by a loop but in the case of the task we would have to write out the various entry bodies which would be irksome.

As an alternative, we could try checking each queue in turn thus

```
task body Controller is
begin
   loop
      for P in Priority loop
         select
            accept Request(P) (D: Data) do
               Action(D);
            end;
            exit;
         else
            null;
         end select;
      end loop;
   end loop;
end Controller;
```

Sadly, this is not satisfactory since the task Controller continuously polls when all the queues are empty. We need a mechanism whereby the task can wait for the first of any requests. This can be done by a two-stage process; the calling task first 'signs in' by calling a common entry and then requeues on the required entry of the family. The details are left as an exercise for the reader.

Exercise 20.9

1. In the protected object Event for the broadcast signal would it be sensible for the requeue on Reset to be **with abort**?
2. Write the protected object Event without the use of requeue. Hint: the last task out switches off the light.

3 Rewrite the task Controller in a way which avoids continuous polling. The specification should be

```
task Controller is
   entry Sign_In(P: Priority; D: Data);
private
   ...
end;
```

where the private part contains the entry family.

4 Show how a protected object could be used as the interface to a task which performs some laborious operation and can be used to monitor the time of calls on the task. Use a requeue from the protected object to the task.

20.10 Summary of structure

In this final section on tasking we briefly summarize the main differences between packages, tasks, and protected objects.

Tasks, protected objects, and packages have a superficial lexical similarity – they all have specifications (with private parts) and bodies. However, there are many differences

- A task is an active construction whereas a package or protected object is passive.
- A task can only have entries in its specification and aspect clauses associated with the entries. A package can have anything except entries. A protected object can have entries and subprograms as well as the data.
- Task types and protected types are types. Packages are not types. Both task types and protected types can have access discriminants; packages cannot.
- A package can be generic but a task type or protected type cannot.
- A package can appear in a use clause but a task or protected object cannot.
- A package can be a library unit but a task or protected object cannot. However, a task or protected body can be a subunit.

The overall distinction is that the package should be considered to be the main tool for structuring purposes whereas the task and protected object provide concurrency. Thus typical subsystems will consist of a (possibly generic) package containing one or more tasks interacting through protected objects.

Note that a task can have a private part but like the visible part this can only contain entries; such entries can only be called by local tasks.

Another closely related but somewhat different matter is the introduction of parallel blocks and loops into Ada 2022. These are discussed in Section 22.8.

Our final example illustrates the use of protected types as private types by the following generic package which provides a general type Buffer.

```
generic
   type Item is private;
package Buffers is
   type Buffer(N: Positive) is limited private;
   procedure Put(B: in out Buffer; X: in Item);
   procedure Get(B: in out Buffer; X: out Item);
private
   type Item_Array is array (Integer range <>) of Item;

   protected type Buffer(N: Positive) is
      entry Put(X: in Item);
      entry Get(X: out Item);
   private
      A: Item_Array(1 .. N);
      In_Ptr, Out_Ptr: Integer := 1;
      Count: Integer := 0;
   end Buffer;

end;

package body Buffers is

   protected body Buffer is
      entry body Put(X: in Item) when Count < N is
      begin
         A(In_Ptr) := X;  In_Ptr := In_Ptr mod N + 1;  Count := Count + 1;
      end Put;

      entry body Get(X: out Item) when Count > 0 is
      begin
         X := A(Out_Ptr); Out_Ptr := Out_Ptr mod N + 1; Count := Count - 1;
      end Get;
   end Buffer;

   procedure Put(B: in out Buffer; X: in Item) is
   begin
      B.Put(X);
   end Put;

   procedure Get(B: in out Buffer; X: out Item) is
   begin
      B.Get(X);
   end Get;

end Buffers;
```

The buffer is implemented as a protected object which contains the storage for the buffer. (Remember that a protected type is limited and can thus be the full type corresponding to a limited private type.) Calls of the procedures Put and Get access the buffer by calling the entries of the protected object.

The length of the buffer has been passed as a discriminant rather than as a generic parameter to avoid unnecessary instantiations. In any event we can no longer use a modular type for the indexing because N is not static (nor would it be if a generic parameter). Also, the bounds of the array have to be 1 and N because a discriminant used as a bound must be used directly and not as part of a larger expression. And moreover, we cannot use the discriminant to give range constraints for In_Ptr, Out_Ptr, and Count (see Section 18.2).

Other arrangements are possible. For example the data could be outside the protected object itself and the protected object could then access the data by using a self-referential access discriminant as we shall see in Section 22.4.

Exercise 20.10

1 A boot repair shop has one man taking orders and three others actually repairing the boots. The shop has storage for 100 boots awaiting repair. The person taking the orders notes the address of the owner and this is attached to the boots; he then puts the boots in the store. Each repairman takes boots from the store when he is free, repairs them and then mails them.

Write a package Cobblers whose specification is

```
package Cobblers is
   procedure Mend(A: Address; B: Boots);
end;
```

The package body should contain four tasks representing the various men. Use an instantiation of the package Buffers to provide a store for the boots and agent objects as mailboxes to deliver them.

Checklist 20

A task is active whereas a package or protected object is passive.

A task specification can contain only entries.

A protected body cannot contain data.

A task or protected object cannot be generic.

A task or protected object name cannot appear in a use clause.

Entries may be overloaded and renamed as procedures.

The Count attribute can only be used inside the task or protected object owning the entry.

An accept statement must not appear in a subprogram.

The order of evaluation of guards is not defined.

A select statement can have just one of an else part, a single terminate alternative, one or more delay alternatives.

A terminate or delay alternative can be guarded.

Several alternatives can refer to the same entry.

Beware of the Count attribute in guards; it is quite safe in barriers.

Task and protected types are inherently limited.

A task declared as an object is dependent on a block, subprogram or task body but not an inner package.

A task created by an allocator is dependent on the block, subprogram or task body containing the access type definition.

A task declared as an object is made active at the following (possibly notional) **begin**.

A task created by an allocator is made active at once.

Do not use **abort** without good reason.

A requeue statement can only occur in an entry body or accept statement.

Requeue must be to a parameterless entry or an entry with the same parameter profile.

The value of Year_Number'Last in package Ada.Calendar was changed from 2099 to 2399 in Ada 2005.

The child packages of Calendar were added in Ada 2005.

Timed and conditional calls could not be done on an entry renamed as a procedure in Ada 95.

The mechanism of extending tasks and protected objects from interfaces was added in Ada 2005.

Entries can have pre- and postconditions in a similar manner to subprograms in Ada 2012.

Requeue is permitted on an entry renamed as a procedure in Ada 2012.

New in Ada 2022

There are restrictions on internal calls of a protected function to prevent race conditions, see AI12-166. This can only apply in a precondition or a default expression for a parameter of a protected operation.

The aspect Nonblocking is introduced and the aspect Max_Entry_Queue_Length is introduced for entries.

Aspect specifications are also allowed on entry bodies.

Conflict checks are introduced, see Section 22.9.

Parallel blocks and loops are introduced to enable greater use of multicore architectures, see Section 22.8.

21 Object Oriented Techniques

21.1 Extension and composition
21.2 Using interfaces
21.3 Mixin inheritance
21.4 Linked structures
21.5 Iterators
21.6 Generalized iteration
21.7 Procedural iterators
21.8 Object factories
21.9 Controlling abstraction

We have now introduced all the main features of Ada 2022 and illustrated them with many examples. An important topic which has arisen in various forms is object oriented programming. Chapter 14 introduced type extension and dynamic polymorphism using tagged types. Chapter 18 discussed type parameterization by discriminants. Chapter 19 covered static polymorphism through genericity. And finally Chapter 20 on tasking discussed concurrent objects which some might feel are the only true objects anyway.

However, the linear exposition has prevented us from exploring a number of important interactions between these features. Accordingly this chapter addresses a number of examples which further illustrate the various techniques and show how they fit together. It also introduces two interesting features, a mechanism for generalized iteration widely used by the containers in Chapter 24 and procedural iterators which provide a convenient shorthand.

The next chapter addresses further examples of tasking.

21.1 Extension and composition

We start this chapter on object oriented techniques with some general remarks about type composition.

There are a number of ways in which types can be put together. An array type is composed of two other types, the index type and the component type. A record type is composed of the several types of its components. Type extension is a special form of composition in which a record type has the same components as another type plus additional components at the same level.

As an example consider the following two declarations of the type Circle

```
type Circle is new Object with                -- type extension
  record
    Radius: Float;
  end record;

type Circle is                                -- type composition
  record
    Obj: Object;
    Radius: Float;
  end record;
```

Although both have three eventual components, the structure and naming are quite different. The fundamental difference is between 'X is a Y' and 'X has a Y'. The circle *is* an object and so the type extension is the appropriate form.

It is interesting to contrast this with

```
type Cylinder is new Circle with
  record
    Height: Float;
  end record;
```

briefly introduced in Section 14.1. This type Cylinder is foolish; a cylinder is *not* a circle although it does have a circle as part of its structure.

In deciding the correct approach, a good test is to check that class wide operations still apply – especially since they cannot be overridden. But clearly Distance is curious because the distance to the centre of the cylinder will depend upon whether we are thinking of the cylinder as standing on the plane or not. It also inherits Area which is obviously wrong since it is a rather different concept in three dimensions. Even if we do provide a definition of Area that is sensible (the area of the surface), the application of the class wide function Moment of Section 3.3 will give a silly answer. The general point is that we have extended out of the domain in which we were originally thinking. A typical pitfall of OOP if misused. Clearly we should write

```
type Cylinder is tagged
  record
    Base: Circle;
    Height: Float;
  end record;
```

which reflects that a cylinder *has* a circle.

Language philosophers get very excited about multiple inheritance. As mentioned in Section 14.8, multiple inheritance can cause problems. The general idea would be to allow one type to have two (or more) parents of equal status. This gets into trouble if the parents have a common ancestor as illustrated in Figure 21.1.

The awkward issues to be resolved are whether the type Child has two copies of the components of the type Ancestor and what happens if the same operation is inherited from both parents and possibly derived from Ancestor and overridden differently on the way.

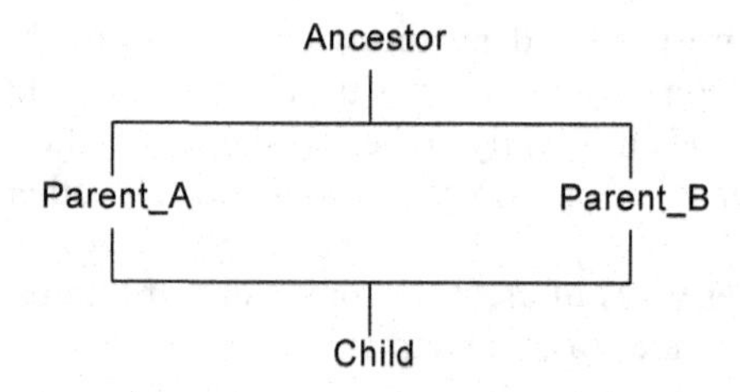

Figure 21.1 The problems of multiple inheritance.

The basic problem is the symmetry and this has to be broken in some way. Various mechanisms are available.

As we saw in Section 14.8, the main Ada solution is to permit multiple inheritance but to break the symmetry by allowing only one ancestor to have components and to allow only one ancestor to have concrete operations. This is done by the introduction of interfaces which in essence are abstract tagged types without components or concrete operations. Only one ancestor (the parent) is allowed to be a normal tagged type and all others must be interfaces (the progenitors).

The symmetry can be broken in other ways using a mixture of inheritance and composition. We inherit from one type and then add other properties as inner components. Consider

```
type Tried_And_Trusted is tagged ...

type Wizard_Stuff is ...
procedure Magic(W: Wizard_Stuff);

type Adventure is new Tried_And_Trusted with
  record
    Spice: Wizard_Stuff;
  end record;
```

where the new type Adventure has all the operations of Tried_And_Trusted but none of those of Wizard_Stuff. However, we can effectively give it the magic operation by simply writing

```
procedure Magic(A: Adventure) is
begin
  Magic(A.Spice);
end Magic;
```

This may seem tedious but in fact often both relationships are 'has a' relationships in which case type extension is probably not appropriate anyway. Composition through explicit components is very clear and the operations then have an explicit ancestry visible at each call.

Other examples are provided by the two versions of the type Linked_Set of Section 14.9. One version directly inherits the abstract set property by being derived from the type Set and indirectly inherits the controlled property since the component Inner is derived from the type Controlled. The other version uses explicit multiple inheritance.

We will see further ways of creating composite structures and explicit multiple inheritance in the remainder of this chapter.

Exercise 21.1

1 Declare appropriate functions Area and Volume for the second type Cylinder. What happens if we attempt to apply the function Moment of Section 3.3 to an object of type Cylinder?

21.2 Using interfaces

The general ideas of interfaces and some simple examples were described in Chapter 14. We will now discuss a more elaborate example.

Before doing so it is important to emphasize that interfaces cannot have components and therefore if we are to perform multiple inheritance then we should think in terms of abstract operations to read and write components rather than the components themselves. This is standard OO thinking anyway because it preserves abstraction by hiding implementation details.

Thus rather than having a component such as Comp it is better to have a pair of operations. The function to read the component can simply be called Comp. A procedure to update the component might be Set_Comp.

Suppose now that we want to print images of geometrical objects. We will assume that the root type is declared as

```
package Geometry is
   type Object is abstract tagged private;
   procedure Move(O: in out Object'Class; X, Y: in Float);
   ...
private
   type Object is abstract tagged
      record
         X_Coord: Float := 0.0;
         Y_Coord: Float := 0.0;
      end record;
   ...
end;
```

The type Object is private and by default both coordinates have the value of zero. The procedure Move, which is class wide, enables any object to be moved to the location specified by the parameters.

Suppose also that we have a line drawing package with the following specification

```
package Line_Draw is
   type Printable is interface;
   type Colour is ... ;
   type Points is ... ;
   procedure Set_Hue(P: in out Printable; C: in Colour) is abstract;
   function Hue(P: Printable) return Colour is abstract;
   procedure Set_Width(P: in out Printable; W: in Points) is abstract;
   function Width(P: Printable) return Points is abstract;
   type Line is ... ;
   type Line_Set is ... ;
   function To_Lines(P: Printable) return Line_Set is abstract;
   procedure Print(P: in Printable'Class);
private
   procedure Draw_It(L: Line; C: Colour; W: Points);
end Line_Draw;
```

The idea of this package is that it enables the drawing of an image as a set of lines. The attributes of the image are the hue (of type Colour) and the width of the lines (in Points) and there are pairs of subprograms to set and read these properties of any object of the interface Printable and its descendants. These operations are of course abstract.

In order to prepare an object in a form that can be printed it has to be converted to a set of lines. The function To_Lines converts an object of the type Printable into a set of lines; again it is abstract. The details of various types such as Line and Line_Set are not shown.

Finally, the package Line_Draw declares a concrete procedure Print which takes an object of type Printable'Class and does the actual drawing using the slave procedure Draw_It declared in the private part. Note that Print is class wide and is concrete. This is an important point. Although all primitive operations of an interface must be abstract this does not apply to class wide operations since these are not primitive.

The body of the procedure Print could take the form

```
procedure Print(P: in Printable'Class) is
   L: Line_Set := To_Lines(P);
   A_Line: Line;
begin
   loop
      -- iterate over the Line_Set and extract each line
      A_Line := ...
      Draw_It(A_Line, Hue(P), Width(P));  -- dispatch to get hue and width
   end loop;
end Print;
```

but this is all hidden from the user. Note that the procedure Draw_It is declared in the private part since it need not be visible to the user.

One reason why the user has to provide To_Lines is that only the user knows about the details of how best to represent the object. For example the poor circle will have to be represented crudely as a polygon of many sides, perhaps a hectogon of 100 sides.

We can now take at least two different approaches. We can for example write

```
with Geometry, Line_Draw;
package Printable_Geometry is
   type Printable_Object is abstract new Geometry.Object
                                and Line_Draw.Printable with private;

   procedure Set_Hue(P: in out Printable_Object; C: in Colour);
   function Hue(P: Printable_Object) return Colour;
   procedure Set_Width(P: in out Printable_Object; W: in Points);
   function Width(P: Printable_Object) return Points;
   function To_Lines(P: Printable_Object) return Line_Set is abstract;
private
   ...
end Printable_Geometry;
```

The type Printable_Object is a descendant of both Object and Printable and all concrete types descended from Printable_Object will therefore have all the operations of both Object and Printable. Note carefully that we have to put Object first in the declaration of Printable_Object and that the following would be illegal

```
type Printable_Object is abstract new Line_Draw.Printable
                      and Geometry.Object with private;    -- illegal
```

This is because of the rule that only the first type in the list can be a normal tagged type; any others must be interfaces. Remember that the first type is always known as the parent type and so the parent type in this case is Object.

The type Printable_Object is declared as abstract because we do not want to implement To_Lines at this stage. Nevertheless we can provide concrete subprograms for all the other operations of the interface Printable. We have given the type a private extension and so in the private part of its containing package we might have

```
private
   type Printable_Object is abstract new Geometry.Object
                                and Line_Draw.Printable with
      record
         Hue: Colour := Black;
         Width: Points := 1;
      end record;
end Printable_Geometry;
```

Just for way of illustration, the components have been given default values. In the package body the operations such as the function Hue are simply

```
function Hue(P: Printable_Object) return Colour is
begin
   return P.Hue;
end Hue;
```

Luckily, the visibility rules are such that this does not do an infinite recursion! This is because the prefix notation does not take preference.

Note that the information containing the style components is in the record structure following the geometrical properties. This is a simple linear structure since interfaces cannot add components. However, since the type Printable_Object has all the operations of both an Object and a Printable, this adds a small amount of complexity to the arrangement of dispatch tables. But this detail is hidden from the user.

The key point is that we can now pass any object of the type Printable_Object or its descendants to the procedure

```
procedure Print(P: in Printable'Class);
```

and then (as outlined above) within Print we can find the colour to be used by calling the function Hue and the line width to use by calling the function Width and we can convert the object into a set of lines by calling the function To_Lines.

And now we can declare the various types Circle, Triangle, Square and so on by making them descendants of the type Printable_Object and in each case we have to implement the function To_Lines.

The unfortunate aspect of this approach is that we have to move the geometry hierarchy. For example the triangle package might now be

```
package Printable_Geometry.Triangles is
   type Printable_Triangle is new Printable_Object with
      record
         A, B, C: Float;
      end record;
   ...   -- functions Area, To_Lines etc.
end;
```

We can now declare a Printable_Triangle thus

```
A_Triangle: Printable_Triangle :=
      (Printable_Object with A => 4.0, B => 4.0, C => 4.0);
```

This declares an equilateral triangle with sides of length 4.0. Its private Hue and Width components are set by default. Its coordinates which are also private are by default set to zero so that it is located at the origin. (The reader can improve the example by making the components A, B, and C private as well.)

We can conveniently move it to wherever we want by using the procedure Move which being class wide applies to all types derived from Object. So we can write

```
A_Triangle.Move(1.0, 2.0);
```

And now we can make a red sign

```
Sign: Printable_Triangle := A_Triangle;
```

Having declared the object Sign, we can give it width and hue and print it

```
Sign.Set_Hue(Red);
Sign.Set_Width(3);
Sign.Print;                              -- print thick red triangle
```

As we observed earlier this approach has the disadvantage that we had to move the geometry hierarchy. A different approach which avoids this is to declare printable objects of just the kinds we want as and when we want them.

So assume now that we have the package Line_Draw as before and the original package Geometry and its child packages. Suppose we want to make printable triangles and circles. We could write

```
with Geometry, Line_Draw;  use Geometry;
package Printable_Objects is

   type Printable_Triangle is new Triangles.Triangle and
                              Line_Draw.Printable with private;
   procedure Set_Hue(P: in out Printable_Triangle; C: in Colour);
   function Hue(P: Printable_Triangle) return Colour;
   procedure Set_Width(P: in out Printable_Triangle; W: in Points);
   function Width(P: Printable_Triangle) return Points;
   function To_Lines(T: Printable_Triangle) return Line_Set;

   type Printable_Circle is new Circles.Circle and
                              Line_Draw.Printable with private;
   procedure Set_Hue(P: in out Printable_Circle; C: in Colour);
   function Hue(P: Printable_Circle) return Colour;
   procedure Set_Width(P: in out Printable_Circle; W: in Points);
   function Width(P: Printable_Circle) return Points;
   function To_Lines(C: Printable_Circle) return Line_Set;

private
   type Printable_Triangle is new Triangles.Triangle and
                              Line_Draw.Printable with
      record
         Hue: Colour := Black;
         Width: Points := 1;
      end record;

   type Printable_Circle is new Circles.Circle and
                              Line_Draw.Printable with
      record
         Hue: Colour := Black;
         Width: Points := 1;
      end record;

end Printable_Objects;
```

and the body of the package will provide the various subprogram bodies.

Now suppose we already have a normal triangle thus

```
A_Triangle: Geometry.Triangles.Triangle := ... ;
```

In order to print A_Triangle we first have to declare a printable triangle thus

```
Sign: Printable_Triangle;
```

and now we can set the triangle components of it using a view conversion thus

```
Triangle(Sign) := A_Triangle;
```

And then as before we write

```
Sign.Set_Hue(Red);
Sign.Set_Width(3);
Sign.Print_It;                              -- print thick red triangle
```

This second approach is probably better since it does not require changing the geometry hierarchy. The downside is that we have to declare the boring hue and width subprograms repeatedly. We can make this much easier by declaring a generic package thus

```
with Line_Draw;  use Line_Draw;
generic
   type T is abstract tagged private;
package Make_Printable is
   type Printable_T is abstract new T and Printable with private;
   procedure Set_Hue(P: in out Printable_T; C: in Colour);
   function Hue(P: Printable_T) return Colour;
   procedure Set_Width(P: in out Printable_T; W: in Points);
   function Width(P: Printable_T) return Points;
private
   type Printable_T is abstract new T and Printable with
      record
         Hue: Colour := Black;
         Width: Points := 1;
      end record;
end;
```

This generic can be used to make any type printable. We simply write

```
package P_Triangle is new Make_Printable(Triangle);
type Printable_Triangle is new P_Triangle.Printable_T with null record;
function To_Lines(T: Printable_Triangle) return Line_Set;
```

The instantiation of the package creates a type Printable_T which has all the hue and width operations and the required additional components. However, it simply inherits the abstract function To_Lines and so itself has to be an abstract type. Note that the function To_Lines has to be especially coded for each type anyway unlike the hue and width operations which can be the same.

We now do a further derivation largely in order to give the type Printable_T the required name Printable_Triangle and at this stage we provide the concrete function To_Lines.

We can then proceed as before. Thus the generic makes the whole process very easy – any type can be made printable by just writing three lines plus the body of the function To_Lines.

Hopefully this example has illustrated a number of important points about the use of interfaces. The key thing perhaps is that we can use the procedure Print to print anything that implements the interface Printable.

21.3 Mixin inheritance

In the previous section we found the generic package Make_Printable very convenient. Its essence was

```
generic
   type T is abstract tagged private;
package Make_Printable is
   type Printable_T is abstract new T and Printable with private;
   ...  -- operations of Printable_T
private
   type Printable_T is abstract new T and Printable with
      record
         ...  -- additional components of Printable_T
      end record;
end Make_Printable;
```

This is an example of what is often known as mixin inheritance. It enables certain properties to be added to the formal type T. In that case it was the operations for manipulating hue and width.

The basic model is

```
generic
   type T is tagged private;
package P is
   type TT is new T with private;
   ...  -- operations of TT
private
   type TT is new T with
      record
         ...  -- additional components of TT
      end record;
end P;
```

where the specification exports the extended type TT plus various operations on TT which are implemented in the body.

We can then use an instantiation of P to add the operations of TT to any existing tagged type and the resulting type will of course still be in the class of the type passed as actual parameter. Thus if we write

```
package Mine is new P(My_Type);
```

then the type Mine.TT is in My_Type'Class and also has all the operations of P.TT.

This approach allows us to extend a type with generic operations that might or might not be made visible to the client. For example, we recall from Section 14.6 that the full type corresponding to

```
type Shape is new Object with private;
```

need not be directly derived from Object. We can therefore write

```
private
   package Q is new P(Object);
   type Shape is new Q.TT with null record;
   ...
end;
```

and then the type Shape will also have all the components and properties of the type TT in the generic package. As written, these are not visible to the client but subprograms in the visible part of the package in which Shape is declared could provide access to them. It is wise to make both T and TT abstract (as in the case of Make_Printable) since this enables the actual generic parameter to be abstract.

A variation is where the extension is only valid for types in a certain class presumably because properties of the class are required for the extension. Thus

```
generic
   type S is new Object with private;
package Colour_Mixin is
   type Coloured_Object is new S with private;
   ...
end;
```

can only be applied to types in Object'Class. Existing inherited operations could be replaced and others added. It is important to note that operations of the exported type can often be provided by renamings (in the package body) of operations inherited from an ancestor as illustrated in one of the exercises below.

Exercise 21.3

1 Write a generic package whose visible part is

```
generic
   type Raw_Type is tagged private;
package Tracking is
   type Tracked_Type is new Raw_Type with private;
   function Identity(TT: Tracked_Type) return Integer;
private ...
```

The purpose is to extend the Raw_Type so that objects of the resulting type Tracked_Type contain an identity number as in the controlled type Thing of Section 14.7. Although Tracked_Type will not be controlled by being directly descended from the type Controlled, it will have a component of a controlled type and this will give the effect of controlling objects of Tracked_Type. The function Identity returns the identity number of the object TT.

2 Using the generic package Tracking of the previous exercise, write a package whose visible part is

```
package Hush_Hush is
   type Secret_Shape is new Object with private;
   function Shape_Identity(SS: Secret_Shape) return Integer;
   ...   -- other operations on a secret shape
private ...
```

The intent is that objects of the type Secret_Shape should be controlled and have an identity number which is returned by a call of Shape_Identity plus other hidden components which can be manipulated by the other operations in the visible part of the package.

21.4 Linked structures

We have encountered a number of linked data structures from time to time. The simplest is the pair Cell and Cell_Ptr which provide a simple list. The pair Node and Node_Ptr from Section 11.2 provide a binary tree. In this section we will use named access types throughout and leave the reader to consider how anonymous access types might best be used.

In Section 3.3 we showed an example where the elements of a list were access to class wide values thereby allowing the list to be indirectly heterogeneous.

An interesting variation was shown in Section 14.2 where various reservations were chained together using a pointer in a root type from which they were all derived; the main advantage of this approach is that the number of pointers is reduced. This mechanism was then encapsulated in the package Queues which we then used to create a queue containing reservations.

However, we saw that it was not possible to prevent any type being placed on the queue of reservations since the queue was completely heterogeneous. We can overcome this by making the package Queues generic thus

```
generic
   type Data is abstract tagged private;
package Queues is
   Queue_Error: exception;
   type Queue is limited private;
   type Element is abstract new Data with private;
   type Element_Ptr is access all Element'Class;
   procedure Join(Q: access Queue; E: in Element_Ptr);
   function Remove(Q: access Queue) return Element_Ptr;
   function Length(Q: Queue) return Integer;
private
   type Element is abstract new Data with
      record
         Next: Element_Ptr;
      end record;
   type Queue is limited
      record
         First, Last: Element_Ptr;
         Count: Integer := 0;
      end record;
end Queues;
```

The generic parameter Data is used as a token to identify the type of queue as we shall see in a moment. It is made abstract so that the actual type can also be abstract. The type Element is now an extension of Data; it is also abstract.

The base of the reservation system can now become (using a formulation similar to that of Section 14.3)

```
with Queues;
package Reservation_System is
   type Root_Resvn is interface;
   procedure Make(R: in out Root_Resvn) is abstract;
   package Resvn_Queues is new Queues(Root_Resvn);
   subtype R_Queue is Resvn_Queues.Queue;
   subtype Reservation is Resvn_Queues.Element;
   subtype Reservation_Ptr is Resvn_Queues.Element_Ptr;
end Reservation_System;
```

with the rest of the system as before. Note the use of subtypes to give relevant names to the types exported from the instantiated package. The actual reservation types are then all derived from Reservation (that is Resvn_Queues.Element) which is abstract as in the formulation of Section 14.3. We can then write

```
type R_Queue_Ptr is access R_Queue;
The_Queue: R_Queue_Ptr := new R_Queue;
...
Join(The_Queue, New_Resvn);
```

and then only reservations can be placed on the queue. (Strictly only types in the class Resvn_Queues.Element'Class.) So although the queue is still heterogeneous it is constrained to accept only objects of the appropriate class and there is no risk of placing the wrong type on the queue. And moreover, no conversion is required when removing an object from the queue

```
Next_Resvn: Reservation_Ptr := Remove(The_Queue);
```

because Reservation_Ptr is correctly an access to the queue element class.

This example also illustrates the use of a series of abstract types. We start with Root_Resvn which is abstract (actually an interface) and only exists in order to characterize the queues; we then add the queue element property and so export Element which is also abstract. Using a subtype we rename that exported type as Reservation and only then do we develop the specific types for the various reservations.

Observe that Make is a primitive operation of Root_Resvn; we could not make it a primitive operation of Reservation since that is really Resvn_Queues.Element and a primitive operation cannot be declared outside the package Resvn_Queues where it is declared.

This is really another example of mixin inheritance discussed in the previous section. The reservations have the properties of Root_Resvn (such as Make) and also the queuing properties of the type Element.

But of course although we have got rid of the conversion problem and ensured that the queue can only have elements of the correct class, nevertheless this approach of extension from a root type is somewhat inflexible. If we really want a reservation to be on two different queues at the same time or perhaps on a tree structure then we are in difficulties.

The most flexible and simplest technique of all is that of Chapter 3 where we simply put the pointers in the queue. We can use the original package Queues of Exercise 12.5(**3**) but make it generic so it starts

```
generic
   type Data is private;
package Queues is ...
```

which can then be instantiated with the original Reservation_Ptr which is a definite type. Of course, we cannot put the reservations themselves on the queue because that would require a class wide actual parameter and hence an indefinite formal and the implementation would then require another level of indirection. But we probably wouldn't want to put the reservations themselves on a queue anyway so this is not a problem.

The type Queue was actually implemented in terms of a singly linked list. For some applications it is convenient to use a lower level of abstraction where the individual cells are directly accessible; this allows items to be added or removed from any point of the list. Thus consider

```
package Lists is
   List_Error: exception;
   type Cell is tagged limited private;
   type Cell_Ptr is access all Cell'Class;
   procedure Insert(After: Cell_Ptr; Item: Cell_Ptr);
   function Remove(After: Cell_Ptr) return Cell_Ptr;
   function Next(After: Cell_Ptr) return Cell_Ptr;
private
   type Cell is tagged limited
      record
         Next: Cell_Ptr;
      end record;
end;
```

In this example we designate a list as a whole, a place in it, an item to be added and an item removed all by values of type Cell_Ptr. Iteration over a list can be performed by the function Next. Singly linked lists are a bit awkward because we cannot go backwards without scanning the list. Thus we designate the place to insert an item by pointing to the cell one before where it is to go; similarly for removing an item. This makes things difficult for the head since there is not one before it and so we cannot use this technique to add an item at the beginning or remove the first one. There are various solutions to this such as having extra procedures, or an extra parameter designating the head of the list, or always having a dummy cell at the start, or using a doubly linked list, or using a circular list. We leave these details to the reader. However, the key point is that the type Cell can be extended with the data to be put on the list just as we did for the queue elements.

Another common structure is the binary tree and a doubly linked form of this might be built around

```
type Node;
type Node_Ptr is access all Node'Class;
```

```
type Node is tagged limited
  record
    Parent: Node_Ptr;
    Left, Right: Node_Ptr;
  end record;
```

These list and tree structures can of course be made generic with respect to some abstract root data type just as for the package Queues.

In practice, it is unlikely that we would want to create these linked structures from scratch; we should simply use the containers provided as part of the standard library in Ada.Containers described in Chapter 24. However, it is instructive to understand the mechanisms involved.

Exercise 21.4

1 Write an appropriate body for the package Lists. Keep a dummy item at the head.

2 Reconsider the previous exercise using null exclusions for the parameters of Insert, Remove, and Next.

21.5 Iterators

A common requirement is to apply an action to all the objects in some sort of structure, for example all the nodes in a tree or all the items in a list. The predefined container library contains a number of examples of the use of iterators. In this section we look at some of the underlying ideas and in the next section we outline the clever features introduced in Ada 2012 which enable the fine details to be hidden from the user.

Iterators can be approached in a number of ways and in order not to confuse the reader with complex actions we will consider two very simple ones. We will suppose that the elements of the list or tree are simply values of an enumeration type such as Colour; we might consider our structures as representing a list or tree of coloured balls. The two actions will be a function which simply counts the number of entities in the structure and a procedure which turns all Green ones into Red ones.

We first consider the approach of a so-called active iterator applied to the simple case of a list. Consider

```
package Lists is
  type List is limited private;
  ...
  type Iterator(L: access List) is limited private;
  procedure Start(It: in out Iterator);
  function Done(It: Iterator) return Boolean;
  procedure Next(It: in out Iterator);
  function Get_Colour(It: Iterator) return Colour;
  procedure Set_Colour(It: in Iterator; C: in Colour);
private
```

```
    type Cell is
      record
        Next: access Cell;                  -- anonymous access type
        C: Colour;
      end record;
    type List is access all Cell;           -- full type

    type Iterator(L: access List) is limited
      record
        This: access Cell;
      end record;

  end;

  package body Lists is
    ...
    procedure Start(It: in out iterator) is
    begin
      It.This := It.L.all;
    end Start;

    function Done(It: Iterator) return Boolean is
    begin
      return It.This = null;
    end Done;

    procedure Next(It: in out Iterator) is
    begin
      It.This := It.This.Next;
    end Next;

    function Get_Colour(It: Iterator) return Colour is
    begin
      return It.This.C;
    end Get_Colour;

    procedure Set_Colour(It: in Iterator; C: in Colour) is
    begin
      It.This.C := C;
    end Set_Colour;

  end Lists;
```

The general idea is that Start, Done, and Next enable us to move over the structure, and the component This of the type Iterator gives access to the current item at each stage.

The subprograms to count the number of items and change the green ones to red are

```
  function Count(L: access List) return Natural is
    It: Iterator(L);
    Result: Natural := 0;
  begin
```

```
      Start(It);
      while not Done(It) loop
         Result := Result + 1;
         Next(It);
      end loop;
      return Result;
   end Count;

   procedure Green_To_Red(L: access List) is
      It: Iterator(L);
      C: Colour;
   begin
      Start(It);
      while not Done(It) loop
         C := Get_Colour(It);
         if C = Green then
            Set_Colour(It, Red);
         end if;
         Next(It);
      end loop;
   end Green_To_Red;
```

Observe how the objects of type Iterator are declared inside Count and Green_To_Red with an access discriminant identifying the List. Using access discriminants is not really necessary for this simple structure since we could have declared Iterator and Start as

```
   type Iterator is
      record
         This: access Cell;
      end record;

   procedure Start(L: in List; It: in out Iterator) is
   begin
      It.This := L;
   end Start;
```

However, for more complex structures the access discriminant enables the iterator to hold on to a pointer to the structure as a whole which would be useful if we wanted to remove items from the list rather than just traverse it.

It should also be noted that the procedure Start could be omitted by declaring the type Iterator as

```
   type Iterator(L: access List) is limited
      record
         This: access Cell := L.all;
      end record;
```

so that initialization occurs automatically.

It is clear that the actions Count and Green_To_Red are quite independent of the structure over which they are being applied. Moreover, we should arrange that the iteration mechanism is itself distinct from other aspects of the structure.

One approach is to put the type List and its other operations in a parent package and then put the iterator operations in a child (which would of course have access to the private part of the parent)

```
package Lists is
   type List is limited private;
   ...
private
   ...
end;

package Lists.Iterators is
   type Iterator(L: access List) is limited private;
   procedure Start(It: in out Iterator);
   ...   -- and Done, Next, Get_Colour
   procedure Set_Colour(It: in Iterator; C: in Colour);
private
   ...
end;
```

If we had several such structures then we could make the subprograms such as Count generic with respect to the various entities. If we had several subprograms to parameterize then we could put them all together in a single generic package. Alternatively, we could keep them distinct and use the package parameter or signature technique to bundle the parameters together

```
generic
   type Structure is limited private;
   type Iterator(S: access Structure) is limited private;
   with procedure Start(It: in out Iterator) is <>;
   ...   -- and Done, Next, Get_Colour
   with procedure Set_Colour(It: in Iterator; C: in Colour) is <>;
package Iteration_Stuff is end;

generic
   with package I_S is new Iteration_Stuff(<>);
   use I_S;
procedure Generic_Green_To_Red(S: access Structure) is
   It: Iterator(S);
   C: Colour;
begin
   Start(It);
   while not Done(It) loop
      C := Get_Colour(It);
      if C = Green then
         Set_Colour(It, Red);
      end if;
      Next(It);
   end loop;
end Generic_Green_To_Red;
```

And now all we have to do is instantiate Iteration_Stuff with the items from Lists.Iterators and then in turn instantiate Generic_Green_To_Red with the resulting package. Note the default parameters in Iteration_Stuff. Using default parameters might be unwise because of the risk of picking up something irrelevant by mistake.

We now turn to a different approach which in a sense is a complete inversion of what we have done so far. The idea here is to hide the iteration loop once and for all inside a fixed procedure and then for that procedure to be called from within the loop to do whatever is required. This perhaps safer approach using an access to subprogram parameter is known as a passive iterator.

The child package Lists.Iterators is simply

```
package Lists.Iterators is
   procedure Iterate(L: List; Action: access procedure (C: in out Colour));
end;

package body Lists.Iterators is
   procedure Iterate(L: List;
                         Action: access procedure (C: in out Colour)) is
      This: access Cell := L;
   begin
      while This /= null loop
         Action(This.C);         -- indirect call of Action
         This := This.Next;
      end loop;
   end Iterate;
end Lists.Iterators;
```

and the subprograms Count and Green_To_Red are then

```
function Count(L: List) return Natural is
   Result: Natural := 0;

   procedure Count_Action(C: in out Colour) is
   begin
      Result := Result + 1;
   end Count_Action;

begin
   Iterate(L, Count_Action'Access);
   return Result;
end Count;

procedure Green_To_Red(L: List) is

   procedure GTR_Action(C: in out Colour) is
   begin
      if C = Green then C := Red; end if;
   end GTR_Action;

begin
   Iterate(L, GTR_Action'Access);
end Green_To_Red;
```

This passive approach seems much easier to understand. Examples of both the active approach where the user calls procedures such as Start and Done and this passive approach using an access to subprogram parameter occur in the predefined package Ada.Directories, see Section 23.10.

But it should be noted that the active approach has the merit that it is much easier to exit from the explicit loop early than is the case with the passive iterator where the loop is hidden from the user.

There is another form of passive iterator which uses dispatching rather than access to subprogram parameters. This is not for the faint-hearted but is interesting and is described in Section 21.5a, Double Dispatching, on the website.

Exercise 21.5

1. Make the appropriate instantiations of the generic package Iteration_Stuff and generic procedure Green_To_Red for the type List.

2. Consider how to generalize the passive iterator approach to work on any structure by declaring a package containing an interface Structure and a procedure Iterate whose parameters are a Structure and an access to subprogram being the action to be performed. Apply the generalization to a binary tree by declaring a type Tree as an extension of Structure. Then declare a function that counts the number of balls of a given colour in any structure and apply it to determine how many Green balls are in a tree.

21.6 Generalized iteration

Iteration is a fundamental activity in programming. The idea of iterating over an array has been a primary concern in programming for at least 60 years. More recently, the development of generalized data structures and more specifically, the introduction of containers in Ada 2005 made prominent the concept of iterating over structures of other kinds as well. The previous section showed the gory details of various approaches to iteration.

This section shows how iteration over an array or container was performed in Ada 2005 and then shows how it can be abbreviated in Ada 2012 and Ada 2022.

As an example suppose we are going to manipulate objects of a type Twin thus

```
type Twin is
   record
      P, Q: Integer;
   end record;
```

Now consider an array of such objects

```
type Artwin is array (1 .. N) of Twin;

The_Array: Artwin;
```

and suppose that we wish to iterate over the array and change all the objects for which the component P is prime by adding some value X to the component Q. We might traditionally write

```
for K in The_Array'Range loop
   if Is_Prime(The_Array(K).P) then
      The_Array(K).Q := The_Array(K).Q + X;
   end if;
end loop;
```

This is laborious and as mentioned in Section 8.1 can be rewritten in Ada 2012 as

```
for E: Twin of The_Array loop
   if Is_Prime(E.P) then
      E.Q := E.Q + X;
   end if;
end loop;
```

This uses a generalized form of iteration where the controlled variable E of subtype Twin takes the identity of each component of The_Array in turn. Note the use of **of** rather than **in** in the loop specification. Incidentally, we do not have to specify the subtype of E because it can be taken from the array but doing so provides a useful documentation aid.

Now suppose we have a list of the type Twin using a container (see Section 24.2) then to manipulate every element of the list in Ada 2005, we had to write something tedious like

```
C := The_List.First;              -- C declared as of type Cursor
loop
   exit when C = No_Element;
   E := Element(C);                            -- E is of type Twin
   if Is_Prime(E.P) then
      Replace_Element(The_List, C, (E.P, E.Q + X));
   end if;
   C := Next(C);
end loop;
```

This reveals the gory details of the iterative process whereas all we want to say is 'add X to the component Q for all members of the list whose component P is prime'. This corresponds to the active approach described in the previous section.

Alternatively, we can use the passive approach in Ada 2005 by using the procedure Iterate. In that case the details of what we are doing have to be placed in a distinct subprogram called perhaps Do_It. Thus we can write

```
declare
   procedure Do_It(C: in Cursor) is
   begin
      E := Element(C);                         -- E is of type Twin
      if Is_Prime(E.P) then
         Replace_Element(The_List, C, (E.P, E.Q + X));
      end if;
   end Do_It;
begin
   The_List.Iterate(Do_It'Access);
end;
```

This avoids the fine detail of calling First and Next but uses what some consider to be a heavy infrastructure.

However, in Ada 2012 and Ada 2022 we can simply say

```
for E: Twin of The_List loop
   if Is_Prime(E.P) then
      E.Q := E.Q + X;
   end if;
end loop;
```

Not only is this just five lines of text rather than nine or eleven, but the possibility of making various errors of detail is completely removed.

Note also the beautiful similarity between iteration over a loop and iteration over a list. The only difference is that The_Array is replaced by The_List.

The mechanisms by which this magic abstraction is achieved are somewhat laborious and it is anticipated that users will take a cookbook approach. Accordingly, we concentrate on how to do it rather than how it works.

Suffice it to say here that it uses the magic package Ada.Iterator_Interfaces and various aspects Implicit_Dereference, Constant_Indexing, Variable_Indexing, Default_Iterator, and Iterator_Element. Just for the record the specification of the package is as follows

```
generic
   type Cursor;
   with function Has_Element(Position: Cursor) return Boolean;
package Ada.Iterator_Interfaces
     with Pure, Nonblocking => False is
   type Forward_Iterator is limited interface;
   function First(Object: Forward_Iterator) return Cursor is abstract;
   function Next(Object: Forward_Iterator;
                 Position: Cursor) return Cursor is abstract;
   type Reversible_Iterator is limited interface and Forward_Iterator;
   function Last(Object: Reversible_Iterator) return Cursor is abstract;
   function Previous(Object: Reversible_Iterator;
                     Position: Cursor) return Cursor is abstract;

   -- new in Ada 2022 from here on

   type Parallel_Iterator is limited interface and Forward_Iterator;

   subtype Chunk_Index is Positive;
   function Is_Split(Object: Parallel_Iterator) return Boolean is abstract;
   procedure Split_Into_Chunks(Object: in out Parallel_Iterator;
                               Max_Chunks: in Chunk_Index) is abstract;
   function Chunk_Count(Object: Parallel_Iterator)
      return Chunk_Index is abstract;
   function First(Object: Parallel_Iterator;
                  Chunk: Chunk_Index) return Cursor is abstract;
   function Next(Object: Parallel_Iterator;
                 Position: Cursor;
                 Chunk: Chunk_Index) return Cursor is abstract;
```

```
type Parallel_Reversible_Iterator is limited interface
                         and Parallel_Iterator and Reversible_Iterator;
end Ada.Iterator_Interfaces;
```

This generic package is used by the various container packages such as Ada.Containers.Doubly_Linked_Lists. Its actual parameters corresponding to the formal parameters Cursor and Has_Element come from the container which includes an instantiation of Ada.Iterator_Interfaces. The instantiation then exports the various required types and functions. An interesting point is that the formal parameter Cursor is incomplete.

The concept of chunks concerns the use of parallel blocks and loops especially with containers that use Ada.Iterator_Interfaces. See Section 22.8.

The type Parallel_Iterator is used for Hashed Maps, Hashed Sets, and Trees. The type Parallel_Reversible_Iterator is used for Lists, Vectors, Ordered Maps, Ordered Sets, and Trees.

We will now briefly summarize the various possible forms of iterators. They are

- A form using **of** with arrays,
- A form using **of** with containers,
- A form using **in** with containers.

The first two have already been illustrated thus

```
for E of The_Array loop
   ...                    -- do something to E
end loop;
for E of The_List loop
   ...                    --do something to E
end loop;
```

and we have already noted the similarity between them. In both cases we can usually add **reverse** in which case iteration goes backwards. So we might have

```
for E of reverse The_List loop ...
```

The container form with **of** depends upon the existence of aspects Default_ Iterator and Iterator_Element. In the case of the type List in Doubly_Linked_Lists we find that the type List has aspects

```
Default_Iterator => Iterate,
Iterator_Element => Element_Type;
```

These aspects indicate that the function Iterate is to be called and that iteration is over the objects of the list which are of type Element_Type. The magic here is that the function Iterate returns a value of the type Reversible_Iterator'Class from (an instantiation of) the package Ada.Iterator_Interfaces and thereby defines the iteration process. However, the user need not understand the fine details.

The third new form is essentially the container analogy of the traditional form for arrays. In the container case we have a cursor which refers to the components of the container much as an index refers to the components of an array.

So we have

```
for K in A'Range loop
   ...                      -- do something via index K
end loop;

for C in The_List.Iterate loop
   ...                      -- do something via cursor C
end loop;
```

Here we have an explicit call of the magic function Iterate that returns a value of the type Reversible_Iterator'Class. In this case iteration occurs over values of the type Cursor which was passed to Ada.Iterator_Interfaces when it was instantiated in the package Doubly_Linked_Lists.

Iteration can be done in reverse by writing

```
for C in reverse The_List.Iterate loop ...
```

Note that only some containers can be iterated in reverse. The details are given in Chapter 24.

A further variation is that writing

```
for C in The_List.Iterate(S) loop
   ...                      -- do something via cursor C
end loop;
```

starts the iteration with the cursor value S. We can of course always stop the iteration at a given cursor value by simply using **exit**.

We now turn to considering the bodies of the loops, that is the code marked 'do something via cursor C' or 'do something to E'.

In the Ada 2005 example we wrote

```
if Is_Prime(E.P) then
   Replace_Element(The_List, C, (E.P, E.Q + X));
end if;
```

It is tedious having to write Replace_Element when using a container whereas in the case of an array we just write

```
if Is_Prime(A(I).P) then
   A(I).Q := A(I).Q + X;
end if;
```

Note that Replace_Element copies the whole new element whereas we just update the one component in the array. This would be a problem if the components were large. To overcome this we can use the procedure Update_Element thus

```
procedure Update_Element(Container: in out List; Position: in Cursor;
                         Process: not null access procedure
                                    (Element: in out Element_Type));
```

To use this we have to write a slave procedure Do_It perhaps thus

```
procedure Do_It(E: in out Twin) is
begin
   E.Q := E.Q + X;
end Do_It;
```

and then

```
if Is_Prime(E.P) then
   Update_Element(The_List, C, Do_It'Access);
end if;
```

This works fine because E is passed by reference and no giant copying occurs. However, the downside is that the distinct procedure Do_It has to be written so that the overall text is more than a bit tedious.

However, the text in the body of Do_It is precisely what we want to say. Using the concepts of left and right hand values, the problem is that The_List(C).Element cannot be used as a left hand value by writing for example

```
The_List(C).Element.Q := ...
```

The problem is overcome in Ada 2012 and Ada 2022 using more magic by the introduction of generalized reference types and further aspects. In particular we find that the containers now include a type Reference_Type and a function Reference which in the case of the list containers are

```
type Reference_Type(Element: not null access Element_Type) is private
   with Implicit_Dereference => Element;

function Reference(Container: aliased in out List; Position: in Cursor)
                                                   return Reference_Type;
```

Note the aspect Implicit_Dereference applied to the type Reference_Type with discriminant Element. There is also a type Constant_Reference_Type and function Constant_Reference for use when the context demands read-only access.

Note the inclusion of **aliased** for the parameter Container of the function Reference. This ensures that the parameter is passed by reference and also permits us to apply 'Access to the parameter Container within the function and to return that access value.

It might be helpful to say a few words about the possible implementation of Reference and Reference_Type. The important part of the type Reference_Type is its access discriminant. The private part might contain housekeeping stuff but we can ignore that. So in essence it is simply a record with just one component being the access discriminant

```
type Reference_Type(E: not null access Element_Type) is null record;
```

and the body of the function might be

```
function Reference(Container: aliased in out List; Position: in Cursor)
                                                   return Reference_Type is
begin
   return (E => Container.Element(Position)'Access);
end Reference;
```

The type List also has other aspects thus

```
Constant_Indexing => Constant_Reference,
Variable_Indexing => Reference,
```

The important aspect here is Variable_Indexing. If this aspect is supplied then in essence an object of the type can be used in a left hand context by invoking the function given as the value of the aspect. In the case of The_List this is the function Reference which returns a value of type Reference_Type. Moreover, this reference type has a discriminant which is of type **access** Element_Type and the aspect Implicit_Dereference with value Element and so gives direct access to the value of type Element.

We can now by stages transform the raw text. So using the cursor form we can start with

```
for C in The_List.Iterator loop
  if Is_Prime(The_List.Reference(C).Element.P) then
    The_List.Reference(C).Element.Q :=
            The_List.Reference(C).Element.Q + X;
  end if;
end loop;
```

Since the aspect Implicit_Dereference applies to the type Reference_Type and has value Element, we can omit the explicit mention of the discriminant Element returned by the call of the function Reference to give

```
for C in The_List.Iterator loop
  if Is_Prime(The_List.Reference(C).P) then
    The_List.Reference(C).Q := The_List.Reference(C).Q + X;
  end if;
end loop;
```

Now because the aspect Variable_Indexing for the type List has value Reference, the explicit calls of Reference can be omitted to give

```
for C in The_List.Iterator loop
  if Is_Prime(The_List(C).P) then
    The_List(C).Q := The_List(C).Q + X;
  end if;
end loop;
```

It should now be clear that the cursor C is simply acting as an index into The_List. We can compare this text with

```
for C in The_Array'Range loop
  if Is_Prime(The_Array(C).P) then
    The_Array(C).Q := The_Array(C).Q + X;
  end if;
end loop;
```

which shows that 'Range is analogous to .Iterator.

Finally, to convert to the element form using E we just replace The_List(C) by E to give

```
for E of The_List loop
   if Is_Prime(E.P) then
      E.Q := E.Q + X;
   end if;
end loop;
```

This underlying technique which transforms the sequence of statements of the container element iterator can be used quite generally. For example, we might not want to iterate over the whole container but just manipulate a particular element given by a cursor C. Rather than calling Update_Element, we can just write

```
The_List(C).Q := ...
```

Although the various aspects were introduced into Ada 2012 primarily to simplify the use of containers they can be used quite generally. These examples and the description given here should be enough for the user to apply the facilities successfully. For more details consult the *Ada 2012 Rationale* and the *ARM*.

21.7 Procedural iterators

A frequently occurring situation is where we need to do some operations on a number of items where what is to be done is described by a subprogram which is then passed as an access to subprogram parameter to some sort of iteration. Thus in Section 23.10 there is an example of printing the state of all environment variables by first declaring a subprogram Print_One thus

```
procedure Print_One(Name, Value: in String) is
begin
   Put_Line(Name & "=" & Value);
end Print_One;
```

and then calling the procedure Iterate in Ada.Environment_Variables whose specification is

```
procedure Iterate(Process: not null access procedure
                                        (Name, Value: in String));
```

by writing

```
Iterate(Print_One'Access);
```

The interesting point is that generally the reason for writing a subprogram is so that it can then be used in various contexts but in this case the subprogram is only ever used once and maybe that seems a bit strange.

In Ada 2022 the explicit subprogram can be avoided by simply writing

```
for (Name, Value) of Iterate(<>) loop
   Put_Line(Name & "=" & Value);
end loop;
```

So the body of the loop is just the body of the procedure Print_One and the parameters of the procedure appear in the loop after **for**.

This new form of loop is known as a procedural iterator. The two parts of the iterator (that is the text between **for** and **loop**) are separated by **of** rather than **in** as occurred in the other more general forms of iteration introduced in Ada 2012.

The part before **of** is known as the iterator parameter specification and is either a normal parameter list (with types and modes) or it can take the simple abbreviated form as in this example which is simply a list of identifiers separated by commas.

The part after **of** is known as the iterator procedure call. This typically consists of the name of a procedure (or entry) followed by some parameters. In the example the parameters just consist of the box symbol <> in the usual parentheses.

The box symbol is familiar as an indication of standing for something to be supplied. Thus it occurred in array type declarations standing for the range (Section 8.2) and in generic formal parameters (Sections 19.2 and 19.3) standing for the details of the actual parameter. In this case it essentially stands for the 'Access of the call in the original text.

In the above example, the procedure Iterate only had one parameter being the access to subprogram. In the general case there are two sets of parameters thus

```
procedure Q(X: T1; Y: T2; ZP: access procedure (A: T3; B: T4);
```

and then the call in a procedural iterator could be of the verbose form

```
for (AA:T3; BB: T4) of Q(X => Val1, Y => Val2, ZP => <>) loop
   -- AA and BB can be used here
end loop;
```

Alternatively, we can write it in an abbreviated form where the formal name and => in the call of Q can be omitted in the normal way. So we end up with

```
for (AA, BB) of Q(Val1, Val2, <>) loop
   -- AA and BB can be used here
end loop;
```

Note that the parameters for the access procedure normally written as <> can be omitted if it is the last parameter.

It is also possible to add a filter (see Section 7.3) of the form **when** condition just before the reserved word **loop**. And we can also put **parallel** in front of **for** to indicate that it is a parallel loop (see Section 22.8).

A minor concern when using a subprogram via an access parameter is the question of whether the subprogram can be left other than through the normal return mechanism. In Ada 2022 it is possible to indicate whether this is permitted or not by the aspect Allows_Exit. And indeed the specification of the procedure Iterate in Ada.Environment_Variables in Ada 2022 is

```
procedure Iterate(Process: not null access procedure
                                        (Name, Value: in String));
with Allows_Exit;
```

Other examples of the use of this aspect are the procedure Search in Ada.Directories (see Section 23.10) and the procedures Iterate and Reverse_Iterate for containers and Iterate_Children and Reverse_Iterate_Children for trees.

21.8 Object factories

There is an interesting lack of symmetry between input and output. It is easy to write a heterogeneous mixture by statements such as

```
Put("John is "); Put(21); Put(" years old");
```

but it is not so easy to read the text back unless we know the order of the items. If we do not then we have to read the text a line at a time and then decode it.

An example of this sort of problem occurs in Program 1 where we wish to read in the values of various properties of a geometrical object. The values are prefixed by a code letter indicating whether it is a Circle, Triangle, or Square. The code is

```
Get(Code_Letter);
case Code_Letter is
   when 'C' => Object_Ptr := Get_Circle;
   when 'T' => Object_Ptr := Get_Triangle;
   when 'S' => Object_Ptr := Get_Square;
   ...
end case;
```

The types Circle, Triangle, and Square are all tagged types derived from the root type Object and Object_Ptr is of the type **access** Object'Class. The function Get_Circle reads the value of the radius from the keyboard and then creates a circle, the function Get_Triangle reads the values of the lengths of the three sides from the keyboard and so on.

The trouble with this formulation is that if we add another type then we have to amend the case statement and this brings us back to variant programming with all its maintenance risks.

What we would like to do is to dispatch to an appropriate procedure to read the values according to the specific type. Now we know that all tagged types carry a tag (hence the name) which identifies the type so somehow we need to be able to create an object with the appropriate tag. We can do this with an amazing generic function called Generic_Dispatching_Constructor whose specification is as follows

```
generic
   type T (<>) is abstract tagged limited private;
   type Parameters (<>) is limited private;
   with function Constructor(Params: not null access Parameters)
                                                    return T is abstract;
function Ada.Tags.Generic_Dispatching_Constructor
      (The_Tag: Tag; Params: not null access Parameters) return T'Class
      with Preelaborate, Convention => Intrinsic, Nonblocking,
            Global => in out synchronized;
```

This generic function is a child function of the package Ada.Tags which we briefly met in Section 14.4.

Note carefully the formal function Constructor which has **is abstract** in its specification. As described in detail in Section 19.3 this means that the actual function must be a dispatching operation of a tagged type. The actual operation can be concrete or abstract. Remember that the overriding rules ensure that the specific operation for any concrete type will always have a concrete body. Note also that since the operation is abstract it can only be called through dispatching.

In this example it therefore has to be a dispatching operation of the type T since that is the only tagged type involved in the profile of Constructor. We say that T is the controlling type. In the general case, the controlling type does not itself have to be a formal parameter of the generic unit but usually will be as here. Moreover, note that although the operation has to be a dispatching operation, it is not primitive and so if we derive from the type T, it will not be inherited.

So the idea is that we instantiate the generic function with a (root) tagged type T, some type Parameters and the dispatching function Constructor. The type Parameters provides a means whereby auxiliary information can be passed to the function Constructor.

The generic function Generic_Dispatching_Constructor takes two parameters, one is the tag of the type of the object to be created and the other is the auxiliary information to be passed to the dispatching function Constructor.

Note that the type Parameters is used as an access parameter in both the generic function and the formal function Constructor. This is so that it can be matched by the profile of the attribute Input which we will meet when we discuss streams in Section 23.7.

Suppose we instantiate Generic_Dispatching_Constructor to give a function Make_T. A call of Make_T takes a tag value, dispatches to the appropriate Constructor which creates a value of the specific tagged type corresponding to the tag and this is finally returned as the value of the class wide type T'Class as the result of Make_T.

In the case of the geometrical example we must first change the individual functions Get_Circle, Get_Triangle, and Get_Square into specific overloadings of a primitive operation Get_Object of the root type Object so that we can dispatch to the appropriate one. In order to match the profile of the formal parameter Constructor they have to have a parameter of the type **access** Parameters.

In some cases there might be no auxiliary information to pass so we could declare

```
type Params is null record;
Aux: aliased Params := (null record);
```

But for illustration we can pass a file of the type Text_IO.File_Type. So we would get

```
function Get_Object(F: not null access File_Type) return Circle is
begin
   return C: Circle do
      Get(F, C.X_Coord);  Get(F, C.Y_Coord);  Get(F, C.Radius);
   end return;
end Get_Object;
```

and so on for the other types. We can now instantiate the generic to give a function for making geometrical objects thus

```
function Make_Object is
    new Generic_Dispatching_Constructor(Object, File_Type, Get_Object);
```

This function takes two parameters, the tag of the object to be created and the auxiliary parameters (in this case the file object); it then dispatches to the appropriate function Get_Object according to the tag and finally returns the newly created object of the appropriate specific type.

What we now need is a way to turn the code letters such as C and T into the corresponding tag. Perhaps the simplest way to do this is to invent some sort of registration system to make a map to convert a letter into a tag. We might have a package

```
with Ada.Tags; use Ada.Tags;
package Tag_Registration is
   procedure Register(The_Tag: Tag; Code: Character);
   function Decode(Code: Character) return Tag;
end;
```

and then we can write

```
Register(Circle'Tag, 'C');
Register(Triangle'Tag, 'T');
Register(Square'Tag, 'S');
```

And now the program to read the code and then make the object becomes simply

```
Get(Code_Letter);
Object_Ptr := new Object'Class'(Make_Object(Decode(Code_Letter),
                                              My_File'Access));
```

and there are no case statements to maintain. Phew!

The really important point about this example is that if we decide at a later date to add more types such as 'P' for Pentagon and 'H' for Hexagon then all we have to do is register the new code letters thus

```
Register(Pentagon'Tag, 'P');
Register(Hexagon'Tag, 'H');
```

and nothing else needs changing. This registration can conveniently be done when the types are declared.

The package Tag_Registration could be implemented trivially as follows by

```
package body Tag_Registration is
   Table: array (Character range 'A' .. 'Z') of Tag := (others => No_Tag);

   procedure Register(The_Tag: Tag; Code: Character) is
   begin
      Table(Code) := The_Tag;
   end Register;

   function Decode(Code: Character) return Tag is
   begin
      return Table(Code);
   end Decode;
end Tag_Registration;
```

The constant No_Tag is a value of the type Tag which does not represent an actual tag. If we forget to register a type then No_Tag will be returned by Decode and this will cause Make_Object to raise Tag_Error.

Note that any instance of Generic_Dispatching_Constructor checks that the tag passed as parameter is indeed that of a type descended from the root type T and raises Tag_Error if it is not.

In simple cases we could in fact perform that check for ourselves by writing something like

```
   Trial_Tag: Tag := The_Tag;
loop
   exit when Trial_Tag = T'Tag;
   Trial_Tag := Parent_Tag(Trial_Tag);
   if Trial_Tag = No_Tag then raise Tag_Error; end if;
end loop;
```

The function Parent_Tag and the constant No_Tag are items in the package Ada.Tags whose full specification is

```
package Ada.Tags
     with Preelaborate , Nonblocking, Global => in out synchronized is
   type Tag is private
     with Preelaborable_Initialization;
   No_Tag: constant Tag;
   function Expanded_Name(T: Tag) return String;
   ...                          -- also Wide and Wide_Wide versions
   function External_Tag(T: Tag) return String;
   function Internal_Tag(External: String) return Tag;
   function Descendant_Tag(External: String; Ancestor: Tag) return Tag;
   function Is_Descendant_At_Same_Level(Descendant, Ancestor: Tag)
                                                          return Boolean;
   function Parent_Tag(T: Tag) return Tag;
   type Tag_Array is (Positive range <>) of Tag;
   function Interface_Ancestor_Tags(T: Tag) return Tag_Array;
   function Is_Abstract(T: Tag) return Boolean;
   Tag_Error: exception;
private
   ...
end Ada.Tags;
```

The function Parent_Tag returns No_Tag if the parameter T of type Tag has no parent which will be the case if it is the ultimate root type of the class. There is also a function Interface_Ancestor_Tags which returns the tags of all those interfaces which are ancestors of T as an array. This includes the parent if it is an interface, any progenitors and all their ancestors which are interfaces as well – but it excludes the type T itself.

If we wish to create an object using Generic_Dispatching_Constructor and the tag passed as a parameter represents an abstract type then Tag_Error is raised because we are not allowed to create an object of an abstract type. In Ada 2012 and

Ada 2022 we can check whether a tag represents an abstract type by calling the function Is_Abstract.

It is instructive to consider another way of using the object constructor for the geometrical objects. We can actually specify the external form of a tag by writing

```
for Circle'External_Tag use "CIRCLE";
for Triangle'External_Tag use "TRIANGLE";
```

and so on. And now we can use these strings rather then the code letters C and T. If we have a function Get_String which reads a string and stops on a blank then we could write

```
Object_Ptr := new Object'Class'(Make_Object(Internal_Tag(Get_String),
                                            My_File'Access));
```

where the function Internal_Tag in Ada.Tags converts the external form which we specified using the attribute External_Tag to the internal form as required by the function Make_Object.

Further uses of the generic function Generic_Dispatching_Constructor and most of the other functions in the package Ada.Tags will be mentioned when we discuss streams in Section 23.9. The function Expanded_Name was explained in Section 14.4.

Incidentally, if we are going to use a registration scheme then an alternative would be to register the functions such as Get_Circle directly. Thus

```
package Registration is
   type Get_Function is access function return Object'Class;
   procedure Register(Func: Get_Function; Code: Character);
   function Decode(Code: Character) return Get_Function;
end;
...
Register(Get_Circle'Access, 'C');
...
Get(Code_Letter);
Object_Ptr := new Object'Class'(Decode(Code_Letter).all);
```

However, the whole purpose of this section was to describe the generic dispatching constructor which is necessary when dealing with tags so this is off the topic!

21.9 Controlling abstraction

We conclude this chapter by considering how we might control access to an abstract model of a system using various techniques. An interesting example is provided by the family saga of Section 18.5. Two formulations were considered, one using discriminants where the discriminant indicated the sex, and the other using tagged types where the tag indicated the sex. That section concluded by observing that the types should be private so that the correct interrelationships are always maintained.

However, as presented, both formulations suffer from the flaw that we can declare persons (objects of type Man or Woman) which are not properly initialized although we have prevented the copying of persons by making them limited. In fact there is no need for the user to actually declare person objects at all; they can all be allocated inside the package and all the user need do is declare names.

As far as names are concerned it would be desirable if all names were initialized. At the moment they are null by default and indeed unset name components are also null by default. As a consequence any misuse of the system gives rise to Constraint_Error inside the various subprograms. Clearly we can arrange that all internal subprograms check properly and raise a specific exception. However, we still need some way of designating nobody when a function such as Spouse is applied to an unmarried person. We could use null for this but it would be better practice to use a Nobody value. It probably does no harm to copy names provided they are always complete (this gives rise to aliasing of course).

In conclusion, we want to be able to declare initialized names but not persons. We would also like to be able to distinguish sex at compile time if possible (so that, for example, any attempt at a wrong marriage is caught by the compiler). Remember that this is a very traditional society.

We can prevent objects from being declared by making them abstract or by making them have unknown discriminants and not providing any means of giving an initial value such as an appropriate constructor function.

Consider the variant formulation. If we give the type Person unknown discriminants then we cannot see the discriminants in the visible part and so cannot declare Mans_Name and Womans_Name in the visible part either. We end up with

```
type Person(<>) is limited private;
type Person_Name is not null access Person;
```

but this is silly since we might as well hide Person anyway.

Let us try

```
package People is
   type Person_Name(<>) is private;
   Nobody: constant Person_Name;          -- deferred constant
   procedure Marry(Bride, Groom: Person_Name);
   function Spouse(P: Person_Name) return Person_Name;
   ...
private
```

All declarations of objects of the type Person_Name now have to be initialized by for example

```
Helen: Person_Name := New_Child( ... );
```

or we could get a new handle on someone else's spouse by

```
Friend: Person_Name := Spouse( ... );
```

Sadly we have lost the ability to distinguish externally between the sexes when calling Marry. (Remember that this is a very traditional society.) We will of course now need various selector subprograms and the type Gender will need to be visible.

We now turn to the tagged type formulation. The first thing to note is that it could match the visible part shown above in which case the tagged nature of the implementation would not be visible.

We could also try

```
package People is
   type Person is abstract tagged private;
   type Man is abstract new Person with private;
   type Woman is abstract new Person with private;

   type Person_Name is not null access all Person'Class;
   type Mans_Name is not null access all Man;
   type Womans_Name is not null access all Woman;
   ...
private
```

We are trying to prevent the external declaration of objects of the types Man and Woman by making them abstract while still retaining the external relationship between the access types. But this does not really work. Internally the true types will have to be extensions of Man and Woman so that objects can be declared and then the access types will have to be **access all** Man'Class and so on. In any event the user can declare concrete types that are extended from Man and Woman so external declarations cannot be prevented. So we give up on visibly tagged types.

Yet another approach is

```
package People is
   type Person_Name(<>) is private;
   type Mans_Name(<>) is private;
   type Womans_Name(<>) is private;
   ...
private
```

but this loses the external relationship between the types and so we have to provide conversion functions between them. We leave consideration of this as an exercise.

As the reader will appreciate, there is a limit to how much support the language can give in defining the relationships between structures. When the requirements reach a certain level it is necessary to program it ourselves even though that might mean extra complexity in the implementation or that the detection of errors is deferred to run time.

We now turn to consider some possible internal details of how our model could be controlled and monitored. For simplicity we will suppose that the external view just provides the type Person_Name as in the first package People and that the internal formulation uses the tagged mechanism.

We might want to iterate over the population so they ought to be linked together in some simple way – other than through the family relationship which is complex and easily leads to double accounting.

So using a list model we might have

```
type Person is abstract new Cell with
   record
      Birth: Date;
      ...
   end record;
```

and then every time a new person is created they are linked onto one unique list. So

```
Everybody: Cell_Ptr;

function New_Child( ... ) return Person_Name is
   ...
   Insert(Everybody, Child);
   return Child;
end New_Child;
```

The total population could be counted using an iterator exactly as in the package Lists.Iterators of Section 21.5. Other queries could be handled in the same way. Indeed, the external user might want to be able to perform such computations and so sufficient mechanism must be available externally for relevant queries to be formulated. So we might provide

```
package People.Iterators is
   procedure Iterate(Action: access procedure (P: Person_Name));
end;
```

Note that there is no parameter designating which list to iterate over since there is only one. To count the number of single persons the corresponding actual procedure might be

```
procedure Count_Single(P: Person_Name) is
begin
   if Spouse(P) = Nobody then
      Result := Result + 1;
   end if;
end Count_Single;
```

A final touch might be to animate the system. Each object of type Person could contain a task which outputs a message at regular intervals giving their name and indicating that they are alive. Other important events might also be announced. The task would have an access discriminant so

```
type Person is tagged;

task type Monitor(My: access Person'Class);

type Person is abstract new Cell with
   record
      ...
      Name: String_Ptr;
      Ego: Monitor(Person'Access);
   end record;
```

```
task body Monitor is
   ...
begin
   loop
      delay ...
      Put_Line("Hello world. It is I, " & My.Name.all & '.');
      Today := Clock;
      if Month_Name'Val(Month_Number(Today)-1) = My.Birth.Month
         and Day_Number(Today) = My.Birth.Day then
            Put_Line("And today is my Birthday.");
      end if;
   end loop;
end Monitor;
```

The internal task automatically becomes active when the object in which it is embedded is created. Clearly some sort of queuing system is needed for the messages otherwise the chatter could become garbled with interleaved messages. We leave the reader to contemplate other possibilities in this world simulation.

This example has shown how all objects of the type Person are created within the system but nevertheless the user has been able to create and manipulate them through the type Person_Name. It is possible to be even more restrictive. For example the coloured balls of Section 21.5 might be kept hidden from the user within a bag; the user might be able to command that another ball of a given colour be created and to institute searches using an iterator. Thus

```
package Balls is
   type Ball(<>) is private;
   procedure Create_Ball(C: Colour);
   function Colour_Of(B: Ball) return Colour;
   procedure Set_Colour(B: in out Ball; C: Colour);
   procedure Remove_Ball(B: in out Ball);
private
   ...
end;

package Balls.Iterators is
   procedure Iterate(Action: access procedure (B: in out Ball));
end;
```

This is weird. We cannot externally declare or get hold of a ball at all since it has unknown discriminants and so needs to be initialized – but we have provided no way to initialise it. We can command that a ball be created but it is then hidden from us. The subprograms that take a ball as a parameter can only be called from within an iterator.

Exercise 21.9

1 Explore the possibility of providing the three external types

```
type Person_Name(<>) is private;
type Mans_Name(<>) is private;
type Womans_Name(<>) is private;
```

together with appropriate conversion functions for both formulations.

Checklist 21

Carefully distinguish the concepts 'is a' and 'has a'.

Beware multiple inheritance.

Distinguish active and passive iterators.

Generalized iterators and references were added in Ada 2012 largely for use with containers.

The function Is_Abstract was added to Ada.Tags in Ada 2012.

New in Ada 2022

Parallel iterators and subprograms for manipulating chunks are added to the package Ada.Iterator_Interfaces.

Procedural iterators are added in Section 21.7.

22 Tasking Techniques

22.1 Dynamic tasks
22.2 Multiprocessors
22.3 Synchronized interfaces
22.4 Discriminants
22.5 Task termination
22.6 Clocks and timers
22.7 Profiles
22.8 Parallel blocks and loops
22.9 Conflict checking

Chapter 20 covered the basic concepts concerning tasking. This chapter provides a number of further examples of tasks and protected objects in order to illustrate how the various facilities work together.

An important topic which was only briefly alluded to earlier is the use of synchronized interfaces which pull together two very important features of programming namely OOP and concurrency.

Some topics covered in this chapter are in fact defined in the Systems Programming or Real-Time Systems annexes and thus might not be available in all implementations. However, they seem worthy of more detail than would otherwise be given in the brief notes on the specialized annexes in Chapter 26.

22.1 Dynamic tasks

The first example is an interesting demonstration of the dynamic creation of task objects. The objective is to find and display the first few prime numbers using the Sieve of Eratosthenes. (Eratosthenes was a Greek mathematician in the 3rd century BC.) This ancient algorithm works using the observation that if we have a list of the primes below *N*, then *N* is also prime if none of these divide exactly into it. So we try the existing primes in turn and as soon as one divides *N* we discard *N* and try again with *N* set to *N*+1. If we get to the end of the list of primes without dividing *N*, then *N* must be prime, so we add it to the list and start again with *N*+1.

A separate task is used for each prime P and is linked (via an access value) to the previous prime task and next prime task as shown in Figure 22.1. Its role is to take a trial number N from the previous task and to check whether it is divisible by its prime P. If it is, the number is discarded; if it is not, the number is passed to the

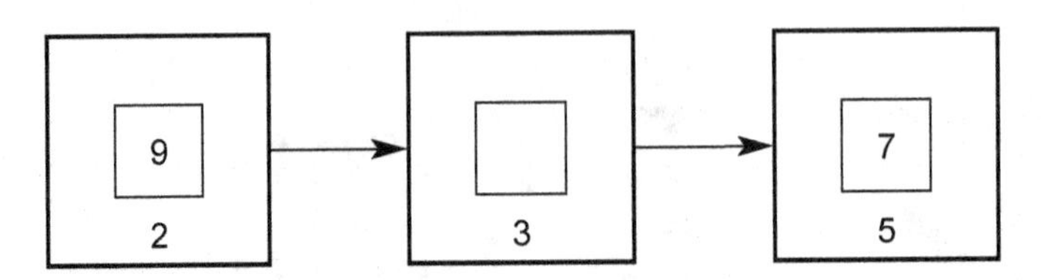

Figure 22.1 The Sieve of Eratosthenes.

next prime task. If there is no next prime task then P was the largest prime so far and N is a newly found prime; the P task then creates a new task whose duty is to check for divisibility by N and links itself to it. Each task thus acts as a filter removing multiples of its own prime value.

On the screen each task displays a frame containing its own prime and an inner box which displays the trial value currently being tested, if any. Figure 22.1 shows the situation when the primes 2, 3, and 5 have been found. The 5 task is testing 7 (which will prove to be a new prime), the 3 task is resting and waiting to receive another number from the 2 task, the 2 task (having just discarded 8 as well as 4 and 6 previously) is testing 9 (which it will pass to the 3 task in a moment).

The program comprises a package Frame containing subprograms for the display (the body of this package is not shown), a task type Filter which describes the activities of the prime tasks, a function Make_Filter which creates new prime tasks as necessary, and a main procedure Sieve.

```
package Frame is
   type Position is private;
   function Make_Frame(Prime: Integer) return Position;
   procedure Write_To_Frame(Value: in Integer; Where: in Position);
   procedure Clear_Frame(Where: in Position);
private
   ...
end Frame;

use Frame;

task type Filter(P: Integer) is
   entry Input(Number: in Integer);
end Filter;

function Make_Filter(N: Integer) return access Filter is
begin
   return new Filter(N);
end Make_Filter;

task body Filter is              -- discriminant P not repeated in body
   N: Integer;                   -- trial number
   Here: Position := Make_Frame(P);
   Next: access Filter;
begin
   loop
```

```
            accept Input(Number: in Integer) do
               N := Number;
            end;
            Write_To_Frame(N, Here);
            if N mod P /= 0 then
               if Next = null then
                  Next := Make_Filter(N);
               else
                  Next.Input(N);
               end if;
            end if;
            Clear_Frame(Here);
         end loop;
      end Filter;

      procedure Sieve is
         First: access Filter := new Filter(2);
         N: Integer := 3;
      begin
         loop
            First.Input(N);
            N := N + 1;
         end loop;
      end Sieve;
```

The subprograms in the package Frame behave as follows. The function Make_Frame draws a new frame on the screen and permanently writes the new prime number (that is P for the task calling it) into the frame; the inner box is left empty. The function returns a value of the private type Position which identifies the position of the frame; this value is later passed as a parameter to the two other procedures which manipulate the inner box in order to identify the frame concerned. The procedure Write_To_Frame has a further parameter giving the value to be written in the inner box; the procedure Clear_Frame wipes the inner box clean.

The task type Filter is fairly straightforward. It has a discriminant P which is its prime number and a single entry Input which is called by the preceding task to give it the next trial divisor. It starts by creating its frame (noting the position in Here) and then enters the loop and awaits a number N to test. Having collected N it displays it in the inner box and then tests for divisibility by P. If it is divisible, it clears the inner box and goes to the beginning of the loop for a new value of N. If N is not divisible by P, it makes a successor task if necessary with discriminant N and otherwise passes the value of N to it by calling its entry Input. Only after the successor task has taken the value does it clear the inner box and go back to the beginning.

The driving procedure Sieve makes the first task with discriminant 2, sets N to 3 and then enters an endless loop giving the task successive integer values until the end of time (or some other limitation is reached).

The reader may wonder why we declared the function Make_Filter rather than simply writing

```
      Next := new Filter(N);
```

within the body of the task Filter. The reason is that within a task body the name of the task normally refers to the current execution of the task and cannot be used as a subtype mark. Thus we could abort the current task or pass it as a parameter to a procedure by using the name Filter and this would refer to the task object currently executing the body. But after **access** the name does denote the type.

22.2 Multiprocessors

In recent years the cost of processors has fallen dramatically and for many applications it is now more sensible to use several individual processors rather than one high performance processor.

Moreover, society has got accustomed to the concept that computers keep on getting faster. This makes them applicable to more and more high volume but low quality applications. But this cannot go on. The finite value of the velocity of light means that increase in processor speed can only be achieved by using devices of ever smaller size. But here we run into problems concerning the nonzero value of Planck's constant. When devices get very small, quantum effects cause problems with reliability.

No doubt, in due course, genuine quantum processors will emerge based perhaps on attributes such as spin. But meanwhile, the current approach is to use multiprocessors to gain extra speed.

In this section we look at facilities introduced in Ada 2012 concerning the numbering of processors and their allocation to tasks. The section concludes by looking at a mathematical example which has been in previous versions of this book for many years and consider how the tasks can be mapped to processors. Lightweight multitasking is discussed in Section 22.8 when we discuss parallel blocks and loops introduced in Ada 2022.

A key feature is a child package of System thus

```
package System.Multiprocessors
    with Preelaborate, Nonblocking, Global => in out synchronized is
  type CPU_Range is range 0 .. implementation-defined;
  Not_A_Specific_CPU: constant CPU_Range := 0;
  subtype CPU is CPU_Range range 1 .. CPU_Range'Last;
  function Number_Of_CPUs return CPU;
end System.Multiprocessors;
```

Note that this is a child of System rather than a child of Ada. This is because System is generally used for hardware related features.

Processors are given a unique positive integer value from the subtype CPU. This is a subtype of CPU_Range which also includes zero; zero is reserved to mean not allocated or unknown and is the value of the constant Not_A_Specific_CPU.

The total number of CPUs is determined by calling the function Number_Of_CPUs. This is a function rather than a constant because there could be several partitions with a different number of CPUs on each partition.

Note that a CPU cannot be used by more than one partition. The allocation of CPU numbers to partitions is not defined; each partition could have a set starting at 1, but they might be numbered in some other way.

Tasks can be allocated to processors by an aspect specification. If we write

```
task My_Task
   with CPU => 10;
```

then My_Task will be executed by processor number 10. The expression giving the processor for a task can be dynamic.

Moreover, in the case of a task type, the CPU can be given by a discriminant. So we can have

```
task type Slave(N: CPU_Range)
   with CPU => N;
```

and then we can declare

```
Tom: Slave(1);
Dick: Slave(2);
Harry: Slave(3);
```

and Tom, Dick, and Harry are then assigned CPUs 1, 2, and 3 respectively. We could also have

```
Fred: Slave(0);
```

and Fred could then be executed by any CPU since 0 is Not_A_Specific_CPU.

The aspect CPU can also be given to the main subprogram in which case the expression must be static.

The child package System.Multiprocessors.Dispatching_Domains whose specification is shown below provides further facilities. We have added use clauses to save space and a subtype D_D as an abbreviation for Dispatching_Domain.

```
with Ada.Real_Time; with Ada.Task_Identification;
use Ada.Real_Time; use Ada.Task_Identification;
package System.Multiprocessors.Dispatching_Domains
      with Nonblocking, Global => in out synchronized is
   Dispatching_Domain_Error: exception;

   type Dispatching_Domain(<>) is limited private;
   subtype D_D is Dispatching_Domain;
   System_Dispatching_Domain: constant D_D;
   function Create(First: CPU; Last: CPU_Range) return D_D;
   function Get_First_CPU(Domain: D_D) return CPU;
   function Get_Last_CPU(Domain: D_D) return CPU_Range;

   type CPU_Set is array (CPU range <>) of Boolean;
   function Create(Set: CPU_Set) return D_D;
   function Get_CPU_Set(Domain: D_D) return CPU_Set;

   function Get_Dispatching_Domain(T: Task_Id := Current_Task)
                                                          return D_D;
   procedure Assign_Task(Domain: in out Dispatching_Domain;
                         CPU: in CPU_Range := Not_A_Specific_CPU;
                         T: in Task_Id := Current_Task);
```

```
   procedure Set_CPU(CPU: in CPU_Range;
                     T: in Task_Id := Current_Task);
   function Get_CPU(T: in Task_Id := Current_Task) return CPU_Range;
   procedure Delay_Until_And_Set_CPU(Delay_Until_Time: in Time;
                                     CPU: in CPU_Range);
private
   ...
end System.Multiprocessors.Dispatching_Domains;
```

The idea is that processors are grouped together into dispatching domains. A task may then be allocated to a domain and it will be executed on one of the processors of that domain.

Domains are of the type Dispatching_Domain. This has unknown discriminants and so uninitialized objects of the type cannot be declared. But initialized objects can be created by the functions Create. One such function works on a contiguous range and the other is more flexible and uses sets. So to declare My_Domain covering processors from 10 to 20 inclusive we can write

```
My_Domain: Dispatching_Domain := Create(10, 20);
```

and to create a domain of just processors 2, 4, 6 we can write

```
My_Small_Domain: Dispatching_Domain :=
                     Create(CPU_Set'(2 | 4 | 6 => true, 3 | 5 => false));
```

All CPUs are initially in the System_Dispatching_Domain. Creating a domain removes those assigned to the domain from the System_Dispatching_Domain. Note carefully that a CPU can only be in one domain. If we attempt to do something silly such as create overlapping domains then Dispatching_Domain_Error is raised.

The environment task is always executed on a CPU in the System_Dispatching_Domain. Clearly we cannot move all the CPUs from this domain otherwise the environment task would be left high and dry. Again an attempt to do so would raise Dispatching_Domain_Error.

A very important rule is that we cannot create domains once the main subprogram is called. Moreover, there is no operation to remove a CPU from a domain once the domain has been created. So the general approach is to create all domains during library package elaboration. This then sets a fixed arrangement for the program as a whole and we can then call the main subprogram.

Tasks can be assigned to a domain in two ways. One way is to use an aspect

```
task My_Task
   with Dispatching_Domain => My_Domain;
```

If we give both the domain and an explicit CPU thus

```
task My_Task
   with CPU => 10, Dispatching_Domain => My_Domain;
```

then they must be consistent. That is the CPU given must be in the domain given. If it is not then task activation fails.

The other way to assign a task to a domain is by calling the procedure Assign_Task. Thus the above example could be written as

```
Assign_Task(My_Domain, 10, My_Task'Identity);
```

Similarly, we can assign a CPU to a task by calling the function Set_CPU thus

```
Set_CPU(A_CPU, My_Task'Identity);
```

So, a task can be assigned a domain and possibly a specific CPU in that domain. If no specific CPU is given then the scheduling algorithm is free to use any CPU in the domain for that task and to change from time to time; moreover, the function Get_CPU applied to the task will return zero meaning not a specific CPU.

If a task is not assigned to a specific domain then it will execute in the domain of its activating task. In the case of a library task the activating task is the environment task and since this executes in the System_Dispatching_Domain, this will be the domain of the library task.

The domain and any specific CPU assigned to a task can be set at any time by calls of Assign_Task and Set_CPU. But note carefully that once a task is assigned to a domain other than the system dispatching domain then it cannot be assigned to a different domain. But the CPU within a domain can be changed at any time; from one specific value to another specific value or maybe to zero indicating no specific CPU. It is also possible to change CPU but for the change to be delayed by the aptly named function Delay_Until_And_Set_CPU.

There are various functions for interrogating the situation regarding domains whose behaviour should be obvious.

As an example we will look at how we might solve a differential equation using several tasks on a computer with several processors with a common address space. Thus when several tasks are active they really will be active and we assume that there are enough processors for all the tasks in the program to truly run in parallel.

Suppose we wish to solve the differential equation

$$\partial^2 P/\partial x^2 + \partial^2 P/\partial y^2 = F(x, y)$$

over a square region. The value of P is given on the boundary and the value of F is given throughout. The problem is to find the value of P at internal points of the region. This equation arises in many physical situations. One example might concern the flow of heat in a thin sheet of material; $P(x, y)$ would be the temperature of point (x, y) in the sheet and $F(x, y)$ would be the external heat flux applied at that point. However, the physics doesn't really matter.

The standard approach is to consider the region as a grid and to replace the differential equation by a corresponding set of difference equations. For simplicity we consider a square region of side N with unit grid. We end up with something like having to solve

$$4P(i, j) = P(i-1, j) + P(i+1, j) + P(i, j-1) + P(i, j+1) - F(i, j) \qquad 0 < i, j < N$$

This equation gives a value for each point in terms of its four neighbours. Remember that the values on the boundary are known and fixed. We use an iterative approach (Gauss–Seidel) and allocate a task to each point (i, j). The tasks then repeatedly compute the value of their point from the neighbouring points until the values cease to change. The function F could be of arbitrary complexity. A possible program is as follows

```
procedure Gauss_Seidel is
   N: constant := 5;
   Next_CPU: CPU := 2;
   subtype Full_Grid is Integer range 0 .. N;
   subtype Grid is Full_Grid range 1 .. N-1;
   type Real is digits 7;                       -- declare our own type
   type Matrix is array (Integer range <>, Integer range <>) of Real
      with Atomic_Components;

   P: Matrix(Full_Grid, Full_Grid);
   Delta_P: Matrix(Grid, Grid);
   Tolerance: constant Real := 0.0001;
   Error_Limit: constant Real := Tolerance * Real(N-1)**2;
   Converged: Boolean := False
      with Atomic;
   Error_Sum: Real;

   function F(I, J: Grid) return Real is separate;

   task type Iterator is
      entry Start(I, J: in Grid);
   end;

   Process: array (Grid, Grid) of Iterator;

   task body Iterator is
      I, J: Grid;
      New_P: Real;
   begin
      accept Start(I, J: in Grid) do
         Iterator.I := Start.I;
         Iterator.J := Start.J;
      end Start;

      loop
         New_P := 0.25 * (P(I-1, J) + P(I+1, J) + P(I, J-1)
                                  + P(I, J+1) - F(I, J));
         Delta_P(I, J) := New_P - P(I, J);
         P(I, J) := New_P;
         exit when Converged;
      end loop;
   end Iterator;

begin                          -- main subprogram, Iterator tasks now active
   ...                         -- initialize P and Delta_P

   Set_CPU(1);                 -- set CPU of environment task

   -- now iterate over all the iterator tasks in a double loop
```

```
      -- in order to tell each task who it is and to allocate a CPU
      for I in Grid loop
         for J in Grid loop
            Process(I, J).Start(I, J);                 -- tell them who they are and
            Set_CPU(Next_CPU, Process(I, J)'Identity);          -- allocate CPU
            Next_CPU := Next_CPU + 1;
         end loop;
      end loop;

      loop
         Error_Sum := 0.0;
         for I in Grid loop
            for J in Grid loop
               Error_Sum := Error_Sum + Delta_P(I, J)**2;
            end loop;
         end loop;

         Converged := Error_Sum < Error_Limit;
         exit when Converged;
      end loop;

      ...                           -- output results

   end Gauss_Seidel;
```

In this simple example we have just used the system dispatching domain. The environment task which is executing the main subprogram, Gauss_Seidel, assigns itself CPU number 1 and then using a double loop tells the iterator tasks who they are through the call of the entry Start and then assigns the next free CPU to each task in turn. Thereafter the individual tasks execute independently and communicate through shared variables. The Iterator tasks continue until the Boolean variable Converged is set by the main task; they then exit their loop and terminate. The main task repeatedly computes the sum of squares of the errors from Delta_P (set by the Iterator tasks) and sets Converged accordingly. When stability is reached the main task outputs the results.

This example illustrates the use of shared variables; they are the Boolean variable Converged and the two arrays P and Delta_P. The aspect Atomic applied to the variable Converged ensures that all accesses to the variable are done atomically and cannot interfere. In the case of the arrays, all we need is to ensure that the individual components are treated atomically and this is ensured by applying the aspect Atomic_Components to the array type.

It is somewhat irritating to have to tell the tasks who they are. We would like to avoid the entry Start and the corresponding accept statement. One approach is to use an allocator in a loop and to provide the tasks with discriminants thus

```
      task type Iterator(I, J: Grid);             -- I and J are discriminants
      An_It: access Iterator;                     -- dummy variable
      ...
   begin                        -- of main subprogram
      ...
      ...
```

```
for I in Grid loop
   for J in Grid loop
      An_It := new Iterator(I, J);
   end loop;
end loop;
```

This creates the tasks dynamically and sets them active one by one. Note that we have to declare a useless variable An_It which feels awkward. And we might feel uncomfortable with using a storage pool for the tasks anyway. So we seek an alternative mechanism which does not involve allocators.

We could laboriously declare all the tasks thus

```
Process_11: Iterator(1, 1);
Process_12: Iterator(1, 2);
...
```

but that is disgusting. Another attempt is to write

```
task type Iterator(I: Grid := Next_I; J: Grid := Next_J);
Process: array (Grid, Grid) of Iterator;
```

where the functions Next_I and Next_J are designed to deliver all the required pairs of values. But that is irritating since the pairs of discriminants might not match the index pairs because the order of evaluation of components of an array is not defined by the language; in a sense that does not matter but it feels uncomfortable. We also have to take care that the pairs of calls of Next_I and Next_J might also be in any order. It is all rather unpleasant.

Nevertheless using discriminants is the key to the solution although we have to beware that a discriminant of a task cannot be changed because tasks are limited. Moreover, aggregates pose a problem because of the undefined order of evaluation and so to ensure proper matching the values have to be set in a loop where the order is under our control.

The crafty solution below involves access discriminants and a self-referential structure as described in Section 18.7 plus the use of the extended return statement in which the loop can be written. The task is restructured as follows

```
type Iterator;
task type Task_Iterator(D: not null access Iterator);

type Iterator is limited
   record
      I, J: Grid;
      T: Task_Iterator(D => Iterator'Access);
   end record;
```

So the task itself is wrapped in a record also containing I and J. The access discriminant then refers to the record and thus gives access to I and J. Although the actual discriminant cannot be changed the values of I and J can be changed via the discriminant. In a sense we have extended the task with additional components which are visible both to the task itself and externally. This is a standard technique that is especially valuable with tasks and protected objects.

The next step is to declare a magic function Make_Process with an extended return in which the array of iterators can be initialized in the required order. To do this we need to declare an array type as well

```
type Process_Type is array (Grid, Grid) of Iterator;

function Make_Process return Process_Type is
begin
   return R: Process_Type do
      for I in Grid loop
         for J in Grid loop
            R(I, J).I := I;
            R(I, J).J := J;
         end loop;
      end loop;
   end return;
end Make_Process;
```

Note that the double loop is setting the values of the components I and J of the individual records of the array R which is the result. We could also set the CPUs in the same loop.

Finally we declare the array Process as follows

```
Process: Process_Type := Make_Process;          -- call magic function
```

Remember that the mechanism of the extended return (see Sections 10.1 and 12.5) for limited types means that the object Process is passed as a hidden parameter to the function so that within the function we are actually manipulating the object Process itself and no copying ever occurs.

The reader is recommended to study this example carefully. There are three key points

- The access discriminant enables the task to be extended with visible writable components.
- The function using the extended return enables arbitrary computation to be done essentially within the declaration of the object Process itself.
- The tasks are not active before the function returns.

This last point is vital. If the tasks became active before the assignments were made to I and J then the whole scheme would fall apart. In this case the tasks only become active at the **begin** of the subprogram Gauss_Seidel and so no problem could arise.

However, if the function were used with an allocator as in

```
P := new Process_Type'(Make_Process);
```

then it is vital that the tasks do not become active before the function returns.

The program as shown (with $N = 5$) requires 17 tasks and thus 17 processors. The reader might like to consider how they could be arranged in different dispatching domains. In order to do this we have to have a library package whose elaboration sets up the domains. Remember that the domains cannot be changed after the main subprogram has started.

There are possibly a number of flaws in the program. The convergence criterion is a bit suspect. It might be possible for waves of divergence to slurp around the grid in a manner which escapes the attention of the asynchronous main task – but this is unlikely. Another point is that the Iterator tasks might still be computing one last iteration while the main task is printing the results. It would be better to add a Stop entry so that the main task can wait until the Iterator tasks have finished their loops. Alternatively, the array of tasks could be declared in an inner block and the main task could do the printing outside that block where it would know that the other tasks must have terminated. Thus

```
begin                          -- main subprogram
   ...                         -- initialize arrays
  declare
     Process: array (Grid, Grid) of Iterator;
  begin                        -- Iterator tasks active
     ...
  end;                         -- wait for Iterator tasks to terminate
   ...                         -- output results
end Gauss_Seidel;
```

Some multiprocessor systems do not have a shared memory in which case a different approach is necessary. A naïve first attempt might be to give each Iterator an entry which when called delivers the current value of the corresponding point of the grid. Direct *ad hoc* calls from one task to another in a casual design would quickly lead to deadlock. A better approach is to have a task and a protected object at each point; the task does the computation and the protected object provides controlled access to the point.

It is hoped that this simple example has given the reader some glimpse of how tasking may be distributed over a multiprocessor system. Of course the example has a ludicrously trivial computation in each task and the system will spend much of its time on communication rather than computation. Nevertheless the principles should be clear.

Exercise 22.2

1 Sketch a solution of the differential equations using a task and a protected object for each point and no shared variables. Evaluate and store Delta_P in the protected object. The main subprogram should use the same convergence rule as before. You will not need arrays (other than for the tasks and protected objects) since the data will be distributed in the protected objects. Also make it apply to a rectangular grid whose size is not statically known.

22.3 Synchronized interfaces

An important aspect of Ada is the interaction between the object oriented and tasking features of the language through inheritance.

Recall from Chapter 14 that we can declare an interface thus

```
type Int is interface;
```

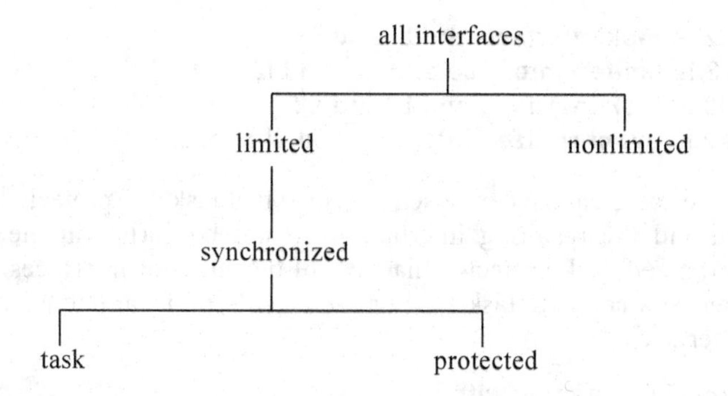

Figure 22.2 Interface hierarchy.

An interface is essentially an abstract tagged type that cannot have any components but can have abstract operations and null procedures. We can then derive other interfaces and tagged types by inheritance such as

```
type Another_Int is interface and Int1 and Int2;
type T is new Int1 and Int2 with ... ;
type TT is new T and Int3 and Int4 with ... ;
```

Remember that a tagged type can be derived from at most one other normal tagged type but can also be derived from several interfaces. In the list, the first is called the parent (it can be a normal tagged type or an interface) and any others (which can only be interfaces) are called progenitors.

There are other categories of interfaces, namely synchronized, protected, and task interfaces which form a hierarchy as shown in Figure 22.2. A synchronized interface can be implemented by either a task or protected type (a concurrent type), a protected interface can only be implemented by a protected type, and a task interface can only be implemented by a task type.

A nonlimited interface can only be implemented by a nonlimited type. However, a limited interface can be implemented by any tagged type (limited or not) or by a protected or task type. Remember that task and protected types are inherently limited and are also treated as tagged types when they implement one or more interfaces.

So we can write

```
type LI is limited interface;            -- similarly type LI2
type SI is synchronized interface;
type TI is task interface;
type PI is protected interface;
```

and we can also provide operations but they must be abstract or null.

We can compose these interfaces provided that no conflict arises. The following are all permitted:

```
type TI2 is task interface and LI and TI;
type LI3 is limited interface and LI and LI2;
type TI3 is task interface and LI and LI2;
type SI2 is synchronized interface and LI and SI;
```

Two or more interfaces can be composed provided that task and protected interfaces are not mixed and the resulting interface must not be earlier in the sequence: limited, synchronized, task/protected than any of the ancestor interfaces.

We can derive a concrete task type or protected type from one or more of the appropriate interfaces

```
task type TT is new TI with
   ...                          -- here we give entries as usual
end TT;
```

or

```
protected type PT is new LI and SI with
   ...
end PT;
```

Unlike tagged record types we cannot derive a task or protected type from another task or protected type. So the derivation hierarchy can only be one level deep once we declare a concrete task or protected type.

The operations of these various interfaces are declared in the usual way and an interface composed of several interfaces has the operations of all of them with the same rules regarding duplication and overriding of an abstract operation by a null one and so on as for normal tagged types.

When we declare a concrete task or protected type then we must implement all of the operations of the interfaces concerned. This can be done in two ways, either by declaring an entry or protected operation in the specification of the task or protected object or by declaring a distinct subprogram in the same list of declarations (but not both). If an operation is a null procedure then it can be inherited or overridden as usual.

Thus the interface

```
package Pkg is
   type TI is task interface;
   procedure P(X: in TI) is abstract;
   procedure Q(X: in TI; I: in Integer) is null;
end Pkg;
```

could be implemented by

```
package PT1 is
   task type TT1 is new TI with
      entry P;                  -- P and Q implemented by entries
      entry Q(I: in Integer);
   end TT1;
end PT1;
```

or by

```
package PT2 is
   task type TT2 is new TI with
      entry P;              -- P implemented by an entry
   end TT2;
                            -- Q implemented by a procedure
   procedure Q(X: in TT2; I: in Integer);
end PT2;
```

or even by

```
package PT3 is
   task type TT3 is new TI with end;
                            -- P implemented by a procedure
                            -- Q inherited as a null procedure
   procedure P(X: in TT3);
end PT3;
```

In this last case there are no entries and so we have the juxtaposition **with end** which is somewhat similar to the juxtaposition **is end** that occurs with generic packages used as signatures. Actually it is unlikely that TT3 would have no entries since the procedure P will almost inevitably be implemented by calling some entries of the task as illustrated by one of the examples below.

Observe how the first parameter which denotes the task is omitted if it is implemented by an entry. This echoes the prefixed notation for calling operations of tagged types in general. Remember that rather than writing

```
Op(X, Y, Z, ...);
```

we can write

```
X.Op(Y, Z, ...);
```

provided certain conditions hold such as that X is of a tagged type and that Op is a primitive operation of that type.

In order for the implementation of an interface operation by an entry of a task type or a protected operation of a protected type to be possible some fairly obvious conditions must be satisfied.

In all cases the first parameter of the interface operation must be of the task type or protected type (it may be an access parameter). In addition, in the case of a protected type, the first parameter of an operation implemented by a protected procedure or entry must have mode **out** or **in out** (and in the case of an access parameter it must be an access to variable parameter).

If the operation does not fit these rules then it has to be implemented as a normal subprogram (as opposed to a protected subprogram). An important example is that a function has to be implemented as a function in the case of a task type because there is no such thing as a function entry. However, a function can often be directly implemented as a protected function in the case of a protected type.

Entries and protected operations which implement inherited operations may be in the visible part or private part of the task or protected type in the same way as for tagged record types.

It may seem rather odd that an operation can be implemented by a subprogram that is not part of the task or protected type itself – it seems as if it might not be task safe in some way. But a common paradigm is where an operation as an abstraction has to be implemented by two or more entry calls. An example occurs in some implementations of the classic readers and writers problem as we shall see in a moment.

A task or protected type which implements an interface can have additional entries and operations as well, just as a derived tagged type can have more operations than its parent.

The overriding indicators **overriding** and **not overriding** can be applied to entries as well as to procedures and functions. Thus the package PT2 above could be written as

```
package PT2 is
   task type TT2 is new TI with
      overriding                -- P implemented by an entry
      entry P;
   end TT2;

   overriding                   -- Q implemented by a procedure
   procedure Q(X: in TT2; I: in Integer);
end PT2;
```

We will now explore a simple readers and writers example in order to illustrate various points. We start with the following interface

```
package RWP is
   type RW is limited interface;
   procedure Write(Obj: out RW; X: in Item) is abstract;
   procedure Read(Obj: in RW; X: out Item) is abstract;
end RWP;
```

The interface RW describes the simple abstraction of providing an encapsulation of a hidden location and a means of writing a value (of some type Item) to it and reading a value from it.

We could implement this in a nonsynchronized manner thus

```
type Simple_RW is new RW with
   record
      V: Item;
   end record;

overriding
procedure Write(Obj: out Simple_RW; X: in Item);
overriding
procedure Read(Obj: in Simple_RW; X: out Item);
...
```

```
procedure Write(Obj: out Simple_RW; X: in Item) is
begin
   Obj.V := X;
end Write;

procedure Read(Obj: in Simple_RW; X: out Item) is
begin
   X := Obj.V;
end Read;
```

This implementation is not task safe (task safe is sometimes known as thread-safe). If a task calls Write and the type Item is a composite type and the writing task is interrupted in the middle of writing, then a task which calls Read might get a curious result consisting of part of the new value and part of the old value.

For illustration we could then derive a synchronized interface

```
type Sync_RW is synchronized interface and RW;
```

This interface can only be implemented by a task or protected type. For a protected type we might have (see Section 20.4)

```
protected type Prot_RW is new Sync_RW with
   overriding
   procedure Write(X: in Item);
   overriding
   procedure Read(X: out Item);
private
   V: Item;
end;

protected body Prot_RW is
   procedure Write(X: in Item) is
   begin
      V := X;
   end Write;

   procedure Read(X: out Item) is
   begin
      X := V;
   end Read;
end Prot_RW;
```

Again observe how the first parameter of the interface operations is omitted when they are implemented by protected operations.

This implementation is perfectly task safe. However, one of the characteristics of the readers and writers example is that it is quite safe to allow multiple readers since they cannot interfere with each other. But the type Prot_RW does not allow multiple readers because protected procedures can only be executed by one task at a time.

Now consider

```
protected type Multi_Prot_RW is new Sync_RW with
   overriding
   procedure Write(X: in Item);
   not overriding
   function Read return Item;
private
   V: Item;
end;

overriding
procedure Read(Obj: in Multi_Prot_RW; X: out Item);
...

protected body Multi_Prot_RW is
   procedure Write(X: in Item) is
   begin
      V := X;
   end Write;

   function Read return Item is
   begin
      return V;
   end Read;
end Multi_Prot_RW;

procedure Read(Obj: in Multi_Prot_RW; X: out Item) is
begin
   X := Obj.Read;
end Read;
```

In this implementation the procedure Read is implemented by a procedure outside the protected type and this procedure then calls the function Read within the protected type. This allows multiple readers because one of the characteristics of protected functions is that multiple execution is permitted (but of course calls of the protected procedure Write are locked out while any calls of the protected function are in progress). The structure is emphasized by the use of overriding indicators.

This example illustrates the point made earlier that the juxtaposition **with end** is rather unlikely to occur in practice.

A simple tasking implementation (see Sections 20.5 and 20.8) might be as follows

```
task type Task_RW is new Sync_RW with
   overriding
   entry Write(X: in Item);
   overriding
   entry Read(X: out Item);
end;

task body Task_RW is
   V: Item;
begin
   loop
```

```
      select
         accept Write(X: in Item) do
            V := X;
         end Write;
      or
         accept Read(X: out Item) do
            X := V;
         end Read;
      or
         terminate;
      end select;
   end loop;
end Task_RW;
```

Finally, as at the end of Section 20.5, we can have a tasking implementation which allows multiple readers and ensures that an initial value is set by only allowing a call of Write the first time.

```
task type Multi_Task_RW(V: access Item) is new Sync_RW with
   overriding
   entry Write(X: in Item);
   not overriding
   entry Start;
   not overriding
   entry Stop;
end;

overriding
procedure Read(Obj: in Multi_Task_RW; X: out Item);
...

task body Multi_Task_RW is
   Readers: Integer := 0;
begin
   accept Write(X: in Item) do
      V.all := X;
   end Write;
   loop
      select
         when Write'Count = 0 =>
         accept Start;
         Readers := Readers + 1;
      or
         accept Stop;
         Readers := Readers - 1;
      or
         when Readers = 0 =>
         accept Write(X: in Item) do
            V.all := X;
         end Write;
```

```
      or
         terminate;
      end select;
   end loop;
end Multi_Task_RW;

overriding
procedure Read(Obj: in Multi_Task_RW; X: out Item) is
begin
   Obj.Start;
   X := Obj.V.all;
   Obj.Stop;
end Read;
```

In this case the data being protected is accessed via the access discriminant of the task. It is structured this way so that the procedure Read can read the data directly. Note also that the procedure Read (which is the implementation of the procedure Read of the interface) calls two entries of the task.

It should be observed that this last example is by way of illustration only. Remember that the Count attribute used in tasks (as opposed to protected objects) can be misleading if tasks are aborted or if entry calls are timed out. Moreover, it would be gruesomely slow.

So we have seen that a limited interface such as RW might be implemented by a normal tagged type (plus its various operations) and by a protected type and also by a task type. We could then dispatch to the operations of any of these according to the tag of the type concerned. Remember that task and protected types that implement interfaces are also forms of tagged types and so we have to be careful to say tagged record type (or informally, normal tagged type) where appropriate.

In the above example the types Simple_RW, Prot_RW, Multi_Prot_RW, Task_RW, and Multi_Task_RW all implement the interface RW.

So we might have

```
RW_Ptr: access RW'Class := ...
...
RW_Ptr.Write(An_Item);                    -- dispatches
```

and according to the value in RW_Ptr this will call the appropriate entry or procedure of an object of any of the types implementing the interface RW.

However if we have

```
Sync_RW_Ptr: access Sync_RW'Class := ...
```

then we know that any implementation of the synchronized interface Sync_RW will be task safe because it can only be implemented by a task or protected type. So the dispatching call

```
Sync_RW_Ptr.Write(An_Item);               -- task safe dispatching
```

will be task safe.

An interesting point is that because a dispatching call might be to an entry or to a procedure, what appear to be procedure calls are permitted in timed entry calls if they might dispatch to an entry.

So we could have

```
select
  RW_Ptr.Read(An_Item);                    -- dispatches
or
  delay Seconds(10);
end select;
```

Thus it might dispatch to the procedure Read if the type concerned turns out to be Simple_RW in which case a time-out could not occur. But if it dispatched to the entry Read of the type Task_RW then it could time out.

On the other hand, we are not allowed to use a timed call if it is statically known to be a procedure. So

```
A_Simple_Object: Simple_RW;
...
select
  A_Simple_Object.Read(An_Item);           -- illegal
or
  delay Seconds(10);
end select;
```

is not permitted.

A note of caution is in order. Remember that the time-out is to when the call gets accepted. If it dispatches to Multi_Task_RW.Read then time-out never happens because the Read itself is a procedure and gets called at once. However, behind the scenes it calls two entries and so could take a long time. But if we called the two entries directly with timed calls then we would get a time-out if there were a lethargic writer in progress. So the wrapper distorts the abstraction. This is not much worse than the problem we have anyway that a time-out is to when a call is accepted and not to when it returns – it could hardly be otherwise.

The same rules apply to conditional entry calls and also to asynchronous select statements where the triggering statement can be a dispatching call.

Another important point to note is that we can as usual assume the common properties of the class concerned. Thus in the case of a task interface we know that it must be implemented by a task and so operations such as **abort** and the attributes Callable and so on can be applied. If we know that an interface is synchronized then we know that it will be implemented by a task or a protected type and so is task safe.

Private extensions can also be marked as **synchronized**. In this case the full type must be a concurrent type or interface – and vice versa. The general rule for tagged types is that the synchronized property cannot be hidden. But for untagged types this is not so as was shown by the type Buffer of Section 20.10.

Typically an interface is implemented by a task or protected type but it can also be implemented by a singleton task or protected object despite the fact that singletons have no type name. Thus we might have

```
protected An_RW is new Sync_RW with
  procedure Write(X: in Item);
  procedure Read(X: out Item);
end;
```

with the obvious body. However we could not declare a single protected object similar to the type Multi_Prot_RW above. This is because we need a type name in order to declare the overriding procedure Read outside the protected object. So singleton implementations are possible provided that the interface can be implemented directly by the task or protected object without needing external subprograms.

As another example consider the stack protocol defined by

```
package Abstract_Stacks is
   type Stack is interface;
   procedure Push(S: in out Stack; E: in Element) is abstract;
   procedure Pop(S: in out Stack; E: out Element) is abstract;
   function Is_Empty(S: Stack) return Boolean is abstract;
   function Is_Full(S: Stack) return Boolean is abstract;
end Abstract_Stacks;
```

This is similar to that in Section 14.9 with additional functions Is_Empty and Is_Full. A concrete type such as Linked_Stack would then be derived from the interface and the inherited abstract operations replaced by concrete ones.

We could also write

```
type Protected_Stack is protected interface and Stack;
```

A concrete type derived from Protected_Stack would then have to be a protected type such as

```
protected type Protected_Array_Stack is new Protected_Stack with
   entry Push(X: in Element);
   entry Pop(X: out Element);
   function Is_Empty return Boolean;
   function Is_Full return Boolean;
private
   ...      -- data implementing the stack
end;

protected body Protected_Array_Stack is
   entry Push(X: in Element) when not Is_Full is
   begin
      ...
   end Push;
   entry Pop(X: out Element) when not Is_Empty is
   begin
      ...
   end Pop;
   ...      -- functions Is_Full and Is_Empty
   ...
end Protected_Array_Stack;
```

Note how the functions Is_Empty and Is_Full which were originally provided to enable the user of the nontasking protocol to guard against misuse of the stack are now also used as barriers which automatically prevent such misuse.

We could now write a procedure to copy between two stacks by dispatching as in the answer to Exercise 14.9(**3**) and this would work between stacks of either type. An alternative approach to this sort of problem using discriminants will be seen in the next section.

We noted above that if a timed call dispatches to a procedure then it simply calls it at once and there is no possibility of a time-out. A related problem arises with requeue. In this case we cannot allow dispatching to a procedure because there is no queue on which to hang it. So we have to ensure that dispatching will be to an entry. This can be done using the aspect Synchronization. If we write

```
procedure Q(S: in out Server; X: in Item) is abstract;
  with Synchronization => By_Entry;
```

then we are assured that we are permitted to do a requeue on Q for any implementation of Server. The other possible values for the aspect Synchronization are By_Protected_Procedure and Optional.

In summary, if the property is By_Entry then the procedure must be implemented by an entry, if the property is By_Protected_Procedure then the procedure must be implemented by a protected procedure, and if the property is Optional then it can be implemented by an entry, procedure, or protected procedure. Naturally enough, the aspect cannot be given for a function.

There are a number of rules regarding consistency. The aspect Synchronization can be applied to a task interface or protected interface as well as to a synchronized interface. However, if it is applied to a task interface then the aspect cannot be specified as By_Protected_Procedure for obvious reasons. If a type or interface is created by inheritance from other interfaces then any Synchronization properties are also inherited and must be consistent. Thus if one is By_Entry then the others must also be By_Entry or Optional.

An example of the use of this aspect will be found in Section 24.8 when we discuss the queue containers.

Exercise 22.3

1 Given

```
type Map is protected interface;
procedure Insert(M: in Map; K: in Key; V: in Value) is abstract;
procedure Find(M: in Map; K: in Key; V: out Value) is abstract;
```

declare the specification of a single protected object A_Map which implements the interface Map.

22.4 Discriminants

In Section 22.2 we saw how access discriminants were essential to the function Make_Process. We will now look at other applications of access discriminants which reference the data being controlled by a task or protected object.

Consider the following which has a similar structure to that of the example at the beginning of Section 18.7

```
type Task_Data is
  record
    X: T;
    ...
  end record;

task type Worker(D: access Task_Data) is
  ...
end Worker;
```

Within the task type the component X of the record being used by the current task instance is denoted by D.X. We can then declare a particular task and its data thus

```
Freds_Data: aliased Task_Data := (X => ... );
Fred: Worker(Freds_Data'Access);
```

This technique only applies in very simple situations. More elaborate examples require type extension.

In many systems it is often the case that a number of activities follow a cyclic pattern. Each activity might have a start time, an end time, and an interval as well as auxiliary activities to be executed when it finishes or if something goes wrong. We can write a task type which acts as a general template with the details provided by dispatching calls or through data referenced by the discriminant. We might have

```
task type Control(Activity: not null access Descriptor'Class);

task body Control is
  Next_Time: Calendar.Time := Activity.Start_Time;
begin
  loop
    delay until Next_Time;
    Action(Activity);                        -- dispatches
    Next_Time := Next_Time + Activity.Interval;
    exit when Next_Time > Activity.End_Time;
  end loop;
  Last_Wishes(Activity);                     -- dispatches
exception
  when Event: others =>
    Handle(Activity, Event);                 -- dispatches
end Control;
```

It is very important to note that the access discriminant is class wide. This is essential so that dispatching to the various operations can occur. The root type would be abstract and contain the timing information thus

```
package Cyclic_Activity is
  type Descriptor is abstract tagged
    record
```

```
         Start_Time, End_Time: Calendar.Time;
         Interval: Duration;
      end record;

   procedure Action(D: access Descriptor) is abstract;
   procedure Last_Wishes(D: access Descriptor) is null;
   procedure Handle(D: access Descriptor; E: in Exception_Occurrence);
end;

package body Cyclic_Activity is
   procedure Handle(D: access Descriptor;
                    E: in Exception_Occurrence) is
   begin
      Put_Line("Unhandled exception");
      Put_Line(Exception_Information(E));
   end Handle;
end Cyclic_Activity;
```

where we have provided a null procedure for last wishes and a default handler for exceptions but the central procedure Action is abstract and thus has to be provided on extension.

A cyclic activity such as one to fire a cannon at noon each day could then be created as follows

```
use Cyclic_Activity;

type Cannon_Data is new Descriptor with
   record
      Pounds_Of_Powder: Integer;
   end record;

procedure Action(CD: access Cannon_Data) is
begin
   Load_Cannon(CD.Pounds_Of_Powder);
   Fire_Cannon;
end Action;

The_Data: aliased Cannon_Data :=
      (Start_Time => High_Noon;
       End_Time => When_The_Stars_Fade_And_Fall;
       Interval => 24*Hours;
       Pounds_Of_Powder => 100);
Cannon_Task: Control(The_Data'Access);
```

Note how the specific data is passed to the procedure Action through the type extension. The timing data itself could have been provided by dispatching functions but it seems best to put such standard information in the root type. The task type Control could conveniently be declared in the same package as the type Descriptor.

As another example consider the package Buffers of Section 20.10. The protected data was actually inside the protected objects. But we could have placed the data outside and referred to it through an access discriminant

```
type Buffer(N: Positive) is
  record
    A: Item_Array(1 .. N);
    In_Ptr, Out_Ptr: Integer := 1;
    Count: Integer := 0;
  end record;

protected type Guardian(B: access Buffer) is
  entry Put(X: in Item);
  entry Get(X: out Item);
end;

protected body Guardian is

  entry Put(X: in Item) when B.Count < B.N is
  begin
    B.A(B.In_Ptr) := X;
    B.In_Ptr := B.In_Ptr mod N + 1;  B.Count := B.Count + 1;
  end Put;
  ...
end Guardian;
```

and now we can declare the buffer and the protected object separately

```
A_Buffer: aliased Buffer(100);
Guarded_A_Buffer: Guardian(A_Buffer'Access);
```

although they are permanently bound together.

Moreover, if we allocate the buffer thus

```
A_Guarded_Buffer: Guardian(new Buffer(100));
```

then the two objects are again permanently bound together. They are allocated together from the same storage pool and finalized and deallocated together. We say that the Buffer is a coextension of the Guardian. The effect of a coextension is to provide a discriminant of a composite type as if it were a genuine component.

Coextensions can also be used with nonlimited types. If the type Guardian were nonlimited then assignment could not be made to a longer lived object and it would not make a copy of the buffer. Consequently, the copied Guardian would not have the tight coupling of the original since it would share the buffer with the original and not have a coextension of its own at all.

In order to create a type suitable as the full type corresponding to the private type Buffer of Section 20.10 we can use the self-referential technique and put the protected object into the buffer alongside the data thus

```
type Buffer(N: Positive) is limited
  record
    A: Item_Array(1 .. N);
    In_Ptr, Out_Ptr: Integer := 1;
    Count: Integer := 0;
    The_Guardian: Guardian(Buffer'Access);
  end record;
```

This curious structure might have some merit if we wanted to bypass the protected object in some circumstances and access the data directly. It is especially curious that the Guardian has access to itself through the buffer.

Another possible arrangement (using the original type Buffer and making it limited) is as follows

```
protected type Guardian(B: access Buffer'Class) is ...

type Guarded_Buffer is new Buffer with
   record
      The_Guardian: Guardian(Guarded_Buffer'Access);
   end record;
```

Note that the discriminant is now class wide so that it can refer to the Guarded_ Buffer. However, from within the Guardian only the root part can be accessed (unless we do an explicit downward conversion).

As a final example we revisit the stack at the end of the previous section. Again we start with

```
package Abstract_Stacks is
   type Stack is interface;
   procedure Push(S: in out Stack; E: in Element) is abstract;
   procedure Pop(S: in out Stack; E: out Element) is abstract;
   function Is_Empty(S: Stack) return Boolean is abstract;
   function Is_Full(S: Stack) return Boolean is abstract;
end Abstract_Stacks;
```

Protected access to any such stack type can be obtained by using

```
protected type Safe_Stack(S: access Stack'Class) is
   entry Push(X: in Element);
   entry Pop(X: out Element);
end;
protected body Safe_Stack is
   entry Push(X: in Element) when not Is_Full(S.all) is
   begin
      Push(S.all, X);
   end Push;

   entry Pop(X: out Element) when not Is_Empty(S.all) is
   begin
      Pop(S.all, X);
   end Pop;
end Safe_Stack;
```

A protected object of this type provides access to the associated unprotected stack through the access discriminant. Note again how the functions Is_Empty and Is_Full are used as barriers which automatically prevent misuse. (We could write S rather than S.**all** by using access parameters for the underlying operations.)

We might now write

```
A_Raw_Stack: aliased Linked_Stack;
A_Safe_Stack: Safe_Stack(A_Raw_Stack'Access);
...
A_Safe_Stack.Push(An_Element);
```

This same technique could be applied to other structures such as queues.

Exercise 22.4

1 Provide a procedure Handle which takes some appropriate action if the cannon should explode and thereby raise the exception Bang from within Fire_Cannon.

2 Restructure the cyclic controller in order to add secret monitoring facilities so that we can find out how many times the loop in the control task has been executed: (a) for each instance, (b) over all instances, by calling functions No_Of_Cycles. Data relating to each instance can be kept in a private root type from which Descriptor is derived in a child package.

3 Could we do the cannon example without dispatching by placing access to subprogram values in the type Descriptor and using indirect calls?

22.5 Task termination

In Section 20.8 we mentioned the problem of how tasks could have a silent death. This happens if a task raises an exception which is not handled by the task itself. Tasks may also terminate because of going abnormal as well as by terminating normally. The detection of task termination and its causes can be monitored by the package Ada.Task_Termination (this is in the Systems Programming annex). Its specification is essentially

```
with Ada.Task_Identification;  use Ada.Task_Identification;
with Ada.Exceptions;  use Ada.Exceptions;
package Ada.Task_Termination
    with Preelaborable, Nonblocking, Global => in out synchronized is

  type Cause_Of_Termination is
                                  (Normal, Abnormal, Unhandled_Exception);
  type Termination_Handler is access protected procedure
                              (Cause: in Cause_Of_Termination;
                               T: in Task_Id; X: in Exception_Occurrence);
  procedure Set_Dependents_Fallback_Handler
                                      (Handler: in Termination_Handler);
  function Current_Task_Fallback_Handler return Termination_Handler;
  procedure Set_Specific_Handler(T: in Task_Id;
                                      Handler: in Termination_Handler);
  function Specific_Handler(T: in Task_Id) return Termination_Handler;
end Ada.Task_Termination;
```

Note that the above includes use clauses in order to simplify the presentation as we did for System.Multiprocessors.Dispatching_Domains in Section 22.2. We will use a similar approach for the other predefined packages described in this chapter.

The general idea is that we can associate a protected procedure with a task. The protected procedure is then invoked when the task terminates with an indication of the reason passed via its parameters. The protected procedure is identified by using the type Termination_Handler which is an access type referring to a protected procedure. The type Task_Id is declared in the package Ada.Task_Identification – see Section 26.1.

The association can be done in two ways. Thus we might declare a protected object Grim_Reaper

```
protected Grim_Reaper is
   procedure Last_Gasp(C: Cause_Of_Termination; T: Task_Id;
                                        X: Exception_Occurrence);
end Grim_Reaper;
```

which contains the protected procedure Last_Gasp whose parameters match those of the access type Termination_Handler.

We can then nominate Last_Gasp as the protected procedure to be called when the specific task T dies by

```
Set_Specific_Handler(T'Identity, Grim_Reaper.Last_Gasp'Access);
```

Alternatively we can nominate Last_Gasp as the protected procedure to be called when any of the tasks dependent on the current task becomes terminated by

```
Set_Dependents_Fallback_Handler(Grim_Reaper.Last_Gasp'Access);
```

Note that a task is not dependent upon itself and so this does not set a handler for the current task.

Thus a task can have two handlers: a fallback handler and a specific handler and either or both of these can be null. When a task terminates (that is after any finalization but just before it vanishes), the specific handler is invoked if it is not null. If the specific handler is null, then the fallback handler is invoked unless it too is null. If both are null then no handler is invoked.

The body of protected procedure Last_Gasp might then log various diagnostic messages

```
procedure Last_Gasp(C: Cause_Of_Termination; T: Task_Id;
                                        X: Exception_Occurrence) is
begin
   case C is
      when Normal => null;
      when Abnormal =>
         Put_Log("Something nasty happened to task ");
         Put_Log(Image(T));
      when Unhandled_Exception =>
         Put_Log("Unhandled exception occurred in task ");
         Put_Log(Image(T));
         Put_Log(Exception_Information(X));
   end case;
end Last_Gasp;
```

We should not call potentially blocking operations such as Put to a file within a protected operation so we call some procedure Put_Log which buffers the messages for later analysis.

There are three possible reasons for termination, it could be normal, abnormal (caused by abort), or because of the propagation of an unhandled exception. In the last case the parameter X gives details of the exception occurrence whereas in the other cases X has the value Null_Occurrence.

Initially, both specific and fallback handlers are null for all tasks. However, note that if a fallback handler has been set for all dependent tasks of T then the handler will also apply to any task subsequently created by T or one of its descendants. Thus a task can be born with a fallback handler already in place.

If a new handler is set then it replaces any existing handler of the appropriate kind. Calling either setting procedure with null for the handler naturally sets the appropriate handler to null.

The current handlers can be found by calling the functions Current_Task_Fallback_Handler or Specific_Handler; they return null if the handler is null.

It is important to realize that the fallback handlers for the tasks dependent on T need not all be the same since one of the dependent tasks of T might set a different handler for its own dependent tasks. Thus the fallback handlers for a tree of tasks can be different in various subtrees. This structure is reflected by the fact that the determination of the current fallback handler of a task is in fact done by searching recursively the tasks on which it depends.

Note that we cannot directly interrogate the fallback handler of a specific task but only that of the current task. Moreover, if a task sets a fallback handler for its dependents and then enquires of its own fallback handler it will not in general get the same answer because it is not one of its own dependents.

It is important to be aware of the environment task (see Sections 20.7 and 27.3). This unnamed task is the task that elaborates the library units and then calls the main subprogram. Library tasks (that is tasks declared at library level) are activated by the environment task before it calls the main subprogram.

Suppose the main subprogram calls the setting procedures as follows

```
procedure Main is
   protected RIP is
      procedure One( ... );
      procedure Two( ... );
   end;
   ...
begin
   Set_Dependents_Fallback_Handler(RIP.One'Access);
   Set_Specific_Handler(Current_Task, RIP.Two'Access);
   ...
end Main;
```

The specific handler for the environment task is then set to Two (because Current_Task is the environment task at this point) but the fallback handler for the environment task is null. On the other hand, the fallback handler for all other tasks in the program including any library tasks is set to One. Note that it is not possible to set the fallback handler for the environment task.

The astute reader will note that there is actually a race condition here since a library task might have terminated before the handler gets set. Preventing this is left as an exercise for the reader.

Some minor points are that if we try to set the specific handler for a task that has already terminated then Tasking_Error is raised. And if we try to set the specific handler for the null task, that is call Set_Specific_Handler with parameter T equal to Null_Task_Id, then Program_Error is raised. These exceptions are also raised by calls of the function Specific_Handler in similar circumstances.

Exercise 22.5

1 Show how to overcome the race condition when the environment task sets the fallback handler. Declare a library package Start_Up so that the elaboration of its body sets the handler. Declare the library tasks in a package Library_Tasks. Use aspects Elaborate and Elaborate_Body to ensure that things happen in the correct order.

22.6 Clocks and timers

As mentioned in Section 20.3, for some applications the timing facilities of the package Calendar are not adequate since the associated clock might not be monotonic. The package Ada.Real_Time (in the Real-Time Systems annex) defines a type Time and function Clock which is monotonic and thus not subject to such difficulties. Time intervals for this clock are defined in terms of a type Time_Span. There are also appropriate operators for manipulating times and time spans and constants such as Time_Span_Zero. The functions Nanoseconds, Microseconds, Milliseconds, Seconds, and Minutes which take an integer parameter and return a time span will be found useful for creating exact regular intervals. Two other important subprograms are

```
procedure Split(T: in Time; SC: out Seconds_Count; TS: out Time_Span);
function Time_Of(SC: Seconds_Count; TS: Time_Span) return Time;
```

These convert a count of a number of seconds plus a residual Time_Span into a Time and vice versa.

The delay until statement can take a value of the type Calendar.Time or Real_ Time.Time. The relative delay statement always takes an interval of the type Duration. This is because durations are intrinsic and not defined by the apparent value of a worldly clock. Boiling an egg takes four minutes and does not suddenly require an extra hour if being cooked when the clocks change (or even a whole day if we cross the International Date Line)!

The Real-Time Systems annex also defines three different kinds of timers. Two are concerned with monitoring the CPU time of tasks – one applies to a single task and the other to groups of tasks. The third timer measures real time rather than execution time and can be used to trigger events at specific real times. We will look first at the CPU timers.

The execution time of a task, or CPU time as it is commonly called, is the time spent by the system executing the task and services on its behalf. CPU times are represented by the type CPU_Time. This type and various subprograms are declared in the package Ada.Execution_Time which is very similar to the package Ada.Real_Time. It defines a function Clock whose specification is

```
function Clock(T: Task_Id := Current_Task) return CPU_Time;
```

which takes a task identity (see Section 26.1) as parameter and returns the CPU_Time that the task has consumed since it was created. There are also appropriate operators for manipulating CPU times and time spans. And again there are subprograms

```
procedure Split(T: in CPU_Time; SC: out Seconds_Count;
                                              TS: out Time_Span);
function Time_Of(SC: Seconds_Count;
                 TS: Time_Span := Time_Span_Zero) return CPU_Time;
```

These convert a count of a number of seconds plus a residual Time_Span into a CPU_Time and vice versa. Note the default value of Time_Span_Zero for the second parameter of Time_Of – this enables times of exact numbers of seconds to be given more conveniently thus

```
Four_Secs: CPU_Time := Time_Of(4);
```

The package Ada.Execution_Time also has facilities to monitor how much time is spent in interrupts. There are two constants and a function thus

```
Interrupt_Clocks_Supported: constant Boolean := implementation-defined;
Separate_Interrupt_Clocks_Supported: constant Boolean := impl-defined;
function Clocks_For_Interrupts return CPU_Time;
```

The constant Interrupt_Clocks_Supported indicates whether the time spent in interrupts is accounted for separately and Separate_Interrupt_Clocks_Supported indicates whether the time is accounted for each interrupt individually. The function Clocks_For_Interrupts returns the CPU_Time used over all interrupts. It is initialized to zero.

Time accounted for in interrupts is not also accounted for in individual tasks. In other words there is never any double accounting.

Calling the function Clocks_For_Interrupts if Interrupt_Clocks_Supported is false raises Program_Error. Note that the function Clock has a parameter giving the task concerned whereas Clocks_For_Interrupts does not since it covers all interrupts.

A child package of Ada.Execution_Time is provided for monitoring the time spent in individual interrupts. This package always exists even if the Boolean constant Separate_Interrupt_Clocks_Supported is false. Its specification is

```
package Ada.Execution_Time.Interrupts
    with Nonblocking, Global => in out synchronized is
  function Clock(Interrupt: Ada.Interrupts.Interrupt_Id) return CPU_Time;
  function Supported(Interrupt: Ada.Interrupts.Interrupt_Id)
                                                  return Boolean;
end Ada.Execution_Time.Interrupts;
```

The function Supported indicates whether the time for a particular interrupt is being monitored. If it is then Clock returns the accumulated CPU_Time spent in that interrupt handler (otherwise it returns zero). However, if the overall constant Separate_Interrupt_Clocks_Supported is false then calling this function Clock for any particular interrupt raises Program_Error.

We now consider timers. To find out when a task reaches a particular CPU time we can use the child package Ada.Execution_Time.Timers whose specification is

```
with System; use System;
package Ada.Execution_Time.Timers
    with Nonblocking, Global => in out synchronized is
  type Timer(T: not null access constant Task_Id) is
                                        tagged limited private;
  type Timer_Handler is access protected procedure(TM: in out Timer)
    with Nonblocking => False;
  Min_Handler_Ceiling: constant Any_Priority := implementation-defined;
  procedure Set_Handler(TM: in out Timer;
                        In_Time: Time_Span; Handler: Timer_Handler);
  procedure Set_Handler(TM: in out Timer;
                        At_Time: CPU_Time; Handler: Timer_Handler);
  function Current_Handler(TM: Timer) return Timer_Handler;
  procedure Cancel_Handler(TM: in out Timer; Cancelled: out Boolean);
  function Time_Remaining(TM: Timer) return Time_Span;
  Timer_Resource_Error: exception;
private
  ...  -- not specified by the language
end Ada.Execution_Time.Timers;
```

The general idea is that we declare an object of type Timer whose discriminant identifies the task to be monitored – note the use of **not null** and **constant** in the discriminant. We also declare a protected procedure which takes the timer as its parameter and which performs the actions required when the CPU_Time of the task reaches some value. Thus to take some action (perhaps abort for example although that would be ruthless) when the CPU_Time of the task My_Task reaches 2.5 seconds we might first declare

```
My_Timer: Timer(My_Task'Identity'Access);
Time_Max: CPU_Time := Time_Of(2, Milliseconds(500));
```

and then

```
protected Control is
  procedure Alarm(TM: in out Timer);
end;
...
protected body Control is
  procedure Alarm(TM: in out Timer) is
  begin
    Abort_Task(TM.T.all);                -- abort the task
  end Alarm;
end Control;
```

Finally, we set the timer in motion by calling the procedure Set_Handler which takes the timer, the time value and (an access to) the protected procedure thus

```
Set_Handler(My_Timer, Time_Max, Control.Alarm'Access);
```

and then when the CPU time of the task reaches Time_Max, the protected procedure Control.Alarm is executed. Note how the timer object incorporates the information regarding the task concerned using an access discriminant T and that this is passed to the handler via its parameter TM.

Aborting the task is perhaps a little violent. Another possibility is simply to reduce its priority so that it is no longer troublesome, thus

```
    -- cool that task
    Set_Priority(Priority'First, TM.T.all);
```

Another version of Set_Handler enables the timer to be set for a given interval (of type Time_Span).

The handler associated with a timer can be found by calling the function Current_Handler. This returns null if the timer is not set in which case we say that the timer is clear.

When the timer expires, and just before calling the protected procedure, the timer is set to the clear state. One possible action of the handler, having perhaps made a note of the expiration of the timer, it to set the handler again or perhaps another handler. So we might have

```
protected Control is
  procedure Alarm(TM: in out Timer);
private
  procedure Kill(TM: in out Timer);
end;

protected body Control is
  procedure Alarm(TM: in out Timer) is
  begin
    Log_Overflow(TM);   -- note that timer had expired
    -- and then reset it for another 500 milliseconds
    Set_Handler(TM, Milliseconds(500), Kill'Access);
  end Alarm;

  procedure Kill(TM: in out Timer) is
  begin
    -- expired again so kill it
    Abort_Task(TM.T.all);
  end Kill;
end Control;
```

In this scenario we make a note of the fact that the task has overrun and then give it another 500 milliseconds but with the handler Control.Kill so that the second time is the last chance.

It is important that the procedure Kill is declared in the private part of the protected object and not just in the body. This ensures that Kill has convention

Protected so that the Access attribute can be applied. Subprograms just declared in the body have convention Intrinsic and Access cannot be applied to them.

Setting the value of 500 milliseconds directly in the call is a bit crude. It might be better to parameterize the protected type thus

```
protected type Control(MS_To_Kill: Integer) is ...
...
My_Control: Control(500);
```

and then the call of Set_Handler in the protected procedure Alarm would be

```
Set_Handler(TM, Milliseconds(MS_To_Kill), Kill'Access);
```

Observe that overload resolution neatly distinguishes whether we are calling Set_Handler with an absolute time or a relative time.

The procedure Cancel_Handler can be used to clear a timer. The out parameter Cancelled is set to True if the timer was in fact set at the time of the call and False if it was clear. The function Time_Remaining returns Time_Span_Zero if the timer is not set and otherwise the time remaining.

Note also the constant Min_Handler_Ceiling. This is the minimum ceiling priority that the protected procedure should have to ensure that ceiling violation cannot occur (see Section 26.2).

This timer facility might be implemented on top of a POSIX system. There might be a limit on the number of timers that can be supported and an attempt to exceed this limit will raise Timer_Resource_Error.

We conclude by summarizing the general principles. A timer can be set or clear. If it is set then it has an associated (non-null) handler which will be called after the appropriate time. The key subprograms are Set_Handler, Cancel_Handler, and Current_Handler. The protected procedure has a parameter which identifies the event for which it has been called. The same protected procedure can be the handler for many events. The same general structure applies to the other kinds of timers.

In order to program various so-called aperiodic servers it is necessary for tasks to share a CPU budget. This can be done using the child package Execution_Time.Group_Budgets. This has much in common with its sibling package Timers but there are a number of important differences.

The first difference is that we are here considering a CPU budget shared among a group of several tasks. A type Group_Budget identifies a group and tasks can be added to or removed from a group by procedures Add_Task and Remove_Task. The members of a group can be identified by functions Is_Member, Is_A_Group_Member, and Members.

The type Group_Budget has a discriminant giving the CPU thus

```
type Group_Budget(CPU: System.Multiprocessors.CPU :=
                System.Multiprocessors.CPU'First) is tagged limited private;
```

This means that a group budget only applies to a single processor. If a task in a group is executed on another processor then the budget is not consumed. Note that the default value for CPU is CPU'First which is always 1.

The value of the budget (initially Time_Span_Zero) can be loaded by a procedure Replenish and increased by a procedure Add. Whenever a budget is nonzero it is counted down as the tasks in the group execute and so consume CPU time. Whenever a budget goes to Time_Span_Zero it is said to have become exhausted. When this happens a handler is called if one has been set. A handler is a protected procedure as before and procedures Set_Handler, Cancel_Handler, and function Current_Handler are much as expected. But a major difference is that Set_Handler does not set the time value of the budget since that is done by Replenish and Add. For further details the reader should consult the *ARM*.

The final kind of timer concerns real time rather than CPU time and so is provided by a child package of Ada.Real_Time. Its specification is

```
package Ada.Real_Time.Timing_Events
    with Nonblocking, Global => in out synchronized is
  type Timing_Event is tagged limited private;
  type Timing_Event_Handler
    is access protected procedure(Event: in out Timing_Event)
    with Nonblocking => False;
  procedure Set_Handler(Event: in out Timing_Event;
                        At_Time: Time; Handler: Timing_Event_Handler);
  procedure Set_Handler(Event: in out Timing_Event;
                   In_Time: Time_Span; Handler: Timing_Event_Handler);
  function Current_Handler(Event: Timing_Event)
                                       return Timing_Event_Handler;
  procedure Cancel_Handler(Event: in out Timing_Event;
                                           Cancelled: out Boolean);
  function Time_Of_Event(Event: Timing_Event) return Time;
private
  ... -- not specified by the language
end Ada.Real_Time.Timing_Events;
```

This package has a very similar pattern to the package Execution_Time.Timers. A handler can be set by Set_Handler and again there are two versions one for a relative time and one for absolute time. Times are of course specified using the type Real_Time rather than CPU_Time.

As a simple example suppose we wish to ring a pinger when our egg is boiled after four minutes. The protected procedure might be

```
protected body Egg is
  procedure Is_Done(Event: in out Timing_Event) is
  begin
    Ring_The_Pinger;
  end Is_Done;
end Egg;
```

and then

```
Egg_Done: Timing_Event;
Four_Min: Time_Span := Minutes(4);
Put_Egg_In_Water;
Set_Handler(Event => Egg_Done, In_Time => Four_Min,
                            Handler => Egg.Is_Done'Access);

-- now read newspaper whilst waiting for egg
```

This is unreliable because if we are interrupted between the calls of Put_Egg_In_Water and Set_Handler then the egg will be boiled for too long. We can overcome this by adding a further procedure to the protected object so that it becomes

```
protected Egg is
   procedure Boil(For_Time: in Time_Span);
private
   procedure Is_Done(Event: in out Timing_Event);
   Egg_Done: Timing_Event;
end Egg;

protected body Egg is
   procedure Boil(For_Time: in Time_Span) is
   begin
      Put_Egg_In_Water;
      Set_Handler(Egg_Done, For_Time, Is_Done'Access);
   end Boil;

   procedure Is_Done(Event: in out Timing_Event) is
   begin
      Ring_The_Pinger;
   end Is_Done;
end Egg;
```

This is much better. The timing mechanism is now completely encapsulated in the protected object and the procedure Is_Done is no longer visible outside. So all we have to do is

```
Egg.Boil(Minutes(4));
-- now read newspaper whilst waiting for egg
```

Of course if the telephone rings as the pinger goes off and before we have a chance to eat the egg then it still gets overdone. One solution is to eat the egg within the protected procedure Is_Done as well. We should never let a telephone call disturb our breakfast.

One protected procedure could be used to respond to several events. In the case of the CPU timer the discriminant of the parameter identifies the task; in the case of the group and real-time timers, the parameter identifies the event.

If we want to use the same timer for several timing events then various techniques are possible. Note that the timers are limited so we cannot test for them directly. However, they are tagged and so can be extended. Moreover, we know that they are passed by reference and that the parameters are considered aliased.

Suppose we are boiling six eggs in one of those French breakfast things with a different coloured holder for each egg. We can write

```
type Colour is (Black, Blue, Red, Green, Yellow, Purple);
Eggs_Done: array (Colour) of aliased Timing_Event;
```

We can then set the handler for the egg in the red holder by something like

```
Set_Handler(Eggs_Done(Red), For_Time, Is_Done'Access);
```

and then the protected procedure might be

```
procedure Is_Done(E: in out Timing_Event) is
begin
   for C in Colour loop
      if E'Access = Eggs_Done(C)'Access then
         -- egg in holder colour C is ready
         ...
         return;
      end if;
   end loop;
   -- falls out of loop – unknown event!
   raise Not_An_Egg;
end Is_Done;
```

Although this does work it is more than a little distasteful to compare access values in this way and moreover requires a loop to see which event occurred.

A much better approach is to use type extension and view conversions. First we extend the type Timing_Event to include additional information about the event (in this case the colour) so we can identify the particular event from within the handler

```
type Egg_Event is new Timing_Event with
   record
      Event_Colour: Colour;
   end record;
```

We then declare an array of these extended events (they need not be aliased) and set the Event_Colour component thus

```
Eggs_Done: array (Colour) of Egg_Event;
...
for C in Colour loop
   Eggs_Done(C).Event_Colour := C;
end loop;
```

We can now call Set_Handler for the egg in the red holder

```
Set_Handler(Eggs_Done(Red), For_Time, Is_Done'Access);
```

This is actually a call on the Set_Handler for the type Egg_Event inherited from Timing_Event. But it is the same code anyway.

Recall that values of tagged types are always passed by reference. This means that from within the procedure Is_Done we can recover the underlying type and so discover the information in the extension. This is done by using view conversions.

In fact we need two view conversions, first we convert to the class wide type Timing_Event'Class and then to the specific type Egg_Event. And then we select the component Event_Colour. We can do these operations in one expression thus

```
procedure Is_Done(E: in out Timing_Event) is
   C: constant Colour := Egg_Event(Timing_Event'Class(E)).Event_Colour;
begin
   -- egg in holder colour C is ready
   ...
end Is_Done;
```

Note that there is a check on the conversion from the class wide type Timing_Event'Class to the specific type Egg_Event to ensure that the object passed as parameter is indeed of the type Egg_Event (or a further extension of it). If this fails then Tag_Error is raised. In order to avoid this possibility we can use a membership test. For example

```
procedure Is_Done(E: in out Timing_Event) is
   C: Colour;
begin
   if Timing_Event'Class(E) in Egg_Event then
      C := Egg_Event(Timing_Event'Class(E)).Event_Colour;
      -- egg in holder colour C is ready
      ...
   else
      -- unknown event – not an egg event!
      raise Not_An_Egg;
   end if;
end Is_Done;
```

The membership test ensures that the event is of the specific type Egg_Event. We could avoid the double conversion to the class wide type by using a local variable for Timing_Event'Class(E).

It is important to appreciate that no dispatching is involved in these operations at all – everything is static apart from the membership test.

Timers can also be used with the Asynchronous Transfer of Control (ATC) mechanism described in Section 20.8. Suppose we have some procedure Compute which we wish to abandon if it takes too much CPU time. We declare a protected object such as

```
protected Control
   procedure Alarm(TM: in out Timer);
   entry Timer_Expired;
private
   ...
end Control;
```

and then a select statement

```
select
   Control.Timer_Expired;
   ...                        -- run out of time
   ...                        -- do cleanup etc.
then abort
   Compute( ... );            -- some time-limited computation
end select;
```

The general idea is that the protected procedure Alarm is set as the handler to be called when the timer expires. This in turn causes the entry Timer_Expired to be accepted which causes the computation to be abandoned.

22.7 Profiles

A profile is a set of restrictions imposed on a program so that it meets certain criteria. A restriction is specified using the configuration pragma Restrictions. Thus we might write

```
pragma Restrictions(No_Task_Hierarchy);
```

and this indicates that all tasks in the program are at library level.

Two key profiles are defined in the *ARM*, the Ravenscar profile and the Jorvik profile. The Ravenscar profile came first and is equivalent to a large set of restrictions and also certain scheduling pragmas. We specify that a program conforms to the Ravenscar profile by writing

```
pragma Profile(Ravenscar);
```

The purpose of the Ravenscar profile is to restrict the use of many tasking facilities so that the effect of the program is predictable. The restrictions are

```
No_Abort_Statements, No_Dynamic_Attachment,
No_Dynamic_CPU_Assignment, No_Dynamic_Priorities,
No_Implicit_Heap_Allocations, No_Local_Protected_Objects,
No_Local_Timing_Events, No_Protected_Type_Allocators,
No_Relative_Delay, No_Requeue_Statements, No_Select_Statements,
No_Specific_Termination_Handlers, No_Task_Allocators,
No_Task_Hierarchy, No_Task_Termination, Simple_Barriers.
Max_Entry_Queue_Length => 1, Max_Protected_Entries => 1,
Max_Task_Enries => 0.
```

plus No_Dependence on

```
Ada.Asynchronous_Task_Control, Ada.Calendar,
Ada.Execution_Time.Group_Budgets, Ada.Execution_Time.Timers,
Ada.Synchronous_Barriers, Ada.Task_Attributes,
System.Multiprocessors.Dispatching_Domains.
```

For details consult the appropriate references in the Bibliography. It is called the Ravenscar profile because it was defined by a workshop held in the remote village of Ravenscar on the North-East coast of England.

The Ravenscar profile has proved very valuable for many critical systems. However, there are applications that do not need its full rigidity. Accordingly, the Jorvik profile is introduced in Ada 2022 in which a number of restrictions imposed by Ravenscar are relaxed.

The following restrictions do not apply to Jorvik

```
No_Implicit_Heap_Allocations, No_Relative_Delay, Max_Entry_Queue_
Length => 1, Max_Protected_Entries => 1, No_Dependence =>
Ada.Calendar, No_Dependence => Ada.Synchronous_Barriers
```

In addition, the restriction Simple_Barriers is replaced by Pure_Barriers.

Both the Ravenscar and Jorvik profiles have pragmas that state that the Task_Dispatching_Policy is FIFO_Within_Priorities, the Locking_Policy is Ceiling_Locking, and Detect_Blocking is required. See Section 26.2.

See also AI12-291 for more details. Incidentally, Jorvik is the Viking name for the city of York in England near the coastal hamlet of Ravenscar.

22.8 Parallel blocks and loops

An important new feature in Ada 2022 is the introduction of lightweight tasking to enable the greater use of multicore architectures. This is distinct from the traditional tasking involving task types, protected types, and entries. The main new features are two parallel constructs, namely parallel blocks and parallel loops which are discussed in this section (AI12-119). There are also related parallel constructs concerning iterating over containers which are mentioned in Chapter 24.

A parallel block comprises two or more sequences of statements separated by **and**. It commences with **parallel do** and is terminated with **end do** as expected. Note that **parallel** is a new reserved word. Thus we might have

```
parallel do
   This(...); That(...);
and
   The_Other(...);
end do;
```

where This, That, and The_Other are typical subprogram calls. The effect is that This and That are executed in that order but in parallel with The_Other. We say that there are two distinct *logical threads of control*. For brevity we can just say thread.

There are no particular restrictions on the statements in the branches. They could be subprogram calls, assignments, and so on. They could include inner parallel blocks or parallel loops.

If local declarations or exception handlers are necessary then they can be included by declaring an inner block; thus one branch might be

```
and
   declare
      ...                          -- various declarations
   begin
      Some_Other(...);
      ...
   end;
and
```

A parallel block as a whole normally completes when all the individual branches have completed. Processing then continues as a single thread in the normal way with the statements following the parallel block. However, there are a number of other ways in which the parallel block can be completed.

If the parallel block is within a subprogram and one of the branches executes a return statement in order to complete a call of the subprogram then the other branches are cancelled. Execution continues as a single thread by returning from the subprogram.

If a branch raises an exception which is not handled locally then all the branches are cancelled. Execution continues as a single thread seeking an appropriate handler.

If a branch executes a goto statement and the label is outside the parallel block then all the branches are cancelled and execution continues as a single thread at the label concerned. Note that a goto cannot go from one branch to another.

When branches are cancelled due allowance is made for any region that is abort deferred (see Section 20.8). Note that the end of a parallel construct and a point within a parallel construct where a new thread is created are also abort completion points.

A parallel block can also include a chunk specification in parentheses between **parallel** and **do**. The chunk specification can be simply an integer or an integer expression. Thus we might have

```
parallel (N) do
   ...
and
   ..
end do;
```

The value of N is the maximum number of chunks that can be processed in parallel. A chunk is simply a collection of branches of the parallel block. So if the computer has 20 processors and we specify N to be 3 then no more than 3 processors can be used to deal with the block.

The other key new feature is the parallel loop. The simplest form is simply a traditional loop prefixed by parallel thus

```
parallel for J in 1 .. 40 loop
   X(J) := Nasty_Problem(J);
end loop;
```

In this case we have notionally forty threads running in parallel. If the hardware has say just twenty real processors then the implementation is free to allocate them as it wishes among the forty threads. A parallel loop cannot include **reverse** because that would imply an order and so would be against the spirit of doing the activities in parallel and thus not in a particular order.

The form of loop with **of** rather than **in** can also be parallel. Thus we could assign zero to all the elements of a massive three dimensional real array AAA by

```
parallel for E of AAA loop
   E := 0.0;
end loop;
```

Parallel loops can also be used with reduction expressions which were described in Section 9.6. Thus the twenty-fourth pyramidal number could be computed by

```
P24 := ([parallel for J in 1 .. 24 => J**2]'Reduce("+", 0));
```

but it seems unlikely that the ability to do the squarings using many processors would save much. Incidentally, P_{24} is the only pyramidal number (apart from 1) that is also a square number. Its value is $70^2 = 4900$.

But it might make sense with the hypothetical nasty problem mentioned above. If we just wanted to calculate the sum of the X(J) we could write

```
Xsum := ([parallel for J in 1 .. 40 => Nasty_Problem(J)]'Reduce("+", 0));
```

Moreover, if we just wanted to add together all the elements of a giant real array G we could use the attribute Parallel_Reduce and write

```
Gsum := G'Parallel_Reduce("+", 0.0);
```

Further control over the degree of parallelism for parallel loops is provided by the use of chunks. Note that chunks for loops are not quite the same as chunks for blocks. The simplest situation is where we want to specify a maximum for the number of threads to be used. In the example of the nasty problem if we want the threads to be limited to 10 then we write

```
parallel (10) for J in 1 .. 40 loop
   X(J) := Nasty_Problem(J);
end loop;
```

A more general case which applies to loops but not to blocks is where we introduce a chunk identifier such as C by writing for example

```
parallel (C in Integer range 1 ..10) ...
```

and then within the body of the loop we can specify what is being done by the various threads according to the value of C.

Two important aspects are Parallel_Calls and Parallel_Iterator. The aspect Parallel_Calls can be applied to a subprogram and specifies that it can be safely called in parallel such as in a parallel block. Similarly Parallel_Iterator can be applied to a loop, and specifies that it can be safely processed by parallel threads.

Readers might care to contemplate the use of parallel blocks and loops and in particular the technique known as divide and conquer. There is a simple example in the *ARM* at section 5.6.1 para 9/5. The general idea is to divide a problem into two or more parts and then solve each part separately.

As another example one might consider the Fermat method of seeking the factors of a large number n.

The general idea is to find x and y such that $n = x^2 - y^2$, in which case we immediately have $n = (x + y) \times (x - y)$.

We start by finding the smallest k for which $k^2 > n$ and then examine $k^2 - n$, $(k+1)^2 - n$, $(k+2)^2 - n$, $(k+3)^2 - n$ and so on until we find one that is a perfect square, say $(k+l)^2 - n = y^2$. So taking $x = k+l$ we have n is $(x + y) \times (x - y)$.

Having found two factors we can then in parallel and recursively apply the same algorithm to the individual factors.

Some interesting problems involve the Mersenne numbers M_n which are defined to be $M_n = 2^n - 1$. If M_n is prime then it is called a Mersenne prime. Father Merin Mersenne (1588–1648) was a French monk and also a famous mathematician. There are some notes on the website regarding example programs involving Mersenne numbers.

Mersenne numbers have some interesting properties. If a Mersenne number M_n is prime then it is always the case that n is itself prime. But the reverse is not true. For example M_{11} which is $2^{11}-1 = 2048-1 = 2047 = 23 \times 89$. For a long time it was known that M_{67} was not prime but the factors were unknown until it was shown by Frederick Nelson Cole in 1903 that

$$M_{67} = 2^{67}-1 = 147{,}573{,}952{,}589{,}676{,}412{,}927 = 193{,}707{,}721 \times 761{,}838{,}257{,}287$$

It is often the case that the largest known prime number is a Mersenne prime. Before the advent of computers the largest known prime was M_{127} which has 39 digits. At the time of writing the largest known prime is $M_{82,589,933}$ which has 24,862,048 digits. See also Section 23.6 on Big Numbers.

Exercise 22.8

1 Write a program using Fermat's method and recursive parallel blocks or loops to find the prime factors of a big number. Use it to find the factors of 969,969 and the Mersenne number M_{55} = 36,028,797,018,963,967.

22.9 Conflict checking

This section looks in a little more detail at the two main issues with tasking mentioned in Section 20.2: checking that a task is not blocked and checking that operations do not interfere with each other. Two areas of the language are involved, one is the original traditional tasking and the other is that involving the new parallel blocks and loops described in Section 22.8.

The approach with regard to traditional tasking is that the programmer should be wary of the risks involved and should write a program so that conflicts do not arise by judicious use of protected types and entry calls. However, this is traditionally outside the checking provided by the language.

On the other hand the new parallel blocks and loops are intended to be such that programmers can easily ensure that conflicts are not possible.

Accordingly, there are separate levels of checking for tasking and parallel operations. In both cases three levels of checking are defined: No, Known, All, where the midway state Known should only disallow conflicts that are clearly unsafe (according to AI12-267).

There are in fact nine possible policy identifiers (where C_C is short for the laborious Conflict_Checks) thus

```
No_Parallel_C_C,     Known_Parallel_C_C,     All_Parallel_C_C,
No_Tasking_C_C,      Known_Tasking_C_C,      All_Tasking_C_C,
No_C_C,              Known_C_C,              All_C_C
```

The new pragma Conflict_Check_Policy gives the user control of the level of checking for both traditional tasking and parallel operations.

For example, if the user wants midway checking for tasking but maximum for parallel then the pragma would be

```
pragma Conflict_Check_Policy(Known_Tasking_Conflict_Checks,
                             All_Parallel_Conflict_Checks);
```

If we want the same level for both then we can use just a single parameter which omits Tasking and Parallel such as

```
pragma Conflict_Check_Policy(Known_Conflict_Checks);
```

and so on. The default is of course

```
pragma Conflict_Check_Policy(No_Tasking_Conflict_Checks,
                             All_Parallel_Conflict_Checks);
```

Note that if we only give one parameter then it will apply to both Tasking and Parallel. For further details of this curious topic see Section 9.10.1 of the *ARM.*

Exercise 22.9

1 Write the pragmas (or pragmata) appropriate for an optimistic programmer and a pessimistic programmer.

Checklist 22

The predefined packages described in this chapter are mostly in the Systems Programming or Real-Time annexes and so may not be available on every implementation.

The facilities for defining CPUs and dispatching domains were added in Ada 2012.

The facilities for monitoring the time spent in interrupts were added in Ada 2012.

The type Group_Budget has a discriminant identifying the CPU concerned.

The aspect Synchronized was added in Ada 2012.

New in Ada 2022

The Jorvik profile which is similar to the Ravenscar profile but less restrictive is added.

The definitions of Time_Offset and related items are improved.

Parallel blocks and loops are added.

Program 4

Super Sieve

This program is a variation of the Sieve of Eratosthenes for finding prime numbers described in Section 22.1. The concept of primality applies to many algebraic structures and not just to the integers and so this program includes a generic form of the sieve and three instantiations.

One instantiation gives the normal primes and so is based on the type Integer.

The second is based on complex or Gaussian integers of the form $p + iq$ where p and q are integers and i is of course the square root of minus one.

The third is based on polynomials whose coefficients are integers modulo 2, that is can only be zero or one. Examples of such polynomials are $x + 1$ and $x^3 + x + 1$.

The program asks the user to identify the kind of primes to be searched for and how many values to test. It prints the primes as they are found and concludes by noting the total number found.

The generic package Eratosthenes contains the various tasks. It has a number of generic parameters defining the algebraic structure. There is also a private generic package Eratosthenes.Frame containing the interface to the display mechanism. In this demonstration exercise the display mechanism is simplified to just printing out each new prime on a new line but the interface is retained for compatibility with the example in Section 22.1.

There are then three packages Integer_Stuff, Complex_Stuff, and Poly_Stuff containing the types and subprograms required for the three algebraic structures. These are followed by three corresponding instantiations of the generic package Eratosthenes which are Integer_Sieve, Complex_Sieve, and Poly_Sieve.

Finally, the main subprogram calls the procedure Do_It of the instantiation of Eratosthenes according to the algebraic structure chosen by the user.

```
generic
   type Element is private;
   Unit: in Element;
   with function Succ(E: Element) return Element;
   with function Is_Factor(E, P: Element)
                                        return Boolean;
   with procedure Put(E: in Element);
package Eratosthenes is
   procedure Do_It(To_Try: in Integer;
                          Found: out Integer);
end;
---------------------------------------

private generic
package Eratosthenes.Frame is
   type Position is private;
   procedure Make_Frame(Prime: in Element;
                                     Where: out Position);
   procedure Write_To_Frame(Value: in Element;
                                     Where: in Position);
   procedure Clear_Frame(Where: in Position);
private
   type Position is null record;
end;

with Ada.Text_IO;
package body Eratosthenes.Frame is
   procedure Make_Frame(Prime: in Element;
                                     Where: out Position) is
   begin
      Ada.Text_IO.New_Line;
      Put(Prime);
      Where := (null record);   -- null aggregate
   end Make_Frame;

   procedure Write_To_Frame(Value: in Element;
                                     Where: in Position) is
   begin null; end;

   procedure Clear_Frame(Where: in Position) is
   begin null; end;
end Eratosthenes.Frame;
---------------------------------------
```

```
with Eratosthenes.Frame;
package body Eratosthenes is
   package Inner_Frame is new Frame;
   use Inner_Frame;

   Primes_Found: Integer := 0;

   protected Finished is
      entry Wait;
      procedure Signal;
   private
      Occurred: Boolean := False;
   end;

   protected body Finished is
      entry Wait when Occurred is
      begin
         null;
      end Wait;

      procedure Signal is
      begin
         Occurred := True;
      end Signal;
   end Finished;

   task type Filter is
      entry Input(Number: in Element);
      entry Stop;
   end Filter;

   function Make_Filter return access Filter is
   begin
      return new Filter;
   end Make_Filter;

   task body Filter is
      P: Element;       -- prime divisor
      N: Element;       -- trial element
      Here: Position;
      Next: access Filter;
   begin
      accept Input(Number: in Element) do
         P := Number;
      end;
      Make_Frame(P, Here);
      Primes_Found := Primes_Found + 1;
      loop
         select
            accept Input(Number: in Element) do
               N := Number;
            end;
         or
            accept Stop;
            if Next = null then
               Finished.Signal;
            else
               Next.Stop;
            end if;
            exit;
         end select;
         Write_To_Frame(N, Here);
         if not Is_Factor(N, P) then
            if Next = null then
               Next := Make_Filter;
            end if;
            Next.Input(N);
         end if;
         Clear_Frame(Here);
      end loop;
   end Filter;

   procedure Do_It(To_Try: in Integer;
                   Found: out Integer) is
      First: access Filter;
      E: Element:= Unit;
   begin
      First := new Filter;
      for I in 1 .. To_Try loop
         E := Succ(E);
         First.Input(E);
      end loop;
      First.Stop;
      Finished.Wait;
      Found := Primes_Found;
   end Do_It;

end Eratosthenes;
----------------------------------------

package Integer_Stuff is
   function Is_Factor(N, P: Integer)
                                    return Boolean;
   procedure Put(N: in Integer);
end;

with Ada.Integer_Text_IO;
package body Integer_Stuff is

   function Is_Factor(N, P: Integer)
                                    return Boolean is
   begin
      return N mod P = 0;
   end Is_Factor;

   procedure Put(N: in Integer) is
   begin
      Ada.Integer_Text_IO.Put(N, 0);
   end Put;

end Integer_Stuff;
----------------------------------------

package Complex_Stuff is
   type Complex is
      record
         Re, Im: Integer;
      end record;

   function Succ(C: Complex) return Complex;
   function Is_Factor(N, P: Complex)
                                    return Boolean;
```

```
   procedure Put(C: Complex);
end;

with Ada.Text_IO, Ada.Integer_Text_IO;
use Ada.Text_IO, Ada.Integer_Text_IO;
package body Complex_Stuff is
   function Succ(C: Complex) return Complex is
      Re: Integer := C.Re;
      Im: Integer := C.Im;
   begin
      if Im = 0 then
         return (1, Re);
      end if;
      return (Re+1, Im-1);
   end Succ;

   function Is_Factor(N, P: Complex)
                                  return Boolean is
      P_Mod: Integer := P.Re*P.Re + P.Im*P.Im;
   begin
    return (N.Re*P.Re + N.Im*P.Im) mod P_Mod = 0
      and (N.Im*P.Re - N.Re*P.Im) mod P_Mod = 0;
   end Is_Factor;

   procedure Put(C: in Complex) is
   begin
      Put(C.Re, 0);
      if C.Im /= 0 then
         Put(" + "); Put(C.Im, 0); Put('i');
      end if;
   end Put;

end Complex_Stuff;
---------------------------------------

package Poly_Stuff is
   Max: constant Integer := 100;
   type Coeff is mod 2;
   subtype Index is Integer range 0 .. Max;
   type Coeff_Vector is
         array (Integer range <>) of Coeff;
   type Polynomial(N: Index := 0) is
      record
         A: Coeff_Vector(0 .. N);
      end record;

   function Succ(P: Polynomial) return Polynomial;
   function Is_Factor(N, P: Polynomial)
                                  return Boolean;
   procedure Put(P: in Polynomial);
end;

with Ada.Text_IO, Ada.Integer_Text_IO;
use Ada.Text_IO, Ada.Integer_Text_IO;
package body Poly_Stuff is

   function Succ(P: Polynomial) return Polynomial is
      N: Integer := P.N;
      A: Coeff_Vector := P.A;
   begin
      for I in 0 .. N-1 loop
         A(I) := A(I) + 1;
         if A(I) = 1 then
            return (N, A);
         end if;
      end loop;
      return (N+1, (0 .. N => 0) & (N+1 => 1));
   end Succ;

   function Is_Factor(N, P: Polynomial)
                                  return Boolean is
      Q: Polynomial := N;
   begin
      while Q.N >= P.N loop
         if Q.A(Q.N) = 1 then
            for I in 1 .. P.N loop
               Q.A(Q.N-I) := Q.A(Q.N-I) - P.A(P.N-I);
            end loop;
         end if;
         Q := (Q.N-1, Q.A(0 .. Q.N-1));
      end loop;
      return Q.A = (0 .. Q.N => 0);
   end Is_Factor;

   procedure Put(P: in Polynomial) is
   begin
      Put('x'); Put(P.N, 0);
      for I in reverse 0 .. P.N-1 loop
         if P.A(I) /= 0 then
            Put(" + "); Put('x'); Put(I, 0);
         end if;
      end loop;
   end Put;

end Poly_Stuff;
---------------------------------------

with Eratosthenes;
with Integer_Stuff; use Integer_Stuff;
package Integer_Sieve is
   new Eratosthenes(Integer, 1, Integer'Succ,
                                  Is_Factor, Put);
---------------------------------------

with Eratosthenes;
with Complex_Stuff; use Complex_Stuff;
package Complex_Sieve is
   new Eratosthenes(Complex, (1, 0), Succ,
                                  Is_Factor, Put);
---------------------------------------

with Eratosthenes;
with Poly_Stuff; use Poly_Stuff;
package Poly_Sieve is
   new Eratosthenes(Polynomial, (0, (0 => 1)),
                             Succ, Is_Factor, Put);
---------------------------------------
```

```
with Integer_Sieve;
with Complex_Sieve;
with Poly_Sieve;
with Ada.Text_IO, Ada.Integer_Text_IO;
use Ada.Text_IO, Ada.Integer_Text_IO;
procedure Super_Sieve is
  Char: Character;
  To_Try: Integer;
  Found: Integer := 0;
begin
  Put("Welcome to the Super Sieve");
  New_Line(2);
  Put("Type I for Integers, C for Complex, " &
      "P for Polynomials  ");
  Get(Char);
  Put("How many do you want to try?  ");
  Get(To_Try);
  New_Line;
  case Char is
    when 'I' | 'i' =>
      Integer_Sieve.Do_It(To_Try, Found);
    when 'C' | 'c' =>
      Complex_Sieve.Do_It(To_Try, Found);
    when 'P' | 'p' =>
      Poly_Sieve.Do_It(To_Try, Found);
    when others =>
      Put("No such sieve");
  end case;
  New_Line(2);
  Put("Number of primes found was ");
  Put(Found, 0);
  New_Line(2);
  Put_Line("Finished");
  Skip_Line(2);
end Super_Sieve;
```

The child package Eratosthenes.Frame is private since it is only relevant to use within the body of the root package. Note how it is instantiated as Inner_Frame within the root body.

The parameters of Eratosthenes give the type of element, the value of its unit (1 for type Integer), a function Succ, a function Is_Factor, and the procedure Put. The function Succ has to deliver the successive values of the type and is called from within Do_It starting from Unit. The function Is_Factor simply checks whether the second parameter is a factor of the first. The procedure Put is used by the child package (if it were a parameter of the child the parent would still have to pass it).

Although the function Succ is obvious for the type Integer (the attribute Integer'Succ is used for the instantiation) it is not so clear what is meant by such a function in the other cases. However, all that matters is that the values are delivered in an order such that each one follows at least one of its factors. Thus if we gave the system the integer 9 before 3 then it would claim that 9 was a prime.

In the complex case 2 is not a prime but is actually $(1 + i)(1 - i)$. This can equally be written as $-i(1 + i)^2$. In fact, factorization is only unique to within factors of i and as a consequence we only need to search for primes with positive real and imaginary parts; moreover, we can omit those on the imaginary axis as being equivalent to those on the real axis. The successor function delivers values on successive lines running NW to SE starting with that with real part 1. The first to be tried is thus the prime $1 + i$ followed by 2 and then $1 + 2i$ and so on.

In the polynomial case the polynomials are generated in ascending order and within each order with coefficients increasing as the binary representation of the integers (the coefficients can only be 0 or 1 because they are of the modular type Coeff). The first few polynomials are thus x, $x + 1$, x^2, $x^2 + 1$, $x^2 + x$ and $x^2 + x + 1$. Note that $x^2 + 1$ is not prime since it is $(x + 1)^2$; the term $2x$ vanishes because $2 \equiv 0 \bmod 2$. The algorithm can easily be extended to coefficients of any modular type but it is not so easy to generate them in a helpful order if the coefficients are ordinary unbounded integers.

The tasking part of the program is much as in Section 22.1. However, the prime divisor cannot be a discriminant of the task type since it is not generally discrete; it is therefore given to each new task on the first call of the entry Input. Stopping the system needs some care. It is no good assuming that the primes have all been found just because the loop in Do_It has been completed since we have to wait for the values in the pipeline to travel to the end. Synchronization is performed using the protected object Finished. The procedure Do_It calls the entry Stop of the first task and then calls the entry Finished.Wait. Each task calls Stop of the next task and then terminates itself. The final task calls Finished.Signal which allows Do_It to exit.

The tally of primes found is kept in the variable Primes_Found. We leave the reader to show that the structure of the program is such that this can only be accessed by one task at a time.

The reader might also care to consider how else this program could be arranged. For example, the algebraic structures could be given by a generic package Integral_Domain defining the arithmetic operations much like the generic package Group of Exercise 19.4(**3**).

Other ways of communicating between the tasks and especially of stopping the system might be considered. For example, it is possible to arrange the structure so that the terminate alternative makes the subsidiary tasks terminate automatically when the main task has finished.

Part 4

Completing the Story

Chapter 23	**Predefined Library**	651
Program 5	**Wild Words**	727
Chapter 24	**Container Library**	731
Chapter 25	**Interfacing**	817
Program 6	**Playing Pools**	839
Chapter 26	**The Specialized Annexes**	843
Chapter 27	**Finale**	861

This final part completes the story and largely describes the predefined library and other material which enable the Ada program to communicate with various aspects of its environment.

In essence the predefined environment comprises the four packages Standard, Ada, System, and Interfaces. Chapters 23 and 24 cover Standard and Ada which are essentially independent of the underlying system whereas Chapter 25 covers System and Interfaces which very much depend upon the underlying system. The discussions in Chapters 23 and 24 are reasonably comprehensive and cover all the points necessary for normal use; the container library is rather large and has been factored off into Chapter 24. The discussion in Chapter 25 is less thorough because it depends upon the implementation.

Chapters 23 and 25 are each followed by a complete program. The first illustrates the use of the string handling and random number generation features provided by the predefined library and the second illustrates how users can provide their own storage pools.

Chapter 26 is an overview of the scope of the six specialized annexes. Although this book does not attempt serious coverage of these annexes nevertheless a brief overview of their scope seems appropriate. In particular enough information is given to use the complex number and vector and matrix facilities of the Numerics annex. Some material on task termination and timers from the Systems Programming and Real-Time annexes was covered in detail in Chapter 22.

The final Chapter 27 pulls together a small number of loose threads, considers the problems of portability and some thoughts on matters such as design and the construction of programs from components and concludes with an introduction to SPARK.

There are then two appendices. Appendix 1 lists the reserved words, attributes, aspects, pragmas, and restrictions; Appendix 2 contains a glossary.

Note that the answers to all the exercises and other supplementary material including the syntax for Ada 2022 and a table of all operations in the containers will be found on the website.

23 Predefined Library

23.1 The package Standard
23.2 The package Ada
23.3 Characters and strings
23.4 Text buffers and images
23.5 Numerics
23.6 Big numbers
23.7 Input and output
23.8 Text input–output
23.9 Streams
23.10 Environment commands

The previous chapters have covered the intrinsic aspects of the Ada language; that is all the various syntactic forms. This chapter and the next two chapters describe the predefined library which comes with every implementation of Ada. As outlined in Chapter 4, the predefined library is structured as three packages: Ada, System, and Interfaces with numerous child packages. These three packages are themselves (like all root library units) considered to be child units of Standard. This chapter is largely about the packages Standard and Ada, apart from the huge package Ada.Containers which is described in Chapter 24; System and Interfaces are discussed in Chapter 25.

The main facilities provided by the package Ada are string and character handling, the mathematical library including elementary functions and random number generation, various forms of input–output, and communication with the operational environment. The most exciting new item in Ada 2022 is the inclusion of a package for manipulating big numbers.

23.1 The package Standard

As mentioned earlier, certain entities are predefined through their declaration in the special package Standard. It should not be thought that this package necessarily actually exists; it is just that the compiler behaves as if it does. Indeed, some entities notionally declared in Standard cannot be truly declared in Ada at all. The general effect of Standard is indicated by the outline specification below.

The predefined operators are given in comments since they are implicitly declared. Italics are used for identifiers not available to users (such as *universal_real*) and for implementation-defined information.

```
package Standard
    with Pure is

  type Boolean is (False, True);

  -- function "=" (Left, Right: Boolean'Base) return Boolean;
  -- similarly "/=", "<", "<=", ">", ">="

  -- function "and" (Left, Right: Boolean'Base) return Boolean'Base;
  -- similarly "or", "xor"

  -- function "not" (Right: Boolean'Base) return Boolean'Base;

  -- The types root_integer and universal_integer are predefined.

  type Integer is range implementation-defined;

  subtype Natural is Integer range 0 .. Integer'Last;
  subtype Positive is Integer range 1 .. Integer'Last;

  -- function "=" (Left, Right: Integer'Base) return Boolean;
  -- similarly "/=", "<", "<=", ">", ">="

  -- function "+" (Right: Integer'Base) return Integer'Base;
  -- similarly "-", "abs"

  -- function "+" (Left, Right: Integer'Base) return Integer'Base;
  -- similarly "-", "*", "/", "rem", "mod"

  -- function "**" (Left: Integer'Base; Right: Natural) return Integer'Base;

  -- And similarly for root_integer and any other predefined integer
  -- types with Integer replaced by the name of the type.
  -- The right operand of "**" remains as Natural.

  -- The types root_real and universal_real are predefined.

  type Float is digits implementation-defined;

  -- function "=" (Left, Right: Float) return Boolean;
  -- similarly "/=", "<", "<=", ">", ">="

  -- function "+" (Right: Float) return Float;
  -- similarly "-", "abs"

  -- function "+" (Left, Right: Float) return Float;
  -- similarly "-", "*", "/"

  -- function "**" (Left: Float; Right: Integer'Base) return Float;

  -- And similarly for root_real and any other predefined floating point
  -- types with Float replaced by the name of the type.

  -- The following also apply to root_integer and root_real

  function "*" (Left: root_integer; Right: root_real) return root_real;
  function "*" (Left: root_real; Right: root_integer) return root_real;
  function "/" (Left: root_real; Right: root_integer) return root_real;

  -- The type universal_fixed is predefined.
```

```
function "*" (Left, Right: universal_fixed) return universal_fixed;
function "/" (Left, Right: universal_fixed) return universal_fixed;

-- The type universal_access is predefined.

function "=" (Left, Right: universal_access) return Boolean;
function "/=" (Left, Right: universal_access) return Boolean;

-- The type Character is based on ISO 8859-1. There are no literals
-- for control characters shown in italics.

type Character is
(nul,  soh,  stx,  etx,   eot,  enq,  ack,  bel,   --   0 ..   7
 bs,   ht,   lf,   vt,    ff,   cr,   so,   si,    --   8 ..  15
 dle,  dc1,  dc2,  dc3,   dc4,  nak,  syn,  etb,   --  16 ..  23
 can,  em,   sub,  esc,   fs,   gs,   rs,   us,    --  24 ..  31
 ' ',  '!',  '"',  '#',   '$',  '%',  '&',  ''',   --  32 ..  39
 '(',  ')',  '*',  '+',   ',',  '-',  '.',  '/',   --  40 ..  47
 '0',  '1',  '2',  '3',   '4',  '5',  '6',  '7',   --  48 ..  55
 '8',  '9',  ':',  ';',   '<',  '=',  '>',  '?',   --  56 ..  63
 '@',  'A',  'B',  'C',   'D',  'E',  'F',  'G',   --  64 ..  71
 'H',  'I',  'J',  'K',   'L',  'M',  'N',  'O',   --  72 ..  79
 'P',  'Q',  'R',  'S',   'T',  'U',  'V',  'W',   --  80 ..  87
 'X',  'Y',  'Z',  '[',   '\',  ']',  '^',  '_',   --  88 ..  95
 '`',  'a',  'b',  'c',   'd',  'e',  'f',  'g',   --  96 .. 103
 'h',  'i',  'j',  'k',   'l',  'm',  'n',  'o',   -- 104 .. 111
 'p',  'q',  'r',  's',   't',  'u',  'v',  'w',   -- 112 .. 119
 'x',  'y',  'z',  '{',   '|',  '}',  '~',  del,   -- 120 .. 127

 128,  129,  bph,  nbh,   132,  nel,  ssa,  esa,   -- 128 .. 135
 hts,  htj,  vts,  pld,   plu,  ri,   ss2,  ss3,   -- 136 .. 143
 dcs,  pu1,  pu2,  sts,   cch,  mw,   spa,  epa,   -- 144 .. 151
 sos,  153,  sci,  csi,   st,   osc,  pm,   apc,   -- 152 .. 159
 ' ',  '¡',  '¢',  '£',   '¤',  '¥',  '¦',  '§',   -- 160 .. 167
 '¨',  '©',  'ª',  '«',   '¬',  '-',  '®',  '¯',   -- 168 .. 175
 '°',  '±',  '²',  '³',   '´',  'µ',  '¶',  '·',   -- 176 .. 183
 '¸',  '¹',  'º',  '»',   '¼',  '½',  '¾',  '¿',   -- 184 .. 191
 'À',  'Á',  'Â',  'Ã',   'Ä',  'Å',  'Æ',  'Ç',   -- 192 .. 199
 'È',  'É',  'Ê',  'Ë',   'Ì',  'Í',  'Î',  'Ï',   -- 200 .. 207
 'Ð',  'Ñ',  'Ò',  'Ó',   'Ô',  'Õ',  'Ö',  '×',   -- 208 .. 215
 'Ø',  'Ù',  'Ú',  'Û',   'Ü',  'Ý',  'Þ',  'ß',   -- 216 .. 223
 'à',  'á',  'â',  'ã',   'ä',  'å',  'æ',  'ç',   -- 224 .. 231
 'è',  'é',  'ê',  'ë',   'ì',  'í',  'î',  'ï',   -- 232 .. 239
 'ð',  'ñ',  'ò',  'ó',   'ô',  'õ',  'ö',  '÷',   -- 240 .. 247
 'ø',  'ù',  'ú',  'û',   'ü',  'ý',  'þ',  'ÿ');  -- 248 .. 255

-- The operators for Character are as any enumeration type.

-- The type Wide_Character is based on the ISO/IEC 10646:2020BMP
-- set. The first 256 positions are as for Character.

type Wide_Character is (nul, soh, ..., FFFE, FFFF);
```

```
-- The type Wide_Wide Character is based on the full ISO/IEC
-- 10646:2020 set. The first 65536 positions are as for
-- Wide_Character.

type Wide_Wide_Character is (nul, soh, ..., 7FFFFFFE, 7FFFFFFF);
for Wide_Wide_Character'Size use 32;

package ASCII is ... -- obsolescent, see Ada.Characters.Latin_1

type String is array (Positive range <>) of Character
   with Pack;

-- function "=" (Left, Right: String) return Boolean;
-- similarly "/=", "<", "<=", ">", ">="

-- function "&" (Left: String; Right: String) return String;
-- function "&" (Left: Character; Right: String) return String;
-- function "&" (Left: String; Right: Character) return String;
-- function "&" (Left: Character; Right: Character) return String;

type Wide_String is array (Positive range <>) of Wide_Character
   with Pack;

type Wide_Wide_String is array (Positive range <>) of
                                          Wide_Wide_Character
   with Pack;

-- The operators for Wide_ and Wide_Wide_String are as for String.

type Duration is delta implementation-defined
                           range implementation-defined;

-- The operators for Duration are as any fixed point type.

Constraint_Error, Program_Error, Storage_Error,
                              Tasking_Error: exception;

end Standard;
```

The above specification is not complete. For example, although the type Boolean can be written showing the literals False and True, the short circuit control forms cannot be expressed explicitly. Moreover, each further type definition implicitly introduces new declarations of some operators. All types, except limited types, introduce new declarations of = and /=. All scalar types and discrete one-dimensional array types introduce new declarations of & and those with Boolean components also introduce new declarations of **and**, **or**, **xor**, and **not**. All fixed point types introduce new declarations of +, –, *, /, and **abs**.

The parameters and results for Integer use the unconstrained subtype Integer'Base as explained in Section 17.1. On the other hand, the subtype Float is already unconstrained and so is used directly. The parameters of the operators for the type Boolean are Boolean'Base; all relational operators return a result of subtype Boolean whereas **and**, **or**, **xor**, and **not** return Boolean'Base (see also the note at the end of Section 12.3).

Note that character 173 is *soft_hyphen* in Ada 2012 and Ada 2022 for obscure reasons and does not have the literal shown above as it did in Ada 2005. But changing it would have upset the layout!

23.2 The package Ada

Most of the predefined library is structured as child units of Ada. The package Ada itself is empty except for the aspect Pure.

Table 23.1 summarizes all the child units and indicates where a more detailed discussion can be found. Some of those in the core language were described in earlier chapters and the remainder are described in detail in later sections in this chapter or the next chapter. Those in the specialized annexes are shown in italics; some were described in Chapter 22 and most of the remainder are mentioned in Chapter 26 (the annex is indicated by a code such as RT for Real-Time).

Note that there are wide versions of many units which use wide characters and wide strings. In all cases there are also wide wide versions.

Table 23.1 Child units of Ada.

Name	*Contents and purpose*	*See*
Assertions	declares procedures Assert and exception Assertion_Error	15.5
Asynchronous_Task_Control	enables tasks to be held and allowed to continue	26.2 RT
Calendar	type Time and its operations	20.3
Arithmetic	arithmetic on days and leap seconds	20.3
Formatting	formatting and other operations involving times	20.3
Time_Zones	time zone manipulation	20.3
Characters	empty except for its child units	23.3
Conversions	like Handling for wide and wide wide characters and strings	23.3
Handling	functions for classification and conversion of characters and strings	23.3
Latin_1	constants of type Character giving names to the characters	23.3
Command_Line	provides access to command line parameters and exit status	23.10
Complex_Text_IO	to output complex values based on type Float	26.5 N
Containers	types common to all containers	24.1
Doubly_Linked_Lists	generic package for manipulating linked lists	24.2
Generic_Array_Sort	for sorting constrained arrays	24.11
Generic_Constrained_Array_Sort	for sorting unconstrained arrays	24.11
Generic_Sort	for sorting general structures	24.11
Hashed_Maps	generic package for hashed maps	24.4
Hashed_Sets	generic package for hashed sets	24.5
Multiway_Trees	generic package for trees	24.6
Ordered_Maps	generic package for ordered maps	24.4
Ordered_Sets	generic package for ordered sets	24.5
Vectors	generic package for vectors	24.3
Indefinite_Holders	generic package to wrap an indefinite type	24.7
Indefinite_Doubly_Linked_Lists, Indefinite_Hashed_Maps	etc. for indefinite types	24.10
Bounded_Doubly_Linked_Lists, Bounded_Hashed_Maps	etc for definite types	24.9
Synchronized_Queue_Interfaces	generic package for queue interfaces	24.8
Unbounded_Synchronized_Queues	generic package for synchronized queues	24.8
Unbounded_Priority_Queues	generic package for priority queues	24.8
Bounded_Synchronized_Queues, Bounded_Priority_Queues	for bounded situations	24.8
Decimal	contains special facilities for decimal arithmetic	26.4 IS
Direct_IO	facilities for indexed I/O	23.7

Table 23.1 Child units of Ada (continued).

Unit	Description	Section
Directories	facilities for manipulating files and directories	23.10
Hierarchical_File_Names	optional for relative file names	23.10
Information	optional for details of files	23.10
Dispatching	root package for dispatching policies	26.2 RT
EDF	for controlling earliest deadline first dispatching	26.2 RT
Round_Robin	for controlling round robin dispatching	26.2 RT
Non_Preemptive	for permitting preemption	26.2 RT
Dynamic_Priorities	enables task priorities to be manipulated dynamically	26.2 RT
Environment_Variables	facilities for accessing variables in operating system	23.10
Exceptions	type Exception_Occurrence and related types and operations	15.4
Execution_Time	the type CPU_Time and related facilities	22.6 RT
Group_Budgets	timers for sharing a CPU time budget	22.6 RT
Interrupts	for accounting interrupts	22.6 RT
Timers	timers for an individual CPU time budget	22.6 RT
Finalization	types Controlled, Limited_Controlled; also Initialize, Adjust, Finalize	14.7
Interrupts	facilities for attaching and detaching interrupts	26.1 SP
Names	constants giving names of interrupt handlers	26.1 SP
Float_Text_IO	to output values of type Float	23.8
Integer_Text_IO	to output values of type Integer	23.8
IO_Exceptions	exceptions such as Data_Error used by all I/O packages	23.7
Iterator_Interfaces	magic package giving syntactic sugar for iterators	21.6
Locales	provides country and language codes for current locale	23.10
Long_Complex_Text_IO	to output complex values based on type Long_Float	26.5 N
Long_Integer_Text_IO	to output values of type Long_Integer	23.8
Long_Float_Text_IO	to output values of type Long_Float	23.8
Numerics	numbers Pi (π) and e and exception Argument_Error	23.5
Discrete_Random	function Random etc. for random numbers of a discrete type	23.5
Float_Random	function Random etc. for random numbers of type Float	23.5
Generic_Elementary_Functions	functions Sqrt, Log, Exp etc.	23.5
Elementary_Functions	nongeneric form for type Float	23.5
Long_Elementary_Functions	nongeneric form for type Long_Float	23.5
Big_Numbers	packages Big_Integers and Big_Reals	23.6
Generic_Complex_Arrays	types and operations for complex arrays	26.5 N
Complex_Arrays	nongeneric form based on type Float	26.5 N
Long_Complex_Arrays	nongeneric form based on type Long_Float	26.5 N
Generic_Complex_Types	type Complex and operations for complex numbers	26.5 N
Complex_Types	nongeneric form based on type Float	26.5 N
Long_Complex_Types	nongeneric form based on type Long_Float	26.5 N
Generic_Complex_Elementary_Functions	functions Sqrt, Log etc. for type Complex	26.5 N
Complex_Elementary_Functions	nongeneric form based on type Float	26.5 N
Long_Complex_Elementary_Functions	nongeneric form based on type Long_Float	26.5 N
Generic_Real_Arrays	types and operations for real arrays	26.5 N
Real_Arrays	nongeneric form for type Float	26.5 N
Real_Time	the monotonic type Time and related facilities	22.6 RT
Timing_Events	for setting timers using the real-time clock	22.6 RT
Sequential_IO	facilities for sequential I/O	23.7
Storage_IO	facilities for I/O into a memory buffer	23.7
Streams	provides root facilities for all streams	23.9
Stream_IO	facilities for I/O using heterogeneous files	23.9

Table 23.1 Child units of Ada (continued).

Unit	Description	Section	
Strings	constants, exceptions and enumeration types used by child units	23.3	
Equal_Case_Insensitive	case insensitive string equality	23.3	
Less_Case_Insensitive	case insensitive string comparison	23.3	
Hash	hash function for use with containers	24.4	
Hash_Case_Insensitive	case insensitive hash function	24.4	
Fixed	subprograms for manipulating objects of type String	23.3	
Equal_Case_Insensitive	case insensitive string equality	23.3	
Less_Case_Insensitive	case insensitive string comparison	23.3	
Hash	renames Strings.Hash for use with containers	24.4	
Hash_Case_Insensitive	case insensitive version	24.4	
Bounded	type Bounded_String and similar subprograms as for Fixed	23.3	
Equal_Case_Insensitive and Less_Case_Insensitive	as for Fixed	23.3	
Hash and Hash_Case_Insensitive	as for Fixed	24.4	
Unbounded	type Unbounded_String and similar subprograms as for Fixed	23.3	
Equal_Case_Insensitive and Less_Case_Insensitive	as for Fixed	23.3	
Hash and Hash_Case_Insensitive	as for Fixed	24.4	
Maps	types Character_Mapping and Character_Set and operations	23.3	
Constants	constants defining standard maps and sets such as Letter_Set	23.3	
Text_Buffers	package with child packages Bounded and Unbounded	23.4	
Wide_Equal_Case_Insensitive	similar to Equal_Case_Insensitive for wide etc	23.3	
Wide_Hash and Wide_Hash_Case_Insensitive	for wide and also wide wide versions	24.4	
Wide_Fixed	similar to Fixed but for Wide_String and Wide_Character	23.3	
Wide_Bounded, Wide_Unbounded, Wide_Maps etc.	ditto and also wide wide versions	23.3	
UTF_Encoding	function Encoding to convert a String to UTF types	23.3	
Conversions	five functions Convert between UTF schemes	23.3	
Strings	functions Encode and Decode between String and UTF schemes	23.3	
Wide_Strings, Wide_Wide_Strings	similar functions for wide and wide wide strings	23.3	
Synchronous_Barriers	enables several tasks to be released together	26.2	RT
Synchronous_Task_Control	enables tasks to suspend on state of suspension objects	26.2	RT
EDF	for deadlines	26.2	RT
Tags	type Tag and related operations	21.7	
Generic_Dispatching_Constructor	generic function for creating objects	21.7	
Task_Attributes	enables attributes to be associated with tasks on a per-task basis	26.1	SP
Task_Identification	type Task_Id and operations to identify and recognize tasks	26.1	SP
Task_Termination	facilities for finding reason for task termination	22.5	RT
Text_IO	facilities for traditional textual I/O such as procedures Get and Put	23.8	
Text_Streams	function Stream for interworking text and binary in one file	23.9	
Bounded_IO	simple text I/O for bounded strings	23.3	
Unbounded_IO	simple text I/O for unbounded strings	23.3	
Complex_IO	for text I/O of values of complex types	26.5	N
Editing	for text I/O of decimal types using picture formats	26.4	IS
Unchecked_Conversion	enables the strong typing to be circumvented	25.2	
Unchecked_Deallocation	provides the ability to release heap storage	25.2	
Unchecked_Deallocate_Subpool	for releasing storage from a subpool	25.4	
Wide_Text_IO	similar to Text_IO but for Wide_Character and Wide_String	23.8	
Text_Streams, Bounded_IO etc.	similarly and also wide wide versions	23.9	
Wide_Characters	empty except for its child units	23.3	
Handling	similar to Characters.Handling but for wide characters	23.3	
Wide_Command_Line	as for Command_Line but for wide text	23.10	
Wide_Directories	as for Directories but for wide text	23.10	
Wide_Environment_Variables	as for Environment_Variables but for wide text	23.10	

23.3 Characters and strings

There are a number of packages for handling characters, wide characters and wide wide characters as follows

Ada.Characters
Ada.Characters.Handling
Ada.Characters.Conversions
Ada.Characters.Latin_1
Ada.Wide_Characters
Ada.Wide_Characters.Handling
Ada.Wide_Wide_Characters
Ada.Wide_Wide_Characters.Handling

However, the parent packages Ada.Characters, Ada.Wide_Characters, and Ada.Wide_Wide_Characters themselves are empty (apart from the aspect Pure).

The package Characters.Handling declares functions for the classification and conversion of characters and strings. Its specification takes the form

```
package Ada.Characters.Handling
    with Pure is

  function Is_Control(Item: in Character) return Boolean;
  ...
  function To_Lower(Item: in Character) return Character;
  ...
  subtype ISO_646 is
        Character range Character'Val(0) .. Character'Val(127);
  function Is_ISO_646(Item: in String) return Boolean;
  ...
end Ada.Characters.Handling
```

The function Is_Control indicates whether the parameter is a control character (that is, in the ranges 0..31 or 127..159). Other similar functions are Is_Graphic, Is_Letter, Is_Lower, Is_Upper, Is_Basic, Is_Digit, Is_Decimal_Digit, Is_Hexadecimal_Digit, Is_Alphanumeric, Is_Special, Is_ISO_646, Is_Line_Terminator, Is_Mark, Is_Other_Format, Is_Punctuation_Connector and Is_Space.

The function Is_NFKC is added in Ada 2022. It stands for Is Normalization Form KC. See the *ARM* for details.

Readers might feel that Is_Mark is a foolish waste of time. However, it is introduced because the corresponding functions in the packages for wide and wide wide characters can return True. These classifications enable a compiler to analyze Ada source code without direct reference to the definition of ISO/IEC 10646. Note further that case insensitive text comparison which is useful for the analysis of identifiers is provided by functions Equal_Case_Insensitive and Less_Case_Insensitive described below.

The type Character covers the full Latin-1 set and thus includes accented letters and other less familiar characters. The upper case letters are A to Z and those in 192..222 (excluding 215 which is multiplication); the lower case are a to z and those in 223..255 (excluding 247 which is division). The basic letters are A to Z plus AE Diphthong (198), Icelandic Eth (208), Icelandic Thorn (222), and their lower case

Table 23.2 Classification of characters.

Classification	*Values in set*
Is_Control	0..31, 127..159
Is_Graphic	**not** Is_Control
Is_Letter	Is_Lower **or** Is_Upper
Is_Lower	97..122, 223..246, 248..255
Is_Upper	65..90, 192..214, 216..222
Is_Basic	65..90, 97..122, 198, 208, 222, 223, 230, 240, 254
Is_Digit	48..57
Is_Decimal_Digit	48..57
Is_Hexadecimal_Digit	48..57, 65..70, 97..102
Is_Alphanumeric	Is_Letter **or** Is_Digit
Is_Special	Is_Graphic **and not** Is_Alphanumeric
Is_ISO_646	0..127
Is_Line_Terminator	10..13, 133
Is_Mark	none
Is_Other_Format	173
Is_Punctuation_Connector	95
Is_Space	32, 160

versions (a to z, 230, 240, 254) plus German Sharp S which only has a lower case form (223). The general rule is that the lower case letter is 32 more than the upper case one. An oddity is lower case y Diaeresis (255) which does not have an upper case form. The various categories are summarized in Table 23.2.

The conversion functions

```
function To_Lower(Item: in Character) return Character;
function To_Lower(Item: in String) return String;
```

convert an individual character or a whole string to lower case (characters not originally upper case are unchanged). There are also functions To_Upper and To_Basic. To_Basic converts to characters without diacritical marks.

There are also functions to check for and convert to the 7-bit set ISO 646

```
function Is_ISO_646(Item: in String) return Boolean;
function To_ISO_646(Item: in String;
                    Substitute: in ISO_646 := ' ') return String;
```

where ISO_646 is a subtype of Character covering 0..127. The conversion function replaces all characters not in ISO 646 by the substitute character. There are similar functions with the same identifiers for individual characters.

The package Wide_Characters.Handling is very similar to the package Characters.Handling with Character and String everywhere replaced by Wide_Character and Wide_String. However, there are no functions corresponding to Is_ISO_646 and To_ISO_646. Such functions are not really necessary because one could always use a membership test on a variable WC of type Wide_Character thus

```
WC in Wide_Characters'POS(0) .. Wide_Characters'POS(127)
```

The package Wide_Characters.Handling also has the function Character_Set_Version thus

```
function Character_Set_Version return String;
```

The string returned identifies the version of the character set standard being used. Typically it will include either '10646:' or 'Unicode'. This function is useful because the categorization of some wide characters depends upon the version of 10646 or Unicode being used. So rather than specifying that the package uses a particular set (which might be a problem in the future if the character set standard changes), this function enables the program to find out exactly which version is being used.

Note that there is no corresponding function in Ada.Characters.Handling. This is because the set used for the type Character is frozen as at 1995 and the classification functions defined for the type Character are frozen as well (and so do not now exactly match 10646 which has since evolved). It might be that classifications for wide and ever wider characters might change in the future for some obscure characters but the programmer can rest assured that Character is for ever reliable.

The package Wide_Wide_Characters.Handling is the same as Wide_Characters.Handling with Wide_Character and Wide_String replaced by Wide_Wide_Character and Wide_Wide_String throughout.

The package Characters.Conversions declares functions to check for and convert to and from the 16-bit set Wide_Character and the 32-bit set Wide_Wide_Character. For example between Wide_String and String there are

```
function Is_String(Item: in Wide_String) return Boolean;
function To_String(Item: in Wide_String;
                   Substitute: in Character := ' ') return String;
function To_Wide_String(Item: in String) return Wide_String;
```

with similar functions for individual characters obtained by replacing String by Character in both the identifiers and profiles. Six similar functions apply to the types Wide_Wide_Character and Wide_Character and six more apply to Wide_Wide_Character and Character.

The package Characters.Latin_1 takes the form

```
package Ada.Characters.Latin_1
   with Pure is
  NUL              : constant Character := Character'Val(0);
  ...
  Space            : constant Character := ' ';
  Exclamation      : constant Character := '!';
  ...
  LC_A             : constant Character := 'a'
  ...
end Ada.Characters.Latin_1;
```

It consists of the declaration of constants giving names to most of the characters so that they can be referred to as simply Exclamation given an appropriate use clause. The names are listed in Table 23.3.

Table 23.3 The Latin-1 names.

0.. 15	NUL, SOH, STX, ETX,	EOT, ENQ, ACK, BEL,	BS, HT, LF, VT,	FF, CR, SO, SI
16.. 31	DLE, DC1, DC2, DC3,	DC4, NAK, SYN, ETB,	CAN, EM, SUB, ESC,	FS, GS, RS, US
32.. 35	Space,	Exclamation,	Quotation,	Number_Sign
36.. 39	Dollar_Sign,	Percent_Sign,	Ampersand,	Apostrophe
40.. 43	Left_Parenthesis,	Right_Parenthesis,	Asterisk,	Plus_Sign
44.. 47	Comma,	Hyphen,	Full_Stop,	Solidus
48.. 57	(the digits 0 to 9 are not named)			
58.. 61	Colon,	Semicolon,	Less_Than_Sign,	Equals_Sign
62.. 64	Greater_Than_Sign,	Question,	Commercial_At	
65.. 90	(the upper case letters A to Z are not named)			
91.. 94	Left_Square_Bracket,	Reverse_Solidus,	Right_Square_Bracket,	Circumflex
95..122	Low_Line,	Grave,	LC_A, LC_B, ...	LC_Z
123..126	Left_Curly_Bracket,	Vertical_Line,	Right_Curly_Bracket,	Tilde
127..130	DEL,	Reserved_128,	Reserved_129,	BPH
131..140	NBH,	Reserved_132,	NEL, SSA, ESA, HTS,	HTJ, VTS, PLD, PLU
141..153	RI, SS2, SS3, DCS,	PU1, PU2, STS, CCH,	MW, SPA, EPA, SOS,	Reserved_153
154..161	SCI, CSI, ST, OSC,	PM, APC,	No_Break_Space,	Inverted_Exclamation
162..165	Cent_Sign,	Pound_Sign,	Currency_Sign,	Yen_Sign
166..169	Broken_Bar,	Section_Sign,	Diaeresis,	Copyright_Sign
170..173	Feminine_Ordinal_Indicator,	Left_Angle_Quotation,	Not_Sign,	Soft_Hyphen
174..177	Registered_Trade_Mark_Sign,	Macron,	Degree_Sign,	Plus_Minus_Sign
178..181	Superscript_Two,	Superscript_Three,	Acute,	Micro_Sign
182..185	Pilcrow_Sign,	Middle_Dot,	Cedilla,	Superscript_One
186..188	Masculine_Ordinal_Indicator,		Right_Angle_Quotation,	Fraction_One_Quarter
189..191	Fraction_One_Half,	Fraction_Three_Quarters,	Inverted_Question	
192..195	UC_A_Grave,	UC_A_Acute,	UC_A_Circumflex,	UC_A_Tilde
196..199	UC_A_Diaeresis,	UC_A_Ring,	UC_AE_Diphthong,	UC_C_Cedilla
200..203	UC_E_Grave,	UC_E_Acute,	UC_E_Circumflex,	UC_E_Diaeresis
204..207	UC_I_Grave,	UC_I_Acute,	UC_I_Circumflex,	UC_I_Diaeresis
208..211	UC_Icelandic_Eth,	UC_N_Tilde,	UC_O_Grave,	UC_O_Acute
212..215	UC_O_Circumflex,	UC_O_Tilde,	UC_O_Diaeresis,	Multiplication_Sign
216..219	UC_O_Oblique_Strike,	UC_U_Grave,	UC_U_Acute,	UC_U_Circumflex
220..223	UC_U_Diaeresis,	UC_Y_Acute,	UC_Icelandic_Thorn,	LC_German_Sharp_S
224..227	LC_A_Grave,	LC_A_Acute,	LC_A_Circumflex,	LC_A_Tilde
228..231	LC_A_Diaeresis,	LC_A_Ring,	LC_AE_Diphthong,	LC_C_Cedilla
232..235	LC_E_Grave,	LC_E_Acute,	LC_E_Circumflex,	LC_E_Diaeresis
236..239	LC_I_Grave,	LC_I_Acute,	LC_I_Circumflex,	LC_I_Diaeresis
240..243	LC_Icelandic_Eth,	LC_N_Tilde,	LC_O_Grave,	LC_O_Acute
244..247	LC_O_Circumflex,	LC_O_Tilde,	LC_O_Diaeresis,	Division_Sign
248..251	LC_O_Oblique_Strike,	LC_U_Grave,	LC_U_Acute,	LC_U_Circumflex
252..255	LC_U_Diaeresis,	LC_Y_Acute,	LC_Icelandic_Thorn,	LC_Y_Diaeresis

There are also a number of alternative names (introduced by renamings). They include: Minus_Sign for Hyphen, NBSP for No_Break_Space, Ring_Above for Degree_Sign, and Paragraph_Sign for Pilcrow_Sign.

The obsolescent package ASCII internal to Standard also defines names for some characters in the 7-bit set. The names of all the control characters and many of the others are the same as for Latin_1.

String handling is provided by an extensive range of packages

```
Ada.Strings
Ada.Strings.Fixed
Ada.Strings.Bounded
Ada.Strings.Unbounded
Ada.Strings.Maps
Ada.Strings.Maps.Constants
```

The reason for so many packages is that different applications have different needs which could not be met by a single approach. There are thus three main packages for three kinds of strings – fixed length strings whose length is determined when an object is created, bounded strings whose length can vary but has an upper limit, and unbounded strings whose length has no upper limit. We will deal with these in turn.

The package Strings.Fixed deals with objects of the type String which of course all individually have a fixed length. Its specification is

```
with Ada.Strings.Maps;
package Ada.Strings.Fixed
    with Preelaborate, Nonblocking, Global => in out synchronized is
  ...      -- various subprograms
end Ada.Strings.Fixed;
```

The package Strings.Bounded deals with a string type which has a maximum length but of which only part is in use at any time; since the maximum length will depend upon the application everything is in an inner generic package which takes this length as its sole parameter. The type Bounded_String is private and will typically be implemented using a discriminated record. The current length of a bounded string is given by a function Length (the lower bound is one). The length has type Integer just as all indexing operations on all forms of strings (fixed, bounded, unbounded) use values of type Integer. The package specification is

```
with Ada.Strings.Maps;
package Ada.Strings.Bounded
    with Preelaborate, Nonblocking, Global => in out synchronized is

  generic
    Max: Positive;          -- maximum length of a Bounded_String
  package Generic_Bounded_Length is
    Max_Length: constant Positive := Max;

    type Bounded_String is private;
    Null_Bounded_String: constant Bounded_String;

    subtype Length_Range is Natural range 0 .. Max_Length;
    function Length(Source: in Bounded_String) return Length_Range;
    ...      -- various subprograms
  private
    ...
  end Generic_Bounded_Length;
end Ada.Strings.Bounded;
```

The package Strings.Unbounded deals with completely unbounded strings (actually bounded by Integer'Last). The type Unbounded_String is private and clearly has to be implemented using dynamic allocation. Again there is a function Length and the lower bound is one. The package specification is

```
with Ada.Strings.Maps;
package Ada.Strings.Unbounded
   with Preelaborate, Nonblocking, Global => in out synchronized is

  type Unbounded_String is private
     with Preelaborable_Initialization;
  Null_Unbounded_String: constant Unbounded_String;

  function Length(Source: in Unbounded_String) return Natural;
  ...       -- various subprograms
private
  ...
end Ada.Strings.Unbounded;
```

The packages Strings.Maps and Strings.Maps.Constants concern mappings between sets of characters; these enable various operations to be performed as modified or qualified by a mapping. We will come back to these later but first consider examples without such complications.

The overall parent package is as follows

```
package Ada.Strings
   with Pure is

  Space: constant Character := ' ';
  Wide_Space: constant Wide_Character := ' ';
  Wide_Wide_Space: constant Wide_Wide_Character := ' ';

  Length_Error, Pattern_Error, Index_Error, Translation_Error: exception;

  type Alignment is (Left, Right, Center);   -- beware spelling
  type Truncation is (Left, Right, Error);
  type Membership is (Inside, Outside);
  type Direction is (Forward, Backward);
  type Trim_End is (Left, Right, Both);
end Ada.Strings;
```

The enumeration types relate to the control of various options and the exceptions naturally get raised when something goes wrong.

We will illustrate all three packages together by considering various common operations. The general idea is that the three styles of strings should be handled in much the same way as far as possible and so most of the subprograms are common to all three packages. But there are important differences which require additional subprograms in some of the packages and we will look at these first.

The main difference is the problem of varying size. The unbounded strings cause no problems; the bounded strings only have a problem with the limit on the size. Thus, provided the length is not exceeded, the corresponding operations do the same on both and, in particular, the Length function will return the same value.

But the fixed length strings are fixed and there is no Length function (there is the attribute Length of course). In order to enable much of the functionality to correspond for fixed strings the concept of justification and padding is introduced. The default behaviour is that fixed strings are left justified and any unused space is filled with space characters. So provided the strings being manipulated do not have significant trailing spaces then they will behave much the same as bounded and unbounded strings.

A related point is that bounded and unbounded strings have a notional lower bound of one whereas a fixed string need not; but we will use fixed strings with a lower bound of one in our examples.

We will suppose that the bounded package has been instantiated and various objects declared thus

```
use Ada.Strings.Fixed;
use Ada.Strings.Bounded;
use Ada.Strings.Unbounded;
package Bounded_80 is new Generic_Bounded_Length(80);
use Bounded_80;
C: Character;
S: String := ... ;
S5: String(1 .. 5);
S10: String(1 .. 10);
S15: String(1 .. 15);
BS: Bounded_String;        -- from Bounded_80
US: Unbounded_String;
```

Objects such as BS and US which are not explicitly initialized are given the values of the constants Null_Bounded_String and Null_Unbounded_String respectively; these represent null strings.

String literals provide values of fixed strings and there are conversion functions for bounded and unbounded strings thus

```
BS := To_Bounded_String("A Bounded String");
US := To_Unbounded_String("An Unbounded String");
```

There is also a version of To_Unbounded_String which takes an integer parameter and returns an uninitialized string of the given length. We can also use procedures

```
Set_Bounded_String(BS, "A Bounded String");
Set_Unbounded_String(US, "An Unbounded String");
```

and these are likely to be more efficient especially in the case of unbounded strings which will almost inevitably be implemented using controlled types; the procedure call avoids the (hidden) call of Adjust required for the assignment when the function is used.

Bounded and unbounded strings can be converted to a string by the function To_String. Conversion between bounded and unbounded strings has to be via an intermediate string thus

```
US := To_Unbounded_String(To_String(BS));
```

Assignment for bounded and unbounded strings is performed in the usual way using :=. Bounded strings always fit and unbounded strings will be properly copied using controlled types so that the old and new values do not interfere. Copying of fixed strings of the same length can also be done by assignment.

Copying fixed strings of different lengths poses problems and so the package Ada.Strings.Fixed provides a procedure Move which uses padding characters and justification as mentioned above. The default behaviour is for all operations to be left justified with space used as a right padding character; strings can then be copied naturally with extra spaces added on the right or removed from the right as necessary; if the destination is too short so that nonpadding characters would be lost by being trimmed off then the exception Length_Error (in Ada.Strings) is raised. The justification and padding/trimming options are controlled by various parameters with defaults

```
procedure Move(Source: in String;
               Target: out String;
               Drop: in Truncation := Error;
               Justify: in Alignment := Left;
               Pad: in Character := Space);
```

Justification on padding can be to the Right or even to the Center (with any odd extra padding character on the right). For trimming, if Drop is Left or Right, then an exception is not raised and characters are simply lost from the end specified; otherwise if Drop is Error then only padding characters can be removed (from the opposite end to that given by Justify) and Length_Error raised if necessary as mentioned above. So

```
Move("Barbara", S10);              -- "Barbara   "
Move(S10, S5, Drop => Right);      -- "Barba"
```

where there are three spaces at the end of the first string.

Bounded strings can be built up using "&" which is overloaded for a bounded string with a character, string or another bounded string; similarly for unbounded strings. (Other combinations would cause irritating ambiguities at times.)

The mechanism of dropping characters is also relevant for some operations on bounded strings. For example, since "&" will always raise Length_Error if an attempt is made to create an oversize bounded string, alternative subprograms Append are provided which give greater control. They have a third optional parameter Drop which causes extra characters to be lost if its value is Left or Right, or Length_Error to be raised if it is Error (the default). The functions Append return the required result, whereas procedures Append update the first parameter. So assuming BS has the value "Barbara", the following assignments are equivalent

```
BS := BS & ' ' & "Barnes";
BS := Append(BS, ' ' & "Barnes");
Append(BS, " Barnes");
```

and all give Barbara her surname. Similar Append procedures exist for unbounded strings but of course need no Drop parameter.

Incidentally, this mechanism also applies to the subprograms To_Bounded_String and Set_Bounded_String whose specifications are

```
function To_Bounded_String(Source: in String;
                          Drop: in Truncation := Error) return Bounded_String;
procedure Set_Bounded_String(Target: out Bounded_String;
                            Source: in String; Drop: in Truncation := Error);
```

If the string is too long then characters are dropped or Length_Error is raised according to the value of Drop.

Appropriate overloadings of "*" create strings which are replications of characters or strings. The first parameter is the number of replications and the second is a character or string (of type String or of the same type as the result). So

```
BS := 2 * "Bar";            -- "BarBar"
US := 3 * 'A';              -- "AAA"
US := 2 * US;               -- "AAAAAA"
S15 := 5 * "SOS";           -- "SOSSOSSOSSOSSOS"
```

There are also similar functions Replicate for bounded strings with an optional parameter Drop in the same way that Append relates to "&".

The equality operators = and /= and the ordering operators <, <=, > and >= are defined between two bounded strings and between a string and a bounded string. And similarly for unbounded strings but they are not defined between a bounded string and an unbounded string; such comparisons can be performed after first converting both to strings.

Extracting or replacing a single character or part of a string at a known position is easy for fixed strings since normal indexing and slice operations can be used. For bounded and unbounded strings, subprograms are provided. Thus

```
C := Element(US, J);                    -- extract Jth element
Replace_Element(US, J, C);              -- replace Jth element
S(J .. K) := Slice(US, J, K);           -- extract elements J to K
Replace_Slice(US, J, K, S(J .. K));     -- replace elements J to K
```

with appropriate sliding as necessary.

The procedures Replace_Slice (there is also a version for fixed strings) can actually insert either a greater or lesser number of elements than those removed. An unbounded string can always adjust to the required size. A bounded string is limited by Max_Length and the behaviour is then controlled by an additional parameter Drop with default Error. In the case of fixed strings there are two further parameters, Justify and Pad, as for the subprogram Move. Finally, there are equivalent functions Replace_Slice which just return the appropriate result and do not alter the original string; in the case of fixed strings no parameters for padding and trimming are required. So if BS contains the string "Barbara Barnes", we can elevate Barbara to the peerage by

```
Lady: Bounded_String;
...
Lady := Replace_Slice(BS, 4, 7, "oness");   -- "Baroness Barnes"
```

The above all use a fixed string as the slice to be extracted or inserted. In order to avoid unnecessary conversions to the type String there are additional subprograms for extracting a slice of a bounded or unbounded string without converting it. Thus we could extract the Baroness by

```
Title: Bounded_String;
...
Title := Bounded_Slice(Lady, 1, 8);          -- "Baroness"
```

There is also a procedure so we could write Bounded_Slice(Title, Lady, 1, 8); instead. There are similar subprograms Unbounded_Slice.

We will now look at the various searching and similar operations which are provided uniformly for all three kinds of strings.

Searching for patterns is performed by Index which returns the index of the start of the pattern or 0 if no match is found. The direction of search is given by the optional parameter Going which can be Forward from the beginning (by default) or Backward. Thus, continuing the theme, but reverting to a commoner

```
BS := To_Bounded_String("Barbara Barnes");
Index(BS, "bar")                         -- is 4
Index(BS, "Bar")                         -- is 1
Index(BS, "Bar", Backward)               -- is 9
```

The start position of the search can be indicated by an additional parameter From. Thus having found the index of the first instance of "Bar" in Place, the next can be found by writing

```
Place := Index(BS, "Bar", From => Place+3);        -- is 9
```

and this could be the kernel of a loop for finding the locations of all instances. There are versions of Index for fixed, bounded, and unbounded strings.

This is possibly a good moment to introduce the idea of character mappings. These enable searches and other operations to be generalized. The private type Character_Mapping is defined in the package Strings.Maps; a value of the type defines a mapping between characters, and the mapped character is given by the function

```
function Value(Map: Character_Mapping;
               Element: Character) return Character;
```

Standard mappings are defined in Strings.Maps.Constants. The mapping Lower_Case_Map maps upper case characters onto the corresponding lower case character and leaves others unchanged. So Value(Lower_Case_Map, 'B') is 'b'. The other predefined maps are Upper_Case_Map and Basic_Map where the latter corresponds to Characters.Handling.To_Basic.

We can now search a string without concern for case by

```
Index(BS, "bar", Forward, Lower_Case_Map)       -- is 1
```

Note that the mapping applies only to the string BS and not to the pattern "bar".

Another version of Index looks for the first of a set of characters. This uses the type Character_Set also defined in Strings.Maps. Again this type is private and a number of constants are defined in Strings.Maps.Constants with names such as

Control_Set, Graphic_Set, Letter_Set, Lower_Set, and Upper_Set corresponding exactly to the functions Is_Control, Is_Graphic, and so on. (Oddly there is no Digit_Set and we have to use the equivalent Decimal_Digit_Set.) Thus we can search for the first lower case letter by

```
Index(BS, Set => Lower_Set)             -- is 2
```

Additional optional parameters (Test and Going) enable us to search for the first character outside the set and also specify the direction of search. Thus

```
Index(BS, Lower_Set, Outside, Forward)  -- is 1
```

because the first character not in the lower case set has index one.

Other forms of search enable us to count the number of matches so

```
Count(BS, "Bar")                        -- is 2
Count(BS, "BAR", Upper_Case_Map)        -- is 3
Count(BS, Lower_Set)                    -- is 11
```

and we can also look for a sequence all of whose characters satisfy some condition. Thus

```
Find_Token(BS, Lower_Set, Inside, I, J);
```

assigns 2 to I and 7 to J because the slice BS(2) .. BS(7) is the first slice whose characters are all inside the lower case set. The largest such slice which is first is chosen. If there were no such slice then 1 would be assigned to I and 0 to J. (Find_Token for a fixed string S would assign S'First to I which might not be 1.)

Although Replace_Slice is very flexible, some other subprograms are provided for convenience, thus a string can be inserted before a given location by Insert

```
Insert(BS, 9, "W ");                    -- "Barbara W Barnes"
-- same as Replace_Slice(BS, 9, 8, "W ");
```

and removed by Delete

```
Delete(BS, 2, 7);                       -- "B W Barnes"
-- same as Replace_Slice(BS, 2, 7, "");
```

A part can be overwritten by Overwrite

```
Overwrite(BS, 2, "ob");                 -- "Bob Barnes"
-- same as Replace_Slice(BS, 2, 3, "ob");
```

and the new section extends the string if necessary with all the usual options. Functional forms of Insert, Delete, and Overwrite are also provided.

The subprograms Trim enable space characters to be removed from one or both ends. Thus

```
Trim(BS, Side => Left);      -- deletes leading spaces
Trim(BS, Side => Both);      -- deletes leading and trailing spaces
```

Another version of Trim gives the character sets which are to be removed for both ends. Thus

```
Trim(BS, Left => Decimal_Digit_Set, Right => Null_Set);
```

deletes leading digits but deletes no trailing characters. For fixed strings the procedure Trim has the odd effect of typically having to add padding characters after doing any trimming. Functional forms of Trim also exist.

Subprograms Head and Tail produce a string comprising a given number of characters from the head or tail respectively. So

```
Head(BS, 3)                -- "Bob"
Tail(BS, 6)                -- "Barnes"
```

If the original string is not long enough then additional padding characters are added at the end for Head and at the beginning for Tail (this is the one case where padding is required for bounded and unbounded strings; it is normally only needed for fixed strings of course). A final optional parameter Drop exists for bounded strings; if the requested number of characters exceeds Max_Length then superfluous characters are dropped as for Replace_Slice; any such dropping is performed after any padding spaces are added. A curiosity is that the procedural form for fixed strings does not have this optional parameter but always raises Length_Error if nonpadding characters would be dropped. So

```
S5 := "ABCDE";
Head(S5, 6, Justify => Left);                 -- "ABCDE"
Head(S5, 6, Justify => Right);                -- Length_Error
```

In both cases an intermediate string "ABCDE " of length 6 has to be truncated but in the second case the attempt to right justify this string would require the nonpadding character 'A' to be dropped. However, the following would succeed because 'A' is explicitly specified to be the padding character so that the intermediate string is "ABCDEA" and then the leading 'A' can be dropped

```
Head(S5, 6, Justify => Right, Pad => 'A');    -- "BCDEA"
```

Finally, the subprograms Translate enable a string to be converted using a translation defined by a character mapping. As usual there are both functional and procedural forms for the three kinds of strings. So using functions

```
Translate("Barbara", Lower_Case_Map)          -- "barbara"
Translate("Café", Basic_Map)                  -- "Cafe"
```

where in the last case the sophisticated café is degraded into a mundane cafe.

The user will often want to use mappings other than those provided as standard. This can be done by writing a character mapping function whose profile corresponds to

```
type Character_Mapping_Function is
                      access function (From: in Character) return Character;
```

For example, suppose we want to remove all accents from upper case letters but not from lower case letters (a common style in French literature). We would declare

```
function Upper_Basic(C: in Character) return Character is
begin
  if Is_Upper(C) then
    return To_Basic(C);
  else
    return C;
  end if;
end Upper_Basic;
```

and then we can use a version of Translate which takes a parameter of the type Character_Mapping_Function rather than the type Character_Mapping as in the previous examples. So

```
Translate("Été", Upper_Basic'Access)        -- "Eté"
```

Character mapping functions can also be used with the subprograms Index and Count.

Another approach is to create our own values of the private type Character_Mapping. This is done by calling the function

```
function To_Mapping(From, To: Character_Sequence)
                                        return Character_Mapping;
```

where the subtype Character_Sequence is simply a renaming of String for readability. The two parameters must be sequences of the same length and there must be no repetition in the first. Characters in the first sequence map to characters in the corresponding position in the second sequence and other characters remain unchanged. Thus if we wanted a map that just made all curly brackets and square brackets into round brackets then we would write

```
Bracket_Map: Character_Mapping := To_Mapping("{}[]", "()()");
Translate(US, Bracket_Map);
```

which would apply the mapping to the unbounded string US. Reverse functions To_Domain and To_Range return the domain and range of a mapping if desired. There is also a constant Identity giving the identity map which is used as a default parameter.

As noted earlier, the subprograms Count, Find_Token, and Trim can take parameters of the type Character_Set and again we will want to define our own values. Although this type is also private, various subprograms are provided in the package Strings.Maps to enable the general manipulation of objects of this type. (Remember that this type defines a set of characters in the mathematical sense and not a complete set such as Latin-1 in the standard sense.)

A character set can be created by various functions To_Set which can take a single character, a string (a character sequence), a single character range, or finally an array of character ranges. The last are defined by

```
type Character_Range is
  record
    Low, High: Character;
  end record;
```

```
type Character_Ranges is array (Positive range <>) of Character_Range;
```

and so possible calls of functions To_Set are

```
S: Character_Set;
...
S := To_Set(Singleton => '?');                -- a single character
S := To_Set(Sequence => "AEIOU");             -- a string
S := To_Set(Span => ('0', '9'));              -- a range
S := To_Set(Ranges => (('A', 'Z'), ('a', 'z')));  -- an array of two ranges
```

Note that any duplicate characters are ignored.

Operations on character sets are "=", "not", "and", "or", "xor" with obvious meanings; "-" is such that X - Y is the same as X **and not** Y; Is_Subset and its renaming "<=" are such that X <= Y returns true if X is a subset of Y. The function Is_In(C, S) returns true if the character C is in the set S. Finally, there is a constant Null_Set, a function To_Ranges that turns a set back into the ordered minimal array of ranges and a function To_Sequence that turns a set back into an ordered sequence.

As a trivial example, suppose we wish to find the index of the last non-blank character in a string. The character set consisting of just the space character is given by To_Set(Space) and so all we need is

```
Index(BS, To_Set(Space), Outside, Backward)
```

and in fact the function Index_Non_Blank does this directly

```
Index_Non_Blank(BS, Backward)
```

As a more realistic example, to find the first octal digit we would write

```
Index(BS, To_Set(('0', '7')))
```

We conclude this survey of the string handling facilities discussed so far by noting that the various exceptions are raised as follows

Exception	Raised when
Length_Error	Result will not fit (fixed or bounded).
Pattern_Error	Pattern is null for Index or Count.
Index_Error	Index(es) outside string for Element, Replace_Element, Slice, or Replace_Slice.
Translation_Error	Parameters of To_Mapping are incorrect.

Although the survey has been brief it has covered all the key topics and the reader ought to be able to use the packages to the full from the description given here. For convenience the main subprograms (which all apply to fixed, bounded, and unbounded strings) are listed in Table 23.4. This shows the second and other parameters. The first parameter is always called Source and it and the result are of the type String, Bounded_String, or Unbounded_String where appropriate. Find_Token only exists as a procedure. Count, Index, and Index_Non_Blank only exist as functions with result Natural. Some procedures for fixed strings and procedures and

Table 23.4 Auxiliary parameters for string subprograms.

Name	*Second and other parameters*	*Notes*
Index	Pattern: String; Going: Direction := Forward; Mapping: Character_Mapping := Identity	
Index	Pattern: String; Going: Direction := Forward; Mapping: Character_Mapping_Function	
Index	Set: Character_Set; Test: Membership := Inside; Going: Direction := Forward	
Index	Pattern: String; From: Positive; Going: Direction := Forward; Mapping: Character_Mapping := Identity	
Index	Pattern: String; From: Positive: Going: Direction := Forward; Mapping: Character_Mapping_Function	
Index	Set: Character_Set; From: Positive; Test: Membership := Inside; Going: Direction := Forward	
Index_Non_Blank	Going: Direction := Forward	
Index_Non_Blank	From: Positive; Going: Direction := Forward	
Count	Pattern: String; Mapping: Character_Mapping := Identity	
Count	Pattern: String; Mapping: Character_Mapping_Function	
Count	Set: Character_Set	
Find_Token	Set: Character_Set; Test: Membership; First: **out** Positive; Last: **out** Natural	
Find_Token	Set: Character_Set; From: Positive; Test: Membership; First: **out** Positive; Last: **out** Natural	
Translate	Mapping: Character_Mapping	
Translate	Mapping: Character_Mapping_Function	
Replace_Slice	Low: Positive; High: Natural; By: String	1, 2
Insert	Before: Positive; New_Item: String	1
Overwrite	Position: Positive; New_Item: String	1
Delete	From: Positive; Through: Natural	2
Trim	Side: Trim_End	2
Trim	Left, Right: Character_Set	2
Head	Count: Natural; Pad: Character := Space	3, 4
Tail	Count: Natural; Pad: Character := Space	3, 4
Notes	*Extra parameters for fixed procedures & bounded subprograms*	*Types*
1	Drop: Truncation := Error	bnd/fixed
2	Justify: Alignment := Left; Pad: Character := Space	fixed
3	Drop: Truncation := Error (after Pad)	bounded
4	Justify: Alignment := Left (between Count and Pad)	fixed

functions for bounded strings have additional parameters for padding and trimming as shown.

A complete program illustrating the use of various features discussed above will be found following this chapter.

Note that there are versions of all the facilities discussed for wide and wide wide characters and strings. For details consult the *ARM*.

There are also child packages for simple text input and output of bounded and unbounded strings which avoid unnecessary conversions to the type String. The specification of the bounded package is

```
with Ada.Strings.Bounded;
generic
   with package Bounded is
                  new Ada.Strings.Bounded.Generic_Bounded_Length(<>);
package Ada.Text_IO.Bounded_IO
     with Global => in out synchronized is
   procedure Put(File: in File_Type; Item: in Bounded.Bounded_String);
   ...    -- also without file parameter and Put_Line and Get_Line similarly
end Ada.Text_IO.Bounded_IO;
```

This is generic because the bounded string package is itself a generic package. The unbounded package has similar subprograms but of course is not generic. The behaviour of the subprograms is similar to those for the type String in the parent package Ada.Text_IO; see Section 23.8.

There are also child functions of the package String for doing case insensitive comparisons. Thus we have

```
function Ada.Strings.Equal_Case_Insensitive(Left, Right: String)
                                                    return Boolean;
```

This compares the strings Left and Right for equality but ignoring case. Thus

```
Equal_Case_Insensitive("Pig", "PIG")
```

is true.

The function Ada.Strings.Fixed.Equal_Case_Insensitive is a renaming of the above. There are also functions Ada.Strings.Bounded.Equal_Case_Insensitive for bounded strings and Ada.Strings.Unbounded.Equal_Case_Insensitive for unbounded strings. There are similar functions for wide and wide wide versions.

Note that the comparison for strings can be phrased as convert to lower case and then compare. But this does not always work for wide and wide wide strings. The proper terminology is 'apply locale-independent case folding and then compare'.

Although it comes to the same thing for Latin-1 characters there are problems with some character sets where there is not a one-one correspondence between lower case and upper case. This used to apply to English with the two forms of lower case S and still applies to the corresponding letters in Greek where the upper case character is Σ and there are two lower case versions namely σ and ς. So

```
Ada.Strings.Wide_Equal_Case_Insensitive("ΣΟΣ", "σος")
```

returns true. Note that if we just convert to lower case first rather than applying case folding then it would not be true.

Furthermore there is also

```
function Ada.Strings.Less_Case_Insensitive(Left, Right: String)
                                                    return Boolean;
```

which does a lexicographic comparison.

There are also child functions such as Ada.Strings.Hash and Ada.Strings.Hash_Case_Insensitive which are hash functions. These are primarily for use with containers and so are described in Chapter 24.

The final topic in this section concerns the encodings UTF-8 and UTF-16 which are now widely used. There are packages defining mechanisms to convert between these encodings and the types String, Wide_String, and Wide_Wide_String.

The encoding UTF-8 works in terms of raw bytes and is straightforward; it is defined in Annex D of ISO/IEC 10646. However, UTF-16 comes in two forms according to whether the arrangement of two bytes into a 16-bit word uses big-endian or little-endian packing. So there are two forms UTF-16BE and UTF-16LE; they are defined in Annex C of ISO/IEC 10646.

The different encodings can be distinguished by a special value known as a BOM (Byte Order Mark) at the start of the string. So we have BOM_8, BOM_16BE, BOM_16LE, and just BOM_16 (for wide strings).

To support these encodings, Ada includes the following five packages

```
Ada.Strings.UTF_Encoding
Ada.Strings.UTF_Encoding.Conversions
Ada.Strings.UTF_Encoding.Strings
Ada.Strings.UTF_Encoding.Wide_Strings
Ada.Strings.UTF_Encoding.Wide_Wide_Strings
```

The first package declares items that are used by the other packages. It is

```
package Ada.Strings.UTF_Encoding
    with Pure is

  type Encoding_Scheme is (UTF_8, UTF_16BE, UTF_16LE);

  subtype UTF_String is String;
  subtype UTF_8_String is String;
  subtype UTF_16_Wide_String is Wide_String;

  Encoding_Error: exception;

  BOM_8: constant UTF_8_String :=
                          Character'Val(16#EF#) &
                          Character'Val(16#BB#) &
                          Character'Val(16#BF#);

  BOM_16BE: constant UTF_String :=
                          Character'Val(16#FE#) &
                          Character'Val(16#FF#);

  BOM_16LE: constant UTF_String :=
                          Character'Val(16#FF#) &
                          Character'Val(16#FE#);

  BOM_16: constant UTF_16_Wide_String :=
                                   (1 => Wide_Character'Val(16#FEFF#));

  function Encoding(Item: UTF_String;
          Default: Encoding_Scheme := UTF_8) return Encoding_Scheme;

end Ada.Strings.UTF_Encoding;
```

Note that the encoded forms are actually still held in objects of type String or Wide_String. However, in order to aid understanding, the subtypes UTF_String, UTF_8_String, and UTF_16_Wide_String are introduced and these should be used when referring to objects holding the encoded forms.

The type Encoding_Scheme defines the various schemes. Note that an encoded string might or might not start with the identifying BOM; it is optional. The function Encoding takes a UTF_String (that is a plain old string), checks the BOM if present and returns the value of Encoding_Scheme identifying the scheme. If there is no BOM then it returns the value of the parameter Default which itself by default is UTF_8.

Note carefully that the function Encoding does not do any encoding – that is done by functions Encode in the other packages which will be described in a moment. Note also that there is no corresponding function Encoding for wide strings; that is because there is only one relevant scheme corresponding to UTF_16_Wide_String, namely that with BOM_16.

We will now look at the other packages. The package UTF_Encoding.Strings contains functions Encode and Decode which convert between the raw type String and the UTF forms. Similar packages apply to wide and wide wide strings. The package UTF_Encoding.Conversions contains functions Convert which convert between the various UTF forms.

The package for the type String is

```
package Ada.Strings.UTF_Encoding.Strings
    with Pure is

  function Encode(Item: String; Output_Scheme: Encoding_Scheme;
                  Output_BOM: Boolean := False) return UTF_String;
  function Encode(Item: String;
                  Output_BOM: Boolean := False) return UTF_8_String;
  function Encode(Item: String; Output_BOM: Boolean := False)
                                         return UTF_16_Wide_String;

  function Decode(Item: UTF_String; Input_Scheme: Encoding_Scheme)
                                                      return String;
  function Decode(Item: UTF_8_String) return String;
  function Decode(Item: UTF_16_Wide_String) return String;

end Ada.Strings.UTF_Encoding.Strings;
```

The functions Encode take a string and return it encoded. The first function has a parameter Output_Scheme which determines whether the encoding is to be to UTF_8, UTF_16BE, or UTF_16LE. The second function is provided as a convenience for the common case of encoding to UTF_8 and the third function is necessary for encoding to UTF_16_Wide_String. In all cases there is a final optional parameter indicating whether or not an appropriate BOM is to be placed at the start of the encoded string.

The functions Decode do the reverse. Thus the first function takes a value of subtype UTF_String and a parameter Input_Scheme giving the scheme to be used and returns the decoded string. If a BOM is present which does not match the Input_Scheme, then the exception Encoding_Error is raised. The second function is a convenience for the common case of decoding from UTF_8 and the third function is necessary for decoding from UTF_16_Wide_String; again, if a BOM is present that does not match the expected scheme then Encoding_Error is raised. In all cases all the strings returned have a lower bound of 1.

The packages UTF_Encoding.Wide_Strings and UTF_Encoding.Wide_Wide_Strings are identical except that the type String is replaced by Wide_String or Wide_Wide_String throughout.

Finally, the package for converting between the various UTF forms is

```
package Ada.Strings.UTF_Encoding.Conversions
    with Pure is

  function Convert(Item: UTF_String;
                   Input_Scheme: Encoding_Scheme;
                   Output_Scheme: Encoding_Scheme;
                   Output_BOM: Boolean := False) return UTF_String;

  function Convert(Item: UTF_String;
                   Input_Scheme: Encoding_Scheme;
                   Output_BOM: Boolean := False)
                                     return UTF_16_Wide_String;

  function Convert(Item: UTF_8_String;
                   Output_BOM: Boolean := False)
                                     return UTF_16_Wide_String;

  function Convert(Item: UTF_16_Wide_String;
                   Output_Scheme: Encoding_Scheme;
                   Output_BOM: Boolean := False) return UTF_String;

  function Convert(Item: UTF_16_Wide_String;
                   Output_BOM: Boolean := False)
                                          return UTF_8_String;

end Ada.Strings.UTF_Encoding.Conversions;
```

The purpose of these should be obvious. The first converts between encodings held as strings with parameters indicating both the Input_Scheme and the Output_Scheme. If the input string has a BOM that does not match the Input_Scheme then the exception Encoding_Error is raised. The final optional parameter indicates whether or not an appropriate BOM is to be placed at the start of the converted string.

The other functions convert between UTF encodings held as strings and wide strings. Two give the explicit Input_Scheme or Output_Scheme and two are provided for convenience for the common case of UTF_8.

Exercise 23.3

1 Search the string S for the first character which is either a decimal digit or decimal point.

2 Evaluate

```
To_Sequence(To_Set(To_Lower("Ben is singing")) - To_Set(Space))
```

3 A simple encryption algorithm is based on a keyword K ignoring any repeating letters. These letters are placed in alphabetical order followed by the remaining letters of the alphabet also in alphabetical order. This sequence then defines the result of mapping the alphabet. Define such a mapping which works on either case of the usual alphabet and leaves other characters unchanged. Use the mapping to encrypt a string S using the keyword "Byron".

4 Given a mapping M write a function Decode which returns the reverse mapping. Raise Translation_Error if the original mapping was not (1, 1). Use this function to decode the message of the previous exercise.

23.4 Text buffers and images

Universal text buffers provide a convenient way for storing and retrieving strings. They are a bit like streams described in Section 23.9, but only handle characters whereas streams handle values of various types. As a consequence text buffers are simpler and more reliable; they were added in Ada 2022.

There are three packages, the parent main package is Ada.Strings.Text_Buffers which includes various procedures to put strings into a buffer and there are two child packages Text_Buffers.Bounded and Text_Buffers.Unbounded which include the corresponding functions for retrieving strings. The specification of the parent package is

```
with Ada.Strings.UTF_Encoding.Wide_Wide_Strings;
package Ada.Strings.Text_Buffers
    with Pure is

  type Text_Buffer_Count is range 0 .. implementation-defined;
  New_Line_Count: constant Text_Buffer_Count := implementation-defined;

  type Root_Buffer_Type is abstract tagged private
    with Default_Initial_Condition =>
                          Current_Indent(Root_Buffer_Type) = 0;

  procedure Put(Buffer: in out Root_Buffer_Type;
                Item: in String) is abstract;

  procedure Wide_Put(Buffer: in out Root_Buffer_Type;
                     Item: in Wide_String) is abstract;

  procedure Wide_Wide_Put(Buffer: in out Root_Buffer_Type;
                          Item: in Wide_Wide_String) is abstract;

  procedure Put_UTF_8(Buffer: in out Root_Buffer_Type;
                      Item: in UTF_Encoding.UTF_8_String) is abstract;

  procedure Wide_Put_UTF_16(Buffer: in out Root_Buffer_Type;
               Item: in UTF_Encoding.UTF_16_Wide_String) is abstract;

  procedure New_Line(Buffer: in out Root_Buffer_Type) is abstract;

  Standard_Indent: constant Text_Buffer_Count:= 3;

  function Current_Indent(Buffer: Root_Buffer_Type)
                                          return Text_Buffer_Count;
```

```
      procedure Increase_Indent(Buffer: in out Root_Buffer_Type;
                                Amount: in Text_Buffer_Count := Standard_Indent)
        with Post'Class =>
             Current_Indent (Buffer) = Current_Indent(Buffer)'Old + Amount;

      procedure Decrease_Indent(Buffer: in out Root_Buffer_Type;
                                Amount: in Text_Buffer_Count := Standard_Indent)
        with Pre'Class => Current_Indent(Buffer) >= Amount
                                          or else raise Constraint_Error,
           Post'Class => Current_Indent(Buffer) =
                                     Current_Indent(Buffer)'Old - Amount;

   private
      ... -- not specified by the language
   end Ada.Strings.Text_Buffers;
```

The general idea is to allow the creation of a buffer representing a number of lines of text where the lines can have various levels of indentation. A new line is started by a call of New_Line (which adds New_Line_Count characters to the buffer to represent the newline). The current level of indentation is given by the function Current_Indent and is initially zero from the Default_Initial_Condition. The level of indentation can be increased or decreased by calls of Increase_Indent and Decrease_Indent. The amount of change by default is Standard_Indent which is the constant 3. Note from the precondition on Decrease_Indent that attempting to decrease the level by more than the current value raises Constraint_Error.

When a string is added by a call of Put (or Wide_Put etc), if it is the first item on a line then it is preceded by Current_Indent spaces. Otherwise it just goes after existing text on that line.

The corresponding Get functions are in the Bounded and Unbounded child packages. The unbounded package is

```
   package Ada.Strings.Text_Buffers.Unbounded
        with Preelaborate, Nonblocking, Global => null is

      type Buffer_Type is new Root_Buffer_Type with private;

      function Get(Buffer: in out Buffer_Type) return String
        with Post'Class => Get'Result'First = 1
                                    and then Current_Indent(Buffer) = 0;

      ... -- also Wide_Get etc

   private
        -- not specified by the language, but will include nonabstract
        -- overridings of all inherited subprograms that require overriding.
   end Ada.Strings.Text_Buffers.Unbounded;
```

and the bounded package is very similar thus

```
   package Ada.Strings.Text_Buffers.Bounded
        with Pure, Nonblocking, Global => null is

      type Buffer_Type(Max_Characters: Text_Buffer_Count) is
                                    new Root_Buffer_Type with private
        with Default_Initial_Condition => not Text_Truncated(Buffer_Type);
```

```
      function Text_Truncated(Buffer: in Buffer_Type) return Boolean;
      ... -- Get, Wide_Get, etc as in the Unbounded child.
   private
      ... -- not specified by the language, etc
   end Ada.Strings.Text_Buffers.Bounded;
```

A call of Get empties the buffer completely; thus Put and Get do not match. It's the old story mentioned in Section 21.7; we know what we are putting in but it's not so easy to know what we are getting out. If we use Get from Bounded then a call of Text_Truncated tells us whether some text was lost by overzealous use of Put.

We now turn to the other main topic of this section, namely images. Previous versions of Ada had a rather purist view of printing out values especially when debugging. To output the value of some variable N of type Integer one had to write

```
   Put(Integer'Image(N));
```

which was irritating since one had to mention the (sub)type of the object. Moreover, it only applied to scalars, so one was typically forced to write tedious code for arrays and records. There was a slight improvement with the Corrigendum in 2016 which permitted the subtype name to be omitted. so we could write

```
   Put(Image(N)):
```

But it still only applied to scalars. This is all changed in Ada 2022.

The attribute Image now applies to all types and generates a string. The new attribute Put_Image is also defined for all types and sends the string to a text buffer. Its specification is

```
   procedure S'Put_Image(
      Buffer: in out Ada.Strings.Text_Buffers.Root_Buffer_Type'Class;
      Arg: in T);
```

where S is a subtype of T.

Put_Image can be redefined as we shall see in the case of Big_Numbers in Section 23.6. The full details of Image and Put_Image are rather extensive and the reader can browse section 4.10 of the *ARM* at leisure. We will just illustrate some important examples

We start with a brief summary of the form of the string for various scalar types using examples from the given sections (quotes enclosing the string are shown to clarify any leading spaces).

```
   " 38"                   -- P: Integer in Section  6.1
   " 6.7000000E-11"        -- G: Float               6.1
   "RED"                   -- C: Colour              6.3
   "TRUE"                  -- Danger: Boolean        6.7
   "'D'"                   -- Dig: Roman_Digit       8.4
```

However, we cannot apply Image to a constant such as Euler's constant γ (= 0.57721_56649) in Section 6.1 but would have to convert it to a genuine type first.

In the case of arrays, the form is that of a named aggregate using square brackets. Both the index values and the component values are written in the same form as for scalars.

Thus for Work_Day in Section 8.3 it is

```
[MON=>TRUE, TUE=>TRUE, WED=>TRUE, .... SUN=>FALSE]
```

and for the constant Red in Section 8.6 it is

```
[R => TRUE, Y => FALSE, B => FALSE]
```

and for Pascal(0 .. 4) in Section 2.4 it is

```
[ 0=> 1, 1=> 4, 2=> 6, 3=> 4, 4=> 1]
```

Multidimensional arrays use a nested format as expected. Thus for Unit_2 of Section 8.3 it is

```
[ 1=>[ 1=> 1.00000E+00, 2=> 0.00000E+00], [ 1=> 0.00000E+00,...]]
```

The fine detail of the spacing in the layout is implementation-defined.

In the case of records, traditional parentheses are used rather than square brackets. Thus for D in Section 8.7 it is

```
(DAY=> 4, MONTH=>JULY, YEAR=> 1776)
```

and for Fred of type Student in Section 8.7, it is

```
(BIRTH=> (DAY=> 19, MONTH=> AUG, YEAR=> 1984),
 FINALS=> [THEOLOGY=> 5, CLASSICS=> 15, MATHEMATICS=> 99])
```

For an access type, if the value is **null** then the string is simply "NULL", otherwise it has the form "(ACCESS" ....)" where the dots typically denote a string of hexadecimal digits giving the address concerned.

For a null record, the string is "(NULL RECORD)". In the case of tasks and protected types the image includes the words TASK and PROTECTED.

Implementations are permitted to transform the image for a composite type in various ways.

For example if the lower bound of an array is the same as the lower bound of the array type then positional array syntax can be used. Thus for Pascal(0 .. 4) it might generate (1, 4, 6, 4, 1) instead. For further details, the reader should consult the *ARM*.

The attribute Image can also be applied to containers; as will be discussed in Chapter 24 the string is in a form appropriate to the kind of container. Thus a vector will be in the form of the image of an array. An empty container has image [].

23.5 Numerics

The numeric library was introduced in Section 4.3 where we saw that the package Ada.Numerics contains the mathematical constants Pi and e plus the exception Argument_Error.

```
package Ada.Numerics
    with Pure is
  Argument_Error: exception;
  Pi: constant := 3.14159_26535_89793_23846_26433_83279_
                                        50288_41971_69399_37511;
  π: constant := Pi;
  e: constant := 2.71828_18284_59045_23536_02874_71352_
                                        66249_77572_47093_69996;
end Ada.Numerics;
```

Note the constant π which is an alternative to Pi for those who like their mathematical expressions to look nice.

The core child packages of Numerics are

```
Numerics.Discrete_Random
Numerics.Float_Random
Numerics.Generic_Elementary_Functions
```

plus nongeneric forms of the last for the predefined types with names such as

```
Numerics.Elementary_Functions         -- for Float
Numerics.Long_Elementary_Functions    -- for Long_Float
```

The Numerics annex defines additional packages for dealing with complex numbers and for manipulating real and complex vectors and matrices; see Section 26.5.

Although numerical applications are in a minority, we will nevertheless consider the elementary functions package in some detail because it provides a good illustration of the use of generics and other key features of Ada.

As we shall see, the package provides ease of use for simple calculations and the ability to provide the ultimate in accuracy for serious numerical work, as well as portability across different implementations.

The package specification is as follows

```
generic
  type Float_Type is digits <>;
package Ada.Numerics.Generic_Elementary_Functions
    with Pure, Nonblocking is

  -- subtype FTB is Float_Type'Base;

  function Sqrt(X: FTB) return FTB;
  function Log(X: FTB) return FTB;
  function Log(X, Base: FTB) return FTB;
  function Exp(X: FTB) return FTB;
  function "**" (Left, Right: FTB) return FTB;
```

```
      function Sin(X: FTB) return FTB;
      function Sin(X, Cycle: FTB) return FTB;
      ...  -- similarly Cos, Tan, Cot, Arcsin, Arccos

      function Arctan(Y: FTB; X: FTB := 1.0) return FTB;
      function Arctan(Y: FTB; X: FTB := 1.0; Cycle: FTB) return FTB;
      function Arccot(X: FTB; Y: FTB := 1.0) return FTB;
      function Arccot(X: FTB; Y: FTB := 1.0; Cycle: FTB) return FTB;

      function Sinh(X: FTB) return FTB;
      function Arcsinh(X: FTB) return FTB;
      ...  -- similarly Cosh, Tanh, Coth, Arccosh, Arctanh, Arccoth

   end Ada.Numerics.Generic_Elementary_Functions;
```

where we have written FTB as an abbreviation for Float_Type'Base.

The single generic parameter is the floating type. The package might be instantiated with a user's own type such as My_Float

```
   package My_Elementary_Functions is
      new Generic_Elementary_Functions(Float_Type => My_Float);
```

The body could then choose an implementation appropriate to the accuracy of the user's type through the attribute Float_Type'Digits rather than necessarily using the accuracy of the predefined type upon which the user's type has been based. This could have significant efficiency advantages.

The package can also be instantiated with a constrained subtype such as

```
   subtype Normal is Float range -1.0 .. +1.0;
```

which might occur if the user were dealing with data which was known to be in such a range. Note however, that the parameters and results of the various functions use Float_Type'Base and so there are no constraint checks on passing the parameters or on returning results.

Moreover, the body of the package will undoubtedly declare working variables for use in the algorithms and these can similarly be declared

```
   Local: Float_Type'Base;
```

so that constraints on the actual subtype do not apply to internal calculations.

We continue by considering the individual functions in the package. The functions Sqrt and Exp need little comment except perhaps concerning exceptions. Calling Sqrt with a negative parameter will raise Argument_Error whereas calling Exp with a large parameter will raise Constraint_Error.

The principle is that intrinsic mathematical restrictions raise Argument_Error whereas implementation range restrictions raise Constraint_Error.

There are two overloadings of Log. That with a single parameter gives the natural logarithm to base e, whereas that with two parameters allows us to choose any base at all. Thus to find $\log_{10}2$, we write

```
   Log(2.0, 10.0)          -- 0.3010...
```

The reader may wonder why there is not just a single function with a default parameter thus

```
function Log(X: Float_Type'Base;  Base: Float_Type'Base := e)
                                          return Float_Type'Base;
```

which would seem to give the desired result with less fuss. The reason concerns obtaining the ultimate in precision. Passing a default parameter means that the accuracy of the value of *e* used can only be that of the Float_Type. Using a separate function enables the function body to obtain the benefit of the full accuracy of the universal real named number Numerics.e.

The restrictions on the parameters of Log are X > 0.0, Base > 0.0, and also Base /= 1.0. So Base could be 0.5 which is an amusing thought.

The function "**" effectively extends the predefined operator to allow non-integral exponents. Consequently, the formal parameters are Left and Right to match (the other functions follow mathematical convention and use X and Y). The parameter Left must not be negative. Note that 0.0**0.0 raises Argument_Error whereas 0.0**0 is 1.0.

The trigonometric functions Sin, Cos, Tan, and Cot also come in pairs like Log and for a similar reason. The single parameter versions assume the parameter is in radians whereas the second parameter allows the use of any unit by giving the number of units in a whole cycle. Thus to find the sine of 30 degrees, we write

```
Sin(30.0, 360.0)            -- 0.5
```

because there are 360 degrees in a cycle.

In these functions the single parameter versions enable the highly accurate number Numerics.Pi or Numerics.π to be used directly rather than being passed with less accuracy as a parameter.

The inverse functions, Arcsin and Arccos, are straightforward; the result of Arcsin is in the range $-\pi/2$ to $+\pi/2$, and that of Arccos is in the range 0 to π. However, Arctan has a default value of 1.0 for a second parameter (the Cycle then being third). This enables us to call Arctan with two parameters giving the classical *x*- and *y*-coordinates (thus fully identifying the quadrant). So

```
Arctan(Y => -1.0, X => +1.0)             -- -π/4
Arctan(Y => +1.0, X => -1.0)             -- +3*π/4
```

Note carefully that the first parameter of Arctan is Y since it is the *x*-coordinate that is taken to be 1.0 by default. Arccot is very similar except that the parameters are naturally in the other order.

Arctan and Arccot are examples where signed zeros may make a difference (signed zeros were explained in Section 17.4). If Float_Type'Signed_Zeros is true then Arctan(Y, X) and Arccot(X, Y) where X is negative and Y is zero return $+\pi$ if Y is a positive zero and $-\pi$ if Y is a negative zero; if Signed_Zeros is false then the result is always $+\pi$.

There are no obvious comments to make on the hyperbolic functions or their inverses.

The random number packages were briefly introduced in Section 4.3. The full specification of the floating point package is

```
package Ada.Numerics.Float_Random
  with Global => in out synchronized is

  type Generator is limited private;
  subtype Uniformly_Distributed is Float range 0.0 .. 1.0;
  function Random(Gen: Generator) return Uniformly_Distributed
    with Global => overriding in out Gen;

  procedure Reset(Gen: in Generator; Initiator: in Integer)
    with Global => overriding in out Gen;
  procedure Reset(Gen: in Generator)
    with Global => overriding in out Gen;

  type State is private;
  procedure Save(Gen: in Generator; To_State: out State);
  procedure Reset(Gen: in Generator; From_State: in State)
    with Global => overriding in out Gen;
  Max_Image_Width: constant := implementation-defined;
  function Image(Of_State: State) return String;
  function Value(Coded_State: String) return State;
private
  ...
end Ada.Numerics.Float_Random;
```

A pseudorandom sequence of values of the type Float uniformly distributed in the range 0.0 .. 1.0 is obtained by declaring an object of the type Generator and then calling the function Random with that object as parameter much as in the example in Section 4.3 for the generic package Discrete_Random. Since Random is a function, the state of the underlying generator has to be updated indirectly and so the full type Generator is inevitably an access type or uses some trickery. Note especially the Global annotations which were explained in Section 16.5.

Other distributions can be obtained by transforming the uniform distribution with a suitable function. Note that the end values 0.0 and 1.0 might or might not be produced. Thus an exponential distribution with mean and variance 1.0 can be obtained by the transformation

```
-Log(Random(G) + Float'Model_Small)
```

where the addition of the smallest possible number ensures that Log never has the parameter zero and thus never raises an exception.

Different sequences can be obtained by calling Reset. The version with just the generator as parameter initiates a sequence whose starting state depends upon the time and so is not reproducible. The version with the additional integer parameter initiates a reproducible sequence corresponding to the value of the integer. Reproducible sequences are often important for debugging purposes. (Note that if we do not call Reset at all then we always get the same sequence which is boring and statistically unhelpful but useful for trivial programs.)

The internal state of the generator can be preserved and then restored by calling Save and Reset. A sequence can thus be stopped, analysed, and then continued from where it was stopped. The saved value could even be written to a file and used in a different execution of the program. Finally, the functions Image and Value enable such a state value to be converted to and from a string for the ultimate in external representation; the maximum length of string required is given by Max_Image_Width.

The specification of the generic package Discrete_Random is as follows

```
generic
   type Result_Subtype is (<>);
package Ada.Numerics.Discrete_Random
      with Global => in out synchronized is
   type Generator is limited private;

   function Random(Gen: Generator) return Result_Subtype
      with Global => overriding in out Gen;

   function Random(Gen: Generator;
                            First, Last: Result_Subtype) return Result_Subtype;
      with Post => Random'Result in First .. Last,
            Global => overriding in out Gen;

      -- also State, Save, Reset, Max_Image_Width, Image, Value
      -- as for Float_Random

end Ada.Numerics.Discrete_Random;
```

Note especially the second function Random which returns a result uniformly distributed in the range First .. Last. This function is added in Ada 2022 because achieving uniform distribution is not at all straightforward for the user to program and so is best provided by the implementation. Note that the postcondition states that the result is in the required range but is unable to specify that it is uniformly distributed.

Exercise 23.5

1 Write a body for the package Simple_Maths of Exercise 2.2(1) using the package Elementary_Functions. Raise Constraint_Error for all errors.

2 Simulate a game of paper, stone, and scissors between Jack and Jill. Paper wraps stone, stone blunts scissors, scissors cut paper. Give each player their own generator.

23.6 Big numbers

Ada 2022 has facilities for doing arithmetic on very large and very accurate numbers outside the accuracy and range of the hardware. There is a common base package whose specification is

```
package Ada.Numerics.Big_Numbers
      with Pure, Nonblocking, Global => null is
   subtype Field is Integer range 0 .. implementation_defined;
   subtype Number_Base is Integer range 2 .. 16;
end Ada.Numerics.Big_Numbers;
```

and then there are two child packages, one for big integers and one for big reals. We will look at them in some detail since they illustrate a number of new features of Ada 2022. The integer package starts as follows

```
with Ada.Strings.Text_Buffers;
package Ada.Numerics.Big_Numbers.Big_Integers
    with Preelaborate, Nonblocking, Global => in out synchronized is
  type Big_Integer is private
    with Integer_Literal => From_Universal_Image,
        Put_Image => Put_Image;

  function Is_Valid(Arg: Big_Integer) return Boolean
    with Convention => Intrinsic;

  subtype Valid_Big_Integer is Big_Integer
    with Dynamic_Predicate => Is_Valid(Valid_Big_Integer),
        Predicate_Failure => (raise Program_Error);

  function "=" (L, R: Valid_Big_Integer) return Boolean;
  -- similarly "<", "<=", ">", ">="
```

This illustrates a number of features of Ada 2022 that enable the program text to describe many properties that would otherwise be described by English text. A key point is the introduction of the subtype Valid_Big_Integer which ensures that a parameter does have a valid representation of a big integer. Note how the check is performed by the Dynamic_Predicate aspect coupled with Predicate_Failure causing the exception Program_Error to be raised if the check fails; this technique is described in Section 16.5. As a consequence we do not have to have an explicit precondition for functions such as "=" to ensure that the parameters are valid nor does the *ARM* need English text to describe the effect.

Note especially the aspect Put_Image which seems tautological! The effect is to replace the normal predefined aspect which would take a parameter of type Big_Integer by the subprogram Put_Image declared below which takes a parameter of subtype Valid_Big_Integer thereby forcing a check.

The specification continues with various subtypes and conversion functions

```
  function To_Big_Integer(Arg: Integer) return Valid_Big_Integer;

  subtype Big_Positive is Big_Integer
    with Dynamic_Predicate => (if Is_Valid(Big_Positive)
                                    then Big_Positive > 0),
        Predicate_Failure => (raise Constraint_Error);

  subtype Big_Natural is Big_Integer
    with Dynamic_Predicate => (if Is_Valid(Big_Natural)
                                    then Big_Natural >= 0),
        Predicate_Failure => (raise Constraint_Error);

  function In_Range(Arg, Low, High: Valid_Big_Integer) return Boolean is
    (Low <= Arg and Arg <= High);

  function To_Integer(Arg: Valid_Big_Integer) return Integer
    with Pre => In_Range(Arg, Low => To_Big_Integer(Integer'First),
                              High => To_Big_Integer(Integer'Last))
                  or else raise Constraint_Error;
```

There are then generic packages for signed and unsigned conversions to and from any integer type.

```
generic
   type Int is range <>;
package Signed_Conversions is
   function To_Big_Integer(Arg: Int) return Valid_Big_Integer;

   function From_Big_Integer(Arg: Valid_Big_Integer) return Int
      with Pre => In_Range(Arg,
                           Low => To_Big_Integer(Int'First),
                           High => To_Big_Integer(Int'Last))
               or else raise Constraint_Error;
end Signed_Conversions;

generic
   type Int is mod <>;
package Unsigned_Conversions is
   function To_Big_Integer(Arg: Int) return Valid_Big_Integer;

   function From_Big_Integer(Arg: Valid_Big_Integer) return Int
      with Pre => In_Range(Arg,
                           Low => To_Big_Integer(Int'First),
                           High => To_Big_Integer(Int'Last))
               or else raise Constraint_Error;
end Unsigned_Conversions;
```

And then there are conversions to and from strings thus

```
   function To_String(Arg: Valid_Big_Integer;
                      Width: Field := 0;
                      Base: Number_Base := 10) return String
      with Post => To_String'Result'First = 1;

   function From_String(Arg: String) return Valid_Big_Integer;

   function From_Universal_Image(Arg: String) return Valid_Big_Integer
      renames From_String;

   procedure Put_Image
      (Buffer: in out Ada.Strings.Text_Buffers.Root_Buffer_Type'Class;
       Arg: in Valid_Big_Integer);
```

And finally the declaration of the basic operations

```
   function "+" (L: Valid_Big_Integer) return Valid_Big_Integer;
 -- similarly unary "–", "abs"
   function "+" (L, R: Valid_Big_Integer) return Valid_Big_Integer;
 -- similarly binary "–", "*", "/", "mod", "rem", Min, Max
   function ** (L: Valid_Big_Integer; R: Natural) return Valid_Big_Integer;
```

```
   function Greatest_Common_Divisor
                              (L, R: Valid_Big_Integer) return Big_Positive
      with Pre => (L /= 0 and R /= 0) or else raise Constraint_Error;
private
   ... -- not specified by the language
end Ada.Numerics.Big_Numbers.Big_Integers;
```

Note that, as usual, the second parameter of ** is simply of subtype Natural. Although the package seems complex at first sight, its use is straightforward. To do a bit of simple arithmetic we might write

```
declare
   N, M, P: Big_Integer;
begin
   N := 123_456_789;
   M := 987_654_321;
   P := M * N;
   ...
```

Writing N := 123_456_789; is shorthand for N := From_String("123_456_789"); since the type Big_Integer has the aspect Integer_Literal => From_Universal_Image which is itself a renaming of From_String (which returns a value of type Valid_Big_Integer). Moreover, note that From_String behaves much as Get from Text_IO.Integer IO. Subsequent operations such as multiplication work as expected.

There is a similar package for the manipulation of big real numbers. It starts

```
with Ada.Numerics.Big_Numbers.Big_Integers;
   use all type Big_Integers.Big_Integer;
with Ada.Strings.Text_Buffers;
package Ada.Numerics.Big_Numbers.Big_Reals
      with Preelaborate, Nonblocking, Global => in out synchronized is
   type Big_Real is private
      with Real_Literal => From_Universal_Image,
           Put_Image => Put_Image;

   function Is_Valid(Arg: Big_Real) return Boolean
      with Convention => Intrinsic;

   subtype Valid_Big_Real is Big_Real
      with Dynamic_Predicate => Is_Valid(Valid_Big_Real),
           Predicate_Failure => raise Program_Error;
```

This is very similar to the big integer package. But then comes some interesting detail which reveals that big reals are really treated as big rationals. Thus

```
   function "/" (Num, Den: Big_Integers.Valid_Big_Integer)
                                                      return Valid_Big_Real
      with Pre => Den /= 0 or else raise Constraint_Error;
```

```
function Numerator(Arg: Valid_Big_Real)
                                    return Big_Integers.Valid_Big_Integer
  with Post => (if Arg = 0.0 then Numerator'Result = 0);

function Denominator(Arg: Valid_Big_Real)
                                         return Big_Integers.Big_Positive
  with Post => (if Arg = 0.0 then Denominator'Result = 1 else
              Big_Integers.Greatest_Common_Divisor
                           (Numerator(Arg), Denominator'Result) = 1);
```

The function "/" constructs a big real from two big integers which are the numerator and denominator plus two corresponding functions which return them.

There are then various conversion functions and comparison operators; note the use of the constructor function "/" just declared

```
function To_Big_Real(Arg: Big_Integers.Valid_Big_Integer)
                                     return Valid_Big_Real is (Arg / 1);

function To_Real(Arg: Integer) return Valid_Big_Real is
  (Big_Integers.To_Big_Integer(Arg) / 1);

function "=" (L, R: Valid_Big_Real) return Boolean;
-- similarly "<", "<=", ">", ">="

function In_Range(Arg, Low, High: Valid_Big_Real) return Boolean is
  (Low <= Arg and Arg <= High);
```

There are then generic constructor functions from real types including both floating and fixed point types (and decimal types added after standardization by AI22-78) thus

```
generic
  type Num is digits <>;
package Float_Conversions is
  function To_Big_Real(Arg: Num) return Valid_Big_Real;
  function From_Big_Real(Arg: Valid_Big_Real) return Num
    with Pre => In_Range (Arg, Low => To_Big_Real(Num'First),
                              High => To_Big_Real(Num'Last))
               or else (raise Constraint_Error);
end Float_Conversions;

generic
  type Num is delta <>;
package Fixed_Conversions is
  function To_Big_Real(Arg: Num) return Valid_Big_Real;
  function From_Big_Real(Arg: Valid_Big_Real) return Num
    with Pre => In_Range (Arg, Low => To_Big_Real(Num'First),
                              High => To_Big_Real(Num'Last))
               or else (raise Constraint_Error);
end Fixed_Conversions;
```

```
generic
   type Num is delta <> digits <>;
package Decimal_Conversions is
   function To_Big_Real(Arg: Num) return Valid_Big_Real;
   function From_Big_Real(Arg: Valid_Big_Real) return Num
      with Pre => In_Range (Arg,  Low => To_Big_Real(Num'First),
                                  High => To_Big_Real(Num'Last))
                  or else (raise Constraint_Error);
end Decimal_Conversions;
```

and then conversion functions to and from strings

```
function To_String(Arg: Valid_Big_Real; Fore: Field := 2;
                         Aft: Field := 3; Exp: Field := 0) return String
   with Post => To_String'Result'First = 1;

function From_String(Arg: String) return Valid_Big_Real;

function From_Universal_Image(Arg: String) return Valid_Big_Real
   renames From_String;

function From_Universal_Image(Num, Den: String)
                                            return Valid_Big_Real is
   (Big_Integers.From_Universal_Image(Num) /
                        Big_Integers.From_Universal_Image(Den));

function To_Quotient_String(Arg: Valid_Big_Real) return String is
     (To_String(Numerator(Arg)) & " / " & To_String(Denominator(Arg)));

                        -- note the single space before and after /

function From_Quotient_String(Arg: String) return Valid_Big_Real;

procedure Put_Image(Buffer: in out
                    Ada.Strings.Text_Buffers.Root_Buffer_Type'Class;
                    Arg: in Valid_Big_Real);
```

and finally the basic operations

```
   function "+" (L: Valid_Big_Real) return Valid_Big_Real;
  -- similarly "-", "abs" "+", "-", "*", "/", "**", Min, Max
   private
   ... -- not specified by the language
end Ada.Numerics.Big_Numbers.Big_Reals;
```

Again, although it looks complex, its use is straightforward. Note in particular that the type Big_Real has aspect Real_Literal => From_Universal_Image which again is a renaming of From_String; but it is in fact a different From_String. The *ARM* states that this From_String behaves much as Get from Text_IO.Float_IO. As a consequence we can write

```
BR1: Big_Real := 1234.5678;
```

This could initially be held as (12345678/10000) but removing common factors gives (6172839/5000). However, we can also write

```
BR2: Big_Real := 23/37;
```

and this uses the function / declared near the beginning of the package Big_Numbers.Big_Reals and takes two Valid_Big_Integer parameters.

Note that the version of From_Universal_Image with two parameters is used if we declare a named number such as XX: **constant** := 23.0/37.0; followed by a declaration BR3: Big_Real := XX; this preserves the required accuracy for the rational representation.

Note also the pair of functions To_Quotient_String and From_Quotient_String. Thus

```
To_Quotient_String(BR2) = "23 / 37"
```

this uses the function To_String in the package Big_Integers and so takes the default values for Width and Base. The function From_Quotient_String reverses the process.

We finish this section with an example of the RSA algorithm for encryption which illustrates the use of big integers. The mathematics is as follows.

Suppose p and q are typically very large prime numbers. Compute $n = p \times q$ and $m = (p - 1) \times (q - 1)$. Now choose e such that e is less than and relatively prime to m so that $\gcd(e, m) = 1$. It can be shown that there is a unique number d (less than m) such that $e \times d \equiv 1 \bmod m$. The numbers n and e are made public but p, q, and d are kept secret. Note that e is for encrypt and d is for decrypt.

A value v can be encrypted by $c = v^e \bmod n$ and then decrypted (using the secret number d) by the similar formula $v = c^d \bmod n$ which amazingly produces the original number v.

One might think that encryption could be performed by

```
function Encrypt(V, E, N: Big_Integer) return Big_Integer is
   ((V**E) mod N);
```

but that fails to compile because the second parameter of ** has to be of type Integer and not Big_Integer. Another approach might be to expand the exponentiation thus

```
function Encrypt(V, E, N: Big_Integer) return Big_Integer is
   Result: Big_Integer := 1;
begin
   for K in Big_Integer range 1 .. E loop
      Result := Result * M mod N;
   end loop;
   return Result;
end Encrypt;
```

taking the modulus on each iteration prevents numbers getting too large but is slow.

A much better approach is to logically consider the exponent as in binary which means that we can just keep on squaring and only include the terms corresponding to the digits which are one and not zero. For example if N is 53 then we note that

53 = 32 + 16 + 4 + 1. So we compute the powers for 1, 2, 4, 8, 16, 32 and just use those that are required to produce 53. The function then is

```
function Encrypt(V, E, N: Big_Integer) return Big_Integer is
   Result: Big_Integer := 1;
   Exp: Big_Integer := E;
   Term: Big_Integer := V;
begin
   loop
      if Exp mod 2 = 1 then
         Result := Result * Term mod N;          -- include a term
      end if;
      Term := Term * Term mod N;                 -- keep on squaring
      Exp := Exp/2;                              -- and halving
      exit when Exp = 0;
   end loop;
   return Result;
end Encrypt;
```

Decryption uses exactly the same algorithm so we can simply write

```
function Decrypt(C, D, N: Big_Integer) return Big_Integer renames Encrypt;
```

We finish with a couple of examples. As a tiny one suppose $p = 13$ and $q = 23$; then $n = 13\times23 = 299$; $m = 12\times22 = 264$. Now we can take $e = 5$ and then we find $d = 53$. To encrypt the letter B we can take the Latin-1 value which is 66. The encrypted value is 66^5 mod 299 which is 79. To decrypt we compute 79^{53} mod 299 and luckily this is 66, so it works.

As a more realistic example suppose the primes p and q are 131_071 and 524_287 so the public key $n = p \times q$ is 68_718_821_377 and $m = (p - 1) \times (q - 1)$ is 68_718_166_020. We then need to choose e relatively prime to m. One possibilty is e = 1_234_567 and then compute d which is 50_048_155_783. If we now have a numeric message which is 987_654_321, it turns out that the encrypted message is 57_688_571_088.

Note that p and q are the Mersenne primes $M_{17} = 2^{17}-1$ and $M_{19} = 2^{19}-1$. Another interesting example is to use M_{67} which is the product of the primes 193,707,721 and 761,838,257,287. See also Section 22.8 and the examples on the website. For details of the mathematics of the RSA algorithm see *Nice_Numbers* by the author.

23.7 Input and output

Unlike many other languages, Ada does not have any intrinsic features for input–output. Instead existing general features such as subprogram overloading and generic instantiation are used. This has the merit of enabling different input–output packages to be developed for different application areas without affecting the language itself. On the other hand, this approach can lead to a consequential risk of anarchy in this area; the (older) reader may recall that this was one of the reasons for the downfall of Algol 60. In order to prevent such anarchy the *ARM* defines standard packages for input–output.

Two categories of input–output are recognized and we can refer to these as binary and text respectively. As an example consider

```
I: Integer := 75;
```

We can output the binary image of I onto file F by

```
Write(F, I);
```

and the pattern transmitted might be (on a 16-bit machine)

```
0000 0000 0100 1011
```

In fact the file can be thought of as essentially an array of the type Integer. On the other hand we can output the text form of I by

```
Put(F, I);
```

and the pattern transmitted might then be

```
0011 0111 0011 0101
```

which is the representation of the characters '7' and '5'. In this case the file can be thought of as an array of the type Character.

Input–output of the binary category is in turn subdivided into sequential and direct access and is provided by distinct generic packages Ada.Sequential_IO and Ada.Direct_IO respectively. Text input–output (which is always sequential) is provided by the nongeneric package Ada.Text_IO.

These forms of input–output result in uniform files containing objects all of the same type. This is often too restrictive and accordingly a general stream mechanism is defined which allows files of arbitrary types; these are implemented using the facilities of the package Ada.Streams.

There is also a package Ada.IO_Exceptions which contains the declarations of the exceptions used by the other packages. We will deal here first with Sequential_IO and then with Direct_IO and consider Text_IO and Streams in subsequent sections. (There is also a package Ada.Storage_IO; this is rather different since it is concerned with memory buffers rather than files and is discussed at the end of this section.)

The specification of the package Ada.Sequential_IO is as follows

```
with Ada.IO_Exceptions;
generic
   type Element_Type(<>) is private;
package Ada.Sequential_IO
     with Global => in out synchronized is
   type File_Type is limited private;
   type File_Mode is (In_File, Out_File, Append_File);

   -- File management

   procedure Create(File: in out File_Type;
                    Mode: in File_Mode := Out_File;
                    Name: in String := "";
                    Form: in String := "");
```

```
      procedure Open(File: in out File_Type;
                     Mode: in File_Mode;
                     Name: in String;
                     Form: in String := "");
      procedure Close(File: in out File_Type);
      procedure Delete(File: in out File_Type);
      procedure Reset(File: in out File_Type; Mode: in File_Mode);
      procedure Reset(File: in out File_Type);
      function Mode(File: in File_Type) return File_Mode;
      function Name(File: in File_Type) return String;
      function Form(File: in File_Type) return String;
      function Is_Open(File: in File_Type) return Boolean;
      procedure Flush(File: in File_Type)
         with Global => overriding in out File;
      -- Input and output operations
      procedure Read(File: in File_Type; Item: out Element_Type)
         with Global => overriding in out File;
      procedure Write(File: in File_Type; Item: in Element_Type)
         with Global => overriding in out File;
      function End_Of_File(File: in File_Type) return Boolean;
      -- Exceptions
      Status_Error: exception renames IO_Exceptions.Status_Error;
      ...
      Data_Error: exception renames IO_Exceptions.Data_Error;
     -- and nested packages Wide_ and Wide_Wide_File_Names
   private
      ...  -- implementation dependent
   end Ada.Sequential_IO;
```

The package has a generic parameter giving the type of element to be manipulated. Note that limited types cannot be handled since the generic formal parameter is private rather than limited private. However, the formal parameter is indefinite and so the actual type can be an indefinite type such as String or a class wide type.

Externally a file has a name which is a string but internally we refer to a file by using objects of type File_Type. An open file also has an associated value of the enumeration type File_Mode; there are three possible values, In_File gives read-only access, Out_File gives write-only access starting at the beginning of the file, Append_File gives write-only access starting at the end of the file. Read-write access is not allowed for sequential files. The mode of a file is originally set when the file is opened or created but can be changed later by a call of the procedure Reset. Manipulation of sequential files is done using various subprograms whose behaviour is much as expected.

As an example, suppose we have a file containing measurements of various populations and that we wish to compute the sum of these measurements. The populations are recorded as values of type Integer and the name of the file is "Census 47". (The actual conventions for the external file name are dependent upon the implementation.) The computed sum is to be written onto a new file to be called "Total 47". This could be done by the following program

```
with Ada.Sequential_IO; use Ada;
procedure Compute_Total_Population is
   package Integer_IO is new Sequential_IO(Integer);
   use Integer_IO;
   Data_File: File_Type;
   Result_File: File_Type;
   Value: Integer;
   Total: Integer := 0;
begin
   Open(Data_File, In_File, "Census 47");
   while not End_Of_File(Data_File) loop
      Read(Data_File, Value);
      Total := Total + Value;
   end loop;
   Close(Data_File);
   -- now write the result
   Create(Result_File, Name => "Total 47");
   Write(Result_File, Total);
   Close(Result_File);
end Compute_Total_Population;
```

We start by instantiating the generic package Sequential_IO with the actual parameter Integer. It is convenient to add a use clause.

The file with the data to be read is referred to via the object Data_File and the output file is referred to via the object Result_File of the type File_Type. This type is limited private and enables the implementation to use techniques similar to those in Section 12.6 where we discussed the example of the key manager.

The call of Open establishes the object Data_File as referring to the external file "Census 47" and sets its mode as read-only. The external file is then opened for reading and positioned at the beginning.

We then obey the loop statement until the function End_Of_File indicates that the end of the file has been reached. On each iteration the call of Read copies the item into Value and positions the file at the next item. Total is then updated. When all the values on the file have been read, it is closed by a call of Close.

The call of Create creates a new external file named "Total 47" and establishes Result_File as referring to it and sets its mode by default to write-only. We then write our total onto the file and then close it.

The procedures Create and Open have a further parameter Form; this is provided so that auxiliary implementation dependent information can be specified; the default value is a null string so its use is not mandatory. Note that the Name parameter of Create also has a default null value; such a value corresponds to a temporary file. The procedures Close and Delete both close the file and thereby sever the connection between the file variable and the external file. The variable can then be reused for another file. Delete also destroys the external file if the implementation so allows.

The overloaded procedures Reset cause a file to be repositioned as for Open. Reset can also change the access mode so that, for example, having written a file, we can now read it. Or perhaps having read it, we can now write further items to it starting at the end. The procedure Flush can be used to clear internal buffers.

The functions Mode, Name, and Form return the corresponding properties of the file. The function Is_Open indicates whether the file is open; that is, it indicates whether the file variable is associated with an external file or not.

The procedures Read and Write automatically reposition the file ready for a subsequent call so that the file is processed sequentially. The function End_Of_File only applies to an input file and returns true if there are no more elements to be read.

If we do something wrong then one of the exceptions in the package Ada.IO_Exceptions will be raised. This package is as follows

```
package Ada.IO_Exceptions
     with Pure is
   Status_Error, Mode_Error, Name_Error, Use_Error,
              Device_Error, End_Error, Data_Error, Layout_Error: exception;
end Ada.IO_Exceptions;
```

This is an example of a package that does not need a body. The various exceptions are declared in this package rather than in Sequential_IO so that the same exceptions apply to all instantiations of Sequential_IO. If they were inside Sequential_IO then each instantiation would create different exceptions and this would be rather more inconvenient in the case of a program manipulating files of various types since general purpose exception handlers would need to refer to all the instances. The renaming declarations on the other hand enable the exceptions to be referred to without use of the name IO_Exceptions.

The following brief summary gives the general flavour of the circumstances giving rise to each exception

Status_Error	File is open when expected to be closed or vice versa.
Mode_Error	File of wrong mode, for example, In_File when should be Out_File.
Name_Error	Something wrong with Name parameter of Create or Open.
Use_Error	Various, such as unacceptable Form parameter or trying to print on card reader.
Device_Error	Physical device broken or not switched on.
End_Error	Malicious attempt to read beyond end of file.
Data_Error	Read or Get (see next section) cannot interpret data as value of desired (sub)type.
Layout_Error	Something wrong with layout in Text_IO (see next section) or Put overfills string parameter.

Note in particular that Data_Error will be raised if the type is wrong (an integer perhaps when a floating point number was expected) or if the value is just out of range (zero perhaps when a value of subtype Positive was expected).

For fuller details of which exception is actually raised in various circumstances consult the *ARM* and the documentation for the implementation concerned.

We continue by considering the package Ada.Direct_IO which is very similar to Ada.Sequential_IO but gives us more flexibility by enabling us to manipulate the file position directly.

As mentioned earlier a file can be considered as a one-dimensional array. The elements in the file are ordered and each has an associated positive index. This ranges from 1 to an upper value which can change since elements can be added to the end of the file. Not all elements necessarily have a defined value in the case of a direct file as we shall see.

Associated with a direct file is a current index which indicates the position of the next element to be transferred. When a file is opened or created this index is set to 1 so that the program is ready to read or write the first element. The main difference between sequential and direct input–output is that in the sequential case this index is implicit and can only be altered by calls of Read, Write, and Reset whereas in the direct case, the index is explicit and can be directly manipulated.

Note that, unlike Sequential_IO, the generic parameter takes the definite form

```
generic
   type Element_Type is private;
package Ada.Direct_IO
   with Global => in out synchronized is ...
```

and so the actual parameter must also be definite; this ensures that the values occupy the same space so that indexing is straightforward.

The extra facilities of Direct_IO are as follows. The enumeration type File_Mode has a value Inout_File (and not Append_File) so that read–write access is possible; this is also the default mode when a new file is created (thus the Mode parameter of Create has a different default for direct and sequential files). The type and subtype

```
type Count is range 0 .. implementation-defined;
subtype Positive_Count is Count range 1 .. Count'Last;
```

are introduced so that the current index can be referred to and finally there are various extra subprograms whose specifications are as follows

```
procedure Read(File: in File_Type; Item: out Element_Type;
                      From: in Positive_Count);
procedure Write(File: in File_Type; Item: in Element_Type;
                      To: in Positive_Count);
procedure Set_Index(File: in File_Type; To: in Positive_Count);
function Index(File: in File_Type) return Positive_Count;
function Size(File: in File_Type) return Count;
```

The extra overloadings of Read and Write first position the current index to the value given by the third parameter and then behave as before. A call of Index returns the current index value; Set_Index sets the current index to the given value and a call of Size returns the number of elements in the file. Note that a file cannot have holes in it; all elements from 1 to Size exist although some may not have defined values.

Note that the three procedures all have **with** Global => **overriding in out** File.

As an illustration of the manipulation of these positions we can alter our example to use Direct_IO and we can then write the total population onto the end of an existing file called "Totals". The last few statements then become

```
      -- now write the result
      Open(Result_File, Out_File, "Totals");
      Set_Index(Result_File, Size(Result_File) + 1);
      Write(Result_File, Total);
      Close(Result_File);
   end Compute_Total_Population;
```

Note that if we set the current index well beyond the end of the file and then write to it, the result will be to add several undefined elements to the file and then finally the newly written element.

Note also that the language does not define whether it is possible to write a file with Sequential_IO and then read it with Direct_IO or vice versa. This depends upon the implementation.

Finally, there is a generic package Ada.Storage_IO which performs input and output to and from a memory buffer. The one generic parameter Element_Type is definite like that for Direct_IO. The package declares

```
Buffer_Size: Storage_Count := implementation-defined;
subtype Buffer_Type is Storage_Array(1 .. Buffer_Size);

procedure Read(Buffer: in Buffer_Type; Item: out Element_Type);
procedure Write(Buffer: out Buffer_Type; Item: in Element_Type);
```

where the types Storage_Count and Storage_Array are from System.Storage_Elements (see Section 25.3). The value of Buffer_Size gives the size in storage elements of the buffer required to hold one value of the type Element_Type. The procedures Read and Write transfer values of the type to and from the buffer.

Exercise 23.7

1 Write a generic library procedure to copy a file to another file but with the elements in reverse order. Pass the external names as parameters.

23.8 Text input–output

Text input–output, which we met in Chapter 4, is the more familiar form and provides two overloaded procedures Put and Get to transmit values as streams of characters as well as various other subprograms such as New_Line for layout control. In addition, the concept of current default files is introduced so that every call of the various subprograms need not tiresomely repeat the file name. Thus if F is the current default output file, we can write

```
Put("Message");
```

rather than

```
Put(F, "Message");
```

There are two current default files, one of mode Out_File for output, and one of mode In_File for input. Ada.Text_IO is like Ada.Sequential_IO and also has a mode Append_File but not Inout_File.

When we enter our program these two files are set to standard default files which are automatically open; we can assume that these are attached to convenient external files such as keyboard and screen or even a card reader and line printer. If we wish to use other files and want to avoid repeating the file names in the calls of Put and Get then we can change the default files to refer to our other files. We can also set them back to their original values. This is done with subprograms

```
procedure Set_Output(File: in File_Type);
function Standard_Output return File_Type;
function Current_Output return File_Type;

type File_Access is access constant File_Type;
function Standard_Output return File_Access;
function Current_Output return File_Access;
```

with similar subprograms for input. The procedure Set_Output enables us to change the current default output file to the file passed as parameter, the two functions Standard_Output return the initial default output file and the two functions Current_Output return the current default output file (directly or indirectly using the type File_Access respectively).

Thus we could bracket a fragment of program with

```
New_File: File_Type;
...
Open(New_File, ... );
Set_Output(New_File);
... -- use Put
Set_Output(Standard_Output);
```

so that having used the file New_File, we can reset the default file to its standard value.

The more general case is where we wish to reset the default file to its previous value which may, of course, not be the standard value. The reader may recall that the type File_Type is limited private and therefore values cannot be assigned; hence the version of Current_Output returning an access value is useful. We could write

```
Old_File_Ref: constant File_Access := Current_Output;
...
Set_Output(New_File);
... -- use Put
Set_Output(Old_File_Ref.all);
```

An alternative approach is to note that the result of a function can be renamed

```
Old_File: File_Type renames Current_Output;
...
Set_Output(Old_File);
```

although this is less flexible since the value cannot be stored.

For convenience a default error file is also provided. This is intended for the user to output messages; it has an initial standard value and there are similar subprograms to manipulate this error file. A typical use might be in an exception handler

```
exception
  when Event: others =>
    Put(Current_Error, Exception_Message(Event));
```

which thereby avoids cluttering the file being used for the current normal output.

The full specification of Ada.Text_IO is rather long and so only the general form is reproduced here. And to save space rather than spelling out the global etc annotations, brief comments are inserted thus for an initial group of items: GoioF is Global => **overriding in out** File, Gioa is Global => **in out all**, Gia is Global => **in all**, and N is Nonblocking.

```
with Ada.IO_Exceptions;
package Ada.Text_IO
    with Global => in out synchronized is
  type File_Type is limited private;
  type File_Mode is (In_File, Out_File, Append_File);

  type Count is range 0 .. implementation-defined;
  subtype Positive_Count is Count range 1 .. Count'Last;
  Unbounded: constant Count := 0;       -- line and page length

  subtype Field is Integer range 0 .. implementation-defined;
  subtype Number_Base is Integer range 2 .. 16;
  type Type_Set is (Lower_Case, Upper_Case);

  -- File management

  ...   -- Create, Open, Close, Delete, Reset, Mode, Name,
  ...   -- Form and Is_Open as for Sequential_IO

  -- Control of default input and output files

  procedure Set_Output(File: in File_Type);
  function Standard_Output return File_Type;
  function Current_Output return File_Type;

  type File_Access is access constant File_Type;
  function Standard_Output return File_Access;
  function Current_Output return File_Access;

  ...   -- Similarly for input and error file

  -- Buffer control

  procedure Flush(File: in File_Type);      -- GoioF
  procedure Flush;                          -- Gioa

  -- Specification of line and page lengths
  -- also with File parameter

  procedure Set_Line_Length(To: in Count);        -- Gioa
```

```
procedure Set_Page_Length(To: in Count);          -- Gioa
function Line_Length return Count;                -- Gia
function Page_Length return Count;                -- Gia

-- Column, line and page control
-- also with File parameter

procedure New_Line(Spacing: in Positive_Count := 1);    -- Gia
procedure Skip_Line(Spacing: in Positive_Count := 1);   -- Gia
function End_Of_Line return Boolean;
procedure New_Page;                               -- Gioa
procedure Skip_Page;                              -- Gioa
function End_Of_Page return Boolean;
function End_Of_File return Boolean;
procedure Set_Col(To: in Positive_Count);         -- Gioa
procedure Set_Line(To: in Positive_Count);        -- Gioa
function Col return Positive_Count;               -- Gia
function Line return Positive_Count;              -- Gia
function Page return Positive_Count;              -- Gia

-- Character input-output

procedure Get(File: in File_Type; Item: out Character);   -- GoioF
procedure Get(Item: out Character);                       -- Gioa
procedure Put(File: in File_Type; Item: in Character);    -- GoioF
procedure Put(Item: in Character);                        -- Gioa
procedure Look_Ahead(Item: out Character;
                     End_Of_Line: out Boolean);           -- Gioa
procedure Get_Immediate(Item: out Character);             -- Gioa
procedure Get_Immediate(Item: out Character;
                        Available: out Boolean);          -- Gioa

-- String input-output

procedure Get(Item: out String);                          -- Gioa
procedure Put(Item: in String);                           -- Gioa
procedure Get_Line(Item: out String; Last: out Natural);  -- Gioa
function Get_Line return String;                          -- Gioa
procedure Put_Line(Item: in String);                      -- Gioa

-- Generic package for input-output of integer types

generic
  type Num is range <>;
package Integer_IO is
  Default_Width: Field := Num'Width;
  Default_Base: Number_Base := 10;
  procedure Get(Item: out Num; Width: in Field := 0);     -- Gioa
  procedure Put(Item: in Num;
                Width: in Field := Default_Width;
                Base: in Number_Base := Default_Base);    -- Gioa
  procedure Get(From: in String; Item: out Num;
                Last: out Positive);                      -- N
```

```
   procedure Put(To: out String;
                 Item: in Num;
                 Base: in Number_Base := Default_Base);   -- N
end Integer_IO;

generic
   type Num is mod <>;
package Modular_IO is
   ...   -- then as for Integer_IO
end Modular_IO;

-- Generic packages for input-output of real types

generic
   type Num is digits <>;
package Float_IO is
   Default_Fore: Field := 2;
   Default_Aft: Field := Num'Digits-1;
   Default_Exp: Field := 3;

   procedure Get(Item: out Num; Width: in Field := 0);
   procedure Put(Item: in Num;
                 Fore: in Field := Default_Fore;
                 Aft: in Field := Default_Aft;
                 Exp: in Field := Default_Exp);
   procedure Get(From: in String;
                 Item: out Num;
                 Last: out Positive);
   procedure Put(To: out String;
                 Item: in Num;
                 Aft: in Field := Default_Aft;
                 Exp: in Field := Default_Exp);
end Float_IO;

generic
   type Num is delta <>;
package Fixed_IO is
   Default_Fore: Field := Num'Fore;
   Default_Aft: Field := Num'Aft;
   Default_Exp: Field := 0;
   -- then as for Float_IO
end Fixed_IO;

generic
   type Num is delta <> digits <>;
package Decimal_IO is
   ...   -- then as for Fixed_IO
end Decimal_IO;

-- Generic package for input-output of enumeration types

generic
   type Enum is (<>);
```

```
   package Enumeration_IO is
      Default_Width: Field := 0;
      Default_Setting: Type_Set := Upper Case;

      procedure Get(Item: out Enum);                    -- and so on!!
      procedure Put(Item: in Enum;
                    Width: in Field := Default_Width;
                    Set: in Type_Set := Default_Setting);

      procedure Get(From: in String;
                    Item: out Enum;
                    Last: out Positive);

      procedure Put(To: out String;
                    Item: out Enum;
                    Set: in Type_Set := Default_Setting);

   end Enumeration_IO;

   -- Exceptions

   Status_Error: exception renames IO_Exceptions.Status_Error;
   ...
   Layout_Error: exception renames IO_Exceptions.Layout_Error;

    -- and nested packages Wide_ and Wide_Wide_File_Names
private
   ...  -- implementation dependent
end Ada.Text_IO;
```

The types File_Type and File_Mode and the various file management procedures are similar to those for Sequential_IO since text files are of course sequential in nature.

Procedures Put and Get occur in two forms for characters and strings, one with the file and one without; both are shown only for type Character.

In the case of type Character, a call of Put just outputs that character; for type String a call of Put outputs the characters of the string.

A problem arises in the case of numeric and enumeration types since there is not a fixed number of such types. This is overcome by the use of internal generic packages for each category. Thus for integer input–output we instantiate the package Integer_IO with the appropriate type thus

```
type My_Integer is range -1E6 .. +1E6;
...
package My_Integer_IO is new Integer_IO(Num => My_Integer);
use My_Integer_IO;
```

However, as noted in Chapter 4, there are nongeneric equivalents to Integer_IO and Float_IO corresponding to instantiations with the predefined types Integer, Long_Integer, Float and so on. They have names such as

```
Ada.Integer_Text_IO               -- for Integer
Ada.Long_Integer_Text_IO          -- for Long_Integer
Ada.Float_Text_IO                 -- for Float
```

For integer output, Put occurs in three forms, one with the file, one without the file, and one with a string as the destination; only the last two are shown.

In the case of Put to a file, there are two format parameters Width and Base which have default values provided by the variables Default_Width and Default_Base. The default width is initially Num'Width which gives the smallest field which is adequate for all values of the subtype expressed with base 10 (including a leading space or minus). Base 10 also happens to be the initial default base. These default values can be changed by the user by directly assigning new values to the variables Default_Width and Default_Base (they are directly visible); remember that a default parameter is re-evaluated on each call requiring it and so the default obtained is always the current value of these variables. The integer is output as an integer literal without underlines and leading zeros but with a preceding minus sign if negative. It is padded with leading spaces to fill the field width specified; if the field width is too small, it is expanded as necessary. Thus a default width of 0 results in the field being the minimum to contain the literal. If base 10 is specified explicitly or by default, the value is output using the syntax of decimal literal; if the base is not 10, the syntax of based literal is used.

The attribute Width deserves attention. It is a property of the subtype of the actual generic type parameter and not of the type. Thus the default format is appropriate to the range as the user sees it and not to that of the underlying machine type on which it is based. This is important for portability. So in the case of My_Integer, the attribute has the value 8 (7 digits for one million plus space or sign).

The general effect is shown by the following sequence of statements where the output is shown in a comment. The quotes delimit the output and s designates a space. We start with the initial default values for the format parameters.

```
X: My_Integer := 1234;
...
Put(X);                  -- "ssss1234"
Put(X, 5);               -- "s1234"
Put(X, 0);               -- "1234"
Put(X, Base => 8);       -- "s8#2322#"
Put(X, 11, 8);           -- "ssss8#2322#"
Default_Base := 8;
Put(X);                  -- "s8#2322#"
```

In the case of Put to a string, the field width is taken as the length of the string. If this is too small, then Layout_Error is raised. Put to strings is useful for building up strings from various items and perhaps editing them before sending them to a file. It will be found that slices are useful for this sort of manipulation.

Similar techniques are used for real types using packages Float_IO and Fixed_IO. A value is output as a decimal literal without underlines and leading zeros but with a preceding minus sign if negative. If Exp is zero, then there is no exponent and the format consists of Fore characters before the decimal point and Aft after the decimal point. If Exp is nonzero, then a signed exponent in a field of Exp characters is output after a letter E with leading zeros if necessary; the exponent value is such that only one significant digit occurs before the decimal point. If the Fore or Exp parts of the field are inadequate, then they are expanded as necessary. Base 10 is always used and the value is rounded to the size of Aft specified. (Note that if Aft is foolishly given as 0 then it is taken to be 1.)

The initial default format parameters for floating point types are 2, Num'Digits–1, and 3; this gives an exponent form with a space or minus sign plus single digit before the decimal point, Num'Digits–1 digits after the decimal point and a two-digit exponent. The corresponding parameters for fixed point types (including decimal types) are Num'Fore, Num'Aft, and 0; this gives a form without an exponent and the attributes give the smallest field such that all values of the type can be expressed with appropriate precision. (As noted in Section 17.5, the Aft attribute is always determined by Delta even if the default *small* is overridden.)

Enumeration types use Enumeration_IO. The default field width is zero. If the field has to be padded then the extra spaces go after the value and not before as with numeric types. Upper case is normal, but lower case may be specified. A value of a character type which is a character literal is output in single quotes.

Note the subtle distinction between Put defined directly for the type Character and for enumeration values.

```
Text_IO.Put('X');
```

outputs the single character X, whereas

```
package Char_IO is new Text_IO.Enumeration_IO(Character);
...
Char_IO.Put('X');
```

outputs the character X between single quotes.

Input using Get works in an analogous way; a call of Get always skips line and page terminators. In the case of the type Character the next character is read. In the case of the type String, the procedure Get reads the exact number of characters as determined by the actual parameter. In the case of enumeration types, leading blanks (spaces or horizontal tabs) are also skipped; input is terminated by a character which is not part of the value or by a line terminator. Numeric types normally have the same behaviour but they also have an additional and optional Width parameter and if this has a value other than zero, then reading stops after this number of characters including skipped blanks. In the case of Get where the source is a string rather than a file, the value of Last indexes the last character read; the end of the string behaves as the end of a file.

The allowed form of data for reading an enumeration value is an identifier (case of letters being ignored), or a character literal in single quotes. The allowed form for an integer value is first and optionally a plus or minus sign and then according to the syntax of an integer literal which may be a based literal and possibly have an exponent (see Section 5.4). The allowed form for a real value is similarly an optional sign followed by a real literal (one with a radix point in it); however leading or trailing zeros and the point may be omitted and so an integer literal is also an acceptable form. If the data item is not of the correct form or not a value of the subtype Num then Data_Error is raised. The collector of Ada curiosities will note that Put cannot output integer based forms where the base is 10 such as

```
10#41#
```

although Get can read them. Furthermore, Put cannot output any real based forms at all although Get can read them. But Get can read whatever Put can write.

A text file is considered as a sequence of lines. The characters in a line have a column position starting at 1. The line length on output can be fixed or variable. A fixed line length is appropriate for the output of tables, a variable line length for dialogue. The line length can be changed within a single file. It is initially not fixed. The lines in turn are similarly grouped into pages starting at page 1.

On output a call of Put will result in all the characters going on the current line starting at the current position in the line. If, however, the line length is fixed and the characters cannot fit in the remainder of the line, a new line is started and all the characters are placed on that line starting at the beginning. If they still will not fit, Layout_Error is raised. If the length is not fixed, the characters always go on the end of the current line.

The layout may be controlled by various subprograms. In some cases they apply to both input and output files; in these cases if the file is omitted then it is taken to apply to the output case and the default output file is assumed. In most cases, a subprogram only applies to one direction and then omitting the file naturally gives the default in that direction.

The function Col returns the current position in the line and the procedure Set_Col sets the position to the given value. A call of Set_Col never goes backwards. On output extra spaces are produced and on input characters are skipped. If the parameter of Set_Col equals the current value of Col then there is no effect; if it is less then a call of New_Line or Skip_Line is implied.

The procedure New_Line (output only) outputs the given number of newlines (default 1) and resets the current column to 1. Spare positions at the end of a line are filled with spaces. The procedure Skip_Line (input only) similarly moves on the given number of lines (default 1) and resets the current column. The function End_Of_Line (input only) returns True if we have reached the end of a line.

The function Line_Length (output only) returns the current line length if it is fixed and zero if it is not. The procedure Set_Line_Length (output only) sets the line length fixed to the given value; a value of zero indicates that it is not to be fixed.

There are also similar subprograms for the control of lines within pages. These are Line, Set_Line, New_Page, Skip_Page, End_Of_Page, Page_Length, and Set_Page_Length. Finally, the function Page returns the current page number from the start of the file. There is no Set_Page.

The procedure Put_Line and function Get_Line are particularly appropriate for manipulating whole lines. A call of Put_Line outputs the string and then moves to the next line (by calling New_Line). A call of Get_Line reads characters until the end of the line is reached and then moves to the next line (by calling Skip_Line); it returns a string comprising the characters read. Successive calls of Get_Line and Put_Line therefore manipulate whole lines.

To copy a file and add the string "--" to each line we simply write

```
while not End_Of_File loop
   Put_Line("--" & Get_Line);
end loop;
```

which assumes default files throughout.

There is also a procedure Get_Line whose behaviour is somewhat odd. A call of Get_Line reads successive characters into the string until the end of the string or the end of the line is encountered; in the latter case it then moves to the next line (by calling Skip_Line); Last indexes the last character moved into the string. However, the behaviour of Get_Line is curious when the string is exactly the right length to accommodate the remaining characters on the line – it doesn't move to the next line! So, given a series of lines of length 80, successive calls of Get_Line with a string of length 80 (bounds 1 .. 80) return alternately lines of 80 characters and null strings (or in other words the value of Last is alternately 80 and 0). This unhelpful behaviour can be overcome by using a string of length 81 or calling Skip_Line ourselves after each call of Get_Line.

The above simple example using the procedure Get_Line becomes the rather laborious

```
S: String(1 .. 100);
N: Natural;
...
while not End_Of_File loop
   Get_Line(S, N);
   Put_Line("--" & S(1 .. N));
end loop;
```

but this only works if the lines are all less than 100 characters. The function Get_Line was added in Ada 2005 in order to avoid the necessity to write this curious code.

The subprograms Flush, Look_Ahead, and Get_Immediate are useful for interactive applications. Flush applies to an output file and flushes any internal buffer.

Get_Immediate reads the next character (control or graphic). There are two versions: that with the extra parameter Available returns false if a character is not immediately available; that without the extra parameter waits for a character if necessary. (The current column, line, and page number are not affected.)

Look_Ahead gets the next character without consuming it. However, it never moves to the next line and so the parameter End_Of_Line indicates whether the end of the line had been reached; Item is undefined if End_Of_Line is set true. Look_Ahead will be found vital for writing subprograms with behaviour corresponding to the predefined Get for the integer, real, and enumeration types which all look ahead to see if the literal value has finished but do not remove the terminating character. Its use is illustrated in Program 3.

The package Text_IO may seem somewhat elaborate but for simple output all we need is Put and New_Line and these are very straightforward.

The packages Text_IO.Unbounded_IO and Text_IO.Bounded_IO for bounded and unbounded strings were briefly described in Section 23.3. The packages Text_IO.Complex_IO and Text_IO.Editing for complex and decimal types are described in Chapter 26.

Finally, there are also packages Wide_Text_IO and Wide_Wide_Text_IO which are identical to Text_IO except that they work in terms of the types Wide_Character and Wide_String, and Wide_Wide_Character and Wide_Wide_String respectively instead of Character and String.

Exercise 23.8

1 What do the following output? Assume the initial values for the default parameters.

(a) Put("Fred");
(b) Put(120);
(c) Put(120, 8);
(d) Put(120, 0);
(e) Put(-120, 0);
(f) Put(120, 8, 8);
(g) Put(-38.0);
(h) Put(0.07, 6, 2, 2);
(i) Put(3.14159, 1, 4);
(j) Put(9_999_999_999.9, 1, 1, 1);

Assume that the real values are of a type with **digits** = 6 and the integer values are of a type with 16 bits.

2 Write a body for the package Simple_IO of Section 2.2. Ignore exceptions.

23.9 Streams

As we remarked earlier, the packages Sequential_IO and Direct_IO only permit the creation of files whose elements are all of the same type. This is often too restrictive and so additional stream facilities are provided which allow the creation of quite arbitrary heterogeneous files.

The stream facilities are quite general and can be used for other purposes as well as creating files. They are provided by the packages

```
Ada.Streams                  Ada.Streams.Storage
Ada.Streams.Stream_IO        Ada.Streams.Storage.Bounded
Ada.Text_IO.Text_Streams     Ada.Streams.Storage.Unbounded
```

The parent package Ada.Streams is as follows

```
package Ada.Streams
    with Pure, Nonblocking => False is
  type Root_Stream_Type is abstract tagged limited private
    with Preelaborable_Initialization;

  type Stream_Element is mod implementation-defined;
  type Stream_Element_Offset is range implementation-defined;
  subtype Stream_Element_Count is
    Stream_Element_Offset range 0..Stream_Element_Offset'Last;
  type Stream_Element_Array is
    array (Stream_Element_Offset range <>) of aliased Stream_Element;

  procedure Read(Stream: in out Root_Stream_Type;
                 Item: out Stream_Array;
                 Last: out Stream_Element_Offset) is abstract;

  procedure Write(Stream: in out Root_Stream_Type;
                  Item: in Stream_Element_Array) is abstract;

private
  ... -- not specified by the language
end Ada.Streams;
```

It defines an abstract type Root_Stream_Type from which all streams are derived. There are two abstract primitive procedures Read and Write which enable streams to be manipulated in terms of stream elements. The type Stream_Element is a modular type and there are also types Stream_Element_Offset and Stream_Element_Array exactly analogous to the types Storage_Element and so on of the package System.Storage_Elements (see Section 25.3). Indeed stream elements and storage elements may often be the same size but need not. We will not consider the underlying mechanism in more detail but simply show how streams can be used with files.

The packages Ada.Streams.Storage and its two child packages are new in Ada 2022. We will discuss them at the end of this section.

Files declared by the package Streams.Stream_IO may be processed sequentially using the stream mechanism. Such files can be created, opened, and closed in the usual manner much as in Sequential_IO.

Moreover, the package Ada.Text_IO.Text_Streams also declares an access type and the function Stream thus

```
with Ada.Streams;
package Ada.Text_IO.Text_Streams
     with Global => in out synchronized is

   type Stream_Access is access all Root_Stream_Type'Class;
   function Stream(File: in File_Type) return Stream_Access;

end Ada.Text_IO.Text_Streams;
```

The function Stream takes a stream file and returns an access to the stream associated with the file. The reading and writing of streams is done with attributes T'Read, T'Write, T'Input, and T'Output. These attributes are predefined for all nonlimited types. They can be replaced by using an attribute definition clause (see Section 25.1) and can also be supplied for limited types. We will first consider T'Read and T'Write; T'Input and T'Output (which are especially relevant for arrays, tagged records, and discriminated records) will be considered later.

The attributes Read and Write take parameters denoting the stream and the element of type T thus

```
procedure T'Write(Stream: not null access Root_Stream_Type'Class;
                                                     Item: in T);
procedure T'Read(Stream: not null access Root_Stream_Type'Class;
                                                    Item: out T);
```

As a simple example, suppose we wish to write a mixture of integers, month names, and dates where the type Date is the familiar

```
type Date is
   record
      Day: Integer range 1 .. 31;
      Month: Month_Name;
      Year: Integer;
   end record;
```

Having created a file in the normal way, we then call the Write attributes for the values to be written to the stream. Thus

```
use Streams.Stream_IO;
Mixed_File: File_Type;
S: Stream_Access;
...
Create(Mixed_File);
S := Stream(Mixed_File);
...
Date'Write(S, Some_Date);
Integer'Write(S, Some_Integer);
Month_Name'Write(S, This_Month);
```

All files created this way are of the same type. Note also that they are binary files. A file written in this way can be read back in a similar manner, but if we attempt to read things in the wrong order and thus with the inappropriate subprogram then we will get a funny value or Data_Error. The attribute Valid discussed in Section 25.2 is useful in such circumstances.

The predefined Write attribute for a simple record such as Date simply calls the attributes for the components in order. So conceptually it is

```
procedure Date'Write(Stream: not null access Root_Stream_Type'Class;
                          Item: in Date) is
begin
   Integer'Write(Stream, Item.Day);
   Month_Name'Write(Stream, Item.Month);
   Integer'Write(Stream, Item.Year);
end;
```

If we wish to output the month as an integer then we can redefine the attribute thus

```
procedure Date_Write(Stream: not null access Root_Stream_Type'Class;
                          Item: in Date) is
begin
   Integer'Write(Stream, Item.Day);
   Integer'Write(Stream, Month_Name'Pos(Item.Month) + 1);
   Integer'Write(Stream, Item.Year);
end Date_Write;

for Date'Write use Date_Write;
```

and then the statement

```
Date'Write(S, Some_Date);
```

will use the new format for the output of dates. Similar facilities apply to input and so in order to read the file back we would have to declare the corresponding version of Date'Read to read the month as an integer and convert to the corresponding value of Month_Name. If we wish to change the format of all months and not just those in dates, then we simply redefine Month_Name'Write and this has the indirect effect of also changing the output of dates.

We could use aspect specifications and declare Date_Write and Date_Read first and then declare the type Date as

```
type Date is
   record
      Day: Integer range 1 .. 31;
      Month: Month_Name;
      Year: Integer;
   end record
   with Write => Date_Write, Read => Date_Read;
```

but maybe this seems a bit like putting the cart before the horse.

Note that the attributes T'Read and T'Write can only be overridden (or declared in the case of limited types) in the same package specification or declarative part as that containing the declaration of the type T. Consequently, they cannot be changed for the predefined types. But they can be changed for types derived from them.

In the case of the predefined limited types, streaming does not make sense for any of them with the exception (curiously) of the type Exception_Occurrence. Accordingly, the package Ada.Exceptions (see Section 15.4) contains the declaration of procedures Read_Exception_Occurrence and Write_Exception_Occurrence and appropriate attribute definition clauses.

In the case of tagged types, the predefined attributes T'Read and T'Write behave rather like predefined equality and are composable. Thus T'Read for a type extension first performs Read on the parent part and then calls the appropriate Read on the additional components. Similar rules apply to T'Write.

The situation is more complex in the case of arrays and records with discriminants since we have to take account of the 'dope' information represented by the bounds and discriminants. (In the case of a record discriminant with defaults, the discriminant is treated as an ordinary component.) This is done with the additional attributes Input and Output. The general idea is that Input and Output process dope information (if any) and then call Read and Write to process the rest of the value. Their profiles are

```
procedure T'Output(Stream: not null access Root_Stream_Type'Class;
                   Item: in T);
function T'Input(Stream: not null access Root_Stream_Type'Class)
                                                         return T;
```

Note that Input is a function since T may be indefinite (such as String) and we may not know the constraints for a particular call.

So for an array the procedure Output first outputs the bounds of the value and then calls Write to output the value itself. Thus

```
Name: String := "String";
...
String'Output(S, Name);     -- outputs bounds
String'Write(S, Name);      -- does not output bounds
```

Note that the attributes Output and Write belong to the types and so it is immaterial whether we write String'Write or use some subtype String_6'Write.

For a record type with discriminants, if it has defaults (is definite) then Output simply calls Write which treats the discriminants as just other components. If there are no defaults then Output first outputs the discriminants and then calls Write to process the remainder of the record.

The above description of T'Input and T'Output applies to the default attributes. They could be redefined to do anything and not necessarily call T'Read and T'Write. Note moreover that Input and Output also exist for elementary types; their defaults just call Read and Write.

There are also attributes T'Class'Output and T'Class'Input for dealing with class wide types. For output, an external representation of the tag is output and then the procedure Output for the specific type is called (by dispatching) in order to output the specific value (which in turn will call Write). Thus

```
procedure T_Class_Output
            (S: not null access Root_Stream_Type'Class; X: in T'Class) is
begin
   if not Is_Descendant_At_Same_Level(X'Tag, T'Tag) then
      raise Tag_Error;
   end if;
   String'Output(S, External_Tag(X'Tag));
   T'Output(S, X);
end T_Class_Output;

for T'Class'Output use T_Class_Output;
```

It is important to note that, in the case of tagged types, streams are designed to work only with types declared at the same accessibility level as the parent type T. The call of Is_Descendant_At_Same_Level in the package Ada.Tags ensures this. See Section 21.7 for the specification of the package Ada.Tags.

The external representation of a tag mentioned above is a string. The functions External_Tag and Internal_Tag also in the package Ada.Tags perform the required conversions. If we try to convert a string which does not represent a tag then Tag_Error is raised. There is also an attribute External_Tag which is such that T'External_Tag is identical to External_Tag(T'Tag). The attribute enables us to set our own representation using an attribute definition clause as we saw in Section 21.7 when we were creating geometrical objects using the function Generic_ Dispatching_Constructor.

A general principle of all stream operations is, of course, that whatever is written can be read back in again by the appropriate reverse operation. So on input the tag is first read and then, according to its value, the corresponding function Input is called by dispatching using the generic constructor. Thus

```
function Dispatching_Input is new Generic_Dispatching_Constructor
                                   (T, Root_Stream_Type'Class, T'Input);

function T_Class_Input(S: not null access Root_Stream_Type'Class)
                                                          return T'Class is
                              -- read tag as string from stream
   The_String: String := String'Input(S);
                              -- convert to a tag
   The_Tag: Tag := Descendant_Tag(The_String, T'Tag);
```

```
begin
                                -- now dispatch to the appropriate function Input
   return Dispatching_Input(The_Tag, S);
end T_Class_Input;

for T'Class'Input use T_Class_Input;
```

The body could of course be written as one giant statement

```
   return Dispatching_Input(Descendant_Tag(String'Input(S), T'Tag), S);
```

but breaking it down hopefully clarifies what is happening.

Note the use of Descendant_Tag rather than Internal_Tag. Although Internal_Tag will do the required conversion and can be used for most types, Descendant_Tag also checks that the tag does indeed correspond to a type descended from the parent type T.

We can use the generic constructor to create our own stream protocols. We could replace T'Class'Input and T'Class'Output or just create our own distinct subsystem. One reason why we might want to use a different protocol is to interface to an existing system such as XML. If we do devise our own protocols then the attribute S'Stream_Size will be useful. It gives the number of bits used in the stream for values of the subtype S. We are guaranteed that it will be a multiple of Stream_Element'Size. So the number of stream elements required will be

```
S'Stream_Size / Stream_Element'Size
```

The attribute can be set by an attribute definition clause in the normal way.

It is also possible to treat a text file as a stream and so mix binary and text in one file. The child package Text_IO.Text_Streams defines a function Stream which takes a file of the type Text_IO.File_Type and returns an access to the corresponding stream. We could then write binary to the current output file by

```
use Text_IO;
S: Text_Streams.Stream_Access := Text_Streams.Stream(Current_Output);
...
Date'Write(S, Today);
```

That concludes a rather brief discussion of streams. They have many other uses. For example, they underlie the mechanism for remote procedure calls in distributed systems; this is briefly alluded to in Section 26.3.

Another feature we have not discussed is that streams can also be processed in an indexed rather than a sequential manner using the notion of stream elements rather like Direct_IO works in terms of typed elements (but the modes include Append_File rather than Inout_File). It is also possible to create streams by the declaration and manipulation of such stream elements; thus a stream could be mapped onto storage in main memory rather than to a file. For details the reader is referred to the *ARM*. And of course there are wide and wide wide versions of these facilities.

We now turn to consider the three packages Ada.Streams.Storage and the child packages Ada.Streams.Storage.Bounded and Ada.Streams.Storage.Unbounded which are new in Ada 2022. They were introduced as sort of companions to text buffers described in Section 23.4. They provide stream implementations that do not make use of any file operations.

The parent package is

```
package Ada.Streams.Storage
  with Pure, Nonblocking is

  type Storage_Stream_Type is
                          abstract new Root_Stream_Type with private;

  function Element_Count(Stream: Storage_Stream_Type)
    return Stream_Element_Count is abstract;

  procedure Clear(Stream: in out Storage_Stream_Type) is abstract;

private
  -- not specified by the language
end Ada.Streams.Storage;
```

The bounded and unbounded packages are similar. The bounded package starts with

```
package Ada.Streams.Storage.Bounded
with Pure, Nonblocking is
  type Stream_Type(Max_Elements: Stream_Element_Count) is
    new Storage.Stream_Type with private
    with Default_Initial Condition => Element_Count(Stream_Type) = 0;
```

wherea the unbounded package starts

```
package Ada.Streams.Storage.Unbounded
with Preelaborated, Nonblocking, Global => in out synchronized is
  type Stream_Type is
    new Storage.Stream_Type with private
    with Default_Initial Condition => Element_Count(Stream_Type) = 0;
```

Both packages then essentially have

```
  procedure Read(Stream: in out Stream_Type;
                 Item: out Stream_Element_Array;
                 Last: out Stream_Element_Offset)

  procedure Write(Stream: in out Stream_Type;
                  Item: in Stream_Element_Array);

  function Element_Count(Stream: Stream_Type)
                                        return Stream_Element_Count;

  procedure Clear(Stream: in out Stream_Type);
```

The full details of these packages are interesting because they illustrate the use of many features such as **overriding**, the aspect Default_Initial_Condition, and pre- and postconditions. See Section 13.13.1 of the *ARM* and AI12-293.

Exercise 23.9

1 Redefine the attribute Date'Read to read the month name as a number.

23.10 Environment commands

There are four predefined packages for interfacing to the operational environment. They are Ada.Command_Line, Ada.Environment_Variables, Ada.Directories, and Ada.Locales. Their detailed behaviour and indeed whether they exist in a particular implementation depend very much upon the interface presented by the operating system.

The package Ada.Command_Line enables the Ada program to access the arguments of the command which invoked it and to set its exit status if defined for the execution environment. Its specification is

```
package Ada.Command_Line
    with Preelaborate, Nonblocking, Global => in out synchronized is

  function Argument_Count return Natural;
  function Argument(Number: in Positive) return String;
  function Command_Name return String;

  type Exit_Status is implementation-defined integer type;
  Success, Failure: constant Exit_Status;
  procedure Set_Exit_Status(Code: in Exit_Status);

private
  ...  -- not specified by the language
end Ada.Command_Line;
```

The parameterless function Argument_Count returns the number of arguments. The function Argument takes an argument number and returns the argument as a string. The function Command_Name returns the invoking command as a string.

The program can set its exit status by a call of Set_Exit_Status which takes a parameter of the type Exit_Status. Note the standard constants Success and Failure of the type Exit_Status.

The package Ada.Environment_Variables has the following specification

```
package Ada.Environment_Variables
    with Preelaborate, Nonblocking, Global => in out synchronized is
  function Value(Name: String) return String;
  function Value(Name: String; Default: String) return String;
  function Exists(Name: String) return Boolean;
  procedure Set(Name: in String; Value: in String);
  procedure Clear(Name: in String);
  procedure Clear;

  procedure Iterate(Process: not null access procedure
                                          (Name, Value: in String))
    with Allows_Exit;

end Ada.Environment_Variables;
```

This package provides access to environment variables by name. The values of the variables are implementation defined and so also represented by strings.

The behaviour is straightforward. We can check to see if there is a variable named "Ada" and then read and print her value and set it to 2022 if it is not, thus

```
if not Exists("Ada") then
   raise Horror;                          -- gosh golly
end if;
Put("Current value of Ada is "); Put_Line(Value("Ada"));
if Value("Ada") /= "2022" then
   Put_Line("Revitalizing Ada now");
   Set("Ada", "2022");
end if;
```

The procedure Clear with a parameter deletes the variable concerned. Thus Clear("Ada") eliminates her completely so that a subsequent call Exists("Ada") will return False. Note that Set actually clears the variable concerned and then defines a new one with the given name and value. The procedure Clear without a parameter clears all variables.

We can iterate over all the environment variables using the procedure Iterate. For example we can print out the current state by

```
procedure Print_One(Name, Value: in String) is
begin
   Put_Line(Name & "=" & Value);
end Print_One;
...
Iterate(Print_One'Access);
```

The procedure Print_One prints the name and value of the variable passed as parameters. We then pass an access to this procedure as a parameter to the procedure Iterate and Iterate then calls Print_One for each variable in turn.

Note that the slave procedure has both Name and Value as parameters. It might be thought that this was unnecessary since the user can always call the function Value. However, real operating systems can sometimes have several variables with the same name; providing two parameters ensures that the name/value pairs are correctly matched.

It might seem a bit peculiar declaring the procedure Print_One just so that it can be passed as a parameter to Iterate. Generally the reason for writing a subprogram is so that it can be used in various contexts. But in this case it is only used once and maybe that seems a bit strange.

This curious situation can be overcome in Ada 2022 by using the concept of a procedural iterator. The explicit subprogram Print_One can be avoided by writing

```
for (Name, Value) of Iterate (<>) loop
   Put_Line(Name & "=" & Value);
end loop;
```

where what was the body of Print_One is inside the explicit loop. Further details of this mechanism will be found in Section 21.7. Note also that the procedure Iterate has the aspect Allows_Exit. This is also explained in Section 21.7.

Attempting to misuse the environment package such as reading a variable that does not exist raises Constraint_Error or Program_Error.

There are dangers of race conditions because the environment variables are globally shared. They might also be shared with the operating system itself as well as programs in other languages.

In the example above, between determining that Ada does exist some malevolent process (such as another Ada task or an external human agent) might execute Clear("Ada"); and then the call of Value("Ada") will raise Constraint_Error. This can be overcome by the use of the function Value with two parameters added in Ada 2012. Calling this version of Value returns the value of the variable if it exists and otherwise returns the value of Default.

A special point is that we must not call the procedures Set or Clear within a procedure which is a parameter of Iterate.

The third environment package is Ada.Directories which was first introduced in Ada 2005. Its specification in Ada 2022 is

```
with Ada.IO_Exceptions;  with Ada.Calendar;
package Ada.Directories
   with Global => in out synchronized is
   -- Directory and file operations:
   function Current_Directory return String;
   procedure Set_Directory(Directory: in String);
   procedure Create_Directory(New_Directory: in String;
                                          Form: in String := "");
   procedure Delete_Directory(Directory: in String);
   procedure Create_Path(New_Directory: in String; Form: in String := "");
   procedure Delete_Tree(Directory: in String);
   procedure Delete_File(Name: in String);
   procedure Rename(Old_Name: in String; New_Name: in String);
   procedure Copy_File(Source_Name: in String;
                                 Target_Name: in String; Form: in String := "");
   -- File and directory name operations:
   function Full_Name(Name: String) return String
      with Nonblocking;
   function Simple_Name(Name: String) return String
      with Nonblocking;
   function Containing_Directory(Name: String) return String
      with Nonblocking;
   function Extension(Name: String) return String
      with Nonblocking;
   function Base_Name(Name: String) return String
      with Nonblocking;
   function Compose(Containing_Directory: String := "";
                              Name: String; Extension: String := "") return String
      with Nonblocking;
   type Name_Case_Kind :=
         (Unknown, Case_Sensitive, Case_Insensitive, Case_Preserving);
   function Name_Case_Equivalence(Name: String)
                                                              return Name_Case_Kind;
```

```
   -- File and directory queries:
   type File_Kind is (Directory, Ordinary_File, Special_File);
   type File_Size is range 0 .. implementation_defined;
   function Exists(Name: String) return Boolean;
   function Kind(Name: String) return File_Kind;
   function Size(Name: String) return File_Size;
   function Modification_Time(Name: String) return Ada.Calendar.Time;

   -- Directory searching:
   type Directory_Entry_Type is limited private;
   type Filter_Type is array (File_Kind) of Boolean;
   type Search_Type is limited private;
   procedure Start_Search(Search: in out Search_Type;
                          Directory: in String; Pattern: in String;
                          Filter: in Filter_Type := (others => True));
   procedure End_Search(Search: in out Search_Type);
   function More_Entries(Search: Search_Type) return Boolean;
   procedure Get_Next_Entry(Search: in out Search_Type;
                            Directory_Entry: out Directory_Entry_Type);
   procedure Search(Directory: in String; Pattern: in String;
                    Filter: in Filter_Type := (others => True);
                    Process: not null access procedure
                             (Directory_Entry: in Directory_Entry_Type))
     with Allows_Exit;

   -- Operations on Directory Entries:
   function Simple_Name(Directory_Entry: Directory_Entry_Type)
                                                         return String;
   function Full_Name(Directory_Entry: Directory_Entry_Type) return String;
   function Kind(Directory_Entry: Directory_Entry_Type) return File_Kind;
   function Size(Directory_Entry: Directory_Entry_Type) return File_Size;
   function Modification_Time(Directory_Entry: Directory_Entry_Type)
                                             return Ada.Calendar.Time;
   Status_Error: exception renames Ada.IO_Exceptions.Status_Error;
   Name_Error: exception renames Ada.IO_Exceptions.Name_Error;
   Use_Error: exception renames Ada.IO_Exceptions.Use_Error;
   Device_Error: exception renames Ada.IO_Exceptions.Device_Error;
private
   ...  -- not specified by the language
end Ada.Directories;
```

Most operating systems have some sort of tree-structured filing system. The general idea of this package is that it allows the manipulation of file and directory names as far as is possible in a unified manner which is not too dependent on the implementation and operating system.

However, experience with the original Ada 2005 package revealed some difficulties and these are overcome by the addition of the type Name_Case_Kind and the function Name_Case_Equivalence in the original parent package and the addition of a child package to deal with hierarchical file systems. It is probably simplest to describe the parent package first and then discuss the features of the child package added in Ada 2012. (There are a few changes in Ada 2022, mostly adding references to Allows_Exit and Global annotations.)

Files are classified as directories, special files, and ordinary files. Special files are things like devices on Windows and soft links on Unix; these cannot be created or read by the predefined Ada input–output packages.

Files and directories are identified by strings in the usual way. The interpretation is implementation defined.

The full name of a file is a string such as

```
"c:\books\ada95\intro.doc"
```

and the simple name is

```
"intro.doc"
```

At least that is in good old DOS which we will use for the sake of illustration. The current directory is that set by the "cd" command. So assuming we have done

```
c:\>cd books
c:\books>
```

then the function Current_Directory will return the string "c:\books". The procedure Set_Directory sets the current default directory. The procedures Create_Directory and Delete_Directory create and delete a single directory. We can either give the full name or just the part starting from the current default. Thus

```
Create_Directory("c:\books\history");
Delete_Directory("history");
```

will cancel out. The procedure Create_Path creates several nested directories as necessary. Thus starting from the situation above, if we write

```
Create_Path("c:\books\ada2022\notes");
```

then it will first create a directory "ada2022" in "c:\books" and then a directory "notes" in "ada2022". But if we wrote Create_Path("c:\books\ada95"); then it would do nothing since the path already exists. The procedure Delete_Tree deletes a whole tree including subdirectories and files.

The procedures Delete_File, Rename, and Copy_File behave as expected. Note in particular that Copy_File can be used to copy any file that could be copied using a normal input–output package such as Text_IO. For example, it is really tedious to use Text_IO to copy a file intact including all line and page terminators. It is a trivial matter using Copy_File.

Note also that the procedures Create_Directory, Create_Path, and Copy_File have an optional Form parameter. Like similar parameters in the input–output packages the meaning is implementation defined.

The next group of six functions, Full_Name, Simple_Name, Containing_Directory, Extension, Base_Name, and Compose just manipulate strings representing file names and do not in any way interact with the actual external file system. Moreover, of these, only the behaviour of Full_Name depends upon the current directory.

The function Full_Name returns the full name of a file. Thus assuming the current directory is still "c:\books"

```
Full_Name("ada2022\intro.doc")
```

returns "c:\books\ada2022\intro.doc" and

```
Full_Name("intro.doc")
```

returns "c:\books\intro.doc". The fact that such a file does not exist is irrelevant. We might be making up the name so that we can then create the file. If the string were malformed in some way (such as "66##77") so that the corresponding full name if returned would be nonsense then Name_Error is raised. But Name_Error is never raised just because the file does not exist.

On the other hand

```
Simple_Name("c:\books\ada2022\intro.doc")
```

returns "intro.doc" and not "ada2022\intro.doc". We can also apply Simple_Name to a string that does not go back to the root. Thus

```
Simple_Name("ada2022\intro.doc");
```

is allowed and also returns "intro.doc".

The function Containing_Directory removes the simple name part of the parameter. We can even write

```
Containing_Directory("..\ada2022\intro.doc")
```

and this returns "..\ada2022"; note that it also removes the separator "\".

The functions Extension and Base_Name return the corresponding parts of a file name thus

```
Base_Name("ada2022\intro.doc")          -- "intro"
Extension("ada2022\intro.doc")          -- "doc"
```

Note that they can be applied to a simple name or to a full name or, as here, to something midway between.

The function Compose can be used to put the various bits together, thus

```
Compose("ada2022", "intro", "doc")
```

returns "ada2022\intro.doc". The default parameters enable bits to be omitted. In fact if the third parameter is omitted then the second parameter is treated as a simple name rather than a base name. So we could equally write

```
Compose("ada2022","intro.doc")
```

We now come to the type Name_Case_Kind and the function Name_Case_Equivalence added in Ada 2012, the function returns the file name equivalence rule for the directory containing Name. It raises Name_Error if Name is not a Full_Name.

It returns Case_Sensitive if file names that differ only in the case of letters are considered to be different. If file names that differ only in the case of letters are considered to be the same, then it returns Case_Preserving if the name has the case of the file name used when the file was created and Case_Insensitive otherwise. It returns Unknown if the name equivalence rule is not known.

We thus see that Unix and Linux are Case_Sensitive, Windows is Case_Preserving, and historic systems such as CP/M and early MS/DOS were Case_Insensitive.

The next group of functions, Exists, Kind, Size, and Modification_Time act on a file name (that is the name of a real external file) and return the obvious result. (The size is measured in stream elements – usually bytes.)

Various types and subprograms are provided to support searching over a directory structure for entities with appropriate properties. This can be done in two ways, either as a loop under the direct control of the programmer (sometimes called an active iterator) or via an access to subprogram parameter (often called a passive iterator). We will look at the active iterator approach first.

The procedures Start_Search, End_Search, and Get_Next_Entry and the function More_Entries control the search loop. The general pattern is

```
Start_Search( ... );
while More_Entries( ... ) loop
   Get_Next_Entry( ... );
   ...                            -- do something with the entry found
end loop;
End_Search( ... );
```

Three types are involved. The type Directory_Entry_Type is limited private and acts as a handle to the entries found. Valid values of this type can only be created by a call of Get_Next_Entry whose second parameter is an **out** parameter of the type Directory_Entry_Type. The type Search_Type is also limited private and contains the state of the search. The type Filter_Type provides a means of identifying the kinds of file to be found. It has three components corresponding to the three values of the enumeration type File_Kind and is used by the procedure Start_Search.

Suppose we want to look for all ordinary files with extension "doc" in the directory "c:\books\ada2022". We could write

```
Book_Search: Search_Type;
Item: Directory_Entry_Type;
Filter: Filter_Type := (Ordinary_File => True, others => False);
...
Start_Search(Book_Search, "c:\books\ada2022", "*.doc", Filter);
while More_Entries(Book_Search) loop
   Get_Next_Entry(Book_Search, Item);
   ...                                          -- do something with Item
end loop;
End_Search(Book_Search);
```

The third parameter of Start_Search (which is "*.doc" in the above example) represents a pattern for matching names and thus provides further filtering of the search. The interpretation is implementation defined except that a null string means match everything. However, we would expect that writing "*.doc" would mean search only for files with the extension "doc".

The alternative mechanism using a passive iterator is as follows. We first declare a subprogram such as

```
procedure Do_It(Item: in Directory_Entry_Type) is
begin
   ...                                          -- do something with Item
end Do_It;
```

and then declare a filter and call the procedure Search thus

```
Filter: Filter_Type := (Ordinary_File => True, others => False);
...
Search("c:\books\ada2022", "*.doc", Filter, Do_It'Access);
```

The parameters of Search are the same as those of Start_Search except that the first parameter of type Search_Type is omitted and a final parameter which identifies the procedure Do_It is added. The variable Item which we declared in the active iterator is now the parameter Item of the procedure Do_It.

Each approach has its advantages. The passive iterator has the merit that we cannot make mistakes such as forget to call End_Search. But some find the active iterator easier to understand and it can be easier to use for parallel searches.

The final group of functions enables us to do useful things with the results of a search. Thus Simple_Name and Full_Name convert a value of Directory_Entry_Type to the corresponding simple or full file name. Having obtained the file name we can do everything we want but for convenience the functions Kind, Size, and Modification_Time are provided which also directly take a parameter of Directory_Entry_Type.

To complete this example we can print out a table of the files found giving their simple name, size, and modification time. The loop might then become

```
while More_Entries(Book_Search) loop
   Get_Next_Entry(Book_Search, Item);
   Put(Simple_Name(Item));  Set_Col(15);
   Put(Size(Item)/1000);  Put(" KB");  Set_Col(25);
   Put_Line(Image(Modification_Time(Item)));
end loop;
```

This might produce a table such as

```
intro.doc              152 KB      2022-08-05  09:03:10
simple.doc             372 KB      2022-09-14  21:39:05
abstraction.doc        281 KB      2022-10-03  08:43:15
...
```

Note that the function Image is from the package Ada.Calendar.Formatting discussed in Section 20.3.

Observe that the search is carried out on the directory given and does not look at subdirectories. If we want to do that then we can use the function Kind to identify subdirectories and then search recursively.

It has to be emphasized that the package Ada.Directories is very implementation dependent and might not be supported by some implementations at all. Implementations are expected to provide any additional useful functions concerning retrieving other information about files (such as the name of the owner or the original creation date) in a child package Ada.Directories.Information.

Finally, note that misuse of the various operations will raise one of the exceptions Status_Error, Name_Error, Use_Error, or Device_Error from the package IO_Exceptions.

We will now consider the problems encountered with the Ada 2005 package which resulted in the addition of the child package in Ada 2012.

One problem found was that it is not possible to concatenate a root directory such as "/tmp" with a relative pathname such as "public/file.txt" using the procedure Compose thus

```
The_Path: String := Compose("/tmp", "public/file.txt");
```

This is because the second parameter of Compose has to be a simple name such as just "file" if there is no extension parameter. If we supply the extension parameter

```
The_Path: String := Compose("/tmp", "public/file", "txt");
```

then the second parameter has to be just a base name such as "public".

Another problem with the Ada 2005 package is that there is no sensible way to check for a root directory. Thus suppose the string S is a directory name and we want to see whether it is just a root such as "/" in Unix then the only thing that we can do is write

```
Containing_Directory(S)
```

which will raise Use_Error which is somewhat ugly.

We could write **if** S = "/" **then** but this would not be portable from Unix to other systems. Indeed, the whole purpose of providing file name operations in Ada.Directories was so that file names can be manipulated in an abstract manner without fiddling with text strings.

These problems were solved in Ada 2012 by the introduction of an optional child package for dealing with systems with hierarchical file names. Its specification is

```
package Ada.Directories.Hierarchical_File_Names
    with Nonblocking, Global => in out synchronized is
  function Is_Simple_Name(Name: String) return Boolean;
  function Is_Root_Directory_Name(Name: String) return Boolean;
  function Is_Parent_Directory_Name(Name: String) return Boolean;
  function Is_Current_Directory_Name(Name: String) return Boolean;
  function Is_Full_Name(Name: String) return Boolean;
  function Is_Relative_Name(Name: String) return Boolean;

  function Simple_Name(Name: String) renames
                                          Ada.Directories.Simple_Name;

  function Containing_Directory(Name: String) renames
                                   Ada.Directories.Containing_Directory;
```

```
      function Initial_Directory(Name: String) return String;
      function Relative_Name(Name: String) return String;
      function Compose(Directory: String := "";
                       Relative_Name: String;
                       Extension: String := "") return String;

   end Ada.Directories.Hierarchical_File_Names;
```

Note that the six functions, Full_Name, Simple_Name, Containing_Directory, Extension, Base_Name, and Compose in the parent package Ada.Directories just manipulate strings representing file names and do not in any way interact with the actual external file system. The same applies to many of the new functions such as Is_Simple_Name.

In particular, Is_Root_Directory_Name returns true if the string is syntactically a root and so cannot be decomposed further. It therefore solves the second problem mentioned above. Thus

```
   Is_Root_Directory_Name("/")
```

returns true for Unix. In the case of Windows, "C:\" and "\\Computer\Share" are roots.

The function Is_Parent_Directory_Name returns true if and only if the Name is ".." for both Unix and Windows.

The function Is_Current_Directory_Name returns true if and only if Name is "." for both Unix and Windows.

The function Is_Full_Name returns true if the leftmost part of Name is a root whereas Is_Relative_Name returns true if Name allows identification of an external file but is not a full name. Note that relative names include simple names as a special case.

The functions Simple_Name and Containing_Directory are just renamings of those in the parent package and are provided for convenience.

Finally, the functions Initial_Directory, Relative_Name, and Compose provide the ability to manipulate relative file names and so solve the problem with Compose mentioned above.

Thus Initial_Directory returns the leftmost directory part of Name and Relative_Name returns the entire full name apart from the initial directory portion.

If we apply Relative_Name to a string that is just a single part of a name then Name_Error is raised. In particular, this happens if Relative_Name is applied to a name which is a Simple Name, a Root Directory Name, a Parent Directory Name, or a Current Directory Name.

The function Compose is much like Compose in the parent package except that it takes a relative name rather than a simple name. It therefore allows us to write

```
   The_Path: String := Compose("/tmp", "public/file.txt");
```

as required. The result of calling the function Compose is a full name if Is_Full_Name(Directory) is true and otherwise is a relative name.

Packages Ada.Wide_Directories and Ada.Wide_Wide_Directories and corresponding child packages are added in Ada 2022.

The final package for interfacing to the environment is Ada.Locales. When writing portable software it is often necessary to know the locality in which the software is to be run. Two key items are the country and the language (human language that is, not programming language).

To enable this to be done, Ada 2012 included the package Ada.Locales whose specification is

```
package Ada.Locales
    with Preelaborate, Remote_Types is
  type Language_Code is new String(1 .. 3)
    with Dynamic_Predicate =>
                        (for all E of Language_Code => E in 'a' .. 'z');
  type Country_Code is new String(1 .. 2)
    with Dynamic_Predicate =>
                        (for all E of Country_Code => E in 'A' .. 'Z');
  Language_Unknown: constant Language_Code := "und";
  Country_Unknown: constant Country_Code := "ZZ";
  function Language return Language_Code;
  function Country return Country_Code;
end Ada.Locales;
```

Note the use of dynamic predicates for constraining the range of the characters and the aspect Remote_Types mentioned in Section 26.3. The various country codes and language codes are defined in ISO/IEC 3166-1:2020 and ISO/IEC 639-3:2007 respectively.

Knowledge of the locale is important for writing programs where the convention for certain information varies. Thus in giving a date we might want to add the name of the day of the week and clearly in order to do this we need to know what language to use.

Canada is interesting in that it has just one country code ("CA") but two language codes ("eng" and "fra"). In Quebec, a decimal value for a million dollars and one cent is written as $1.000.000,01 whereas in English language parts it is written as $1,000,000.01 with the comma and stop interchanged.

Checklist 23

The packages for wide wide character and string handling were added in Ada 2005.

The package Ada.Characters.Conversions was added in Ada 2005. The conversion functions involving Wide_String and Wide_Character were in the package Ada.Characters.Handling in Ada 95.

The functions Index and Index_Non_Blank with parameter From were added in Ada 2005.

The subprograms Set_Bounded_String, Set_Unbounded_String, Bounded_Slice, and Unbounded_Slice were added in Ada 2005.

The packages Text_IO.Bounded_IO and Text_IO.Unbounded_IO were added in Ada 2005.

The functions Get_Line in Text_IO were added in Ada 2005.

The packages Environment_Variables and Directories were added in Ada 2005.

Packages for handling wide and wide wide characters were added in Ada 2012.

The versions of Find_Token with the parameter From were added in Ada 2012.

The functions for case insensitive comparison and hashing were added in Ada 2012.

The packages for conversions to and from UTF-8 and UTF-16 encodings were added in Ada 2012.

An additional function Value for avoiding race conditions with environment variables was added in Ada 2012.

The child package.Ada.Directories.Hierarchical_File_Names was added in Ada 2012.

The package Ada.Locales was added in Ada 2012.

New in Ada 2022

There are better facilities for outputting images of all types.

There are new packages Ada.Strings.Text_Buffers, and Bounded and Unbounded child packages for use as universal text buffers.

There are new packages Ada.Numerics.Big_Numbers and child packages Big_Integers and Big_Reals for manipulating large and accurate numbers.

The generic random number package Discrete_Random has an additional function Random including parameters giving the range of the required result.

There are variants of all IO packages where the file names are given by wide or wide wide strings.

There are new packages Ada.Streams.Storage and child packages Bounded and Unbounded.

There are variants of packages Command_Line and Environment_Variables where various parameters are given by wide or wide wide strings.

Program 5

Wild Words

This program simulates the familiar attempt by a chimpanzee to type one of the works of Shakespeare. It illustrates a number of aspects of the use of the predefined library for string handling and generating random numbers.

The idea is that the chimp types away basically at random. However, the magic word processor knows of certain characteristics of English text such as the frequency of letters.

Various checks ensure that only certain one- and two-letter words are possible; and of course certain four-letter words are forbidden. Words are also rejected if they start with an unacceptable pattern, have no vowels or have no consonants.

The text is processed as a series of unjustified lines of words with parameters such as a maximum line length of 70 characters. The text is structured into sentences and paragraphs. The stop is the only punctuation mark.

After a set number of lines have been typed, the program searches for names that might occur in a play such as Hamlet, Lear, Puck and so on. If successful it types a friendly message for the chimp and then prints the story, otherwise it types a less friendly message and tries again.

Characters are drawn from a conceptual infinite bag containing characters in the correct proportion. Word, sentence, and paragraph length are given by rolling an associated pair of conceptual dice as required. After a word is completed it is checked for feasibility. If accepted, it is added to the story.

The program has two main packages, Bag which contains the notional stock of characters, which are accessed by calling the procedure Draw, and Story which contains the text being created and also the constant Max_Lines.

Other packages are Lines which contains the instantiation for bounded string handling and Rules which contains information regarding word formation such as the letter frequencies and a list of forbidden words.

```
with Ada.Strings.Bounded;
use Ada.Strings.Bounded;
package Lines is
   Line_Length: constant := 70;
   package Bounded_Lines is
      new Generic_Bounded_Length(Line_Length);
   use Bounded_Lines;

   subtype B_String is Bounded_String;
   type BS_Array is
      array (Positive range <>) of B_String;

   function "+" (S: String) return B_String;
end;

package body Lines is
   function "+" (S: String) return B_String is
   begin
      return To_Bounded_String(S);
   end "+";
end Lines;
---------------------------------------------

with Lines;
use Lines;
package Rules is
   use Bounded_Lines;

   Letter_Freq: constant
        array (Character) of Integer :=
        ('a' => 9,  'b' => 2,  'c' => 2,  'd' => 4,
         'e' =>12,  'f' => 2,  'g' => 3,  'h' => 2,
         'i' => 9,  'j' => 1,  'k' => 1,  'l' => 4,
        'm' => 2,  'n' => 6,  'o' => 8,  'p' => 2,
         'q' => 1,  'r' => 6,  's' => 4,  't' => 6,
         'u' => 4,  'v' => 2,  'w' => 2,  'x' => 1,
         'y' => 2,  'z' => 1,          others => 0);
                          -- frequencies as in Scrabble

   function Letter_Tot return Integer;
   function Word_OK(W: B_String) return Boolean;

end Rules;
```

```
with Ada.Strings.Maps;
use Ada.Strings.Maps;  use Ada.Strings;
package body Rules is
   Vowels: constant String := "aeiouy";
   Consonants: constant String :=
                             "bcdfghjklmnpqrstvwxyz";
   Short_Words: constant BS_Array :=
  (+"a",   +"i",    +"o",
   +"am", +"an", +"as", +"at", +"be", +"by", +"do",
   +"eh", +"go", +"he", +"if", +"in", +"is", +"it",
   +"me", +"my", +"no", +"of", +"on", +"or", +"ox",
   +"so", +"up", +"us", +"we", +"ye");
   Four_Letters: constant
       array (Positive range <>) of String(1 .. 4) :=
     ("blow", "dash", "drat");  -- add your own!
   Starts: constant BS_Array :=
  (+"bl", +"br", +"ch", +"cl", +"cr", +"dr", +"dw",
   +"fl", +"fr", +"gh", +"gl", +"gn", +"gr", +"kl",
   +"kn", +"kr", +"mn", +"ph", +"pl", +"pn", +"pr",
   +"ps", +"pt", +"rh", +"sc", +"sh", +"sk", +"sl",
   +"sm", +"sn", +"sp", +"sq", +"st", +"th", +"tr",
   +"wh", +"wr",
   +"chr", +"phl", +"phr", +"sch", +"scl", +"scr",
   +"shr", +"sph", +"spl", +"spr", +"str", +"thr");
   function Letter_Tot return Integer is
      Total: Integer := 0;
   begin
      for C in Character loop
         Total := Total + Letter_Freq(C);
      end loop;
      return Total;
   end Letter_Tot;
   function Word_OK(W: B_String) return Boolean is
      First_Seq, Last_Seq: Integer;
   begin
      if Length(W) <= 2 then
         for I in Short_Words'Range loop
            if W = Short_Words(I) then
               return True;
            end if;
         end loop;
         return False;    -- non-English word
      end if;
      for I in Four_Letters'Range loop
         if W = Four_Letters(I) then
            return False;    -- coarse slang
         end if;
      end loop;
      if Count(W, "q") /= Count(W, "qu") then
         return False;    -- q without u
      end if;
      if Count(W, To_Set(Vowels)) = 0 then
         return False;    -- no vowels
      end if;
      if Count(W, To_Set(Consonants)) = 0 then
         return False;    -- no consonants
      end if;
      Find_Token(W, To_Set(Consonants) -
           To_Set('y'), Inside, First_Seq, Last_Seq);
      if First_Seq = 1 and Last_Seq > 1 then
         for I in Starts'Range loop
            if Slice(W, First_Seq, Last_Seq) =
                                          Starts(I) then
               return True;
            end if;
         end loop;
         return False;    -- illegal start
      end if;
      return True;
   end Word_OK;
end Rules;
----------------------------------------
package Bag is
   procedure Start;
   procedure Draw(C: out Character);
   function Dice return Integer;
end;
with Ada.Numerics.Discrete_Random;
use Ada.Numerics;
with Rules;  use Rules;
package body Bag is
   subtype Letter_Range is
                     Integer range 1 .. Letter_Tot;
   Letters: array (Letter_Range) of Character;
   package Letter_Random is
         new Discrete_Random(Letter_Range);
   Letter_Gen: Letter_Random.Generator;
   subtype Die_Range is Integer range 1 .. 6;
   package Die_Random is
         new Discrete_Random(Die_Range);
   Die_Gen: Die_Random.Generator;
   procedure Start is
      L: Integer := 0;
   begin
      Letter_Random.Reset(Letter_Gen);
      Die_Random.Reset(Die_Gen);
      for C in Character loop
         for I in 1 .. Letter_Freq(C) loop
            L := L + 1;  Letters(L) := C;
         end loop;
      end loop;
   end Start;
   procedure Draw(C: out Character) is
      use Letter_Random;
   begin
      C := Letters(Random(Letter_Gen));
   end Draw;
```

```
   function Dice return Integer is
      use Die_Random;
   begin
      return Random(Die_Gen) + Random(Die_Gen);
   end Dice;

end Bag;
----------------------------------------

with Lines; use Lines;
package Story is
   use Bounded_Lines;

   Done: exception;

   procedure Start;
   procedure Add_Word(A_Word: in B_String);
   procedure Add_Stop;
   procedure New_Para;
   function Is_A_Good_One return Boolean;
private
   Para_Indent: constant := 3;
   Max_Lines: constant := 1000;
   Text: array (1 .. Max_Lines) of B_String;
   Line_No: Integer;
   New_Sentence: Boolean;
end Story;

with Ada.Characters.Handling;
use Ada.Characters.Handling;
with Ada.Strings.Maps;
use Ada.Strings.Maps;
use Ada.Strings;
package body Story is

   procedure Start is
   begin
      Text := (others => Null_Bounded_String);
      Line_No := 1;
      New_Sentence := True;
   end Start;

   procedure Increment_Line is
      Last_Stop: Integer;
   begin
      if Line_No < Max_Lines then
         Line_No := Line_No + 1;
      else
         -- discard partial sentence if any
         loop
            Last_Stop := Index(Text(Line_No),
                   To_Set('.'), Going => Backward);
            if Last_Stop /= 0 then
               if Last_Stop /=
                       Length(Text(Line_No)) then
                  Delete(Text(Line_No), Last_Stop+1,
                              Length(Text(Line_No)));
               end if;
               raise Done;
            end if;
            Text(Line_No) := Null_Bounded_String;
            Line_No := Line_No - 1;
         end loop;
      end if;
   end Increment_Line;

   procedure Add_Word(A_Word: in B_String) is
      W: B_String := A_Word;
      This_Line: B_String renames Text(Line_No);
   begin
      if New_Sentence or W = "i" or W = "o" then
         Replace_Element(W, 1,
                        To_Upper(Element(W, 1)));
      end if;
      if Index_Non_Blank(This_Line) = 0 then
         Append(This_Line, W);
      elsif Length(This_Line) + Length(W) <
                                   Line_Length then
         Append(This_Line, Space & W);
      else
         Increment_Line;
         Text(Line_No) := W;
      end if;
      New_Sentence := False;
   end Add_Word;

   procedure Add_Stop is
      This_Line: B_String renames Text(Line_No);
      Last_Space: Integer;
   begin
      if Length(This_Line) < Line_Length then
         Append(This_Line, '.');
      else
         Last_Space := Index(This_Line,
                 To_Set(Space), Going => Backward);
         Increment_Line;
         Text(Line_No) := +Slice(This_Line,
              Last_Space+1, Length(This_Line)) & '.';
         Delete(This_Line, Last_Space,
                                Length(This_Line));
      end if;
      New_Sentence := True;
   end Add_Stop;

   procedure New_Para is
   begin
      Increment_Line;
      Text(Line_No) := Para_Indent * Space;
   end New_Para;

   function Is_A_Good_One return Boolean
                                   is separate;
end Story;
----------------------------------------

with Ada.Strings.Maps.Constants;
use Ada.Strings.Maps.Constants;
with Ada.Text_IO; use Ada.Text_IO;
separate(Story)
```

```
function Is_A_Good_One return Boolean is
   Names: constant BS_Array :=
      (+"Lear", +"Puck", +"Hamlet", +"Belch");

   I, J: Integer;  -- start and end of sequence
begin
   for L in Text'Range loop
      for K in Names'Range loop
         I := Index(Text(L), To_String(Translate
                   (Names(K), Lower_Case_Map)),
                        Forward, Lower_Case_Map);
         if I > 0 then
               -- now check not part of larger word
            J := I + Length(Names(K)) - 1;
            if (I = 1 or else not Is_Letter(Element
                                 (Text(L), I-1))) and
               (J = Length(Text(L)) or else
                     not Is_Letter(Element
                                 (Text(L), J+1))) then
               Put_Line(To_String(Names(K)) &
                                       " found in");
               Put_Line(To_String(Text(L)));
               return True;
            end if;
         end if;
      end loop;
   end loop;
   return False;
end Is_A_Good_One;
```

```
with Ada.Text_IO;  use Ada.Text_IO;
procedure Story.Print is
begin
   for I in Text'Range loop
      Put_Line(To_String(Text(I)));
   end loop;
end Story.Print;
```

```
with Rules, Bag, Story, Lines;  use Lines;
procedure Process_Paragraph is
   use Bounded_Lines;
   The_Word: B_String;
   The_Char: Character;
   Chars_In_Word: Integer;
begin
   for S in 1 .. Bag.Dice/2 loop
      for W in 1 .. Bag.Dice+2 loop
         Chars_In_Word := Bag.Dice-1;
         loop
            The_Word := Null_Bounded_String;
            for C in 1 .. Chars_In_Word loop
               Bag.Draw(The_Char);
               Append(The_Word, The_Char);
            end loop;
            exit when Rules.Word_OK(The_Word);
         end loop;
         Story.Add_Word(The_Word);
      end loop;
      Story.Add_Stop;
   end loop;
   Story.New_Para;
end Process_Paragraph;
```

```
with Bag, Story.Print, Process_Paragraph;
with Ada.Text_IO;  use Ada.Text_IO;
procedure Wild_Words is
begin
   Put_Line("Welcome to the Wild Word Processor");
   Bag.Start;
   loop
      Story.Start;
      loop
         begin
            Process_Paragraph;
         exception
            when Story.Done => exit;
         end;
      end loop;
      exit when Story.Is_A_Good_One;
      Put_Line("Wretched animal - try again.");
   end loop;
   Put_Line("Well done - have a banana!");
   Skip_Line;  Story.Print;  Skip_Line;
end Wild_Words;
```

A further discussion will be found on the website but the following general points should be noted.

The function "+" in the package Lines cannot be a renaming because To_Bounded_String has a default parameter Drop to control truncation. In Increment_Line, when the line limit is reached, any partial sentence is removed by scanning back for a stop and deleting all subsequent text; it then raises the exception Done. In Add_Word, a space is added if the word fits unless the line currently has no text; otherwise another line is started. In Add_Stop, if the current line is full, then the last word is moved to the next line before adding the stop. In Is_A_Good_One, the search pattern is translated to lower case.

The text produced is not a bit like English. A better approach might be to construct words out of tokens comprising groups of letters such as ck, ing, sh, and so on. The length distribution produced by the two dice is also somewhat wrong since it does not produce enough short words. Another problem is that the rejection strategy distorts the distribution of letters. For example most instances of 'q' are rejected since they are not followed by a 'u'.

It will be found that the word Lear occurs sufficiently frequently to give the chimp the odd banana. The others are quite rare. It is clear that the chance of producing a complete play is very small!

24 Container Library

24.1 Organization of library
24.2 Doubly linked lists
24.3 Vectors
24.4 Maps
24.5 Sets
24.6 Trees
24.7 Holders
24.8 Queues
24.9 Bounded containers
24.10 Indefinite containers
24.11 Sorting
24.12 Summary table

A major feature of the predefined library is the container library. This is rather extensive and merits this separate chapter on its own.

The main part of the library comprises several generic packages for the manipulation of various kinds of important data structures. These structures are vectors, doubly linked lists, maps and sets (both hashed and ordered), multiway trees, holders (or wrappers), and various task safe queues. There are also generic packages for sorting arrays and other general structures.

There are a number of improvements to the existing containers and one new kind of container in Ada 2022. The overall improvements are designed to increase the efficiency of container operations and to make their use more convenient and flexible.

One interesting change is that the behaviour of the various container operations is now largely described by pre- and postconditions rather than by English text. This is hopefully less liable to misinterpretation provided the reader can understand the detail which in some cases is tricky.

Efficiency is increased by introducing a stable view of the containers which eliminates a number of checks. This is done by a nested package called Stable which contains the relevant operations which thereby avoids many tampering checks; see the end of Section 16.5.

Additional subprograms are added so that all commonly used operations can be done using prefixed notation as well as classical notation. Another advantage is that the Global aspect can avoid having to say **in out all** or **in all** but instead can rely on the mode of the Container parameter to allow the Global aspect to be **null**.

Another helpful innovation is the ability to declare a container aggregate which is useful for the initialization of a container in one step. Note that container aggregates are enclosed in square brackets rather than parentheses. Container aggregates can also use appropriate new aggregate features as for arrays and records described in Section 8.8.

All containers now have a function Empty. If the container has a concept of capacity (vectors and hashed maps and sets) then the capacity is a parameter. In other cases it simply returns the constant such as Empty_List.

The various iteration functions now use the parallel iterators from Ada.Iterator_Interfaces so that one can take advantage of multiple processors.

The one new kind of container is Bounded_Indefinite_Holders which is therefore added to Table 24.1. For convenience it is outlined in Section 24.7 along with the existing definite holder container.

24.1 Organization of library

The container library comprises a root package Ada.Containers plus various child generic packages. The main generic packages form three groups. The first group concerns objects of definite types; the second group concerns indefinite types, and the third group has bounded capacity for objects of definite types.

It will be remembered that an indefinite (sub)type is one for which we cannot declare an object without giving a constraint (either directly or indirectly from its initial value).

The reason for distinct containers for definite and indefinite types concerns efficiency. It is much easier to manipulate definite types and although the packages for indefinite types can be used for definite types, this would be rather inefficient.

Another consideration is reliability. For high integrity applications it is generally unacceptable to use dynamic storage allocation. Accordingly, a separate group of containers with bounded capacity is provided for definite types.

Note that there are no bounded containers for indefinite types because objects of indefinite types (such as the type String) by their very nature are of variable size and thus inevitably need dynamic storage management.

The containers can also be categorized in other ways.

Sequence containers – these hold sequences of elements. There are packages for manipulating vectors and for manipulating linked lists. These packages have much in common. But they have different behaviours in terms of efficiency according to the pattern of use. In general (with some planning) it should be possible to change from one to the other with little effort.

Multiway tree containers – these hold elements as a multiway tree. They have much in common with the sequence containers.

Associative containers – these associate a key with each element and then store the elements in order of the keys. There are packages for manipulating hashed maps, ordered maps, hashed sets, and ordered sets. These packages also have much in common and changing between hashed and ordered versions is usually feasible.

Holder containers – these enable a single object of an indefinite type to be encapsulated so that it appears to be of a definite type.

Table 24.1 Generic child units of Ada.Containers.

Vectors	Indefinite_Vectors	Bounded_Vectors
Doubly_Linked_Lists	Indefinite_Doubly_Linked_Lists	Bounded_Doubly_Linked_Lists
Multiway_Trees	Indefinite_Multiway_Trees	Bounded_Multiway_Trees
Hashed_Maps	Indefinite_Hashed_Maps	Bounded_Hashed_Maps
Ordered_Maps	Indefinite_Ordered_Maps	Bounded_Ordered_Maps
Hashed_Sets	Indefinite_Hashed_Sets	Bounded_Hashed_Sets
Ordered_Sets	Indefinite_Ordered_Sets	Bounded_Ordered_Sets
	Indefinite_Holders	Bounded_Indefinite_Holders
Synchronized_Queue_Interfaces		
	Unbounded_Synchronized_Queues	Bounded_Synchronized_Queues
	Unbounded_Priority_Queues	Bounded_Priority_Queues
Generic_Sort	Generic_Array_Sort	Generic_Constrained_Array_Sort

Queue containers – these are particularly relevant to tasking applications and enable objects of a definite type to be held on various forms of queues. There are no queue containers for indefinite types; however, the holder container can be used to wrap an indefinite type.

The names of all the containers are shown in Table 24.1.

- In the case of vectors, lists, and so on there are three as explained above. There are two holder containers.
- The queue containers are structured slightly differently. The package Ada.Containers.Synchronized_Queue_Interfaces exports an interface which is imported by the others as will be explained later.
- Finally, there are three generic procedures for sorting.

The root package is

```
package Ada.Containers
    with Pure is

  type Hash_Type is mod implementation-defined;
  type Count_Type is range 0 .. implementation-defined;
  Capacity_Error: exception;

end Ada.Containers;
```

The type Hash_Type is used by the associative containers and Count_Type is used generally for the number of elements in a container. Note that we talk about elements in a container rather than the components in a container – components is the Ada term for the items of an array or record as an Ada type and it is convenient to use a different term since in the case of containers the actual data structure is hidden. The exception Capacity_Error is used by the bounded containers.

Perhaps a remark about using containers from a multitasking program would be helpful. The queue containers which were new in Ada 2012 are explicitly designed to enable several tasks to add items to or remove items from a queue without interference. However, there is nothing special to say about the others. So we have to protect ourselves by using the normal techniques such as protected objects when container operations are invoked concurrently on the same object from multiple tasks even if the operations are only reading from the container.

24.2 Doubly linked lists

We will first consider the list container since in some ways it is the simplest. Here is the specification of the unbounded version for components of a definite type interspersed with some explanation

```
with Ada.Iterator_Interfaces;
generic
  type Element_Type is private;
  with function "=" (Left, Right: Element_Type) return Boolean is <>;
package Ada.Containers.Doubly_Linked_Lists
    with Preelaborate, Remote_Types,
         Nonblocking, Global => in out synchronized is     -- new

  type List is tagged private
    with Constant_Indexing => Constant_Reference,
         Variable_Indexing => Reference,
         Default_Iterator => Iterate,
         Iterator_Element => Element_Type

         Iterator_View => Stable.List,                 -- new from here
         Aggregate => (Empty => Empty,
                       Add_Unnamed => Append),
         Stable_Properties => (Length,
                               Tampering_With_Cursors_Prohibited,
                               Tampering_With_Elements_Prohibited),
         Default_Initial_Condition =>
             Length(List) = 0 and then
             (not Tampering_With_Cursors_Prohibited(List)) and then
             (not Tampering_With_Elements_Prohibited(List)),
         Preelaborable_Initialization;

  type Cursor is private
    with Preelaborable_Initialization;
```

The context clause refers to the magic package Ada.Iterator.Interfaces described in Section 21.6 which enables iteration to be performed in a simple and elegant manner. The two generic parameters are the type of the elements in the list and the definition of equality for comparing elements. The equality relation must be such that x = y and y = x always have the same value. Note that the default is used if the parameter is omitted.

A list container is an object of the type List. It is tagged since it will inevitably be implemented as a controlled type. The fact that it is visibly tagged means that all the advantages of object oriented programming are available. For one thing it enables the use of the prefixed notation so that we can write operations such as

```
My_List.Append(Some_Value);
```

rather than

```
Append(My_List, Some_Value);
```

The type List has aspects Constant_Indexing, Variable_Indexing, Default_Iterator, and Iterator_Element. These concern the magic referencing and iteration processes mentioned in Section 21.6.

We will now look at the new aspects introduced in Ada 2022 in turn

```
Iterator_View => Stable.List,
```

states that iterators from Ada.Iterator_Interfaces will use the type Stable.List rather than List for all operations. The package Stable is internal to Doubly_Linked_Lists; we will see its declaration later.

```
Aggregate => (Empty => Empty,
              Add_Unnamed => Append),
```

states that container aggregates are applicable and in particular an empty aggregate can be established by calling the new function Empty and that when using unnamed notation in the aggregate, an item is introduced by calling the function Append. Note carefully that Ada 2022 has two procedures Append, one has two parameters and the other has three thus

```
procedure Append(Container: in out List; New_Item: in Element_Type);
procedure Append(Container: in out List; New_Item: in Element_Type;
                                                   Count: in Count_Type);
```

The aggregate refers to the version with two parameters.

An empty list can now be created by using aggregate syntax thus

```
L: List := [];                    -- same as L: List := Empty;
```

where Empty is a new function that returns an empty list thus

```
function Empty return List is
  (Empty_List);
```

In the case of the Float_Container example at the end of this section, a stack of items of type Float could be initialized with several values in place by

```
The_Stack: List := [1.0, 37.5, 99.0];
```

and this would be equivalent to

```
The_Stack := Empty_List;
Append(The_Stack, 1.0);                  -- or The_Stack.Append(1.0);
Append(The_Stack, 37.5);                 -- and so on
Append(The_Stack, 99.0);                 -- and on
```

The aspect

```
Stable_Properties => (Length,
                      Tampering_With_Cursors_Prohibited,
                      Tampering_With_Elements_Prohibited),
```

indicates that the Length of a container is a stable property (see Section 16.5). Remember that Length is an existing function in Ada 2012 but in Ada 2022 it also has aspects Nonblocking etc, so the specification now is

```
function Length(Container: List) return Count_Type;
  with Nonblocking, Global => null, Use_Formal => null;
```

In Ada 2022 there are two similar functions concerning tampering thus

```
function Tampering_With_Cursors_Prohibited(Container: List)
                                                    return Boolean;
  with Nonblocking, Global => null, Use_Formal => null;
function Tampering_With_Elements_Prohibited(Container: List)
                                                    return Boolean;
  with Nonblocking, Global => null, Use_Formal => null;
```

and these give the state with regard to tampering.

The final aspect is

```
Default_Initial_Condition =>
    Length(List) = 0 and then
    (not Tampering_With_Cursors_Prohibited(List)) and then
    (not Tampering_With_Elements_Prohibited(List)),
```

which simply states that when a List is declared its default condition is that its length is zero and both sorts of tampering are permitted. The concept of tampering is described at the end of this section.

The type Cursor is a vital concept. It provides the means of access to individual elements in the container. Not only does it contain a reference to an element but it also identifies the container as well. This enables various checks to be made to ensure that we do not accidentally meddle with an element in the wrong container.

We now have

```
Empty_List: constant List;
No_Element: constant Cursor;
function Has_Element(Position: Cursor) return Boolean;
  with Nonblocking, Global => in all, Use_Formal => null;
function Has_Element(Container: List; Position: Cursor) return Boolean;
  with Nonblocking, Global => null, Use_Formal => null;  -- Ada 2022
```

The constants Empty_List and No_Element are as expected and also provide default values for objects of types List and Cursor respectively. The functions Has_Element return False if the cursor does not identify an element; for example if it is No_Element.

Note that there are two functions Has_Element. The version with the Container parameter is added in Ada 2022 so that prefixed notation can be used everywhere as mentioned in the introduction to this chapter. The original version without the Container parameter is left for compatibility.

Both versions have the aspects Nonblocking etc as shown. This is a common list of aspects and applies to many subprograms. Note that global is **in all** in one case and **null** in the other. It is better to be truthful and mention the Container explicitly.

We will now survey the outline of the remaining declarations in order. However, we will omit most of the aspects in order to save space. Note in particular that the meanings of most declarations are now given by pre- and postconditions rather than English text. The reader should consult the *ARM* for details.

```
package List_Iterator_Interfaces is
   new Ada.Iterator_Interfaces(Cursor, Has_Element);
```

This is an instantiation of the package Ada.Iterator_Interfaces. See Section 21.6.

```
function "=" (Left, Right: List) return Boolean;
function Empty return List is (Empty_List);          -- added in 2022
function Length(Container: List) return Count_Type;
function Is_Empty(Container: List) return Boolean;
procedure Clear(Container: in out List);
```

The function "=" compares two lists. It only returns true if both lists have the same number of elements and corresponding elements have the same value as determined by the generic parameter "=" for comparing elements. The function Empty is added for uniformity. The subprograms Length, Is_Empty, and Clear are as expected.

Note that A_List = Empty_List, Is_Empty(A_List), and Length(A_List) = 0 all have the same value.

```
function Element(Position: Cursor) return Element_Type;
function Element(Container: List;                         -- added in
                 Position: Cursor) return Element_Type;   -- Ada 2022

procedure Replace_Element(Container: in out List;
                          Position: in Cursor;
                          New_Item: in Element_Type);
```

The function Element takes a cursor and returns the value of the corresponding element (remember that a cursor identifies the list as well as the element itself); the version with Container parameter is added in Ada 2022. The procedure Replace_Element replaces the value of the element identified by the cursor by the value given.

Note carefully that Replace_Element has both the list and cursor as parameters. There are two reasons for this concerning correctness. One is to enable a check that the cursor does indeed identify an element in the given list (if this fails then Program_Error is raised). The other is to ensure that we do have write access to the container (the parameter has mode **in out**). Otherwise it would be possible to modify a container even though we only had a constant view of it.

So the general principle in Ada 2012 was that any operation that modifies a container must have the container as a parameter whereas an operation that only reads it such as the function Element need not. But in Ada 2022 we prefer the additional version with the Container parameter in all cases.

In future we will only give the version with the Container parameter but in those cases where the version without still exists for compatibility it will be marked by three asterisks on the left of the page alongside the Ada 2022 version as below.

```
***   procedure Query_Element(Container: in List;
                              Position: in Cursor;
                              Process: not null access procedure
                                          (Element: in Element_Type));
      procedure Update_Element(Container: in out List;
                               Position: in Cursor;
                               Process: not null access procedure
                                          (Element: in out Element_Type));
```

These procedures provide *in situ* access to an element. One parameter is the cursor identifying the element and another is an access to a procedure to be called with that element as parameter. In the case of Query_Element, we can only read the element whereas in the case of Update_Element we can change it as well since the parameter mode of the access procedure is **in out**.

To read a component Q of an element of My_List using the cursor C we write

```
X := Element(C).Q;
```

or we can first declare a slave procedure

```
procedure Get_Q(E: in Element_Type) is
begin
   X := E.Q;
end Get_Q;
```

and then call Query_Element thus

```
Query_Element(C, Get_Q'Access);          -- two parameter version
```

The advantage of the former is that it is easy but it could be slow because the function Element returns a copy of the whole element which could be enormous. The advantage of the latter is that it does not copy the element; its disadvantage is that it is laborious.

In Ada 2012 and Ada 2022, we can do much better as explained in Section 21.6. There are types and functions thus

```
type Constant_Reference_Type
         (Element: not null access constant Element_Type) is private
   with Implicit_Dereference => Element;

type Reference_Type
                  (Element: not null access Element_Type) is private
   with Implicit_Dereference => Element;

function Constant_Reference
                  (Container: aliased in List; Position: in Cursor)
                                 return Constant_Reference_Type;
```

```
function Reference(Container: aliased in out List; Position: in Cursor)
                                              return Reference_Type;
```

These types and functions relate to the aspects Constant_Indexing and Variable_Indexing which apply to the type List itself as shown in its declaration above. They enable updates and queries to be done very simply; the subprograms Query_Element and Replace_Element are more or less redundant but have to be retained for compatibility. Thus we can now write

```
X := My_List.Constant_Reference(C).Q;
```

This works because the function Constant_Reference returns a value of Constant_Reference_Type and this moreover has aspect Implicit_Dereference whose value is Element.

However, we can simplify this even more because the type List has aspects Constant_Indexing and Variable_Indexing which refer to the functions Constant_Reference and Reference. The result is that we can simply write

```
X := My_List(C).Q;                          -- much better
```

which is a lot better than calling Query_Element.

Similarly, if we just want to update the component Q of some element given by a cursor C, then in Ada 2005 we either had to create a whole new element with the new value for Q and then use Replace_Element or declare a slave procedure and use Update_Element. However, in Ada 2012 and Ada 2022 we simply write

```
My_List(C).Q := X;                          -- better again
```

which implicitly uses the aspect Variable_Indexing to call the function Reference which gives access to the element.

```
procedure Assign(Target: in out List; Source: in List);
function Copy(Source: List) return List;
procedure Move(Target, Source: in out List);
```

The procedure Assign simply copies the elements of Source to Target as for an assignment statement. The function Copy returns a list whose elements match those of Source. The procedure Move is equivalent to Assign(Target, Source) followed by Clear(Source).

```
procedure Insert(Container: in out List;
                 Before: in Cursor;
                 New_Item: in Element_Type;
                 Count: in Count_Type := 1);

procedure Insert(Container: in out List;
                 Before: in Cursor;
                 New_Item: in Element_Type;
                 Position: out Cursor;
                 Count: in Count_Type := 1);
```

```
procedure Insert(Container: in out List;
                 Before: in Cursor;
                 Position: out Cursor;
                 Count: in Count_Type := 1);
```

These three procedures enable one or more identical elements to be added anywhere in a list. The place is indicated by the parameter Before – if this is No_Element, then the new elements are added at the end. The second procedure is similar to the first but also returns a cursor to the first of the added elements. The third is like the second but the new elements take their default values. Note the default value of one for the number of elements.

```
procedure Prepend(Container: in out List;
                  New_Item: in Element_Type;
                  Count: in Count_Type := 1);
procedure Append(Container: in out List;
                 New_Item: in Element_Type;
                 Count: in Count_Type);
procedure Append(Container: in out List;
                 New_Item: in Element_Type);
```

These add one or more new elements at the beginning or end of a list respectively. Clearly these operations can be done using Insert but they are sufficiently commonly needed that it is convenient to provide them specially. For curious historic reasons there are two versions of Append, one for a single item and a distinct one for multiples.

```
procedure Delete(Container: in out List;
                 Position: in out Cursor;
                 Count: in Count_Type := 1);
procedure Delete_First(Container: in out List;
                       Count: in Count_Type := 1);
procedure Delete_Last(Container: in out List;
                      Count: in Count_Type := 1);
```

These delete one or more elements at the appropriate position. In the case of Delete, the parameter Position is set to No_Element. If there are not as many as Count elements to be deleted at the appropriate place then it just deletes as many as possible (this clearly results in the container becoming empty in the case of Delete_First and Delete_Last).

```
procedure Reverse_Elements(Container: in out List);
```

This does the obvious thing. It would have been nice to call this procedure Reverse but sadly that is a reserved word.

```
procedure Swap(Container: in out List; I, J: in Cursor);
procedure Swap_Links(Container: in out List; I, J: in Cursor);
```

These handy procedures swap the values in the two elements denoted by the two cursors. The elements must be in the given container otherwise Program_Error is raised. Note that the cursors are **in** parameters in both procedures and do not change but remain pointing at the same elements. However, for Swap, the values of the elements are interchanged, whereas for Swap_Links, the values of the elements stay the same but the elements swap position in the list. Swap_Links is faster if the elements are large but has no analogy in the vectors package.

```
procedure Splice(Target: in out List;
                 Before: in Cursor;
                 Source in out List);
procedure Splice(Target: in out List;
                 Before: in Cursor;
                 Source: in out List;
                 Position: in out Cursor);
procedure Splice(Container: in out List;
                 Before: in Cursor;
                 Position: in Cursor);
```

These three procedures enable elements to be moved (without copying). The place is indicated by the parameter Before – if this is No_Element, then the elements are added at the end. The first moves all the elements of Source into Target just before the position given by Before; as a consequence, like the procedure Move, after the operation the source is empty and Length(Source) is zero. The second moves a single element at Position from the list Source to Target and so the length of target is incremented whereas that of source is decremented; Position is updated to its new location in Target. The third moves a single element within a list and so the length remains the same (the formal parameter is Container rather than Target in this case). There are no corresponding operations in the vectors package because, like Swap_Links, we are just moving the links and not copying the elements.

```
    function First(Container: List) return Cursor;
    function First_Element(Container: List) return Element_Type;
    function Last(Container: List) return Cursor;
    function Last_Element(Container: List) return Element_Type;

*** function Next(Container: List; Position: Cursor) return Cursor;   -- new
*** function Previous(Container: List; Position: Cursor) return Cursor;-- new
*** procedure Next(Container: in List; Position: in out Cursor);    -- new
*** procedure Previous(Container: in List; Position: in out Cursor);  -- new
```

Here we find that new versions of Next and Previous are introduced in Ada 2022 which explicitly give the Container as a parameter.

```
function Find(Container: List; Item: Element_Type;
              Position: Cursor := No_Element) return Cursor;
function Reverse_Find(Container: List; Item: Element_Type;
                      Position: Cursor := No_Element) return Cursor;
function Contains(Container: List; Item: Element_Type) return Boolean;
```

Hopefully the purpose of these is almost self-evident. The function Find searches for an element with the given value starting at the given cursor position (or at the beginning if the position is No_Element); if no element is found then it returns No_Element. Reverse_Find does the same but backwards. Note that equality used for the comparison in Find and Reverse_Find is that defined by the generic parameter "=" ; the default will be used if no actual parameter was supplied.

```
procedure Iterate(Container: in List;
                  Process: not null access procedure (Position: in Cursor))
                     with Allows_Exit;

procedure Reverse_Iterate(Container: in List;
                  Process: not null access procedure (Position: in Cursor));
                     with Allows_Exit;
```

These apply the procedure designated by the parameter Process to each element of the container in turn in the appropriate order. Note the addition of Allows_Exit (see Section 21.6).

```
function Iterate(Container: in List)
      return List_Iterator_Interfaces.Parallel_Reversible_Iterator'Class;

function Iterate(Container: in List; Start: in Cursor)
            return List_Iterator_Interfaces.Reversible_Iterator'Class;
```

These functions are invoked when we write loops of the forms

```
for C in My_List.Iterate loop ...
for C in My_List.Iterate(S) loop ...
```

as explained in Section 21.6.

Note that the function Iterate without the parameter Start returns values of Parallel_Reversible_Iterator'Class in Ada 2022 and so the iterations can be done using a parallel loop thus

```
parallel for C in My_List.Iterate loop ...
```

However, remember that parallel loops cannot specify **reverse**. On the other hand if we use the Iterate with parameter Start then we cannot use a parallel loop, but we can specify **reverse**.

There are no material changes to the package Generic_Sorting other than the addition of pre- and postconditions which we have generally ignored. However, this is perhaps an opportunity to illustrate their use. So here is the package specification in full as it is in Ada 2022

```
generic
   with function "<" (Left, Right: Element_Type) return Boolean is <>;
package Generic_Sorting
     with Nonblocking, Global => null is
   function Is_Sorted(Container: List) return Boolean;
   procedure Sort(Container: in out List)
     with Pre => not Tampering_With_Cursors_Prohibited(Container)
                        or else raise Program_Error;
```

```
   procedure Merge(Target, Source: in out List)
      with Pre => (not Tampering_With_Cursors_Prohibited(Target)
                              or else raise Program_Error) and then
                  (not Tampering_With_Elements_Prohibited(Source)
                              or else raise Program_Error) and then
                  (Length(Target) <= Count_Type'Last - Length(Source)
                              or else raise Constraint_Error) and then
                  ((Length(Source) = 0 or else
                                     not Target'Has_Same_Storage(Source))
                              or else raise Constraint_Error),
           Post => (declare
                          Result_Length: constant Count_Type :=
                              Length(Source)'Old + Length(Target)'Old;
                     begin
                              (Length(Source) = 0 and then
                                             Length(Target) = Result_Length));
   end Generic_Sorting;
```

The pre- and postconditions generally make good use of **or else** and **and then**. These were described at the end of Section 6.9. A recap is perhaps in order. In the case of

```
X or else Y
```

X is evaluated first. If X is true, the answer is true and Y is not evaluated. If X is false then Y has to be evaluated and the value of Y is the answer.

An important point is that Y can be a raise expression (see the end of Section 15.2) and that is widely used in preconditions.

In the case of Sort the precondition is

```
      with Pre => not Tampering_With_Cursors_Prohibited(Container)
                                  or else raise Program_Error;
```

So if tampering with cursors is prohibited (that is True) then not tampering etc is False, that is X is False and so Y has to be evaluated and so Program_Error is raised. On the other hand if tampering is not prohibited, then not tampering etc is True and the precondition is true so all is well. No postcondition is given for Sort, it is described by English text as in Ada 2012.

Turning now to Merge, the precondition consists of four subconditions separated by **and then**. These subconditions all have to be true if the precondition as a whole is going to be true. As soon as one is found to be false then the precondition as a whole is false and the others are not evaluated.

The first two subconditions concerning tampering and as in the case of Sort, if tampering is prohibited then the precondition is false and Program_Error is raised. If both types of tampering are permitted then the next subcondition is evaluated. Consider

```
Length(Target) <= Count_Type'Last – Length(Source)
```

The type Count_Type is declared in the parent package Ada.Containers. It is an integer type and is used for counting in containers in general. In particular the function Length returns a value of Count_Type. So this checks that the length of Target and Source together do not exceed Count_Type'Last. Hence this subcondition checks that there is enough space and raises Constraint_Error if there is not.

The final subcondition is

```
((Length(Source) = 0 or else
                not Target'Has_Same_Storage(Source))
    or else raise Constraint_Error),
```

and this ensures that Source and Target do not overlap.

Finally, consider the postcondition for Merge which is

```
Post => (declare
             Result_Length: constant Count_Type :=
                 Length(Source)'Old + Length(Target)'Old;
         begin
             (Length(Source) = 0 and then
                           Length(Target) = Result_Length));
```

This takes the form of a declare expression (see Section 9.5) and ensures that the Source is empty and that the length of the Target is the sum of the original lengths of Source and Target.

And finally we have the new nested package Stable; it starts as follows

```
package Stable is

   type List(Base: not null access Doubly_Linked_Lists.List) is
      tagged limited private
      with Constant_Indexing => Constant_Reference,
           Variable_Indexing => Reference,
           Default_Iterator => Iterate,
           Iterator_Element => Element_Type,
           Stable_Properties => (Length),
           Global => null,
           Default_Initial_Condition => Length(List) = 0,
           Preelaborable_Initialization;
```

A few points are worth noting. The stable version is limited and has an access discriminant linking it to the normal version. There are no global side effects. Aggregates are not available and tampering is not mentioned (because it cannot happen). It then continues thus as before

```
   type Cursor is private
      with Preelaborable_Initialization;

   Empty_List: constant List;
   No_Element: constant Cursor;

   function Has_Element(Position: Cursor) return Boolean
      with Nonblocking, Global => in all, Use_Formal => null;
```

```
package List_Iterator_Interfaces is
   new Ada.Iterator_Interfaces (Cursor, Has_Element);
```

Now come two new subprograms Assign and Copy.

```
procedure Assign
            (Target: in out Doubly_Linked_Lists.List; Source: in List)
   with Post => Length(Source) = Length(Target);
function Copy(Source: Doubly_Linked_Lists.List) return List
   with Post => Length(Copy'Result) = Length(Source);
```

Note carefully that the parameter Target of Assign and Source of Copy are not of the type List declared in the package Stable but of the type List in the outer package. They can therefore be used to convert a stable view of a list to a normal view and vice versa.

Next are new versions of the types Constant_Reference_Type and Reference_Type thus

```
type Constant_Reference_Type
         (Element: not null access constant Element_Type) is private
   with Implicit_Dereference => Element,
        Nonblocking, Global => null, Use_Formal => null,
        Default_Initial_Condition => (raise Program_Error);
type Reference_Type
                  (Element: not null access Element_Type) is private
   with Implicit_Dereference => Element,
        Nonblocking, Global => null, Use_Formal => null,
        Default_Initial_Condition => (raise Program_Error);
```

Note the aspect Default_Initial_Condition which ensures that if we declare an object of either of these types without giving an explicit initial value then the exception Program_Error is raised

And finally

```
   ... -- and other subprograms.
private
   ... -- not specified by the language
end Stable;
```

The *other subprograms* are essentially the same as in the main package with identical specifications except that those that might change the size of a list are omitted. Hence Assign, Move, Insert, Append, Prepend, Clear, Delete, Delete_First, Slice, Swap_Links, and Reverse_Elements are not provided but First, Last, Replace and so on are provided.

Thus we see that we can have two views of a list and can use the stable view for quick safe operations and the full view for those which are perhaps slow and vulnerable to tampering. For example, suppose we declare a list container representing a list of animals. We can extend the type Animal of Section 9.3 thus

```
type Animal is (Ape, Bear, Cat, Dog, Elephant, Fox, Giraffe, Horse, ...)
```

and then declare a container

```
package Zoo is new Doubly_Linked_Lists(Animal);
use Zoo;
The_Zoo: aliased List := [Cat, Dog];
```

This initializes The_Zoo with a Cat and a Dog. To add a Fox and Bear to the zoo we can write

```
The_Zoo.Append(Fox);
The_Zoo.Append(Bear);
```

and so on. Now suppose we want to change the Fox to a Vixen. This can be done using the stable view of the zoo since the size (that is Length) will not be changed. To get a stable view we write

```
Stable_Zoo: Stable.List(The_Zoo'Access);
```

Note that this uses the access discriminant of the type Stable.List.

Assuming that we have lost track of where the Fox is in the list (that is the cursor value) we can find the Fox and then replace him by the Vixen as follows

```
C := Stable_Zoo.Find(Fox);              -- find that Fox
Stable_Zoo.Replace(C, Vixen);           -- and replace him
```

where C is of type Stable.Cursor. Of course we might have put several foxes into the zoo. The function Find finds the first one starting from the start of the list.

An interesting consequence of having both a stable view and a full view is that at any point where both views are visible, the full view prohibits tampering and so one cannot use the full view perhaps to change the length.

Another way of getting a stable view is to declare a stable list without an explicit discriminant but with an initial value (which thereby provides the discriminant) thus

```
Stable_Zoo: Stable.List := Copy(The_Zoo);
```

which uses the function Copy declared above. Note that the postcondition of Copy shows that the length of the copy is the same as that of the source. But note also that we do now have two zoos!

And at long last we have

```
private
   ...      -- not specified by the language
end Ada.Containers.Doubly_Linked_Lists;
```

We finish this discussion by first looking at a simple example and then conclude with some thoughts on the curious topic of tampering.

As a very simple illustration of the use of a container here is a simple stack of floating point numbers. Note that in this example, the use of the container is completely hidden.

```
package Stack is
   procedure Push(X: in Float);
   function Pop return Float;
   function Size return Integer;
   Stack_Empty: exception;
end;

with Ada.Containers.Doubly_Linked_Lists;
use Ada.Containers;
package body Stack is
   package Float_Container is new Doubly_Linked_Lists(Float);
                                    -- not necessary to say use Float_Container
   The_Stack: Float_Container.List;

   procedure Push(X: in Float) is
   begin
      The_Stack Append(X);
   end Push;

   function Pop return Float is
      Result: Float;
   begin
      if The_Stack.Is_Empty then
         raise Stack_Empty;
      end if;
      Result := The_Stack.Last_Element;
      The_Stack.Delete_Last;
      return Result;
   end Pop;

   function Size return Integer is
   begin
      return Integer(The_Stack.Length);
   end Size;
end Stack;
```

This barely needs any explanation. The lists package is instantiated in the package Stack and the object The_Stack is the list container. The rest is straightforward. Note that we did not provide an actual parameter for equality so the predefined equality on Float will be used if needed.

We have used the prefixed notation throughout. We can do this because the type List is tagged since it is derived from the tagged type List in the container package. The prefixed notation is considered more elegant and does not need the use clause. However, it can only be used with subprograms whose first parameter is of the type List. Note that in Ada 2012 it could not be applied to calls of Element, Query_Element, Next, and Previous where the first parameter was of type Cursor rather than List. However, Ada 2022 includes new versions with the extra parameter as mentioned earlier so all is well. So we do not have to say **use** Float_Container.

An important point about cursors is that if a cursor C is referencing some element E and we add or delete other elements, then C will continue to refer to E even if the other elements are earlier in the list. This property also applies to multiway trees but not to the other containers such as vectors.

Another point concerning lists (and containers in general) is that attempts to do foolish things typically result in Constraint_Error or Program_Error being raised. This especially applies to the procedures Process in Query_Element, Update_Element, Iterate, and Reverse_Iterate. The concepts of tampering with cursors and elements are introduced in order to dignify a general motto of 'Thou shalt not violate thy container'.

Tampering with cursors occurs when elements are added to or deleted from a container (by calling Insert and so on) whereas tampering with elements is an action that could replace an element with one of a different size (by calling Replace_Element for example).

The procedure Process in Query_Element and Update_Element must not tamper with elements and the procedure Process in the other cases must not tamper with cursors. It might be thought odd that Update_Element should not be allowed to tamper with elements since the whole purpose is to update the element; this comes back to the point mentioned earlier that update element gives access to the existing element *in situ* via the parameter of Process and that is allowed – calling Replace_Element within Process would be tampering. Tampering causes Program_Error to be raised.

A major innovation in Ada 2022 is the introduction of stable properties. This greatly reduces the risk of tampering but does not entirely eliminate it.

Exercise 24.2

1 Implement the package Queues of Exercise 12.5(**3**) using a list container.

24.3 Vectors

The generic package Ada.Containers.Vectors has much in common with the package Ada.Containers.Doubly_Linked_Lists that we have just been discussing. Its specification starts

```
with Ada.Iterator_Interfaces;
generic
   type Index_Type is range <>;
   type Element_Type is private;
   with function "=" (Left, Right: Element_Type) return Boolean is <>;
package Ada.Containers.Vectors
     with Preelaborate, Remote_Types
          Nonblocking, Global => in out synchronized is    -- new
```

This is similar to the lists package except for the additional generic parameter Index_Type (note that this is an integer type and not a discrete type). This additional parameter reflects the idea that a vector is essentially an array and we can index directly into an array. Thus the vectors package enables us to access elements either by using an index or by using a cursor. And another important difference is that vectors have a concept of Capacity as well as Length.

The improvements to the package Ada.Containers.Vectors for Ada 2022 are much the same as for Doubly_Linked_Lists.

The additional aspects for vectors are

```
type Vector is tagged private
  with ...
        Iterator_View => Stable.Vector,                  -- new from here
        Aggregate => (Empty => Empty,
                      Add_Unnamed => Append,
                      New_Indexed => New_Vector,
                      Assign_Indexed => Replace_Element),
        Stable_Properties => (Length, Capacity,
                                  Tampering_With_Cursors_Prohibited,
                                  Tampering_With_Elements_Prohibited),
        Default_Initial_Condition =>
            Length(Vector) = 0 and then
            (not Tampering_With_Cursors_Prohibited(Vector)) and then
            (not Tampering_With_Elements_Prohibited(Vector)),
        Preelaborable_Initialization;
```

These are much as for lists. However, the aggregate aspect has additional items reflecting that vectors can also be indexed. And the stable properties have Capacity as well as Length as might be expected. The default initial condition is the same as for lists. Note that the functions Empty and New_Vector are new in Ada 2022 as follows

```
function Empty(Capacity: Count_Type := implm-defined) return Vector;

function New_Vector(First, Last: Index_Type) return Vector is
  (To_Vector(Count_Type(Last–First+1)));
```

Note that the existing function To_Vector simply returns a new vector with empty elements.

Aggregates for vectors can take various forms and deserve investigation in some detail. The aspect is

```
Aggregate => (Empty => Empty,
              Add_Unnamed => Append,
              New_Indexed => New_Vector,
              Assign_Indexed => Replace_Element),
```

The entries for Empty and Add_Unnamed are the same as for lists and have a similar meaning.

As in the case of lists, Ada 2022 has two procedures Append for vectors, one has two parameters and the other has three thus

```
procedure Append(Container: in out Vector; New_Item: in Element_Type);
procedure Append(Container: in out Vector; New_Item: in Element_Type;
                                              Count: in Count_Type);
```

The aggregate refers to the version with two parameters. Note also that originally in Ada 2012, there were four versions of Append, one had a default of 1 for the Count parameter and the other three appended a vector rather than a single element, these are now renamed Append_Vector to avoid confusion.

An empty vector can be created by using aggregate syntax thus

```
V: Vector := [];                 -- same as V: Vector := Empty;
```

and if we implemented the Float_Container using vectors rather than lists we could also initialize it by

```
The_Stack: Vector := [1.0, 37.5, 99.0];
```

Alternatively we could use a named aggregate thus

```
The_Stack: Vector := [0 => 1.0, 1 => 37.5, 2 => 99.0];
```

where we have assumed that the container was instantiated with the generic parameters (Natural, Float) and so the index starts at zero. The same vector could have been created in stages thus (using the last two entries in the aspect Aggregate)

```
The_Stack := New_Vector(0, 2);
The_Stack.Replace_Element(0, 1.0);
The_Stack.Replace_Element(1, 37.5);
The_Stack.Replace_Element(2, 99.0);
```

Aggregates for containers can also use other new aggregate features such as iterators and filters.

The dual nature of Vectors means that many operations are replicated such as

```
function Element(Container: Vector; Index: Index_Type)
                                                  return Element_Type;
function Element(Position: Cursor) return Element_Type;

function Element(Container: Vector;                          -- added
                 Position: Cursor) return Element_Type;      -- in 2022

procedure Replace_Element(Container: in out Vector;
                          Index: in Index_Type;
                          New_Item: in Element_Type);
procedure Replace_Element(Container: in out Vector;
                          Position: in Cursor;
                          New_Item: in Element_Type);
```

If we use an index then there is always a distinct parameter identifying the vector as well. If we use a cursor then the vector parameter was omitted in Ada 2012 if the vector was unchanged as in the case with the function Element. But in Ada 2022 there is also a version with the vector parameter so that prefixed notation can be used.

There are also functions First_Index and Last_Index thus

```
function First_Index(Container: Vector) return Index_Type;
function Last_Index(Container: Vector) return Extended_Index;
```

These return the values of the index of the first and last elements respectively. The function First_Index always returns Index_Type'First whereas Last_Index will return No_Index if the vector is empty. The function Length returns Last_Index-First_Index+1 which is zero if the vector is empty.

Note the irritating subtype Extended_Index thus

```
subtype Extended_Index is Index_Type'Base range
Index_Type'First-1 .. Index_Type'Min(index_Type'Base'Last-1, Index_Type'Last)+1;
```

which has to be introduced to cope with end values. The constant No_Index has the value Extended_Index'First which is equal to Index_Type'First-1.

There are operations to convert between an index and a cursor thus

```
function To_Cursor(Container: Vector; Index: Extended_Index)
                                                    return Cursor;
function To_Index(Position: Cursor) return Extended_Index;
```

It is perhaps slightly messier to use the index and vector parameters because of questions concerning the range of values of the index but probably slightly faster and maybe more familiar. And sometimes of course using an index is the whole essence of the problem. For example we could easily use the procedure Update_Element to double the values of those elements of a vector whose index was an even number. This would be somewhat tedious with cursors.

But an advantage of using cursors is that (provided certain operations are avoided) it is easy to replace the use of vectors by lists. For example here is the package Stack rewritten to use vectors

```
with Ada.Containers.Vectors;                              -- changed
use Ada.Containers;
package body Stack is

   package Float_Container is new Vectors(Natural, Float); -- changed
   The_Stack: Float_Container.Vector;                       -- changed

   procedure Push(X: in Float) is
   begin
      The_Stack.Append(X);
   end Push;
   ...      -- etc. exactly as before
end Stack;
```

So the changes are very few indeed and can be quickly done with a simple edit. Note that the index parameter has been given as Natural rather than Integer. Using Integer will not work since attempting to elaborate the subtype Extended_Index would raise Constraint_Error when evaluating Integer'First-1. But in any event it is more natural for the index range of the container to start at 0 (or 1) rather than a large negative number such as Integer'First.

The type Vector has aspects Constant_Indexing, Variable_Indexing, Default_Iterator, and Iterator_Element similar to the type List. It also has an instantiation of Ada.Iterator_Interfaces and types and functions for the magic manipulation of references (the functions are duplicated to allow manipulation via both indexes and cursors).

There are other important properties of vectors that should be mentioned. One is that there is a concept of capacity. Vectors are adjustable and will extend if necessary when new items are added. However, this might lead to lots of extensions and copying and so we can set the capacity of a container by calling

```
procedure Reserve_Capacity(Container: in out Vector;
                           Capacity: in Count_Type);
```

There is also

```
function Capacity(Container: Vector) return Count_Type;
```

which naturally returns the current capacity. Note that Length(V) cannot exceed Capacity(V) but might be much less.

If we add items to a vector whose length and capacity are the same then no harm is done. The capacity will be expanded automatically by effectively calling Reserve_Capacity internally. So the user does not need to set the capacity although not doing so might result in poorer performance.

There is also the concept of 'empty elements'. These are elements whose values have not been set. There is no corresponding concept with lists. It is a bounded error to read an empty element. Empty elements can be created in a number of ways.

The first way in which empty elements arise is if we declare a vector by calling

```
function To_Vector(Length: Count_Type) return Vector;
```

as in

```
My_Vector: Vector := To_Vector(100);
```

Note that there is also the much safer

```
function To_Vector(New_Item: Element_Type; Length: Count_Type)
                                                  return Vector;
```

which sets all the elements to the value New_Item.

Empty elements can also arise by calling

```
procedure Set_Length(Container: in out Vector; Length: in Count_Type);
```

This changes the length of a vector. This may require elements to be deleted (from the end) or to be added – in which case the new elements are empty.

The final way to get an empty element is by calling one of the procedures Insert_Space which are

```
procedure Insert_Space(Container: in out Vector;
                       Before: in Extended_Index;
                       Count: in Count_Type := 1);

procedure Insert_Space(Container: in out Vector;
                       Before: in Cursor;
                       Position: out Cursor;
                       Count: in Count_Type := 1);
```

These insert the number of empty elements given by Count at the place indicated. Existing elements are slid along as necessary. These should not be confused with the versions of Insert which do not provide an explicit value for the elements – in those cases the new elements take their default values.

Care needs to be taken if we use empty elements. For example, we should not compare two vectors using "=" if they have empty elements because this implies reading them. But the big advantage of empty elements is that they provide a quick way to make a large lump of space in a vector which can then be filled in with appropriate values. One big slide is a lot faster than lots of little ones.

For completeness, we briefly mention the remaining few subprograms that are unique to the vectors package. There are functions Insert_Vector (they were just Insert in Ada 2012) thus

```
procedure Insert_Vector(Container: in out Vector; Before: in
                        Extended_Index; New_Item: in Vector);
procedure Insert_Vector(Container: in out Vector; Before: in Cursor;
                        New_Item: in Vector);
procedure Insert_Vector(Container: in out Vector; Before: in Cursor;
                        New_Item: in Vector; Position: out Cursor);
```

These insert copies of a vector into another vector (rather than just single elements).

There are also corresponding versions of Prepend_Vector and Append_Vector (they were Prepend and Append in Ada 2012) thus

```
procedure Prepend_Vector(Container: in out Vector;
                         New_Item: in Vector);
procedure Append_Vector(Container: in out Vector;
                        New_Item: in Vector);
```

Finally, there are four functions "&" which concatenate vectors and elements by analogy with those for the type String. Their specifications are

```
function "&" (Left, Right: Vector) return Vector;
function "&" (Left: Vector; Right: Element_Type) return Vector;
function "&" (Left: Element_Type; Right: Vector) return Vector;
function "&" (Left, Right: Element_Type) return Vector;
```

Note the similarity between

```
Append(V1, V2);
V1 := V1 & V2;
```

The result is the same but using "&" is less efficient because of the extra copying involved. But "&" is a familiar operation and so is provided for convenience.

It will be recalled that in the case of a list if a cursor C denotes an element E and we add or delete other elements, then the cursor continues to denote E. This stability does not apply to vectors. The underlying reason is that vectors can also be indexed. Note however, that the stack example above is satisfactory.

We conclude by summarizing the subprograms that are added in Ada 2022 to enable prefixed notation to be used.

```
function Has_Element(Container: Vector; Position: Cursor) return Boolean;
function To_Index(Container: Vector; Position: Cursor)
                                          return Extended_Index;
function Element(Container: Vector; Position: Cursor) return Element_Type;
procedure Query_Element(Container: in Vector; Position: In Cursor;
        Process: not null access procedure (Element in Element_Type));
function Next(Container: Vector; Position: Cursor) return Cursor;
procedure Next(Container: in Vector; Position: in out Cursor);
function Previous(Container: Vector; Position: Cursor) return Cursor;
procedure Previous(Container: in Vector; Position: in out Cursor);
```

As in the case of lists, if the Container parameter does not match that identified by the Cursor then Program_Error is raised.

Exercise 24.3

1. Repeat Exercise 24.2(1) but use a vector rather than a list. Would this be a good idea?

2. Write a generic function to convert a vector into a list. The generic parameters will be instantiations of Doubly_Linked_Lists and Vectors.

3. Write a generic function to compare a vector and a list. Assume that there is no duplication of elements.

24.4 Maps

Remember that a map is just a means of getting from a value of one type (the key) to another type (the element). This is not a one-one relationship. Given a key there is a unique element (if any), but several keys may correspond to the same element. A simple example is an array. This is a map from the index type to the component type. Thus if we have

```
S: String := "animal";
```

then this provides a map from integers in the range 1 to 6 to some values of the type Character. Given an integer such as 4 there is a unique character 'm' but given a character such as 'a' there might be several corresponding integers (in this case both 1 and 5). But no element corresponds to a key value of 7.

More interesting examples are where the set of used key values is quite sparse. For example we might have a store where various spare parts are held. The parts have a five-digit part number and there are perhaps twenty racks where they are held identified by a letter. However, only a handful of the five-digit numbers are in use so it would be very wasteful to use an array with the part number as index. What we want is a container which holds just the pairs that matter such as (34618, 'F'), (27134, 'C') and so on. We can do this using a map. We usually refer to the pairs of values as nodes of the map.

There are two maps packages with much in common. One keeps the keys in order and the other uses a hash function. We will look at the specification of the ordered maps package showing just those facilities common to both.

```
with Ada.Iterator_Interfaces;
generic
   type Key_Type is private;
   type Element_Type is private;
   with function "<" (Left, Right: Key_Type) return Boolean is <>;
   with function "=" (Left, Right: Element_Type) return Boolean is <>;
package Ada.Containers.Ordered_Maps
      with Preelaborate, Remote_Types,
            Nonblocking, Global => in out synchronized is

   function Equivalent_Keys(Left: Right: Key_Type) return Boolean is
      (not ((Left < Right) or (Right < Left)));
```

The generic parameters include the ordering relationship "<" on the keys and equality for the elements.

It is assumed that the ordering relationship is well behaved in the sense that if x < y is true then y < x is false. We say that two keys x and y are equivalent if both x < y and y < x are false. In other words this defines an equivalence class on keys. The relationship must also be transitive, that is if x < y and y < z are both true then x < z must also be true.

This concept of an equivalence relationship occurs throughout the various maps and sets. Sometimes, as here, it is defined in terms of an order but in other cases, as we shall see, it is defined by an equivalence function.

It is absolutely vital that the equivalence relations are defined properly and meet the above requirements. It is not possible for the container packages to check this and if the operations are wrong then peculiar behaviour is almost inevitable.

Note how the function Equivalent_Keys is declared explicitly as an expression function.

The equality operation on elements is not so demanding. It must be symmetric so that x = y and y = x are the same but transitivity is not required (although cases where it would not automatically be transitive are likely to be rare). The operation is only used for the function "=" on the containers as a whole.

Note that Find and similar operations for maps and sets work in terms of the equivalence relationship rather than equality as was the case with lists and vectors.

```
   type Map is tagged private
      with Constant_Indexing => Constant_Reference,
            Variable_Indexing => Reference,
            Default_Iterator => Iterate,
            Iterator_Element => Element_Type,
            Iterator_View => Stable.Map,                    -- new from here
            Aggregate => (Empty => Empty,
                                 Add_Named => Insert),
            Stable_Properties => (Length,
                                          Tampering_With_Cursors_Prohibited,
                                          Tampering_With_Elements_Prohibited),
```

```
Default_Initial_Condition =>
   Length(Map) = 0 and then
   (not Tampering_With_Cursors_Prohibited(Map)) and then
   (not Tampering_With_Elements_Prohibited(Map)),
Preelaborable_Initialization;
```

The most notable thing here is that the Aggregate aspect is

```
Aggregate => (Empty => Empty,
              Add_Named => Insert),
```

which uses Add_Named unlike Lists and Vectors. The stock control system which we will meet later could be initialized with two items in it thus

```
The_Store := [((34618,1998) => ('F', 25)), ((27134, 2022) => ('C', 45))];
```

The declarations continue as follows

```
type Cursor is private
   with Preelaborable_Initialization;

Empty_Map: constant Map;
No_Element: constant Cursor;

function Has_Element(Position: Cursor) return Boolean;
function Has_Element(Container: Map                    -- added in
                     Position: Cursor) return Boolean;  -- Ada 2022
```

The types Map and Cursor and constants Empty_Map and No_Element and the functions Has_Element are similar to the corresponding entities in the lists and vectors containers. Note the additional Has_Element in Ada 2022 once more.

```
package Map_Iterator_Interfaces is
   new Ada.Iterator_Interfaces(Cursor, Has_Element);
```

is the expected instantiation of Ada.Iterator_Interfaces. This is followed by

```
function Tampering_With_Cursors_Prohibited(Container: Map)
                                                     return Boolean;
   with Nonblocking, Global => null, Use_Formal => null;

function Tampering_With_Elements_Prohibited(Container: Map)
                                                     return Boolean;
   with Nonblocking, Global => null, Use_Formal => null;

function Empty return Map is
   (Empty_Map);

function "=" (Left, Right: Map) return Boolean;
function Length(Container: Map) return Count_Type;
function Is_Empty(Container: Map) return Boolean;
procedure Clear(Container: in out Map);
```

These are again similar to the corresponding entities for lists. Note that two maps are said to be equal if they have the same number of nodes with equivalent keys (as defined by "<") whose corresponding elements are equal (as defined by "=").

Again there are new versions of many subprograms with a Map parameter so that prefixed notation can be used. As earlier we just give the new version and mark with *** to indicate that the version without Map parameter still exists.

```
***   function Key(Container: Map; Position: Cursor) return Key_Type;
***   function Element(Container: Map; Position: Cursor) return Element_Type;

      procedure Replace_Element(Container: in out Map;
                                Position: in Cursor;
                                New_Item: in Element_Type);
***   procedure Query_Element(Container: in Map;
                              Position: in Cursor;
                              Process: not null access procedure
                            (Key: in Key_Type; Element: in Element_Type));
      procedure Update_Element(Container: in out Map;
                               Position: in Cursor;
                               Process: not null access procedure
                     (Key: in Key_Type; Element: in out Element_Type));
```

In this case there is a function Key as well as a function Element. But there is no procedure Replace_Key since it would not make sense to change a key without changing the element as well and this really comes down to deleting the whole node and then inserting a new one.

The procedures Query_Element and Update_Element are slightly different in that the procedure Process also takes the key as parameter as well as the element to be read or updated. Note again that the key cannot be changed. Nevertheless the value of the key is given since it might be useful in deciding how the update should be performed. Remember that we cannot get uniquely from an element to a key but only from a key to an element.

In Ada 2012 and Ada 2022, the above subprograms are again somewhat redundant because we can use the magic types and functions for references which in this case are

```
type Constant_Reference_Type
        (Element: not null access constant Element_Type) is private
   with Implicit_Dereference => Element;

type Reference_Type
                     (Element: not null access Element_Type) is private
   with Implicit_Dereference => Element;

function Constant_Reference
                 (Container: aliased in Map; Position: in Cursor)
                                   return Constant_Reference_Type;

function Reference(Container: aliased in out Map; Position: in Cursor)
                                              return Reference_Type;

function Constant_Reference
                 (Container: aliased in Map; Key: in Key_Type)
                                   return Constant_Reference_Type;

function Reference(Container: aliased in out Map; Key: in Key_Type)
                                              return Reference_Type;
```

Note that there are two pairs of functions Constant_Reference and Reference because we can use either the cursor or the key to identify the element concerned.

```
procedure Assign(Target: in out Map; Source: in Map);
function Copy(Source: Map) return Map;
procedure Move(Target, Source: in out Map);
```

These are similar to the corresponding subprograms operating on lists.

```
procedure Insert(Container: in out Map;
                 Key: in Key_Type;
                 New_Item: in Element_Type;
                 Position: out Cursor;
                 Inserted: out Boolean);

procedure Insert(Container: in out Map;
                 Key: in Key_Type;
                 Position: out Cursor;
                 Inserted: out Boolean);

procedure Insert(Container: in out Map;
                 Key: in Key_Type;
                 New_Item: in Element_Type);
```

These insert a new node into the map unless a node with an equivalent key already exists. If it does exist then the first two return with Inserted set to False and Position indicating the node whereas the third raises Constraint_Error (the element value is not changed). If a node with an equivalent key is not found then a new node is created with the given key, the element value is set to New_Item when that is given and otherwise it takes its default value (if any), and Position is set when given.

An important property of keys is therefore that they are unique; two nodes cannot have the same (that is, equivalent) key.

Unlike vectors and lists, we do not have to say where the new node is to be inserted because of course this is an ordered map and it just goes in the correct place according to the order given by the generic parameter "<".

```
procedure Include(Container: in out Map;
                  Key: in Key_Type;
                  New_Item: in Element_Type);
```

This is somewhat like the last Insert except that if an existing node with an equivalent key is found then it is replaced (rather than raising Constraint_Error). Note that both the key and the element are updated. This is because equivalent keys might not be totally equal.

For example the key part might be a record with part number and year of introduction, thus

```
type Part_Key is
   record
      Part_Number: Integer;
      Year: Integer;
   end record;
```

and we might define the ordering relationship to be used as the generic parameter simply in terms of the part number

```
function "<" (Left, Right: Part_Key) return Boolean is
begin
   return Left.Part_Number < Right.Part_Number;
end "<";
```

In this situation, the keys could match without the year component being the same and so it would need to be updated. In other words with this definition of the ordering relation, two keys are equivalent provided just the part numbers are the same.

```
procedure Replace(Container: in out Map;
                  Key: in Key_Type;
                  New_Item: in Element_Type);
```

In this case, Constraint_Error is raised if the node does not already exist. On replacement both the key and the element are updated as for Include.

Perhaps a better example of equivalent keys not being totally equal is if the key were a string. We might decide that the case of letter did not need to match in the test for equivalence but nevertheless we would probably want to update with the string as used in the parameter of Replace.

```
procedure Exclude(Container: in out Map; Key: in Key_Type);
```

If there is a node with an equivalent key then it is deleted. If there is not then nothing happens.

```
procedure Delete(Container: in out Map; Key: in Key_Type);
procedure Delete(Container: in out Map; Position: in out Cursor);
```

These delete a node. In the first case if there is no such equivalent key then Constraint_Error is raised (by contrast to Exclude which remains silent in this case). In the second case if the cursor is No_Element then again Constraint_Error is raised – there is also a check to ensure that the cursor otherwise does designate a node in the correct map (remember that cursors designate both an entity and the container); if this check fails then Program_Error is raised.

Perhaps it is worth observing that Insert, Include, Replace, Exclude, and Delete form a sort of progression from an operation that will insert something, through operations that might insert, will neither insert nor delete, might delete, to the final operation that will delete something. Note also that Include, Replace, and Exclude do not apply to lists and vectors.

```
    function First(Container: Map) return Cursor;
    function Last(Container: Map) return Cursor;
*** function Next(Container: Map; Position: Cursor) return Cursor;
*** procedure Next(Container: Map; Position: in out Cursor);
    function Find(Container: Map; Key: Key_Type) return Cursor;
    function Element(Container: Map; Key: Key_Type) return Element;
    function Contains(Container: Map; Key: Key_Type) return Boolean;
```

These should be self-evident. Unlike the operations on vectors and lists, Find logically searches the whole map and not just starting at some point (and since it searches the whole map there is no need for Reverse_Find). (In implementation terms it won't actually search the whole map because it will be structured in a way that makes this unnecessary – as a balanced tree perhaps.) Moreover, Find uses the equivalence relation based on the "<" parameter so in the example it only has to match the part number and not the year. The function call Element(My_Map, My_Key) is equivalent to Element(Find(My_Map, My_Key)).

```
procedure Iterate(Container: in Map;
                  Process: not null access procedure(Position: in Cursor));
                       with Allows_Exit;
```

This is also as for other containers. Note that there is also Reverse_Iterate for ordered maps but not for hashed maps; note again Allows_Exit which is added in Ada 2022.

There are also two functions Iterate for ordered maps which are

```
function Iterate(Container: in Map)
      return Map_Iterator_Interfaces.Parallel_Reversible_Iterator'Class;

function Iterate(Container: in Map; Start: in Cursor)
            return Map_Iterator_Interfaces.Reversible_Iterator'Class;
```

These are similar to those for list containers and are invoked if we write a loop of the form

```
for C in My_Map.Iterate loop ...
for C in My_Map.Iterate(S) loop ...
```

Again the function without Start returns the parallel iterator so we can use a parallel loop thus

```
parallel for C in My_Map.Iterate loop ...
```

Again remember that parallel loops cannot specify **reverse**. But if we use the iterate with the parameter Start then we can specify **reverse**.

However, in the case of hashed maps there is only one function (the start cannot be specified since we have no concept of ordering) and the one function returns a value of Forward_Iterator'Class and so the iterations cannot be done in reverse.

And at last (ignoring the stable which is much as expected)

```
private
   ...      -- not specified by the language
end Ada.Containers.Ordered_Maps;
```

We have omitted to mention quite a few operations that have no equivalent in hashed maps – we will come back to these in a moment.

As an example we can make a container to hold the information concerning spare parts. We can use the type Part_Key and the function "<" as above. We can suppose that the element type is

```
type Stock_Info is
  record
    Shelf: Character range 'A' .. 'T';
    Stock: Integer;
  end record;
```

This gives both the shelf letter and the number in stock. We can then declare the container thus

```
package Store_Maps is
      new Ordered_Maps(Key_Type => Part_Key,
                       Element_Type => Stock_Info,
                       "<" => "<",
                       "=" => "=");
The_Store: Store_Maps.Map;
use Store_Maps;
```

The last parameter could be omitted since the formal has <> default for it and the actual is a record which does have predefined equality. If Part_Key were of a scalar type then the penultimate parameter could also be omitted. But it is also a record type and only = and /= are predefined for records, not <.

We can now add items to our store by calling Insert (using named parameters for clarity) thus

```
The_Store.Insert(Key => (34618, 1998), New_Item => ('F', 25));
The_Store.Insert(Key => (27134, 2022), New_Item => ('C', 45));
```

We might now have a procedure which, given a part number, checks to see if it exists and that the stock is not zero, and if so returns the shelf letter and year number and decrements the stock count.

```
procedure Request(Part: in Integer; OK: out Boolean;
                  Year: out Integer; Shelf: out Character) is
  C: Cursor;
  K: Part_Key;
  E: Stock_Info;
begin
  C := The_Store.Find((Part, 0));
  if C = No_Element then
    OK := False;  return;                    -- no such key
  end if;
  E := Element(C);  K := Key(C);
  Year := K.Year;  Shelf := E.Shelf;
  if E.Stock = 0 then
    OK := False;  return;                    -- out of stock
  end if;
  Replace_Element(The_Store, C, (Shelf, E.Stock-1));
  OK := True;
end Request;
```

Note that we had to put a dummy year number in the call of Find. We could of course use the <> notation for this

```
C := The_Store.Find((Part, others => <>));
```

In Ada 2012 and Ada 2022, rather than using Replace_Element, we can simply write

```
The_Store(C) := (Shelf, E.Stock-1);
```

As another example suppose we wish to check all through the stock looking for parts whose stock is low, perhaps less than some given parameter. We first illustrate this using the Ada 2005 mechanisms and then show the neater form in Ada 2012 and Ada 2022.

We can use the procedure Iterate for this as follows

```
procedure Check_Stock(Low: in Integer) is

   procedure Check_It(C: in Cursor) is
   begin
      if Element(C).Stock < Low then
         Put("Low stock of part ");          -- print a message perhaps
         Put(Key(C).Part_Number);  New_Line;
      end if;
   end Check_It;

begin
   The_Store.Iterate(Check_It'Access);
end Check_Stock;
```

Note that this uses a so-called downward closure. The procedure Check_It has to be declared locally to Check_Stock in order to access the parameter Low. An alternative approach is to use First and Next and so on thus

```
procedure Check_Stock(Low: in Integer) is
   C: Cursor := The_Store.First;
begin
   loop
      exit when C = No_Element;
      if Element(C).Stock < Low then
         Put("Low stock of part ");          -- print a message perhaps
         Put(Key(C).Part_Number);  New_Line;
      end if;
      C := Next(C);
   end loop;
end Check_Stock;
```

In Ada 2012 and Ada 2022 we can more simply use the new function Iterate as follows

```
procedure Check_Stock(Low: in Integer) is
begin
   for C in The_Store.Iterate loop
      if Element(C).Stock < Low then
         Put("Low stock of part ");          -- print a message perhaps
```

```
        Put(Key(C).Part_Number); New_Line;
      end if;
    end loop;
  end Check_Stock;
```

We will now consider hashed maps. The trouble with ordered maps in general is that searching can be slow when the map has many entries. Techniques such as a binary tree can be used but even so the search time will increase at least as the logarithm of the number of entries. A better approach is to use a hash function. This will be familiar to many readers (especially those who have written compilers). The general idea is as follows.

We define a function which takes a key and returns some value in a given range. In the case of the Ada containers it has to return a value of the modular type Hash_Type which is declared in the root package Ada.Containers. We could then convert this value onto a range representing an index into an array whose size corresponds to the capacity of the map. This index value is the preferred place to store the entry. If there already is an entry at this place (because some other key has hashed to the same value) then a number of approaches are possible. One way is to create a list of entries with the same index value (often called buckets); another way is simply to put it in the next available slot. The details don't matter. But the overall effect is that provided the map is not too full and the hash function is good then we can find an entry almost immediately more or less irrespective of the size of the map.

Note that because of the pseudo-random nature of hashing in general, there is no obvious relation between the order of the elements, the order of the keys, or the order in which they were inserted.

So as users all we have to do is to define a suitable hash function. It should give a good spread of values across the range of Hash_Type for the population of keys, it should avoid clustering and above all for a given key it must always return the same hash value.

Defining good hash functions needs care. In the case of the part numbers we might multiply the part number by some obscure prime number and then truncate the result down to the modular type Hash_Type. The author hesitates to give an example but perhaps

```
function Part_Hash(P: Part_Key) return Hash_Type is
   M31: constant := 2**31-1;      -- a nice Mersenne prime
begin
   return Hash_Type(P.Part_Number) * M31;
end Part_Hash;
```

On reflection that's probably a very bad prime to use because it is so close to half of 2**32 a typical value of Hash_Type'Last+1. Of course it doesn't have to be prime but simply relatively prime to it such as 5**13. Knuth (see Bibliography) suggests dividing the range by the golden number $\tau = (\sqrt{5}+1)/2 = 1.618...$ and then taking the nearest number relatively prime which is in fact simply the nearest odd number (in this case it is 2654435769).

The specification of the hashed maps package is very similar to that for ordered maps. It starts

```
with Ada.Iterator_Interfaces;
generic
   type Key_Type is private;
   type Element_Type is private;
   with function Hash(Key: Key_Type) return Hash_Type;
   with function Equivalent_Keys(Left, Right: Key_Type) return Boolean;
   with function "=" (Left, Right: Element_Type) return Boolean is <>;
package Ada.Containers.Hashed_Maps
     with Preelaborate, Remote_Types
          Nonblocking, Global => in out synchronized is
```

The differences from the ordered maps package are that there is an extra generic parameter Hash giving the hash function and the ordering parameter "<" has been replaced by the function Equivalent_Keys. It is this function that defines the equivalence relationship for hashed maps; it is vital that Equivalent_Keys(X, Y) is always the same as Equivalent_Keys(Y, X). Moreover, if X and Y are equivalent and Y and Z are equivalent then X and Z must also be equivalent.

Note that the function Equivalent_Keys in the ordered maps package discussed above corresponds to the formal generic parameter of the same name in this hashed maps package. This should make it easier to convert between the two forms of packages.

Returning to our example, if we now write

```
function Equivalent_Parts(Left, Right: Part_Key) return Boolean is
begin
   return Left.Part_Number = Right.Part_Number;
end Equivalent_Parts;
```

then we can instantiate the hashed maps package as follows

```
package Store_Maps is
        new Hashed_Maps(Key_Type => Part_Key,
                        Element_Type => Stock_Info,
                        Hash => Part_Hash,
                        Equivalent_Keys => Equivalent_Parts);
The_Store: Store_Maps.Map;
```

and then the rest of our example will be exactly as before. It is thus easy to convert from an ordered map to a hashed map and vice versa provided of course that we only use the facilities common to both.

We will finish this discussion of maps by briefly considering the additional facilities in the two packages.

The ordered maps package has the following additional subprograms

```
procedure Delete_First(Container: in out Map);
procedure Delete_Last(Container: in out Map);
function First_Element(Container: Map) return Element_Type;
```

```
        function First_Key(Container: Map) return Key_Type;
        function Last_Element(Container: Map) return Element_Type;
        function Last_Key(Container: Map) return Key_Type;
***     function Previous(Container: Map; Position: Cursor) return Cursor;
***     procedure Previous(Container: Map; Position: in out Cursor);
        function Floor(Container: Map; Key: Key_Type) return Cursor;
        function Ceiling(Container: Map; Key: Key_Type) return Cursor;
        function "<" (Left, Right: Cursor) return Boolean;
        function ">" (Left, Right: Cursor) return Boolean;
        function "<" (Left: Cursor; Right: Key_Type) return Boolean;
        function ">" (Left: Cursor; Right: Key_Type) return Boolean;
        function "<" (Left: Key_Type; Right: Cursor) return Boolean;
        function ">" (Left: Key_Type; Right: Cursor) return Boolean;
        procedure Reverse_Iterate(Container: in Map;
                                  Process: not null access procedure
                                                       (Position: in Cursor));
```

These are again largely self-evident. The functions Floor and Ceiling are interesting. Floor searches for the last node whose key is not greater than Key and similarly Ceiling searches for the first node whose key is not less than Key – they return No_Element if there is no such element. The subprograms Previous are of course the opposite of Next and Reverse_Iterate is like Iterate only backwards as mentioned earlier.

The functions "<" and ">" are mostly for convenience. Thus the first is equivalent to

```
function "<" (Left, Right: Cursor) return Boolean is
begin
   return Key(Left) < Key(Right);
end "<";
```

Clearly these additional operations must be avoided if we wish to retain the option of converting to a hashed map later.

Hashed maps have a very important facility not in ordered maps which is the ability to specify a capacity as for the vectors package. (Underneath their skin the hashed maps are a bit like vectors whereas the ordered maps are a bit like lists.) Thus we have

```
procedure Reserve_Capacity(Container: in out Map;
                           Capacity: in Count_Type);
function Capacity(Container: Map) return Count_Type;
```

The behaviour is much as for vectors. We don't have to set the capacity ourselves since it will be automatically extended as necessary but it might significantly improve performance to do so. In the case of maps, increasing the capacity requires the hashing to be redone which could be quite time consuming, so if we know that our map is going to be a big one, it is a good idea to set an appropriate capacity right from the beginning. Note again that Length(M) cannot exceed Capacity(M) but might be much less.

The other additional subprograms for hashed maps are

```
function Equivalent_Keys(Left, Right: Cursor) return Boolean;
function Equivalent_Keys(Left: Cursor; Right: Key_Type) return Boolean;
function Equivalent_Keys(Left: Key_Type; Right: Cursor) return Boolean;
```

These (like the additional "<" and ">" for ordered maps) are again mostly for convenience. The first is equivalent to

```
function Equivalent_Keys(Left, Right: Cursor) return Boolean is
begin
  return Equivalent_Keys(Key(Left), Key(Right));
end Equivalent_Keys;
```

Before moving on to sets it should be noticed that there are also some useful functions in the string packages discussed in Section 23.3. The main one is

```
with Ada.Containers;
function Ada.Strings.Hash(Key: String) return Containers.Hash_Type
  with Pure;
```

There is a similar function Strings.Unbounded.Hash where the parameter Key has type Unbounded_String. It simply converts the parameter to the type String and then calls Strings.Hash. There is also a generic function for bounded strings which again calls the basic function Strings.Hash. For completeness the function Strings.Fixed.Hash is a renaming of Ada.Strings.Hash.

These are provided because it is often the case that the key is a string and they save the user from devising good hash functions for strings which might cause a nasty headache.

We could for example save ourselves the worry of defining a good hash function in the above example by making the part number into a five-character string. So we might write

```
function Part_Hash(P: Part_Key) return Hash_Type is
begin
  return Ada.Strings.Hash(P.Part_Number);
end Part_Hash;
```

and if this doesn't work well then we can blame the vendor.

For the wide enthusiast there are also functions such as Ada.Strings.Wide_Hash and Ada.Strings.Wide_Wide_Hash for basing a hash on wide or wide wide characters. And there are also functions Ada.Strings.Hash_Case_Insensitive and so on for use when we don't want to be bothered by the case of characters.

Exercise 24.4

1 Implement the package Tag_Registration of Section 21.6 using an ordered map.

24.5 Sets

Sets, like maps, come in two forms: hashed and ordered. Sets are of course just collections of values and there is no question of a key (we can perhaps think of the value as being its own key since a set is simply a map from the value to Boolean). Thus in the case of an ordered set the values are stored in order whereas in the case of a map, it is the keys that are stored in order. As well as the usual operations of inserting elements into a set and searching and so on, there are also many operations on sets as a whole that do not apply to the other containers – these are the familiar set operations such as union and intersection.

Here is the specification of the ordered sets package giving just those facilities that are common to both kinds of sets.

```
with Ada.Iterator_Interfaces;
generic
   type Element_Type is private;
   with function "<" (Left, Right: Element_Type) return Boolean is <>;
   with function "=" (Left, Right: Element_Type) return Boolean is <>;
package Ada.Containers.Ordered_Sets
      with Preelaborate, Remote_Types,
           Nonblocking, Global => in out synchronized is

   function Equivalent_Elements(Left, Right: Element_Type)
                                                        return Boolean;

   type Set is tagged private
      with Constant_Indexing => Constant_Reference,
           Default_Iterator => Iterate,
           Iterator_Element => Element_Type,
           Iterator_View => Stable.Set,                -- new from here
           Aggregate => (Empty => Empty,
                         Add_Unnamed => Include),
           Stable_Properties => (Length,
                                 Tampering_With_Cursors_Prohibited);
           Default_Initial_Condition =>
               Length(Set) = 0 and then
               (not Tampering_With_Cursors_Prohibited(Set)),
           Preelaborable_Initialization;
```

Note that Tampering_With_Elements_Prohibited does not appear in these aspects for the type Set.

The Aggregate aspect uses Add_Unnamed like lists (and unlike maps) thus

```
Aggregate => (Empty => Empty,
              Add_Unnamed => Include),
```

where Add_Unnamed uses Include in the case of Sets rather than Append in the case of lists.

The specification then continues much as expected

```
type Cursor is private
   with Preelaborable_Initialization;

Empty_Set: constant Set;
No_Element: constant Cursor;
function Has_Element(Position: Cursor) return Boolean;
function Has_Element(Container: Set; Position: Cursor) return Boolean;

package Set_Iterator_Interfaces is
   new Ada.Iterator.Interfaces(Cursor, Has_Element);
```

Again there is an additional Has_Element in Ada 2022.

One difference from the maps package (apart from the identifiers) are that there is no key type and both "<" and "=" apply to the element type (whereas in the case of maps, the operation "<" applies to the key type). Thus the ordering relationship "<" defined on elements defines equivalence between the elements whereas "=" defines equality. Another difference is that the type Map does not have the aspect Variable_Indexing; the reason is given below when we discuss the absence of Update_Element.

It is possible for two elements to be equivalent but not equal. For example if they were strings then we might decide that the ordering (and thus equivalence) ignored the case of letters but that equality should take the case into account. (They could also be equal but not equivalent but that is perhaps less likely.)

And as in the case of the maps package, the equality operation on elements is only used by the function "=" for comparing two sets.

Again we have the usual rules as explained for maps. Thus if x < y is true then y < x must be false; x < y and y < z must imply x < z; and x = y and y = x must be the same.

For the convenience of the user the function Equivalent_Elements is declared explicitly. It is equivalent to

```
function Equivalent_Elements(Left, Right: Element_Type)
                                                   return Boolean is
begin
   return not(Left < Right) and not(Right < Left);
end Equivalent_Elements;
```

This function Equivalent_Elements corresponds to the formal generic parameter of the same name in the hashed sets package discussed below. This should make it easier to convert between the two forms of packages.

Then we have

```
function "=" (Left, Right: Set) return Boolean;
function Equivalent_Sets(Left, Right: Set) return Boolean;

function Tampering_With_Cursors_Prohibited(Container: Set)
                                                   return Boolean;
with Nonblocking, Global => null, Use_Formal => null;

function Empty return Set is
   (Empty_Set);
```

```
function To_Set(New_Item: Element_Type) return Set;
function Length(Container: Set) return Count_Type;
function Is_Empty(Container: Set) return Boolean;
procedure Clear(Container: in out Set);
```

Note the addition of Equivalent_Sets and To_Set. Two sets are equivalent if they have the same number of elements and the pairs of elements are equivalent. This contrasts with the function "=" where the pairs of elements have to be equal rather than equivalent. Remember that elements might be equivalent but not equal (as in the example of a string mentioned above). The function To_Set takes a single element and creates a set. It is particularly useful when used in conjunction with operations such as Union described below. The other subprograms are as in the other containers.

```
***   function Element(Container: Set; Position: Cursor) return Element_Type;

      procedure Replace_Element(Container: in out Set;
                                Position: in Cursor;
                                New_Item: in Element_Type);
***   procedure Query_Element(Container: in Set;
                              Position: in Cursor;
                              Process: not null access procedure
                                          (Element: in Element_Type));
```

Again these are much as expected except that there is no procedure Update_ Element. This is because the elements are arranged in terms of their own value (either by order or through the hash function) and if we just change an element *in situ* then it might become out of place (this problem does not arise with the other containers). This is also why the type Set does not have the aspect Variable_ Indexing. Note also that Replace_Element has to ensure that the value New_Item is not equivalent to an element in a different position; if it is then Program_Error is raised. Beware when using Replace_Element not to introduce a new value that changes the order since this might cause a loop to miss an element or visit it twice. We will return to the problem of the missing Update_Element later.

```
type Constant_Reference_Type
          (Element: not null access constant Element_Type) is private
   with Implicit_Dereference => Element;

function Constant_Reference
               (Container: aliased in Set; Position: in Cursor)
                              return Constant_Reference_Type;
```

In the case of sets only the constant versions of these magic facilities are available; again because we cannot update an element *in situ*.

```
procedure Assign(Target: in out Set; Source: in Set);
function Copy(Source: Set) return Set;
procedure Move(Target, Source: in out Set);
```

These are just as for the other containers.

```
procedure Insert(Container: in out Set;
                 New_Item: in Element_Type;
                 Position: out Cursor;
                 Inserted: out Boolean);

procedure Insert(Container: in out Set;
                 New_Item: in Element_Type);
```

These insert a new element into the set unless an equivalent element already exists. If it does exist then the first one returns with Inserted set to False and Position indicating the element whereas the second raises Constraint_Error (the element value is not changed). If an equivalent element is not in the set then it is added and Position is set accordingly.

```
procedure Include(Container: in out Set; New_Item: in Element_Type);
```

This is somewhat like the last Insert except that if an equivalent element is already in the set then it is replaced (rather than raising Constraint_Error).

```
procedure Replace(Container: in out Set; New_Item: in Element_Type);
```

In this case, Constraint_Error is raised if an equivalent element does not already exist.

```
procedure Exclude(Container: in out Set; Item: in Element_Type);
```

If an element equivalent to Item is already in the set, then it is deleted.

```
procedure Delete(Container: in out Set; Item: in Element_Type);
procedure Delete(Container: in out Set; Position: in out Cursor);
```

These both delete an element. In the first case, if there is no such equivalent element then Constraint_Error is raised. In the second case, if the cursor is No_Element then again Constraint_Error is raised – there is also a check to ensure that the cursor otherwise does designate an element in the correct set (remember that cursors designate both an entity and the container); if this check fails then Program_Error is raised.

And then we have the usual set operations.

```
procedure Union(Target: in out Set; Source: in Set);
function Union(Left, Right: Set) return Set;
function "or" (Left, Right: Set) return Set renames Union;

procedure Intersection(Target: in out Set; Source: in Set);
function Intersection(Left, Right: Set) return Set;
function "and" (Left, Right: Set) return Set renames Intersection;

procedure Difference(Target: in out Set; Source: in Set);
function Difference(Left, Right: Set) return Set;
function "–" (Left, Right: Set) return Set renames Difference;

procedure Symmetric_Difference(Target: in out Set; Source: in Set);
function Symmetric_Difference (Left, Right: Set) return Set;
function "xor" (Left, Right: Set) return Set renames Symmetric_Difference;
```

These all do exactly what one would expect using the equivalence relation on the elements.

```
function Overlap(Left, Right: Set) return Boolean;
function Is_Subset(Subset: Set; Of_Set: Set) return Boolean;
```

These are self-evident as well.

```
    function First(Container: Set) return Cursor;
    function Last(Container: Set) return Cursor;
*** function Next(Container: Set; Position: Cursor) return Cursor;
*** procedure Next(Container: Set; Position: in out Cursor);
    function Find(Container: Set; Item: Element_Type) return Cursor;
    function Contains(Container: Set; Item: Element_Type) return Boolean;
```

These should be self-evident and are very similar to the corresponding operations on maps. Again unlike the operations on vectors and lists, Find logically searches the whole set and not just starting at some point (there is also no Reverse_Find). Moreover, Find uses the equivalence relation based on the "<" parameter.

And then we have

```
procedure Iterate(Container: in Set;
                  Process: not null access procedure
                                          (Position: in Cursor))
                  with Allows_Exit;
```

This is also as for other containers. Note that there is also Reverse_Iterate for ordered sets but not for hashed sets; note again Allows_Exit which is added in Ada 2022.

There also have two functions Iterate for ordered sets which are

```
function Iterate(Container: in Set)
        return Set_Iterator_Interfaces.Parallel_Reversible_Iterator'Class;
function Iterate(Container: in Set; Start: in Cursor)
                  return Set_Iterator_Interfaces.Reversible_Iterator'Class;
```

These are similar to those for map containers and are invoked if we write a loop of the form

```
for C in A_Set.Iterate loop ...
for C in A_Set.Iterate(S) loop ...
```

Again the function without Start returns the parallel iterator so we can use a parallel loop thus

```
parallel for C in My_Set.Iterate loop ...
```

Again remember that parallel loops cannot specify **reverse**. But if we use the iterator with the parameter Start then we can specify **reverse**.

However, in the case of hashed sets there is only one function (the start cannot be specified since we have no concept of ordering) and the one function returns a value of Forward_Iterator'Class and so the iterations cannot be done in reverse. These rules are exactly the same as for maps.

The sets packages have an internal generic package called Generic_Keys. This generic package enables some set operations to be performed in terms of keys where the key is a function of the element. Note carefully that in the case of a map, the element is defined in terms of the key whereas here the situation is reversed. An equivalence relationship is defined for these keys as well; this is defined by a generic parameter "<" for ordered sets and Equivalent_Keys for hashed sets.

In the case of ordered sets the formal parameters are

```
generic
   type Key_Type(<>) is private;
   with function Key(Element: Element_Type) return Key_Type;
   with function "<" (Left, Right: Key_Type) return Boolean is <>;
package Generic_Keys
   with Nonblocking, Global => null is
```

The following are then common to the package Generic_Keys for both hashed and ordered sets.

```
***   function Key(Container: Set; Position: Cursor) return Key_Type;
      function Element(Container: Set; Key: Key_Type) return Element_Type;

      procedure Replace(Container: in out Set;
                        Key: in Key_Type; New_Item: in Element_Type);

      procedure Exclude(Container: in out Set; Key: in Key_Type);
      procedure Delete(Container: in out Set; Key: in Key_Type);

      function Find(Container: Set; Key: Key_Type) return Cursor;
      function Contains(Container: Set; Key: Key_Type) return Boolean;

      procedure Update_Element_Preserving_Key
                             (Container: in out Set; Position: in Cursor;
                              Process: not null access procedure
                                          (Element: in out Element_Type));

      type Reference_Type(Element: not null access Element_Type) is private
         with Implicit_Dereference => Element;

      function Reference_Preserving_Key(Container: aliased in out Set;
                                        Position: in Cursor)
                                                return Reference_Type;

      function Constant_Reference(Container: aliased in Set;
                                  Key: in Key_Type)
                                          return Constant_Reference_Type;

      end Generic_Keys;
```

and finally (ignoring the stable which is much as expected)

```
private
   ...      -- not specified by the language
end Ada.Containers.Ordered_Sets;
```

It is expected that most users of sets will use them in a straightforward manner and that the operations specific to sets such as Union and Intersection will be dominant.

However, sets can be used as sort of economy class maps by using the inner package Generic_Keys. Although this is certainly not for the novice we will illustrate how this might be done by reconsidering the stock problem using sets rather than maps. We declare

```
type Part_Type is
   record
      Part_Number: Integer;
      Year: Integer;
      Shelf: Character range 'A' .. 'T';
      Stock: Integer;
   end record;
```

Here we have put all the information in the one type. We then declare "<" much as before

```
function "<" (Left, Right: Part_Type) return Boolean is
begin
   return Left.Part_Number < Right.Part_Number;
end "<";
```

and then instantiate the package thus

```
package Store_Sets is new Ordered_Sets(Element_Type => Part_Type);

The_Store: Store_Sets.Set;
use Store_Sets;
```

We have used the default generic parameter mechanism for "<" this time by way of illustration.

In this case we add items to the store by calling

```
The_Store.Insert(New_Item => (34618, 1998, 'F', 25));
The_Store.Insert(New_Item => (27134, 2004, 'C', 45));
...
```

The procedure for checking the stock could now become

```
procedure Request(Part: in Integer; OK: out Boolean;
                          Year: out Integer; Shelf: out Character) is
   C: Cursor;
   E: Part_Type;
begin
   C := The_Store.Find((Part, others => <>));
   if C = No_Element then
      OK := False;  return;                -- no such item
   end if;
   E := Element(C);
   Year := E.Year;  Shelf := E.Shelf;
```

```
   if E.Stock = 0 then
      OK := False; return;                -- out of stock
   end if;
   -- now update the stock level
   Replace_Element(The_Store, C,
                  (E.Part_Number, Year, Shelf, E.Stock-1));
   OK := True;
end Request;
```

This works but is somewhat unsatisfactory. For one thing we have had to make up dummy components in the call of Find (using <>) and also we have had to replace the whole of the element although we only wanted to update the Stock component. Rather than using Replace_Element we could write

```
The_Store(C) := (E.Part_Number, Year, Shelf, E.Stock-1);
```

Moreover, we cannot use Update_Element because it is not defined for sets at all. Remember that this is because it might make things out of order; that wouldn't be a problem in this case because we don't want to change the part number and our ordering is just by the part number.

A better approach is to use the part number as a key. We define

```
type Part_Key is new Integer;

function Part_No(P: Part_Type) return Part_Key is
begin
   return Part_Key(P.Part_Number);
end Part_No;
```

and then

```
package Party is new Generic_Keys(Key_Type => Part_Key,
                                  Key => Part_No);
use Party;
```

Note that we do not have to define "<" on the type Part_Key because it already exists since Part_Key is an integer type. And the instantiation then uses it by default.

And now we can rewrite the Request procedure as follows

```
procedure Request(Part: in Part_Key; OK: out Boolean;
                  Year: out Integer; Shelf: out Character) is
   C: Cursor;
   E: Part_Type;
begin
   C := Find(The_Store, Part);
   if C = No_Element then
      OK := False; return;                -- no such item
   end if;
   E := Element(C);
   Year := E.Year; Shelf := E.Shelf;
```

```
    if E.Stock = 0 then
      OK := False; return;                       -- out of stock
    end if;
    -- now update the stock level
    declare
      procedure Do_It(E: in out Part_Type) is
      begin
        E.Stock := E.Stock - 1;
      end Do_It;
    begin
      Update_Element_Preserving_Key(The_Store, C, Do_It'Access);
    end;
    OK := True;
  end Request;
```

This seems hard work but has a number of advantages. The first is that the call of Find is more natural and only involves the part number (the key) – note that this is a call of the function Find in the instantiation of Generic_Keys and takes just the part number. And the other is that the update only involves the component being changed. We mentioned earlier that there was no Update_Element for sets because of the danger of creating a value that was in the wrong place. In the case of the richly named Update_Element_Preserving_Key it also checks to ensure that the element is indeed still in the correct place (by checking that the key is still the same); if it isn't it removes the element and raises Program_Error.

In Ada 2012 and Ada 2022 we can avoid using the tedious Update_Element_Preserving_Key because of the magic function Reference_Preserving_Key also declared in the package Generic_Keys. So instead of the block containing Do_It and the update call we can just write

```
E.Stock := E.Stock - 1;
```

The user is warned to take care when using the package Generic_Keys. It is absolutely vital that the relational operation and the function (Part_No) used to instantiate Generic_Keys are compatible with the ordering used to instantiate the parent package Containers.Ordered_Sets itself. If this is not the case then the sky might fall in.

Incidentally, the procedure for checking the stock which previously used the maps package now becomes

```
procedure Check_Stock(Low: in Integer) is
begin
  for C in The_Store.Iterate loop
    if Element(C).Stock < Low then
      Put("Low stock of part ");           -- print a message perhaps
      Put(Element(C).Part_Number);  New_Line;    -- changed
    end if;
  end loop;
end Check_Stock;
```

The only change is that the call of Key in

```
Put(Key(C).Part_Number);
```

when using the maps package has been replaced by Element. A minor point is that we could avoid calling Element twice by declaring a constant E thus

```
E: constant Part_Type := Element(C);
```

and then writing E.Stock < Low and calling Put with E.Part_Number.

A more important point is that if we have instantiated the Generic_Keys inner package as illustrated above then we can leave it unchanged to call Key. But it is important to realize that we are then calling the function Key internal to the instantiation of Generic_Keys (flippantly called Party) and not that from the instantiation of the parent ordered sets package (Store_Sets) because that has no such function. This illustrates the close affinity between the sets and maps packages.

And finally there is a hashed sets package which has strong similarities to both the ordered sets package and the hashed maps package. We can introduce this much as for hashed maps by giving the differences between the two sets packages, the extra facilities in each and the impact on the part number example.

The specification of the hashed sets package starts

```
with Ada.Iterator_Interfaces;
generic
   type Element_Type is private;
   with function Hash(Element: Element_Type) return Hash_Type;
   with function Equivalent_Elements(Left, Right: Element_Type)
                                                    return Boolean;
   with function "=" (Left, Right: Element_Type) return Boolean is <>;
package Ada.Containers.Hashed_Sets
     with Preelaborate, Remote_Types
          Nonblocking, Global => in out synchronized  is
```

The only differences from the ordered sets package are that there is an extra generic parameter Hash and the ordering parameter "<" has been replaced by the function Equivalent_Elements.

To illustrate the similarity between maps and sets note that we can rewrite the functions Equivalent_Parts and Part_Hash of the previous section as follows

```
function Equivalent_Parts(Left, Right: Part_Type) return Boolean is
begin
   return Left.Part_Number = Right.Part_Number;
end Equivalent_Parts;

function Part_Hash(P: Part_Type) return Hash_Type is
   M31: constant := 2**31-1;      -- a nice Mersenne prime
begin
   return Hash_Type(P.Part_Number) * M31;
end Part_Hash;
```

These are very similar to the hashed map example – the only changes are that the parameter type is now Part_Type rather than Part_Key. And now we can instantiate the hashed sets package as follows

```
package Store_Sets is
   new Hashed_Sets(Element_Type => Part_Type, Hash => Part_Hash,
                   Equivalent_Elements => Equivalent_Parts);
The_Store: Store_Sets.Set;
```

and then the rest of our example will be exactly as before. It is thus easy to convert from an ordered set to a hashed set and vice versa provided of course that we only use the facilities common to both.

It should also be mentioned that the inner package Generic_Keys for hashed sets has the following formal parameters

```
generic
   type Key_Type(<>) is private;
   with function Key(Element: Element_Type) return Key_Type
   with function Hash(Key: Key_Type) return Hash_Type;
   with function Equivalent_Keys(Left, Right: Key_Type) return Boolean;
package Generic_Keys
   with Nonblocking, Global => null is
```

The differences from that for ordered sets are the addition of the function Hash and the replacement of the comparison operator "<" by Equivalent_Keys. (Incidentally the package Generic_Keys for ordered sets also exports a function Equivalent_Keys for uniformity with the hashed sets package.)

Although our example itself is unchanged we do have to change the instantiation of Generic_Keys thus

```
type Part_Key is new Integer;
function Part_No(P: Part_Type) return Part_Key is
begin
   return Part_Key(P.Part_Number);
end Part_No;
function Part_Hash(P: Part_Key) return Hash_Type is
   M31: constant := 2**31-1;     -- a nice Mersenne prime
begin
   return Hash_Type(P) * M31;
end Part_Hash;
```

and then

```
package Party is
      new Generic_Key(Key_Type => Part_Key,
                      Key => Part_No,
                      Hash => Part_Hash,
                      Equivalent_Keys => "=");
use Party;
```

The hash function is similar to that used with hashed maps. The type Part_Key and function Part_No are the same as for ordered sets. Note the use of "=" as the actual parameter for Equivalent_Keys – this is the predefined "=" for the type Part_Key which is just an integer type.

We will finish this discussion of sets by briefly considering the additional facilities in the two sets packages (and their inner generic keys packages) just as we did for the two maps packages (the discussion is almost but not quite identical).

The ordered sets package has the following additional subprograms

```
    procedure Delete_First(Container: in out Set);
    procedure Delete_Last(Container: in out Set);
    function First_Element(Container: Set) return Element_Type;
    function Last_Element(Container: Set) return Element_Type;
*** function Previous(Container: Set; Position: Cursor) return Cursor;
*** procedure Previous(Container: Set; Position: in out Cursor);
    function Floor(Container: Set; Item: Element_Type) return Cursor;
    function Ceiling(Container: Set; Item: Element_Type) return Cursor;

    function "<" (Left, Right: Cursor) return Boolean;
    function ">" (Left, Right: Cursor) return Boolean;
    function "<" (Left: Cursor; Right: Element_Type) return Boolean;

    function ">" (Left: Cursor; Right: Element_Type) return Boolean;
    function "<" (Left: Element_Type; Right: Cursor) return Boolean;
    function ">" (Left: Element_Type; Right: Cursor) return Boolean;

    procedure Reverse_Iterate(Container: in Set;
                              Process: not null access procedure
                                                  (Position: in Cursor));
```

These are again largely self-evident. The functions Floor and Ceiling are similar to those for ordered maps – Floor searches for the last element which is not greater than Item and Ceiling searches for the first element which is not less than Item – they return No_Element if there is not one.

The functions "<" and ">" are very important for ordered sets. The first is equivalent to

```
function "<" (Left, Right: Cursor) return Boolean is
begin
   return Element(Left) < Element(Right);
end "<";
```

There is a general philosophy that the container packages should work efficiently even if the elements themselves are very large – perhaps even other containers. We should therefore avoid copying elements. (Passing them as parameters is of course no problem since they will be passed by reference if they are large structures.) So in this case the built-in comparison is valuable because it can avoid the copying which would occur if we wrote the function ourselves with the explicit internal calls of the function Element.

On the other hand, there is a general expectation that keys will be small and so there is no corresponding problem with copying keys. Thus such built-in functions are less important for maps than sets but they are provided for maps for uniformity.

The following are additional in the package Generic_Keys for ordered sets

```
function Equivalent_Keys(Left, Right: Key_Type) return Boolean;
```

This corresponds to the formal generic parameter of the same name in the package Generic_Keys for hashed sets as mentioned earlier.

```
function Floor(Container: Set; Key: Key_Type) return Cursor;
function Ceiling(Container: Set; Key: Key_Type) return Cursor;
```

These are much as the corresponding functions in the parent package except that they use the formal parameter "<" of Generic_Keys for the search.

Hashed sets, like hashed maps also have the facility to specify a capacity as for the vectors package. Thus we have

```
procedure Reserve_Capacity(Container: in out Set;
                                        Capacity: in Count_Type);
function Capacity(Container: Set) return Count_Type;
```

The behaviour is much as for vectors and hashed maps. We don't have to set the capacity ourselves since it will be automatically extended as necessary but it might significantly improve performance to do so. Note again that Length(S) cannot exceed Capacity(S) but might be much less.

The other additional subprograms for hashed sets are

```
function Equivalent_Elements(Left, Right: Cursor) return Boolean;
function Equivalent_Elements(Left: Cursor; Right: Element_Type)
                                                    return Boolean;
function Equivalent_Elements(Left: Element_Type; Right: Cursor)
                                                    return Boolean;
```

Again, these are very important for sets. The first is equivalent to

```
function Equivalent_Elements(Left, Right: Cursor) return Boolean is
begin
   return Equivalent_Elements(Element(Left), Element(Right));
end Equivalent_Elements;
```

and once more we see that the built-in functions can avoid the copying of the type Element that would occur if we wrote the functions ourselves.

Exercise 24.5

1 Implement the interface Set of Section 14.9 using an ordered set.

24.6 Trees

Another form of container is for manipulating multiway trees. It has all the operations required to operate on a tree structure where each node can have multiple child nodes to any depth. Thus there are operations on subtrees, the ability to find siblings, to insert and remove children and so on. It will be noted that many operations on trees are similar to corresponding operations on lists. Its specification starts with the usual generic parameters.

```
with Ada.Iterator_Interfaces;
generic
   type Element_Type is private;
   with function "=" (Left, Right: Element_Type) return Boolean is <>;
package Ada.Containers.Multiway_Trees
     with Preelaborate, Remote_Types,
          Nonblocking, Global => in out synchronized is

   type Tree is tagged private
     with Constant_Indexing => Constant_Reference,
          Variable_Indexing => Reference,
          Default_Iterator => Iterate,
          Iterator_Element => Element_Type,
          Preelaborable_Initialization;
          Iterator_View => Stable.Tree,            -- new from here
          Stable_Properties => (Node_Count,
                               Tampering_With_Cursors_Prohibited,
                               Tampering_With_Elements_Prohibited),
          Default_Initial_Condition =>
               Node_Count(Tree) = 1 and then
               (not Tampering_With_Cursors_Prohibited(Tree)) and then
               (not Tampering_With_Elements_Prohibited(Tree)),
          Preelaborable_Initialization;
```

Note that there is no aspect Aggregate. That is because container aggregates do not apply to trees – they would be cumbersome.

```
   type Cursor is private
     with Preelaborable_Initialization;
   Empty_Tree: constant Tree;
   No_Element: constant Cursor;
   function Equal_Element(Left, Right: Element_Type)
                                        return Boolean renames "=";
   function Has_Element(Position: Cursor) return Boolean;
   function Has_Element(Container: Tree; Position: Cursor)
                                                      return Boolean;
   package Tree_Iterator_Interfaces is
     new Ada.Iterator_Interfaces(Cursor, Has_Element);
```

This is much as expected and follows the same pattern as the start of the other containers. Note the additional Has_Element as usual. Equal_Element is also new in Ada 2022 with the obvious meaning.

We then have

```
function Equal_Subtree(Left_Position: Cursor;
                       Right_Position: Cursor) return Boolean;

function "=" (Left, Right: Tree) return Boolean;
```

and then the following are added in Ada 2022 as expected

```
function Tampering_With_Cursors_Prohibited(Container: Tree)
                                                    return Boolean;
  with Nonblocking, Global => null, Use_Formal => null;

function Tampering_With_Elements_Prohibited(Container: Tree)
                                                    return Boolean;
  with Nonblocking, Global => null, Use_Formal => null;

function Empty return Tree is
  (Empty_Tree);
```

and then it continues with the usual duplication indicated by three asterisks

```
    function Is_Empty(Container: Tree) return Boolean;
    function Node_Count(Container: Tree) return Count_Type;
*** function Subtree_Node_Count(Container: Tree;
                                Position: Cursor) return Count_Type;
*** function Depth(Container: Tree; Position: Cursor) return Count_Type;
*** function Is_Root(Container: Tree; Position: Cursor) return Boolean;
*** function Is_Leaf(Container: Tree; Position: Cursor) return Boolean;

    function Root(Container: Tree) return Cursor;
    procedure Clear(Container: in out Tree);
```

A tree consists of a set of nodes linked together in a hierarchical manner. Nodes are identified as usual by the value of a cursor. Nodes can have one or more child nodes; the children are ordered so that there is a first child and a last child. Nodes with the same parent are siblings. One node is the root of the tree. If a node has no children then it is a leaf node.

All nodes other than the root node have an associated element whose type is Element_Type. The whole purpose of the tree is of course to give access to these element values in a structured manner.

The function "=" compares two trees and returns true if and only if they have the same structure of nodes and corresponding nodes have the same values as determined by the generic parameter "=" for comparing elements. Similarly, the function Equal_Subtree compares two subtrees.

The function Node_Count gives the number of nodes in a tree. All trees have at least one node, the root node. The function Is_Empty returns true only if the tree consists of just this root node. Note that the expressions A_Tree = Empty_Tree, Node_Count(A_Tree) = 1 and Is_Empty(A_Tree) always have the same value. The function Subtree_Node_Count returns the number of nodes in the subtree identified by the cursor. If the cursor value is No_Element then the result is zero.

The functions Is_Root and Is_Leaf indicate whether a node is the root or a leaf respectively. If a tree is empty and so consists of just a root node then that node is both the root and a leaf so both functions return true.

The function Depth returns 1 if the node is the root, and otherwise indicates the number of ancestor nodes. Thus a node which is an immediate child of the root has depth equal to 2. The function Root returns the cursor designating the root of a tree. The procedure Clear removes all elements from the tree so that it consists just of a root node.

Two new functions in Ada 2022 are

```
function Is_Ancestor_Of(Container: Tree:
                        Parent: Cursor;
                        Position: Cursor) return Boolean;

function Meaningful_For(Container: Tree; Position Cursor) return Boolean is
                        (Position = No_Element or else
                        Is_Root(Container, Position) or else
                        Has_Element(Container, Position));
```

The function Is_Ancestor_Of is added for obvious convenience. The expression function Meaningful_For may seem strange. However, it is used in many preconditions; remember that the descriptions of operations in the *ARM* for Ada 2022 are largely given by pre- and postconditions rather than English text (as an example see the illustration of the package Generic_Sorting in Section 24.2). Regarding Meaningful_For the *ARM* remarks as follows. When this function is true the Position can be meaningfully used with operations for Container. We define this because many operations allow the root (which does not have an element, so Has_Element returns False), so many preconditions get unwieldy. We allow No_Element as it is allowed by many queries, and for existing routines, it raises a different exception (Constraint_Error rather than Program_Error) than a cursor for the wrong container does)

We then have

```
*** function Element(Container: Tree; Position: Cursor) return Element_Type;

    procedure Replace_Element(Container: in out Tree;
                              Position: in Cursor;
                              New_Item: in Element_Type);

*** procedure Query_Element(Container: in Tree;
                            Position: in Cursor;
                            Process: not null access procedure
                                          (Element: in Element_Type));

    procedure Update_Element(Container: in out Tree;
                             Position: in Cursor;
                             Process: not null access procedure
                                          (Element: in out Element_Type));
```

These subprograms have the expected behaviour similar to other containers.

```
type Constant Reference_Type
          (Element: not null access constant Element_Type) is private
   with Implicit_Dereference => Element;

type Reference_Type
                (Element: not null access Element_Type) is private
   with Implicit_Dereference => Element;

function Constant_Reference
             (Container: aliased in Tree; Position: in Cursor)
                            return Constant_Reference_Type;
function Reference
             (Container: aliased in out Tree; Position: in Cursor)
                                    return Reference_Type;
```

These magic types and functions are similar to those for the other containers.

```
procedure Assign(Target: in out Tree; Source: in Tree);
function Copy(Source: Tree) return Tree;
procedure Move(Target, Source: in out Tree);
```

Assign simply assigns Source to Target thereby overriding what was in Target. Copy does the obvious thing. Move is subtle; if Source is the same as Target then nothing happens; otherwise it does Clear(Target) and then the nodes of Source (other than the root) are moved to Target. At the end Node_Count(Source) is 1.

```
procedure Delete_Leaf(Container: in out Tree;
                      Position: in out Cursor);

procedure Delete_Subtree(Container: in out Tree;
                         Position: in out Cursor);

procedure Swap(Container: in out Tree; I, J: in Cursor);
```

The procedures Delete_Leaf and Delete_Subtree check that the cursor value designates a node of the container and raise Program_Error if it does not. Program_Error is also raised if Position designates the root node and so cannot be removed. In the case of Delete_Leaf, if the node has any children then Constraint_Error is raised. The appropriate nodes are then deleted and Position is set to No_Element.

The procedure Swap interchanges the values in the two elements denoted by the two cursors. The elements must be in the given container (and must not denote the root) otherwise Program_Error is raised.

```
    function Find(Container: Tree; Item: Element_Type) return Cursor;

*** function Find_In_Subtree(Container: Tree;
                             Position: Cursor;
                             Item: Element_Type) return Cursor;

*** function Ancestor_Find(Container: Tree;
                           Position: Cursor;
                           Item: Element_Type) return Cursor;

    function Contains(Container: Tree; Item: Element_Type) return Boolean;
```

These search for an element in the container with the given value Item. The function Contains returns false if the item is not found; the other functions return No_Element if the item is not found. The function Find searches the whole tree starting at the root node, Find_In_Subtree searches the subtree rooted at the node given by Position including the node itself; these searches are in depth-first order. The function Ancestor_Find searches upwards through the ancestors of the node given by Position including the node itself.

Depth-first order is explained at the end of the section.

```
procedure Iterate(Container: in Tree;
                  Process: not null access procedure
                                        (Position: in Cursor))
                  with Allows_Exit;
```

```
***  procedure Iterate_Subtree(Container: in Tree;
                               Position: in Cursor;
                               Process: not null access procedure
                                             (Position: in Cursor))
                               with Allows_Exit;
```

These apply the procedure designated by the parameter Process to each element of the whole tree or the subtree. This includes the node at the subtree but not at the root; iteration is in depth-first order. Note Allows_Exit added in Ada 2022.

```
function Iterate(Container: in Tree)
                 return Tree_Iterator_Interfaces.Parallel_Iterator'Class;
```

```
***  function Iterate_Subtree(Container: in Tree; Position: in Cursor)
                 return Tree_Iterator_Interfaces.Parallel_Iterator'Class;
```

The first of these is called if we write

```
for C in The_Tree.Iterate loop
   ...                       -- do something via cursor C
end loop;
```

and iterates over the whole tree in the usual depth-first order. Note that we can use a parallel loop in Ada 2022. In order to iterate over a subtree we write

```
for C in Iterate_Subtree(S) loop
   ...                       -- do something via cursor C
end loop;
```

and this iterates over the subtree rooted at the cursor position given by S.

If we use the other form of loop using **of** thus

```
for E of The_Tree loop
   ...                       -- do something to element E
end loop;
```

then this also calls Iterate since the aspect Default_Iterator of the type Tree (see above) is Iterate. However, we cannot iterate over a subtree using this mechanism.

```
***  function Child_Count(Container: Tree;
                          Parent: Cursor) return Count_Type;

***  function Child_Depth(Container: Tree;
                          Parent, Child: Cursor) return Count_Type;
```

The function Child_Count returns the number of child nodes of the node denoted by Parent. This count covers immediate children only and not any grandchildren. The function Child_Depth indicates how many ancestors there are from Child to Parent. If Child is an immediate child of Parent then the result is 1; if it is a grandchild then 2 and so on.

```
procedure Insert_Child(Container: in out Tree;
                       Parent: in Cursor;
                       Before: in Cursor;
                       New_Item: in Element_Type;
                       Count: in Count_Type := 1);

procedure Insert_Child(Container: in out Tree;
                       Parent: in Cursor;
                       Before: in Cursor;
                       New_Item: in Element_Type;
                       Position: out Cursor;
                       Count: in Count_Type := 1);

procedure Insert_Child(Container: in out Tree;
                       Parent: in Cursor;
                       Before: in Cursor;
                       Position: out Cursor;
                       Count: in Count_Type := 1);
```

These three procedures enable one or more new child nodes to be inserted. The parent node is given by Parent. If Parent already has children then the new nodes are inserted before the child node identified by Before; if Before is No_Element then the new nodes are inserted after all existing children. The second procedure is similar to the first but also returns a cursor to the first of the added nodes. The third is like the second but the new elements take their default values. Note the default value of one for the number of new nodes.

Note that trees are like lists in that cursors are preserved when elements are added or deleted.

```
procedure Prepend_Child(Container: in out Tree;
                        Parent: in Cursor;
                        New_Item: in Element_Type;
                        Count: in Count_Type := 1);

procedure Append_Child(Container: in out Tree;
                       Parent: in Cursor;
                       New_Item: in Element_Type;
                       Count: in Count_Type := 1);
```

These insert the new children before or after any existing children.

```
procedure Delete_Children(Container: in out Tree;
                          Parent: in Cursor);
```

This procedure simply deletes all the children, grandchildren, and so on of the node designated by Parent.

```
procedure Copy_Subtree(Target: in out Tree;
                       Parent: in Cursor;
                       Before: in Cursor;
                       Source: in Cursor);
```

This copies the complete subtree rooted at Source into the tree denoted by Tree as a subtree of Parent at the place denoted by Before using the same rules as Insert_Child. Note that this makes a complete copy and creates new nodes with values equal to the corresponding existing nodes. Note also that Source might be within Tree but might not. There are the usual various checks.

```
procedure Copy_Local_Subtree(Target: in out Tree;
                             Parent: in Cursor;
                             Before: in Cursor;
                             Source: in Cursor);
procedure Copy_Subtree(Target: in out Tree;
                       Parent: in Cursor;
                       Before: in Cursor;
                       Source: in Tree;
                       Subtree: in Cursor);
```

These are new in Ada 2022. Copy_Local_Subtree is essentially the original Copy_Subtree (with four parameters) but restricted to Source being within the Target. The new Copy_Subtree applies when the source is a Subtree of Source outside the Target.

```
procedure Splice_Subtree(Target: in out Tree;
                         Parent: in Cursor;
                         Before: in Cursor;
                         Source: in out Tree;
                         Position: in out Cursor);
procedure Splice_Subtree(Container: in out Tree;
                         Parent: in Cursor;
                         Before: in Cursor;
                         Position: in Cursor);
procedure Splice_Children(Target: in out Tree;
                          Target_Parent: in Cursor;
                          Before: in Cursor;
                          Source: in out Tree;
                          Source_Parent: in Cursor);
procedure Splice_Children(Container: in out Tree;
                          Target_Parent: in Cursor;
                          Before: in Cursor;
                          Source_Parent: in Cursor);
```

These are similar to the procedures Splice applying to lists. They enable nodes to be moved without copying. The destination is indicated by Parent or Target_Parent together with Before as usual indicating where the moved nodes are to be placed with respect to existing children of Parent or Target_Parent.

The first Splice_Subtree moves the subtree rooted at Position in the tree Source to be a child of Parent in the tree Target. Note that Position is updated to be the appropriate element of Target. We can use this procedure to move a subtree within a tree but an attempt to create circularities raises Program_Error.

The second Slice_Subtree is similar but only moves a subtree within a container. Again, circularities cannot be created.

The procedures Splice_Children are similar but move all the children and their descendants of Source_Parent to be children of Target_Parent.

```
***   function Parent(Container: Tree; Position: Cursor) return Cursor;
***   function First_Child(Container: Tree; Parent: Cursor) return Cursor;
***   function First_Child_Element(Container: Tree;
                                   Parent: Cursor) return Element_Type;
***   function Last_Child(Container: Tree; Parent: Cursor) return Cursor;
***   function Last_Child_Element(Container: Tree;
                                   Parent: Cursor) return Element_Type;
***   function Next_Sibling(Container: Tree; Position: Cursor) return Cursor;
***   function Previous_Sibling(Container: Tree;
                                Position: Cursor) return Cursor;
***   procedure Next_Sibling(Container; in Tree; Position: in out Cursor);
***   procedure Previous_Sibling(Container: in Tree; Position: in out Cursor);
```

Hopefully, the purpose of these is self-evident.

```
      procedure Iterate_Children(Parent: in Cursor;
                   Process: not null access procedure (Position: in Cursor));

      procedure Reverse_Iterate_Children(Parent : in Cursor;
                   Process: not null access procedure (Position: in Cursor));
```

These apply the procedure designated by the parameter Process to each child of the node given by Parent. The procedure Iterate_Children starts with the first child and ends with the last child whereas Reverse_Iterate_Children starts with the last child and ends with the first child. Note that these do not iterate over grandchildren.

```
      function Iterate_Children(Container: in Tree; Parent: in Cursor)
            return Tree_Iterator_Interfaces.Parallel_Reversible_Iterator'Class;
```

This is called if we write

```
      for C in The_Tree.Iterate_Children(P) loop
         ...                          -- do something via cursor C
      end loop;
```

and iterates over all the children (using cursor C) of the parent node designated by cursor P from P.First_Child to P.Last_Child. We could also insert **reverse** thus

```
for C in reverse The_Tree.Iterate_Children(P) loop
   ...                              -- do something via cursor C
end loop;
```

in which case the iteration goes in reverse from P.Last_Child to P.First_Child. Note that this function returns Parallel_Reversible_Iterator'Class and so can go in either direction and use parallel loops whereas the functions Iterate and Iterate_Subtree described earlier use Parallel_Iterator'Class.

But note that we cannot go backwards and be parallel at the same time. So we could instead write

```
parallel for C in The_Tree.Iterate_Children(P) loop
   ...                              -- do something via cursor C
end loop;
```

and so deal with the children in parallel.

And then at last (skipping over the package Stable) we come to

```
private
   ...                              -- not specified by the language
end Ada.Containers.Multiway_Trees;
```

The above descriptions have not described all the situations in which something can go wrong and so raise Constraint_Error or Program_Error. Generally, the former is raised if a source or target is No_Element; the latter is raised if a cursor does not belong to the appropriate tree. In particular, as mentioned above, an attempt to create an illegal tree such as one with circularities using Splice_Subtree raises Program_Error. Remember also that every tree has a root node but the root node has no element value; attempts to remove the root node or read its value or assign a value similarly raise Program_Error.

As an example consider a tree representing a simple algebraic expression involving just the binary operations of addition, subtraction, multiplication, and division applied to simple variables and real literals. Nodes are of three kinds, those representing operations have two children giving the two operands, and those representing variables and literals have no children and so are leaf nodes.

We can declare the element type thus

```
type Operator is ('+', '–', '×', '/');
type Kind is (Op, Var, Lit);

type El(K: Kind := Op) is        -- note default to make type definite
   record
      case K is
         when Op =>
            Fn: Operator;
         when Var =>
            V: Character;
         when Lit =>
            Val: Float;
      end case;
   end record;
```

Note that the variables are (as typically in mathematics) represented by single letters. So the expression

$(x + 3) \times (y - 4)$

is represented by nodes with elements such as

```
(Op, '×')
(Var, 'x')
(Lit, 3.0)
```

So now we can declare a suitable tree thus

```
package Expression_Trees is
  new Ada.Containers.Multiway_Trees(El);
use Expression_Trees;
My_Tree: Tree := Empty_Tree;
C: Cursor;
```

and then build it by the following statements

```
C := Root(My_Tree);
Insert_Child(Container => My_Tree,
             Parent => C,
             Before => No_Element,
             New_Item => (Op, '×'),
             Position => C);
```

This puts in the first real node as a child of the root which is designated by the cursor C. There are no existing children so Before is No_Element. The New_Item is the node for multiply. Finally, the cursor C is changed to designate the position of the newly inserted node.

We can then insert the two children of this node which represent the mathematical operations + (plus) and – (minus).

```
Insert_Child(My_Tree, C, No_Element, (Op, '+'));
Insert_Child(My_Tree, C, No_Element, (Op, '–'));
```

These calls are to a different overloading of Insert_Child and have not changed the cursor. The second call also has Before equal to No_Element and so the second child goes after the first child.

We now change the cursor to that of the first newly inserted child and then insert its children which represent x and 3. Thus

```
C := First_Child(C);
Insert_Child(My_Tree, C, No_Element, (Var, 'x'));
Insert_Child(My_Tree, C, No_Element, (Lit, 3.0));
```

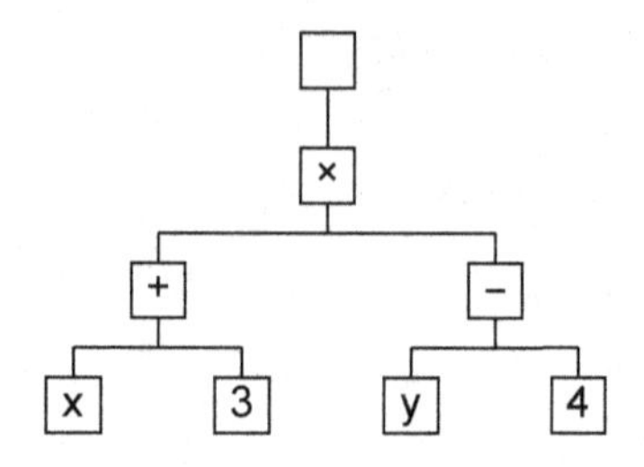

Figure 24.1 The expression tree.

And then we can complete the tree by inserting the final two nodes thus

```
C := Next_Sibling(C);
Insert_Child(My_Tree, C, No_Element, (Var, 'y'));
Insert_Child(My_Tree, C, No_Element, (Lit, 4.0));
```

Of course a compiler will do all this recursively and keep track of the cursor rather more neatly than we have in this manual illustration. The resulting tree should be as in Figure 24.1.

We will now assume that the variables are held in an array as follows

```
subtype Variable_Name is Character range 'a' .. 'z';
Variables: array (Variable_Name) of Float;
```

We can then evaluate the tree by a recursive function such as

```
function Eval(C: Cursor) return Float is
  E: El := Element(C);
  L, R: Float;
begin
  case E.K is
    when Op =>
      L := Eval(First_Child(C));
      R := Eval(Last_Child(C));
      case E.Fn is
        when '+' => return (L+R);
        when '–' => return (L–R);
        when '×' => return (L*R);
        when '/' => return (L/R);
      end case;
    when Var =>
      return Variables(E.V);
    when Lit =>
      return E.Val;
 end case;
end Eval;
```

Finally, we obtain the value of the tree by

```
X := Eval(First_Child(Root(My_Tree)));
```

Remember that the node at the root has no element so hence the call of First_Child.

An alternative approach would be to use tagged types with a different type for each kind of node rather than the variant record. This would be much more flexible but would have required the use of the unbounded indefinite container Ada.Containers.Indefinite_Multiway_Trees.

Finally, an explanation of depth-first order. The general principle is that child nodes are visited in order before their parent. We can symbolically write this as

```
procedure Do_Node(N: Node) is
begin
   for C in N.First_Child .. N.Last_Child loop
      Do_Node(C);
   end loop;
   if not N.Is_Root then
      Do_Element(N);
   end if;
end Do_Node;
```

and the whole thing is triggered by calling Do_Node(Root). Remember that the root node has no element. The result is that the first element to be processed is that of the leftmost leaf. (Note that the above symbolic code is not legal Ada; the reader might care to write it properly as an exercise.)

Thus in the tree illustrated below in Figure 24.2, the elements are visited in order A, B, C, D, and so on. Note that the root has no element and so is not visited.

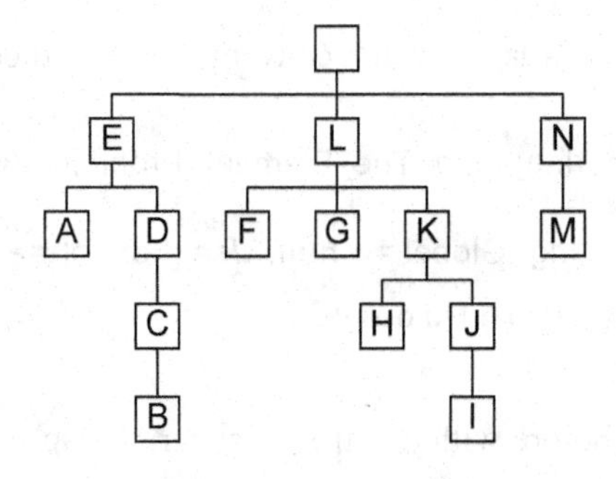

Figure 24.2 A tree showing depth-first order.

24.7 Holders

It is not possible to declare an object of an indefinite type that can hold any value of that type since the object becomes constrained by the mandatory initial value. Thus if we write

```
Pet: String := "dog";
```

then we can assign "cat" to Pet but we cannot assign "rabbit" because it is too long.

This is overcome by the introduction of the indefinite holder container which can hold a single indefinite object. Its specification is

```
generic
   type Element_Type(<>) is private;
   with function "=" (Left, Right: Element_Type) return Boolean is <>;
package Ada.Containers.Indefinite_Holders
      with Preelaborate, Remote_Types,
         Nonblocking, Global => in out synchronized is

   type Holder is tagged private
      with Stable_Properties => (Is_Empty,
                                 Tampering_With_The_Element_Prohibited),
         Default_Initial_Condition => Is_Empty(Holder),
         Preelaborable_Initialization;
```

Note the quaintly named Tampering_With_The_Element_Prohibited. It continues thus

```
   Empty_Holder: constant Holder;
   function Equal_Element(Left, Right: Element_Type)
                                       return Boolean renames "=";
   function "=" (Left, Right: Holder) return Boolean;
```

Equal_Element is new in Ada 2022 for convenience and then the following are added as expected

```
   function Tampering_With_The_Element_Prohibited(Container: Holder)
                                                      return Boolean
      with Nonblocking, Global => null, Use_Formal => null;

   function Empty return Holder is
      (Empty_Holder);
```

and it then continues as before with just the procedure Swap added at the end

```
   function To_Holder(New_Item: Element_Type) return Holder;
   function Is_Empty(Container: Holder) return Boolean;
   procedure Clear(Container: in out Holder);

   function Element(Container: Holder) return Element_Type;
   procedure Replace_Element(Container: in out Holder;
                             New_Item: in Element_Type);
```

```
      procedure Query_Element(Container: in Holder;
                              Process: not null access procedure
                                         (Element: in Element_Type));
      procedure Update_Element(Container: in out Holder;
                               Process: not null access procedure
                                         (Element: in out Element_Type));
      type Constant_Reference_Type
            (Element: not null access constant Element_Type) is private
         with Implicit_Dereference => Element;
      type Reference_Type
                     (Element: not null access Element_Type) is private
         with Implicit_Dereference => Element;
      function Constant_Reference(Container: aliased in Holder)
                                        return Constant_Reference_Type;
      function Reference(Container: aliased in out Holder)
                                                 return Reference_Type;
      procedure Assign(Target: in out Holder; Source: in Holder);
      function Copy(Source: Holder) return Holder;
      procedure Move(Target: in out Holder; Source: in out Holder);
      procedure Swap(Left, Right: in out Holder);
   private
      ...                    -- not specified by the language
   end Ada.Containers.Indefinite_Holders;
```

Note that there is no concept of a cursor since by its nature the holder contains just a single element. Similarly, there is no possibility of iteration so the type Holder does not have the various aspects that apply to the other containers. But it does have the magic types and functions such as Reference_Type.

Hopefully, the purpose of the facilities provided by this container are obvious given an understanding of the use of the other containers. It would be possible to use a list container with just a single element to act as a holder but it seems better to have an explicit container with probably less overhead and risk of confusion.

A trivial example of its use might be to provide a holder for pets. We write

```
   package Strings is
      new Ada.Containers.Indefinite_Holders(String);
   Kennel: Strings.Holder := Strings.To_Holder("cat");
```

This declares an object Kennel which is a wrapper for a string and initializes it with the string "cat". Assuming that we have written **use** Strings; we can now replace the cat with a rabbit by writing

```
   Kennel := To_Holder("rabbit");
```

However, using To_Holder in this way could be a bit slow since this will create a new object which has to be destroyed after the assignment. It is better to write

```
Replace_Element(Kennel, "rabbit");
```

In order to print out the contents of the kennel we can just write

```
Put(Element(Kennel));
```

and this will output "rabbit".

Operations such as Update_Element are provided partly for uniformity but also because the object might be large so that it is better to update it *in situ*. However, beware that although Update_Element can change just part of the element, it cannot change its length. If we attempt to do so then the procedure Process written by the user will inevitably raise Constraint_Error. We have to use Element to read its value and then Replace_Element to put the new value back.

So we can use Update_Element to change "kitten" to "mitten" but we cannot use it to change "robin" to "bobbin". See also the more detailed note in the last paragraph of Section 24.10.

Ada 2022 also has a bounded holder for use with high integrity systems. It is somewhat different to the existing bounded containers for lists, vectors and so on. Its specification starts

```
with System.Storage_Elements; use System.Storage_Elements;
generic
   type_Element_Type is private;
   Max_Element_Size_In_Storage_Elements: Storage_Count;
   with function "=" (Left, Right: Element_Type) return Boolean is <>;
package Ada.Containers.Bounded_Indefinite_Holders is
```

Note the use of Max_Element_Size rather than Capacity which would always be one since it is just a holder for one element. To_Holder and Replace_Element raise Program_Error if the object will not fit.

For further details see AI12-254 which notes: The Bounded_Indefinite_Holder container is added to allow the use of class-wide objects in safety critical environments where dynamic allocation is not allowed.

24.8 Queues

The final type of container to be introduced is that for queues. These are quite different from the other containers. They do not have cursors since by their very nature only certain elements in a queue can be accessed. In particular, we can only take elements out of a queue or add elements to a queue but we cannot meddle with elements already in the queue.

Accordingly, there are no subprograms such as Replace_Element and Update_Element with their fears of tampering. And there are no magic types and functions for dereferencing and no need for iteration.

Queues are amenable to controlled access by several tasks. One or more tasks can place objects on a queue and one or more can remove them.

There are four different queue containers. These are all for elements of a definite type. Two are bounded and two are unbounded, two have priority and two do not. The names are

```
Unbounded_Synchronized_Queues        -- without priority
Bounded_Synchronized_Queues          -- without priority
Unbounded_Priority_Queues            -- with priority
Bounded_Priority_Queues              -- with priority
```

Note that there are no queue containers for indefinite types. This is because there would be significant problems with the Dequeue operation to remove an indefinite object related to the fact that Ada does not have entry functions. The absence of queues of indefinite types is easily overcome by making the elements of the queue a holder container as will be illustrated later.

The four different queue containers are all derived from a single synchronized interface declared in a generic package whose specification is as follows

```
generic
   type Element_Type is private;                              -- definite
package Ada.Containers.Synchronized_Queue_Interfaces
     with Pure, Nonblocking, Global => null  is

   type Queue is synchronized interface;

   procedure Enqueue(Container: in out Queue;
                     New_Item: in Element_Type) is abstract
      with Synchronization => By_Entry,
           Nonblocking => False,
           Global'Class => in out synchronized;

   procedure Dequeue(Container: in out Queue;
                     Element: out Element_Type) is abstract
      with Synchronization => By_Entry,
           Nonblocking => False,
           Global'Class => in out synchronized;

   function Current_Use(Container: Queue) return Count_Type is abstract
      with Nonblocking, Global'Class => null, Use_Formal => null;

   function Peak_Use(Container: Queue) return Count_Type is abstract
      with Nonblocking, Global'Class => null, Use_Formal => null;

end Ada.Containers.Synchronized_Queue_Interfaces;
```

This generic package declares the synchronized interface Queue and four operations on queues. These are the procedures Enqueue and Dequeue to add items to a queue and remove items from a queue respectively; note the aspect Synchronization which ensures that all implementations of these abstract procedures must be by an entry. There are also functions Current_Use and Peak_Use which can be used to monitor the number of items on a queue.

The four queue containers are generic packages which themselves declare a type Queue derived in turn from the interface Queue declared in the package above. We will look first at the synchronized queues and then at the priority queues.

The package for unbounded synchronized queues is as follows

```
with System;  use System;
with Ada.Containers.Synchronized_Queue_Interfaces;
generic
   with package Queue_Interfaces is
      new Ada.Containers.Synchronized_Queue_Interfaces(<>);
   Default_Ceiling: Any_Priority := Priority'Last;
package Ada.Containers.Unbounded_Synchronized_Queues
      with Preelaborate,
            Nonblocking, Global => in out synchronized is

   package Implementation is
      ...                           -- not specified by the language
   end Implementation;

   protected type Queue(Ceiling: Any_Priority := Default_Ceiling)
      with Priority => Ceiling
                                    is new Queue_Interfaces.Queue with

      overriding
      entry Enqueue(New_Item: in Queue_Interfaces.Element_Type);
      overriding
      entry Dequeue(Element: out Queue_Interfaces.Element_Type);

      overriding
      function Current_Use return Count_Type
         with Nonblocking, Global => null, Use_Formal => null;

      overriding
      function Peak_Use return Count_Type
         with Nonblocking, Global => null, Use_Formal => null;

   private
      ...                           -- not specified by the language
   end Queue;

private
   ...                              -- not specified by the language
end Ada.Containers.Unbounded_Synchronized_Queues;
```

Note that there are two generic parameters. The first (Queue_Interfaces) has to be an instantiation of the interface generic Synchronized_Queue_Interfaces; remember that the parameter (<>) means that any instantiation will do. The second parameter concerns priority and has a default value so we ignore it for the moment.

Inside this package there is a protected type Queue which controls access to the queues via its entries Enqueue and Dequeue. This protected type is derived from Queue_Interfaces.Queue and so promises to implement the operations Enqueue, Dequeue, Current_Use, and Peak_Use of that interface. And indeed it does implement them and moreover implements Enqueue and Dequeue by entries as required by the aspect Synchronization.

As an example suppose we wish to create a queue of some records such as

```
type Rec is record ... end record;
```

First of all we instantiate the interface package (using named notation for clarity) thus

```
package Rec_Interface is
   new Ada.Containers.Synchronized_Queue_Interfaces
                                          (Element_Type => Rec);
```

This creates an interface from which we can create various queuing mechanisms for dealing with objects of the type Rec.

Thus we might write

```
package Unbounded_Rec_Package is
   new Ada.Containers.Unbounded_Synchronized_Queues
                              (Queue_Interfaces => Rec_Interface);
```

Finally, we can declare a protected object, My_Rec_UQ which is the actual queue, thus

```
My_Rec_UQ: Unbounded_Rec_Package.Queue;
```

To place an object on the queue we can write (assuming a use clause for the package Unbounded_Rec_Package)

```
Enqueue(My_Rec_UQ, Some_Rec);
```

or perhaps more neatly (and no need for that use clause)

```
My_Rec_UQ.Enqueue(Some_Rec);
```

And to remove an item from the queue we can write

```
My_Rec_UQ.Dequeue(The_Rec);
```

where The_Rec is an object of type Rec which thereby is given the value removed.

The statement

```
N := Current_Use(My_Rec_UQ);
```

assigns to N the number of items on the queue when Current_Use was called (it could be obsolete by the time it gets into N) and similarly Peak_Use(My_Rec_UQ) gives the maximum number of items that have been on the queue since it was declared.

This is all task safe because of the protected type; several tasks can place items on the queue and several, perhaps the same, can remove items from the queue without interference.

It should also be noticed that since the queue is unbounded, we never get blocked by Enqueue since extra storage is allocated as required just as for the other unbounded containers (perhaps we might get Storage_Error).

The observant reader will note the mysterious local package called Implementation. This enables the implementation to declare local types to be used by the protected type. It will be recalled that there is an old rule that one cannot declare a type within a type. These local types really ought to be within the private part of the protected type; maybe this is something for a future version of Ada.

The package for bounded synchronized queues is very similar. The only differences (apart from its name) are that it has an additional generic parameter Default_Capacity and the protected type Queue has an additional discriminant Capacity. So its specification is

```
with System;  use System;
with Ada.Containers.Synchronized_Queue_Interfaces;
generic
   with package Queue_Interfaces is
      new Ada.Containers.Synchronized_Queue_Interfaces(<>);
   Default_Capacity: Count_Type;
   Default_Ceiling: Any_Priority := Priority'Last;
package Ada.Containers.Bounded_Synchronized_Queues
      with Preelaborate,
            Nonblocking, Global => in out synchronized is

   package Implementation is
      ...                          -- not specified by the language
   end Implementation;

   protected type Queue(Capacity: Count_Type := Default_Capacity,
                        Ceiling: Any_Priority := Default_Ceiling)
      with Priority => Ceiling is
            new Queue_Interfaces.Queue with

   ...              -- Enqueue, Dequeue, etc as for the unbounded one

end Ada.Containers.Bounded_Synchronized_Queues;
```

So using the same example, we can use the same interface package Rec_Interface. Now to declare a bounded queue with capacity 1000, we can write

```
package Bounded_Rec_Package is
   new Ada.Containers.Bounded_Synchronized_Queues
         (Queue_Interfaces => Rec_Interface, Default_Capacity => 1000);
```

Finally, we can declare a protected object, My_Rec_BQ which is the actual queue, thus

```
My_Rec_BQ: Bounded_Rec_Package.Queue;
```

And then we can use the queue as before. To place an object on the queue we can write

```
My_Rec_BQ.Enqueue(Some_Rec);
```

And to remove an item from the queue we can write

```
My_Rec_BQ.Dequeue(The_Rec);
```

The major difference is that if the queue becomes full then calling Enqueue will block the calling task until some other task calls Dequeue. Thus, unlike the other bounded containers which are discussed in the next section, Capacity_Error is never raised.

Note that having given a value for Default_Capacity, it can be overridden when the queue is declared, perhaps

```
My_Rec_Giant_BQ: Bounded_Rec_Package.Queue(Capacity => 100000);
```

These packages also provide control over the ceiling priority of the protected type. By default it is Priority'Last. This default can be overridden by our own default when the queue package is instantiated and can be further specified as a discriminant when the actual queue object is declared.

So we might write

```
My_Rec_Ceiling_BQ: Bounded_Rec_Package.Queue(Ceiling => 10);
```

In the case of the bounded queue, if we do not give an explicit capacity then the ceiling has to be given using named notation. This does not apply to the unbounded queue which only has one discriminant, so to give that a ceiling priority we can just write

```
My_Rec_Ceiling_UQ: Unbounded_Rec_Package.Queue(10);
```

But clearly the use of the named notation is advisable.

Being able to give default discriminants is very convenient. In Ada 2005, this was not possible if the type was tagged. However, in Ada 2012 and Ada 2022, it is permitted in the case of limited tagged types (see Section 18.2) and a protected type implementing an interface is considered to be limited and tagged (see Section 22.3).

If we wanted to make a queue of indefinite objects, then as mentioned above, there is no special container for this because Dequeue would be difficult to use since it is a procedure and not a function. So the actual parameter would have to be constrained which means knowing before the call the value of the discriminant, tag, or bound of the object which is unlikely. However, we can use the holder container to wrap the indefinite type so that it looks definite.

So to create a queue for strings, using the example of the previous section, we can write

```
package Strings is
   new Ada.Containers.Indefinite_Holders(String);
package Strings_Interface is
   new Ada.Containers.Synchronized_Queue_Interfaces
                              (Element_Type => Strings.Holder);
package Unbounded_Strings_Package is
   new Ada.Containers.Unbounded_Synchronized_Queues
                         (Queue_Interfaces => Strings_Interface);
```

and then finally declare the actual queue

```
My_Strings_UQ: Unbounded_Strings_Package.Queue;
```

To put some strings on this queue, we write

```
My_Strings_UQ.Enqueue(To_Holder("rabbit"));
My_Strings_UQ.Enqueue(To_Holder("horse"));
```

or even

```
My_Strings_UQ.Enqueue(Element(Kennel));
```

We now turn to considering the two other forms of queue which are the unbounded and bounded priority queues.

Here is the specification of the unbounded priority queue

```
with System;  use System;
with Ada.Containers.Synchronized_Queue_Interfaces;
generic
   with package Queue_Interfaces is
      new Ada.Containers.Synchronized_Queue_Interfaces(<>);

   type Queue_Priority is private;
   with function Get_Priority(Element: Queue_Interfaces.Element_Type)
                                             return Queue_Priority is <>;
   with function Before(Left, Right: Queue_Priority) return Boolean is <>;
   Default_Ceiling: Any_Priority := Priority'Last;
package Ada.Containers.Unbounded_Priority_Queues
      with Preelaborate,  Nonblocking, Global => in out synchronized is

   package Implementation is
      ...                        -- not specified by the language
   end Implementation;

   protected type Queue(Ceiling: Any_Priority := Default_Ceiling)
      with Priority => Ceiling is
                              new Queue_Interfaces.Queue with

      overriding
      entry Enqueue(New_Item: in Queue_Interfaces.Element_Type);
      overriding
      entry Dequeue(Element: out Queue_Interfaces.Element_Type);

      not overriding
      procedure Dequeue_Only_High_Priority(At_Least: in Queue_Priority
                         Element: in out Queue_Interfaces.Element_Type;
                         Success: out Boolean);

      overriding
      function Current_Use return Count_Type
         with Nonblocking, Global => null, Use_Formal => null;

      overriding
      function Peak_Use return Count_Type;
         with Nonblocking, Global => null, Use_Formal => null;

   private
      ...                        -- not specified by the language
   end Queue;

private
   ...                           -- not specified by the language
end Ada.Containers.Unbounded_Priority_Queues;
```

The differences from the synchronized bounded queue are that there are several additional generic parameters, namely the private type Queue_Priority and the two functions Get_Priority and Before which operate on objects of the type Queue_Priority, and also that the protected type Queue has an additional operation, the protected procedure Dequeue_Only_High_Priority.

The general idea is that elements have an associated priority which can be ascertained by calling the function Get_Priority. The meaning of this priority is given by the function Before.

When we call Enqueue, the new item is placed in the queue taking due account of its priority with respect to other elements already on the queue. So it will go before all less important elements as defined by Before. If existing elements already have the same priority then it goes after them.

As expected Dequeue just returns the first item on the queue and will block if the queue is empty.

The procedure Dequeue_Only_High_Priority (note that it is marked as **not overriding** unlike the other operations) is designed to enable us to process items only if they are important enough as defined by the parameter At_Least. The priority of the first element E on the queue is P as given by Get_Priority(E). And so if Before(At_Least, P) is false, then the item on the queue is indeed important enough and so is removed from the queue and the Boolean parameter Success is set to true. On the other hand if Before(At_Least, P) is true then the item is not removed and Success is set to false. Note especially that Dequeue_Only_High_Priority never blocks. If the queue is empty, then Success is just set to false; it never waits for an item to be put on the queue.

As an (unrealistic) example, suppose we decide to make the queue of strings into a priority queue and that the priority is given by their length so that "rabbit" takes precedence over "horse". Remember that the type of the elements is Strings.Holder. We can define the priority as given by the attribute Length so we might as well make the actual (sub)type corresponding to Queue_Priority as simply Natural. Then we define

```
function S_Get_Priority(H: Strings.Holder) return Natural is
  (H.Element'Length);

function S_Before(L, R: Natural) return Boolean is
  (L > R);
```

Note the convenient use of expression functions for this sort of thing.

The instantiation now becomes

```
package Unbounded_Priority_Strings_Package is
  new Ada.Containers.Unbounded_Priority_Queues
                          (Queue_Interfaces => Strings_Interface,
                           Queue_Priority => Natural,
                           Get_Priority => S_Get_Priority,
                           Before => S_Before);
```

and we then declare a queue thus

```
My_Strings_UPQ: Unbounded_Priority_Strings_Package.Queue;
```

To put some strings on this queue, we write

```
My_Strings_UPQ.Enqueue(To_Holder("rabbit"));
My_Strings_UPQ.Enqueue(To_Holder("horse"));
My_Strings_UPQ.Enqueue(To_Holder("donkey"));
My_Strings_UPQ.Enqueue(To_Holder("gorilla"));
```

The result is that "gorilla" will have jumped to the head of the queue despite having been put on last. It will be followed by "rabbit" and "donkey" (these are both the same length but the rabbit got there first) and the "horse" is last.

If we do

```
My_Strings_UPQ.Dequeue_Only_High_Priority(7, Kennel, OK);
```

then the "gorilla" will be taken from the queue and placed in the Kennel and OK will be true. But if we then do it again, nothing will happen because the resulting head of the queue (the "rabbit") is not long enough.

Finally, we need to consider bounded priority queues. They are exactly like the unbounded priority queues except that they have the same additional features regarding capacity as found in the bounded synchronized queues. Thus the only differences (apart from the name) are that there is an additional generic parameter Default_Capacity and the protected type Queue has an additional discriminant Capacity.

As a final example we will do a bounded priority queue of records. Suppose the records concern requests for servicing a dishwasher. They might include usual information such as the model number, name and address of owner and so on. They might also have a component indicating degree of urgency, such as

Urgent – machine has vomited dirty water all over floor; spouse having a tantrum,

Major – machine won't do anything; children refuse to help with washing up,

Minor – machine leaves some dishes unclean, parents are coming next week,

Routine – machine needs annual service.

So we might have

```
type Degree is (Urgent, Major, Minor, Routine);

type Dish_Job is
   record
      Urgency: Degree;
      Name: ...
      ...
   end record;
```

First we declare the interface for this type

```
package Dish_Interface is
   new Ada.Containers.Synchronized_Queue_Interfaces
                                          (Element_Type => Dish_Job);
```

and then we declare the two slave functions for the priority mechanism thus

```
function W_Get_Priority(X: Dish_Job) return Degree is
   (X.Urgency);

function W_Before(L, R: Degree) return Boolean is
   (Degree'Pos(L) < Degree'Pos(R));          -- or simply L < R
```

The instantiation is then

```
package Washer_Package is
   new Ada.Containers.Bounded_Priority_Queues
                                  (Queue_Interfaces => Dish_Interface,
                                   Queue_Priority => Degree,
                                   Get_Priority => W_Get_Priority,
                                   Before => W_Before,
                                   Default_Capacity => 100);
```

and we declare the queue of waiting calls thus

```
Dish_Queue: Washer_Package.Queue;
```

which gives a queue with the default capacity of 100.

The staff taking requests then place the calls on the queue by

```
Dish_Queue.Enqueue(New_Job);
```

To cope with the possibility that the queue is full, they can do a timed entry call; remember that this is possible because the procedure Enqueue in the interface package has Synchronization => By_Entry.

And then general operatives checking in and taking the next job do

```
Dish_Queue.Dequeue(Next_Job);
```

However, at weekends we can suppose that just one operative (old George) is on call and deals with only Urgent and Major calls. He might check the queue from time to time by calling

```
Dish_Queue.Dequeue_Only_High_Priority(Major, My_Job, Got_Job);
```

and if Got_Job is false, old George can relax and go back to resting in the garden.

24.9 Bounded containers

As mentioned in Section 24.1, bounded containers were introduced so that high integrity systems could use containers without dynamic storage mechanisms. Queues with bounded storage were described in the previous section and we have already noted that by their very nature the holder containers only ever have one element anyway. So all that remains to be discussed are the bounded forms for lists, vectors, maps and sets (both ordered and hashed), and multiway trees.

The packages for bounded containers are almost identical to the unbounded ones. The main differences are that the types have discriminants giving the capacity and we have to consider the behaviour when a container becomes full.

So (denoting the various aspects by ...) the type Vector in the bounded case is

```
type Vector(Capacity: Count_Type) is tagged private ...
```

whereas in the unbounded case the type Vector is simply

```
type Vector is tagged private ...
```

The other types in the bounded packages are

```
type List(Capacity: Count_Type) is tagged private...
type Map(Capacity: Count_Type; Modulus: Hash_Type) is tagged private...
type Map(Capacity: Count_Type) is tagged private ...
type Set(Capacity: Count_Type; Modulus: Hash_Type) is tagged private ...
type Set(Capacity: Count_Type) is tagged private ...
type Tree(Capacity: Count_Type) is tagged private ...
```

Note that the types for hashed maps and sets have an extra discriminant to set the modulus; this will be explained in a moment.

Remember that the types Count_Type and Hash_Type are declared in the parent package Ada.Containers shown in Section 24.1.

When a bounded container is declared, its capacity is set once and for all by the discriminant and cannot be changed. If we subsequently attempt to add more elements to the container than it can hold then the exception Capacity_Error is raised. (Remember that queue containers are different and never raise Capacity_Error.)

If we are using a bounded container and want to make it larger then we cannot. But what we can do is create another bounded container with a larger capacity and copy the values from the old container into the new one. Remember that we can check the number of items in a container by calling the function Length.

So we might have a sequence such as

```
My_List: List(100);
                                      -- use my list
if Length(My_List) > 90 then          -- it's nearly full

  declare
    My_Big_List: List := Copy(My_List, 200);
  begin
    ...
```

The specification of the function Copy for bounded lists is

```
function Copy(Source: List; Capacity: Count_Type := 0) return List;
```

If the parameter Capacity is not specified (or is given as zero) then the capacity of the copied list is the same as the length of Source.

If the given value of Capacity is larger than (or equal to) the length of the Source (as in our example) then the returned list has this capacity and the various elements are copied. If we foolishly supply a value which is less than the length of Source then Capacity_Error is naturally raised. Remember that a discriminant can be set by an initial value.

Note that if we write

```
My_Copied_List: List := My_List;
```

then My_Copied_List will have the same capacity as My_List because discriminants are copied as well as the contents.

Copy is very valuable and indeed almost vital for operations on bounded sets such as Union and Intersection which return a set whose size is somewhat unpredictable. Thus we can write

```
Set_3 := Copy(Union(Set_1, Set_2), Capacity_Of_Set_3);
```

In order to make it easier to move from the bounded form to the unbounded form, a function Copy was added in Ada 2012 to the unbounded containers as well although it does not need a parameter Capacity in the case of lists and ordered maps and sets. So in the case of the list container it is simply

```
function Copy(Source: List) return List;          -- unbounded
```

Similar unification between bounded and unbounded forms occurs with assignment. Thus in both cases we have

```
procedure Assign(Target: in out List; Source: in List);
```

In the bounded case, if the length of Source is greater than the capacity of Target, then Capacity_Error is raised. In the unbounded case, the structure is automatically extended.

Lists, ordered maps and sets, and trees do not explicitly have a notion of capacity. It is in their very nature that they automatically extend as required. However, in the case of vectors and hashed maps and sets (which have a notion of indexing) taking a purely automatic approach could lead to lots of extensions and copying so the notion of capacity was introduced. As mentioned in Section 24.3, the capacity can be set by calling

```
procedure Reserve_Capacity(Container: in out Vector;
                           Capacity: in Count_Type);
```

and the current value of the capacity can be ascertained by calling

```
function Capacity(Container: Vector) return Count_Type;
```

which naturally returns the current capacity. Remember that Length(V) cannot exceed Capacity(V) but might be much less.

If we add items to a vector whose length and capacity are the same then no harm is done. The capacity will be expanded automatically by effectively calling Reserve_Capacity internally. So the user does not need to set the capacity although not doing so might result in poorer performance.

For uniformity, the bounded vectors and hashed maps and sets also declare a procedure Reserve_Capacity. However, since the capacity cannot be changed for the bounded forms it just checks that the value of the parameter Capacity does not exceed the actual capacity of the container; if it does then Capacity_Error is raised and otherwise it does nothing. There is also a function Capacity for bounded vectors and hashed maps and sets which simply returns the fixed value of the capacity.

There are other differences between unbounded and bounded containers. Unbounded containers have aspect Preelaborate whereas bounded containers have aspect Pure. They both have aspect Remote_Types, see Section 26.3.

The bounded containers for hashed maps and hashed sets are treated somewhat differently to those for the corresponding unbounded containers regarding hashing.

In the case of unbounded containers, the hashing function to be used is left to the user and is provided as an actual generic parameter. For example, in the case of hashed sets, the package specification begins

```
with Ada.Iterator_Interfaces;
generic
   type Element_Type is private;
   with function Hash(Element: Element_Type) return Hash_Type;
   with function Equivalent_Elements(Left, Right: Element_Type)
                                                   return Boolean;
   with function "=" (Left, Right: Element_Type) return Boolean is <>;
package Ada.Containers.Hashed_Sets is ...
```

What the implementation actually does with the hash function is entirely up to the implementation. The value returned is in the range of Hash_Type which is a modular type declared in the root package Ada.Containers. The implementation will typically then map this value onto the current range of the capacity in some way. If the unbounded container becomes nearly full then the capacity will be automatically extended and a new mapping will be required; this in turn is likely to require the existing contents to be rehashed. None of this is visible to the user.

In the case of the bounded containers, these problems do not arise since the capacity is fixed. Moreover, the modulus to be used for the mapping is given when the container is declared since the type has discriminants thus

```
type Set(Capacity: Count_Type; Modulus: Hash_Type) is tagged private;
```

The user can then choose the modulus explicitly or alternatively can use the additional function Default_Modulus whose specification is

```
function Default_Modulus(Capacity: Count_Type) return Hash_Type;
```

This returns an implementation defined value for the number of distinct hash values to be used for the given capacity. Thus we can simply write

```
My_Set: Set(Capacity => My_Cap,
            Modulus => Default_Modulus(My_Cap));
```

Moreover, for these bounded hashed maps and sets, the function Copy has an extra parameter thus

```
function Copy(Source: Set;
              Capacity: Count_Type := 0;
              Modulus: Hash_Type := 0) return Set;
```

If the capacity is given as zero then the newly returned set has the same capacity as the length of Source as mentioned above. If the modulus is given as zero then the value to be used is obtained by applying Default_Modulus to the new capacity.

Finally, note that there are no bounded containers for indefinite types. This is because the size of an object of an indefinite type is generally not known and so indefinite types need some dynamic storage management. However, the whole point of introducing bounded containers was to avoid such management.

24.10 Indefinite containers

As mentioned earlier, the holder container is always for indefinite types and the queue containers cannot be used directly with indefinite types because of problems with entries; however, an indefinite type can always be wrapped in a holder and then becomes definite thereby catering for queues of indefinite types. So all that remains to be discussed are the explicit versions of the container packages for lists, vectors, maps, sets, and trees for indefinite types.

Remember that an indefinite (sub)type is one for which we cannot declare an object without giving a constraint (either explicitly or though an initial value). Moreover, we cannot have an array of an indefinite subtype. The type String is a good example. Thus we cannot declare an array of the type String because the components might not all be the same size and indexing would be a pain. Class wide types are also indefinite.

In the case of the indefinite container for lists the specification starts

```
with Ada.Iterator_Interfaces;
generic
   type Element_Type(<>) is private;
   with function "=" (Left, Right: Element_Type) return Boolean is <>;
package Ada.Containers.Indefinite_Doubly_Linked_Lists is ...
```

where we see that the formal type Element_Type has unknown discriminants and so permits the actual type to be any indefinite type (and indeed a definite type as well). So if we want to manipulate lists of strings where the individual strings can be of any length then we declare

```
package String_Lists is
            new Ada.Containers.Indefinite_Doubly_Linked_Lists(String);
```

In the case of ordered maps we have

```
with Ada.Iterator_Interfaces;
generic
   type Key_Type(<>) is private;
   type Element_Type(<>)  is private;
   with function "<" (Left, Right: Key_Type) return Boolean is <>;
   with function "=" (Left, Right: Element_Type) return Boolean is <>;
package Ada.Containers.Indefinite_Ordered_Maps is ...
```

showing that both Element_Type and Key_Type can be indefinite.

There are two other differences from the definite versions which should be noted.

One is that the Insert procedures for Vectors, Lists, and Maps and the Insert_Child procedure for multiway trees which insert an element with its default value are omitted (because there is no way to create a default initialized object of an indefinite type anyway).

The other is that the parameter Element of the access procedure Process of Update_Element (or the garrulous Update_Element_Preserving_Key in the case of sets) can be constrained even if the type Element_Type is unconstrained.

As an example of the use of an indefinite container consider the problem of creating an index. For each word in a text file we need a list of its occurrences. The individual words can be represented as just objects of the type String. It is perhaps convenient to consider strings to be the same irrespective of the case of characters and so we define

```
function Same_Strings(S, T: String) return Boolean is
begin
   return To_Lower(S) = To_Lower(T);
end Same_Strings;
```

where the function To_Lower is from the package Ada.Characters.Handling. Or we can use the function Ada.Strings.Equal_Case_Insensitive added in Ada 2012 or perhaps rename it. See Section 23.3.

We can suppose that the positions of the words are given by a type Place thus

```
type Place is
   record
      Page: Text_IO.Positive_Count;
      Line: Text_IO.Positive_Count;
      Col: Text_IO.Positive_Count;
   end record;
```

The index is essentially a map from values of the type String to lists of values of type Place. We first create a definite list container for handling the lists thus

```
package Places is new Doubly_Linked_Lists(Place);
```

We then create an indefinite map container from the type String to the type List thus

```
package Indexes is
      new Indefinite_Hashed_Maps(Key_Type => String,
                                 Element_Type => Places.List,
                                 Hash => Ada.Strings.Hash,
                                 Equivalent_Keys => Same_Strings,
                                 "=" => Places."=");
```

An index is then declared by writing

```
An_Index: Indexes.Map;
```

Note that this example illustrates the use of nested containers since the elements in the map are themselves containers (lists of places).

It might be useful for the index to contain information saying which file it refers to. We can extend the type Map thus (remember that container types are tagged)

```
type Text_Map is new Indexes.Map with
   record
      File_Ref: Text_IO.File_Access;
   end record;
```

We also have to declare a corresponding function Copy; this is because if a type is extended then a primitive function returning the type must be overridden as explained in Section 14.3. We can use an expression function thus

```
overriding
function Copy(Source: Text_Map;
              Capacity: Count_Type := 0) return Text_Map is
(Indexes.Copy(Indexes.Map(Source), Capacity) with Source.File_Ref);
```

Now we can more usefully declare

```
My_Index: Text_Map := (Indexes.Empty_Map with My_File'Access);
```

which creates an empty map plus the extra component giving the file reference.

We can now declare various subprograms to manipulate our map. For example to add a new item we have first to see whether the word is already in the index – if it is not then we add the new word to the map and set its list to a single element whereas if it is already in the index then we add the new place entry to the corresponding list. Thus

```
procedure Add_Entry(The_Index: in out Text_Map; Word: String; P: Place) is
   use type Indexes.Cursor;
   M_Cursor: Indexes.Cursor;
   A_List: Places.List;                          -- empty list of places
begin
   M_Cursor := The_Index.Find(Word);
   if M_Cursor = Indexes.No_Element then         -- it's a new word
      A_List.Append(New_Item => P);
      The_Index.Insert(Key => Word, New_Item => A_List);
   else                                          -- it's an old word
      A_List := Indexes.Element(M_Cursor);       -- get old list
      A_List.Append(P);                          -- add to it
      The_Index.Replace_Element(M_Cursor, New_Item => A_List);
   end if;
end Add_Entry;
```

The type Text_Map being derived from Indexes.Map inherits all the map operations and so we can write either of

```
The_Index.Find(Word)                 -- prefixed notation
Indexes.Find(The_Index, Word)        -- classical notation
```

On the other hand, auxiliary entities such as the type Cursor and the constant No_Element are of course in the package Indexes and have to be referred to as Indexes.Cursor and so on.

Note the use type clause for Indexes.Cursor so the comparison of M_Cursor with Indexes.No_Element can be done in the ordinary infixed way.

A big problem with the procedure as written however is that it uses Element and Replace_Element. This means that it copies the whole of the existing list, adds the new item to it, and then copies it back. We can of course use Update_Element thus

```
procedure Add_Entry(The_Index: in out Text_Map; Word: String; P: Place) is
   use type Indexes.Cursor;
   M_Cursor: Indexes.Cursor;
   A_List: Places.List;                            -- empty list of places
begin
   M_Cursor := The_Index.Find(Word);

   if M_Cursor = Indexes.No_Element then           -- it's a new word
      A_List.Append(New_Item => P);
      The_Index.Insert(Key => Word, New_Item => A_List);
   else                                            -- it's an old word
      declare                      -- this procedure adds to the list in situ
         procedure Add_It(The_Key: in String;
                          The_List: in out Places.List) is
         begin
            The_List.Append(New_Item => P);
         end Add_It;
      begin
                                   -- and here we call it via Update_Element
         The_Index.Update_Element(M_Cursor, Add_It'Access);
      end;
   end if;
end Add_Entry;
```

However, this is truly tedious. It is far better to use the magic references thus

```
procedure Add_Entry(The_Index: in out Text_Map; Word: String; P: Place) is
   use type Indexes.Cursor;
   M_Cursor: Indexes.Cursor;
   A_List: Places.List;                            -- empty list of places
begin
   M_Cursor := The_Index.Find(Word);
   if M_Cursor = Indexes.No_Element then           -- it's a new word
      A_List.Append(New_Item => P);
      The_Index.Insert(Key => Word, New_Item => A_List);
   else                                            -- it's an old word
                                                   -- add to the list
      The_Index(M_Cursor).Append(New_Item => P);
   end if;
end Add_Entry;
```

This is still somewhat untidy. In the case of a new word we might as well make the new map entry with an empty list and then update it thereby sharing the calls of Append. We get

```
procedure Add_Entry(The_Index: in out Text_Map; Word: String; P: Place) is
   use type Indexes.Cursor;
   M_Cursor: Indexes.Cursor := The_Index.Find(Word);
   OK: Boolean;
begin
   if M_Cursor = Indexes.No_Element then          -- it's a new word
      The_Index.Insert(Word, Places.Empty_List, M_Cursor, OK);
                  -- M_Cursor now refers to new position and OK will be True
   end if;
   The_Index(M_Cursor).Append(New_Item => P);   -- add to the list
 end Add_Entry;
```

It will be recalled that there are various versions of Insert. We have used that which has two out parameters being the position where the node was inserted and a Boolean parameter indicating whether a new node was inserted or not. In this case we know that it will be inserted and so the final parameter is a nuisance (but sadly we cannot default out parameters). Note also that we need not give the parameter Places.Empty_List because another version of Insert will do that automatically since that is the default value of a list anyway.

Yet another approach might be not to use Find but just call Insert. If the type of Key were definite we could use the definite version of the container and write

```
procedure Add_Entry(The_Index: in out Text_Map; Word: String; P: Place) is
   M_Cursor: Indexes.Cursor;
   Inserted: Boolean;
begin
   The_Index.Insert(Word, M_Cursor, Inserted);
                                  -- M_Cursor now refers to position of node
                                  -- and Inserted indicates whether it was added
   The_Index(M_Cursor).Append(New_Item => P);   -- add to the list
end Add_Entry;
```

This aims to call the defaulted version of Insert. If the word is not present then a new node is made with an empty list. The position parameter indicates where it is.

However, this is not possible for the indefinite version of the container as it is not possible to create a default initialized object of an indefinite type and hence that version of Insert does not exist as was mentioned earlier. So we leave the reader to decide which of the four legal approaches is best.

We can now do some queries on the index. For example we might want to know how many different four-lettered words there are in the text. We can do this by iterating various ways. We can do it ourselves using Next, or we can use the procedure Iterate or we can use the function Iterate introduced in Ada 2012 thus

```
function Four_Letters(The_Index: Text_Map) return Integer is
   Count: Integer := 0;
begin
   for M_Cursor in The_Index.Iterate loop
      if Indexes.Key(Position => M_Cursor)'Length = 4 then
         Count := Count + 1;
      end if;
```

```
    end loop;
    return Count;
  end Four_Letters;
```

We might finally wish to know how many four-lettered words there are on a particular page. (This is just an exercise – it would clearly be simplest to search the original text!) We use the functions Iterate this time both to scan the map for the words and then to scan each list for the page number.

```
  function Four_Letters_On_Page(The_Index: Text_Map;
                         The_Page: Text_IO.Positive_Count) return Integer is
    Count: Integer := 0;
  begin
    for M_Cursor in The_Index.Iterate loop
      if Indexes.Key(Position => M_Cursor)'Length = 4 then

        for L_Cursor in Indexes.Element(M_Cursor).Iterate loop
          if Places.Element(L_Cursor).Page = The_Page then
            Count := Count + 1;
          end if;
        end loop;

      end if;
    end loop;

    return Count;
  end Four_Letters_On_Page;
```

The inner loop could be written as

```
        for E of Indexes.Element(M_Cursor) loop
          if E.Page = The_Page then
            Count := Count + 1;
          end if;
        end loop;
```

which uses the form of loop with **of** rather than **in**. Note that we cannot use this form for the outer loop because we need to use the cursor explicitly in order to get hold of the key. Remember that in a map we cannot get from an element to its key and that the **of** form of loop iterates over the elements and not over the keys.

An important point to note regarding indefinite containers is that Update_Element can be unhelpful if the element is something like a string rather than a nested container as in the example above. The problem is that the procedure Process which is written by the user applies the usual run-time checks and so if we try to change the length of a string it will raise Constraint_Error. So the irony is that we can have a container whose individual elements are of different lengths (so it has to be an indefinite container), but Update_Element cannot be used to change the lengths of individual elements. We have to use Element to read the value and can then use Replace_Element to put the new value back. And similarly, we cannot use Reference to change the length of an individual element either.

Exercise 24.10

1 Write a function Most that takes a Text_Map and returns the word that has the most entries.

24.11 Sorting

The final facilities in the container library are generic procedures for sorting. There are three versions, one for unconstrained arrays, one for constrained arrays and an ingenious one for general sorting over arbitrary structures.

That for unconstrained arrays is

```
generic
   type Index_Type is (<>);
   type Element_Type is private;
   type Array_Type is array (Index_Type range <>) of Element_Type;
   with function "<" (Left, Right: Element_Type) return Boolean is <>;
procedure Ada.Containers.Generic_Array_Sort
                                        (Container: in out Array_Type)
   with Pure;
```

and that for constrained arrays is

```
generic
   type Index_Type is (<>);
   type Element_Type is private;
   type Array_Type is array (Index_Type) of Element_Type;
   with function "<" (Left, Right: Element_Type) return Boolean is <>;
procedure Ada.Containers.Generic_Constrained_Array_Sort
                                        (Container: in out Array_Type)
   with Pure;
```

These generic procedures do the obvious thing. They sort the array Container into order as defined by the generic parameter "<". The emphasis is on speed.

Note that the procedure for unconstrained arrays is identical in form to the generic procedure Sort discussed in Section 19.3.

As an example we could sort the letters in a string into alphabetical order. We would declare

```
procedure String_Sort is
   new Ada.Containers.Generic_Array_Sort(Positive, Character, String);
```

and then if we had a string such as

```
My_Pet: String := "rabbit";
```

we could apply String_Sort to it thus

```
String_Sort(My_Pet);
```

and the value in My_Pet will now be "abbirt".

However, sorting doesn't just apply to arrays and Ada 2012 introduced another container which takes a more general approach. Its specification is

```
generic
   type Index_Type is (<>);
   with function Before(Left, Right: Index_Type) return Boolean;
   with procedure Swap(Left, Right: in Index_Type);
procedure Ada.Containers.Generic_Sort(First, Last: Index_Type'Base)
   with Pure;
```

This can be used to sort any indexable structure and not just arrays. The generic parameters define the required ordering through the parameter Before much as expected. The cunning trick however, is that the means of interchanging two items in the structure is provided by the parameter Swap.

As an illustration we can use this on the array My_Pet. We can use an expression function for MP_Before and so we write

```
function MP_Before(L, R: Positive) return Boolean is
   (My_Pet(L) < My_Pet(R));

procedure MP_Swap(L, R: in Positive) is
   Temp: Character;
begin
   Temp := My_Pet(L);  My_Pet(L) := My_Pet(R);  My_Pet(R) := Temp;
end MP_Swap;

procedure MP_Sort is
   new Ada.Containers.Generic_Sort(Positive, MP_Before, MP_Swap);
```

and then we actually do the sort by

```
MP_Sort(My_Pet'First, My_Pet'Last);
```

That may seem to be rather a struggle but the key point is that the technique can be used to sort items in any indexable structure such as a vector container.

Suppose we have a number of records of a type Score which might be

```
type Score is
   record
      N: Natural := 0;
      OS: Other_Stuff;
   end record;
```

and we declare a vector container to hold such objects thus

```
package Scores is
   new Ada.Containers.Vectors(Natural, Score);
My_Vector: Scores.Vector;
```

Now assume that we have added various objects of the type Score to our vector and that we decide that we would like them sorted into the order determined by their component N. We write

```
function MV_Before(L, R: Natural) return Boolean is
   (Scores.Element(My_Vector, L).N < Scores.Element(My_Vector, R).N);
procedure MV_Swap(L, R: in Natural) is
begin
   Scores.Swap(My_Vector, L, R);
end MV_Swap;
procedure MV_Sort is
   new Ada.Containers.Generic_Sort(Natural, MV_Before, MV_Swap);
```

and then we do the sort by

```
MV_Sort(Scores.First_Index(My_Vector), Scores.Last_Index(My_Vector));
```

Note that the vectors container package conveniently already has a procedure Swap. This vector example is not very exciting because it might be recalled that the vectors containers already have their own internal generic sort. To use it on this example we would have to write

```
package MV_Sorting is
   new Scores.Generic_Sorting(MV_Before);
MV_Sorting.Sort(My_Vector);
```

which is somewhat simpler. However, note that this sorts the whole vector. If we only wanted to sort part of it, say from elements in index range P to Q then it cannot be used. But that would be easy with the new one since we would simply write

```
MV_Sort(P, Q);
```

Note that curiously this does not need to mention My_Vector.

24.12 Summary table

The various facilities in the vectors, lists, maps, sets, and tree containers are summarized in a comprehensive table which will be found on the website. The queue containers do not fit the same pattern so are excluded. The holder containers are also omitted because they are so simple.

Checklist 24

The containers for Lists, Vectors, Maps and Sets, and Trees come in three forms: definite unbounded, definite bounded and indefinite unbounded.

Maps and Sets can be ordered or hashed.

There was only one form of holder container originally: indefinite unbounded.

There are four forms of queues, unbounded/bounded, priority/synchronized.

There are three forms of sort, for definite arrays, indefinite arrays, and general.

Cursors do not change if new items are added or deleted in the case of lists and trees; but can change for vectors and other containers.

Keys for maps are never duplicated.

Beware that using Replace_Element with Sets may cause the set to be reordered.

The containers for multiway trees, holders, and queues were added in Ada 2012.

The bounded forms were also new in Ada 2012.

The use of Ada.Iterator_Interfaces and the types and functions concerning references were introduced in Ada 2012.

The general form of the sort container was added in Ada 2012.

New in Ada 2022

There is one new container Ada.Containers.Bounded.Indefinite.Holder.

There are better facilities for preventing and detecting tampering. Most containers include a nested package Stable for this purpose.

Additional subprograms are added to allow operations to be done using prefixed as well as classical notation.

The subprograms Append, Prepend, and Insert that add a Vector in Ada 2012 are Append_Vector, Prepend_Vector, and Insert_Vector respectively in Ada 2022.

Containers can be initialized with a form of aggregate using square brackets. Such aggregates can also include loops.

25 Interfacing

25.1 Representations
25.2 Unchecked programming
25.3 The package System
25.4 Storage pools and subpools
25.5 Other languages

This chapter considers various aspects of how an Ada program interfaces to the outside world. One area is the mapping of the abstract Ada program onto the computer; how the data structures are represented and so on. Another important aspect is communication with programs in other languages. (Communication with the outside world through interrupts is dealt with in Section 26.1.) However, the discussion in this chapter cannot be exhaustive because many details of this area will depend upon the implementation. The intent, therefore, is to give the reader a general overview of the facilities available.

25.1 Representations

When compiling a program, the compiler needs to decide how the various data items are to be represented. Certain uses or occurrences of entities require that their representation be known; such occurrences are called freezing points. We can provide explicit information about certain aspects of an entity and that of course determines that aspect. If no such information is provided then the compiler has to make some default decision at the first freezing point.

An obvious example is that declaring an object requires that its type be frozen. This does not apply to deferred constants since in the case of private types the freezing is deferred until the full type declaration. If we wish to provide our own representation then this must be done before the entity is otherwise frozen. Note that the end of a declarative part freezes all the items declared in it anyway (except for incomplete types) and so any representation information must be in the same declarative part as the entity to which it applies.

In most cases the freezing rules are very much a fringe aspect of the language and intuitively obvious and have little impact on the normal programmer. One important interaction with other rules was however noted in Section 14.6 when we discussed type extension and primitive operations. We noted that we cannot add further primitive operations after a tagged type is frozen and moreover that a tagged type is frozen when a type is derived from it.

In summary, a type must be fully defined, including all aspect clauses and dispatching operations, before it is frozen. Declaring an object of a type or extending a type freezes it. A couple of minor points are that a null procedure never freezes anything (AI12-157) and that an expression function acting as a completion just freezes the expression (AI12-103).

We can provide our own representations using aspect clauses which can take various forms or in most cases by using aspect specifications.

For example we can specify the amount of storage to be allocated for objects of a type, for the storage pool of an access type, and for the working storage of a task type. This can done by an attribute definition clause thus

```
type Byte is range 0 .. 255;
for Byte'Size use 8;
```

and indicates that objects of the type Byte should occupy only 8 bits. Alternatively we can use an aspect specification thus

```
type Byte is range 0 .. 255
   with Size => 8;
```

The size of individual objects can also be specified.

The topic of Size is quite complex since it concerns not only single objects but also arrays and records containing components of the type. In addition the aspect Pack can be given for a composite type and is a broad hint to the compiler to squeeze things up. An example occurs in Standard where we find

```
type String is array (Positive range <>) of Character
   with Pack;
```

The attribute Size may be applied to an object or subtype whether we have supplied a value or not. For a single object it gives the actual size occupied. For a subtype it gives the value that will be used in a packed record. We are assured that Boolean'Size = 1 and so in a packed record any Boolean components will be squeezed right up. But individual objects would typically occupy a word or byte for addressability reasons.

The attribute Size is subtype-specific – this means that different subtypes of the same type may have different values of Size. Confusion over Size is overcome in Ada 2022 by the introduction of the new attribute Object_Size which when applied to a definite subtype returns the size in bits of a stand-alone aliased object of the subtype. It also gives the size of a component provided that no representation item is applied. The other subtype-specific attribute is Alignment.

The alignment of a subtype or object may be specified in a similar way. Thus we might decide that all objects of the type My_Float should be on 8-byte boundaries (assuming that the storage unit is a byte). We could write

```
for My_Float'Alignment use 8;
```

or we could use an aspect specification with the declaration of My_Float.

There is also an aspect Component_Size which applies to array types and objects and indicates the size for the components. Thus for speed we might want a particular array of Booleans to be packed loosely, thus

```
Loose_Bits: array (1 .. 10) of Boolean
  with Component_Size => 4;
```

and the compiler should then pack the bits two to a byte which we assume is easily addressed on the hardware concerned.

The space for standard storage pools and tasks is indicated using the aspect Storage_Size. In these cases the unit is not bits but storage elements. The number of bits in a storage element is implementation dependent and is given by the named number Storage_Unit in the package System. Thus if we wanted to ensure that the pool for the type Cell_Ptr will accommodate 500 cells then we can write

```
type Cell_Ptr is access Cell
  with Storage_Size => 500 * Cell'Size / System.Storage_Unit;
```

which assumes that Cell'Size is an exact multiple of Storage_Unit and that there are no hidden overheads. Alternatively we can write

```
type Cell_Ptr is access Cell
  with Storage_Size => 500 * Cell'Max_Size_In_Storage_Elements;
```

which does any rounding up and is especially appropriate if the accessed type is indefinite (such as the discriminated record Person of Section 18.5) since it allows for the maximum possible size of object. In that example, an object of subtype Woman would typically take more space than one of subtype Man.

The data space for tasks can similarly be set and can thus depend upon a discriminant. This enables individual tasks of a task type to have different amounts of storage. Thus we might have

```
task type T(Work_Space: Integer)
  with Storage_Size => Work_Space is
  ...
end;
```

The value of *small* for a fixed point type can also be indicated as was discussed in Section 17.5.

An enumeration representation clause can be used to specify, as an aggregate, the internal integer codes for the literals of an enumeration type. We might have a status value transmitted into our program as single bit settings, thus

```
type Status is (Off, Ready, On);
for Status use (Off => 1, Ready => 2, On => 4);
```

There is a constraint that the ordering of the values must be the same as the logical ordering of the literals. However, despite the holes, the functions Succ, Pred, Pos, and Val always work in logical terms. If a variable S of type Status has value Ready then Status(S) is in fact 1 rather than 2 which is perhaps surprising.

Ada 2022 has two new attributes Enum_Rep and Enum_Val which work on the representation values rather than the position. So Status'Enum_Rep(S) is 2 and

Status'Enum_Rep(Off) is 1. And similarly Status'Enum_Val(1) is Off and Status'Enum_Val(2) is Ready. And, as expected, Status'Enum_Val(3) causes Constraint_Error to be raised.

If these single bit values were autonomously loaded into our machine at location octal 100 then we could conveniently access them in our program by declaring a variable of type Status thus

```
S: Status
  with Address => 8#100#;
```

However, if by some hardware mishap a value which is not 1, 2, or 4 turns up then the program will clearly misbehave. We will see how to overcome this difficulty in the next section.

The final form of aspect clause is used to indicate the layout of a record type. So if we have

```
type Register is range 0 .. 15;
type Opcode is ( ... );

type RR is
  record
    Code: Opcode;
    R1: Register;
    R2: Register;
  end record;
```

which represents a machine instruction of the RR format in the classic IBM System 370, then we can specify the exact mapping by a record representation clause thus

```
for RR'Alignment use 2;

for RR use
  record
    Code at 0 range 0 .. 7;
    R1 at 1 range 0 .. 3;
    R2 at 1 range 4 .. 7;
  end record;
```

The first aspect clause indicates that the record is to be aligned on a double byte boundary; the alignment is given in terms of the number of storage elements and in the case of the 370 a storage element would naturally be an 8-bit byte.

The position and size of the individual components are given relative to the start of the record. The value after **at** gives a storage element and the range is in terms of bits. However, it is not always clear which way round the bits are numbered and different compilers might choose to number them the different way by default. We can specify the ordering for a type. Thus

```
for RR'Bit_Order use High_Order_First;
```

states that the first bit (bit 0) of a storage element is the most significant (often called big endian as opposed to little endian). This just concerns the logical numbering of the bits and does not involve any run-time conversion. Its main purpose is so that record representation clauses can be given consistently regardless of the default order. The details are defined in terms of so-called machine scalars. The type Bit_Order and its default value are declared in System.

The bit number can extend outside the storage element (such as R1 **at** 0 **range** 8 .. 11) but beware that the meaning will depend on the bit order. We are not assured that bit 8+*b* of byte *a* is the same as bit *b* of byte *a*+1 whatever the order.

If we do not specify the location of every component, the compiler is free to juggle the rest as best it can. However, we must allow enough space for those we do specify and they must not overlap unless they are in different alternatives of a variant. There may also be hidden components (array dope information for example) and this may interfere with our freedom. The position of a component is given by attributes First_Bit, Last_Bit, and Position (see Appendix 1 for details).

Note that both the alignment and the bit order could have been given by aspect specifications with the type declaration thus

```
type RR is
   record
      ...
   end record
      with Alignment => 2, Bit_Order => High_Order_First;
```

It is a general principle that all subtypes of a type have the same representation. Accordingly, the attributes can only be specified for the first subtype of a type. However, the compiler may choose different values for Size and Alignment of subtypes from those of the type. Derived types inherit the representation of the parent but can generally override it.

The functional attributes Has_Same_Storage and Overlaps_Storage are useful for ensuring that objects do not occupy the same space in whole or in part. Thus if

```
X'Has_Same_Storage(Y)
```

is true then the objects X and Y have all bits in common. If there are zero bits then Has_Same_Storage is false. On the other hand if

```
X'Overlaps_Storage(Y)
```

is true then the objects X and Y have at least one bit in common. X and Y can be of any type. For example, we might have a procedure Exchange and wish to ensure that the parameters do not overlap in any way by writing

```
procedure Exchange(X, Y: in out T)
   with Pre => not X'Overlaps_Storage(Y);
```

The balance between space and time can be set by the pragma Optimize (see Section 5.6) and the aspect Inline. The aspect Inline indicates that all calls of the subprogram concerned should be expanded inline. This can be given with the declaration of individual subprograms or might be done by the obsolete pragma Inline. A good example would be the short subprograms for manipulating complex numbers. So in the specification of the package Complex_Numbers in Section 12.2, we might write

```
pragma Inline("+", "-", "*", "/");
```

25.2 Unchecked programming

Sometimes the strict integrity of a strongly typed language is a nuisance. This particularly applies to system programs where, in different parts of a program, an object is thought of in different terms. This can be overcome by the use of a generic function called Ada.Unchecked_Conversion. Its specification is

```
generic
   type Source(<>) is limited private;
   type Target(<>) is limited private;
function Ada.Unchecked_Conversion(S: Source) return Target
   with Pure, Nonblocking, Convention => Intrinsic;
```

As an example, we can overcome our problem with possible unexpected values of S of the type Status of the previous section. We can receive the values into our program as values of type Byte, check their validity in numeric terms and then convert the values to type Status for the remainder of the program. In order to perform the conversion we first instantiate the generic function thus

```
function Byte_To_Status is new Unchecked_Conversion(Byte, Status);
```

and we can then write

```
B: Byte
   with Address => 8#100#;                  -- same address as S
...
case B is
   when 1 | 2 | 4 =>
      null;
   when others =>
      raise Bad_Data;
end case;
S := Byte_To_Status(B);
```

The effect of the unchecked conversion is typically nothing; the bit pattern of the source type is merely passed on unchanged and reinterpreted as the bit pattern of the target type. Clearly, certain conditions must be satisfied for this to be possible; an obvious one is that the size of the two (sub)types must be the same; another is that the alignment of the source must be the same as or an integer multiple of that of the target. Misuse of unchecked conversion can give rise to an erroneous program.

The Valid attribute is useful for checking that the value of a scalar object is sensible. Thus we can rewrite the above example as simply

```
S: Status
   with Address => 8#100#;
...
if not S'Valid then raise Bad_Data; end if;
```

Another good example of unchecked conversion is converting between signed integer and modular types. Thus following Section 17.2 we might write

```
function Convert_Byte is
     new Unchecked_Conversion(Signed_Byte, Unsigned_Byte);
function Convert_Byte is
     new Unchecked_Conversion(Unsigned_Byte, Signed_Byte);
```

where we have used the same name in both directions. For this to work it is important that the size of the two types be the same.

Another area of unchecked programming occurs with the deallocation of objects created by an allocator. We can do this with a generic procedure called Ada.Unchecked_Deallocation. Its specification is

```
generic
   type Object(<>) is limited private;
   type Name is access Object;
procedure Ada.Unchecked_Deallocation(X: in out Name)
   with Preelaborate, Nonblocking, Convention => Intrinsic,
        Global => in out Name'Storage_Pool;          -- note this
```

So we might write something like

```
type Cell_Ptr is access Cell;
...
procedure Free is new Unchecked_Deallocation(Cell, Cell_Ptr);
...
List: Cell_Ptr;
...
Free(List);
```

After calling Free, the value of List will be **null** (Constraint_Error would be raised if null excluding) and the cell will have been returned to free storage. Of course, if we mistakenly still had another variable referring to the cell then we would be in a mess; the program would be erroneous. If we use unchecked deallocation then the onus is on us to get it right. Note that it can also be used with a class wide type; thus we could replace Cell by Cell'Class in the above example. This topic is considered in more detail in Section 25.4 when we discuss how to create our own storage pools.

An interesting point is that if we use anonymous access types then it is impossible to use unchecked deallocation because the access type has no name to use in the instantiation of Unchecked_Deallocation. Thus if we have

```
type Cell is
   record
      Next: access Cell;
      Value: Integer;
   end record;
```

then although we can write

```
L: access Cell := new Cell( ... );
```

and thereby allocate storage, it is not possible to free it. In order to prevent this difficulty we can write

```
pragma Restrictions(No_Anonymous_Allocators);
```

and all such allocation is thereby prevented.

Unchecked programming needs care and it is wise to restrict its use to privileged parts of a program. However, since the generic subprograms are library units, any compilation unit using them must mention them in a with clause. This makes it fairly straightforward for a tool (and our manager) to check for their use. Unfortunately, it is not quite so easy to monitor the use of the attribute Unchecked_Access described in Section 11.5.

25.3 The package System

The package System is concerned with the fine details of the control of storage and related issues. It has a number of child packages as follows

```
System.Address_To_Access_Conversions
System.Machine_Code
System.RPC
System.Storage_Elements
System.Storage_Pools
```

System.RPC concerns the Distributed Systems annex (see Section 26.3). The others are discussed in this chapter.

The specification of the parent package System is as follows

```
package System
    with Pure is
  type Name is implementation-defined-enumeration-type;
  System_Name: constant Name := implementation-defined;

  -- system-dependent named numbers

  Min_Int: constant := root_integer'First;
  Max_Int: constant := root_integer'Last;
  Max_Binary_Modulus: constant := implementation-defined;
  Max_Nonbinary_Modulus: constant := implementation-defined;
  Max_Base_Digits: constant := root_real'Digits;
  Max_Digits: constant := implementation-defined;
  Max_Mantissa: constant := implementation-defined;
  Fine_Delta: constant := implementation-defined;
  Tick: constant := implementation-defined;

  -- storage-related declarations

  type Address is implementation-defined;
                         -- pragma Preelaborable_Initialization(Address);
  Null_Address: constant Address;

  Storage_Unit: constant := implementation-defined;
  Word_Size: constant := implementation-defined * Storage_Unit;
  Memory_Size: constant := implementation-defined;
```

```
    function "<" (Left, Right: Address) return Boolean
       with Convention => Intrinsic;
    -- similarly "<=", ">", ">=" and "="

    type Bit_Order is (High_Order_First, Low_Order_First);
    Default_Bit_Order: constant Bit_Order := implementation-defined;

    subtype Any_Priority is Integer range implementation-defined;
    subtype Priority is
               Any_Priority range Any_Priority'First .. implementation-defined;
    subtype Interrupt_Priority is
                     Any_Priority range Priority'Last+1 .. Any_Priority'Last;
    Default_Priority: constant Priority := (Priority'First+Priority'Last)/2;

private
    ...
end System;
```

Min_Int and Max_Int give the most negative and most positive values of a signed integer type, Max_Binary_Modulus and Max_Nonbinary_Modulus give the maximum supported modulus of a modular type, Max_Base_Digits is the largest number of decimal digits of a floating type where a range is given (Max_Digits if no range given, this will not be more than Max_Base_Digits), and Max_Mantissa is the largest number of binary digits of a fixed type; they are all of type *universal_integer*. Fine_Delta is a bit redundant since it has the value 2.0**(–Max_Mantissa) and Tick is the clock period in seconds; they are both of type *universal_real*.

The type Address is that given by the corresponding attribute; it might be an integer type or possibly a record type. Whatever it is, we are guaranteed that it has preelaborable initialization. Note that the pragma is given as a comment because the pragma strictly applies only to private types and Address might not be private.

The numbers Storage_Unit, Word_Size, and Memory_Size give the number of bits in a storage element, the number of bits in a word, and the memory size in storage elements; they are of type *universal_integer*.

The child package System.Storage_Elements defines facilities for messing about with addresses and offsets. Its specification is

```
package System.Storage_Elements
     with Pure is
   type Storage_Offset is range implementation-defined;
   subtype Storage_Count is Storage_Offset
                                       range 0 .. Storage_Offset'Last;
   type Storage_Element is mod implementation-defined;
   for Storage_Element'Size use Storage_Unit;
   type Storage_Array is array (Storage_Offset range <>)
                                         of aliased Storage_Element;
   for Storage_Array'Component_Size use Storage_Unit;

   function "+" (Left: Address; Right: Storage_Offset) return Address
      with Convention => Intrinsic;
   function "+" (Left: Storage_Offset; Right: Address) return Address
      with Convention => Intrinsic;
```

```
      function "-" (Left: Address; Right: Storage_Offset) return Address
         with Convention => Intrinsic;
      function "-" (Left, Right: Address) return Storage_Offset
         with Convention => Intrinsic;
      function "mod" (Left: Address; Right: Storage_Offset)
                                                   return Storage_Offset
         with Convention => Intrinsic;
      type Integer_Address is implementation-defined;
      function To_Address(Value: Integer_Address) return Address
         with Convention => Intrinsic;
      function To_Integer(Value: Address) return Integer_Address
         with Convention => Intrinsic;
   end System.Storage_Elements;
```

The type Storage_Element is unsigned and represents a storage element and the type Storage_Array represents a contiguous lump of store. The various operations "+" and "–" enable addresses and offsets to be added and subtracted. The functions To_Address and To_Integer enable an Address to be converted to the integer type Integer_Address and vice versa.

The generic child package

```
   generic
      type Object(<>) is limited private;
   package System.Address_To_Access_Conversions
         with Preelaborate, Nonblocking, Global => in out synchronized is
      type Object_Pointer is access all Object;
      function To_Pointer(Value: Address) return Object_Pointer
         with Convention => Intrinsic;
      function To_Address(Value: Object_Pointer) return Address
         with Convention => Intrinsic;
   end System.Address_To_Access_Conversions;
```

enables naughty peeking and poking to be done. Note in particular that X'Unchecked_Access can also be written as To_Pointer(X'Address) and that To_Address(Y) returns Y.**all**'Address if Y is not null and Null_Address if it is.

The use of these packages clearly needs care and can easily give rise to erroneous situations. Note finally that all subprograms in these three packages have convention Intrinsic.

25.4 Storage pools and subpools

Recall that the storage for allocated objects associated with an access type is termed a storage pool. We can use standard storage pools or provide our own using another child package of System. Before looking at this in detail, we will first consider some general aspects of the deallocation of storage.

Implementations of Ada typically do not have a garbage collector and so we need to deallocate storage as appropriate in order to prevent so-called storage leaks. This is very important with long running programs – it is not unknown for a program to run satisfactorily for many weeks and then suddenly raise Storage_

Error. Such behaviour is almost always caused by a failure to deallocate storage somewhere so that the available storage slowly diminishes.

Storage can be deallocated by calling a procedure obtained by an instantiation of Unchecked_Deallocation as explained in Section 25.2. Such procedures traditionally have the identifier Free.

In simple cases such as the various examples of a stack it is perfectly clear that the storage can be released in the body of the subprogram Pop. So following the example in Section 12.5 we might write

```
procedure Pop(S: in out Stack; X: out Integer) is
   Old_S: Stack := S;
begin
   X := S.Value;
   S := S.Next;
   Free(Old_S);
end Pop;
```

Note that we have to make a local copy of S first otherwise we lose the ability to free the unused storage. It is important to appreciate that the call of Free is safe because the user's view of the type Stack is limited and so the user cannot have made a copy of the structure. This means that it is not possible for there to be any other reference to the deallocated storage.

An abstract data type should always be constructed so that any deallocation is safe. The key is to avoid shallow copying. One approach is to make the type limited so that any copy has to be through a procedure defined as part of the abstraction; another is to make it controlled so that Adjust can ensure that a deep copy is always made as in the type Linked_Set of Section 14.9.

Another problem is that an object such as a stack or queue might go out of scope when it is not logically empty in which case the allocated storage will be lost. This can be overcome by making the type controlled so that Finalize can deallocate any residual storage. So for the type Linked_Set we might write

```
procedure Finalize(Object: in out Inner) is
   P: Cell_Ptr := Inner.The_Set;
   Old_P: Cell_Ptr;
begin
   while P /= null loop
      Old_P := P;
      P := P.Next;
      Free(Old_P);
   end loop;
end Finalize;
```

In more complex examples of processing lists it is sometimes hard to know when storage can be released. In some applications it might be possible to keep a reference count in an object and then free it when the last reference is removed. However, this fails for applications with cyclic data structures since inaccessible islands can result.

Another example might be in resource management where we are keeping track of some objects of a type T by chaining them together. We might have

```
package P is
   type T is limited private;
   ...                              -- operations visible to the user
private
   type T is new Limited_Controlled with
      record
         ...
         Next: access T;
      end record;
   procedure Initialize(X: in out T);
   procedure Finalize(X: in out T);
end;

package body P is
   List: access T := null;

   procedure Initialize(X: in out T) is
   begin
      X.Next := List;
      List := X'Unchecked_Access;
   end Initialize;

   procedure Finalize(X: in out T) is
   begin
      pragma Assert(List = X'Unchecked_Access);
      List := List.Next;
   end Finalize;

   ...                              -- other operations
end P;
```

The objects of the type T are all chained together through each other; the head of the chain is List. When an object is declared Initialize links it onto the chain and similarly Finalize removes it. All this is invisible to the user. Operations visible to the user might search the chain. We have to use Unchecked_Access to put the objects on the chain and Finalize ensures that no dangerous dangling references are left around. This version assumes that objects are finalized in reverse order to initialization (note the Assert pragma). It will need modifying if the objects of type X are allocated rather than declared so that this is no longer true.

In principle the complete storage pool for an access type can be released at the end of the lifetime of the type itself (bearing in mind that an access type shares its pool with any type derived from it). We might therefore be able to make individual access types localized so that the associated pools are relinquished from time to time. However, if we allow the compiler to use standard storage pools then it might take a crude approach and use just one global storage pool although we are assured that if we specify the size of the pool by setting the attribute Storage_Size as explained in Section 25.1 then indeed the pool will be released at the end of the lifetime of the access type.

We can gain more control by declaring our own storage pools as will now be described. All storage pools are objects of a type derived from the abstract type

Root_Storage_Pool declared in the package System.Storage_Pools whose specification is

```
with Ada.Finalization;
with System.Storage_Elements;
package System.Storage_Pools
    with Pure, Nonblocking => False is
  type Root_Storage_Pool is abstract
                  new Ada.Finalization.Limited_Controlled with private
    with Preelaborable_Initialization;

  procedure Allocate(Pool: in out Root_Storage_Pool;
                  Storage_Address: out Address;
                  Size_In_Storage_Elements, Alignment:
                      in Storage_Elements.Storage_Count) is abstract;

  procedure Deallocate ... ;   -- similar with in for Storage_Address

  function Storage_Size(Pool: Root_Storage_Pool)
                  return Storage_Elements.Storage_Count is abstract;
private
  ...
end System.Storage_Pools;
```

The general idea is that we define a type derived from Root_Storage_Pool and provide concrete subprograms for Allocate, Deallocate, and Storage_Size. We then declare an object of the type to be used as the pool and then associate this object with an access type using the attribute Storage_Pool. Thus

```
with System.Storage_Pools;
with System.Storage_Elements;
use System;
package My_Pools is
  type Pond(Size: Storage_Elements.Storage_Count) is
                  new Storage_Pools.Root_Storage_Pool with private;
  procedure Allocate(Pool: in out Pond; ... );
  ...   -- also Deallocate, Storage_Size, (maybe Initialize, Finalize)
end My_Pools;
...
Cell_Ptr_Pool: My_Pools.Pond(Size => 5000);
for Cell_Ptr'Storage_Pool use Cell_Ptr_Pool;
```

We could alternatively have used an aspect specification giving the pool when declaring the type Cell_Ptr thus

```
type Cell_Ptr is access Cell;
  with Storage_Pool => Cell_Ptr_Pool;
```

The implementation then automatically calls Allocate and Deallocate whenever storage is required (by an allocator) or released (by unchecked deallocation). This is similar to the automatic calling of Initialize and Finalize for controlled types. Note that the storage pool types are controlled anyway and so the user may provide procedures Initialize and Finalize as well.

The parameters for Allocate give the identity of the pool, the address of the allocated object, the space required for the object and its alignment. Thus given the above pool, calling an allocator as in X: Cell_Ptr := **new** Cell; implies a call of Allocate thus

```
Allocate(Cell_Ptr_Pool, The_Address, The_Size, Cell'Alignment);
```

where The_Address returns the address of the space allocated and The_Size gives the size of the cell in storage elements (not the same as the attribute Size which gives it in bits). Note that we are guaranteed that the size demanded will never be more than Cell'Max_Size_In_Storage_Elements and that the alignment demanded will never be more than Cell'Max_Alignment_For_Allocation. Deallocate is similar except that The_Address is that of the storage to be released.

The function Storage_Size is automatically called when the attribute Storage_Size is evaluated and so should be defined appropriately. Note how in this example the size of the pool is given through the discriminant (in some appropriate units). Of course, if we do declare our own storage pool for an access type then we should not also set the Storage_Size attribute for it – that facility is just for standard pools.

If we declare our own storage pool type then we can add other functionality. We might declare a function giving the amount of space still available in the pool. Another possibility might be to have a facility to release all storage allocated since some previously indicated event – this is often called a Mark/Release capability. It is clear that care is needed in implementing our own storage pools. An example will be found in Program 6 which follows this chapter and the reader might also care to consider how Program 2 should be reconstructed to avoid Storage_Error.

A major new facility in Ada 2012 was the introduction of subpools. The general idea is that one wants to manage storage collections with different lifetimes. It is often the case that an access type is declared at library level but various groups of objects of the type are declared and so could be reclaimed at a more nested level. This is done by splitting a pool into separately reclaimable subpools using a child package of System.Storage_Pools with the following specification

```
package System.Storage_Pools.Subpools
    with Preelaborate, Global => in out synchronized is

  type Root_Storage_Pool_With_Subpools is abstract
    new Root_Storage_Pool with private
    with Preelaborable_Initialization;

  type Root_Subpool is abstract tagged limited private
    with Preelaborable_Initialization;

  type Subpool_Handle is access all Root_Subpool'Class;
    for Subpool_Handle'Storage_Size use 0;

  function Create_Subpool
                 (Pool: in out Root_Storage_Pool_With_Subpools)
                        return not null Subpool_Handle is abstract;
```

```
   function Pool_Of_Subpool (Subpool: not null Subpool_Handle)
                  return access Root_Storage_Pool_With_Subpools'Class;

   procedure Set_Pool_Of_Subpool(Subpool: not null Subpool_Handle;
                To: in out Root_Storage_Pool_With_Subpools'Class)
      with Global => overriding in out Subpool;

   procedure Allocate_From_Subpool
                      (Pool: in out Root_Storage_Pool_With_Subpools;
                       Storage_Address: out Address;
                       Size_In_Storage_Elements: in Storage_Count;
                       Alignment: in Storage_Count;
                       Subpool: in not null Subpool_Handle) is abstract
      with Pre'Class => Pool_Of_Subpool(Subpool) = Pool'Access,
           Global => overriding in out Subpool;

   procedure Deallocate_Subpool
                      (Pool: in out Root_Storage_Pool_With_Subpools;
                       Subpool: in out Subpool_Handle) is abstract
      with Pre'Class => Pool_Of_Subpool(Subpool) = Pool'Access;

   function Default_Subpool_For_Pool
                      (Pool: in out Root_Storage_Pool_With_Subpools)
                                       return not null Subpool_Handle;

   overriding
   procedure Allocate(Pool: in out Root_Storage_Pool_With_Subpools;
                      Storage_Address: out Address;
                      Size_In_Storage_Elements: in Storage_Count;
                      Alignment: in Storage_Count);

   overriding
   procedure Deallocate( ... ) is null;

   overriding
   function Storage_Size(Pool: Root_Storage_Pool_With_Subpools)
                                          return Storage_Count is
      (Storage_Count'Last);

private
   ...                          -- not specified by the language
end System.Storage_Pools.Subpools;
```

If we wish to declare a storage pool that can have subpools then rather than declare an object of the type Root_Storage_Pool in the package System.Storage_Pools we have to declare an object of the derived type Root_Storage_Pool_With_Subpools declared in the child package.

The type Root_Storage_Pool_With_Subpools inherits operations Allocate, Deallocate, and Storage_Size from the parent type. Remember that Allocate and Deallocate are automatically called by the compiled code when items are allocated and deallocated. In the case of subpools we don't need Deallocate to do anything so it is null. The function Storage_Size determines the value of the attribute Storage_Size and is given by a function expression.

Subpools are separately reclaimable parts of a storage pool and are identified and manipulated by objects of the type Subpool_Handle (these are access values). We can create a subpool by a call of Create_Subpool. So we might have (assuming appropriate with and use clauses)

```
package My_Pools is
   type Pond(Size: Storage_Count) is
      new Root_Storage_Pool_With_Subpools with private;
   subtype My_Handle is Subpool_Handle;
   ...
```

and then

```
My_Pool: Pond(Size => 1000);
Puddle: My_Handle := Create_Subpool(My_Pool);
```

The implementation of Create_Subpool should call

```
Set_Pool_Of_Subpool(Puddle, My_Pool);
```

before returning the handle. This enables various checks to be made.

In order to allocate an object of type T from a subpool, we have to use a new form of allocator. But first we must ensure that T is associated with the pool itself. So we might write

```
type T_Ptr is access T;
   for T_Ptr'Storage_Pool use My_Pool;
```

And then to allocate an object from the subpool identified by the handle Puddle we write

```
X := new (Puddle) T'( ... );
```

where the subpool handle is given in parentheses following **new**.

Of course we don't have to allocate all such objects from a specified subpool since we can still write

```
Y := new T'( ... );
```

and the object will be allocated from the parent pool My_Pool. It is actually allocated from a default subpool in the parent pool and this is determined by writing a suitable body for the function Default_Subpool_For_Pool and this is called automatically by the allocation mechanism. Note that in effect the whole pool is divided into subpools one of which may be the default subpool. If we don't provide an overriding body for Default_Subpool_For_Pool then Program_Error is raised.

The implementation carries out various checks. For example, it will check that a handle refers to a subpool of the correct pool by calling the function Pool_Of_Subpool. Both this function and Set_Pool_Of_Subpool are provided by the Ada implementation and typically do not need to be overridden by the implementer of a particular type derived from Root_Storage_Pool_With_Subpools.

In the case of allocation from a subpool, the procedure Allocate_From_Subpool rather than Allocate is automatically called. Note the precondition to check that all is well.

It will be recalled that for normal storage pools, Deallocate is automatically called from an instance of Unchecked_Deallocation. In the case of subpools the general idea is that we get rid of the whole subpool rather than individual items in it. Accordingly, Deallocate does nothing as mentioned earlier and there is no Deallocate_From_Subpool. Instead we have to write a suitable implementation of Deallocate_Subpool. Note again the precondition to check that the subpool belongs to the pool.

Deallocate_Subpool is called automatically as a consequence of calling the following library procedure

```
with System.Storage_Pools.Subpools;
use System.Storage_Pools.Subpools;
procedure Ada.Unchecked_Deallocate_Subpool
                                    (Subpool: in out Subpool_Handle);
```

So when we have finished with the subpool Puddle we can write

```
Unchecked_Dellocate_Subpool(Puddle);
```

and the handle becomes null. Appropriate finalization also takes place.

In summary, the writer of a subpool implementation typically only has to provide Create_Subpool, Allocate_From_Subpool, and Deallocate_Subpool since the other subprograms are provided by the Ada implementation of the package System.Storage_Pools.Subpools and can be inherited unchanged.

An example of an implementation will be found in subclause 13.11.6 of the *ARM*. This shows an implementation of a Mark/Release pool in a package MR_Pool.

Further control over the use of storage pools (nothing to do with subpools) is provided by the ability to define our own default storage pool. Thus we can write

```
pragma Default_Storage_Pool(Master_Pool);
```

and then all allocation within the scope of the pragma will be from Master_Pool unless a different specific pool is given for a type. A pragma Default_Storage_Pool can be overridden by another one so that for example all allocation in a package (and its children) is from another pool.

The default pool can be specified as **null** thus

```
pragma Default_Storage_Pool(null);
```

which prevents allocation from standard pools. We can revert to standard pools by

```
pragma Default_Storage_Pool(Standard);
```

Allocation normally occurs from the default pool unless a specific pool has been given for a type. But there are two exceptions, one concerns access parameter allocation and the other concerns coextensions; in these cases allocation uses a pool that depends upon the context. There are also subtleties regarding the interaction between specifying Storage_Size and the Storage_Pool. For details see the *ARM* and AI12-43.

Exercise 25.4

1 Rewrite the package Queues of Exercise 12.5(**3**) so that the procedure Remove deallocates unused storage.

25.5 Other languages

Another possible form of communication between an Ada program and the outside world is interfacing to other languages. These could be machine languages or other high level languages such as Fortran or C. In order for such communication to be possible it is clear that the data must conform to appropriate conventions. This can be ensured by various aspects.

For example, suppose we wish to communicate with some mouse handling program written in C. We wish the C program to call our Ada procedure Action when the mouse is clicked. We therefore need to tell the C program which procedure to call. Suppose this is done by calling the C function Set_Click with the address of our Ada procedure as its parameter. We might write

```
type Response is access procedure (D: in Data)
  with Convention => C;

procedure Set_Click(P: in Response)
  with Import, Convention => C;

procedure Action(D: in Data) is separate
  with Convention => C;

...
Set_Click(Action'Access);
```

The aspect Import indicates that the body of Set_Click is external to the Ada program. There is also a similar aspect Export which makes a subprogram visible externally (we did not need it in this example because the address was passed indirectly via the access value). The convention can be Intrinsic or Ada as well as the name of some foreign language such as C.

A deferred constant can be completed using the aspect Import and also Link_Name and External_Name. The *ARM* has the following example

```
CPU_Identifier: constant String(1 .. 8)
  with Import, Convention => Assembler, Link_Name => "CPU_ID";
```

An interesting feature that is important for interfacing to C is the aspect Unchecked_Union. This applies to an (unconstrained) Ada discriminated record with a variant part and specifies that space is not to be allocated for a discriminant to be stored. Since C never stores a discriminant, this enables the Ada structure to be mapped directly to a C union. Consider

```
type Number(Kind: Precision) is
  record
    case Kind is
      when Single_Precision =>
        SP_Value: Long_Float;
```

```
        when Multiple_Precision =>
           MP_Value_Length: Integer;
           MP_Value_First: access Long_Float;
      end case;
   end record
      with Unchecked_Union;
```

The aspect ensures that no space is allocated for discriminants, that discriminant checks are suppressed and that the convention of the type is C.

The above Ada text provides a mapping of the following C union

```
union {
   double spvalue;
   struct {
      int length;
      double* first;
      } mpvalue;
} number;
```

The general idea is that the C programmer has created a type which can be used to represent a floating point number in one of two ways according to the precision required. One way is just as a double length value (a single item) and the other way is as a number of items considered juxtaposed to create a multiple precision value. This latter is represented as a structure consisting of an integer giving the number of items followed by a pointer to the first of them. These two different forms are the two alternatives of the union.

In the Ada text the choice of precision is governed by the discriminant Kind which is of an enumeration type as follows

```
type Precision is (Single_Precision, Multiple_Precision);
```

In the single precision case the component SP_Value of type Long_Float maps onto the C component spvalue of type double.

The multiple precision case is somewhat troublesome. The Ada component MP_Value_Length maps onto the C component length and the Ada component MP_Value_First of type **access** Long_Float maps onto the C component first of type double*.

In our Ada program we can declare a variable thus

```
X: Number(Multiple_Precision);
```

and we then obtain a value in X by calling some C subprogram. We can then declare an array and map it onto the C sequence of double length values thus

```
A: array (1 .. X.MP_Value_Length) of aliased Long_Float
   with Import, Convention => C, Address =>
                                     X.MP_Value_First.all'Address;
```

The elements of A are now the required values. Note the use of the aspect Import – this ensures that no additional space is allocated for A and that no further initialization takes place (since it is in fact part of the C structure) and that the convention is C. Note also that we do not use an Ada array in the declaration of Number because there might be problems with dope information.

The Ada type can also have a non-variant part preceding the variant part and variant parts can be nested. It may have several discriminants.

When an object of an unchecked union type is created, values must be supplied for all its discriminants even though they are not stored. This ensures that appropriate default values can be supplied and that an aggregate contains the correct components. However, since the discriminants are not stored, they cannot be read. So we can write

```
X: Number := (Single_Precision, 45.6);
Y: Number(Single_Precision);
...
Y.SP_Value := 55.7;
```

The nominal subtype of X is unconstrained whereas the nominal subtype of Y is constrained. (The concept of nominal subtype was mentioned in Section 8.2.) As a consequence, the variable Y is said to have an inferable discriminant whereas X does not. Although it is clear that playing with unchecked unions is potentially dangerous, nevertheless Ada imposes certain rules that avoid some dangers. One rule is that predefined equality can only be used on operands of a subtype with inferable discriminants; Program_Error is raised otherwise. This only applies to objects as a whole and not to the individual components. So

```
if X.SP_Value = 45.7 then          -- OK, but False
if X = Y then                      -- raises Program_Error
```

It is important to be aware that unchecked union types are for the sole purpose of interfacing to C programs and not for living dangerously. Thus consider

```
type T(Flag: Boolean := False) is
  record
    case Flag is
      when False =>
        F1: Float := 0.0;
      when True =>
        F2: Integer := 0;
    end case;
  end record
    with Unchecked_Union;
```

The type T can masquerade as either type Integer or Float. But we should not use unchecked union types as an alternative to unchecked conversion. Thus consider

```
X: T;                     -- Float by default
Y: Integer := X.F2;       -- erroneous
```

The object X has discriminant False by default and thus has the value zero of type Float. In the absence of the aspect Unchecked_Union, the attempt to read X.F2 would raise Constraint_Error because of the discriminant check. The use of Unchecked_Union suppresses the discriminant check and so the assignment will occur. But, as mentioned in Section 15.5, if a check is suppressed and the corresponding error situation arises then the program is erroneous.

However, assigning a Float value to an Integer object using Unchecked_Conversion is not erroneous providing certain conditions hold such as that Float'Size and Integer'Size are equal.

The package Interfaces was mentioned in Chapter 17 when we noted that it contained the declarations of the various machine integer types plus the shift and rotate functions for modular types.

The package Interfaces also has a number of child packages namely

```
Interfaces.C
Interfaces.C.Strings
Interfaces.C.Pointers
Interfaces.COBOL
Interfaces.Fortran
```

which contain other facilities for communication with the languages concerned. We leave the reader to consult the *ARM* for details.

Finally, it may be possible to use machine code by writing a subprogram consisting of code statements which take the unlikely form of an aggregate of a type defined in System.Machine_Code. Such a subprogram is typically called inline. For details the reader is referred to the implementation concerned. Another approach is to call an external subprogram written in assembler.

Checklist 25

Avoid allocation with anonymous access types.

Take care when using Unchecked_Conversion and Unchecked_Deallocation.

The pragma Unchecked_Union did not exist in Ada 95.

Many pragmas are replaced by aspect specifications in Ada 2012.

The attribute Max_Alignment_For_Allocation were new in Ada 2012.

Subpools were new in Ada 2012.

New in Ada 2022

A new attribute Object_Size is defined to overcome confusion with the existing attribute Size; also attributes Enum_Rep and Enum_Val are defined.

In a record representation clause, **end record** may be optionally followed by the type name.

Program 6

Playing Pools

The purpose of this program is to illustrate the declaration of storage pools. It implements the well known Tower of Hanoi in which a tower of graduated discs is moved from one pole to another pole using a third pole as 'temporary storage' without any disc ever being on a smaller disc.

Each pole is represented by a stack and, in order to add variety, two kinds of stack are implemented much as in Exercise 14.9(**3**). One is a linked list and the other is an array. The choice of stack is made by the user and the program dispatches to the relevant subprograms Push and Pop as necessary.

The package Pools declares the type Pond and its associated operations Allocate, Deallocate, Initialize, Finalize, and Storage_Size. The package Heap then declares the one storage pool used in this example with the discriminant giving its size.

There is then a root package Stacks defining the limited interface Stack and abstract operations Push and Pop. The child packages Stacks.Linked and Stacks.Vector then declare the derived types L_Stack and V_Stack which implement the two kinds of stacks.

The package Tower declares the three poles and procedures Start and Move. The procedure Move implements the familiar recursive algorithm.

Two access types are involved, Stack_Ptr in the package Tower for the class wide pointers to the three stacks and Cell_Ptr in the package Stacks.Linked for the cells making up the linked stack. Both use the same storage pool, The_Pool, declared in the package Heap.

Finally, the main subprogram interacts with the user and asks for the size of the stack and the kinds of stack for the three poles; it then calls Start and Move as requested.

```
with System.Storage_Pools;
use System.Storage_Pools;
with System.Storage_Elements;
use System.Storage_Elements;
use System;
package Pools is
   type Pond(Size: Storage_Count) is
         new Root_Storage_Pool with private;

   procedure Allocate(Pool: in out Pond;
         Storage_Address: out Address;
         SISE: in Storage_Count;
         Align: in Storage_Count);

   procedure Deallocate(Pool: in out Pond;
         Storage_Address: in Address;
         SISE: in Storage_Count;
         Align: in Storage_Count);

   function Storage_Size(Pool: Pond)
                             return Storage_Count;
   procedure Initialize(Pool: in out Pond);
   procedure Finalize(Pool: in out Pond);
   Error: exception;
private
   type Integer_Array is
      array (Storage_Count range <>) of Integer;
   type Boolean_Array is
      array (Storage_Count range <>) of Boolean;
   type Pond(Size: Storage_Count) is
         new Root_Storage_Pool with
      record
         Monitoring: Boolean;
         Free: Storage_Count;
         Count: Integer_Array(1 .. Size);
         Used: Boolean_Array(1 .. Size);
         Store: Storage_Array(1 .. Size);
      end record;
end;

with Ada.Text_IO, Ada.Integer_Text_IO;
use Ada.Text_IO, Ada.Integer_Text_IO;
package body Pools is
   procedure Put_Usage(Pool: in Pond) is
      NoI: Integer := 0;
      Mark: constant
                  array (Boolean) of Character := ".*";
```

```
begin
   if not Pool.Monitoring then return; end if;
   for I in 1 .. Pool.Size loop
      Put(Mark(Pool.Used(I)));
      NoI := NoI + 1;
      if NoI = 64 then
         New_Line;  NoI := 0;
      end if;
   end loop;
   Skip_Line;
end Put_Usage;

procedure Allocate(Pool: in out Pond;
         Storage_Address: out Address;
         SISE: in Storage_Count;
         Align: in Storage_Count) is
   Index: Storage_Offset;
begin
   if Pool.Monitoring then
      Set_Col(40);  Put("Allocating     ");
      Put(Integer(SISE), 2);  Put(" , ");
      Put(Integer(Align), 2);  New_Line;
   end if;
   if Pool.Free < SISE then
      raise Error with "Not enough space";
   end if;
   Index := Align -
            Pool.Store(Align)'Address mod Align;
   while Index <= Pool.Size-SISE+1 loop
      if Pool.Used(Index .. Index+SISE-1) =
                        (1 .. SISE => False) then
         for I in Index .. Index+SISE-1 loop
            Pool.Used(I) := True;
            Pool.Count(I) := Pool.Count(I) + 1;
         end loop;
         Pool.Free := Pool.Free - SISE;
         Storage_Address :=
                        Pool.Store(Index)'Address;
         Put_Usage(Pool);
         return;
      end if;
      Index := Index + Align;
   end loop;
   raise Error with "Pool fragmented";
end Allocate;

procedure Deallocate(Pool: in out Pond;
         Storage_Address: in Address;
         SISE: in Storage_Count;
         Align: in Storage_Count) is
   Index: Storage_Offset;
begin
   if Pool.Monitoring then
      Set_Col(40);  Put("Deallocating  ");
      Put(Integer(SISE), 2);  New_Line;
   end if;
   Index := Storage_Address -
                              Pool.Store(1)'Address;
   for I in 1 .. SISE loop
      Pool.Used(I+Index) := False;
   end loop;
   Pool.Free := Pool.Free + SISE;
   Put_Usage(Pool);
end Deallocate;

function Storage_Size(Pool: Pond)
                          return Storage_Count is
begin
   return Pool.Size;
end Storage_Size;

procedure Initialize(Pool: in out Pond) is
   Char: Character;
begin
   Put("Initializing pool of type Pond ");
   Put(" Pool size is ");
   Put(Integer(Pool.Size), 0);
   New_Line;
   loop
      Put("Is pool monitoring required, Y or N?  ");
      Get(Char);
      case Char is
         when 'y' | 'Y' =>
            Pool.Monitoring := True;  exit;
         when 'n' | 'N' =>
            Pool.Monitoring := False;  exit;
         when others =>
            null;
      end case;
      New_Line;
   end loop;
   New_Line;
   Pool.Free := Pool.Size;
   for I in 1 .. Pool.Size loop
      Pool.Count(I) := 0;  Pool.Used(I) := False;
   end loop;
end Initialize;

procedure Finalize(Pool: in out Pond) is
   NoI: Integer := 0;
begin
   Put_Line("Finalizing pool – usages were");
   for I in 1 .. Pool.Size loop
      Put(Pool.Count(I), 4);  NoI := NoI + 1;
      if NoI = 16 then New_Line; NoI := 0; end if;
   end loop;
   New_Line;  Skip_Line;
end Finalize;
end Pools;
---------------------------------------

with Pools;
package Heap is
   The_Pool: Pools.Pond(Size => 128);
end Heap;
---------------------------------------
```

```
package Stacks is
   type Stack is limited interface;
   procedure Push(S: in out Stack; X: in Integer)
                                          is abstract;
   procedure Pop(S: in out Stack; X: out Integer)
                                          is abstract;
end Stacks;
----------------------------------------

with Heap; use Heap;
package Stacks.Linked is
   type L_Stack is
                  limited new Stack with private;
   procedure Push(S: in out L_Stack; X: in Integer);
   procedure Pop(S: in out L_Stack; X: out Integer);
private
   type Cell;
   type Cell_Ptr is access Cell;
   for Cell_Ptr'Storage_Pool use The_Pool;

   type Cell is
      record
         Next: Cell_Ptr;
         Value: Integer;
      end record;

   type L_Stack is limited new Stack with
      record
         Head: Cell_Ptr;
      end record;

end;

with Ada.Unchecked_Deallocation; use Ada
package body Stacks.Linked is

   procedure Free is
      new Unchecked_Deallocation(Cell, Cell_Ptr);

   procedure Push(S: in out L_Stack; X: in Integer) is
   begin
      S.Head := new Cell'(S.Head, X);
   end Push;

   procedure Pop(S: in out L_Stack; X: out Integer) is
      Old_Head: Cell_Ptr := S.Head;
   begin
      X := S.Head.Value;
      S.Head := S.Head.Next;
      Free(Old_Head);
   end Pop;

end Stacks.Linked;
----------------------------------------

package Stacks.Vector is
   type V_Stack(Max: Integer) is
                  limited new Stack with private;
   procedure Push(S: in out V_Stack; X: in Integer);
   procedure Pop(S: in out V_Stack; X: out Integer);
private
   type Integer_Vector is
      array (Integer range <>) of Integer;
   type V_Stack(Max: Integer) is
                           limited new Stack with
      record
         V: Integer_Vector(1 .. Max);
         Top: Integer;
      end record;
end;

package body Stacks.Vector is

   procedure Push(S: in out V_Stack; X: in Integer) is
   begin
      S.Top := S.Top + 1;
      S.V(S.Top) := X;
   end Push;

   procedure Pop(S: in out V_Stack; X: out Integer) is
   begin
      X := S.V(S.Top);
      S.Top := S.Top - 1;
   end Pop;

end Stacks.Vector;
----------------------------------------

package Tower is
   type Kind is (Linked, Vector);
   type Kinds is array (1 .. 3) of Kind;
   procedure Start(K: Kinds; N, On: in Integer);
   procedure Move(N, From, To: in Integer);
end;

with Heap; use Heap;
with Stacks; use Stacks;
with Stacks.Linked, Stacks.Vector;
use Stacks.Linked, Stacks.Vector;
with Ada.Text_IO, Ada.Integer_Text_IO;
use Ada.Text_IO, Ada.Integer_Text_IO;
package body Tower is

   type Stack_Ptr is access Stack'Class;
   for Stack_Ptr'Storage_Pool use The_Pool;

   Pole: array (1 .. 3) of Stack_Ptr;

   procedure Start(K: Kinds; N, On: in Integer) is
   begin
      for I in 1 .. 3 loop
         case K(I) is
            when Linked =>
               Pole(I) := new L_Stack;
            when Vector =>
               Pole(I) := new V_Stack(N);
         end case;
      end loop;
      for I in reverse 1 .. N loop
         Push(Pole(On).all, I);
      end loop;
   end Start;
```

```
   procedure Move(N, From, To: in Integer) is
      The_Disc: Integer;
   begin
      if From = To then
         null;          -- nothing to do!
      elsif N = 1 then
         Pop(Pole(From).all, The_Disc);
         Put("Moving disc number ");
         Put(The_Disc, 0);
         Put(" from pole ");  Put(From, 0);
         Put(" to pole ");  Put(To, 0);
         New_Line;
         Push(Pole(To).all, The_Disc);
      else
         Move(N-1, From, 6-From-To);
         Move(1, From, To);
         Move(N-1, 6-From-To, To);
      end if;
   end Move;

end Tower;
---------------------------------------

with Tower, Pools;
with Ada.Exceptions;  use Ada.Exceptions;
with Ada.Text_IO, Ada.Integer_Text_IO;
use Ada.Text_IO, Ada.Integer_Text_IO;
procedure Tower_Of_Hanoi is
   Size: Positive;
   Lets: String(1 .. 3);
   Patt: Tower.Kinds;
begin
   Put_Line("Welcome to the Tower of Hanoi");
   Put("Tower size please  ");
   Get(Size);
<<Again>>
   New_Line;
   Put("Give three letters L or V for types of poles  ");
   Get(Lets);  Skip_Line;
   for I in 1 .. 3 loop
      case Lets(I) is
         when 'l' | 'L' => Patt(I) := Tower.Linked;
         when 'v' | 'V' => Patt(I) := Tower.Vector;
         when others => goto Again;
      end case;
   end loop;
   New_Line;
   Tower.Start(Patt, Size, 1);
   Tower.Move(Size, 1, 2);
   New_Line(2);
   Put_Line("Finished");  Skip_Line;
exception
   when Event: Pools.Error =>
      Put_Line(Exception_Information(Event));
      Put_Line("Tower toppled - tough");
      Skip_LIne;
end Tower_Of_Hanoi;
```

The storage pool is implemented as a discriminated record with the discriminant giving the size of the pool. The storage space itself is provided by the array component Store of the type Storage_Elements.Storage_Array. Other components of the record concern usage and monitoring and are implemented crudely in order to simplify the illustration. The array Count indicates how many times the corresponding component of Store has been allocated and the array Used indicates which components are in use.

The alignment calculation in Allocate is interesting. No attempt is made to align the pool itself; instead the index of the first appropriately aligned component is computed in Index using the **mod** operation between types Address and Storage_Offset – this operation is declared in the package System.Storage_Elements.

The search is implemented extremely crudely in order to fit within the space here available. Other algorithms can obviously be devised.

The user can switch on monitoring when the declaration of the pool is elaborated and so calls Initialize. If monitoring is switched on, a record of the calls of Allocate and Deallocate is output after each call together with a usage map thus

```
************************************************
********........********........................
```

where an asterisk indicates that a storage element is in use and a dot indicates that it is not; Skip_Line is then called to allow the user to view the situation before the program continues. On termination the call of Finalize prints out a table showing how many times each storage element had been allocated.

The choice of stack implementation is controlled by the user giving a string of three characters such as VLV which would indicate that the middle pole is to be a linked stack and the others vector stacks. (For brevity, a wicked label is used in the code checking the string!) The various stacks are established by Tower.Start which also calls Push to set up the tower on the initial pole.

The transfer between poles is then performed by Tower.Move using the traditional recursive algorithm. Note that the calls of Push and Pop dispatch to the relevant implementations.

It will be found that the program as shown (with a pool of 128 storage elements) on a typical byte machine copes with towers of size from 4 to 7 according to the kinds of poles but raises the exception Pool.Error for larger towers. For some combinations of kinds of poles, holes do appear in the storage pool area but fragmentation does not occur in this example. However, the more elaborate version on the web does produce fragmentation.

26 The Specialized Annexes

26.1 Systems Programming
26.2 Real-Time Systems
26.3 Distributed Systems
26.4 Information Systems
26.5 Numerics
26.6 High Integrity Systems

This chapter contains a survey of the scope of the six specialized annexes. These annexes are optional and many implementations will not support them in full.

It is a general principle that the annexes contain no new syntax but only additional packages, attributes, aspects, and pragmas. They are thus not intrinsically hard to understand especially if the reader is familiar with the application area concerned. We thus briefly catalogue the principles and main facilities with just a few examples and refer the reader to the *ARM* for the full details. But note that some of the more interesting tasking features were described in Chapter 22.

The annexes are generally independent but the Real-Time Systems annex requires that the Systems Programming annex be supported.

26.1 Systems Programming

This annex covers access to machine code, interrupt handling, some extra requirements on representations and preelaboration, aspects for discarding names at run time, aspects for shared variables, and packages for general task identification, per task attributes, and the detection of task termination.

Interrupt handling is important. The general idea is that an interrupt handler is provided by a parameterless protected procedure which is called by some mythical external task. The protected procedure can be attached statically to the interrupt by the aspect Attach_Handler or it can be attached dynamically by a procedure of the same name in the package Ada.Interrupts. This package also defines the type Interrupt_Id which is used to identify the interrupts which are declared as constants in the package Ada.Interrupts.Names.

Thus we might write a protected object Contact_Handler as follows

```
protected Contact_Handler is
   use Ada.Interrupts;
   procedure Response
      with Attach_Handler => Names.Contact_Int;
end;

protected body Contact_Handler is
   procedure Response is
      ...       -- the interrupt handling code
   end Response;
end Contact_Handler;
```

The aspect specification statically attaches the procedure Response to the interrupt identified by the constant Contact_Int. Dynamic attachment is performed by using the aspect Interrupt_Handler to indicate that the protected procedure is to be used as a handler and then calling the procedure Attach_Handler thus

```
protected Contact_Handler is
   procedure Response
      with Interrupt_Handler;
end;
...
use Ada.Interrupts;
...
Attach_Handler(Contact_Handler.Response'Access, Names.Contact_Int);
```

where the first parameter of Attach_Handler is of the access type

```
type Parameterless_Handler is access protected procedure;
```

Other subprograms in Ada.Interrupts allow a handler to be detached or exchanged. There is also a function Get_CPU which returns the CPU number on which the handler for an interrupt is executed. It might be zero meaning not specific. See Section 22.2.

The aspect Discard_Names may be used to indicate that various voluminous tables of names associated with a type or exception need not be retained at run time. Such tables are required by Image and Value, Enumeration_IO, Exception_Name, and Tags.Expanded_Name. Omitting these tables might be vital for saving space in an embedded application. Thus we might have

```
type Greek is (alpha, beta, gamma, ..., omega)
   with Discard_Names;
```

The use of the aspects Atomic and Atomic_Components with shared variables was illustrated in Section 22.2. They ensure that all reads and writes to the variables are atomic. The aspects Volatile and Volatile_Components ensure that accesses to an object are always seen in the same order by all tasks; this certainly applies if the object is always in memory. We should really have written

```
S: Status
   with Address => 8#100#, Volatile;
```

when we were discussing the status value in Section 25.1. There are also aspects Independent and Independent_Components which require that an object be independently addressable. Note that atomic implies both volatile and independent.

Sometimes it is convenient to be able to refer to a task without it having to be of a specific task type. A server task might wish to keep a record of past callers so that they can be recognized in the future. This can be done using the package Ada.Task_Identification which defines a type Task_Id and a few operations upon tasks in general. Its specification is as follows

```
package Ada.Task_Identification
      with Preelaborate, Nonblocking, Global => in out synchronized is
   type Task_Id is private
      with Preelaborable_Initialization;
   Null_Task_Id: constant Task_Id;
   function "=" (Left, Right: Task_Id) return Boolean;
   function Current_Task return Task_Id;
   function Environment_Task return Task_Id;
   procedure Abort_Task(T: in Task_Id)
      with Nonblocking => False;
   function Is_Terminated(T: Task_Id) return Boolean;
   function Is_Callable(T: Task_Id) return Boolean;
   function Activation_Is_Complete(T: Task_Id) return Boolean;
private
   ...                              -- not specified by the language
end Ada.Task_Identification;
```

The parameterless function Current_Task returns the identity of the currently executing task (but should not be called from within an entry body); the attribute E'Caller may be applied inside an accept statement or entry body and gives the task currently calling the entry E. The task identity of a task T is denoted by T'Identity. Task identities are particularly useful for manipulating priorities as discussed in the next section. The function Environment_Task returns the identity of the environment task. The function Activation_Is_Complete returns true if the task concerned has finished activation. Moreover, if Activation_Is_Complete is applied to the environment task then it indicates whether all library items of the partition have been elaborated. This can be useful in a start-up situation.

The generic package Ada.Task_Attributes enables attributes to be identified with tasks on a per-task basis using task identities. Finally, the package Ada.Task_Termination can be used to monitor the termination of tasks; it was discussed in detail in Section 22.5.

Various atomic operations are provided in Ada 2022 for applications where synchronization can be performed through the use of shared memory. There are several packages as follows. First there is a parent package whose specification is

```
package System.Atomic_Operations
      with Pure, Nonblocking is
end System.Atomic_Operations;
```

Note that it has no body. There are four child packages, we first consider the one for simple test and set operations thus

```
package System.Atomic_Operations.Test_And_Set
    with Pure, Nonblocking is

  type Test_And_Set_Flag is mod implementation-defined
    with Atomic, Default_Value => 0, Size => implementation-defined;

  function Atomic_Test_And_Set(Item: aliased in out Test_And_Set_Flag)
    return Boolean with Convention => Intrinsic;

  procedure Atomic_Clear(Item: aliased in out Test_And_Set_Flag)
    with Convention => Intrinsic;

  function Is_Lock_Free(Item: aliased Test_And_Set_Flag) return Boolean
    with Convention => Intrinsic;

end System.Atomic_Operations.Test_And_Set;
```

The type Test_And_Set_Flag gives the state of some atomic object whose value can be zero or nonzero. The procedure Atomic_Clear sets a flag to zero. The function Atomic_Test_And_Set sets a flag to nonzero and returns true if its previous value was nonzero and false if it was not.

There are also generic child packages Exchange, Integer_Arithmetic, and Modular_Arithmetic for doing other appropriate atomic operations. The single generic parameter is **private**, **range** <>, **mod** <> respectively. In each case it also has the aspect **with** Atomic. Here is the specification of that for integer arithmetic.

```
generic
  type Atomic_Type is range <> with Atomic;
package System.Atomic_Operations.Integer_Arithmetic
  with Pure, Nonblocking is

  procedure Atomic_Add(Item: aliased in out Atomic_Type;
                       Value: Atomic_Type)
    with Convention => Intrinsic;

  procedure Atomic_Subtract(Item: aliased in out Atomic_Type;
                            Value: Atomic_Type)
    with Convention => Intrinsic;

  function Atomic_Fetch_And Add
              (Item: aliased in out Atomic_Type;
               Value: Atomic_Type) return Atomic_Type;
    with Convention => Intrinsic;

  function Atomic_Fetch_And Subtract
              (Item: aliased in out Atomic_Type;
               Value: Atomic_Type) return Atomic_Type
    with Convention => Intrinsic;

  function Is_Lock_Free(Item: aliased Atomic_Type) return Boolean
    with Convention => Intrinsic;

end System.Atomic_Operations.Integer_Arithmetic;
```

Note that this only does addition and subtraction, nothing hard like multiplication. For more details see the *ARM*.

26.2 Real-Time Systems

This annex covers priorities and scheduling, detailed requirements on the immediacy of the abort statement, restrictions enabling simplified run-time systems, a monotonic time package, various timers, and direct task control.

The core language says little about scheduling and priorities. It does however state that entry queues are serviced in order of arrival.

This annex defines various scheduling policies in terms of priorities. A general problem is the risk of priority inversion; this is where a high priority task is held up by a lower priority task using some resource. In order to define appropriate behaviour a task has a base priority and an active priority. The active priority is the one that is used to decide which tasks get the processors and is never less than the base priority; indeed, unless a task is involved in some interaction with another task or protected object, the active priority is the same as the base priority. These priority rules are slightly modified in the case of EDF scheduling mentioned below.

An obvious example of interaction is the rendezvous in which case the called task takes the active priority of the caller if that is higher; this is called priority inheritance. Similarly, when a task is being activated, it inherits the active priority of its activator if it is higher.

The base priority of a task can be set by the aspect Priority thus

```
task type T(P: Priority)
  with Priority => P is
  ...
```

This shows the priority being set to the value of the discriminant; but it could of course be a constant.

The subtype Priority is a subtype of Any_Priority which itself is a subtype of Integer. The subtypes Priority and Interrupt_Priority together cover the full range of Any_Priority without overlapping; they are all declared in System. The range of values of Priority is always at least 30 and there is always at least one value for Interrupt_Priority which is of course higher. The aspect Interrupt_Priority can be used to set the priority level of a protected object used as an interrupt handler.

Base priorities can be manipulated dynamically using subprograms Set_Priority and Get_Priority in the package Ada.Dynamic_Priorities. For example

```
Fred: T(10);                        -- initial base priority 10
...
Set_Priority(15, Fred'Identity);    -- change to 15
```

The second parameter can be omitted and then by default is the result of calling Current_Task in Ada.Task_Identification.

Various pragmas enable different dispatching policies to be specified; some policies are standard and others may be defined by the implementation. The standard policies are

FIFO_Within_Priorities – Within each priority level to which it applies tasks are dealt with on a first-in–first-out basis. Moreover, a task may preempt a task of a lower priority.

Non_Preemptive_FIFO_Within_Priorities – Within each priority level to which it applies tasks run to completion or until they are blocked or execute a delay statement. A task cannot be preempted by one of higher priority. This sort of policy is widely used in high integrity applications.

Round_Robin_Within_Priorities – Within each priority level to which it applies tasks are time-sliced with an interval that can be specified. This is a very traditional policy widely used since the earliest days of concurrent programming.

EDF_Within_Priorities – This provides Earliest Deadline First dispatching. The general idea is that across a range of priority levels, each task has a deadline and the one with the earliest deadline is processed first. This policy has mathematically provable advantages with respect to resource utilization.

A single policy can be selected for a whole partition by for example

```
pragma Task_Dispatching_Policy(Round_Robin_Within_Priorities);
```

whereas to mix different policies across different priority levels we can write

```
pragma Priority_Specific_Dispatching(Round_Robin_Within_Priorities, 1, 1);
pragma Priority_Specific_Dispatching(EDF_Within_Priorities, 2, 10);
pragma Priority_Specific_Dispatching(FIFO_Within_Priority, 11, 24);
```

This sets Round Robin at priority level 1, EDF at levels 2 to 10, and FIFO at levels 11 to 24.

Note that in Ada 2022, the policy which was known as EDF_Across_Priorities has been replaced by EDF_Within_Priorities. The background to this important change is described in detail in AI12-230.

The fine details of the various policies can be controlled by the package Ada.Dispatching and its children Non_Preemptive, Round_Robin, and EDF.

The parent package includes a procedure Yield which can be used by a task in the FIFO regime to yield to another task and so go to the end of the ready queue for its active priority (this can also be done by writing **delay** 0.0;).

The package Non_Preemptive includes a procedure Yield_To_Higher which can be called to indicate that a task is willing to be preempted by a task of higher but not equal priority; it also includes a procedure Yield_To_Same_Or_Higher which is simply a renaming of Yield.

The package Round_Robin includes procedures Set_Quantum for setting the quantum for time slicing for an individual priority level or a range of levels.

The deadline of a task for EDF is a property similar to priority and can be set when a task is created using the aspect and attribute Relative_Deadline thus

```
task T
   with Relative_Deadline => RD is
   ...
```

This ensures that the absolute deadline of the task when it is created is equal to RD (of type Real_Time.Time_Span, see Section 22.6) after its time of creation. Deadlines can be manipulated by calling procedures such as Set_Deadline and Delay_Until_And_Set_Deadline in the package Ada.Dispatching.EDF.

Protected objects can also be given priorities thus

```
protected Object
  with Priority => 20 is
  entry E ...
```

If the pragma Locking_Policy is used to specify Ceiling_Locking by writing

```
pragma Locking_Policy(Ceiling_Locking);
```

then the priority of the protected object is known as its ceiling priority. A task calling the entry E will then inherit the ceiling priority while executing the protected operation. A calling task with higher active priority than the ceiling priority is not permitted to execute the protected operation; it receives Program_Error instead. This ensures a degree of predictability for the other tasks. A protected object can change its ceiling priority dynamically by assigning to the attribute Priority within a protected operation. For example

```
protected type PT is
  procedure Change_Priority(Change: in Integer);
  ...
end;

protected body PT is
  procedure Change_Priority(Change: in Integer) is
  begin
    ...                         -- PT'Priority has old value here
    PT'Priority := PT'Priority + Change;
    ...                         -- PT'Priority has new value here
    ...
  end Change_Priority;
  ...
end PT;
```

Changing the ceiling priority is thus done while mutual exclusion is in force. Although the value of the attribute itself is changed immediately the assignment is made, the actual ceiling priority of the protected object is only changed when the protected operation (in this case the call of Change_Priority) is finished. Note the unusual syntax in which an attribute is permitted as the destination of an assignment statement. This happens nowhere else in the language.

The servicing of entry queues can be controlled by the pragma Queuing_Policy. There are three policies, namely FIFO_Queuing, Ordered_FIFO_Queuing, and Priority_Queuing.

FIFO_Queuing is the default core policy for servicing entry queues which is in order of arrival irrespective of priorities. Priority_Queuing on the other hand indicates that entry queues are to be serviced according to the active priorities of the queued tasks (note that the Resource_Allocator example of Section 20.9 assumes FIFO_Queuing). Ordered_FIFO_Queuing (which is new in Ada 2022) is a sort of halfway house.

Another new feature in Ada 2022 is the introduction of the concept of an Admission Policy. One such policy is defined which is FIFO_Spinning. For the background and details, see AI12-276.

Also new in Ada 2022 is the pragma Generate_Deadlines. If in effect, the deadline of a task is recomputed each time it becomes ready. See AI12-230.

Many of the pragmas that we have been discussing are of a category known as configuration pragmas and generally apply throughout a partition.

Another configuration pragma is the pragma Restrictions; this enables the programmer to assert that the program does not use certain features of the language at all or does not exceed some limits. For example

```
pragma Restrictions(No_Task_Hierarchy, Max_Tasks => 100);
```

indicates that all tasks are at the library level and that there are no more than 100. Judicious use of this pragma may enable a program to use a specially small or fast version of the run-time system.

The package Ada.Synchronous_Task_Control enables tasks to suspend according to the state of suspension objects; these objects can be set true and false and a task can suspend itself until an object is true. Its specification is

```
package Ada.Synchronous_Task_Control
      with Preelaborate, Nonblocking, Global => in out synchronized is
   type Suspension_Object is limited private;
   procedure Set_True(S: in out Suspension_Object);
   procedure Set_False(S: in out Suspension_Object);
   function Current_State(S: Suspension_Object) return Boolean;
   procedure Suspend_Until_True(S: in out Suspension_Object)
      with Nonblocking => False;
private
   -- not specified by the language
end Ada.Synchronous_Task_Control;
```

It has a child package EDF which enables a task to suspend itself until an object is true and then set a deadline.

```
with Ada.Real_Time:
package Ada.Synchronous_Task_Control.EDF is
      with Nonblocking, Global => in out synchronized is
   procedure Suspend_Until_True_And_Set_Deadline
      (S: in out Suspension_Object; TS: in Ada.Real_Time.Time_Span)
         with Nonblocking => False;
end Ada.Synchronous_Task_Control.EDF
```

The package Ada.Asynchronous_Task_Control enables tasks to be held and then allowed to continue; these essentially work by reducing the priority to an idle level; the task is identified by its task identity. The package Ada.Synchronous_Barriers enables tasks to wait until a specified given number of them are ready to be released together.

The Real-Time Systems annex also defines a real-time clock and various timers. These are very important and were described in some detail in Section 22.6. The equally important Ravenscar and Jorvik profiles were outlined in Section 22.7.

26.3 Distributed Systems

This annex defines facilities for splitting a program up into a number of partitions and for communication between the partitions.

Partitions may be active or passive. An active partition is like a traditional complete program and has its own environment task, copy of the run-time system and so on. A passive partition has no threads of control and thus may only include certain kinds of library units; it has no main subprogram and no library level elaboration.

The general idea is that the partitions execute independently other than when communicating. However, unlike a set of quite distinct programs, the strong typing is imposed rigorously between the partitions. Incidentally, an active partition need not have a main subprogram since its activity could be entirely within library package elaboration.

Library units are categorized into a hierarchy by a number of aspects thus

```
Pure
Shared_Passive
Remote_Types
Remote_Call_Interface
```

Each category imposes restrictions on what the unit can contain. An important rule is that a unit can only depend on (via with clauses) units in the same or higher categories (the bodies of the last two are not restricted).

A pure unit cannot contain any state; since it has no state a distinct copy can be placed in each partition. This is important since it reduces the need for communication between partitions.

A shared passive unit can have visible state but no tasks or protected objects with entries. It must be preelaborable. There are also restrictions on access types. The idea is to avoid references to active units. It requires no run-time system.

A remote types unit defines types used for communication between partitions. Its specification must be preelaborable but its body need not. The only visible access types permitted are so-called remote access types; these are essentially access to subprogram or access to class wide limited private types. Remote access to subprogram types must not be nonblocking.

A remote call interface (RCI) unit cannot have visible state; its main purpose is to define the subprograms to be called remotely from other partitions. Its specification also has to be preelaborable but its body need not.

Incidentally, although many language defined packages have the aspect Pure the main ones to have the aspect Remote_Types are the various containers.

A passive partition can only contain pure and shared passive units; as a consequence it does not need a copy of a run-time system. Communication via a passive partition is normally through the use of protected objects declared in shared passive units in the partition. But note that an object in a passive partition can be read directly.

Communication between active partitions is via remote procedure calls on RCI units. Such remote calls are processed by stubs at each end of the communication; parameters and results are passed as streams. This is all done automatically by the partition communication subsystem (PCS) and need not concern the user.

Thus to do a remote call from planet Earth to *Starship Enterprise* we might have one partition containing

```
package Enterprise                          -- in space
     with Remote_Call_Interface is
   procedure Command(S: in String);
end;
```

and then in a different partition

```
with Enterprise;
package Earth_Station is                    -- on terra firma
   ...
     Enterprise.Command("Invade Jupiter");
   ...
end;
```

so that the call of Enterprise.Command is all done by magic. Note incidentally that each active partition has its own package Calendar so there are no relativistic problems regarding simultaneity.

The means of dividing the system up into partitions is not defined by the language and is done by some appropriate post-compilation tools.

26.4 Information Systems

This brief annex addresses material typically associated with COBOL programs. The main topics are decimal types and output formats.

The rules about decimal types follow those for fixed point types in general. In addition, the package Ada.Decimal declares a number of constants giving limits on the values of delta and digits supported plus a generic procedure for doing division giving both quotient and remainder in one operation. The constant Max_Decimal_Digits gives the largest digits value supported by the implementation; at least 18.

A major concern in information system processing is the format of human-readable output. Extensive facilities are provided by the predefined package Ada.Text_IO.Editing. This defines a private type Picture for the control of format and subprograms to manipulate and create pictures from strings. There is also a generic package Decimal_Output for the output of decimal values according to the format specified by a picture. Means are provided for localizing the currency symbol, filler character, digits separator, and radix mark.

26.5 Numerics

This annex contains the definitions of packages for complex arithmetic, the manipulation of both real and complex vectors and matrices, accuracy requirements for floating and fixed point arithmetic (the former based on the concept of model numbers) and accuracy requirements for the various predefined packages.

The predefined generic packages for complex arithmetic are

```
Numerics.Generic_Complex_Types
Numerics.Generic_Complex_Elementary_Functions
Text_IO.Complex_IO
```

plus nongeneric forms such as

```
Numerics.Complex_Types                          -- for Float
Numerics.Long_Complex_Types                     -- for Long_Float
Numerics.Complex_Elementary_Functions           -- for Float
Numerics.Long_Complex_Elementary_Functions      -- for Long_Float
Complex_Text_IO                                 -- for Float
Long_Complex_Text_IO                            -- for Long_Float
```

In earlier chapters in this book we have discussed a hypothetical package Generic_Complex_Numbers in which the type Complex was private plus various child packages. This was introduced purely as an illustration largely because the concept of complex numbers and the possibility of implementing them in different ways is common knowledge. However, in practice, complex numbers are inevitably implemented in cartesian form and the package Generic_Complex_Types defined in this annex thus declares the type Complex as a visible record type with two components giving the real and imaginary parts as follows

```
type Complex is
  record
    Re, Im: Real'Base;
  end record;
```

where Real is the formal generic parameter. This has the advantage that record aggregates can be used in a natural way to provide complex literals. Moreover, there is also a type Imaginary which is private and whose full declaration is

```
type Imaginary is new Real'Base;
```

and this type has values representing pure imaginary numbers. Deferred constants i (and j for electrical engineers) having the appropriate value are also defined. This enables complex expressions to be written in a very natural way such as

```
X, Y: Real;
Z: Complex;
...
Z := X + i*Y;
```

The four operators "+", "–", "*", "/" are provided for all combinations of the types Real'Base, Imaginary, and Complex. The operator "**" is provided for a first parameter of the type Imaginary or Complex and a second parameter of type Integer as usual.

Constructor and selector functions are also provided with names Re, Im, Compose_From_Cartesian and Modulus, Argument, Compose_From_Polar. Although somewhat redundant since the type Complex is not private they might be useful as generic actual parameters. Compose_From_Cartesian takes one or two real parameters or one imaginary parameter. Compose_From_Polar and Argument take an optional Cycle parameter like the trigonometric functions. There are also functions "abs" and Conjugate with appropriate specifications.

The specification of the generic package for complex elementary functions has the form

```
with Ada.Numerics.Generic_Complex_Types;
generic
   with package Complex_Types is
        new Ada.Numerics.Generic_Complex_Types(<>);
   use Complex_Types;
package Ada.Numerics.Generic_Complex_Elementary_Functions
     with Pure, Nonblocking is

   function Sqrt(X: Complex) return Complex;
   ...   -- similarly for Log, Exp, Sin, Cos, Tan, Cot,
   ...   -- Arcsin, Arccos, Arctan, Arccot,
   ...   -- Sinh, Cosh, Tanh, Coth, Arcsinh, Arccosh, Arctanh, Arccoth
end;
```

and also includes Exp taking an imaginary parameter and "**" taking combinations of real and complex parameters.

Note that this package has a formal package parameter which has to be an instantiation of Generic_Complex_Types. This technique was discussed in Section 19.4. However, unlike the hypothetical package Generic_Complex_Functions of that section, this annex package does not also import (an instantiation of) the real elementary functions since it could well use different algorithms to those we discussed; moreover, they are not needed for the specification of the package and can always be imported by the body.

The package Complex_IO also takes an instantiation of Generic_Complex_Types as its only parameter. It provides various Put and Get procedures with similar specifications to those in Text_IO.Float_IO except that they take a parameter of type Complex rather than Num. Thus Put outputs the values of the real and imaginary parts as a pair of real values in the corresponding format separated by a comma and all enclosed in parentheses. And Get accepts the same format but the parentheses and/or comma may be omitted.

The Numerics annex also includes packages for real and complex vectors and matrices. The real package takes the form

```
generic
   type Real is digits <>;
package Ada.Numerics.Generic_Real_Arrays
     with Pure, Nonblocking is

   -- Types
   type Real_Vector is array (Integer range <>) of Real'Base;
   type Real_Matrix is array (Integer range <>,
                              Integer range <>) of Real'Base;

   -- Real_Vector operations
   ... -- unary and binary "+" and "-" giving a vector; also inner product
```

```
... -- and two versions of "abs" – one returns a vector and the other a
... -- value of Real'Base; and operations "*" and "/" to multiply and
... -- divide a vector by a scalar

function Unit_Vector(Index: Integer; Order: Positive; First: Integer := 1)
                                                      return Real_Vector;

-- Real_Matrix operations
... -- unary "+", "-", "abs", binary "+", "-" giving a matrix
... -- "*" on two matrices giving a matrix, on a vector
... -- and a matrix giving a vector, outer product of two
... -- vectors giving a matrix
... -- operations "*" and "/" to multiply and divide a matrix by a scalar

function Transpose(X: Real_Matrix) return Real_Matrix;
function Solve(A: Real_Matrix; X: Real_Vector) return Real_Vector;
function Solve(A, X: Real_Matrix) return Real_Matrix;
function Inverse(A: Real_Matrix) return Real_Matrix;

function Determinant(A: Real_Matrix) return Real'Base;
function Eigenvalues(A: Real_Matrix) return Real_Vector;
procedure Eigensystem(A: in Real_Matrix;
                      Values: out Real_Vector; Vectors: out Real_Matrix);
function Unit_Matrix(Order: Positive; First_1, First_2: Integer := 1)
                                                      return Real_Matrix;
end Ada.Numerics.Generic_Real_Arrays;
```

Many of these operations are quite self-evident. The general idea as far as the usual arithmetic operations are concerned is that we just write an expression in the normal way. But the following points should be noted.

There are two operations "abs" applying to a Real_Vector thus

```
function "abs"(Right: Real_Vector) return Real_Vector;
function "abs"(Right: Real_Vector) return Real'Base;
```

One returns a vector each of whose elements is the absolute value of the corresponding element of the parameter and the other returns a scalar which is the norm of the vector (this is the square root of the inner product of the vector with itself). This is provided as a distinct operation in order to avoid any intermediate overflow that might occur if the user were to compute it directly using the inner product "*".

There are two functions Solve for solving one and several sets of linear equations respectively. Thus if we have the single set of n equations

$$Ax = y$$

then we might write

```
X, Y: Real_Vector(1 .. N);
A: Real_Matrix(1 .. N, 1 .. N);
...
X := Solve(A, Y);
```

and if we have *m* sets of *n* equations we might write

```
XX, YY: Real_Matrix(1 .. N, 1 .. M)
A: Real_Matrix(1 .. N, 1 .. N);
...
XX := Solve(A, YY);
```

The functions Inverse and Determinant are provided for completeness although they should be used with care. Remember that it is foolish to solve a set of equations by writing X := Inverse(A) * Y; because it is both slow and prone to errors. The main problem with Determinant is that it is liable to overflow or underflow even for moderate sized matrices. Thus if the elements are of the order of a thousand and the matrix has order 10, then the magnitude of the determinant will be of the order of 10^{30}. The user may therefore have to scale the data.

Two subprograms are provided for determining the eigenvalues and eigenvectors of a symmetric matrix. These are commonly required in many calculations in domains such as elasticity, moments of inertia, confidence regions and so on. The function Eigenvalues returns the eigenvalues (which will be non-negative) as a vector with them in decreasing order. The procedure Eigensystem computes both eigenvalues and vectors; the parameter Values is the same as that obtained by calling the function Eigenvalues and the parameter Vectors is a matrix whose columns are the corresponding eigenvectors in the same order. The eigenvectors are mutually orthonormal (that is, of unit length and mutually orthogonal) even when there are repeated eigenvalues. These subprograms apply only to symmetric matrices and if the matrix is not symmetric then Argument_Error is raised. Other errors such as the mismatch of array bounds raise Constraint_Error by analogy with built-in array operations.

The reader will observe that the facilities provided here are rather humble and presented in a simple black-box style. It is important to appreciate that the Ada predefined numerics library is not in any way in competition with or a substitute for professional libraries such as the renowned BLAS (Basic Linear Algebra Subprograms, see www.netlib.org/blas). Indeed the overall goal is to provide commonly required simple facilities for the user who is not a numerical professional, and to provide a baseline of types and operations that forms a firm foundation for binding to more general facilities such as the BLAS.

The complex package is very similar and will not be described in detail. However, the generic formal parameters are interesting. They are

```
with Ada.Numerics.Generic_Real_Arrays,
                               Ada.Numerics.Generic_Complex_Types;
generic
   with package Real_Arrays is
                         new Ada.Numerics.Generic_Real_Arrays(<>);
   use Real_Arrays;
   with package Complex_Types is
                    new Ada.Numerics.Generic_Complex_Types(Real);
   use Complex_Types;
package Ada.Numerics.Generic_Complex_Arrays
      with Pure, Nonblocking is ...
```

Thus we see that it has two formal packages which are the corresponding real array package and the complex types package. The formal parameter of the first is <> and that of the second is Real which is exported from the first package and ensures that both are instantiated with the same floating point type.

As well as the obvious array and matrix operations, the complex package also has operations for composing complex arrays from cartesian and polar real arrays, and computing the conjugate array by analogy with scalar operations in the complex types package. There are also mixed real and complex array operations but not mixed imaginary, real, and complex array operations.

By analogy with real symmetric matrices, the complex package has subprograms for determining the eigensystems of Hermitian matrices. A Hermitian matrix is one whose complex conjugate equals its transpose; such matrices have real eigenvalues and are well behaved.

Floating point is defined in terms of model numbers as outlined in Section 17.4. The model numbers are zero plus all values of the canonical form where the mantissa has Model_Mantissa digits and the exponent is greater than or equal to Model_Emin. Note that there is no upper bound and so no attribute Model_Emax. The general principle is that associated with each value is a model interval. If a value is a model number then the associated model interval is simply the model number. Otherwise it is the interval defined by the two model numbers surrounding the value.

When an operation is performed, the bounds of the result are given by the smallest model interval that can arise as a consequence of operating upon any values in the model intervals of the operands. The relational operators are also defined in terms of model intervals. If the result is the same, whatever values are chosen in the intervals, then its value is clearly not in dispute. If, however, the result depends upon which values in the intervals are chosen then the result is undefined.

Some care is needed in the interpretation of these principles. Although we may not know where a value lies in a model interval, nevertheless it does have a specific value and should not be treated in a stochastic manner. For example X = X is always true even if we do not know the specific value of X. There is perhaps some philosophical analogy here with Quantum Mechanics – the execution of the program equates to performing an observation on X and the knowing of the specific value is the collapse of the wave packet!

One tiny example must suffice. Consider a hypothetical binary machine with a type Rough that has just 5 bits in the mantissa, 3 bits in the exponent and a sign bit (we can squeeze this into a byte because the leading mantissa bit need not be stored). The model numbers around 1 are

$$\ldots, {}^{30}/_{32}, {}^{31}/_{32}, 1, 1^{1}/_{16}, 1^{2}/_{16}, 1^{3}/_{16}, \ldots$$

Now suppose that R of type Rough has the value $1^{1}/_{8}$ and consider the values of ((R*R)*R)*R and (R*R)*(R*R). The final model intervals are $[1^{1}/_{2}, 1^{11}/_{16}]$ and $[1^{9}/_{16}, 1^{3}/_{4}]$ respectively. The upper bound of the latter interval is outside the former interval and so the optimization of computing R**4 by repeated squaring actually gives a different result. However, the language definition allows this optimization anyway since it does not prescribe any order to the association of the multiplications for exponentiation as mentioned in Section 6.5.

This annex thus prescribes the accuracy of floating point operations in terms of model numbers; it also gives accuracy and performance requirements for the various packages. Some awkward hardware might not be able to meet these requirements for built-in implementations of certain functions (such as Sin) and rather than exclude such implementations from conformance to this annex altogether, it is instead possible to conform in either a relaxed or strict mode. In the relaxed mode, the accuracy requirements do not have to be met. For further details consult the *ARM*.

26.6 High Integrity Systems

This annex provides a number of pragmas which can be used to aid the analysis of a program to ensure that it is correct.

The configuration pragma Normalize_Scalars ensures that all otherwise uninitialized objects have an initial value. This reduces the dangers of bounded errors.

The configuration pragma Reviewable directs the compiler to generate code and listings enabling an understandable mapping between the source and object code.

The pragma Inspection_Point takes a list of object names and ensures that at each point where it is written the named objects are in sensible places from which their values can be obtained for analysis or debugging.

The configuration pragma Partition_Elaboration_Policy can be used to select a policy from the predefined policies which are Sequential or Concurrent or an implementation defined policy. In the case of Sequential we also have to specify the restriction No_Task_Hierarchy. The policy Sequential ensures that library tasks are not activated until all library units are elaborated and interrupts are attached.

The configuration pragma Detect_Blocking ensures that an implementation will detect a potentially blocking operation within a protected operation.

The pragma Restrictions has additional arguments defined in this annex which can be used to ensure that the program uses only simple and consequently formally analysable features of the language.

Examples of such restrictions are No_Protected_Types, No_Floating_Point, No_Recursion, and No_Exceptions.

A number of new aspects and restrictions are added in Ada 2022. They include Use_Formal and Dispatching. The *ARM* (para H.7.1) explains that they are provided to more precisely describe the use of generic formal parameters and dispatching calls within the execution of an operation. This enables more precise checking of conformance with the Nonblocking and Global aspects that apply at the point of invocation of the operation. See also the discussion on global state in Section 16.5.

Checklist 26

The packages Ada.Dispatching, Ada.Task_Termination, Ada.Execution_Time, and Ada.Real_Time.Timing_Events were added in Ada 2005.

The pragmas Priority_Specific_Dispatching and Relative_Deadline were added in Ada 2005.

The policies Non_Preemptive_FIFO_Within_Priorities, Round_Robin_Within_Priorities, and EDF_Across_Priorities were added in Ada 2005.

The pragma Profile and the Ravenscar profile were added in Ada 2005.

It was not possible to change the priority of a protected object in Ada 95.

The packages Ada.Numerics.Generic_Real_Arrays and Ada.Numerics.Generic_Complex_Arrays were added in Ada 2005.

The High Integrity Systems annex was known as the Safety and Security annex in Ada 95.

The pragmas Partition_Elaboration_Policy and Detect_Blocking were added in Ada 2005.

Various pragmas were replaced by aspect specifications in Ada 2012.

The functions Environment_Task and Activation_Is_Complete in the package Ada.Task_Identification were new in Ada 2012.

The procedure Yield was added to Ada.Dispatching and a child package Ada.Dispatching.Non_Preemptive containing a procedure Yield_To_Higher.

The packages Ada.Synchronous_Task_Control.EDF and Ada.Synchronous_Barriers were also new in Ada 2012.

New in Ada 2022

The package System.Atomic_Operations for manipulating items in an atomic manner is added to the Systems Programming annex.

Various aspects such as Use_Formal and Dispatching are added in the High Integrity Systems annex.

There are a number of changes and additions to the Real-Time Systems annex as follows:

The policy EDF_Across_Priorities is replaced by EDF_Within_Priorities.

An additional queuing policy is defined: Ordered_FIFO_Queuing.

The concept of an admission policy is introduced. The policy is either FIFO_Spinning or an implementation defined policy.

27 Finale

27.1 Names and expressions
27.2 Type equivalence
27.3 Overall program structure
27.4 Portability
27.5 Penultimate thoughts
27.6 SPARK

This final chapter summarizes various overall aspects of Ada. The first three sections cover some general topics which have been introduced in stages throughout the book. There is then a section on the important issue of portability. The penultimate section airs a few philosophical thoughts about programming in general and Ada in particular. And then the last section is an introduction to the SPARK language which is based on Ada and is used for those essential high integrity systems where life and limb and the environment are at risk if the software is incorrect.

27.1 Names and expressions

The idea of a name should be carefully distinguished from that of an identifier. An identifier is a syntactic form such as Fred which is used for various purposes including introducing entities when they are declared. An operator has a similar status to an identifier; operators and identifiers are collectively referred to as direct names in the syntax. A name, on the other hand, may be more complex and is the form used to denote entities in general. The reader might find it helpful to consult the syntax (which is on the website) and especially items 42 (name), 43 (direct_name), and 56 (primary).

Syntactically, a name typically starts with an identifier such as Fred or an operator symbol such as "+" and can then be followed by one or more of the following in an arbitrary order

- one or more index expressions in parentheses; this denotes a component of an array,
- a discrete range in parentheses; this denotes a slice of an array,
- an apostrophe plus an identifier, possibly indexed; this denotes an attribute,

- a dot followed by an identifier, operator or **all**; this denotes a record component, an object designated by an access value, or an entity in a package, task, protected object, subprogram, block, or loop,
- an actual parameter list in parentheses; this denotes a function call.

Type conversions, qualified expressions, and character literals are also treated as names in the syntax. Character literals are considered to be parameterless functions and can be renamed. In addition, generalized references and generalized indexing, which were discussed in Section 21.6, are also considered to be names. In Ada 2022 the target name symbol @ is also a name.

Names are just one of the primary components of an expression. The others are literals (numeric literals, strings and **null**), aggregates, and allocators as well as quantified expressions, conditional expressions, declare expressions, and expressions in parentheses. Expressions involving scalar operators were summarized in Section 6.9. Note that declare expressions are new in Ada 2022.

For convenience, all the operators and their predefined uses are shown in Table 27.1. They are grouped according to precedence level. The table also includes the short circuit forms **and then** and **or else** and the membership tests **in** and **not in** although these are not technically classed as operators (they cannot be overloaded).

Note the careful distinction between Boolean which means the predefined type and 'Boolean' which means Boolean or any type derived from it. Similarly Integer means the predefined type and 'integer' means any integer type (including *root_integer*). Also 'floating' means any floating type plus *root_real*.

Remember that & can take either an array or a component for both operands so four cases arise.

Observe that the membership tests apply to any type and not just scalar types which were discussed in Section 6.9. Thus we can check whether an array or record has a particular subtype by using a membership test rather than testing the bounds or discriminant. So we can write

```
V in Vector_5              -- true, see Section 8.2
John in Woman              -- false, see Section 18.5
```

rather than

```
V'First = 1 and V'Last = 5
John.Sex = Female
```

which are equivalent. Furthermore, as explained in Section 9.1, membership tests were greatly extended in Ada 2012 and enable a value of any type to be tested against combinations of subtypes, single values, and ranges.

From time to time we have referred to the need for certain expressions to be static. This means that they have to be evaluated at compile time. An expression is static if every constituent is one of the following

- a numeric, enumeration, character, or string literal,
- a named number,
- a constant initialized by a static expression,
- a predefined operator with scalar parameters and result,

Table 27.1 Predefined operators.

Operator	*Operand(s)*		*Result*
and or xor	Boolean		same
	one-dimensional Boolean array		same
	modular		same
and then or else	Boolean		same
= /=	any, not limited		Boolean
< <= > >=	scalar		Boolean
	one-dimensional discrete array		Boolean
in not in	any appropriate combination		Boolean
+ - (binary)	numeric		same
&	one-dimensional array | component		same array
+ - (unary)	numeric		same
*	integer	integer	same
	fixed	Integer	same fixed
	Integer	fixed	same fixed
	univ fixed	univ fixed	univ fixed
	floating	floating	same
	root real	root integer	root real
	root integer	root real	root real
/	integer	integer	same
	fixed	Integer	same fixed
	univ fixed	univ fixed	univ fixed
	floating	floating	same
	root real	root integer	root real
mod rem	integer	integer	same
**	integer	Natural	same integer
	floating	Integer	same floating
not	Boolean		same
	one-dimensional Boolean array		same
	modular		same
abs	numeric		same

The mathematical operators also apply to the types Big_Integer and Big_Real in the package Ada.Numerics.Big_Numbers and its child packages. See Section 23.6.

They also apply to the types Real, Imaginary, and Complex in the package Ada.Numerics.Complex_Types and various generic forms in the Numerics Annex. They also apply to the types Real_Vector and Real_Matrix in the generic package Ada.Numerics.Generic_Real_Arrays in the Numerics Annex. See Section 26.5.

- a call of a static function with static parameters,
- a predefined concatenation operator returning a string,
- a short circuit form with static parts,
- a membership test whose simple expression and choices are all static,
- a static attribute or a functional attribute with static parameters,
- a type conversion provided that any constraint involved is static,
- a qualified static expression provided that any constraint involved is static,
- a conditional expression provided all parts are static,
- a declare expression provided all parts are static.

Note that renaming preserves staticness so a renaming of one of the above (for which renaming is allowed) is also an allowed constituent of a static expression.

There are a number of impovements in Ada 2022 to allow more expressions where the value has to be known statically.

An obvious case is that a declare expression is somewhat like a conditional expression and is static if all its parts are static. (This is the last item in the list above.)

The shift and rotate functions in the package Interfaces are now permitted in static expressions.

The new aspect Static can be applied to an expression function. This ensures that a call of the function will be considered to be static provided various obvious conditions are satisfied such as all the actual parameters are static. And it must not be recursive.

The equality and relational operators are now permitted in static string expressions. Thus S = "abc" is now static if S is itself static. And of course "dog" = "cat" is also static (and indeed false).

Note however, that if a non-numeric type uses the aspects Integer_Literal or Real_Literal then the literals are not static.

It is an important point that staticness is an intrinsic property of an expression and does not depend upon its context. Thus 2 + 3 is always statically evaluated to give 5 irrespective of whether the context demands a static expression or not (provided of course that the "+" is predefined and not a user-defined operation). Moreover, the value of a static expression must not be outside the base range of the expected type – this is also checked statically.

Exercise 27.1

1 Given

```
L: Integer := 6;
M: constant Integer := 7;
N: constant := 8;
```

then classify the following as static or dynamic expressions and give their type.

(a) L + 1 (b) M + 1 (c) N + 1

27.2 Type equivalence

It is perhaps worth emphasizing the rules for type equivalence. The basic rule is that every type definition introduces a new type. Remember the difference between a type definition and a type declaration. A type definition introduces a type whereas a type declaration also introduces an identifier referring to it. Thus

```
type T is (A, B, C);
```

is a type declaration whereas

```
(A, B, C)
```

is a type definition.

The language rules are such that, strictly speaking, types never have names; it is subtypes that have names. Thus T is the name of a subtype of an unnamed type; T is known as the first subtype. This nicety rarely need concern the user but explains why the *ARM* usually refers to subtypes and never has to say 'type or subtype'. We should thus really say 'the type of T' meaning the type of which T is the name of the first subtype. But of course we just loosely say 'the type T'. Note that the attribute Base does not give the type but simply an unconstrained subtype of the type.

Most subtypes have names but in a few cases even all subtypes may be anonymous. The obvious cases occur with the declarations of arrays, tasks, protected objects, and objects of anonymous access types. Thus

```
A: array (I range L .. R) of C;
```

is essentially short for

```
type anon is array (I range <>) of C;
A: anon(L .. R);
```

and

```
task T is ...
```

is essentially short for

```
task type anon is ...
T: anon;
```

and

```
P: access T;
```

is essentially short for

```
type anon is access T;
P: anon;
```

In such cases we say that A, T, and P are of anonymous types.

In some cases the first subtype is constrained. This occurs with array types, derived types, and numeric types. Thus

```
type T is array (I range L .. R) of C;
```

is essentially short for

```
subtype index is I range L .. R;
type anon is array (index range <>) of C;
subtype T is anon(L .. R);
```

and

```
type S is new T constraint;
```

is essentially short for

```
type anon is new T;
subtype S is anon constraint;
```

and

```
type T is range L .. R;
```

is essentially short for

```
type anon is new some_integer_type;
subtype T is anon range L .. R;
```

where *some_integer_type* corresponds to an underlying implemented type. Similar expansions apply to floating types.

When interpreting the rule that each type definition introduces a new type, remember that generic instantiation is equivalent to text substitution in this respect. Thus each instantiation of a package with a type definition in its specification introduces a distinct type. It will be remembered that a similar rule applies to the identification of different exceptions. Each textually distinct exception declaration introduces a new exception; an exception in a recursive procedure is the same for each incarnation but generic instantiation introduces different exceptions.

Remember also that multiple declarations are equivalent to several single declarations written out explicitly. Thus if we have

```
A, B: array (I range L .. R) of C;
```

then A and B are of different anonymous types.

In summary then, Ada has named equivalence rather than the weaker structural equivalence used for some types in C++. As a consequence Ada gives greater security in the sense that more errors can be found during compilation. However, the overzealous use of lots of different types can lead to trouble and there are stories of programs that could never be got to compile.

An obvious area of caution is with numeric types (a novice programmer often uses lots of numeric types with great glee). Using different types can prevent errors as we saw when counting apples and oranges in Section 12.3. However, attempts to use different numeric types to separate different units of measurement (for example the lengths and areas of Exercise 12.3(**1**)) can lead to messy situations where either lots of overloadings of operators have to be introduced or so many type conversions are required that the clarity sought is lost by the extra clutter. Another problem is that each different numeric type will require a separate instantiation of the relevant package in Ada.Text_IO if input–output is required. An example of possible overuse of numeric types is in Ada.Text_IO itself where the distinct integer type Count (used for counting characters, lines, and pages) is a frequent irritant.

So too many types can be unwise sometimes. However, the use of appropriate constraints (as explicit subtypes or directly) always seems to be a good idea. Remember that subtypes are merely shorthand for a type plus constraint with possible predicates and as a consequence have structural equivalence. Thus, recalling an example in Section 6.4, we can declare

```
subtype Day_Number is Integer range 1 .. 31;
subtype Feb_Day is Day_Number range 1 .. 29;
D1: Integer range 1 .. 29;
D2: Day_Number range 1 .. 29;
D3: Feb_Day;
```

and then D1, D2, and D3 all have exactly the same subtype.

When to use a subtype and when to use a new type is a matter of careful judgement. The guidelines must be the amount of separation between the abstract concepts. If the abstractions are truly distinct then separate types are justified but if there is much overlap and thus much conversion then subtypes are probably appropriate. Thus we could make a case for Day_Number being a distinct type

```
type Day_Number is range 1 .. 31;
```

but we would find it hard to justify making Feb_Day not simply a subtype of Day_Number.

Another important distinction between types and subtypes is in their representation. The basic rule is that a subtype has the same representation (apart perhaps from Size and Alignment) as the base type whereas a derived type can have a different representation. Of course, the compiler can still optimize, but that is another matter.

It should also be remembered that checking subtype properties is generally a run-time matter. Thus

```
S: String(1 .. 4) := "ABC";
```

raises Constraint_Error although we can expect that any reasonable compiler would pick this up during compilation.

However, although checking that a value lies within a subtype is a run-time matter, static matching is generally required between two subtypes. This means that either all constraints must be static and the same, or that both subtypes are the same. This was explained in more detail in Section 8.2 when we discussed matching the component subtypes in array conversions. Static matching also occurs in other contexts such as with access types (Section 11.4), deferred constants (Section 12.2), renaming subprogram bodies (Section 13.7), and discriminants in generics (Section 19.2).

27.3 Overall program structure

Ada has five structural units in which declarations can occur; these are blocks, subprograms, packages, tasks, and protected units (function expressions in Ada 2012 and declare expressions in Ada 2022 are somewhat of a special nature and so we shall ignore them in this analysis). They can be classified in various ways. First of all packages, tasks, and protected units have separate specifications and bodies; for subprograms this separation is optional; for blocks it is not possible or relevant since a block has no specification. We can also consider separate compilation: packages, tasks, protected units, and subprogram bodies can all be subunits but only packages and subprograms can be library units including child library units. Note also that only packages and subprograms can be generic. Finally, tasks, subprograms, and blocks can have dependent tasks; packages cannot since they are only passive scope control units and declaring a task in a protected unit is not allowed since it is a potentially blocking operation (see Section 20.4). These various properties of units are summarized in Table 27.2.

We can also consider the scope and nesting of these five structural units. (Note that a block is a statement whereas the others are declarations.) Each unit can generally appear inside any of the other units and this lexical nesting can in principle go on indefinitely, although in practical programs a depth of three will not often be exceeded. The only restrictions to this nesting are that a block, being a statement, cannot appear in a package specification but only in its body (and directly only in the initialization sequence); none of these units can appear in a task specification but only in its body; similarly none can appear in a protected specification although they can appear inside the bodies of entries and subprograms of the unit. In practice, however, some of the combinations will arise rarely. Blocks will usually occur inside subprograms and task bodies and occasionally inside other blocks. Subprograms will occur as library units and inside packages and protected units and less frequently inside tasks and other subprograms. Packages will usually be library units or inside other packages. Tasks will nearly always be inside packages and occasionally inside other tasks or subprograms. Protected units will usually be inside packages.

The *ARM* does not prescribe how a complete program is to be started but, as discussed in Section 13.1, we can imagine that the main subprogram is called by some magic outside the language itself. Moreover, we must imagine that this originating flow of control is associated with an anonymous task; this is called the environment task. The priority of this task can be set by the pragma Priority in the outermost declarative part of this main subprogram and the identity of the environment task can be found by calling Ada.Task_Identification.Environment_Task (see Section 26.1).

A further point is that whether a main subprogram can have parameters or not or whether there are restrictions on their types and modes or indeed whether the main subprogram can be a function is dependent on the implementation. It may indeed be very convenient for a main subprogram to have parameters, and for the calling and parameter passing to be performed by the magic associated with the interpretation of a statement in some non-Ada command language. On the other hand any such parameters may be accessible through the package Command_Line discussed in Section 23.10.

Table 27.2 Properties of units.

Property	*Blocks*	*Subprograms*	*Packages*	*Tasks*	*Protected*
Separation	no	optional	yes	yes	yes
Subunits	no	yes	yes	yes	yes
Library units	no	yes	yes	no	no
Generic units	no	yes	yes	no	no
Dependent tasks	yes	yes	no	yes	no

The main subprogram is usually a library unit but could be local to some other unit such as a library package. It could also be a child subprogram or even an instance of a generic subprogram. It is also possible for there to be no main subprogram at all and for all execution to be performed as a consequence of the elaboration of library packages.

The general effect is illustrated by the following simple model

```
task Environment_Task;

task body Environment_Task is
   -- all the library units, including Standard are declared here in
   -- some order consistent with the elaboration requirements
begin
   -- call the main subprogram, if any
end;
```

This model also illustrates that

- Delay statements executed by the environment task during the elaboration of a library package delay the environment task.
- Library tasks (that is task objects declared in library packages) are started before any main subprogram is called.
- After normal termination of the main subprogram the environment task must wait for all library tasks to terminate. If all library tasks terminate, then the program as a whole terminates.

Special rules apply if the pragma Partition_Elaboration_Policy is applied with the parameter Sequential – see Section 26.6. As mentioned in Section 26.1, we can use the function Activation_Is_Complete to determine whether a task has been activated and whether all library items have been elaborated.

If the environment task terminates abnormally (perhaps because of an exception during library elaboration or propagated from the main subprogram), then the implementation should wait for library tasks to terminate. However, it is permitted to abort such tasks.

After the main subprogram and all library tasks have terminated (or if execution of the main subprogram is abandoned because of an unhandled exception), any effects on any external files are not specified; in particular, any files that have been left open may (but need not) be closed.

Table 27.3 Compilation units.

Unit	*Dependency*
package spec	may need a body
package body	depends on: [generic] package spec
subprogram spec	needs a body
subprogram body	may depend on: [generic] subprogram spec
gen package spec	may need a body
gen subprogram spec	needs a body
subunit	depends on: package body | subprogram body | subunit
child spec	depends on: package spec
gen package instance	all in one lump
gen subprogram instance	all in one lump

The model just discussed corresponds to an active partition. A passive partition has no environment task and so all library units are preelaborated.

We continue by reconsidering the rules for order of compilation and recompilation of the units in a program library. The various different units are summarized in Table 27.3 which also shows their basic interrelationships. Remember also that renaming is permitted at the library level so that a library item can simply be the renaming of another library item.

The reader will recall from Chapter 13 that the compilation order is determined by the dependency relationships. A unit cannot be compiled unless all the units on which it depends have already been entered into the library. And similarly, if a unit is changed then all units depending upon it also have to be recompiled. The basic rules for dependency are

- A body is dependent on its specification.
- A child is dependent upon its parent's specification.
- A subunit is dependent on its parent body or subunit.
- A unit is dependent on the specifications of units given in its nonlimited with clauses.

In addition, an implementation may impose the following auxiliary rule

- If a subprogram call is inlined using the aspect Inline (see Appendix 1) then the calling unit will be dependent upon the called subprogram body (as well as the specification).

The language definition is very careful not to impose unnecessary restrictions on how the compilation process actually works. At the end of the day there is really only the one requirement that a linked partition be consistent. The discussion here must thus be interpreted as the sort of way in which a library environment might work rather than how it must work.

The basic rules for compilation are as follows. A new library unit will replace an existing library unit (of any sort) with the same name. A unit will be rejected unless there already exist all those units on which it depends; thus a body is rejected unless its specification exists and a subunit is rejected unless its parent body or subunit exists. If a unit is replaced then all units dependent on it need to be (re)compiled.

These fairly simple rules are complicated by the fact that a subprogram need not have a distinct specification and a package may not need a body.

If we start with an empty library and compile a procedure body P, then it will be accepted as a library unit (and not needing a distinct specification). If we subsequently compile a new version of the body then it will replace the previous library unit. If, however, we subsequently enter and compile just the specification of P, then it will make the old body obsolete and we must then compile a new body which will now be classed as dependent upon the separate specification. In other words we cannot add the specification as an afterthought and then provide a new body perhaps in the expectation that units dependent just on the specification could avoid recompilation. Moreover, once we have a distinct specification and body we cannot join them up again – if we provide a new body which matches the existing distinct specification then it will replace the old body, if it does not match the specification then it will be rejected.

A library package cannot have a body unless it needs one to satisfy language rules. This prevents the inadvertent loss of a body when a specification is replaced since otherwise the system would link without the body. The pragma Elaborate_Body can be used to force a body to be required as mentioned in Section 13.8.

Generics also have to be considered and we need to take care to distinguish between generic units and their instantiations. A generic subprogram always has a distinct specification but a generic package may not need a body. Note carefully that the body of a generic package or subprogram looks just like the body of a plain package or subprogram. It will be classed as one or the other according to the category of the existing specification (a body is rejected if there is no existing specification except in the case of a subprogram which we discussed above). An instantiation however is all in one lump, it is classed as a single library unit in its own right and the separation of specification and body does not occur. The notional body of a generic instance cannot be replaced by a newly compiled plain body. For example, suppose we first compile

```
generic
procedure GP;

procedure GP is
begin ... end GP;
```

and then separately compile

```
with GP;
procedure P is new GP;
```

and then submit

```
procedure P is
begin ... end P;
```

The result is that the new unit P will be accepted. However, it will be classed as a nongeneric library unit and completely replace the existing instantiation. It cannot be taken as a new body for the generic instance since the instantiation is treated as one lump.

We conclude by emphasizing that we have been discussing a possible implementation of the program library. Some implementations may insist that compilation is only possible if all units on which a unit depends have already been compiled; other implementations may merely insist that just the source of the units be present. The above discussion must thus be interpreted accordingly. We can expect an implementation to provide utility programs which manipulate the library in additional ways. Such facilities are outside the scope of this book and so the reader must refer to the documentation for the implementation concerned.

27.4 Portability

An Ada program may or may not be portable. In many cases a program will be intimately concerned with the particular hardware on which it is running; this is especially true of embedded applications. Such a program cannot be transferred to another machine without significant alteration. On the other hand it is highly desirable to write portable software components so that they can be reused in different applications. In some cases a component will be totally portable; more often it will make certain demands on the implementation or be parameterized so that it can be tailored to its environment in a straightforward manner. This section contains some thoughts on the writing of portable Ada programs.

One thing to avoid is programs with bounded errors and those whose behaviour depends upon some order of evaluation which is not specified. They can be insidious. A program may work quite satisfactorily on one implementation and may seem superficially to be portable. However, if it happens to depend upon some unspecified feature then its behaviour on another implementation cannot be guaranteed. A common example in most programming languages occurs with variables which accidentally are not initialized. It is often the case that the intended value is zero and furthermore some operating systems clear the program area before loading the program. Under such circumstances the program will behave correctly but may give surprising results when transferred to a different implementation. In the previous chapters we have mentioned various causes of such doubtful programs. For convenience we summarize them here.

An important group of situations concerns the order of evaluation of expressions. Since an expression can include a function call and a function call can have side effects, it follows that different orders of evaluation can sometimes produce different results. The order of evaluation of the following is not specified

- the operands of a binary operator,
- the destination and value in an assignment,
- the components in an aggregate,
- the parameters in a subprogram or entry call,
- the index expressions in a multidimensional name,
- the expressions in a range,

- the barriers in a protected object,
- the guards in a select statement.

Moreover, we recall from Section 13.8 that the order of elaboration of library units is not fully determined.

There is an important situation where a junk value can arise

- reading an uninitialized variable before assigning to it.

Such reading results in a bounded error.

There are also situations where the language mechanism is not specified

- the passing of array and record parameters,
- the algorithm for choosing a branch of a select statement or between several queues in protected units.

But note that the last can be specified by the pragma Queuing_Policy if the Real-Time Systems annex is implemented.

There are also situations where the programmer is given extra freedom to overcome the stringency of the type model; abuse of this freedom can lead to erroneous programs. Examples are: suppressing exceptions, using unchecked access, unchecked deallocation, and unchecked conversion.

Numeric types are another important source of portability problems. The reason is the compromise necessary between achieving absolutely uniform behaviour on all machines and maximizing efficiency. Ada gives the user various options in choosing an appropriate compromise as discussed in Section 17.1.

There are various attributes which, if used correctly, can make programs more portable. Thus we can use Base to find out what is really going on and Machine_Overflows to see whether Constraint_Error will occur or not. But the misuse of these attributes can obviously lead to very nonportable programs.

It is good practice to declare our own floating point types and not directly use the predefined types such as Float. A similar approach should be taken with integer types, although the language does encourage us to assume that the predefined type Integer has a sensible range.

Another area to consider is tasking. Any program that uses tasking is likely to suffer from portability problems because instruction execution times vary from machine to machine. In some cases a program may not be capable of running at all on a particular machine because it does not have adequate processing power. Hard guidelines are almost impossible to give but the following points should be kept in mind.

Take care that the type Duration is accurate enough for the application. Remember that regular loops (that is, those executed at uniform intervals) cannot easily be achieved if the interval required is not a multiple of *small*.

Avoid the unsynchronized use of shared variables as far as possible. Protected objects should be used but sometimes timing considerations demand quick and dirty techniques. In simple cases the aspect Atomic may be able to prevent interference between tasks.

The use of the abort statement and asynchronous transfer of control might also give portability problems because of their asynchronous nature.

And finally, different machines have different sizes and so a program that runs satisfactorily on one machine might raise Storage_Error on another. Moreover, different implementations may use different storage allocation strategies for access types and task data. We have seen how aspect clauses and aspect specifications can be used to give control of storage allocation and thereby reduce portability problems.

27.5 Penultimate thoughts

In early versions of this book, this section was entitled Program Design. However, that seemed too specific a title and it was called Final Thoughts in the 2005 version thereby enabling the author to ruminate somewhat. In this version it is called Penultimate Thoughts so that the book can conclude with a further section giving an introduction to SPARK.

It is some seventy years since programming started in earnest and forty years since the birth of Ada. Ada 2022 in particular includes all those concepts that appear to have gained favour with practitioners. Nevertheless, it seems clear that we do not have much agreement on what a universal programming language should look like or perhaps even what a language is for.

Is the prime purpose of the program to tell the computer what to do or is the computer a means of executing an abstract program? The control program in an airplane falls into the former category whereas the typical numerical program for solving some differential equations fits the latter model. This dichotomy seems to lie at the heart of many difficulties we have in deciding upon what we are doing. Ada aims to be helpful in both situations and should enable the programmer to divide the program into parts that are purely algorithmic or abstract and parts that communicate with the outside world. Ada also recognizes that other languages exist and provides means of communicating with them so that if perhaps it is more appropriate to use some other language for a part of a program then this can be done. Most programming languages are very insular in this respect.

Program design is a thorny subject. Indeed, program design is still largely an art and the value of different methods of design is often a matter of opinion rather than a matter of fact. We have therefore tried to stick to the facts of Ada and to remain neutral regarding design.

An important aspect of design is ensuring that software can be reused. In a broad sense this includes maintenance. This means that the external behaviour should be clearly specified and that the internal structure should be understandable. It is not possible to say much in the space available but the following thoughts might be helpful.

With regard to the internal structure, there are various low level and stylistic issues which are perhaps obvious. Identifiers should be meaningful. The program should be laid out neatly. Useful comments should be added. The block structure should be used to localize declarations to their use. Whenever possible a piece of information should only be written once; thus number and constant declarations should be used rather than explicit literals. And so on.

The decomposition of a program into various component parts is a familiar and vital concept. There is a strong analogy with the decomposition of mechanical and electronic devices into components. A component should have a well specified interface which defines its external behaviour. Essentially there exists a contract between the component and its users.

This idea of a contract is not new. If we look at the programming libraries developed in the early 1960s, particularly in mathematical areas and perhaps written in Algol 60 (a language favoured for the publication of such material in respected journals such as the *Communications of the ACM* and the *Computer Journal*), we find that the manual tells us what parameters are required, any constraints on their range and so on. In essence there is a contract between the writer of the subroutine and the user. The user promises to hand over suitable parameters and the subroutine promises to produce the correct answer.

A key aspect of the programming process is therefore to define what the component parts do and hence the interfaces between them. This enables the parts to be developed independently of each other. If we write each part correctly (that is so that it satisfies its side of the contract implied by its interface) and if we have defined the interfaces correctly then we are assured that when we put the parts together to create the complete system, it will work correctly. An important addition in Ada 2012 was Chapter 16 on Contracts with pre- and postconditions which strengthen the definition of interfaces; Ada 2022 continues this with Global state.

One of Ada's strengths is its ability to separate the definition of an interface from its implementation. The primary means of doing this is through packages. The specification defines what the component does and the body says how it does it. No other practically used language seems to separate these concerns so clearly.

Designing an Ada program is therefore largely concerned with designing groups of packages and the interfaces between them. A group can often be organized as a child hierarchy. Often we will hope to use existing packages. For this to be possible it is clear that they must have been designed with consistent, clean, and sufficiently general interfaces.

An important use of packages is to provide abstract data types where the use of private types enables us to separate the representation of the type from the operations upon it. In a way the new types can be seen as natural extensions to the language. Indeed, the whole essence of object oriented programming is the reuse and composition of abstract data types through extension and inheritance.

Object oriented programming raises interesting issues. It is clear that object oriented design is an excellent concept but it is not at all clear that the programming techniques of inheritance, polymorphism, and dispatching are the panacea some would claim. It is clear that the overuse of inheritance can lead to very obscure programs. Luckily, Ada very clearly separates the concept of a specific type from a class of types with the result that one can always find out the specific type of an object. Thus in Section 14.5 we were able to ensure that the Posh passenger always gets a decent meal.

The problem with inheritance is that it provides properties that are not directly stated. In fact it is almost the opposite of information hiding. Good abstraction is about hiding irrelevant detail; the trouble with inheritance is that it hides (or at least obscures) *relevant* detail.

Another major design area concerns the use of tasks. Ada is unusual in having tasking embedded within the language itself. However, this brings major benefits of reliability and predictability. An important development is the Ravenscar profile which by restricting the user to a number of carefully defined features ensures that a program is highly predictable and thus trustworthy for high integrity applications.

High integrity applications are very important. Software now pervades all aspects of modern society and there are serious programs that, if incorrect, cause real difficulties. These range from safety critical programs such as engine controllers to communications systems where security is a major concern.

Ada is widely used for high integrity applications for various reasons. One is that it aims to encourage the writing of a program correctly in the first place. Strong static typing and related features enable the compiler to detect errors early in the development process. And the separation of components by package interfaces encourages the construction of well-defined components. For the most critical applications the SPARK language should be considered. This is a subset of Ada but also adds extra information that strengthens the definition of interfaces and enables programs to be proved to be correct. A brief introduction to SPARK will be found in the next and last section of this book.

Although dividing a program into components is very well established, a market place for software components does not seem to have emerged in the same way that it has for electronic and other components. When I wrote the first version of this book, I foresaw a conversation in a local software store in which I predicted the use of a common European currency. But it looks like it will take much longer for a universal market in components to emerge. Accordingly, here is that conversation updated to look a little further ahead.

Customer:	Could I have a look at the reader–writer package you have in the window?
Server:	Certainly. Would you be interested in this robust version – proof against abort? Or we have this slick version for trusty callers. Just arrived by Cool Comet Courier through our private wormhole from our subsidiary on Sirius Three.
Customer:	Well – it's for a cooperating system so the new one sounds good. How much is it?
Server:	It's 250 Galactic Guineas but as it's new there is a special offer with it – a free copy of this random number generator and 10% off your next certification.
Customer:	Great. Is it validated?
Server:	All our products conform to the highest standards. The parameter mechanism conforms to MP13-8191 and it has the usual intergalactic multitasking certificate.
Customer:	OK, I'll take it.
Server:	Will you take it as it is or shall I instantiate it for you?
Customer:	As it is please. I prefer to do my own instantiation.

On this fantasy note we almost come to the end of this book. It is hoped that the reader will have gained some general understanding of the principles of Ada as well as a lot of the detail. Further understanding will come with use and the author hopes that he has in some small way prepared the reader for the successful use of a very good programming language.

But finally we will look at some key ideas behind SPARK which has proved to be highly successful in many critical applications.

27.6 SPARK

SPARK is essentially a subset of Ada with additional features which enable a program to be analysed for correctness *before* the program is executed.

Most software is produced in a somewhat haphazard way with emphasis on time-to-market rather than correctness. There is typically a feeling, well if it's not quite correct we can release an update. So it is tested and if it seems OK, it is released. But as we know, testing can never prove that a program is correct. Testing can find the presence of errors but cannot show that there are none (except perhaps in the most trivial cases.) The goal of SPARK is to provide a development approach in which it is feasible to show conclusively that a program has no errors.

Over the last twenty or so years, software has become widely used in systems throughout almost all areas of society. Banking, transport, medical, and industrial control systems all depend upon the functioning of software. As a consequence the safety of many human lives and the security of much property and the environment now depend upon the correctness of software.

There are now many serious programs that, if incorrect, cause real difficulties. These range from safety critical programs such as aircraft control systems to programs in financial systems where security is a major concern. For safety critical programs the consequence of an error can be loss of life or environmental damage. For secure programs the consequence of an error may be equally catastrophic such as loss of national security or commercial reputation or just plain theft.

So it is important that these high integrity systems are programmed with great care so that maximum confidence in their correctness is achieved.

SPARK has played a major role in the development of many critical systems since it first emerged in around 1988. The first version of SPARK was based on Ada 83. A major upgrade was made when Ada itself was upgraded to give Ada 95 and later to Ada 2005. A further version which incorporates the Ravenscar tasking profile (see Section 22.7) is called RavenSPARK.

The key differences between Ada and SPARK are that

- Certain features of Ada are omitted from SPARK on the grounds that programs using them are difficult to show to be correct. Thus the use of access types is forbidden because the misuse of pointers is the source of many software evils.
- Features are added to SPARK to enable the proof of program correctness to be feasible. These largely concern the definition of interfaces between different components of a program.

The original versions of SPARK added this extra information through the medium of Ada comments. Thus these comments were ignored by the Ada compiler but were processed by separate SPARK tools comprising the Examiner, the Simplifier, and the Proof Checker.

In the later version of SPARK known as SPARK 2014 the additional information is added by using aspect specifications. There are still restrictions on the use of certain features of Ada such as access types but the important thing is that the analysis of the extra information is now done by tools which are more integrated with the Ada compiler. This makes things much easier for the user. Incidentally, in the discussion that follows we will refer to the original SPARK as SPARK 2005 because it is based on Ada 2005.

Important features introduced explicitly in Ada 2012 include those for what we might call 'programming by contract' which were described in Chapter 16. This idea of a contract is not new as mentioned in the previous section.

The decomposition of a program into various component parts is very familiar and the essence of the programming process is to define what these parts do and therefore the interfaces between them. This enables the parts to be developed independently of each other. If we write each part correctly (that is, so that it satisfies its side of the contract implied by its interface) and if we have defined the interfaces correctly then we are assured that when we put the parts together to create the complete system, it will work correctly. In Ada the key component parts are packages and subprograms and the interfaces between them are given by their specifications.

Bitter experience shows that life is not quite like that. Two things go wrong: on the one hand the interface definitions are not usually complete (the package and subprogram specifications do not say everything, so there are holes in the contracts) and on the other hand, the individual components are not correct or are used incorrectly (the contracts are violated). And of course the contracts might not say what we meant to say anyway.

The key purpose of SPARK is to prevent these difficulties by improving the detail of the description of an interface and ensuring that the code implements the interface correctly.

As a simple example of an interface definition consider the interface to a subprogram. The interface should describe the full contract between the user and the implementor. The details of how the subprogram is implemented should not concern us. In order that these two concerns be clearly distinguished it is helpful to use a programming language in which they are lexically distinct.

As we have seen, Ada has such a structure separating interface (the specification) from the implementation (the body). This applies both to individual subprograms (procedures or functions) and to groups of entities encapsulated into packages.

As mentioned above, SPARK requires additional information to be provided and this is done through the mechanism of annotations or aspect specifications which increase the amount of information about the interface without providing unnecessary information about the implementation. In fact SPARK allows such information to be added at various levels of detail as appropriate to the needs of the application.

Consider the information given by the following Ada specification

```
procedure Add(X: in Integer);
```

Frankly, it tells us very little. It just says that there is a procedure called Add and that it takes a single parameter of type Integer whose formal name is X. This is enough to enable the compiler to generate code to call the procedure. But it says nothing about what the procedure does. It might do anything at all. It certainly doesn't have to add anything nor does it have to use the value of X. It could for example subtract two unrelated global variables and print the result to some file. But now consider what happens when we add the lowest level of SPARK annotation. In SPARK 2005, the specification might become

```
procedure Add(X: in Integer);
--# global in out Total;
```

This states that the only global variable that the procedure can access is that called Total. Moreover, the mode information tells us that the initial value of Total must be used (**in**) and that a new value will be produced (**out**). The SPARK rules also say more about the parameter X. Although in Ada a parameter need not be used at all, nevertheless an **in** parameter must be used in SPARK.

So now we know rather a lot. We know that a call of Add will produce a new value of Total and that it will use the initial value of Total and the value of X. We also know that Add cannot affect anything else. It certainly cannot print anything or have any other malevolent side effect.

Of course, the information regarding the interface is not complete since nowhere does it require that addition be performed in order to obtain the new value of Total. In order to do this we can add optional annotations which concern proof and obtain

```
procedure Add(X: in Integer);
--# global in out Total;
--# post Total = Total~ + X;
```

The postcondition explicitly says that the final value of Total is the result of adding its initial value (distinguished in SPARK 2005 by ~) to that of X.

It is also possible to provide preconditions. Thus we might require X to be positive so that the specification would become

```
procedure Add(X: in Integer);
--# global in out Total;
--# pre X > 0;
--# post Total = Total~ + X;
```

So now the specification is complete.

In SPARK 2014 this specification is written entirely in Ada style as follows

```
procedure Add(X: in Integer)
   with Global => (in_out => Total),
        Pre => X > 0,
        Post => Total = Total'Old + X;
```

The pre- and postconditions use precisely the syntax described in Chapter 16. Note that 'Old is used instead of the tilde used in SPARK 2005. The global annotation has now become the new Global aspect. This is an example of an implementation-defined aspect especially added for SPARK 2014. The meaning of the aspect is as before. It ensures that the body of Add cannot manipulate any global variables other then Total. It also ensures that it both reads the original value of Total and writes a new value to it. As mentioned above, modes in Ada (which only apply to parameters) give access permission, whereas in SPARK, modes (which apply to both parameters and globals) impose a requirement to read and/or update as applicable. (See Section 16.5 for the more consistent style of global annotation in Ada 2022.)

It is important to note that many annotations/aspects are optional in SPARK. Thus the global annotations are traditionally mandatory but the pre- and postconditions are optional. Analysis can be performed at various levels. At the very lowest level, we can omit all annotations including the global ones but provided that we stay within the SPARK subset of Ada, we can obtain useful simple data flow analysis.

By adding global annotations we get more comprehensive data flow analysis and so on. At the highest level, by adding proof annotations, we can do formal proof that shows that the body of the subprogram does indeed perform the operations indicated by the pre- and postconditions.

In SPARK 2014 we have even more flexibility. The pre- and postconditions can be used to control a formal proof or they can simply be used as in Ada 2012 to perform dynamic checks and so raise Assertion_Error if they fail.

In Chapter 16 we saw how in Ada 2012 we can give preconditions and postconditions for a subprogram. The idea is that if the preconditions are satisfied then we should be assured that the postconditions will also be satisfied; the intent therefore is that the subprogram body should be written so that if the preconditions are satisfied then the postconditions will also be satisfied. However, in Ada this is basically a question of dynamic testing.

In SPARK, the same general approach is adopted. But the goal is that we should be able to prove mathematically that if the preconditions are satisfied then the postconditions will also be satisfied. There is no question of testing here at all but (simply!) formal proof before the program is executed. As mentioned above, testing can never show the absence of an error but only the presence of one; the SPARK approach is to show that there can be no errors whatever the inputs.

Here is a trivial example of proof. Consider a procedure Exchange which aims to interchange the values of two actual parameters. It might be

```
procedure Exchange(X, Y: in out Float) is
   T: Float;
begin
   T := X;  X := Y;  Y := T;
end Exchange;
```

The intent is that this will work for any input values so there is no need for a precondition. In Ada 2012, we can supply an appropriate postcondition as follows

```
procedure Exchange(X, Y: in out Float)
   with Post => X = Y'Old and Y = X'Old;
```

This is satisfactory as far as it goes. We will find that whatever values we pass as parameters, the postcondition will be satisfied and Assertion_Error will not be raised. But this is not very satisfactory; all we can show is that we have not found a pair of values for which the postcondition is not satisfied but we cannot show that no such pair exists.

If we ask for formal analysis, that is, request that verification conditions (VCs) be generated then we might obtain an internal file containing something like

```
H1:   true .
      ->
C1:   y = y .
C2:   x = x .
```

The notation used is that first we have a number of hypotheses (H1, H2, ...) and these are then followed by a number of conclusions (C1, C2, ...) which have to be verified using the hypotheses. The whole verification condition is therefore

```
H1 and H2 and ...   ->   C1 and C2 and ...
```

where the arrow -> means implies.

In this example there is no precondition and so effectively no hypotheses (this is represented as the single hypothesis H1 which is true). The two conclusions to be proved are that y = y and x = x which as pure mathematical expressions are reasonably self-evident and so it is pretty clear that the procedure Exchange is indeed correct. Proving that a verification condition is correct is known as discharging the verification condition.

The reader might wonder why the verification condition takes the form shown. Roughly speaking what happens is that the analysis takes the postcondition (which is X = Y'Old **and** Y = X'Old) required at the end of the subprogram and determines what would have to hold just before the last assignment which is Y := T; a simple mechanical transformation shows that it would be

```
X = Y'Old and T = X'Old
```

The analysis then considers the situation progressively until we find that at the beginning we need

```
Y = Y'Old and X = X'Old
```

In doing these transformations we have to remember that X'Old means the initial value of X and so is not changed by the transformations – the initial value is just the initial value whereas X means the current value at a particular point.

So we end up with the condition that at the beginning the current value of X must equal the initial value of X and similarly for Y. Of course at the beginning the current value *is* the initial value and so the 'Old can be dropped resulting in the final condition

```
Y = Y and X = X
```

We note that this corresponds exactly to the VCs generated by the analysis

```
H1:   true .
      ->
C1:   y = y .
C2:   x = x .
```

where the hypothesis H1 is just true because there is no precondition in this case. This analysis incidentally reveals why the analysis gives the y conclusion first (a human being would surely give them in alphabetical order!).

Much can be learnt about a program by simple data and information flow analysis without the use of proof. SPARK does not allow us to create information that is not used; it does not allow us to read from variables that have not been written to and so on. For example, suppose we accidentally make an error in the procedure Exchange and write

```
procedure Exchange(X, Y: in out Float) is
   T: Float;
begin
   T := X;  X := Y;  Y := X;
end Exchange;
```

where we have assigned X to Y rather than T to Y in the last statement. As an Ada program this is fine (although silly). However, SPARK is very grumpy about this and even without any consideration of proof it reports errors such as

- the initial value of X is not used (that is does not affect any external state),
- there is an unused assignment (this refers to the assignment to T which is then neither referenced nor exported).

Given these messages it is pretty obvious that something is wrong. Although this example is trivial, this kind of analysis in a large program is very helpful in discovering errors which a lenient traditional compiler will overlook.

Another form of annotation which gives extra information regarding flow analysis concerns the dependency of exported variables on imported variables. In the Exchange example we could add the aspect Depends thus

```
procedure Exchange(X, Y: in out Float)
   with Depends => (X => Y,              -- read as X depends on Y
                    Y => X);             -- and Y depends on X
```

In this simple example, this does not add very much but in an elaborate subprogram with a number of parameters and globals, it can reveal many errors such as inadvertent cross-coupling of relationships.

The two important features of programming that make the computer so successful are

- The ability of subprograms to call other subprograms thus making a nested hierarchy of instructions.
- The ability to iterate with usually some parameterization over a series of instructions many times using the notion of a loop.

In the case of nested calls suppose we have a procedure P that calls a procedure Q. And suppose that the specifications of both P and Q have some pre- and post-conditions thus

```
procedure P( ... )
   with Pre => PreP,
        Post => PostP;

procedure Q( ... )
   with Pre => PreQ,
        Post => PostQ;
```

Now, the body of P calls Q and so has the form

```
procedure P( ... ) is
   ...                          -- 1  start of P, assume PreP
begin
   ...
   Q( ... );                    -- 2  call of Q, check PreQ
                                -- 3  return from Q, assume PostQ
   ...
   return;                      -- 4  end of P, check PostP
end P;
```

The call of Q breaks the body of P into two pieces. We have to take account of the fact that the code of P must be such that it satisfies the precondition for Q at the point where Q is called. And when Q returns, we can assume that Q has done its job correctly so that PostQ is true. So we have a section of code from *1* to *2* whose precondition is PreP and whose postcondition is PreQ. And there is a section of code from *3* to *4* whose precondition is PreP **and** PostQ and whose postcondition is PostP. (In fact this second path is often treated as the overall path from *1* to *4*.)

Anyway, the important point is that we don't have just one VC for P but at least two and both of these have to be discharged if we wish to prove that the program is correct.

Loops similarly cause one section of code to be split into several pieces for the purpose of analysis. Consider a procedure R with pre- and postconditions PreR and PostR whose body contains a loop thus

```
procedure R( ... ) is
   ...                          -- 1  start of R, assume PreR
begin
   ...
   loop                         -- 2  start of loop, check L
      pragma Loop_Invariant => L;
      ...                       -- 3  another iteration, assume L
         exit when C;
      ...
   end loop;
   ...
   return;                      -- 4  end of R, check PostR
end R;
```

It is an important rule that all loops have to be broken by a loop invariant, that is an expression L which we have to show will be true each time around the loop.

So the procedure R is split into three pieces. There is a section of code from *1* to *2* whose precondition is PreR and whose postcondition is L; this corresponds to entering R and stops at the loop invariant in the loop the first time it is encountered. There is another section of code from *3* to *2* whose precondition is L and whose postcondition is also L; this corresponds to going around the loop and not taking the exit path. At first sight it may seem strange that the pre- and postconditions appear to be the same; however, L will inevitably be an expression involving variables whose values change each time around the loop. Finally, there is a path from *3* to *4* which corresponds to taking the exit path from the loop and eventually returns from the procedure R, the precondition is L and the postcondition is PostR.

So we find that there are typically three VCs to be discharged for the procedure R because of the inner loop.

The reader might wonder about the form of a typical Loop_Invariant. Because it concerns loops it will often take the form of a quantified expression. This is because a quantified expression is rather like a loop expression and so is the expression analogy of a loop statement. For example, suppose we have a simple procedure Idarr whose purpose is to set each element of an integer array to its index value, that is A(1) to 1, A(2) to 2 and so on. Its specification might be

```
type Atype is array(Natural range <>) of Integer;

procedure Idarr(A: out Atype)
   with Depends => (A => null),
        Post => (for all M in A'Range => (A(M) = M));
```

where we have added a dependency aspect for illustration (it shows that the **out** parameter A does not depend on any input) and then a postcondition using a quantified expression. There is no need for a precondition. The body of Idarr might then be

```
procedure Idarr(A: out Atype) is
begin
   for K in A'Range loop
      A(K) := K;
      pragma Loop_Invariant => (for all M in A'First .. K => (A(M) = M));
   end loop;
end Idarr;
```

Again, three VCs will be generated. The first is from the beginning of the procedure to the type invariant in the loop the first time the loop is executed; this will be satisfied because the only component of A that has been set will be A(A'First) and that exactly matches the quantified expression in the invariant which the first time traverses only one component. Each time around the loop, another component is set and the quantified expression is extended to take account of the extra component.So we see how the loop invariant is both the pre- and postcondition for the loop; although its form remains the same, the values involved change on each iteration. At the end of the loop, all components are set and so the third VC which corresponds to leaving the subprogram will simply find that the final version of the loop invariant exactly matches the postcondition.

Note that we used a for loop in this example rather than the simple loop with an exit statement in the case of the procedure R. For loops are very useful because they always terminate. It is an important feature of proving that a program is correct to show that it terminates and does not loop for ever. Remember that SPARK only proves partial correctness, that is correctness on the assumption of termination.

A typical program will comprise many procedures and involve many loops. As a consequence, many VCs will be generated and these all have to be discharged.

Many VCs will be straightforward such as that for Exchange discussed above and are eliminated automatically as part of program analysis. Others can be tackled by SMT solvers such as CVC4. SMT is short for Satisfiability Modulo Theory; suffice it to say here that much progress has been made in the last decade on theorem proving.

An indication of the size of problem that can be tackled these days is given by the example of iFACTS (see below) which generates over 120,000 VCs relating to freedom from exceptions.

It is especially important to note that in SPARK the pre- and postconditions are checked before the program executes. If they were only checked when the program executes then it would be a bit like bolting the door after the horse has bolted (which reveals a nasty pun caused by overloading in English!). We don't really want to be told that the conditions are violated as the program runs. For example, we might have a precondition for landing an aircraft

```
procedure Touchdown( ... )
  with Pre => Undercarriage_Down;
```

It is pretty unhelpful to be told that the undercarriage is not down as the plane lands; we really want to be assured that the program has been analysed to show that the situation will not arise. And certainly the last thing we want is for the pilot to be told Assertion_Error! This thought reminds one of that wonderful film Airplane! – at one point the pilot exclaims 'That's strange', and we see a shot of an instrument showing STRANGE. It could so easily have been Assertion_Error.

We will conclude this brief overview of SPARK by considering the various ways in which bugs are found. They are

(1) By the compiler or other static tools such as those for SPARK. These are usually easy to fix because the tool tells us exactly what is wrong.

(2) At run time by a language check. This applies in languages which carry out checks that, for example, ensure that we do not write outside an array. Typically we obtain an error message saying which structure was violated and whereabouts in the program this happened.

(3) By testing. This means running various examples and poring over the (un)expected results and wondering where it all went wrong.

(4) By the program crashing. In olden days this resulted in a nice coredump which you could take home and browse over in the middle of the night. A similar modern effect is when your computer becomes remarkably silent because some application has written all over the operating system and usually destroyed the evidence. Reboot and try again!

Clearly these four ways provide a progression of difficulty. Errors are easier to locate and correct if they are detected early. Good programming tools are those which move bugs from one category to a lower numbered category. Thus good programming languages are those which provide facilities enabling one to protect oneself against errors that are hard to find. Strong typing is one example and the enumeration type is a simple feature which correctly used makes hard bugs of type 3 into easy bugs of type 1.

As we have seen, a major goal of SPARK is to enable the strengthening of interface definitions (the contracts) and thus move errors to a low category and ideally to type 1 so that they are all found before the program executes. The global annotations do this because they prevent us writing a program that accidentally changes the wrong global variables. Similarly, detecting the violation of pre- and postconditions in SPARK results in a type 1 error.

A very important point is that we cannot prove a verification condition if it does not match the program. If the proof process fails then it could be that the program is incorrect or that some pre- or postcondition or invariant is incorrect. If the program itself is incorrect then no amount of processing by the most sophisticated proof solvers is likely to help.

An important approach to this problem is provided by the fact that a SPARK 2014 program can also be viewed as an Ada 2012 or Ada 2022 program. So we can execute the program on various sets of data in which case the pre- and postconditions will be treated as dynamic tests and hopefully the bug will be found by the raising of Assertion_Error. In other words we can resort to Testing to find a counter example.

So we can run the system in Test mode to shake out the bugs and then switch to Proof mode to prove that it is correct.

On the other hand, if we believe that the program is correct, we can start in Proof mode and then if any proofs remain elusive, we can use Test mode to demonstrate that all remaining parts also appear to be correct. Being able to switch between the two modes of operation is a huge advantage of the unification of the syntax with Ada in SPARK 2014.

Although the writing of appropriate pre- and postconditions can be tedious, they are not necessary for all forms of proof. An important feature of SPARK is the ability to show that a program cannot raise any exceptions and so is free from run-time errors. This can be done even if all annotations, including the global annotations, are omitted. Information inherent in the Ada program, such as information about ranges, enables the automatic generation of appropriate hypotheses and typically results in verification conditions that can be discharged in a straightforward manner. Lesser languages that do not have even simple things such as strong typing are really not fit for purpose.

In this short introduction, it has not been possible to describe many features of SPARK. However, we will conclude by mentioning a few important applications which have used or are using SPARK.

- The Lockheed Martin C130J shares the familiar Hercules airframe with its predecessors but features entirely new engines, propellors, and avionics. The Mission Computer controls most of the aircraft's major systems. It comprises over 100,000 lines of code, most of which is written in SPARK.
- The first and possibly largest SPARK project in the world is the EuroFighter Typhoon aircraft where nearly all safety critical systems are programmed in SPARK.
- The MULTi-application Operating System (Multos) is a smartcard OS that allows several applications to reside on a single card. Multos applications can be loaded and deleted dynamically, so a major security concern is the prevention of forged applications. To this end, a Multos application is accompanied by a digital certificate that is signed by the Multos Certification Authority (CA).

The CA was built as far as was practicable to meet the highest standard of the UK ITsec scheme. SPARK therefore seemed appropriate for the most security critical aspects of the system. The static analysis offered by SPARK proved to be most effective. Data flow errors can cause subtle security problems – for example, an uninitialized variable might just acquire an initial value which happens to be a piece of cryptographic key material 'left over' on the stack from the execution of another subprogram. Knowing that the use of SPARK prevented such problems was important.

- NATS, the UK's leading air traffic services provider, has pioneered research and development into new predictive (look-ahead) tools for air traffic controllers. iFACTS – the interim Future Area Control Tool Support – provides tools for trajectory prediction, conflict detection, and monitoring aids, and replaces traditional paper information strips with electronic data and new displays. iFACTS enables air traffic controllers to increase the amount of air traffic they can handle, providing capacity increases and enhancing safety.

  The iFACTS source code consists of over 200 KLOC of RavenSPARK source code, from which over 120,000 verification conditions are generated to prove exception-freedom. iFACTS was successfully deployed to the live operational environment at Swanwick's en-route centre between July and November 2011, and is now fully operational.

- A CubeSat is a small satellite, nominally a 10cm cube, that is typically launched into a low Earth orbit by piggybacking on a launch vehicle that is carrying a much larger satellite. CubeSats are usually built using commodity parts and are vey inexpensive to build and launch. As a result they have garnered much interest among academic and commercial satellite builders.

  Because of their small size and limited resources, CubeSats typically do not support any ability to modify the flight software after launch. Thus it is essential that the software be highly reliable. Unfortunately, many early CubeSat missions failed due to software problems. Vermont Technical College has been using SPARK to write CubeSat flight software since 2013 and has had success building reliable systems even using inexperienced student developers and very limited budgets.

Appendix 1

Reserved Words, etc.

This appendix lists the reserved words and predefined attributes, aspects, pragmas, and restriction identifiers. An implementation may not define additional reserved words but may define additional attributes, aspects, pragmas, and restriction identifiers.

The lists do not include obsolete items such as pragmas which are now replaced by aspect specifications. In particular, in the case of aspects such as Pure and Preelaborate which were given by library unit pragmas, the preferred style now is to use aspect specifications and so they are listed under aspects and not pragmas.

A1.1 Reserved words

The following 74 words are reserved; the index contains references to the places where their use is described.

abort
abs
abstract
accept
access
aliased
all
and
array
at

begin
body

case
constant

declare
delay
delta
digits
do

else
elsif
end
entry
exception
exit

for
function

generic
goto

if
in
interface
is

limited
loop

mod

new
not
null

of
or
others
out
overriding

package
parallel
pragma
private
procedure
protected

raise
range
record
rem
renames
requeue
return
reverse

select
separate
some
subtype
synchronized

tagged
task
terminate
then
type

until
use

when
while
with

xor

The reserved words **access**, **delta**, **digits**, **mod**, and **range** are also used as attributes; there is no conflict.

A1.2 Predefined attributes

This section lists all the predefined attributes. Most attributes are dealt with in detail in the body of this book and so the descriptions are generally brief.

X'Access: Applies to an object or subprogram. Yields an access value designating the entity. (See 11.4, 11.8)

X'Address: Applies to an object, program unit or label. Denotes the address of the first storage element associated with the entity. Of type System.Address. (See 25.1)

S'Adjacent: Applies to a floating point subtype S of a type T and denotes
  function S'Adjacent(X, Towards: T) **return** T;
which returns the machine number adjacent to X in the direction of Towards.

S'Aft: Applies to a fixed point subtype. Yields the number of decimal digits needed after the point to accommodate the subtype S, unless the delta of the subtype S is greater than 0.1, in which case it yields the value one. (S'Aft is the smallest positive integer N for which (10**N)*S'Delta is greater than or equal to one.) Of type *universal_integer.* (See 23.8)

X'Alignment: Applies to a subtype or object. Zero means that an object is not necessarily aligned. Otherwise, the Address of an object is an integer multiple of this attribute. Of type *universal_integer.* (See 25.1)

S'Base: Applies to a scalar subtype S and denotes an unconstrained subtype of its type. (See 27.2, also 17.1, 17.4)

S'Bit_Order: Applies to a record subtype and denotes the bit ordering. Of type System.Bit_Order. (See 25.1)

P'Body_Version: Applies to a program unit and returns a String. Used by the Distributed Systems annex.

T'Callable: Applies to a task. Yields true unless the task is completed, terminated or abnormal. Of type Boolean. (See 20.8)

E'Caller: Applies to an entry. Yields the identity of the task calling the entry body or accept statement. Of type Task_Identification.Task_ID. (See 26.1)

S'Ceiling: Applies to a floating point subtype S of a type T and denotes
  function S'Ceiling(X: T) **return** T;
which returns the algebraically smallest integral value not less than X. (17.4)

S'Class: Applies to a tagged subtype. Denotes its class wide type. (See 14.2)

X'Component_Size: Applies to an array subtype or object. Denotes the size in bits of components of the type. Of type *universal_integer.* (See 25.1)

S'Compose: Applies to a floating point subtype S of a type T and denotes
  function S'Compose(Fraction: T; Exponent: *universal_integer*) **return** T;
which in essence is Fraction but with its exponent replaced by Exponent. (17.4)

A'Constrained: Applies to an object of a discriminated type. Yields true if A is a constant or is constrained. Of type Boolean. (See 18.2)

S'Copy_Sign: Applies to a floating point subtype S of a type T and denotes
  function S'Copy_Sign(Value, Sign: T) **return** T;
which returns the magnitude of Value with the sign of Sign. (See 17.4)

E'Count: Applies to an entry. Yields the number of tasks queued on the entry. Of type *universal_integer*. (See 20.4, 20.5)

S'Definite: Applies to a generic formal indefinite subtype. Yields true if the actual subtype is definite. Of type Boolean. (See 19.2)

S'Delta: Applies to a fixed point subtype. Yields the value of the delta of the subtype. Of type *universal_real*. (See 17.5)

S'Denorm: Applies to a floating point subtype. Yields true if every denormalized number is a machine number. Of type Boolean.

S'Digits: Applies to a floating point or decimal subtype. Yields the requested number of decimal digits. Of type *universal_integer*. (See 17.4, 17.6)

S'Enum_Rep: Applies to a discrete subtype S and denotes

function S'Enum_Rep(Arg: S'Base) **return** *universal_integer*;

which returns the representation value of Arg. (See 25.1)

S'Enum_Val: Applies to a discrete subtype S and denotes

function S'Enum_Val(Arg: *universal_integer*) **return** S'Base;

which returns the value of type S corresponding to representation value Arg. (See 25.1)

S'Exponent: Applies to a floating point subtype S of a type T and denotes

function S'Exponent(X: T) **return** T;

which returns the normalized exponent of X. (See 17.4)

S'External_Tag: Applies to a tagged subtype. Yields an external representation of the tag. Of type String. (See 23.9)

A'First(N): Applies to an array object or constrained subtype. Denotes the lower bound of the Nth index range. Of the type of the bound. (See 8.1, 8.2)

A'First: Same as A'First(1).

S'First: Applies to a scalar subtype. Yields the lower bound of the range of S. Of the same type as S. (See 6.8)

R.C'First_Bit: Applies to a component C of a composite, non-array object R. If the nondefault bit ordering applies to the type and a component clause specifies the placement of C, denotes the value given for the first bit; otherwise denotes the offset, measured in bits, from the start of the first storage element occupied by C, of the first bit occupied by C. The first bit of a storage element is numbered zero. Of type *universal_integer*. (See 25.1)

S'First_Valid: Applies to a static discrete subtype. Denotes the smallest value that satisfies the predicates of S. Of the same type as S. (See 16.4)

S'Floor: Applies to a floating point subtype S of a type T and denotes

function S'Floor(X: T) **return** T;

which returns the algebraically largest integral value not greater than X. (17.4)

S'Fore: Applies to a fixed point subtype. Yields the minimum number of characters needed before the decimal point for the decimal representation of any value of the subtype S, assuming that the representation does not include an exponent, but includes a one-character prefix that is either a minus sign or a space. (This minimum number does not include superfluous zeros or underlines; and is at least two.) Of the type *universal_integer*. (See 23.8)

S'Fraction: Applies to a floating point subtype S of a type T and denotes
```
function S'Fraction(X: T) return T;
```
which in essence is the number X with exponent replaced by zero. (See 17.4)

X'Has_Same_Storage: Applies to an object and denotes
```
function X'Has_Same_Storage(Arg: any type) return Boolean;
```
which returns true if X occupies exactly the same bits as Arg. (See 25.1)

E'Identity: Applies to an exception. Yields the identity of the exception. Of type Exceptions.Exception_ID. (See 15.4)

T'Identity: Applies to a task. Yields the identity of the task. Of type Task_Identification.Task_ID. (See 26.1)

S'Image: Applies to a subtype S and denotes
```
function S'Image(Arg: S'Base) return String;
```
S'Image calls S'Put_Image passing Arg (which stores a sequence of characters in a text buffer) and then returns the result of retrieving the contents with function Get (See 23.4)

X'Image: Applies to a prefix X of type T (not *universal_real* or *universal_fixed*): X'Image denotes the result S'Image where S is nominal subtype of X. (See 23.4)

E'Index: Applies to an entry family E: Within a pre- or postcondition for family E denotes the entry index for the call of E. (See 16.2, 20.9)

S'Class'Input: Applies to a subtype S'Class of a class wide type T'Class and denotes
```
function S'Class'Input(Stream: not null access
                  Ada.Streams.Root_Stream_Type'Class) return T'Class;
```
Reads the external tag from Stream, determines the corresponding internal tag and then dispatches to the Input attribute. (See 23.9)

S'Input: Applies to a subtype S of a specific type T and denotes
```
function S'Input(Stream: not null access
                  Ada.Streams.Root_Stream_Type'Class) return T;
```
Reads and returns one value from the stream. (See 23.9)

A'Last(N): Applies to an array object or constrained subtype. Denotes the upper bound of the Nth index range. Of the type of the bound. (See 8.1, 8.2)

A'Last: Same as A'Last(1).

S'Last: Applies to a scalar subtype. Yields the upper bound of the range of S. Of the same type as S. (See 6.8)

R.C'Last_Bit: Applies to a component C of a composite, non-array object R. If the nondefault bit ordering applies to the type and a component clause specifies the placement of C, denotes the value given for the last bit; otherwise denotes the offset, measured in bits, from the start of the first storage element occupied by C, of the last bit occupied by C. Of type *universal_integer*. (See 25.1)

S'Last_Valid: Applies to a static discrete subtype. Denotes the largest value that satisfies the predicates of S. Of the same type as S. (See 16.4)

S'Leading_Part: Applies to a floating point subtype S of a type T and denotes
```
function S'Leading_Part(X: T; D: universal_integer) return T;
```
which in essence is the number X but with all except the first D digits in the mantissa set to zero.

A'Length(N): Applies to an array object or constrained subtype. Denotes the number of values of the Nth index range (zero for a null range). Of type *universal_integer*. (See 8.1, 8.2)

A'Length: Same as A'Length(1).

S'Machine: Applies to a floating point subtype S of a type T and denotes
function S'Machine(X: T) **return** T;
which is X if it is a machine number and otherwise one adjacent to X.

S'Machine_Emax: Applies to a floating point subtype. Yields the largest value of *exponent* in the canonical representation for which all numbers are machine numbers. Of type *universal_integer*. (See 17.4)

S'Machine_Emin: Applies to a floating point subtype. Yields the smallest value of *exponent* in the canonical representation for which all numbers are machine numbers. Of type *universal_integer*. (See 17.4)

S'Machine_Mantissa: Applies to a floating point subtype. Yields the largest number of digits in the mantissa in the canonical representation for which all numbers are machine numbers. Of type *universal_integer*. (See 17.4)

S'Machine_Overflows: Applies to a floating or fixed point subtype. Yields true if overflow and divide by zero raise Constraint_Error for every predefined operation returning a value of the base type of S. Of type Boolean. (See 17.4, 17.5)

S'Machine_Radix: Applies to a floating or fixed point subtype. Yields the radix used by the machine representation of the base type of S. Of type *universal_integer*. (See 17.4, 17.5, 17.6)

S'Machine_Rounding: Applies to a floating point subtype S of a type T and denotes
function S'Machine_Rounding(X: T) **return** T;
Returns the integral value nearest to X choosing either one if X is midway.

S'Machine_Rounds: Applies to a floating or fixed point subtype. Yields true if rounding is performed on inexact results of every predefined arithmetic operation returning a value of the base type of S. Of type Boolean.

S'Max: Applies to a scalar subtype and denotes
function S'Max(Left, Right: S'Base) **return** S'Base;
The result is the greater of the parameters. (See 6.8)

S'Max_Alignment_For_Allocation: Applies to any subtype. Denotes the maximum value for Alignment that could be requested by Allocate for an access type designating S. Of type System.Storage_Elements.Storage_Count. (See 25.1)

S'Max_Size_In_Storage_Elements: Applies to any subtype. Denotes the maximum value for Size_In_Storage_Elements that could be requested by Allocate for an access type designating S. Of type Storage_Count. (See 25.1)

S'Min: Applies to a scalar subtype and denotes
function S'Min(Left, Right: S'Base) **return** S'Base;
The result is the lesser of the parameters. (See 6.8)

S'Mod: Applies to a modular subtype and denotes
function S'Mod(Arg: *universal_integer*) **return** S'Base;
The result is Arg **mod** S'Modulus. (See 17.2)

S'Model: Applies to a floating point subtype S of a type T and denotes

function S'Model(X: T) **return** T;

which is X if it is a model number and otherwise one adjacent to X. This attribute and Model_Emin, Model_Epsilon, Model_Mantissa and Model_Small concern the floating point model. (See 17.4)

S'Modulus: Applies to a modular subtype. Yields its modulus. Of type *universal_integer*. (See 17.2)

S'Object_Size: Applies to a subtype S. If S is definite denotes the size in bits of a stand-alone aliased object or of a component provided no representation item is applied. Of type *universal_integer*. (See 25.1)

X'Old: Applies to an object of a nonlimited type. Can only be used in a postcondition and denotes the value of X when the subprogram was called. (See 16.2)

S'Class'Output: Applies to a subtype S'Class of a class wide type T'Class and denotes

procedure S'Class'Output(Stream: **not null access**
Ada.Streams.Root_Stream_Type'Class; Item: **in** T'Class);

Writes the external tag of Item to the Stream, and then dispatches to the Output attribute. (See 23.9)

S'Output: Applies to a subtype S of a specific type T and denotes

procedure S'Output (Stream: **not null access**
Ada.Streams.Root_Stream_Type'Class; Item: **in** T);

Writes the value of Item to the stream including bounds and discriminants. (See 23.9)

X'Overlaps_Storage: Applies to an object and denotes

function X'Overlaps_Storage(Arg: *any type*) return Boolean;

which returns true if X shares at least one bit with Arg. (See 25.1)

X'Parallel_Reduce(Reducer, Initial_Value): Applies to an array type X. Is a reduction expression with the value sequence

[**parallel for** Item **of** X => Item]. (See 22.8)

D'Partition_ID: Applies to a library level declaration (not pure). Denotes the partition in which the entity was elaborated. Of type *universal_integer*. Used by the Distributed Systems annex.

S'Pos: Applies to a discrete subtype and denotes

function S'Pos(Arg: S'Base) **return** *universal_integer*;

Returns the position number of Arg. (See 6.8)

R.C'Position: Applies to a component C of a composite, non-array object R. If the nondefault bit ordering applies to the type and a component clause specifies the placement of C, denotes the value given for the position of the component clause; otherwise denotes R.C'Address – R'Address. Of type *universal_integer*. (See 25.1)

S'Preelaborable_Initialization: Applies to a subtype S. Of type Boolean; indicates whether the type of S has preelaborable initialization. (See 13.8)

P'Priority: Applies to a protected object. Denotes the priority of P. (See 26.2)

S'Pred: Applies to a scalar subtype and denotes

```
function S'Pred(Arg: S'Base) return S'Base;
```

For an enumeration type, returns the value whose position number is one less than that of Arg. For an integer type, returns the result of subtracting one from Arg (with wraparound for a modular type). For a fixed type, returns the result of subtracting *small* from Arg. For a floating type, returns the machine number below Arg. The exception Constraint_Error is raised if appropriate. (See 6.8)

S'Put_Image: Applies to subtype S of T but not *universal_real* or *universal_fixed* and denotes

```
procedure S'Put_Image(
   Buffer: in out Ada.Strings.Text_Buffers.Root_Buffer_Type'Class;
   Arg: in T);
```

The default implementation writes an image using Wide_Wide_Put. (See 23.4)

A'Range(N): Applies to an array object or constrained subtype. Is equivalent to A'First(N) .. A'Last(N). (See 8.1, 8.2)

A'Range: Same as A'Range(1).

S'Range: Applies to a scalar subtype. Equivalent to S'First .. S'Last. (See 6.8)

S'Class'Read: Applies to a subtype S'Class of T'Class and denotes

```
procedure S'Class'Read(Stream: not null access
                 Ada.Streams.Root_Stream_Type'Class; Item: out T'Class);
```

Dispatches to the Read attribute according to the tag of Item. (See 23.9)

S'Read: Applies to a subtype S of a specific type T and denotes

```
procedure S'Read(Stream: not null access
                 Ada.Streams.Root_Stream_Type'Class; Item: out T);
```

Reads the value of Item from the stream. (See 23.9)

V'Reduce(Reducer, Initial_Value): Applies to a value sequence V. (See 9.6)

X'Reduce(Reducer, Initial_Value). Applies to an array type or iterable container object. (See 9.6)

P'Relative_Deadline: Applies to a projected object P. Denotes the relative deadline of P. Can only be used within the body of P. (See 26.2)

S'Remainder: Applies to a floating point subtype S of a type T and denotes

```
function S'Remainder(X, Y: T) return T;
```

Essentially returns $V = X - nY$ where n is the integer nearest to X/Y. (n is even if midway.)

F'Result: Applies to a function. Used only in a postcondition for F and denotes the result value returned by the call of F. (See 16.2)

S'Round: Applies to a decimal fixed point subtype and denotes

```
function S'Round(X: universal_real) return S'Base;
```

Returns the value obtained by rounding X (away from zero if X is midway between values of S'Base). (See 17.6)

S'Rounding: Applies to a floating point subtype S of a type T and denotes

```
function S'Rounding(X: T) return T;
```

Returns the integral value obtained by rounding X (away from zero for exact halves). (See 17.4)

S'Safe_First: Applies to a floating point subtype S of a type T. Yields the lower bound of the safe range of T. Of type *universal_real*. (See 17.4)

S'Safe_Last: Applies to a floating point subtype S of a type T. Yields the upper bound of the safe range of T. Of type *universal_real*. (See 17.4)

S'Scale: Applies to a decimal fixed point subtype. Yields N such that S'Delta = 10.00**(-N). Of type *universal_integer*. (See 17.6)

S'Scaling: Applies to a floating point subtype S of a type T and denotes

function S'Scaling(X: T; Adjustment: *universal_integer*) **return** T;

which in essence is the number X but with its exponent increased by Adjustment. (See 17.4)

S'Signed_Zeros: Applies to a floating point subtype S of a type T. Returns true if the hardware representation for T supports signed zeros. Of type Boolean. (See 17.4)

S'Size: Applies to any subtype. For a definite subtype gives the size in bits of a packed record component of the subtype; for an indefinite subtype is implementation defined. Of type *universal_integer*. (See 25.1)

X'Size: Applies to an object. Denotes its size in bits. Of type *universal_integer*. (See 25.1)

S'Small: Applies to a fixed point subtype. Denotes its *small*. Of type *universal_real*. (See 17.5)

S'Storage_Pool: Applies to an access to object subtype. Denotes its storage pool. Of type System.Storage_Pools.Root_Storage_Pool'Class. (See 25.4)

S'Storage_Size: Applies to an access to object subtype. Yields a measure of the number of storage elements reserved for its pool. Of type *universal_integer*. (See 25.1)

T'Storage_Size: Applies to a task. Yields the number of storage elements reserved for the task. Of type *universal_integer*. (See 25.1)

S'Stream_Size: Applies to an elementary subtype. Denotes the number of bits used for items of the subtype. Of type *universal_integer*. (See 23.9)

S'Succ: Applies to a scalar subtype and denotes

function S'Succ(Arg: S'Base) **return** S'Base;

For an enumeration type, returns the value whose position number is one more than that of Arg. For an integer type, returns the result of adding one to Arg (with wraparound for a modular type). For a fixed type, returns the result of adding *small* to Arg. For a floating type, returns the machine number above Arg. The exception Constraint_Error is raised if appropriate. (See 6.8)

S'Tag: Applies to a subtype S of a tagged type T. Denotes the tag of T. Of type Ada.Tags.Tag. (See 14.4)

X'Tag: Applies to an object of a class wide type. Denotes the tag of X. Of type Ada.Tags.Tag. (See 14.4)

T'Terminated: Applies to a task. Yields true if the task is terminated. Of type Boolean. (See 20.8)

S'Truncation: Applies to a floating point subtype S of a type T and denotes

```
function S'Truncation(X: T) return T;
```

Returns the integral value obtained by truncation towards zero. (See 17.4)

S'Unbiased_Rounding: Applies to a floating point subtype S of a type T and denotes

```
function S'Unbiased_Rounding(X: T) return T;
```

Returns the integral value nearest to X choosing the even one if X is midway. (See 17.4

X'Unchecked_Access: Applies to an aliased view of an object. As for X'Access but as if X were at the library level. (See 11.5)

S'Val: Applies to a discrete subtype and denotes

```
function S'Val(Arg: universal_integer) return S'Base;
```

Returns the value of S with position number Arg. (See 6.8)

X'Valid: Applies to a scalar object. Yields true if X is normal and has a valid representation. Of type Boolean. (See 25.2)

S'Value: Applies to a scalar subtype and denotes

```
function S'Value(Arg: String) return S'Base;
```

Returns the value corresponding to the given image, ignoring any leading or trailing spaces.

P'Version: Applies to a program unit and returns a String. Used by the Distributed Systems annex.

S'Wide_Image: Similar to Image but returns a Wide_String.

S'Wide_Value: Similar to Value but applies to a Wide_String.

S'Wide_Wide_Image: Similar to Image but returns a Wide_Wide_String.

S'Wide_Wide_Value: Similar to Value but applies to a Wide_Wide_String.

S'Wide_Wide_Width: Similar to Width for Wide_Wide_Image and Wide_Wide_String.

S'Wide_Width: Similar to Width for Wide_Image and Wide_String.

S'Width: Applies to a scalar subtype. Yields the length of a String returned by S'Image over all values of the subtype. Of type *universal_integer*. (See 23.8)

S'Class'Write: Applies to a subtype S'Class of a class wide type T'Class and denotes

```
procedure S'Class'Write(Stream: not null access
                   Ada.Streams.Root_Stream_Type'Class; Item: in T'Class);
```

Dispatches to the Write attribute according to the tag of Item. (See 23.9)

S'Write: Applies to a subtype S of a specific type T and denotes

```
procedure S'Write(Stream: not null access
                         Ada.Streams.Root_Stream_Type'Class; Item: in T);
```

Writes the value of Item to the stream. (See 23.9)

A1.3 Predefined aspects

The aspects listed here are mostly set by an aspect specification. In early versions of Ada, aspects were usually set by a pragma and these pragmas are now obsolete.

Note that the values of many aspects can be interrogated by a corresponding attribute. Thus the aspect Address can be set by an aspect specification such as

```
X: Integer
  with Address => ... ;
```

and then interrogated by X'Address. Some aspects (indeed such as Address) can also be set by an attribute definition clause thus

```
for X'Address use ... ;
```

To avoid unnecessary repetition aspects such as Address in the list below simply refer to the previous section.

Address: See A1.2.

Aggregate: Applies to containers. Describes how a container aggregate is structured. (See 24.2)

Alignment: See A1.2.

All_Calls_Remote: Applies to a library unit. Of type Boolean. Used by the Distributed Systems annex.

Allows_Exit: Applies to a subprogram. Enables control to exit the subprogram safely by for example a goto or exception. (See 23.10 and procedural iterators)

Asynchronous: Applies to a remote procedure. Of type Boolean. Specifies that the caller need not wait. (See 26.3)

Atomic: Applies to a type, object or component. Of type Boolean. Indicates that access is atomic. (See 22.2)

Atomic_Components: Applies to an array type or object. Of type Boolean. Indicates that access to individual components is atomic. (See 22.2)

Attach_Handler: Applies to a protected procedure. Of type Ada.Interrupts.Interrupt_Id. Indicates that the protected procedure handles the given interrupt. (See 26.1)

Bit_Order: See A1.2.

Component_Size: See A1.2.

Constant_Indexing: Applies to a tagged type. Specifies function(s) to be used to automate indexing. (See 21.6)

Convention: Applies to a type or subprogram. Specifies convention for interfacing. (See 25.5)

CPU: Applies to a task or task type. Of type System.Multiprocessors.CPU_Range. Indicates the CPUs to be used for the tasks. (See 22.2)

Default_Component_Value: Applies to an array subtype with scalar components. Of type of the array components. Specifies default value for the components. (See 8.2)

Default_Initial_Condition: Applies to a private type. Specifies default initial value. (See 16.2)

Default_Iterator: Applies to a container type. Specifies a function to be used for iteration. (See 21.6)

Default_Value: Applies to a scalar subtype. Of same type. Specifies default value for objects of the subtype. (See 6.8)

Discard_Names: Applies to types and exceptions. Allows tables of names to be discarded at run time. (See 26.1)

Dispatching: Applies to declarations in the High Integrity Annex. (See 26.6)

Dispatching_Domain: Applies to a task. Of type System.Multiprocessors. Dispatching_Domains.Dispatching_Domain. Specifies the domain on which a task runs. (See 22.2)

Dynamic_Predicate: Applies to a subtype. Of type Boolean. Indicates a condition that will hold for values of the subtype. (See 16.4)

Elaborate_Body: Applies to a library unit. Of type Boolean. Specifies that the body of the unit is to be elaborated immediately after its declaration (spec). (See 13.8)

Exclusive_Functions: Applies to protected types. Prevents parallel calls to protected functions. This was added by AI12-129. (See 20.4)

Export: Applies to objects and subprograms. Of type Boolean. Makes the Ada entity externally visible. (See 25.5)

External_Name: Applies to entities imported and exported. Specifies the name by which the entity is known externally.

External_Tag: See A1.2.

Full_Access_Only: Applies to a volatile entity. In Systems Programming annex. (See 26.1)

Global: Applies to units. Indicates global state that may be manipulated by a package or subprogram etc. (See 16.5)

Global'Class: As for Global.

Implicit_Dereference: Applies to a discriminated type. Of type of the discriminant. (See 21.6)

Import: Applies to objects and subprograms. Of type Boolean. Makes the external entity visible to the Ada program. (See 25.5)

Independent: Applies to a type, object or component. Of type Boolean. Indicates that access is independently addressable. (See 22.2)

Independent_Components: Applies to an array or record type or an array object. Of type Boolean. Indicates that access to individual components are independently addressable. (See 22.2)

Inline: Applies to subprograms, entries and generic subprograms. Of type Boolean. Specifies that inline expansion is requested for all calls. (See 5.6 and 25.1)

Input: See A1.2.

Integer_Literal: Applies to a type. Specifies a function to create a generalized integer literal. (See 5.6 and 23.6)

Interrupt_Handler: Applies to a protected procedure. Of type Boolean. Indicates that the protected procedure can be used as an interrupt handler. (See 26.1)

Interrupt_Priority: Applies to a task or protected type or object. Of type Integer. Specifies the priority which must be in the range applicable to interrupts. (See 26.2)

Iterator_Element: Applies to a container type. Specifies the type over which iteration occurs. (See 21.6)

Iterator_View: Applies to a container type. Specifies another type to be used for iterators. (See 24.2)

Link_Name: Applies to entities imported and exported. Specifies a name that can be used by link-time tools. (See 25.5)

Machine_Radix: See A1.2.

Max_Entry_Queue_Length. Applies to a task, protected type, or entry. Specifies maximum length of an entry queue. (See 20.2)

No_Controlled_Parts: Applies to a type. States that it and descendants do not have any controlled parts. Used by the High Integrity annex. (See 26.6)

No_Return: Applies to a procedure. Of type Boolean. Indicates that the procedure does not return normally. (See 15.5)

Nonblocking: Applies to packages, subprograms, and entries etc. Specifies that it does not block. (See 20.2)

Output: See A1.2.

Pack: Applies to a composite subtype. Of type Boolean. Specifies that storage minimization should be the main criterion when selecting the representation of the type. For a type extension only applies to the extension part. (See 25.1)

Parallel_Calls: Applies to a subprogram. Specifies that it can be safely called in parallel. (See 22.8)

Parallel_Iterator: Applies to a loop. Specifies that it can be processed by parallel threads. (See 22.8)

Preelaborable_Initialization: Specifies that all objects of the type have preelaborable initialization expressions. (See 13.8)

Preelaborate: Applies to a library unit. Of type Boolean. Specifies that the unit is to be preelaborated. (See 13.8)

Post: Applies to a subprogram. Of type Boolean. Specifies an expression which should be true on return from the subprogram. (See 16.2)

Post'Class: Applies to a subprogram. Specifies a postcondition that is inherited on type derivation. (See 16.2)

Pre: Applies to a subprogram. Of type Boolean. Specifies an expression which should be true before calling the subprogram. (See 16.2)

Pre'Class: Applies to a subprogram. Specifies a precondition that is inherited on type derivation. (See 16.2)

Predicate_Failure: Applies to a predicate and indicates exception to be raised and/or a message. (See 16.5)

Priority: Applies to a task or protected object or type. Of type Integer. Specifies the priority which must not be in the range applicable to interrupts. Note that the value of the priority of a task cannot be interrogated using the attribute Priority. (See 26.2)

Pure: Applies to a library unit. Of type Boolean. Specifies that the unit is pure. (See 13.8)

Put_Image: Applies to a type. Defines a procedure that sends an image to a text buffer. (See 23.4)

Read: See A1.2.

Real_Literal: Applies to a type. Specifies a function to create a generalized real literal. (See 5.6 and 23.6)

Relative_Deadline: Applies to a task. Of type Ada.Real_Time.Time_Span. Specifies task deadline for EDF scheduling. (See 26.2)

Remote_Call_Interface: Applies to a library unit. Of type Boolean. Used by Distributed Systems annex. (See 26.3)

Remote_Types: Applies to a library unit. Of type Boolean. Used by Distributed Systems annex. (See 26.3)

Shared_Passive: Applies to a library unit. Of type Boolean. Used by Distributed Systems annex. (See 26.3)

Size: See A1.2.

Small: See A1.2.

Stable_Properties: Applies to a type. Lists functions that do not change often. (See 16.5 and 24.2)

Stable_Properties'Class:: Ditto.

Static: Applies to an expression function States that it can be used in a static expression. (See 27.1)

Static_Predicate: Applies to a static subtype. Of type Boolean. Specifies a condition that must be true for values of the subtype. (See 16.4)

Storage_Pool: See A1.2.

Storage_Size: See A1.2.

Stream_Size: See A1.2.

String_Literal: Applies to a type. Specifies a function to create a generalized string literal. (See 5.6)

Synchronization: Applies to an operation of a synchronized interface. Specifies whether the operation has to be implemented by an entry or protected procedure. (See 22.3)

Type_Invariant: Applies to a private type. Of type Boolean. Specifies a condition that should be true for all externally visible values of the type. (See 16.3)

Type_Invariant'Class: Applies to a private type. Specifies a condition that is inherited on type derivation. (See 16.3)

Unchecked_Union: Applies to an unconstrained record subtype with a variant part. Of type Boolean. Specifies that objects of the subtype do not include storage space for discriminants. (See 25.5)

Use_Formal: Applies to a generic parameter. In the High Integrity Systems annex. (See 26.6)

Variable_Indexing: Applies to a tagged type. Specifies function(s) to be used to automate indexing. (See 21.6)

Volatile: Applies to a type, object, or component. Of type Boolean. Indicates that it is volatile. (See 22.2)

Volatile_Components: Applies to an array type or object. Of type Boolean. Indicates that individual components are volatile. (See 22.2)

Write: See A1.2.

Yield: Applies to a subprogram etc. Ensures that it has a task dispatching point. (See 26.2)

Some aspects are always set by special syntax. These are the internal representation of enumeration types and the layout of record components. (See 25.1)

A1.4 Predefined pragmas

Pragmas are of various categories. Configuration pragmas appear at the beginning of a compilation and apply to all units in it; configuration pragmas compiled alone apply to all further units compiled into the library and so apply to the remainder (which might be all) of the partition. Program unit pragmas immediately follow or are immediately within the unit to which they apply; in the latter case the argument is optional.

The following pragmas are predefined. The syntax of the parameters is given in the same style as the general syntax in Appendix 3.

pragma Admission_Policy(*policy*_identifier);
Applies to Real_Time Systems annex. (See 26.2)

pragma Assert([Check =>] *Boolean*_expression
[, [Message =>] string_expression]);
If the Boolean expression is False then the exception Assertion_Error is raised with the message if any. (See 15.5, 16.2)

pragma Assertion_Policy(*policy*_identifier);
pragma Assertion_Policy(*assertion*_aspect_mark => *policy*_identifier
{, *assertion*_aspect_mark => *policy*_identifier});
Controls the pragma Assert according to the policy identifier which can be Check, Ignore or implementation-defined. (See 15.5, 16.2)

pragma Conflict_Check_Policy(policy_identifier[, policy_identifier]);
Controls checking of tasking and parallel operations. (See 22.9)

pragma Default_Storage_Pool(storage_pool_indicator);
Specifies storage pool to be used by default. (See 25.4)

pragma Detect_Blocking;
Requires the implementation to detect potentially blocking operations. (See 26.6)

pragma Elaborate(*library_unit*_name {, *library_unit*_name});
This pragma is only allowed in the context clause of a compilation unit. Each argument must be the name of a library unit mentioned by the context clause. This pragma specifies that the corresponding library unit bodies must be elaborated before the current compilation unit. (See 13.8)

pragma Elaborate_All(*library_unit*_name {, *library_unit*_name});
As Elaborate but is transitive and specifies that every unit needed by the named units is elaborated before the current unit. (See 13.8)

pragma Generate_Deadlines;
Causes task deadlines to be recomputed. (See 26.2)

pragma Inspection_Point[(*object*_name {, *object*_name})];
Ensures that the named objects are implemented in a way that permits analysis. (See 26.6)

pragma Linker_Options(*string*_expression);
Applies to the immediately enclosing compilation unit. The string is passed to the system linker for partitions including the unit. The effect is implementation-defined.

pragma List(identifier);
Takes one of the identifiers On or Off as argument. Specifies that listing of the compilation is to be continued or suspended until a List pragma with the opposite argument is given within the same compilation. The pragma itself is always listed if the compiler is producing a listing.

pragma Locking_Policy(*policy*_identifier);
Specifies a policy such as Ceiling_Locking. (See 26.2)

pragma Normalize_Scalars;
Ensures that all scalar objects have an initial value. (See 26.6)

pragma Optimize(identifier);
Takes one of Time, Space, or Off as argument. Specifies whether time or space is the primary optimization criterion or that optimization should be turned off. It applies until the end of the immediately enclosing declarative region or the end of the compilation. (See 5.6)

pragma Page;
Specifies that the program text which follows the pragma should start on a new page (if the compiler is currently producing a listing).

pragma Partition_Elaboration_Policy(*policy*_identifier);
Controls activation of library tasks. (See 26.6)

pragma Priority_Specific_Dispatching(*policy*_identifier,
*first_priority*_expression, *last_priority*_expression);
Specifies dispatching policy to be used for a range of priorities. (See 26.2)

pragma Profile(*profile*_identifier {, *profile*_pragma_argument_association});
Specifies a profile to be enforced; for example the Ravenscar profile. (See 22.7)

pragma Queuing_Policy(*policy*_identifier);
Specifies the policy to be used for entry queues. (See 26.2)

pragma Restrictions(restriction {, restriction});
where:

restriction ::= *restriction*_identifier
| *restriction_parameter*_identifier => restriction_parameter_argument
restriction_parameter_argument ::= name | expression

This is usually a configuration pragma and asserts that the restrictions given apply to the partition concerned. Used mainly by the Real-Time Systems and High Integrity Systems annexes. (See 13.8, 26.2 and 26.6)

pragma Reviewable;
Directs compiler to generate reviewable object code. (See 26.6)

pragma Suppress(identifier);
Applies to the identifier of a check. Allowed as a configuration pragma or immediately within a declarative part or immediately within a package specification. Permission to omit the given check extends from the place of the pragma to the end of the innermost enclosing declarative region. (See 15.5)

pragma Task_Dispatching_Policy(*policy*_identifier);
Specifies the dispatching policy for a whole partition. (See 26.2)

pragma Unsuppress(identifier);
Similar to Suppress but revokes any permission to omit checks. (See 15.5)

A1.5 Predefined restrictions

This section lists the restriction identifiers used with the pragma Restrictions that are predefined. The restriction often applies just to the current compilation not the entire partition.

Restrictions have a number of purposes. They can ensure that a program is portable by forbidding the use of implementation defined features. They can enable a small run-time system to be used for critical applications. They can ensure that a program is simple so that proof tools can be used. These lists should give the reader some appreciation of the scope of the facility. For further details consult the *ARM*.

In the core language there are some blanket restrictions

No_Implementation_Aspect_Specifications, No_Implementation_Attributes,
No_Implementation_Identifiers, No_Implementation_Pragmas,
No_Implementation_Units, No_Obsolescent_Features,
No_Unrecognized_Aspects, No_Unrecognized_Pragmas

and then there are some restrictions on the use of particular entities

No_Dependence => *some_library_unit*,
No_Specification_Of_Aspect => *some_aspect*,
No_Use_Of_Attribute => *some_attribute*,
No_Use_Of_Pragma => *some_pragma*.

Note that writing

```
pragma Profile(No_Implementation_Extensions);
```

is equivalent to

```
pragma Restrictions(No_Implementation_Aspect_Specifications,
                    No_Implementation_Attributes,
                    No_Implementation_Identifiers,
                    No_Implementation_Pragmas,
                    No_Implementation_Units);
```

This profile and the Ravenscar and Jorvik profiles are the only predefined profiles.

Note also that it is illegal to try to be clever and write things like

```
pragma Restrictions(No_Use_Of_Pragma => Restrictions);
```

because of the obvious paradox.

A number of restrictions are defined in the Real-Time systems annex; the blanket ones are

```
No_Abort_Statements, No_Dynamic_Attachment,
No_Dynamic_CPU_Assignment, No_Dynamic_Priorities,
No_Implicit_Heap_Allocations, No_Local_Protected_Objects,
No_Local_Timing_Events, No_Nested_Finalization,
No_Protected_Type_Allocators, No_Relative_Delay,
No_Requeue_Statements, No_Select_Statements,
No_Specific_Termination_Handlers,
No_Standard_Allocators_After_Elaboration, No_Task_Allocators,
No_Task_Hierarchy, No_Tasks_Unassigned_To_CPU, No_Task_Termination,
No_Terminate_Alternatives, Pure_Barriers, Simple_Barriers.
```

and then there are restrictions giving a limit

```
Max_Asynchronous_Select_Nesting => N, Max_Entry_Queue_Length => N,
Max_Protected_Entries => N, Max_Select_Alternatives => N,
Max_Storage_At_Blocking => N, Max_Task_Entries => N, Max_Tasks => N.
```

Finally, a number of blanket restrictions are defined in the High-Integrity systems annex thus

```
No_Access_Parameter_Allocators, No_Access_Subprograms,
No_Anonymous_Allocators, No_Allocators, No_Coextensions, No_Delay,
No_Dispatch, No_Exceptions, No_Fixed_Point, No_Floating_Point,
No_Hidden_Indirect_Globals, No_IO, No_Local_Allocators,
No_Protected_Types, No_Recursion, No_Reentrancy,
No_Unchecked_Access, No_Unspecified_Globals.
```

and then there is one restriction giving a limit

```
Max_Image_Length => N
```

Checklist A1

An implementation may not define additional reserved words.

Three reserved words were added in Ada 2005, namely: **interface**, **overriding**, **synchronized**.

The reserved words **access**, **delta**, **digits**, **mod**, and **range** are also used as attributes.

The word **some** was also reserved in Ada 2012.

Many pragmas such as Inline were made obsolete in Ada 2012 and replaced by aspect specifications.

The following attributes were new in Ada 2012: First_Valid, Has_Same_Storage, Last_Valid, Max_Alignment_For_Allocation, Old, Overlaps_Storage, Result.

The following aspects were new in Ada 2012: Constant_Indexing, CPU, Default_Component_Value, Default_Iterator, Default_Value, Dispatching_Domain, Dynamic_Predicate, Exclusive_Functions, Implicit_Dereference, Iterator_Element, Post, Post'Class, Pre, Pre'Class, Static_Predicate, Synchronization, Type_Invariant, Type_Invariant'Class, Variable_Indexing.

The following pragma was new in Ada 2012: Default_Storage_Pool.

The following restriction identifiers were new in Ada 2012: No_Access_Parameter_Allocators, No_Anonymous_Allocators, No_Coextensions, No_Dynamic_CPU_Assignment, No_Implementation_Aspect_Specifications, No_Implementation_Identifiers, No_Implementation_Units, No_Specification_Of_Aspect, No_Standard_Allocators_After_Elaboration, No_Use_Of_Attribute, No_Use_Of_Pragma.

New in Ada 2022

The word **parallel** is also reserved in Ada 2022.

The following aspects are new: Aggregate, Allows_Exit, Default_Initial_Condition, Dispatching, Full_Access_Only, Global, Global'Class, Integer_Literal, Iterator_View, Max_Entry_Queue_Length, No_Controlled_Parts, Nonblocking, Parallel_Calls, Parallel_Iterator, Preelaborable_Initialization, Put_Image, Real_Literal, Stable_Properties, Stable_Properties'Class, Static, String_Literal, Use_Formal, Yield.

The following attributes are new: Enum_Rep, Enum_Val, Index, Object_Size, Parallel_Reduce, Preelaborable_Initialization, Put_Image, Reduce, Relative_Deadline.

The following pragmas are new: Admission_Policy, Conflict_Check_Policy, Generate_Deadlines.

The following restriction identifiers are new: Max_Image_Length, No_Hidden_Indirect_Globals, No_Tasks_Unassigned_To_CPU, No_Unrecognized_Aspects, No_Unrecognized_Pragmas, No_Unspecified_Globals, Pure_Barriers.

Appendix 2

Glossary

The following glossary of important terms is adapted from the *ARM*.

Abstract type An abstract type is a tagged type intended for use as an ancestor of other types, but which is not allowed to have objects of its own.

Access type An access type has values that designate aliased objects. Access types correspond to 'pointer types' or 'reference types' in some other languages.

Aggregate An aggregate is a construct used to define a value of a composite type by specifying the values of the components of the type.

Aliased An aliased view of an object is one that can be designated by an access value. Objects allocated by allocators are aliased. Objects can also be explicitly declared as aliased with the reserved word **aliased**. The Access attribute can be used to create an access value designating an aliased object.

Ancestor An ancestor of a type is the type itself or, in the case of a type derived from other types, its parent type or one of its progenitor types or one of their ancestors. Note that ancestor and descendant are inverse relationships.

Array type An array type is a composite type whose components are all of the same type. Components are selected by indexing.

Aspect An aspect is a specifiable property of an entity. An aspect may be specified by an aspect specification on the declaration of the entity. Some aspects may be queried via attributes.

Assertion An assertion is a Boolean expression that appears in any of the following: a pragma Assert, a predicate, a precondition, a postcondition, an invariant, a constraint, or a null exclusion. An assertion is expected to be True at run time at certain specified places.

Attribute An attribute is a characteristic or property of an entity that can be queried, and in some cases specified.

Category (of types) A category of types is a set of types with one or more common properties, such as primitive operations. A category of types that is closed under derivation is also known as a class.

Character type A character type is an enumeration type whose values include characters.

Check A check is a test made during execution to determine whether a language rule has been violated.

Class (of types) A class of types is a set of types that is closed under derivation, which means that if a given type is in the class, then all types derived from that type are also in the class. The set of types of a class share common properties, such as their primitive operations.

Compilation unit The text of a program can be submitted to the compiler in one or more compilations. Each compilation is a succession of compilation units. A compilation unit contains either the declaration, the body, or a renaming of a program unit.

Composite type A type which may have components such as an array or record.

Construct A construct is a piece of text (explicit or implicit) that is an instance of a syntactic category.

Container A container is an object that contains other objects all of the same type, which could be class wide. Several predefined container types are provided by the children of the package Ada.Containers.

Controlled type A controlled type supports user-defined assignment and finalization. Objects are always finalized before being destroyed.

Default initial condition A default initial condition is a property that holds for every default initialized object of a given type.

Declaration A declaration is a language construct that associates a name with (a view of) an entity. A declaration may appear explicitly in the program text (an explicit declaration), or may be supposed to occur at a given place in the text as a consequence of the semantics of another construct (an implicit declaration).

Derived type A derived type is a type defined in terms of one or more other types given in a derived type definition. The first of those types is the parent type of the derived type and any others are progenitor types. Each class containing the parent type or a progenitor type also contains the derived type. The derived type inherits properties such as components and primitive operations from the parent and progenitors. A type together with the types derived from it (directly or indirectly) form a derivation class.

Descendant A type is a descendant of itself, its parent and progenitor types, and their ancestors. Note that descendant and ancestor are inverse relationships.

Discrete type A discrete type is either an integer type or an enumeration type. Discrete types may be used, for example, in case statements and as array indices.

Discriminant A discriminant is a parameter for a composite type. It can control, for example, the bounds of a component of the type if the component is an array. A discriminant for a task type can be used to pass data to a task of the type upon its creation.

Elaboration Elaboration is the process by which a declaration has its run-time effect. Elaboration is one of the forms of execution.

Elementary type An elementary type is a type that does not have components.

Enumeration type An enumeration type is defined by an enumeration of its values, which may be named by identifiers or character literals.

Evaluation The process by which an expression has its run-time effect is called evaluation. Evaluation is one of the forms of execution.

Exception An exception represents a kind of exceptional situation; an occurrence of such a situation (at run time) is called an exception occurrence. To raise an exception is to abandon normal program execution so as to draw attention to the fact that the corresponding situation has arisen. Performing some actions in response to the arising of an exception is called handling the exception.

Execution The process by which a construct achieves its run-time effect is called execution. Execution of a declaration is also called elaboration. Execution of an expression is also called evaluation.

Function A function is a form of subprogram that returns a result and can be called as part of an expression.

Generic unit A generic unit is a template for a (nongeneric) program unit; the template can be parameterized by objects, types, subprograms, and packages. An instance of a generic unit is created by a generic instantiation. The rules of the language are enforced when a generic unit is compiled, using a generic contract model; additional checks are performed upon instantiation to verify the contract is met. Generic units can be used to perform the role that macros sometimes play in other languages.

Incomplete type An incomplete type gives a view of a type that reveals only some of its properties. The remaining properties are provided by the full view given elsewhere. Incomplete types can be used for defining recursive data structures.

Indexable container type An indexable container type is one that has user defined behaviour for indexing, via the Constant_Indexing or Variable_Indexing aspects.

Integer type Integer types comprise the signed integer types and the modular types. A signed integer type has a base range that includes both positive and negative numbers, and has operations that may raise an exception when the result is outside the base range. A modular type has a base range whose lower bound is zero, and has operations with 'wraparound' semantics. Modular types subsume what are called 'unsigned types' in some other languages.

Interface type An interface type is a form of abstract tagged type which has no components or concrete operations except possibly null procedures. Interface types are used for composing other interfaces and tagged types and thereby provide multiple inheritance. Only an interface type can be used as a progenitor of another type.

Invariant An invariant is an assertion that is expected to be True for all objects of a given private type when viewed from outside the defining package.

Iterable container type An iterable container type is one that has user defined behaviour for iteration, via the Default_Iterator and Iterator_Element aspects.

Iterator An iterator is a construct that is used to loop over the elements of an array or container. Iterators my be user defined and may perform arbitrary computations to access elements from a container.

Library unit A library unit is a separately compiled program unit, and is always a package, subprogram, or generic unit. Library units may have other (logically nested) library units as children, and may have other program units physically nested within them. A root library unit, together with its children and grandchildren and so on, form a subsystem.

Limited type A limited type is a type for which copying (such as in an assignment statement) is not allowed. A nonlimited type is a type for which copying is allowed.

Object An object is either a constant or a variable. An object contains a value. An object is created by an object declaration or by an allocator. A formal parameter is (a view of) an object. A subcomponent of an object is an object.

Operational aspect An operational aspect is an aspect that indicates a logical property of an entity, such as the precondition of a subprogram, or the procedure used to write a given type of object to a stream.

Overriding operation An overriding operation is one that replaces an inherited primitive operation. Operations may be marked explicitly as overriding or not overriding.

Package Packages are program units that allow the specification of groups of logically related entities. Typically, a package contains the declaration of a type (often a private type or private extension) along with the declarations of primitive subprograms of the type, which can be called from outside the package, while their inner workings remain hidden from outside users.

Parallel construct A parallel construct is an executable construct that defines multiple activities of a single task that can proceed in parallel, via the execution of multiple logical threads of control.

Parent The parent of a derived type is the first type given in the definition of the derived type. The parent can be almost any kind of type, including an interface type.

Partition A partition is a part of a program. Each partition consists of a set of library units. Each partition may run in a separate address space, possibly on a separate computer. A program may contain just one partition, or it can be distributed across multiple partitions, which can execute concurrently.

Postcondition A postcondition is an assertion that is expected to be True when a given subprogram returns normally.

Pragma A pragma is a compiler directive. There are language-defined pragmas that give instructions for optimization, listing control, etc. An implementation may support additional (implementation-defined) pragmas.

Precondition A precondition is an assertion that is expected to be True when a given subprogram is called.

Predicate A predicate is an assertion that is expected to be True for all objects of a given subtype.

Primitive operations The primitive operations of a type are the operations (such as subprograms) declared together with the type declaration. They are inherited by other types in the same class of types. For a tagged type, the primitive subprograms are dispatching subprograms, providing run-time polymorphism. A dispatching subprogram may be called with statically tagged operands, in which case the subprogram body invoked is determined at compile time. Alternatively, a dispatching subprogram may be called using a dispatching call, in which case the subprogram body invoked is determined at run time.

Private extension A private extension is a type that extends another type, with the additional properties hidden from its clients.

Private type A private type gives a view of a type that reveals only some of its properties. The remaining properties are provided by the full view given elsewhere. Private types can be used for defining abstractions that hide unnecessary details from their clients.

Procedure A procedure is a form of subprogram that does not return a result and can only be invoked by a statement.

Progenitor A progenitor of a derived type is one of the types given in the definition of the derived type other than the first. A progenitor is always an interface type. Interfaces, tasks and protected types may also have progenitors.

Program A program is a set of partitions, each of which may execute in a separate address space, possibly on a separate computer. A partition consists of a set of library units.

Program unit A program unit is either a package, a task unit, a protected unit, a protected entry, a generic unit, or an explicitly declared subprogram other than an enumeration literal. Certain kinds of program units can be separately compiled. Alternatively, they can appear physically nested within other program units.

Protected type A protected type is a composite type whose components are accessible only through one of its protected operations which synchronize concurrent access by multiple tasks.

Real type A real type has values that are approximations of the real numbers. Floating point and fixed point types are real types.

Record extension A record extension is a type that extends another type by optionally adding additional components.

Record type A record type is a composite type consisting of zero or more named components, possibly of different types.

Reference type A reference type is one that has user defined behaviour for '**.all**', defined by the Implicit_Dereference aspect.

Renaming A renaming declaration is a declaration that does not define a new entity, but instead defines a view of an existing entity.

Representation aspect A representation aspect is an aspect that indicates how an entity is mapped onto the underlying hardware, for example the size or alignment of an object.

Scalar type A scalar type is either a discrete type or a real type.

Stable property A stable property of a type is a characteristic of objects of the type that is preserved by many of the primitive operations of the type.

Storage pool Each access to object type has an associated storage pool object. The storage for an object created by an allocator comes from the storage pool of the type of the allocator. Some storage pools may be partitioned into subpools in order to support finer grained storage management.

Stream A stream is a sequence of elements that can be used, along with the stream oriented attributes, to support marshalling and unmarshalling of values of most types.

Subprogram A subprogram is a unit of a program that can be executed in various contexts. It is invoked by a subprogram call that may qualify the effect of the subprogram through the passing of parameters. There are two forms of subprogram: functions, which return values, and procedures, which do not.

Subtype A subtype is a type together with optional constraints, null exclusions, and predicates which constrain the values of the subtype to satisfy certain conditions. The values of a subtype are a subset of the values of its type.

Subunit A subunit is a body of a program unit that can be compiled separately from its enclosing program unit.

Suppress To suppress a check is to assert that it cannot fail, and to request that the compiler optimize by disabling the check. The compiler is not required to honour this request. Suppressing checks that can fail can cause programs to behave in arbitrary ways.

Synchronized Informally, a synchronized entity is one that will work safely with multiple tasks at one time. A synchronized interface can be an ancestor of a task or a protected type. Such a task or protected type is called a synchronized tagged type.

Tagged type The objects of a tagged type have a run-time type tag, which indicates the specific type with which the object was originally created. An operand of a class wide tagged type can be used in a dispatching call; the tag indicates which subprogram body to invoke. Nondispatching calls, in which the subprogram body to invoke is determined at compile time, are also allowed. Tagged types may be extended with additional components.

Task type A task type is a composite type used to represent active entities which execute concurrently and which can communicate via queued task entries. The top level task of a partition is called the environment task.

Type Each object has a type. A type has an associated set of values, and a set of primitive operations which implement the fundamental aspects of its semantics. Types are grouped into categories. Most language-defined categories of types are also classes of types.

View A view of an entity reveals some or all of the properties of the entity. A single entity may have multiple views. A partial view is a view that reveals only some of its properties. The full view reveals all of its properties.

Answers to Exercises

Specimen answers to the exercises in Part 1 are given here; answers to them all are on the website. In some cases they do not necessarily represent the best technique for solving a problem but merely one which uses the material introduced at that point in the discussion.

Answers 2

Exercise 2.2

1
```
package Simple_Maths is
   function Sqrt(F: Float) return Float;
   function Log(F: Float) return Float;
   function Ln(F: Float) return Float;
   function Exp(F: Float) return Float;
   function Sin(F: Float) return Float;
   function Cos(F: Float) return Float;
end Simple_Maths;
```

The first few lines of the program Print_Roots could now become

```
with Simple_Maths, Simple_IO;
procedure Print_Roots is
   use Simple_Maths, Simple_IO;
```

Exercise 2.4

1
```
for I in 0 .. N loop
   Pascal(I) := Next(I);
end loop;
```

2
```
for N in 0 .. 10 loop
   Pascal2(N, 0) := 1;
   for I in 1 .. N–1 loop
      Pascal2(N, I) := Pascal2(N–1, I–1) +
                                 Pascal2(N–1, I);
   end loop;
   Pascal2(N, N) := 1;
end loop;
```

3
```
type Month_Name is (Jan, Feb, Mar, Apr, May,
                    Jun, Jul, Aug, Sep, Oct, Nov, Dec);

type Date is
   record
      Day: Integer;
      Month: Month_Name;
      Year: Integer;
   end record;

Today: Date;
...
Today := (24, May, 1819);
```

Answers 3

Exercise 3.1

1
```
package Buffer_System is             -- visible part
   type Buffer is private;
   Buffer_Error: exception;
   procedure Load(B: in out Buffer; S: in String);
   procedure Get(B: in out Buffer;
                        C: out Character);
   function Is_Empty(B: Buffer) return Boolean;
private                              -- private part
   Max: constant Integer := 80;
   type Buffer is
      record
         Data: String(1 .. Max);
         Start: Integer := 1;
         Finish: Integer := 0;
      end record;
end Buffer_System;
```

```
package body Buffer_System is
   procedure Load(B: in out Buffer; S: in String) is
   begin
      if S'Length > Max or B.Start <= B.Finish then
         raise Buffer_Error;
      end if;
      B.Start := 1;
      B.Finish := S'Length;
      B.Data(B.Start .. B.Finish) := S;
   end Load;

   procedure Get(B: in out Buffer;
                 C: out Character) is
   begin
      if B.Start > B.Finish then
         raise Buffer_Error;
      end if;
      C := B.Data(B.Start);
      B.Start := B.Start + 1;
   end Get;

   function Is_Empty(B: Buffer) return Boolean is
   begin
      return B.Start > B.Finish;
   end Is_Empty;
end Buffer_System;
```

The parameter Buffer of Load now has to be **in out** because the original value is read. Also, we could replace the test in Get by

```
      if Is_Empty(B) then
```

Exercise 3.2

1

```
package Objects is
   type Object is tagged
      record
         X_Coord: Float;
         Y_Coord: Float;
      end record;

   function Distance(O: Object) return Float;
   function Area(O: Object) return Float;
end Objects;

package body Objects is
   function Distance(O: Object) return Float is
   begin
      return Sqrt(O.X_Coord**2 + O.Y_Coord**2);
   end Distance;

   function Area(O: Object) return Float is
   begin
      return 0.0;
   end Area;
end Objects;
```

```
with Objects;  use Objects;
package Shapes is
   type Circle is new Object with
      record
         Radius: Float;
      end record;

   function Area(C: Circle) return Float;

   type Point is new Object with null record;

   type Triangle is new Object with
      record
         A, B, C: Float;
      end record;

   function Area(T: Triangle) return Float;
end Shapes;

package body Shapes is
   function Area(C: Circle) return Float is
   begin
      return π * C.Radius**2;
   end Area;

   function Area(T: Triangle) return Float is
      S: constant Float := 0.5 * (T.A + T.B + T.C);
   begin
      return Sqrt(S * (S - T.A) * (S - T.B) * (S - T.C));
   end Area;
end Shapes;
```

Note that we can put the use clause for Objects immediately after the with clause.

Exercise 3.3

1

```
procedure Add_To_List(The_List: in out List;
                      Obj_Ptr: in Pointer) is
   Local: List := new Cell;
begin
   Local.Next := The_List;
   Local.Element := Obj_Ptr;
   The_List := Local;
end Add_To_List;
```

or more briefly using a form of allocation with initial values

```
procedure Add_To_List(The_List: in out List;
                      Obj_Ptr: in Pointer) is
begin
   The_List := new Cell'(The_List, Obj_Ptr);
end Add_To_List;
```

2

```
package body Objects is
   function Distance(O: Object) return Float is
   begin
      return Sqrt(O.X_Coord**2 + O.Y_Coord**2);
   end Distance;
end Objects;
```

3 We have to add the function Area for the type Point.

4 We cannot declare the function Moment for the abstract type Object because it contains a call of the abstract function Area.

5
```
function MO(OC: Object'Class) return Float is
begin
   return MI(OC) + Area(OC) * Distance(OC)**2;
end MO;
```

Answers 4

Exercise 4.2

1 The default field is 6 for a 16-bit type Integer and 11 for a 32-bit type Integer so

```
Put(123);        -- "sss123" and "ssssssss123"
Put(-123);       -- "ss-123" and "sssssss-123"
```

Exercise 4.4

1
```
with Ada.Text_IO, Etc;
use Ada.Text_IO, Etc;
procedure Table_Of_Square_Roots is
   use My_Float_IO, My_Elementary_Functions;
   Last_N: Integer;
   Tab: Count;
begin
   Tab := 10;
   Put("What is the largest value please? ");
   Get(Last_N);
   New_Line(2);
   Put("Number"); Set_Col(Tab);
   Put("Square root");
   New_Line(2);
   for N in 1 .. Last_N loop
      Put(N, 4); Set_Col(Tab);
      Put(Sqrt(My_Float(N)), 3, 6, 0);
      New_Line;
   end loop;
end Table_Of_Square_Roots;
```

2
```
with Ada.Text_IO;
package My_Numerics.My_Float_IO is
      new Ada.Text_IO.Float_IO(My_Float);

with Ada.Text_IO;
package My_Numerics.My_Integer_IO is
      new Ada.Text_IO.Integer_IO(My_Integer);

with Ada.Numerics.Generic_Elementary_Functions;
package My_Numerics.My_Elementary_Functions is
      new Ada.Numerics.
            Generic_Elementary_Functions(My_Float);
```

3
```
package Objects is ...

with Ada.Numerics.Elementary_Functions;
use Ada.Numerics.Elementary_Functions;
package body Objects is ...

with Objects; use Objects;
package Shapes is ...

with Ada.Numerics.Elementary_Functions;
use Ada.Numerics.Elementary_Functions;
package body Shapes is ...

with Shapes; use Shapes;
with Ada.Text_IO, Ada.Float_Text_IO;
use Ada.Text_IO, Ada.Float_Text_IO;
procedure Area_Of_Triangle is
   T: Triangle;
begin
   Get(T.A); Get(T.B); Get(T.C);
   Put(Area(T));
end Area_Of_Triangle;
```

We should really check that the sides do form a triangle, if they do not then the call of Sqrt in Area will have a negative parameter and so raise Ada.Numerics.Argument_Error. See Program 1.

4
```
with Ada.Text_IO, Ada.Integer_Text_IO;
use Ada.Text_IO, Ada.Integer_Text_IO;
with Ada.Numerics.Discrete_Random;
procedure Sundays is
   type Day is (Mon, Tue, Wed, Thu, Fri, Sat, Sun);
   package Random_Day is
         new Ada.Numerics.Discrete_Random(Day);
   use Random_Day;
   G: Generator;
   D: Day;
   Number_Of_Sundays: Integer;
begin
   Number_Of_Sundays := 0;
   for I in 1 .. 100 loop
      D := Random(G);
      if D = Sun then
         Number_Of_Sundays :=
                          Number_Of_Sundays + 1;
      end if;
   end loop;
   Put("Percentage of Sundays in selection was ");
   Put(Number_Of_Sundays);
   New_Line;
end Sundays;
```

5

```
with Ada.Text_IO, Ada.Integer_Text_IO;
use Ada.Text_IO, Ada.Integer_Text_IO;
procedure Triangle is
   Size: Integer;
begin
   Put("Size of triangle please: ");  Get(Size);
   declare
      Pascal: array (0 .. Size) of Integer;
      Tab: Count;        -- indentation at start of row
   begin
      Tab := Count(2*Size + 1);
      Pascal(0) := 1:
      for N in 1 .. Size loop
         Pascal(N) := 1;
         for I in reverse 1 .. N-1 loop
            Pascal(I) := Pascal(I-1) + Pascal(I);
         end loop;
         Tab := Tab - 2;
         New_Line(2);  Set_Col(Tab);
         for I in 0 .. N loop
            Put(Pascal(I), 4);
         end loop;
      end loop;
      New_Line(2);
      if 2*Size > 8 then
         Set_Col(Count(2*Size - 8));
      end if;
      Put("The Triangle of Pascal");
      New_Line(2);
   end;
end Triangle;
```

It is instructive to consider how this should be written to accommodate larger values of Size in a flexible manner and so avoid the confusing repetition of the literal 2. A variable Half_Field might be declared with the value 2 in the above but would need to be 3 for values of Size up to 19 which will go off the screen anyway. Care is needed with variables of type Count which are not allowed to take negative values.

Bibliography

The following selection for further reading comprises a number of books which the author believes will be found helpful. It also includes some interesting historic books on Ada 83 and Ada 95 which are still relevant. And some jolly books on the life and times of Ada Lovelace.

Christine Ausnit-Hood, Kent A Johnson, Robert G Pettit, and Steven B Opdahl (1995). *Ada 95 Quality and Style.* LNCS 1344, Springer

This is a valuable guide to writing good Ada programs. Topics covered include portability and reuse as well as stylistic issues. Most of it also applies to Ada 2012 and Ada 2022.

John Barnes (2008). *Ada 2005 Rationale.* LNCS 5020, Springer-Verlag

John Barnes (2013). *Ada 2012 Rationale.* LNCS 8338, Springer-Verlag

These are the Rationales which were written as part of the standardization process for Ada 2005 and Ada 2012.

John Barnes with Altran Praxis (2012). *SPARK – The proven approach to High Integrity Software.* Altran Praxis

This describes SPARK 2005 which is based on a subset of Ada and includes annotations as Ada comments that enable tools to analyse a program for correctness.

John Barnes with Ben Brosgol (2015). *Safe and Secure Software, an Invitation to Ada 2012.* AdaCore

This handy booklet describes aspects of Ada that contribute to safe programming. Topics range from Safe Syntax to Safe Concurrency and conclude with Certified Safe with Spark.

Grady Booch (1986). *Software Engineering with Ada*, 2nd edn. Benjamin Cummings

This well known classic was one of the first books to discuss how to design programs in Ada; it is especially famed for object oriented programming.

Alan Burns and Andy Wellings (2007). *Concurrent and Real-Time programming in Ada 2005.* Cambridge University Press

This is a very complete account of tasking in Ada 2005 and contains many canonical examples which explore all aspects of concurrency in considerable depth. It includes comprehensive coverage of both the Real-Time Systems annex and the Distributed Systems annex as well as the Ravenscar tasking profile.

Alan Burns and Andy Wellings (2016). *Analysable Real-Time Systems Programmed in Ada.* Amazon Books

This book provides an in-depth analysis of the requirements for designing, verifying, and implementing real-time, embedded, cyber-physical systems; and discusses how these requirements are supported by the Ada programming language.

John McCormick and Peter Chapin (2015). *Building High Integrity Applications with SPARK.* Cambridge University Press

This is a good introduction to SPARK 2014 which is integrated into Ada 2012 using aspects such as pre- and postconditions and other assertions. The book targets students and programmers who are new to formal software verification.

ISO/IEC TR 24718:2004 (2004). *Guide for the use of the Ada Ravenscar profile in high integrity systems.*

This ISO technical report describes the motivation for and how to use the Ravenscar profile.

S. Tucker Taft, Robert A. Duff, Randall L. Brukardt, Erhard Plödereder, Pascal Leroy, Edmond Schonberg eds (2014). *Ada 2012 Reference Manual.* LNCS 8339, Springer

This is the Ada 2012 standard in one volume. A version for Ada 2022 is underway.

The following websites give information on organizations concerned with Ada and standardization.

www.ada-europe.org Ada-Europe is an interest group on Ada. Its activities include an annual conference and the publication of the quarterly *Ada User Journal.* See for example the excellent paper entitled *An Overview of Ada 202x* by Jeff Cousins in the issue for September 2020 which formed the basis for the *Language Enhancement Guide*, the *LEG*, mentioned in the preface.

www.adaic.org This site organized by the Ada Resource Association on behalf of the Ada Information Clearinghouse provides much information concerning companies offering Ada software and services.

www.adacore.com/about-spark This site gives access to a variety of material concerning SPARK and applications.

www.ada-auth.org This site contains various items concerning the standardization of Ada. It includes the *Ada Reference Manual*, the *ARM,* and the *AARM*, the annotated version, and also the various Ada Issues concerning technical details.

www.cambridge.org/barnes22 This site contains auxiliary material regarding this book such as a complete set of answers to all exercises and demonstration programs.

The following books are of a more general nature.

Edwin A Abbot (1884). *Flatland.* Basil Blackwell

This entertaining book is the basis of an example of multiple inheritance in Section 14.8. It is a satire on the structure of society at the time. Various reprints abound. Mine was published by HarperCollins and includes a preface by Isaac Asimov.

William Gibson. and Bruce Sterling (1991). *The Difference Engine.* Bantam Books

This colourful *Steampunk* novel is about a Victorian world where Babbage's engines are widely used, Lord Byron is Prime Minister of Britain and Ada is the Queen of Engines.

Donald E Knuth (1973). *The Art of Computer Programming, vol 3: Searching and Sorting.* Addison-Wesley

These books cover many aspects of the mathematics behind programming. Hash functions are in Section 6.4 of this volume.

Doris Langley Moore (1977). *Ada, Countess of Lovelace.* John Murray

Dorothy Stein (1985). *Ada, A Life and a Legacy.* MIT Press

These classic biographies contain much information about the life of Ada Lovelace.

Betty Alexander Toole (1992). *Ada, the Enchantress of Numbers.* Strawberry Press

This biography captures more of the spirit of Ada as the first programmer.

Robin Hammerman and Andrew L Russell (2016). *Ada's Legacy.* ACM Books

This fascinating book covers cultures of computing from the Victorian to Digital age. Topics include the mathematics of the difference engine, Ada Lovelace's work with Babbage, the background, evolution, and use of the Ada language, and the role of women in computing and other industries.

John Barnes (2012). *Gems of Geometry*, 2nd edn. Springer

John Barnes (2016). *Nice Numbers.* Birkhauser

These books are based on lectures given at Oxford to mature students. Several examples in *Programming in Ada* are based on topics in these books such as: the geometry of the Fano plane where each line has just three points and vice versa, the Tower of Hanoi puzzle, the Sieve of Eratosthenes, Gaussian and polynomial primes, the RSA algorithm for encryption, pyramidal numbers and so on.

Index

General index

Entries to predefined entities are distinguished by being in the program font (such as 'Adjust') and are followed by an indication of their category and, where appropriate, the name of the package containing their declaration. Abbreviations are ch (child), const (constant), el (enumeration literal), ex (exception), f (function), nn (named number), pkg (package), p (procedure), s (subprogram); the prefix g means generic and multiple declarations add s so that gss means that the reference is to several generic subprograms. Reserved words are shown in bold (such as '**abstract**'). The names of important examples will be found in the distinct 'Index to examples' following this general index. References to answers have the section number and exercise number prefixed by Ans.

abort 545
  in asynchronous transfer 548
  with protected objects 552
  in requeue 554
abort completion points 548
abort deferred region 545
abs,
  fixed point 442
  integer and floating 83
  as an operator symbol 173
abstract 250, 319, 485, 500
  in generic parameter 589
abstract data types 27, 41, 238, 827, 875
abstract state machines 233, 236
abstract subprograms 38, 250, 319
abstract types 37, 319
  no private operations 339
  series of 573
abstraction 27, 357, 593
  design as a whole 875
  evolution of 6
  using subprograms 165
  types v subtypes 867
accept statement 514
access 67, 195
  in access discriminants 475
  in access parameters 210
  in generic parameters 483, 490
Access attribute 67, 205
  with access discriminants 475
  applied to subprograms 218, 224
  restriction with components 462
access discriminants 475, 577, 585
  with concurrent objects 608, 618, 621
access parameters 210, 316
  with access discriminants 476
  not of entries 514
access types 21, 193
  anonymous 210, 214
  to class wide types 314
  to discriminated records 466
  general 204
  as generic parameters 483, 490
  pool specific 225
  to protected operations 530
  and renaming 294
  to subprograms 218, 399, 499, 579, 587
  to tasks and protected objects 541
accessibility rules 194, 208, 215, 222
  dynamic 211, 475
    in generic body 494
  for packages 290
  and renaming 294

Ada (ch of Standard) 48, 71, 655
  child packages and subprograms 655
  .Assertions 380
  .Asynchronous_Task_Control 850
  .Calendar 520
  .Characters 658
  .Command_Line 715
  .Containers 733
  .Decimal 852
  .Direct_IO 697
  .Directories 717
  .Dispatching 848
  .Dynamic_Priorities 847
  .Environment_Variables 715
  .Execution_Time 630
  .Exceptions 377
  .Finalization 347
  .Float_Text_IO 51, 703
  .Integer_Text_IO 51, 703
  .Interrupts 843
  .Iterator_Interfaces 582, 734
  .IO_Exceptions 696
  .Locales 725
  .Numerics 52, 681, 685
  .Real_Time 629
  .Sequential_IO 693
  .Storage_IO 698
  .Streams 708
  .Strings 662
  .Synchronous_Barriers 850
  .Synchronous_Task_Control 850
  .Tags 331, 592, 712
  .Task_Attributes 845
  .Task_Identification 845
  .Task_Termination 626
  .Text_IO 49, 699
  .Unchecked_Conversion 822
  .Unchecked_Deallocation 823, 827
  .Wide_Text_IO 707
Ada convention 224, 834
Ada, Countess of Lovelace 3, 918
Ada Issues,
  AI05-220, requirements for static discriminants 460
  AI12-32, type of X'Old 393
  AI12-43, storage pools and sizes 833
  AI12-54-2, addition of Predicate_Failure 422
  AI12-61, loops in aggregates 148
  AI12-71, order of predicate checks 422
  AI12-74, parameter modes and Default_Value 180
  AI12-79-3, global in and out annotations 448
  AI12-86, aggregates and variants 460
  AI12-90, contracts and requeue 554
  AI12-98, abort completion points 548
  AI12-100, checking subtype properties 95
  AI12-103, expression functions and freezing 818
  AI12-119, parallel operations 639
  AI12-125-3, @ as target name symbol 98
  AI12-129, addition of Exclusive_Functions 526
  AI12-157, rules for expression functions 168, 818
  AI12-166, protected function restrictions 560
  AI12-187, stable properties 421
  AI12-212, array and container aggregates 147
  AI12-230, EDF within priorities 848
  AI12-250, iterator filters 113
  AI12-254, Bounded_Indefinite_Holders 794
  AI12-258, null ranges of quantified expressions 159
  AI12-267, data race and conflict checking 642
  AI12-291, the Jorvik profile 639
  AI12-302, default global aspects 418
  AI12-336, meaning of Time_Offset 522
  AI12-345, dynamic accessibility 226
  AI12-362-2, fixed point Floor and Ceiling 447
  AI12-380, improved global aspects 418
  AI12-406, static accessibility 226
  AI22-78, decimal conversions for Big_Real 689
Ada Rationales for 2015, 2012 917
Ada Reference Manual 10
addition,
  fixed point 442
  integer and floating 83
Address attribute 820, 825
Address (type in System) 824
Address_To_Access_Conversions (ch of System) 826
Adjust (p in Ada.Finalization) 346
agent tasks 542
aggregates,
  array 120, 127, 147
    bounds of 127, 128
    mixed 128, 130
    multiple index values 203
    named 127
    with **others** 128
    positional 120, 124, 128
    sliding 128
    string as positional 134
  delta 148
  extension 33, 307, 338, 350
    ancestor type must be specific Ans18.4(3)
  container 147, 732
  record 143, 147
    with discriminants 450, 459
    as initial value of limited type 259
    mixed 144
    with **others** 144
  visibility in named 289
aliased 204, 207, 242, 318, 585
aliasing through parameters 177, 585
aliasing rules 181

Alignment attribute 818, 820
all 159, 197, 200, 220
  in general access type declaration 204
  in generic parameters 490
Allocate (p in System.Storage_Pools) 829
allocator 195
Allows_Exit aspect 588, 716
alphabet 67
ancestor type 246, 306, 353, 915
and,
  in **and then** 96
  Boolean operator 90
  on type Character_Set 671
  on container sets 770
  on modular types 434
  on one-dimensional arrays 138
  as an operator symbol 173
and then 96, 104, 175
annexes 843
anonymous types 865
  access 214, 221, 473
  access parameters 210
  arrays 122, 125, 139
  base type on derivation 249
  floating 438
  integer 428
Argument_Error (ex in Ada.Numerics) 52
array 117, 489
arrays 19, 117
  access types 203
  aggregates, *see* aggregates
  assignment 122, 125
    with slices 137
  attributes 119, 127
  bounds and constraints 117, 123
    with access types 203
    with generic parameters 439
    from initial value 124
    on parameters 168, 178
    on results 170, 178
    sliding 169, 178
  conversion 126, 246
  equality 125, 254, Ans12.4(4)
  generic parameters 489
  null arrays 124, 130
  one-dimensional operations 137, 138
  parameters 165, 174
  ragged, *see* ragged arrays
  slice 137, 456, Ans12.4(4)
  types 122, 865
    component subtype 135, 462
    index subtype 118, 431
  unconstrained, *see* unconstrained array types
ASCII (pkg in Standard) 276, 299, 654, 661
aspect clause 818
aspect specification 72, 77, 389, 818, 898
aspects 72
Assert pragma 380, 392, 828
Assertion_Error (ex in Ada.Assertions) 381, 400
Assertion_Policy pragma 380, 392, 400
assignment 74, 81
  array 122, 125
    with slices 137
  order of evaluation 98
  private 239, 255
  record 143
    deep copy 360
    with discriminants 454, 460
    tagged 328, 333, 336
asynchronous transfer of control 548, 619
at 820
Atomic aspect 607, 844
Atomic_Components aspect 607, 844
Atomic_Operations (ch of System) 845
attribute definition clause 441, 818
attributes 68, 82, 890
  array 119, 127
  subprogram attributes are intrinsic 224

barriers 527, 528, 621
Base attribute 429, 438
base range 429, 435, 437
base type 79, 429
based literal 69
begin,
  of block 76
  of bodies 166, 235, 513
Big_Numbers (ch of Ada.Numerics) 685
Bit_Order attribute 820
block 76
  exception handlers 367
  named 188
  parallel 639
block structure 7, 234
body 235, 512, 525
body stub 271
BOM 674
Boolean (type in Standard) 90
  arrays as basis for sets 139, 490
  in if statement 101
  operations on arrays 138
  special rules for types derived from 251
  use own two-valued type 92, 180
Bounded (ch of Ada.Strings) 662
bounded containers 794, 803
bounded errors 24, 872
  with aliasing 177

blocking operations 531
uninitialized variables 74, 98, 375
Bounded_String (type in Generic_Bounded_Length) 662
Bounded_Priority_Queues (ch of Ada.Containers) 803
Bounded_Synchronized_Queues (ch of Ada.Containers) 798
bounds of aggregates, *see* aggregates
bounds of arrays, *see* arrays, bounds and constraints
Byte Order Mark 674

Callable attribute 547
capacity of containers 421, 752, 765, 779, 863
Capacity_Error (ex in Ada.Containers) 733
case,
in case expression 157
in case statement 105
in variant part 459
case expression 157
case statement 105, 111
cases of alphabet 14, 67
distinct for literals 133, 134
in operators 175
categories of types 41, 93
Character (type in Standard) 133, 653, 658
character handling 640
character literals 133
character set of program text 66
character strings 134
concatenation 141
defining operators 173
character types 132
Character_Mapping (type in Ada.Strings.Maps) 667
Character_Mapping_Function (type in Ada.Strings.Maps) 669
Character_Sequence (subtype in Ada.Strings.Maps) 670
Character_Set (type in Ada.Strings.Maps) 667
Character_Set_Version (f in Ada.Wide_Characters.Handling) 660
checks 374, 384
child units 47, 272
functions 342, 466
generic 508
private 277
choice 106, 461
chunk 583, 640, 641
Class attribute 34, 311
class wide operations 314, 324
in interfaces 565
class wide types 34, 311, 354
are indefinite 474, 485
have no primitive operations 35, 314
classes 34, 41
Clock (p in Ada.Calendar) 520
Clock (p in Ada.Execution_Time) 630
Clock (p in Ada.Real_Time) 629
Close (p in Ada.Sequential_IO etc.) 695
coextension 624
Col (f in Ada.Text_IO) 50, 706
comments 70
communication 12, 817, 834
compilation order 270, 870
compilation unit 267, 870
complete program 54, 296, 868
Complex (type in Numerics.Complex_Types etc.) 853
complex arithmetic 852
Complex_Elementary_Functions (ch of Ada.Numerics) 852
Complex_Types (ch of Ada.Numerics) 852
component,
array 117, 135
record 143, 185, 288, 305
discriminant 449
Component_Size attribute 819
composability,
of equality 254, 329
of stream operations 711
composite types 93, 117
and discriminants 449
composition of interfaces 353
concatenation 141
concrete type 319
conditional entry call 537, 552, 619
Conflict_Check_Policy pragma 642
conformance 293
full conformance 187, 452
mode conformance 292
subtype conformance 220, 293, 308
type conformance 185, 288, 308
constant 75
in access type 205, 216
in generic parameter 490
constant declarations 75
access objects 197
deferred 239, 242
in child package 275
with discriminants 452
external completion 834
Constant_Reference (fs in containers) 738
Constant_Indexing aspect 582, 739
Constants (ch of Ada.Strings.Maps) 667
Constrained attribute 456
Constraint_Error (ex in Standard) 365
with allocators 203
on array assignment 125, 867
on array indexing 117
associated checks 384
on conversion 95
of discriminated types 464

with modular types 434
of tagged types 307, 317
on discriminant checks 451, 460, 469, 837
on dispatching 326
on divide by zero 98
on exponentiation 85
with null value 256, 469
with one-dimensional array operations 139
on overflow in general 168, 429, 437
on parameter evaluation 168, 469
with Pred or Succ 88
on qualification 95, 128
on range violation in general 79, 98
in return statement 167
constraints 79
with access types 202, 466
discriminant, *see* discriminant constraints
index of array, *see* arrays, bounds and constraints
on parameters 169, 178, 456
ignored in renaming 292
range, *see* range constraints
constructor functions 258, 341, 466, 590
context clause 269, 271, 290
contract between components 875
contract model 42, 494
control characters 133, 141, 661
control structures 101
Controlled (type in Ada.Finalization) 347
controlled types 346, 474, 827
controlling operands 325, 359
Convention aspect 834
conventions 224, 292, 530, 834
conversion,
between access types 207, 209, 211, 215
accessibility check 211, 222
access to subprogram types 221, 222
not pool specific 225
to tagged types 318
to types with discriminants 472
array 126, 246
between character and string types 660, 664
between derived types 246
with fixed point 442, 446
between numeric types 83, 94, 430, 438
as a parameter 178
of tagged types 33, 307, 317, 333
implicit from universal types 217, 431, 438, 442
view 178, 333, 636
Conversions (ch of Ada.Characters) 660
Convert (fs in Ada.Strings.UTF_Encoding.Conversions) 676
Count (type in Ada.Text_IO) 50
Count attribute 516, 530
care with 533, 537, 552, 618
Count_Type (type in Ada.Containers) 733
CPU aspect 602
CPU time 629
Create (p in Ada.Sequential_IO etc.) 695
current instance rule 406, 410, 602
cursors for containers 736

data abstraction 6, 12, 825
Data_Error (ex in Ada.IO_Exceptions) 696
Day_Of_Week (f in Ada.Calendar.Formatting) 523
Deallocate (p in System.Storage_Pools) 829
decimal types 446, 689, 852
declarations 13, 73, 538
multiple 74, 122, 866
number 75
order of 237
of units 186, 237
declarative region 246, 273, 287
declare 76
declare expression 161
Decode (fs in Ada.Strings.UTF_Encoding.Strings) 675
default components in aggregates 131, 145
Default_Component_Value aspect 126, 390
default discriminants 453, 456, 460
with allocated objects 468
with limited tagged types 454, 464
default entry parameters 514
default files 699
default formats 704
default generic parameters,
for objects 481, 482
for subprograms 497, 506
for types 494
Default_Initial_Condition aspect 400
Default_Iterator aspect 582
default record components 145, 146, 262
of ancestor type in extension aggregates 307
multiple evaluation 185
using Unchecked_Access attribute 476
Default_Storage_Pool pragma 833
default subprogram parameters 184, 211
no conformance problems 187
renaming can change 292
Default_Value aspect 94, 98, 180, 390
deferred constants, *see* constant declarations
Definite attribute 494
definite subtypes 126, 135, 145
with default discriminants 454
delay alternative 535
as triggering alternative 548
delay statement 519, 521, 629, 869
delta 67, 148, 440, 487
in aggregates 148
in decimal types 446

Delta attribute 67, 446
delta constraints 446
dependency of compilation units 268, 283, 870
  on private part 241
  of subunits 271
dependency of tasks 513, 539
  created by allocators 541
  on library units 539, 869
depth-first order 791
dereferencing 197, 204, 325
derived types 17, 245
  for concurrent types 612
  and discriminants 463
  model for numeric types 431, 490
  and representation 867
  for type extension 31, 305
  versus subtypes 492
digits 67, 437, 487
  in decimal types 446
Digits attribute 67, 439
digits constraint 438
direct name 288, 861
direct visibility 188, 288
Discard_Names aspect 844
discrete choice 106, 461
discrete types 93
Discrete_Random (gch of Ada.Numerics) 53, 684, 685
discriminant constraints 450, 454
  with access types 466, 468
  changing 461, 468
  default 453
  and renaming of components 461, 483
  unknown 473, 594
    in generic parameters 485, 493
discriminant, inferable 836
discriminant use 456
  in protected types 530
  in task types 601, 847
discriminated records 449
  tagged 454, 464
  and Unchecked_Union 835
dispatching 35, 313, 325
  on assignment and equality 328, 334, 336, 360
  redispatching 334
  of tasks 847
Dispatching_Domain (type in System.Multiprocessors.Dispatching_Domains) 603
division,
  fixed point 442, 443
  integer and floating 83
  by zero 98
do 171, 514
Doubly_Linked_Lists (ch of Ada.Containers) 734
downward closures 194, 221
Duration (type in Standard) 519, 629
dynamic, terminology 25
dynamic polymorphism 30, 34, 311
Dynamic_Predicate aspect 410
dynamically tagged 325

e (nn in Ada.Numerics) 52
eigenvalues and vectors 856
Elaborate pragma 296
Elaborate_All pragma 296
Elaborate_Body aspect 297, 871
elaboration, terminology 26
elaboration 76
  of library units 296, 368
Element (fs in containers) 737
elementary functions 52, 681, 854
elementary types 93
Elementary_Functions (ch of Ada.Numerics) 681
else,
  in if statement 102
  in select statement 536
  in short circuit operation 96
elsif 103
embedded systems 3
encapsulation 42
Encode (fs in Ada.Strings.UTF_Encoding.Strings) 675
end,
  of accept statement 514
  of block 76
  of bodies 166, 235, 513
  of control structures 101
  of extended return statement 171
  of record type 143
  of select statement 532
End_Of_File (f in Ada.Sequential_IO etc.) 696
End_Of_Line (f in Ada.Text_IO) 706
entry 514, 526
entry barrier 527, 528, 621
entry call 42, 514, 527, 528
  conditional or timed 536, 537, 552, 619
  pre/post conditions on 518, 554
  as triggering alternative 548
entry family 554
entry queue 516, 528
  leaving on time out 537, 552, 618, 619
enumeration literals 87, 133
  are intrinsic 224
  overloading and subprograms 185
  renaming 293
enumeration representation clause 819
enumeration types 16, 87
  character types 132
  two-valued 92, 180
Enumeration_IO (gpkg in Ada.Text_IO) 703

Enum_Rep attribute 89, 819
Enum_Val attribute 89, 819
environment commands 715
environment task 539, 628, 845, 868
Equal_Case_Insensitive (f in Ada.Strings) 673
equality, *see* =, /=
Equivalent_Keys (f) 755
erroneous programs 24, 873
  because checks suppressed 385, 837
  misuse of unchecked access 209
  misuse of unchecked conversion 822
  misuse of unchecked deallocation 225, 823
errors 23
evaluation, terminology 26
exception 366, 369
exception declaration 369
exception handler 366
exception propagation 367
  during library elaboration 368, 869
  from main subprogram 869
  from rendezvous 548
  out of scope 385
  from task body 548, 626
  from task declaration 540
Exception_Id (type in Ada.Exceptions) 379, 385
Exception_Identity (f in Ada.Exceptions) 379
Exception_Information (f in Ada.Exceptions) 376
Exception_Message (f in Ada.Exceptions) 376, 423
Exception_Name (fs in Ada.Exceptions) 376, 379
Exception_Occurrence (type in Ada.Exceptions) 376
exceptions 12, 23, 365
  distinct on instantiation 482, 866
  in rendezvous 547
  suppressing checks 384
  in task declaration 540
Exclusive_Functions aspect 526
execution, terminology 26, 76
exit statement 109
  from block in loop 190
  in exception handler 367
  in named loop 113
Exp (f in Numerics.Elementary_Functions etc.) 682, 854
Expanded_Name (f in Ada.Tags) 332
exponentiation 83, 85, 682
  for complex types 854
  second operand of type Integer 428
Export aspect 834
expressions 74, 95, 861
expression function 172, 175, 244
extended return, *see* return statement, extended
extension aggregate, *see* aggregates

False (el in Standard) 90

file 694, 697, 869
  and directory names 718
  heterogeneous 708
File_Mode (type in Ada.Sequential_IO etc.) 694
File_Type (type in Ada.Sequential_IO etc.) 694
filter 113, 148, 588
Finalize (p in Ada.Finalization) 346, 830
First attribute,
  on array types and objects 119, 127
  on real types 440, 446
  on scalar types 82, 88, 94
first subtype 79, 865
First_Valid attribute 94, 416, 488
Fixed (ch of Ada.Strings) 662
fixed point types 436, 440, 705
  decimal types 446, 852
Fixed_IO (gpkg in Ada.Text_IO) 704
Float (type in Standard) 13, 81, 437, 873
Float_IO (gpkg in Ada.Text_IO) 40, 50, 704
Float_Random (ch of Ada.Numerics) 684
Float_Text_IO (ch of Ada) 51, 703
floating point types 436, 437, 705
Flush (s in Ada.Sequential_IO etc) 696, 707
for 110
  in aspect clause 818
  in quantified expression 159
for statement 110, 431
formats 50, 705
Forward_Iterator (type in Ada.Iterator_Interfaces) 582
freezing rules 343, 817, Ans19.2(4)
function 165
  as generic parameter 496
function parameters 164
function result 166, 178
  constraints on 178
    ignored in renaming 291
  in handler 367
  null exclusion 217
  as an object 170, 469
  and type extension 322, 340
  with unconstrained array 170
  with unconstrained discriminant 451

garbage collection 225
Generator (type in Discrete_/Float_Random) 54, 684
generic 480
generic instantiation 480, 508, 866
generic parameters 480, 482, 484, 516
  that are packages 502, 578
generic units 12, 40, 479
  child units 508
  for multiple inheritance 570
Generic_Array_Sort (ch of Ada.Containers) 813
Generic_Bounded_Length (gpkg in Strings.Bounded) 662

Generic_Complex_Elementary_Functions (ch of Ada.Numerics) 853
Generic_Complex_Types (ch of Ada.Numerics) 853
Generic_Constrained_Array_Sort (ch of Ada.Containers) 813
Generic_Dispatching_Constructor (gf in Ada.Tags) 589, 712
Generic_Elementary_Functions (ch of Ada.Numerics) 681
Generic_Keys (gp in set containers) 772
Generic_Real_Arrays (ch of Ada.Numerics) 856
Generic_Sort (ch of Ada.Containers) 814
Get (ps in Ada.Text_IO) 51, 703, 705
Get_Immediate (p in Ada.Text_IO) 707
Get_Line (ss in Ada.Text_IO) 706
Global aspect 418, 731, 858
global state 185, 418
goto statement 114
  to end of loop 114
  in exception handler 367
  out of loop or block 190
guards 532

handled, terminology 26
handlers,
  interrupt 843
  task termination 627
  timers 632
Handling (ch of Ada.Characters) 658
hashing 763, 776
Hash (ch of Ada.Strings) 766
Hash_Type (type in Ada.Containers) 733
Has_Same_Storage attribute 821
Hierarchical_File_Names (ch of Ada.Directories) 723
hierarchical library 47, 272, 508
high integrity systems 858, 876
holder container 792

identifiers 13, 67
Identity attributes 375, 845
if,
  if expression 153
    of type Boolean 154
  if statement 101
illegal program 24
Image (f in Ada.Calendar.Formatting) 523
Image attribute 51, 230, 679
Imaginary (type in Numerics.Complex_Types etc.) 853
Implicit_Dereference aspect 582
Import aspect 834, 835
in,
  for statement 110, 583
  generic parameter mode 482
  membership test 95, 175
  parameter mode 176, 184
in out,
  danger of exceptions 387
  generic parameter mode 482
  parameter mode 176, 184
    versus access parameters 212
incomplete declaration 195, 214, 283, 467
  in access to subprogram profile 223
  completed in package body 257, 275
  tagged 315
incomplete generic parameter 486, 507
indefinite containers 807
Indefinite_Holders (ch of Ada.Containers) 792
indefinite subtypes 126
  class wide type 312, 474
  discriminants without defaults 454
  generic type parameters 485, 493
  private type with unknown discriminants 473
Independent aspect 845
Independent_Components aspect 845
index subtype 118, 431
Index_Error (ex in Ada.Strings) 671
inheritance 30, 245, 305
  multiple, *see* multiple inheritance
initial values 73, 375
  access types 195, 202
  of allocated objects 195, 201
  arrays 120, 124, 131
  class wide 312, 328
  of limited types 258
  private types 239, 400
  records 144, 262, 450
initialization of packages 235, 237
  exception raised in 367
Initialize (p in Ada.Finalization) 346, 830
Inline aspect 243, 389, 391, 821, 870
input–output 49, 692
  as streams 708
  of text 698
instantiation 480, 866
  of generic children 508
Integer (type in Standard) 81, 427, 873
  in loops and arrays 112, 118, 431
integer literals 68, 431
Integer_Arithmetic (ch of System.Atomic_Operations) 846
Integer_Literal aspect 71, 864
integer types 93
  modular 433
  signed 427
Integer_16 (type in Interfaces) 430, 434
Integer_IO (gpkg in Ada.Text_IO) 51, 703
Integer_Text_IO (ch of Ada) 51, 703
interface 26, 38, 319
  examples of use 357, 564, 795
  in generic parameters 486

provide multiple inheritance 351
synchronized 610
Interfaces (ch of Standard) 48, 430, 434, 837
interfacing 817, 834
interrupts 843
Intrinsic convention 224, 530, 834, Ans19.2(4)
intrinsic subprograms 224, 308, 826
in protected body 530
shift and rotate operations 434
is,
in case expression 157
in case statement 105
in package 235
in subprogram body 166
in type declaration 78
Is_Abstract (f in Ada.Tags) 593
Iterator_Element aspect 582
iterators 575
active 575
filter 113
generalized 580
passive 579, 581
Iterator_View aspect 735

Jorvik profile 638

keys with containers 754, 755, 767

label 115
scope of 190
Last attribute,
on array types and objects 119, 127
on real types 440, 446
on scalar types 82, 88, 94
Last_Valid attribute 94, 416, 488
Latin_1 (ch of Ada.Characters) 660
leap seconds 523
Length (f in Unbounded etc.) 662
Length attribute 119, 127
Length_Error (ex in Ada.Strings) 671
Less_Case_Insensitive (f in Ada.Strings) 673
lexical element 66
library package body 237, 297, 871
library task 539, 869
library unit 12, 267, 868
elaboration 296, 368, 869
generic 481, 508
hierarchy 47, 272, 508
predefined 48, 651
renaming 291
ligatures 67
limited,
in derivation 355
in generic parameters 485
in private type 255
record type 255
in with clause 283, 286
limited types 255
and access discriminants 475
explicitly 259
important in OOP 315
and interfaces 355, 611
partial and full view 338
Limited_Controlled (type in Ada.Finalization) 347
line length of program text 67
linear elaboration 77, 187
not with aspect specifications 77, 391
Liskov Substitution Principle 400
list processing 22, 734
heterogeneous 36, 314, 572
literals,
character 133
enumeration, *see* enumeration literals
generalized 71
numeric 68
string, *see* string literals
Locking_Policy pragma 849
Log (fs in Numerics.Elementary_Functions etc.) 682, 854
Long_Float (type in Standard) 437
Long_Integer (type in Standard) 427, 430
Look_Ahead (p in Ada.Text_IO) 707
loop parameter 110
hidden in named loop 189
of type Integer 421
loop statement 108
label at end 114
parallel 640
LSP 400

machine attributes 438, 893
machine code 837
Machine_Overflows attribute 437
main subprogram 12, 268, 296
called by environment task 513, 868
exception propagated from 368
main task, *see* environment task
maps,
Hashed_Maps (ch of Ada.Containers) 764
Ordered_Maps (ch of Ada.Containers) 755
Maps (ch of Ada.Strings) 663, 667
mathematical library 52, 681, 853
matrices 854
Max attribute 94, 230
Max_Base_Digits, Max_Digits (nns in System) 825
Max_Binary_Modulus (nn in System) 825
Max_Entry_Queue_Length aspect 518
Max_Int (nn in System) 430, 825
Max_Length (const in Generic_Bounded_Length) 662

Max_Mantissa (nn in System) 825
Max_Nonbinary_Modulus (nn in System) 825
membership tests 95, 151, 331, 862
  purpose 153, 423
  are not operators 175
  before a conversion 152
methods, terminology 41
Mersenne primes 641, 692, 763
Min attribute 94
Min_Int (nn in System) 430, 825
mixed notation,
  in array aggregates 128, 130
  in discriminant constraints 453
  in generic instantiations 484
  in record aggregates 144
  in subprogram calls 184
mod 67
  as generic formal 487
  on integer types 83
  modular types 433
  as an operator symbol 173
  on types Address and Storage_Offset 826
Mod attribute 67, 434
model attributes 439, 894
model numbers 857
modes,
  of files 697, 699
  of generic parameters 482
  of subprogram parameters 175
modular types 433
  as array index type 526
Modulus attribute 435
Move (p in Ada.Strings.Fixed) 665
multiple declarations 74, 122, 866
multiple evaluation 74, 185, 200
multiple inheritance 351, 562
  mixin using generics 570, 573
  avoids a wrapper 362
multiplication,
  fixed point 442, 443
  integer and floating 83
Multiprocessors (ch of System) 602
Multiway_Trees (ch of Ada.Containers) 780
mutability 454, 462, 474

named aggregates,
  of arrays 127
  of records 144
  visibility in 289
named block 188
  scope of 190
named discriminants 453
named loop 113
  scope of 190
named numbers 75
  cannot rename 295
named parameters,
  generic 480
  with if expression 154
  to resolve ambiguities 186
  subprogram 183
  visibility in 289
names 861
Natural (subtype in Standard) 82
new,
  for allocators 195
  for derived types 245
  for generic formals 486, 502
  for instantiation 480
New_Line (p in Ada.Text_IO) 50, 706
No_Dependence restriction identifier 299
No_Implementation_Extensions profile 865
No_Return aspect 382, 389
nominal subtype 124, 206, 836
Nonblocking aspect 517, 858
not,
  Boolean operator 90, 95
  on type Character_Set 671
  on modular types 434
  in null exclusion 202
  on one-dimensional arrays 138
  as an operator symbol 173
  with **overriding** 251
not in 95, 175
null,
  as access value 195, 217
  in null aggregate 145
  in null declaration 145, 460
  in null exclusion 202
  for null procedure 250
  in null statement 105
null aggregate 130, 134, 145
null array 124, 130
null exclusion 79, 166, 202
  and conversions 216
  and deferred constants 242
  in generic parameters 482, 483, 488, 489
  in renaming 292, 330
null generic package 506
null package 281
null procedure 250, 319, 347
null range 82
  for array bounds 124
  of enumeration type 88
  in for statement 111
null record 145, 306, 474
null slice 137
null statement 105

null string 134, 140
Null_Address (const in System) 824
Null_Bounded_String (const in
Generic_Bounded_Length) 662, 664
Null_Id (const in Ada.Exceptions) 377
Null_Occurrence (const in Ada.Exceptions) 377
Null_Unbounded_String (const in
Strings.Unbounded) 663, 664
numbers 68
declaration of 75
are of universal type 431, 436
numeric library 52, 681, 685, 852
numeric literals 68, 71
numeric types 17, 81, 427, 865
Numeric_Error (renaming in Standard) 366, 367

object factories 589
object oriented programming 12, 29, 305, 561
design as a whole 875
and tasking 610
terminology 41
objects 41, 73, 538
Object_Size attribute 818
of,
in arrays 117
in loops 121, 160, 583
Old attribute 393
one-component aggregates 131, 134, 145
one-dimensional array operations 138
Open (p in Ada.Sequential_IO etc.) 695
operators 95, 173, 862
calling as function 189
no default parameters 185
not as expression function 175
not as library units 270
order of operands 98
new overloadings 174
parameters only of mode in 175
renaming 291
not as subunits 271
optimization 71, 429, 821
Optimize pragma 71, 821
or,
Boolean operator 90
on type Character_Set 671
on container sets 770
on modular types 434
on one-dimensional arrays 138
as an operator symbol 173
in **or else** 96
in **or use** 494
in generic parameters 494
in select statements 532
or else 96, 175
order of compilation 270, 870
order of declarations 237
order of elaboration 76, 187
of library units 296, 869
order of evaluation 98, 181, 872
others,
in array aggregate 128
in case statement 105
in exception handler 367
not as procedure parameters 184
in record aggregate 144
in variant 461
out,
danger of exceptions 387
not as generic parameter mode 482
parameter mode 176
overflow 168, 429, 437
Overlaps_Storage attribute 821
overloading 14, 78
characters 133
compilation units 270, 271
entries 516
enumeration literals 87, 112, 186
operators 174
subprograms 185, 247, 480, 484
visibility rules 288
overriding indicators 251, 321, 350
in global annotation 421
are optional 344
with entries 614
with generics 484
with renaming 291

Pack aspect 818
package 28, 235
as generic parameter 502
package 234
design as a whole 875
as generic parameter 502, 578
private part 239, 278
scope and visibility 287
package initialization 235, 237
exceptions in 367
package renaming 294
parallel 588, 639
parallel block 639
parallel loop 588, 640
Parallel_Iterator (type in Ada.Interator_Interfaces) 582
Parallel_Reduce attribute 641
parameterless function versus constant 243
parameterless instantiations 484
parameterless subprogram calls 170, 220
parameters 166, 175
bounds of 168

constraints on 178, 187, 456
ignored in renaming 292
default 183
generic 482, 497
mechanism 177
for access types 198
for concurrent types 537
for elementary types 182
danger of exceptions 387
with derived types 249
generic 483, 497
for limited and/or private types 258
for tagged types 318
mixed 184, 484
modes 175, 482
named 183, 289, 480
positional 183, 480
parent of subunit 271
parent type 245, 305, 352
Parent_Tag (f in Ada.Tags) 332, 592
partitions 296, 851, 870
Pattern_Error (ex in Ada.Strings) 671
Pi (nn in Ada.Numerics) 52
pointers, dangers of 193
policies 847
polymorphism 40, 479
dynamic 30, 34, 311
static 40, 479
portability 24, 872
elaboration order 296
further reading 917
and numeric types 433
order of assignment and operations 98
Pos attribute 88, 94
positional aggregates,
of arrays 120, 124, 128
not of one component 131
of records 144
as a string 134
positional discriminants 453
positional parameters 183, 184, 480
Positive (subtype in Standard) 82
Post aspect 392
Post'Class aspect 393, 396
postcondition 392, 399, 742
pragma 71, 902
precedence 85, 862
of logical operators 90
of membership tests 95
Pre aspect 392
Pre'Class aspect 393, 396
precondition 392, 399, 742
Pred attribute,
on discrete types 88, 94
on modular types 435
on real types 94
predefined library 48, 276, 651
predicates 159, 391
Predicate_Failure aspect 423
Preelaborable_Initialization aspect 298
Preelaborate aspect 297
preference control 554
preference rule 217, 431, 436
prefixed notation 39, 325, 341, 613
primitive operations 31, 245, 305, 322
child subprogram is not 273, 341
not for class wide types 35, 314
in generic units 491
private 339
priorities 519, 847
Priority aspect 847
Priority (subtype in System) 825, 847
Priority_Specific_Dispatching pragma 848
private,
for child units 277
in generic parameters 485
in private types 239
in protected types 525
in task types 557
in with clause 278
private child units 276
private extension 339, 619
private types 28, 238
with discriminants 451, 473
implemented as an access type 255
implemented as a concurrent type 538, 557
implemented as a derived type 248, 257
sharing between packages 272
as a tagged type 306
and type extension 338
and type invariants 404
private with clause 276, 281
procedure 175
see also subprogram
procedural iterators 587, 716
profile, of subprogram 187, 202
Profile pragma 638, 905
progenitor type 352
program, complete 54, 296, 868
program library 12, 56, 268, 870
Program_Error (ex in Standard) 365
access before elaboration 237, 293, 296
with access parameters 211
associated checks 384
at end of function 167
when all guards false 533
programming by contract 5, 8
programming in the large 12

protected 525, 530
protected convention 530
protected operations 518, 525
protected types 43, 524, 527, 865
  and OOP 610
  as parameters 539
public child units 272
Pure aspect 298, 851
Put (ps in Ada.Text_IO) 50, 703
Put_Image attribute 679, 686
Put_Line (p in Ada.Text_IO) 50, 706

qualification,
  of array aggregates 128
  as initial value with allocator 196
  to resolve ambiguities 94
    of character literals 133
    of derived types 247
    of enumeration literals 87, 185
    of strings 135, 140
quantified expression 159, 431
quantifier 159
Query_Element (fs in containers) 738
Queuing_Policy pragma 849
queues 794

ragged arrays 136
  using access types 200, 207
  using discriminants 457
raise expressions 373, 422
raise statement 369, 370, 372
raised, terminology 26
Raise_Exception (p in Ada.Exceptions) 379
Random (f in Discrete_/Float_Random) 53, 421, 685
random numbers 53, 683
range 67
  in array types 117
  in fixed point 440
  in for statement 111
  as generic formal 487
  in integer types 428
  null range 111
  in range constraints 79
  for record representation 820
Range attribute 67
  of array 119, 127
    not in initial value 131
range constraints 79
  in discrete choice 106
  on enumeration types 88
  on floating types 437
  on integer types 429
Ravenscar profile 638, 875
Read (p in Ada.Sequential_IO etc.) 695
Read attribute 709
readability 11
real literals 68, 436
Real_Literal aspect 71, 864
real, terminology 26
real types 93, 436
Real_Matrix (type in Numerics.Generic_Real_Arrays) 854
Real_Vector (type in Numerics.Generic_Real_Arrays) 854
record 143, 820
records,
  aggregates, *see* aggregates
  assignment, *see* assignment
  components, *see* component
  discriminants 450, 464
  extension 306
  representation 820
  tagged 30, 306
  types 20, 143
  variants 459, 594
recursion 167
  accidental with overloading 174, 217, 256
  exceptions with 386
  not with generics 484
  mutually recursive procedures 187
  in tree manipulation 199
redispatching 334
Reduce attribute 162
reduction expression 162, 640
Reference (fs in containers) 739
reference types 193
Relative_Deadline aspect 848
rem,
  as an operator symbol 173
  for remainder 83
remainder 83
Remote_Types aspect 851
renames 291
renaming 161, 291
  as bodies 292, Ans19.2(4)
  of class wide 334
  in declare expression 161
  and dispatching 330
  of entries 516
  of exceptions 385
  like generic parameters 483, 497
  of generic units 484
  of library units 291
  preserves staticness 864
  restriction with components 461, 483
rendezvous 513
Replace_Element (ps in containers) 737
representations 818, 867
requeue statement 551
Reraise_Occurrence (p in Ada.Exceptions) 379

reserved words 13, 67, 889
Reset (p in Ada.Sequential_IO etc.) 696
Reset (ps in Discrete_/Float_Random) 684
resource management 261, 370, 473, 827
  multitasking 552
Restrictions pragma 299, 850, 904
result, *see* function result
Result attribute 394
return 166, 167, 171
return statement 167, 179
  in accept statement 517
  in block or loop 190
  in constructor function 259
  in exception handler 367
  extended 171
    for limited types 259
    with task types 608
    with variants 462
RSA algorithm 691
reverse 110
Reversible_Iterator (type in Ada.Iterator_Interfaces) 582
root_integer 430, 436, 490
root_real 436
root type of class 311
Root_Storage_Pool (type in System.Storage_Pools) 829
rotate operations 434, 864
rounding 84, 439

Save (p in Discrete_/Float_Random) 684
Save_Occurrence (ss in Ada.Exceptions) 378
scalar types 16, 73, 93
scheduling 519, 550, 847
scope 77, 186, 287
  of exceptions 385
select statement 531, 532, 535
  asynchronous 548, 619
semicolons 13, 101, 166
separate 271
separate compilation 267, 870
Set_Col (p in Ada.Text_IO) 50, 706
sets 139, 357, 490, 552
  Hashed_Sets (ch of Ada.Containers) 776
  Ordered_Sets (ch of Ada.Containers) 767
shared variables 607, 845, 873
  via protected objects 525
shift operations 434, 864
short circuit control forms 96, 862
  are not operators 175
side effects 190, 872
signalling 550
signatures 506, 578
Signed_Zeros attribute 439, 683
Size attribute 441, 818
Skip_Line (p in Ada.Text_IO) 706
slice 137, 456, Ans12.4(4)
sliding semantics,
  with array assignment and equality 125
  in aggregates 128
  with one-dimensional operations 139
  with parameters and results 169, 178
Small attribute 441, 446
small for fixed point 440
software components 40, 874, 876
some 159
sorting 742, 813
SPARK language 6, 877
  further reading 918
Split (ps in Ada.Calendar) 520, 523
  in Ada.Execution_Time 630
  in Ada.Real_Time 629
Sqrt (f in Numerics.Elementary_Functions etc.) 682, 854
Stable_Property aspect 421, 736
Stable package (ch of containers) 744
Standard 290, 651, 869
  no with clause 16, 270
statements 13, 101, 115
Static aspect 864
static, terminology 25
static expression 862
  must be in base range 434
  in conditional expression 156
  in discrete choice 107
  generic parameters not static 483
  in named number declaration 431
  in numeric types 429, 434, 437, 440
static matching 126, 867
  with Access attribute 206
  in array conversion 126, 246
  in deferred constants 242, 452
  of generic parameters 485, 489, 502
  on renaming subprogram body 292
static polymorphism 40, 479
Static_Predicate aspect 410
static subtype 107
statically unevaluated 156, 157
storage pools 224, 826, 830
  with derived types 245
Storage_Array (type in System.Storage_Elements) 825
Storage_Count (subtype in System.Storage_Elements) 825
Storage_Element (type in System.Storage_Elements) 825
Storage_Elements (ch of System) 825
Storage_Error (ex in Standard) 365
  on allocation 256, 826
  associated check 384
  on stack overflow 168
Storage_Offset (type in System.Storage_Elements) 825
Storage_Pools (ch of System) 829
Storage_Size (f in System.Storage_Pools) 829

Storage_Size aspect 819
Storage_Unit (nn in System) 819, 825
streams 708, 851
String (type in Standard) 134, 654
string handling 662
String_Literal aspect 71
string literals 134
  as operator 173
  too long for a line 141
strong typing 11, 16
  in assignment 78, 81
  overcoming 822
  overzealous 866
stub 271
Subpools (ch of System.Storage_Pools) 830
subprogram 165
  abstract 38, 250, 319, 339
  body 186, 235
  call 176, 247
  declaration 186, 235
  generic parameters 495, 506
  overloading 174, 185, 484
  as a parameter 219
  parameters 175, 216, 539
  specification 186, 235
subtraction,
  fixed point 442
  integer and floating 83
subtype 79
subtype indication 80
  in derived type 249
  need subtype mark alone,
    as generic actual type parameter 488
    in generic formal type parameter 489
    as index subtype 123
    in membership test 95
  with null exclusion but no constraints,
    in discriminants 453
    in generic formal object parameters 483
    in parameters and result 166, 169, 187
subtype mark, *see* subtype indication
subtype predicates 410
subtypes 25, 79
  with access types 203
  of anonymous types 428, 488, 865
  of arrays 123
  with derived types 248
  of discriminated records 453
  of enumeration types 88
  as renaming 295
subunits 270
Succ attribute,
  on discrete types 88, 94
  on modular types 435
  on real types 94
Suppress pragma 384
Synchronization aspect 621
synchronized 420, 611, 619
Synchronized_Queue_Interfaces (ch of
    Ada.Containers) 795
syntax 65, Web
  examples of 68, 106
System (ch of Standard) 48, 824

tag 30, 306, 311, 331
Tag attribute 331, 712
Tag (type in Ada.Tags) 331, 592
tag indeterminate 327, 359
  in conditional expressions 155
tagged 31, 306, 485
tagged type 30, 306
  concurrent 538, 611
  and discriminants 464
  versus variants 462, 470, 594
tampering 748
target name symbol 98
task 512
Task_Dispatching_Policy pragma 848
Tasking_Error (ex in Standard) 366, 384
  on abort in rendezvous 545
  calling entry of completed task 547
  on communication failure in general 547
  during task activation 540
tasks 12, 42, 511
  activation 513, 539, 541
  attributes 547, 845
  dependency 513, 539
    created by allocator 541
    on library units 539, 869
  design as a whole 875
  further reading 917
  identity 845
  objects 539
  and OOP 610
  as parameters 539
  storage space 819
  termination 513, 544, 626, 869
  types 538, 865
templates, terminology 42
terminate alternative 545
Terminated attribute 547
terminology 25, 41, 907
Text_Buffers (ch of Ada.Strings) 677
then,
  in asynchronous select 548
  in if statement 101
  in short circuit operation 96

threads of control 639
Tick (nn in System) 825
Time (type in Ada.Calendar) 520
Time (type in Ada.Real_Time) 629
time zones 522
Time_Error (ex in Ada.Calendar) 520
Time_Of (fs in Ada.Calendar) 521, 523
  in Ada.Execution_Time 630
  in Ada.Real_Time 629
Time_Offset (type in Ada.Calendar.Time_Zones) 522
timed entry call 536, 552, 618
timers 629
To_Address (f in System.Address_To_Access_Conversions) 826
To_Address (f in System.Storage_Elements) 826
To_Bounded_String (f in Generic_Bounded_Length) 664, 666
To_Integer (f in System.Storage_Elements) 826
To_Pointer (f in System.Address_To_Access_Conversions) 826
To_String (f in Strings.Unbounded etc.) 664
To_Unbounded_String (fs in Strings.Unbounded) 664
Translation_Error (ex in Ada.Strings) 671
trees 780
triggering alternative 548, 619
trigonometric functions 683, 854
True (el in Standard) 90
type,
  in concurrent types 526, 538
  in generic parameters 485
  in type declaration 78
  in use type clause 248
Type_Invariant aspect 404
Type_Invariant'Class aspect 408
types 77
  abstract, *see* abstract types
  access, *see* access types
  anonymous, *see* anonymous types
  array, *see* arrays, types
  categories of 41, 93
  composition 561
  concurrent 537, 611, 865
  controlled 346
  conversion, *see* conversion
  declaration 78, 122, 865
  definition 78, 122, 865
  enumeration, *see* enumeration types
  equivalence 865
    of arrays 122, 139, 168
  extension 30, 305, 570
  generic parameters 485
  incomplete 195, 214, 223, 467
    completed in body 257, 275
  invariant 404
  model 16, 77, 866
  numeric, *see* numeric types
  pool specific 225
  protected 527, 537
  qualification, *see* qualification
  record, *see* records, types
  task 538
Type_Invariant aspect 404
Type_Invariant'Class aspect 408

Unbounded (ch of Ada.Strings) 663
Unbounded_String (type in Ada.Strings.Unbounded) 663
Unbounded_Priority_Queues (ch of Ada.Containers) 800
Unbounded_Synchronized_Queues (ch of Ada.Containers) 796
Unchecked_Access attribute 209, 476, 826
  not for access to subprogram types 222
  avoiding dangling references 828
Unchecked_Conversion (gchf of Ada) 822
Unchecked_Deallocation (gchp of Ada) 823, 827
Unchecked_Union aspect 834
unconstrained array types 123
  with access types 203, 206
  as generic parameter 489
  as result 170, 178
  as subprogram parameter 168, 178
unconstrained discriminant,
  with access types 468
  and defaults 453
  restriction on renaming 461, 483
underline 14, 67, 68
universal_access 217
universal_fixed 442
universal_integer 431
universal_real 436
unknown discriminants, *see* discriminants
unsigned types, *see* modular types
Unsigned_16 (type in Interfaces) 434
Unsuppress pragma 384
until 521
Update_Element (ps in containers) 738
use 236, 248
  in aspect clause 818
  in generic parameter 494
use clause 237, 269, 286, 289
  bad for you 248
  for child units 274
  not for concurrent objects 514, 557
  in context clause 269, 290
  in generic formal list 505
  not with limited/private with clauses 278, 286
  use type clause 248, 290
  use all type clause 248

Use_Formal aspect 421
UTF_Encoding (ch of Ada.Strings) 674

Val attribute 88, 94
Valid attribute 822
Value (fs in Ada.Calendar.Formatting) 523
variables 73
Variable_Indexing aspect 582, 739
variant parts 459, 594
vectors and matrices 854
Vectors (ch of Ada.Containers) 748
views 28
  constant and variable 243
  of discriminants 464
  limited 255, 338
  partial and full 242, 338, 404
  prefixed 39, 325, 341, 613
  view conversions 178, 333, 636
visibility 186, 233
  block structure 7, 77
  and child units 274, 281
  of operators 248
  and packages 287
  of private types 345
  and subprograms 186, 188
Volatile aspect 844
Volatile_Components aspect 844

when,
  in case expression 157
  in case statement 105
  in entry body 527
  in exception handler 366
  in exit statement 109
  in select statement 533
  in variant part 459, 461
while statement 110
Wide_Character (type in Standard) 133, 653, 660
Wide_String (type in Standard) 135, 140, 654
Wide_Wide_Character (type in Standard) 133, 654, 660
Wide_Wide_String (type in Standard) 135, 140, 654
Width attribute 704
with 268
  in extension aggregate 307
  in generic formals 486, 496, 502
  in raise statement 379
  in requeue 554
  in type extension 306
with clause 269
  for child units 274
  limited 283, 286
  private 278, 281, 286
Write (p in Ada.Sequential_IO etc.) 695
Write attribute 709

xor,
  Boolean operator 90
  on type Character_Set 671
  on container sets 770
  on modular types 434
  on one-dimensional arrays 138
  as an operator symbol 173

Yield (p in Ada.Dispatching) 548, 848

+ , – ,
  as arithmetic operators 82
  for complex types 853
  for fixed point 442
  as operator symbols 173
  on types Address and Storage_Offset 826
  on types Time and Duration 521, 523
  – on type Character_Set 671
  – on container sets 770

* , / ,
  as arithmetic operators 83
  for complex types 853
  for fixed point 442
  as operator symbols 173
  * for string construction 666

** ,
  for exponentiation 83, 85, 682
    on complex types 853
  as an operator symbol 173
  second operand of type Integer 428

= , /= ,
  on access values 217
  on arrays 125, 254, Ans12.4(4)
  on type Character_Set 671
  composability 254, 329
  on enumeration types 89
  generic parameter = of containers 734, 755, 767
  on limited types 255
  on numeric types 83
  as operator symbols 173
  on private types 239
  on records 145
  (re)defining 175, 251, 254, 256, 329
  special rules for /= 175, 224
  on strings 666
  on type extension 329

< , <= , > , >= ,
  on enumeration types 89
  generic parameter < of containers 755, 767
  on numeric types 83

on one-dimensional arrays 140
on strings 666
as operator symbols 173
on type Address 825
on type Time 520
<= on type Character_Set 671

<> ,
in aggregates as default 131, 145
for components of concurrent types 538
good for null aggregates 134
with generic formals 487
with generic package parameter 502
as generic subprogram default 497
in procedural iterators 588
for unconstrained array types 123, 125

& ,
for concatenation 141, 665
as an operator symbol 173

@,
target name symbol 98

Index to examples

This index lists the more significant examples. The entries give the section numbers (or program numbers such as P2) where they are declared or referenced; entries followed by the letter e indicate that the reference is to an exercise. Predefined entities such as the type String will be found in the 'General index'.

A_String, type 11.2
Abstract_Sets, package 14.9
Animal, type 9.3, 24.2
Angle, type 17.5
Apples, type 12.3, 17.1
Apply, generic function 19.3, 19.4

Balls, package 21.9
Bank, package 12.6e, 15.6
Boxer, type 18.4
Buffer, task 20.2
Buffer_System, package 3.1
Buffering, protected type 20.4
Buffers, generic package 20.10
Button, type 11.8

Cell, type 11.2, 11.7, P2, 21.4, 25.1

Check_Stock, procedure 24.4, 24.5
Clean_Up, procedure 15.2
Cobblers, package 20.10e
Colour, type 6.6, 8.6
Complex, type 8.7, 12.2
Complex_Numbers, package 12.2, 13.3, 17.5
Compute_Total_Population, procedure 23.7
Contact_Handler, protected object 26.1
Controller, task 20.9
Cow, type 10.6

Dæmon, task type 20.7
Date, type 8.7, 19.3
Day, type 6.6
Decrypt, function 23.6
Diurnal, package 12.1
Dry_Martini, procedure 10.5

Egg, protected object 22.6
Encrypt, function 23.6
Eratosthenes, package P4
Etc, package 4.4
Even, subtype 16.4
Event, protected object 20.9
Exchange, generic procedure 19.1
Expression_Trees, package 24.6

Factorial, function 10.1, 15.1e, 15.2e
Farmyard, array 8.5
Flatlander, type 14.8
Four_Letters, function 24.10
Frac, type 16.5
Frankenstein, function 18.3

G_String, type 11.4
Garment, type 10.6
Gauss_Seidel, procedure 22.2
GCD, function 10.1e, 13.4e, P3
Gender, type 18.3
Generic_Complex_Functions, generic package 19.4, 19.5e
Generic_Complex_Numbers, generic package 19.2, 19.4, 19.5e
Generic_Complex_Vectors, generic package 19.4

Geometry, package 3.2, P1, 14.6, 18.4
Greek, type 26.1
Grim_Reaper, protected object 22.5
Group, generic package 19.4e

Hush_Hush, package 21.3e

Increment, procedure 10.3
Indexes, package 24.10
Inner, function 10.1, 10.2
Inner, type 18.7
Integrate, function 11.8
Integrate, generic function 19.3
Is_Unduplicated, function 16.3

Key, type 12.6, 18.6
Key_Manager, package 12.6, 14.7e, 15.2

Line_Draw, package 21.2
Lists, package P2, 21.4, 21.5

Mailbox, task type 20.7
Make, procedure 14.1, 14.5
Make_Process, function 20.7, 22.2
Make_Unit, function 10.1e, 18.1e
Marry, procedure 18.5, 21.9
Math_Function, type 11.8, 16.2
Matrix, type 8.2, 11.3
Mixture, package 14.9
Moment, function 3.3, P1
Month_Name, type 7.1e, 8.7
Mutant, type 18.3, 18.6, 19.3e

New_Child, function 18.5, 21.9
Next, generic function 19.2
Next_Digit, function 19.2
Next_Work_Day, array 8.2
Next_Work_Day, function 19.2
Node, type 11.2, P2

Object, type 3.2, 14.1, 14.3e, 14.8, 21.1, 21.2
Old_Person, type 18.4
Oranges, type 12.3, 17.1
Order_Meal, procedure 14.1, 14.5
OS, package 13.4
Outer, function 10.1e, 19.2e
Outer, type 18.7

Palette, package 12.3
Pascal, array 2.4
People, array 8.7, 13.7, 19.3e
People, package 21.9
Person, type 14.3, 14.4, 14.8, 18.3, 18.4, 18.5, 21.9
Person_Name, type 18.5, 21.9
Pets, object 8.5, 13.7e
Planet, type 7.3
Polygon, type 18.4
Polynomial, type 18.2, 19.3e, P4
Pond, type 25.4, P6
Pop, function 12.1
Primary, type 8.6, 19.2
Printable_Objects, package 21.2
Print_Roots, procedure 2.2
Process_Reservation, procedure 14.2
Protected_Variable, task 20.5, 20.8
Push, procedure 12.1

Quadratic, procedure 10.3, 15.1e
Queues, generic package 21.4
Queues, package 12.5e, 14.2

Random, package 12.1e, 15.2e, 16.5
Rational, type 12.2e, P3
Rational_Numbers, package 12.2e, 13.4e, P3, 19.2e
Rational_Polynomial, type 18.2
Reader_Writer, package 20.5
Rectangle, type 18.1, 18.4
Regular_Polygon, type 18.4
Reservation, type 14.1, 14.6
Resource_Allocator, protected object 20.9
Rev, function 10.1
Ring_5, subtype 8.2e
Ring_5, type 17.2e
Roman_Digit, type 8.4
Roman_Number, type 8.4
Rough, type 26.5
Row, type 8.2
Royal_Events, array 8.3

Schedule, type 8.3
Score, type 16.4
Semaphore, protected type 20.4
Set, interface 14.9
Set_Of, generic package 19.2, 20.9
Shopping, procedure 20.1, 20.2
Sieve, procedure 22.1
Sign, function 10.1
Signal, type 2.6
Simple_IO, package 2.2, 23.8e
Simple_Maths, package 2.2e, 23.5e
Solve, procedure 16.2
Sort, generic procedure 19.3, 19.4
Sort, procedure 11.2
Spouse, function 18.5, 21.9
Square, type 18.1
Stack, generic package 19.1
Stack, package 12.1, 13.1, 13.4, 15.2, 24.2
Stack, type 12.4, 18.1

Stacks, package 12.4, 18.1, 19.2
Status, type 25.1, 25.2
Store_Maps, package 24.4
Story, package P5
Strange, array 8.1
String_Array, array 8.5
Student, type 8.7
Sum, function 10.1, 10.2
Sum, generic function 19.2
Sum_5, function 10.1
Swap, procedure 14.5, 19.1

Thing, type 14.7
Tomorrow, array 8.1
Tomorrow, function 15.1, 19.2
Tower_Of_Hanoi, procedure P6
Trace, function 18.1
Tracking, generic package 21.3e
Transpose, function 18.1
Trees, package P2
Truncate, procedure 18.2

V_String, type 18.2
Variable, protected object 20.4
Vector, type 8.2
Vector_5, subtype 8.2
Vector_6, type 8.2

Washer_Package, package 24.8
Weekday, subtype 6.6
Wheel_State, type 6.7
Winter, subtype 16.4
Withdraw, procedure 12.6e, 15.6
Work_Day, array 8.1, 8.3
Work_Place, package 13.5

Zoo, array 8.5, 11.2, 11.4e, 18.2
Zoo, package 24.2

Printed in the United States
by Baker & Taylor Publisher Services